Combinatorial Problems and Exercises

Second Edition

COMBINATORIAL PROBLEMS AND EXERCISES

SECOND EDITION

LÁSZLÓ LOVÁSZ

AMS CHELSEA PUBLISHING
American Mathematical Society • Providence, Rhode Island

This book was first published in coedition between North Holland Publishing Company and Akadémiai Kiadó in 1979. It was revised and updated in 1993 in coedition between Elsevier Science Publishers and Akadémiai Kiadó.

2000 *Mathematics Subject Classification*. Primary 05–01.

For additional information and updates on this book, visit
www.ams.org/bookpages/chel-361

Library of Congress Cataloging-in-Publication Data
Lovász, László, 1948–
    Combinatorial problems and exercises / László Lovász.—2nd ed.
      p. cm.
    Includes index.
    ISBN 978-0-8218-4262-1 (alk. paper)
    1. Combinatorial analysis—Problems, exercises, etc.   I. Title.

QA164.L69   2007
511′.6076—dc22
                                                                            2007060765

To Kati

Contents

Preface to the Second Edition

When the publishers of this book asked me to revise and update my problem book for a second edition, I had to decide how much to change, taking into consideration the fast development of the field (but also that the first edition was out of print). Combinatorics has grown a lot in the last decade, especially in those fields interacting with other branches of mathematics, like polyhedral combinatorics, algebraic combinatorics, combinatorial geometry, random structures and, most significantly, algorithmic combinatorics and complexity theory. (The theory of computing has so many applications in combinatorics, and vice versa, that sometimes it is difficult to draw the border between them.) But combinatorics is a discipline on its own right, and this makes this collection of exercises (subject to some updating) still valid.

I decided not to change the structure and main topics of the book. Any conceptual change (like introducing algorithmic issues consistently, together with an analysis of the algorithms and the complexity classification of the algorithmic problems) would have meant writing a new book. I could not resist, however, to working out a series of exercises on random walks on graphs, and their relations to eigenvalues, expansion properties, and electrical resistance (this area has classical roots but has grown explosively in the last few years). So Chapter 11 became substantially longer.

In some other chapters I also found lines of thought that have been extended in a natural and significant way in the last years. Altogether, I have added about 60 new exercises (more if you count subproblems), simplified several solutions, and corrected those errors that I became aware of.

In the preface of the first edition, I said that I plan a second volume on important topics left out, like matroids, polyhedral combinatorics, lattice geometry, block designs, etc. These topics have grown enormously since then, and to cover all of them would certainly need more than a single volume. I still love the procedure of selecting key results in various fields and analyzing them so that their proofs can be broken down to steps adding one idea at a time, thus creating a series of exercises leading up to a main result. (This love was revigorated while working on this new edition.) But writing a new volume is at the moment beyond my time and energy constraints.

In the meanwhile, many monographs were published on these topics, and several of these (in the first line, A. Recski's book *Matroid Theory and its Applications in Electric Network Theory and Statics*, Akadémiai Kiadó–Springer Verlag, 1989) contain extensive and very carefully compiled lists of problems and exercises.

Acknowledgements. I have received many remarks, corrections, and suggestions for improvements from my collegues; many of these were based on experience while teaching a course based on this book. Needless to say how pleased I felt by their interest in my work, and how grateful I am to these collegues for taking the trouble of formulating these remarks and sending them to me. Virtually all of them were right, and I have implemented almost all of these comments while revising the text (a few concerned research results and further topics related to the material, and were beyond the scope of the book). I am particularly grateful to J. Burghduff, A. Frank, F. Galvin, D. E. Knuth, and I. Tomescu for their extensive and thoughtful comments. My special thanks are due to D. E. Knuth and D. Aldous for reflecting on the new series of exercises, and in fact so fast that I could implement their remarks in the final version of the revised manuscript.

I also feel indebted to Ms. K. Fried for the very careful and expert typing into TEX (and for discovering many errors in the first edition while doing so), and to G. Bacsó and T. Csizmazia for their help in the proofreading and their thoughtful observations made during this work.

Budapest, March 1992.

Preface

Having vegetated on the fringes of mathematical science for centuries, combinatorics has now burgeoned into one of the fastest growing branches of mathematics — undoubtedly so if we consider the number of publications in this field, its applications in other branches of mathematics and in other sciences, and also, the interest of scientists, economists and engineers in combinatorial structures. The mathematical world had been attracted by the success of algebra and analysis and only in recent years has it become clear, due largely to problems arising from economics, statistics, electrical engineering and other applied sciences, that combinatorics, the study of finite sets and finite structures, has its own problems and principles. These are independent of those in algebra and analysis but match them in difficulty, practical and theoretical interest and beauty.

Yet the opinion of many first-class mathematicians about combinatorics is still in the pejorative. While accepting its interest and difficulty, they deny its depth. It is often forcefully stated that combinatorics is a collection of problems, which may be interesting in themselves but are not linked and do not constitute a theory. It is easy to obtain new results in combinatorics or graph theory because there are few techniques to learn, and this results in a fast-growing number of publications.

The above accusations are clearly characteristic of any field of science at an early stage of its development — at the stage of collecting data. As long as the main questions have not been formulated and the abstractions to a general level have not been carried through, there is no way to distinguish between interesting and less interesting results — except on an aesthetic basis, which is, of course, too subjective. Those techniques whose absence has been disapproved of above await their discoverers. So underdevelopment is not a case against, but rather for, directing young scientists toward a given field.

In my opinion, combinatorics is now growing out of this early stage. There *are* techniques to learn: enumeration techniques, matroid theory, the probabilistic method, linear programming, block design constructions, etc. There *are* branches which consist of theorems forming a hierarchy and which contain central structure theorems forming the backbone of study: connectivity of graphs (network flows) or factors of graphs, just to pick two examples from graph theory. There *are* notions abstracted from many non-trivial results, which unify large parts of the theory, such as matroids or the concept of good characterization (see below). My feeling that it is no longer possible to obtain significant results without the knowledge of these facts, concepts and techniques. (Of course, exceptions may occur, since the field is destined to cover such a large part of the world of mathematics that entirely new problems may still arise.)

*

The reader will forgive I hope the insertion of some general ideas which tend to play the role of systematizing and unifying concepts. The first of these is the notion of the class NP.[†] A property T of graphs is in NP if we are able to efficiently prove (exhibit) T if it holds. (Technically, "efficiently" means that the length of the proof is bounded by a polynomial in the size of the graph.) For example, if a graph G is Hamiltonian, we can exhibit this by specifying a Hamilton cyclein G. This notion leads us to the notion of a *good characterization*, or — in the language of computational complexity theory — of the class NP∩co-NP, formulated by J. Edmonds. A property T of graphs is in NP∩co-NP if (making use of the different formulations of the definition and the equivalent condition) we are able to efficiently prove T if it holds and to efficiently disprove if it does not. For example, Kuratowski's classical characterization of planar graphs provides a good characterization of planarity: if a graph is planar, we easily establish this by drawing it in the plane; if it is non-planar, we can show this by exhibiting one of Kuratowski's graphs in it (see problem 5.37). A good characterization reflects a deep underlying logical duality of the property and, as the reader may convince himself by comparing good and "non-good" characterizations occurring throughout this book, often amounts to "the" solution of the problem. Of course, this does not mean that "non-good" characterizations may not be deep and useful theorems.

The existence of good characterizations tends to go hand in hand with the existence of good decision algorithms (by a "good" or "efficient" algorithm we mean one whose running time in the worst case is only a polynomial in the input data; this again does not directly affect its practical value). Several combinatorial properties are known for which there exists a polynomial-time algorithm to decide whether a given structure has the property, but its existence is by no means obvious (e.g. having a 1-factor). An interesting theoretical result, due to S. A. Cook, R. M. Karp and L. A. Levin is the following. Several properties of graphs (for example, the existence of Hamiltonian circuits, independence number, chromatic number, the existence of a kernel etc.) are equivalent in the sense that if any of them could be solved by a polynomial-time algorithm, then all of them could, and in this case also the "good" algorithmic solvability of very many other

---

[†] For a detailed description see e.g. A. V. Aho, J. E. Hopcroft, J. D. Ullman, *The Design and Analysis of Computer Algorithms,* Addison–Wesley, 1974, Chap. 10, or M.R. Garey and D.S. Johnson, *Computers and Intractability: A Guide to the Theory of NP-Completeness,* Freeman, San Francisco, 1979.

problems in different fields of mathematics would follow (just to mention a very distant one: testing an n-digit number for primarility). These "most difficult" problems in NP are called NP-*complete*. It is unlikely that all of these could be solved efficiently, but there is as yet no proof of this. This is the famous $P \neq NP$ problem of computer science.

Another idea which has proved very fruitful is that combinatorial optimization problems can generally be formulated as linear programming problems with integrality constraints. If one could disregard the constraints, the Duality Theorem of linear programming would provide the solution. So the solution of these problems is connected to investigating the effect of the integrality constraints on the behavior of optima. For example, if we can show that they do not change the optimum, we obtain a *minimax theorem*. Several instances of this idea can be found in § 13 (Hypergraphs).

Last but not least we mention the use of *linear algebra*. This ranges from applications of matrix calculus to the introduction of homology and cohomology groups. A common background of many applications of linear algebra is *Matroid Theory*, which is now a flourishing branch of combinatorics itself.

*

The main purpose of this book is to provide help in learning existing techniques in combinatorics. The most effective (but admittedly very time-consuming) way of learning such techniques is to solve (appropriately chosen) exercises and problems.

This book presents all the material in the form of problems and series of problems (apart from some general comments at the beginning of each chapter). We hope that it will be useful to those students who intend to start research in graph theory, combinatorics or their applications, and for those who feel that combinatorial techniques might help them with their work in other branches of mathematics, management science, electrical engineering and so on. For background, only the elements of linear algebra, group theory, probability and calculus are needed.

When selecting the material I have had to restrict the topics covered. I feel that a more detailed analysis of a few basic notions is more useful than touching of all possible fields of research. So in this volume only enumeration problems, graphs and set-systems are discussed. Some fields have had to be completely omitted: random structures (here the reader is advised to read the book *Probabilistic Methods in Combinatorics* by P. Erdős and J. Spencer, Akadémiai Kiadó, Budapest and Academic Press, New York–London, 1974), integer programming, matroids (combinatorial geometries), finite geometries, block designs, lattice geometry, etc. I hope eventually to write a sequel to this volume covering some of these latter topics.

The book consists of three major parts: Problems, Hints and Solutions. A reader with less experience may read the hint given to a problem before trying to solve it; those problems with one or two asterisks are thought to be difficult and the reader may read the hint given right away unless he is prepared to sacrifice several days to the problem (some of them are worth it, I venture to say). Even having solved a problem the reader is advised to compare his solution with the one given here: it may be that the idea occurring in our solution will be basic in a later series of problems. Here it should be pointed out that problems come in series and previous problems often serve as steps to the last, deepest result in the series. Also note that the solution often uses notation or properties introduced in the hint.

As references I have preferred to give those where further development of the subject can best be seen. Thus, for those results reproduced in textbooks or monographs, usually the latter are given as reference. No reference means either that the assertion of the exercise is so well known that it would have been impossible or superfluous to trace it back or that the problem is believed to be new. A list of those textbooks and monographs most often cited is given at the end of the volume. A dictionary containing the definitions of combinatorial concepts used in the book and a list of symbols as well as an Author Index and a Subject Index are included.

Acknowledgements. First of all, I wish to express my gratitude to Professors P. Erdős and T. Gallai, through whom I have come to love combinatorics, and from whom in principle I have learnt almost everything included in this book. I have to mention in the same group Professors Vera T. Sós, A. Hajnal and many other graph theorists and combinatorists, who (directly or indirectly) contributed to this book. I am particularly indebted to J. C. Ault, L. Babai, A. Bondy, A. Hajnal, G. Katona, M. D. Plummer and M. Simonovits, who having read the manuscript or parts of it, made many valuable remarks, suggestions and corrections; and to L. Babai and his students, in particular E. Boros and Z. Füredi for their invaluable help in proofreading.

I want to express my thanks to Prof. I Rábai for the idea of such a problem book and for encouraging its writing; also to Mrs. A. Porubszky for the careful typing of the manuscript; and to Akadémiai Kiadó, in particular Mrs. Á. Sulyok for the competent and thorough editorial work.

But above all I have to thank my wife Kati; her unbroken encouragement and professional, technical and moral support have been the firm basis of my work.

I. Problems

§ 1. Basic enumeration

There is no rule which says that enumeration problems, even the simplest ones, must have solutions expressible as closed formulas. Some have, of course, and one important thing to be learnt here is how to recognize such problems. Another approach, avoiding the difficulty of trying to produce a closed formula, is to look for "substitute" solutions in other forms such as formulas involving summations, recurrence relations or generating functions. A typical (but not unique or universal) technique for solving an enumeration problem, in one or more parameters, is to find a recurrence relation, deduce a formula for the generating function (the recurrence relation is usually equivalent to a differential equation involving this function) and finally, where possible, to obtain the coefficients in the Taylor expansion of the generating function.

However, it should be pointed out that, in many cases, elementary transformations of the problems may lead to another problem already solved. For example, it may be possible by such transformations to represent each element to be counted as the result of n consecutive decisions such that there are a_i possible choices at the ith step. The answer would then be $a_1 a_2 \ldots a_n$. This is particularly useful when each decision is independent of all the previous decisions. Finding such a situation equivalent to the given problem is usually difficult and a matter of luck combined with experience.

1. In a shop there are k kinds of postcards. We want to send postcards to n friends. How many different ways can this be done? What happens if we want to send them different cards? What happens if we want to send two different cards to each of them (but different persons may get the same card)?

2. We have k distinct postcards and want to send them all to our n friends (a friend can get any number of postcards, including 0). How many ways can this be done? What happens if we want to send at least one card to each friend?

3. How many anagrams can be formed from the word CHARACTERIZATION? (An anagram is a word having the same letters, each occurring the same number of times; this second word does not need to have a meaning.)

4. (a) How many possibilities are there to distribute k forints[†] among n people so that each receives at least one?

(b) Suppose that we do not insist that each person receives something. What will be the number of distributions in this case?

5. There are k kinds of postcards, but only in a limited number of each, there being a_i copies of the ith one. What is the number of possible ways of sending them to n friends? (We may send more than one copy of the same postcard to the same person).

6. (a) Find recurrence relations for the Stirling partition numbers $\left\{{n \atop k}\right\}$ and the Stirling cycle numbers $\left[{n \atop k}\right]$, and tabulate them for $n \le 6$.

(b) Prove that $\left\{{n \atop n-k}\right\}$ and $\left[{n \atop n-k}\right]$ are polynomials in n for each fixed k.

(c) Show that there is a unique way to extend the definition of $\left\{{n \atop k}\right\}$ and $\left[{n \atop k}\right]$ over all integers n and k so that the recurrence relations in (a) are preserved and the "boundary conditions" $\left\{{0 \atop 0}\right\} = \left[{0 \atop 0}\right] = 1$ and $\left\{{0 \atop m}\right\} = \left[{m \atop 0}\right] = 0$ ($m \ne 0$) are fulfilled.

(d) Prove the duality relation

$$\left\{{n \atop k}\right\} = \left[{-k \atop -n}\right].$$

7. Prove the identities

(a) $\displaystyle\sum_{k=0}^{n} \left\{{n \atop k}\right\} x(x-1)\dots(x-k+1) = x^n,$

(b) $\displaystyle\sum_{k=0}^{n} \left[{n \atop k}\right] x^k = x(x+1)\dots(x+n-1),$

(c) $\displaystyle\sum_{k=0}^{n} (-1)^k \left\{{n \atop k}\right\} \left[{k \atop j}\right] = \begin{cases} 1, & \text{if } j = n, \\ 0, & \text{otherwise.} \end{cases}$

[†] forint: Hungarian currency.

8. Prove that

$$\left\{{n \atop k}\right\} = \frac{1}{k!} \sum_{j=0}^{k} (-1)^{k-j} \binom{k}{j} j^n.$$

In particular, the right-hand side is 0 for $k > n$ (cf. also 2.4).

9. (a) Let B_n denote the nth Bell number, i.e. the number of all partitions of n objects. Prove the formula

$$B_n = \frac{1}{e} \sum_{k=0}^{\infty} \frac{k^n}{k!}.$$

(b)* $\quad B_n \sim \frac{1}{\sqrt{n}} \lambda(n)^{n+1/2} e^{\lambda(n)-n-1},$

where $\lambda(n)$ is defined by $\lambda(n) \log \lambda(n) = n$.

10. Find a recurrence relation for B_n.

11. Let $p(x)$ be the exponential generating function of the sequence B_n, i.e.

$$p(x) = \sum_{n=0}^{\infty} \frac{B_n}{n!} x^n.$$

Determine $p(x)$.

12. (a) Prove that

$$B_n = \sum_{\substack{k_1,\ldots,k_n \geq 0 \\ k_1+2k_2+\ldots+nk_n=n}} \frac{n!}{k_1!(1!)^{k_1} k_2!(2!)^{k_2} \ldots k_n!(n!)^{k_n}}.$$

Deduce the formula for $p(x)$ from this expression.

(b) What do we get (asymptotically) if the summation is extended over all systems $(k_1,\ldots,k_n)$ with $k_1 + k_2 + \ldots + k_n = n$?

13. Give another proof of

$$B_n = \frac{1}{e} \sum_{k=0}^{\infty} \frac{k^n}{k!}.$$

14. (a) Let Q_n denote the number of partitions of an n-element set into an even number of classes ($Q_0 = 1$). Determine

$$q(x) = \sum_{n=0}^{\infty} \frac{Q_n}{n!} x^n,$$

and find an analogue to the formula of 1.9a.

(b) Denote by R_n the number of partitions of an n-set into classes of even cardinality ($R_0 = 1$). Determine

$$r(x) = \sum_{n=0}^{\infty} \frac{R_n}{n!} x^n.$$

15. Let S_n denote the number of all possible final results at a competition where ties are possible. More precisely, S_n is the number of mappings $f : \{1,\ldots,n\} \to \{1,\ldots,n\}$ such that if f takes a value i then it also takes each value j, $1 \le j \le i$. Let $S_0 = 1$.

(a) Prove the identities

$$S_n = \sum_{k=0}^{n} k! \left\{ {n \atop k} \right\} = \sum_{k=0}^{\infty} \frac{k^n}{2^{k+1}}$$

and determine the generating function

$$s_n = \sum_{n=0}^{\infty} \frac{S_n}{n!} x^n.$$

(b) Prove the asymptotics

$$\frac{S_n (\log 2)^{n+1}}{n!} \to \frac{1}{2} \quad (n \to \infty).$$

$*$

16. The number of partitions of the number n into (a sum of) no more than r terms is equal to the number of partitions of n into any number of terms, each at most r.

17. The number of partitions of a number n into exactly m terms is equal to the number of partitions of $n-m$ into no more than m terms. Find a similar identity involving the number of partitions of n into exactly m *distinct* terms.

18*. The number of partitions of n into (any number of) distinct terms is equal to the number of partitions of n into odd terms.

19*. (Pentagon Numbers Theorem) If $n \ne \frac{3k^2 \pm k}{2}$ then the number of partitions of n into an odd number of distinct terms is the same as the number of its partitions into an even number of distinct terms.

20. Let π_n be the number of partitions of the number n. Determine the generating function of the sequence π_n.

21. Prove that the number of partitions of n into distinct terms is equal to the number of partitions of n into odd terms (1.18), by calculating the generating functions of both sides.

22. Which identity follows from problem 1.19?

23*. Prove the following identities by combinatorial considerations:

$$\text{(a)} \quad (1+x)(1+x^3)(1+x^5)\ldots = \sum_{k=0}^{\infty} \frac{x^{k^2}}{(1-x^2)(1-x^4)\ldots(1-x^{2k})},$$

$$\text{(b)} \quad (1+x^2)(1+x^4)(1+x^6)\ldots = \sum_{k=0}^{\infty} \frac{x^{k^2+k}}{(1-x^2)(1-x^4)\ldots(1-x^{2k})}.$$

24. We look for those partitions of the number n which have the property that each number between 1 and n is uniquely representable as a partial sum of the partition. When will the trivial partition $n=1+\ldots+1$ be the only solution?

25. The number of non-congruent triangles with circumference $2n$ and integer sides is equal to the number of non-congruent triangles with circumference $2n-3$ and integer sides. This number is also equal to the number of partitions of n into exactly three terms. Determine this number.

*

26. We have n forints. Every day we buy exactly one of the following products: pretzel (1 forint), candy (2 forints), ice-cream (2 forints). What is the number M_n of possible ways of spending all the money?

27. What is the number A_n of ways of going up n stairs, if we may take one or two steps at a time? Determine

$$\sum_{n=0}^{\infty} A_n x^n.$$

28. (a) We have n forints and each day we make exactly one purchase. There are a_i sorts of goods which can be bought for i forints $(i=1,\ldots,k)$. Suppose that the polynomial $x^k - a_1 x^{k-1} - \ldots - a_k = 0$ has distinct roots $\vartheta_1,\ldots,\vartheta^k$. Prove that the number of possible ways of spending our money is

$$C_n = \frac{\begin{vmatrix} 1 & \vartheta_1 & \cdots & \vartheta_1^{k-2} & \vartheta_1^{k-1+n} \\ \vdots & & & & \\ 1 & \vartheta_k & \cdots & \vartheta_k^{k-2} & \vartheta_k^{k-1+n} \end{vmatrix}}{\begin{vmatrix} 1 & \vartheta_1 & \cdots & \vartheta_1^{k-1} \\ \vdots & & & \\ 1 & \vartheta_k & \cdots & \vartheta_k^{k-1} \end{vmatrix}}.$$

(b) Determine the generating function of C_n, and thereby a formula for C_n.

29. Determine the eigenvalues of the matrix

$$\begin{pmatrix} 0 & 1 & & & & & & 0 \\ 1 & 0 & 1 & & & & & \\ & 1 & \ddots & \ddots & & & & \\ & & \ddots & \ddots & \ddots & & & \\ 0 & & & & \ddots & 0 & 1 \\ & & & & & 1 & 0 \end{pmatrix}}_{n}$$

30. How many sequences of length n can be composed from a, b, c, d in such a way that a and b are never neighboring elements?

31. What is the number of k-tuples chosen from $1, \dots, n$ containing no two consecutive integers?

32. What is the number of monotone increasing functions mapping $\{1, \dots, n\}$ into itself?

33. (a) How many monotone increasing mappings f of $\{1, \dots, n\}$ into itself satisfy the condition $f(x) \leq x$ for every $1 \leq x \leq n$?

(b) What is the number of sequences of n 0's and n 1's such that there are at least as many 0's as 1's among the first k digits for each $1 \leq k \leq 2n$?

34. Prove that the number of sequences $(x_1, \dots, x_r)$ $(1 \leq x_i \leq n)$ containing less than i entries from $\{1, \dots, i\}$ for each $i = 1, \dots, n$ is $(n-r)n^{r-1}$ $(1 \leq r \leq n)$.

*

35. A sequence of partitions of the n-element set S is constructed as follows: We start with S itself and then we obtain the $(i+1)^{\text{st}}$ partition from the ith by splitting any one of its classes having more than one element into two non-empty subsets. Thus the partition into one-element classes will be reached after $n-1$ steps. What is the number of ways of carrying out this procedure?

36. The previous problem is modified to the extent that the $(i+1)^{\text{st}}$ partition is obtained from the ith by splitting all classes having more than one element into two non-empty subsets. What is the number of ways of carrying out the procedure now?

37. We want to break a stick of length n into n pieces of unit length. What is the number of ways of doing so if

(a) at each step, we break one of the pieces with a length greater than 1 into two,

(b) at each step, we break all pieces with a length greater than 1 into two.

38. In how many ways can one put brackets into the product $x_1 \cdot x_2 \cdot \dots \cdot x_r$ (in the sense that any bracket encloses a product of exactly two factors)?

39. What is the number D_n of triangulations of a convex n-gon? (A triangulation is a set of $n-3$ diagonals no two of which intersect internally and which, therefore, divide the n-gon into $n-2$ triangles.)

40. Determine, without using the preceding result, the generating function of D_n and prove the result of the preceding problem from this.

41. In how many ways can one divide a convex n-gon into triangles by $n-3$ non-intersecting diagonals, in such a way that each triangle has an edge in common with the convex n-gon?

*

42. Find a closed formula for each of the following expressions:

(a) $\displaystyle\sum_{k=0}^{\lfloor n/2 \rfloor} \binom{n}{2k}$,

(b) $\displaystyle\sum_{k=0}^{m} \binom{n-k}{m-k}$,

(c) $\displaystyle\sum_{k=0}^{m} \binom{u}{k}\binom{v}{m-k}$,

(d) $\displaystyle\sum_{k=0}^{m} (-1)^k \binom{u}{k}\binom{u}{m-k}$,

(e) $\displaystyle\sum_{k=m}^{n} \binom{k}{m}\binom{n}{k}$,

(f) $\displaystyle\sum_{k=0}^{\lfloor n/7 \rfloor} \binom{n}{7k}$,

(g) $\displaystyle\sum_{k=0}^{\lfloor n/2 \rfloor} \binom{n-k}{k} z^k$,

(h) $\displaystyle\sum_{k=0}^{m} (-1)^k \binom{n}{k}$,

(i) $\displaystyle\sum_{k=0}^{m} \binom{u+k}{k}\binom{v-k}{m-k}$.

43. Prove the following identities:

(a) $\displaystyle\sum_{k=0}^{m}\binom{m}{k}\binom{n+k}{m} = \sum_{k=0}^{m}\binom{m}{k}\binom{n}{k}2^k,$

(b) $\displaystyle\sum_{k=0}^{m}\binom{m}{k}\binom{n+k}{m} = (-1)^m\sum_{k=0}^{m}\binom{m}{k}\binom{n+k}{k}(-2)^k,$

(c) $\displaystyle\sum_{k=0}^{p}\binom{p}{k}\binom{q}{k}\binom{n+k}{p+q} = \binom{n}{p}\binom{n}{q},$

(d) $\displaystyle\sum_{k=0}^{p}\binom{p}{k}\binom{q}{k}\binom{n+p+q-k}{p+q} = \binom{n+p}{p}\binom{n+q}{q},$

(e) $\displaystyle\frac{d}{dx}\binom{x}{n} = \sum_{k=1}^{n}\frac{(-1)^{k+1}}{k}\binom{x}{n-k} = \sum_{k=1}^{n}\frac{1}{k}\binom{x-k}{n-k}.$

44*. Prove the following Abel identities:

(a) $\displaystyle\sum_{k=0}^{n}\binom{n}{k}x(x+k)^{k-1}(y+n-k)^{n-k} = (x+y+n)^n,$

(b) $\displaystyle\sum_{k=0}^{n}\binom{n}{k}(x+k)^{k-1}(y+n-k)^{n-k-1} = \left(\frac{1}{x}+\frac{1}{y}\right)(x+y+n)^{n-1},$

(c) $\displaystyle\sum_{k=1}^{n-1}\binom{n}{k}k^{k-1}(n-k)^{n-k-1} = 2(n-1)n^{n-2},$

45. Let

$$f_n(x) = x(x-1)\dots(x-n+1).$$

Prove that

$$f_n(x+y) = \sum_{k=0}^{n}\binom{n}{k}f_k(x)f_{n-k}(y).$$

Find other examples of such sequences of polynomials.

§ 2. The sieve

One powerful tool in the theory of enumeration as well as in prime number theory is the inclusion-exclusion principle (sieve of Eratosthenes). This relates the cardinality of the union of certain sets to the cardinalities of intersections of some of them, these latter cardinalities often being easier to handle. However, the formula does have some handicaps: it contains terms alternating in sign, and in general it has too many of them!

A natural setting for the sieve is in the language of probability theory. Of course, this only means a division by the cardinality of the underlying set, but it has the advantage that independence of occurring events can be defined. Situations in which events are almost independent are extremely important in number theory and also arise in certain combinatorial applications. Number theorists have developed ingenious methods to estimate the formula when the events (usually divisibilities by certain primes) are almost independent. We give here the combinatorial background of some of these methods. Their actual use, however, rests upon complicated number-theoretic considerations which are here illustrated only by two problems.

It should be emphasized that the sieve formula has many applications in quite different situations. These are scattered throughout the book, but § 15 (Reconstruction) in particular may be pointed out in this connection.

A beautiful general theory of inclusion-exclusion, usually referred to as the theory of the Möbius function, is due to L. Weisner, P. Hall and G.C. Rota (see [St]); and the second half of this section is devoted to this.

1. In a high school class of 30 pupils, 12 pupils like mathematics, 14 like physics and 13 chemistry, 5 pupils like both mathematics and physics, 7 both physics and chemistry, 4 pupils like mathematics and chemistry. There are 3 who like all three subjects. How many pupils do not like any of them?

2. (a) (The Sieve Formula) Let $A_1, \ldots, A_n$ be arbitrary events of a probability space (Ω, P). For each $I \subseteq \{1, \ldots, n\}$, let

$$A_I = \prod_{i \in I} A_i; \quad A_\emptyset = \Omega;$$

and let

$$\sigma_k = \sum_{|I|=k} \mathsf{P}(A_I), \quad \sigma_0 = 1.$$

Then

$$\mathsf{P}(A_1 + \ldots + A_n) = \sum_{j=1}^{n} (-1)^{j-1} \sigma_j.$$

(b) (Inclusion-Exclusion Formula) Let $A_1, \ldots, A_n \subseteq S$ where S is a finite set, and let

$$A_I = \bigcap_{i \in I} A_i; \quad A_\emptyset = S.$$

Then

$$|S - (A_1 \cup \ldots \cup A_n)| = \sum_{I \subseteq \{1, \ldots, n\}} (-1)^{|I|} |A_I|.$$

3. Determine the number $\varphi(n)$ of integers between 1 and n coprime to n, given the prime factorization $p_1^{\alpha_1}, \ldots p_r^{\alpha_r}$ of n.

4. Prove the identity

$$\sum_{i=0}^{n}(-1)^i\binom{n}{i}i^k = \begin{cases} 0 & \text{if } 0 \leq k < n, \\ (-1)^n n! & \text{if } k = n. \end{cases}$$

What do we get if $k > n$?

5. Let $p(x_1, \ldots, x_n)$ be a polynomial of degree m, and denote by $\sigma^k p$ the polynomial obtained by substituting 0's for k of the variables in p in every possible combination and summing the arising $\binom{n}{k}$ polynomials $(\sigma^0 p = p)$. Prove that

$$\sigma^0 p - \sigma^1 p + \sigma^2 p - \ldots = \begin{cases} 0 & \text{if } m < n \\ c \cdot x_1 \ldots x_n & \text{if } m = n. \end{cases}$$

What is c?

6. Let $A_1, \ldots, A_n$ be any events, $B_i = f_i(A_1, \ldots, A_n)$ $(i = 1, \ldots, k)$ polynomials in $A_1, \ldots, A_n$ and $c_1, \ldots, c_n$ reals. Then

$$\sum_{i=1}^{k} c_i P(B_i) \geq 0$$

holds for every $A_1, \ldots, A_k$, provided it holds in those cases when $P(A_j) = 0$ or 1 for $j = 1, \ldots, n$.

7. (a) Let $A_1, \ldots, A_n$ be events as in 2.2a. The probability that exactly q of them occur is

$$\sum_{j=q}^{n}(-1)^{j+q}\binom{j}{q}\sigma_j.$$

(b) Let η denote the number of A_i''s that occur (this is a random variable). Then

$$E(x^\eta) = \sum_{j=0}^{n}(x-1)^j \sigma_j.$$

8. Express by a similar formula the probability of the event D_p that at least p of $A_1, \ldots, A_n$ occur.

9. (Bonferoni's Inequalities) The partial sums of

$$P(\bar{A}_1 \ldots \bar{A}_n) - \sigma_0 + \sigma_1 - \sigma_2 + \ldots$$

are alternating in sign.

10. Prove that

$$\sigma_r \leq \frac{n-r+1}{r}\sigma_{r-1}.$$

Sharpen this inequality if $\sigma_m = 0$ for some $r+1 \leq m \leq n$.

11. For each $I \subseteq \{1,\ldots,n\}$, let a number p_I be given. When can one find events $A_1,\ldots,A_n$ such that

$$P(A_I) = p_I?$$

*

12. Let G be a simple graph on $V(G) = \{1,\ldots,n\}$. Denote by $\mathcal{O}_0$ the collection of odd independent subsets of $V(G)$ and by $\mathscr{E}_1$, the collection of those even subsets which span at most one edge. Then

$$P(\bar{A}_1 \ldots \bar{A}_n) \leq \sum_{I \in \mathscr{E}_1} P(A_I) - \sum_{I \in \mathcal{O}_0} P(A_I).$$

Find an analogous lower estimate. (Note that for $E(G) = \emptyset$ we obtain the sieve formula.)

13. (Brun's Sieve) Let $f(k) \geq 0$ be any integer-valued function defined on $\{1,\ldots,n\}$ and set

$$\mathscr{I} = \{I \subseteq \{1,\ldots,n\} : |I \cap \{1,\ldots,k\}| \leq 2f(k), \ k = 1,\ldots,n\}.$$

Then

$$P(\bar{A}_1 \ldots \bar{A}_n) \leq \sum_{I \in \mathscr{I}}(-1)^{|I|}P(A_I).$$

Find an analogous lower estimate.

14*. (Selberg's Sieve) (a) For each $I \subseteq \{1,\ldots,n\}$, let λ_I be a real number, and $\lambda_\emptyset = 1$. Then

$$P(\bar{A}_1,\ldots\bar{A}_n) \leq \sum_{I,J \subseteq \{1,\ldots,n\}} \lambda_I \lambda_J P(A_{I \cup J}).$$

There is always a choice of the λ_I for which equality holds.

(b) Find an analogous lower estimate.

15*. (Cont'd) Let $p_i = P(A_i)$ and

$$p_I = \prod_{i \in I} p_i.$$

Determine the minimum and the minimizing system $\{\lambda_I\}$ of

$$\sum_{I,J \subseteq \{1,\ldots,n\}} \lambda_I \lambda_J p_{I \cup J}$$

under the constraints

$$\lambda_{\emptyset} = 1, \quad \lambda_I = 0 \text{ for } |I| > k,$$

where $p_1, p_2, \ldots, p_n$ and k are supposed given.

16*. (Cont'd) Let $\varepsilon > 0$, $M \geq 1$, $0 < p_i < 1$ and

$$S = \sum p_1^{l_1} \cdots p_n^{l_n},$$

where the summation extends over those systems $l_1, \ldots, l_n \geq 0$ with $p_i^{l_1} \cdots p_n^{l_n} \geq \frac{1}{M}$. Assume that

$$|P(A_K) - p_K| < \varepsilon \qquad (K \subseteq \{1, \ldots, n\}).$$

Then

$$P(\bar{A}_1 \ldots \bar{A}_n) \leq \frac{1}{S} + \frac{\varepsilon M^2}{(1 - p_1)^2 \cdots (1 - p_n)^2}.$$

17*. Prove that, for large enough x, the number of primes in the sequence $l, k + l, \ldots, k(x-1) + l$ $(0 < l < k)$ is not greater than

$$3 \cdot \frac{k}{\varphi(k)} \cdot \frac{x}{\log x}.$$

*

18*. Let G be a simple graph on $V(G) = \{1, \ldots, n\}$ with all degrees at most d and let an event A_i be associated with each point i. Suppose that

(i) $P(A_i) \leq \dfrac{1}{4d}$, and

(ii) every A_i is independent of the set of all A_j's for which j is not adjacent to i.

Then

$$P(\bar{A}_1 \ldots \bar{A}_n) > 0.$$

19. (Second Moment Method) Prove the inequality

$$P(\bar{A}_1 \ldots \bar{A}_n) \leq \frac{\sigma_1 + 2\sigma_2}{\sigma_1^2} - 1.$$

20. Let $P(A_i) = p$, and assume any two events are independent. Estimate

$$P(\bar{A}_1 \ldots \bar{A}_n)$$

from the above using each of 2.9, Selberg's method (2.14–16, with $k = 2$ in 2.15) and the preceding formula and compare the results.

*

21. Let $V = \{x_1, \ldots, x_n\}$ be a set partially ordered by a relation $\leq$. Call an $n \times n$-matrix (a_{ij}) *compatible* if

$$a_{ij} \neq 0 \Rightarrow x_i \leq x_j.$$

Show that the sum, the product and (if it exists) the inverse of compatible matrices is compatible.

22. Prove that there exists a unique function μ defined on $V \times V$ such that

$$\mu(x, y) = 0 \quad \text{if } x \not\leq y,$$
$$\mu(x, x) = 1,$$
$$\sum_{x \leq y \leq z} \mu(x, y) = 0 \quad (x < z)$$

(the *Möbius function* of V).

23. Evaluate the function $\mu(x, y)$ if V is

(a) the lattice of all subsets of a set S (ordered by inclusion),

(b) an arborescence (here $x \leq y$ means that the unique path going from the root to y passes through x),

(c) the set of integers $1, \ldots, n$ where $x \leq y$ means $x | y$.

24. Turn the partially ordered set "upside down" i.e. consider $(V, \leq^*)$ where $x \leq^* y$ iff $y \leq x$. Let μ^* denote the Möbius function of $(V, \leq^*)$. Then

$(*)$ $$\mu^*(x, y) = \mu(y, x).$$

25. Write M as a polynomial in Z (see the solution of 2.22).

26. (Möbius Inversion Formula) Let $f(x)$ be any function defined on V, and set

$$g(x) = \sum_{z \leq x} f(z).$$

Then

$$f(x) = \sum_{z \leq x} g(z)\mu(z, x).$$

Show that the sieve formula is a special case of this statement.

27. Let V be a lattice, $a \leq b \in V$, and denote by 0 and 1 the minimal and maximal elements of V. Then

(a) $\displaystyle\sum_{x \vee a = b} \mu(0, x) = 0$ (if $a > 0$);

(b) $\displaystyle\sum_{x \wedge b = a} \mu(x, 1) = 0$ (if $b < 1$).

28. Let V be a lattice again and $x \in V$ such that x is not representable as the union of atoms. Then $\mu(0, x) = 0$.

29. Let V be a lattice and $\{A, B, C\}$ a partition of V such that if $x \in A$ and $y \leq x$ then $y \in A$; on the other hand, if $x \in C$ and $x \leq y$ then $y \in C$. Then

$$1 + \sum_{x \in A} \sum_{y \in C} \mu(x, y) = \sum_{x, y \in B} \mu(x, y).$$

30. (a) Let V be the lattice of all partitions of the set $S = \{1, \dots, n\}$, where $\{A_1, \dots, A_p\} \leq \{B_1, \dots, B_q\}$ means that each A_i is contained in some B_j. Denote by 0 the partition into one-element sets and by 1 the partition into a single set. Then

$$\mu(0, 1) = (-1)^{n-1}(n - 1)! .$$

(b) Let V be the lattice of faces of a d-dimensional convex polytope P, ordered by containment. (The smallest face is $\emptyset$, of dimension (-1), the largest, P itself.) Prove that $\mu(0, 1) = (-1)^{d+1}$.

31. Let $(V, \leq)$ be a lattice, $V = \{x_1, \dots, x_n\}$, $f(x)$ any function on V and

$$g(x) = \sum_{y \leq x} f(y).$$

Let $g_{ij} = g(x_i \wedge x_j)$, then

$$\det(g_{ij}) = f(x_1) \dots f(x_n).$$

Generalize this formula for partially ordered sets.

32. Evaluate the determinants

$$\begin{vmatrix} (1,1) & (1,2) & \dots & (1,n) \\ (2,1) & (2,2) & \dots & (2,n) \\ & & & \\ (n,1) & (n,2) & \dots & (n,n) \end{vmatrix} ; \qquad \begin{vmatrix} (2,2) & (2,3) & \dots & (2,n) \\ (3,2) & (3,3) & \dots & (3,n) \\ & & & \\ (n,2) & (n,3) & \dots & (n,n) \end{vmatrix} ,$$

where (i, j) denotes the greatest common divisor of i and j.

33*. Let T be a tree, $V = V(T) = \{x_1, \dots, x_n\}$ and

$$d_{ij} = d(x_i, x_j)$$

the distance of x_i from x_j in T. Then

$$\det(d_{ij})_{i,j=1}^{n} = -(-2)^{n-2}(n - 1).$$

34. Let V be a partially ordered set, $x, y \in V$, and let p_k denote the number of chains $a_0 = x < a_1 < \dots < a_k = y$. Then

$$\mu(x, y) = p_0 - p_1 + p_2 - p_3 + \dots .$$

35. Let V be a lattice and let R be the set of its atoms. Denote by q_k the number of those k-tuples in R whose union is 1 ($q_0 = 0$ except when $|V| = 1$). Then

$$\mu(0,1) = q_0 - q_1 + q_2 - q_3 + \cdots .$$

36*. Let C be a set of pairwise incomparable elements of the lattice V such that every maximal chain (linearly ordered subset) contains an element of C. Denote by q_k the number of those k-tuples from C whose union is 1 and whose intersection is 0. Then

$$\mu(0,1) = q_0 - q_1 + q_2 - q_3 + \cdots .$$

37. Let V be a (finite) geometric lattice of rank r, then $\mu(0,1) \neq 0$ and has sign $(-1)^r$.

§ 3. Permutations

Permutations play a role in combinatorics mainly as symmetries of combinatorial structures. Various properties of the symmetric group and its subgroups are important. We shall focus our attention on a few questions here which do not involve extensive algebraic machinery (for example, group representations). So we discuss simple enumeration problems, and two important tools in handling such problems: the cycle index polynomial and a coding of permutations due to M. Hall and A. and C. Rényi.

When we enumerate certain structures we sometimes want to consider some of them as not being essentially different (for example, they may be isomorphic). This usually means that a certain permutation group acts on the underlying set of structures, and we do not want to distinguish two structures if some element of this group maps one onto the other. Pólya's method, discussed in the second part of this section, is a beautiful way to handle such situations.

Many other combinatorial properties of permutation groups, only touched upon here, can be found, for example, in Wielandt's *Finite Permutation Groups*, Academic Press, 1966.

1. What is the number of conjugacy classes in the symmetric group S_n?

2. Determine the number of the permutations in S_n

(a) having no fixed points,

(b) consisting of one cycle only.

3. We select a permutation of $\{1,\ldots,n\}$ at random. What is the probability that the cycle containing 1 has length k? (In this and all subsequent problems all permutations are supposed to have an equal probability of selection.)

4. What is the probability that in a permutation of $\{1,\ldots,n\}$, chosen at random, 1 and 2 belong to the same cycle?

5. We choose a permutation of $\{1,\ldots,n\}$ at random. What is the expected number of cycles?

6. We have n savings boxes, with different keys. Someone locks the boxes and throws the keys into them at random, putting one key in each box. We break open k boxes. What is the probability that we are now able to open all the rest? (In mathematical form: We choose a permutation π of $\{1,\ldots,n\}$ at random. What is the probability that the cycles of π containing $1,\ldots,k$ cover all points?)

*

7. Denote by $p_n(x_1,\ldots,x_n)$ the cycle index of S_n, the symmetric group on n elements, and put $p_0 = 1$. Prove that

$$\sum_{n=0}^{\infty} p_n(x_1,\ldots,x_n)y^n = \exp\left(x_1 y + \frac{x_2}{2}y^2 + \ldots + \frac{x_k}{k}y^y + \ldots\right).$$

8. Determine the cycle index of

(a) the group of rotations of a regular n-gon;

(b) the group of permutations of nk elements which keep a given partition P into k n-element classes invariant.

Generalize these results.

9. Two permutation groups (acting effectively) have the same cycle index. Are they necessarily isomorphic?

10. Denote by $q_n(x_1,\ldots,x_n)$ the cycle index of the alternating group A_n ($q_0 = 2$). Determine the generating function

$$\sum_{n=0}^{\infty} q_n(x_1,\ldots,x_n)y^n.$$

11. Prove that if all but a finite number of the numbers $x_1, x_2, \ldots$ are equal to 1, then

$$\lim_{n\to\infty} p_n(x_1,\ldots,x_n)$$

exists, and determine its value.

12. (a) Use 3.7 to give a new proof of the identity in 1.7(b):

$$\sum_{k=0}^{n} \begin{bmatrix} n \\ k \end{bmatrix} x^k = x(x+1)\ldots(x+n-1).$$

(b) Use this identity to give a new proof of 3.5.

13. What is the exponential generating function of the number g_n of 2-regular simple graphs on n given points?

14. In how many ways can one partition the $2n$ letters $\{1,1,2,2,\ldots,n,n\}$ into n pairs if

(a) the order of elements in a pair does not count,

(b) the order of elements in a pair does count? (Determine the generating function.)

15. Give a new solution of 3.2, based on the cycle index of S_n.

*

16. Choose a permutation π at random. Denote by $\bar{\pi}(i)$ the number of those integers $1 \leq j \leq i$ with $\pi(j) \geq \pi(i)$. Then $\bar{\pi}(1), \ldots, \bar{\pi}(n)$ are random variables. Prove that they are independent.

17. (a) What is the expected number of (on-going) records at a long jump competition, in which each of the n competitors has one jump, these jumps are all different and the starting order is drawn at random?

(b) What is the probability that there are exactly k records?

18*. Imagine a long jump competition, in which no two of the n competitors jump the same distance. They jump in a drawn order. One can bet according to the following rule: After a jump, one can say "This is best of all." We arrive on the scene after the kth jump.

(a) What is the probability p_k that, following the best strategy, we will guess the winner correctly (information about the previous results is available)?

(b) What strategy should we follow in order to have the largest probability of winning (provided we arrive in time)?

19. What is the number of permutations π of $\{1, \ldots, n\}$ so that there is no triple $i < j < k$ with
$$\pi(j) < \pi(i) < \pi(k)?$$

20. Let $a_{n,k}$ denote the number of those permutations π of $\{1, \ldots, n\}$ which have k inversions (pairs $i < j$ with $\pi(i) > \pi(j)$). Prove that
$$\sum_{k=0}^{\binom{n}{2}} a_{n,k} x^k = (1+x)(1+x+x^2) \ldots (1+x+\ldots+x^{n-1}).$$

21. Along a speed track there are some gas-stations. The total amount of gasoline available in them is equal to what our car (which has a very large tank) needs for going around the track. Prove that there is a gas-station such that if we start there with an empty tank, we shall be able to go around the track without running out of gasoline.

22. (a) Let $x_1, \ldots, x_n$ be real numbers. For each permutation π of $\{1, \ldots, n\}$, define
$$a(\pi) = \max\{0, x_{\pi(1)}, x_{\pi(1)} + x_{\pi(2)}, \ldots, x_{\pi(1)} + \ldots + x_{\pi(n)}\}.$$

Also consider the cycles $C_1, \ldots, C_k$ of π and set

$$b(\pi) = \sum_{l=1}^{k} \max \left(0, \sum_{j \in C_l} x_j \right).$$

Then $\{a(\varrho) : \varrho \in S_n\}$ and $\{b(\pi) : \pi \in S_n\}$ equal *as collections*.

(b) m boys and m girls form "rings" for a dance at random (there may be pairs or even singles: two and one-element "rings"). Prove that the probability that in each "ring" the number of boys is equal to the number of girls is $\frac{1}{m+1}$.

*

23. (a) How many convex k-gons can be formed from the vertices of a regular n-gon, if two k-gons are considered essentially different when they do not arise from each other by rotation?

(b) What is the number of k-colorations of the vertices of a regular n-gon, if two k-colorations are not considered as essentially different when they arise by rotation from each other?

24. (Burnside's Lemma) Let Γ be a group of permutations of $\{1, \ldots, n\}$ with k orbits. Then the average number of fixed points of permutations in Γ is k.

25. Let Γ be a permutation group acting on a set Ω. Suppose that a "weight" $w(x)$ is associated with each $x \in \Omega$ such that $w(x)$ is invariant under Γ, i.e. for any given orbit Θ of Γ, $w(x)$ takes the same value $w(\Theta)$ for all $x \in \Theta$. Prove that

$$\sum_{\Theta} w(\Theta) = \frac{1}{|\Gamma|} \sum_{\pi \in \Gamma} \sum_{\pi(x)=x} w(x).$$

26. (Pólya–Redfield Enumeration Method I) Let Γ be a permutation group with cycle index $F(x_1, \ldots, x_n)$ on a set D and let R be another set. We say that the mappings f, $g : D \to R$ are *essentially different* if no $\pi \in \Gamma$ can be found such that

$$\pi f = g.$$

What is the number of essentially different mappings of D into R?

27* . (Cont'd) What happens if a further permutation group Γ_1 is given on R and two mappings f and g are only considered essentially different if for no $\pi \in \Gamma$, $\varrho \in \Gamma_1$

$$\pi f = g\varrho?$$

(Express this number by the cycle indices F, G of Γ and Γ_1.)

28* . (Cont'd) Count the number of essentially different injections (one-to-one mappings) of D into R (under the actions of the given groups Γ and Γ_1).

29. (Pólya–Redfield Method II) Let D be a finite set, R arbitrary; let Γ be a permutation group acting on D with cycle index F. Assume that each $y \in R$ has an integral weight $w(y) \geq 0$. Denote by r_n the number of elements of R with weight n (suppose that this is finite) and set

$$r(x) = \sum_{n=0}^{\infty} r_n x^n.$$

Let a_n denote the number of essentially different mappings $f : D \to R$ (in the sense of 4.26) such that

$$(*) \qquad \sum_{x \in D} w(f(x)) = n.$$

Prove that

$$\sum_{n=0}^{\infty} a_n x^n = F(r(x), r(x^2), \ldots) .$$

30. Solve the problem of distributing k forints to n persons (problem 1.4(b)) using the above technique.

31. Give a new proof of the formula for the exponential generating function of the number of partitions of a set (1.11), using the Pólya–Redfield–De Bruijn method.

§ 4. Two classical enumeration problems in graph theory

I have long felt it a very surprising fact that the number of trees on n labelled points is n^{n-2}. The reason why this is surprising is that one does not see any direct way to enumerate trees and one would expect a less compact result. The difficulty and beauty of this problem have challenged several authors to find proofs, often through a well-chosen generalization of the problem. Some of these are given here; the reader may test his knowledge of much basic enumeration technique.

Another theory, with less general results — but quite ingenious methods — is the enumeration of 1-factors in graphs. The special case when the graphs are bipartite is equivalent to the problem of "restricted permutations", that is, the problem of enumerating those permutations of $\{1, \ldots, n\}$ which map each element i onto one of the elements of a specified subset A_i. The methods here also use a variety of enumeration techniques but they usually work for certain special cases only.

It is also interesting to note that the theory of enumeration of trees is important in solving the Kirchoff equations in electrical network theory and that enumeration of 1-factors plays an important role in ferromagnetism (see Seshu–Read, *Linear Graphs and Electrical Networks*, Addison–Wesley, 1961; E.W. Montroll, in: *Applied Combinatorial Mathematics*, E. F. Beckenbach, ed. Wiley, 1964; or LP, Chapter 8).

1. Let $v_1, \ldots, v_n$ be given points and $d_1, \ldots, d_n$ be given numbers such that $\sum_{i=1}^{n} d_i = 2n - 2$, $d_i \geq 1$. Prove that the number of trees on the set $\{v_1, \ldots, v_n\}$ in which v_i has degree d_i $(i = 1, \ldots, n)$ is

$$\frac{(n-2)!}{(d_1 - 1)! \ldots (d_n - 1)!} .$$

2. (Cayley Formula) Prove that the number of all trees on n points is n^{n-2}.

3. Let

$$p_n(x_1, \ldots, x_n) = \sum x_1^{d_T(v_1) - 1} \ldots x_n^{d_T(v_n) - 1}$$

where $d_T(v_i)$ denotes the degree of the point v_i in the tree T and the summation extends over all trees on $\{v_1, \ldots, v_n\}$. Prove, without using 4.1, that

$$p_n(x_1, \ldots, x_n) = (x_1 + \ldots + x_n)^{n-2}.$$

Deduce the Cayley formula from this.

4. Let $T_1, \ldots, T_r$ be trees on disjoint sets of points and $V = V(T_1) \cup \ldots \cup V(T_r)$. What is the number of trees on V containing $T_1, \ldots, T_r$?

5. Let T be a tree on points $v_1, \ldots, v_n$. Delete the endpoint having the least index and write down the index of its neighbor. Repeat this procedure with the resulting tree, until a tree with only one point remains. This associates a sequence of $n-1$ numbers with T, called the *Prüfer code* of T. Prove that

    (a) the Prüfer code of T uniquely characterizes T,

    (b) given any sequence $(a_1, \ldots, a_{n-1})$ such that $1 \leq a_i \leq n$, $a_{n-1} = n$, there is a (unique) tree with this Prüfer code.

    (c) Deduce the Cayley formula.

6. Denote by T_n the number of trees on points $v_1, \ldots, v_n$. Prove that

$$(*) \qquad\qquad T_n = \sum_{k=1}^{n-1} k \binom{n-2}{k-1} T_k T_{n-k}$$

and prove Cayley's formula from this identity.

7. (a) The exponential generating function

$$t(x) = \sum_{n=1}^{\infty} \frac{T_n}{(n-1)!} x^n$$

satisfies the functional identity

$$t(x) e^{-t(x)} = x.$$

    (b) Prove the Cayley formula from this identity.

8. What is the number of all trees on n points with exactly $n-l$ endpoints?

9. (a) Let G be a digraph without loops, $V(G) = \{v_1, \ldots, v_n\}$ and $E(G) = \{e_1, \ldots, e_m\}$. Let A be the point-edge incidence matrix of G, i.e. the matrix $A = (a_{ij})$ defined by

$$a_{ij} = \begin{cases} 1 & \text{if } v_i \text{ is the head of } e_j, \\ -1 & \text{if } v_i \text{ is the tail of } e_j, \\ 0 & \text{otherwise.} \end{cases}$$

Remove any row from A and let A_0 be the remaining matrix. Prove that the number $T(G)$ of spanning trees of G is $\det A_0 A_0^T$.

(b) What are the entries of $A_0 A_0^T$?

(c) Deduce the Cayley formula from this.

10. Let

$$p_G(x_1, \ldots, x_n) = \sum x_1^{d_T(v_1)} \ldots x_n^{d_T(v_n)}$$

where G is a graph on $\{v_1, \ldots, v_n\}$ and the summation extends over all spanning trees of G (cf. 4.3). Define

$$a_{ij}(x_1, \ldots, x_n) = \begin{cases} -x_i x_j & \text{if } v_i, v_j \text{ are adjacent,} \\ 0 & \text{if } i \neq j \text{ and } v_i, v_j \text{ are non-adjacent,} \\ x_i \cdot \displaystyle\sum_{v_\nu \in \Gamma(v_j)} x_\nu & \text{if } i = j; \end{cases}$$

and

$$D = (a_{ij})_{i=1\ j=1}^{n-1\,n-1}.$$

Prove that

$$p_G(x_1, \ldots, x_n) = \det D.$$

11. What is the number of trees on $\{v_1, \ldots, v_n; w_1, \ldots, w_m\}$ in which each edge joins a v_i to a w_j?

12. (a) The number $T(G)$ of spanning trees of a graph G is given by

$$T(G) = \sum (-1)^{n-r} T(\bar{G}[X_1]) \ldots T(\bar{G}[X_r])|X_1| \ldots |X_r| n^{r-2}$$

where the summation extends over all partitions $(X_1, \ldots, X_r)$ of $V(G)$ and $n = |V(G)|$.

Determine $T(G)$ when G is

(b) the complement of a graph consisting of q independent edges and $n - 2q$ isolated points,

(c) the complement of a graph consisting of q edges adjacent to a point v and $n - q - 1$ isolated points.

13. (a) What is the number of "binary plane trees", i.e. those trees on $2n$ points which are embedded in the plane, and each point has degree 1 or 3 with some endpoint specified as root? (Two plane trees are the "same" if there is an isomorphism between them which preserves the cyclic ordering of edges at each point.)

(b) What is the number of plane trees on n points rooted at an endpoint?

*

14. What is the number $E(n,k)$ of forests F on points $v_1,\ldots,v_n$ having k components and such that $v_1,\ldots,v_k$ belong to distinct components?

15. Let G be an acyclic digraph with a specified root a and suppose that G has a spanning arborescence rooted at a. What is the number of spanning arborescences of G?

16. Let G be a digraph without loops, $V(G)=\{v_1,\ldots,v_n\}$. Let a_{ij} be the number of edges joining v_i to v_j and let d_i^- be the indegree of v_i.

(a) The number of spanning arborescences of G rooted at v_n is equal to the determinant

$$\Delta(G) = \begin{vmatrix} d_1^- & -a_{12} & \cdots & -a_{1,n-1} \\ -a_{21} & d_2^- & & -a_{2,n-1} \\ \vdots & & & \vdots \\ -a_{n-1,1} & -a_{n-1,2} & \cdots & d_{n-1}^- \end{vmatrix}.$$

(b) Find a similar formula for the polynomial

$$\sum y_1^{d_T^+(v_1)} \ldots y_n^{d_T^+(v_n)},$$

where T ranges over all spanning arborescences of G rooted at v_n (cf. 4.10).

17*. Let n be odd and let π be a permutation of the n-element set V. The number of trees on V admitting π as an automorphism is

$$0 \qquad \text{if } \pi \text{ has no fixed points,}$$

$$k_1^{k_1-2} \prod_{i=2}^n \left(\sum_{d|i} dk_d \right)^{k_i-1} \left(\sum_{\substack{d|i \\ d \neq i}} dk_d \right) \qquad \text{if } \pi \text{ has a fixed point,}$$

where $k_i = k_i(\pi)$ is the number of i-cycles in π.

18. Let W_n denote the number of non-isomorphic trees on n points with a specified point called *root*. Prove that

$$2^n < W_n < 4^n \quad (n \geq 6).$$

19*. Set

$$w(x) = \sum_{n=1}^{\infty} W_n x^n.$$

(a) Prove the identity

$$w(x) = x(1-x)^{-W_1}(1-x^2)^{-W_2} \ldots (1-x^n)^{-W_n} \ldots .$$

(b) Deduce the following identities from both (a) and the Pólya–Redfield method:

$$w(x) = x \cdot \exp\left(\frac{w(x)}{1} + \frac{w(x^2)}{2} + \ldots\right)$$

$$w(x) = x \sum_{n=0}^{\infty} p_n(w(x), w(x^2), \ldots),$$

where p_n is the cycle index of the symmetric group S_n.

20**. (a) Show that $w(x)$ has a radius of convergence $0 < \tau < 1$ and has a unique singularity on the circle $|x| = \tau$, namely $x = \tau$. Moreover, $w(x)$ is an analytic function of $\sqrt{\tau - x}$ near this singularity.

(b) Prove that there is a positive constant c such that

$$W_n \sim cn^{-3/2}\tau^{-n}.$$

$$*$$

21. Let G be a bipartite graph with 2-coloration $\{U, V\}$, $U = \{u_1, \ldots, u_n\}$, $V = \{v_1, \ldots, v_n\}$ and let a_{ij} denote the number of (u_i, v_j)-edges in G. Set

$$A = (a_{ij})_{i=1\,j=1}^{n\ \ n}.$$

Then the number of 1-factors of G is per A.

22. Determine the number of 1-factors in the "ladder" graph shown in Fig. 1.

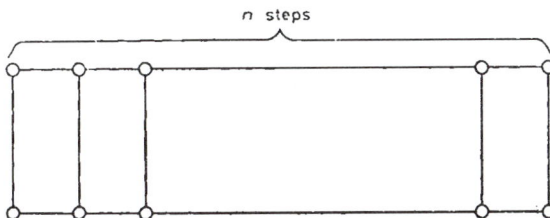

FIGURE 1

23. What is the number of 1-factors in $K_{n,n}$ minus n disjoint edges? (That is, in how many ways can a company consisting of n married couples dance if everyone dances but nobody dances with his spouse?)

24. Let B be any skew symmetric matrix. Then

$$\det B = (\text{Pf } B)^2.$$

25. (a) Let G be a simple oriented graph on $V(G) = \{1, \ldots, \}$ and B be defined by

$$B = (b_{ij}), \quad b_{ij} = \begin{cases} 1 & \text{if } (i,j) \in E(\overrightarrow{G}), \\ -1 & \text{if } (j,i) \in E(\overrightarrow{G}), \\ 0 & \text{otherwise.} \end{cases}$$

Prove that the number of 1-factors of G is at least $|\text{Pf } B|$.

(b) For every oriented graph, the following are equivalent:

(i) $|\text{Pf } B|$ is equal to the number of 1-factors of G;

(ii) every circuit alternating with respect to any 1-factor of G has an odd number of edges oriented in a given direction;

(iii) every circuit alternating with respect to some (fixed) 1-factor of G has an odd number of edges oriented in a given direction.

26. (Cont'd) Let G be a simple graph on $\{1, \ldots, n\}$. Orient each edge of G at random. Prove that the expectation of $\det B = (\text{Pf } B)^2$ is equal to the number of 1-factors of G.

27. Let G be a connected planar graph which has a 1-factor F. Prove that G has an orientation $\overrightarrow{G}$ such that every circuit which alternates with respect to F, contains an odd number of edges oriented in some given direction.

28. Determine the number of 1-factors of the "ladder" (4.19) from the preceding results.

29*. Let a_n be the number of ways to cover a $(2n) \times (2n)$ chessboard by dominoes. Prove the formulas

(a) $$a_n = 2^{2n^2} \prod_{k=1}^{n} \prod_{l=1}^{n} \left(\cos^2 \frac{k\pi}{2n+1} + \cos^2 \frac{l\pi}{2n+1} \right)$$

(b)

$$
a_n = 2^n \left| \begin{array}{cccccc}
\binom{2n}{0} & \binom{2n-2}{2} & \binom{2n-4}{4} & \cdots & & \\
0 & \binom{2n}{0} & \binom{2n-2}{2} & \cdots & & \\
\vdots & & \ddots & & & \\
0 & & \cdots & \binom{2n}{0} & \binom{2n-2}{2} & \cdots \\
\binom{2n-1}{1} & \binom{2n-3}{3} & \cdots & & & \\
0 & \binom{2n-1}{1} & \binom{2n-3}{3} & \cdots & & \\
\vdots & & \ddots & & & \\
0 & & \cdots & & \binom{2n-1}{1} & \binom{2n-3}{3} & \cdots
\end{array} \right|^2
\left. \begin{array}{c} \\ \\ \\ \end{array} \right\} \left\lfloor \frac{n-1}{2} \right\rfloor \text{ rows}
\left. \begin{array}{c} \\ \\ \\ \end{array} \right\} \left\lfloor \frac{n}{2} \right\rfloor \text{ rows}
$$

$$\underbrace{}_{n-1 \text{ columns}}$$

(c) Determine

$$\lim_{n \to \infty} \frac{\log a_n}{n^2}.$$

30. (a) Prove that the number of spanning trees of the $n \times n$ "chessboard" (i.e. the graph whose points are the squares and two of them are adjacent iff they share an edge) is equal to the number of ways of covering a $(2n-1) \times (2n-1)$ chessboard minus its upper left-hand corner by dominoes.

(b) Show that the number of spanning trees of any planar graph G can be expressed as the number of 1-factors of another planar graph.

31. What is the number of k-element matchings in the following bipartite graph:

$$V(G) = \{v_1, \ldots, v_n; u_1, \ldots, u_n\};$$
$$E(G) = \{(u_i, v_j) : i < j\}.$$

32. What is the number of those permutations π of $\{1, \ldots, n\}$ having the property that

$$|\pi(k) - k| \le 1 \quad (k = 1, \ldots, n)?$$

33*. Let a_n denote the number of permutations π of $\{1, \ldots, n\}$ such that

$$|\pi(k) - k| \le 2.$$

Determine

$$f(x) = \sum_{n=0}^{\infty} a_n x^n.$$

34. Determine the number $U_{n,p}$ of those permutations π of $\{1, 2, \ldots, n\}$ which satisfy

$$\pi(k) \le k + p - 1 \quad (k = 1, \ldots, n).$$

35. Let $a(n, p)$ denote the number of permutations π such that

$$|\pi(k) - k| \le n - p \quad (k = 1, \ldots, n).$$

Prove that if p is fixed, then $a(n, p)/(n - 2p)!$ is a polynomial in n.

36. Denote by p_n the number of those permutations π of $\{1, \ldots, 2n\}$ satisfying

$$|\pi(i) - i| < n.$$

Prove that

$$p_n = \sum_{k=0}^{n} S(n, k)^2 (k!)^2.$$

§ 5. Parity and duality

Parity considerations are applied in graph theory quite often. This loosely connected section contains various examples of parity problems, together with questions of different sorts to which these exercises lead. The first graph-theoretic result, Euler's solution of the Kőnigsberg bridge problem, is typical of the role of parity. Studying Euler trails will lead to another classical problem, finding the way out of a labyrinth.

The most important applications of parity considerations in graph theory are linear spaces of circuits and cuts. This is the starting point of matroid theory on the one hand, algebraic topology on the other. Basic results in algebraic topology such as Sperner's lemma (5.29) depend on parity arguments. We also note that certain existence results follow by showing that the number of objects in question is odd (for example, 5.20 and, more convincingly, 5.30). Linear spaces of circuits and cuts play a role in the characterization of planar graphs and in the study of the relationship between a planar graph and its dual.

Having finished with enumeration problems, we shall more frequently encounter the notion of a *good characterization* (see the introduction). Problems 5.3, 5.4, 5.6 are simple examples of well-characterizing necessary and sufficient conditions. A more important example is Kuratowski's criterion for planarity. Note that the MacLane and Whitney criteria are not "good characterizations" in this sense. (This shows at the same time that the name "good characterization" may be misleading; it does not have anything to do with the depth of the theorem.)

1.   (a) Is there a graph with degrees 3, 3, 3, 3, 5, 6, 6, 6, 6, 6, 6?

     (b) Is there a bipartite graph with degrees 3, 3, 3, 3, 3, 5, 6, 6, 6, 6, 6, 6, 6, 6?

     (c) Is there a simple graph with degrees 1, 1, 3, 3, 3, 3, 5, 6, 8, 9?

2. Which numbers occur as orders of k-regular simple graphs?

3. A graph is bipartite iff every circuit of it is even. Is the statement (with cycle instead of circuit) true for digraphs? Or strongly connected digraphs?

4. We associate a value $v(e)$ with each edge of a digraph G. This may be considered as the work needed to go from the tail of e to the head of e. To go conversely we need $-v(e)$ work. We want to find a "potential", i.e. a function $p(x)$ defined on $V(G)$ such that if $e = (x,y)$ then $v(e) = p(y) - p(x)$. Show that this is possible if and only if going around any circuit, the total work needed is 0.

5. The points of a strongly connected digraph G can be 2-colored so that each point is joined to a point of opposite color by a directed edge if and only if G has an even cycle.

6. If G is a weakly connected digraph with $d_G^-(x) = d_G^+(x)$ for each x then G has an Euler trail.

7. Let $G_{k,n}$ ($k \geq 2$) be defined as follows. The points of $G_{k,n}$ are vectors of dimension k composed from $1, \ldots, n$. The vectors $(a_1, \ldots, a_k)$ and $(b_1, \ldots, b_k)$ are joined by a directed edge if $a_2 = b_1$, $a_3 = b_2$, $\ldots$, $a_k = b_{k-1}$. Show that $G_{k,n}$ is

   (a) Eulerian;
   (b) Hamiltonian.

8. Show that we can associate 0 or 1 with each point of a 2^k-cycle ($k \geq 2$) in such a way that all arcs of length k of the cycle give different 01-sequences.

9. If G has a point of outdegree at least 3, then the number of its Euler trails is even. (Two Euler trails are not considered to be different if they give the same cyclic permutation of the edges.)

10. Let G be a digraph in which the outdegree of each point is equal to its indegree. Let T be a spanning inverse arborescence with root x_0. We start from x_0 and walk along the edges of G according to the following rule:

   $(*)$ From each point x we go to the next one along an edge (starting from x) not used previously. We use the edges of T only if we have no other choice.

   Show that we get stuck at x_0 and that, by this time we have traversed an Euler trail.

11. Let $V(G) = \{x_0, \ldots, x_{n-1}\}$, $d_G^+(x_i) = d_G^-(x_i) = d_i$. Show that the number of Euler trails in G is a multiple of

$$(d_0 - 1)! \ldots (d_{n-1} - 1)!$$

12. (Labyrinth Problem). Starting at a point x_0 we walk along the edges of a connected graph G according to the following rules:

   $(*)$ We never use the same edge twice in the same direction.

   $(**)$ Whenever we arrive at a point $x \neq x_0$ not previously visited, we mark the edge along which we entered x. We use the marked edge to leave x only if we must, i.e. if we have used all the other edges before.

Show that we get stuck at x_0 and that, by then, every edge has been traversed in both directions.

13. If G is a graph with even degrees, then G can be oriented in such a way that every point of the resulting digraph $\vec{G}$ has the same outdegree as indegree.

14. (a) A graph G has an Euler trail iff it is connected and has even degrees.

(b) If a connected graph G has $2k$ points of odd degree, then it is the union of k- edge-disjoint open trails ($k \geq 1$).

15. If a planar map has an Euler trail then the map can be drawn without lifting the pencil from the paper, without going over the same line twice and without crossing our line (at most touching it).

16. The number of subgraphs G' of a connected graph G which have even degrees and $V(G') = V(G)$ is equal to 2^{m-n+1} ($m = |E(G)|$, $n = |V(G)|$).

17*. (a) The set of points of any graph G can be partitioned into two classes V_1, V_2 such that both V_1 and V_2 span subgraphs with even degrees.

(b) For any graph G, $V(G)$ can be partitioned into two classes V_1, V_2 such that V_1 spans a subgraph with even degrees and V_2 spans a subgraph with odd degrees.

(c) Assume that at each vertex of a graph there is a light and a botton. At the beginning, all lights are on. Pushing a botton will change the status of the light at the vertex and at its neighbors. Show that one can push some bottons so that all lights will be turned off.

18*. (a) Let G be a graph and denote by $m_r(G)$ the number of r-element matchings of G. Express $m_1(G), m_2(G), \ldots$ in terms of $|V(G)| = n$ and $m_1(\bar{G})$, $m_2(\bar{G})$, $\ldots$.

(b) If $|V(G)|$ is even and $\bar{G}$ has an odd number of matchings then G has a 1-factor.

(c) G has an even number of 1-factors iff there is a non-empty set $S \subseteq V(G)$ such that every point is adjacent to an even number of points of S (e.g. Eulerian graphs).

19. If G a simple digraph and $h(G)$ denotes the number of Hamiltonian paths in G, then

$$h(G) \equiv h(\bar{G}) \qquad (\text{mod } 2).$$

Is this true for undirected graphs?

20. Any tournament has an odd number of Hamiltonian paths.

21*. Every edge e of a 3-regular graph G is contained in an even number of Hamiltonian circuits.

22. A 3-regular bipartite graph with at least 4 points has an even number of Hamiltonian circuits.

*

23. Let G be a connected planar map and G^* its dual. Show that G and G^* have the same number of spanning trees.

24. (Euler's Formula) show that if G is a connected planar map then it has $m - n + 2$ faces, where

$$m = |E(G)|, \quad n = |V(G)|.$$

25. (a) A simple planar graph with n points has at most $3n - 6$ edges.

(b) A simple triangle-free planar graph with n points has at most $2n - 4$ edges.

26. If a planar map has even degrees then the faces can be 2-colored in such a way that faces with a common edge on their boundary have different colors.

27. Let G be a simple 4-regular planar map.

(a) Show that G can be oriented in such a way that each point has two edges coming in, two edges going out, and these two pairs separate each other in the given embedding.

(b) Prove that the edges of G *cannot* be 2-colored in such a way that each point is incident with two red edges and two blue edges, and these two pairs separate each other in the given embedding.

28. (a) Can you draw a planar map in the interior of a pentagon so that the faces are triangles (except, of course, the outermost one), and each point has even degree?

(b) Suppose that we split the interior of a pentagon into triangles by a map such that all points except those on the pentagon have even degree. Which points of the pentagon have odd degree?

29. Let G be a planar map with triangular faces. Suppose that the points of G are 3-colored. Show that the number of faces which get all three colors is even.

30*. Let G be a planar map whose faces are triangles except for the exterior face $abcd$. Let $V(G) = V_1 \cup V_2$, $a, c \in V_1$, $b, d \in V_2$. Then either V_1 contains an (a, c)-path or V_2 contains a (b, d)-path.

*

31. Let V be the space of vectors $(a_1, \ldots, a_n)^T$ where the a_i's are from a field F. For $\mathbf{a} = (a_1, \ldots, a_n)^T$, $\mathbf{b} = (b_1, \ldots, b_n)^T$, define their inner product by $\mathbf{a}^T \mathbf{b} = a_1 b_1 + \ldots + a_n b_n$. Further, let M be a subspace of V and A a linear transformation of V. Decide whether the following assertions are true [†]

[†] This exercise really belongs to linear algebra. Most textbooks, however, do not consider inner products over finite fields which is the important case for us. (Note that,

(a) If A is non-singular, so is its transpose A^T.

(b) More generally, if A is non-singular, so is $A^T A$.

(c) Even more generally, $A^T A$ and A have the same null space.

(d) $M \cap M^\perp = \{0\}$ (where $M^\perp$ is the subspace orthogonal to M i.e. $M^\perp = \{\mathbf{a}: \mathbf{a}^T \mathbf{b} = 0$ for each $\mathbf{b} \in M\}$.

(e) $\langle M \cup M^\perp \rangle = V$ ($\langle X \rangle$ denotes the subspace generated by X).

(f) $\dim M + \dim M^\perp = n$.

(g) $(M^\perp)^\perp = M$.

32. (a) Let V be the n-dimensional vector space over $GF(2)$, and M a subspace of V. Show that
$$\mathbf{j} = (1, \ldots, 1)^T \in \langle M \cup M^\perp \rangle.$$

(b) Let A be a symmetric 0–1 matrix. Prove that the diagonal of A, considered as a row vector, is contained in the row space of A over $GF(2)$. Give a new proof of 5.17, using this fact.

33. Let $E(G) = \{e_1, \ldots, e_m\}$. Represent each subset of $E(G)$ by a 01-vector, where the jth entry is 1 iff e_j belongs to the subset. This way we identify the subsets of $E(G)$ with the elements of an m-dimensional vector space V_G over $GF(2)$.

(a) Determine the subspace U_G generated by the stars, and the orthogonal subspace W_G. Give new proofs of 5.3, 5.16 and 5.17, using these linear spaces.

(b) Prove that the decomposition in 5.17(a) is unique iff G has an odd number of spanning trees.

34. Let G be a 2-connected planar map. Show that the boundaries of all finite faces form a basis for the space W_G.

35*. Let G be a graph, $C_1, \ldots, C_f$ circuits of G and suppose that

(1) each edge of G is contained in at most two of $C_1, \ldots, C_f$;

(2) $C_1, \ldots, C_f$ constitute a basis of W_G.

Show that

(a) If $K = \sum_{i \in I} C_i$ is a circuit and
$$\sum_{i \in J} C_i \subseteq \bigcup_{i \in I} C_i$$
then $J \subseteq I$ ($I, J \subseteq \{1, \ldots, f\}$);

---

if F is the field of complex numbers, our inner product is not the same as usually introduced.) The reader will find it instructive to clear up the similarities and differences through this exercise. (All statements are true for the real field.)

(b) either all blocks of G are single circuits or there are two C_i's, say C_1 and C_2, such that $C_1 + C_2$ is a circuit;

(c) G is planar, and $C_1, \ldots, C_f$ are faces.

(d) (MacLane's Criterion) A graph G is planar iff W_G has a basis such that each edge belongs to at most two elements of it.

36*. (Whitney's Criterion) A connected graph G is planar iff there exist another graph G^* and a one-to-one mapping φ of $E(G)$ onto $E(G^*)$ such that whenever T is a spanning tree of G, $\varphi(E(G) - E(T))$ gives a spanning tree in G^* and vice versa.

37*. Let G be a minimal non-planar graph (i.e. each proper subgraph of G is planar) with all degrees at least 3. Then

(a) G is 3-connected,

(b) G contains a circuit with a chord,

(c) $G \cong K_5$ or $K_{3,3}$,

(d) (Kuratowski's Criterion) A graph is planar if and only if it contains no subdivision of K_5 or $K_{3,3}$.

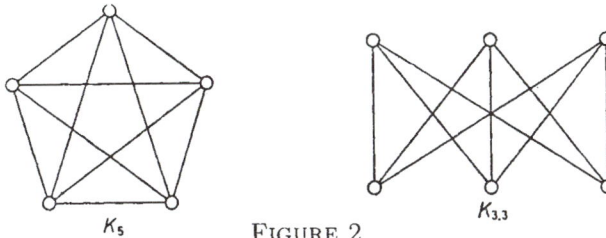

K_5 FIGURE 2 $K_{3,3}$

38*. If G is a simple planar graph, then G has an embedding in the plane such that all edges are straight segments.

§ 6. Connectivity

Several central concepts of graph theory belong to this chapter; separating sets of points and edges, connected components, blocks, paths, circuits, trees and forests. These concepts occur in most fields of graph theory and hence they belong to the basic techniques a graph theorist uses.

The theory of connectivity is one of the most developed branches of graph theory (together with factorization problems, which are discussed in the next chapter). One main reason for this is Menger's theorem, which links connectivity defined in terms of separation to connectivity defined in terms of connecting paths. This gives a "good characterization" of k-connected graphs. Closely related to Menger's theorem is the so-called Max-Flow-Min-Cut theorem, which plays a fundamental role in flow theory. This puts the duality between connection and separation in a very graphic form.

Problems involving connectivity between two specified points are usually settled without difficulty using Menger's theorem. On the other hand, connectivities between more than two points are more difficult to handle and are, to a large extent, independent of Menger's theorem. Such problems arise in the study of minimal k-connected graphs, multicommodity flows, safe communication networks, etc. Their solutions are difficult but some typical manipulations with cuts occur repeatedly and these may lead to ideas for a general approach.

Some of the strongest results in the field are structure theorems, which prove that certain classes of graphs can be constructed by repeated application of some simple transformation, e.g. 2-connected graphs by repeatedly attaching "ears" (cf. 6.27, 6.28, 6.33, 6.52, 6.53, 6.64).

1. Show that
$$c(G) + |E(G)| \geq |V(G)|$$
holds for every graph G.

2. (a) Let G_1, G_2 be two graphs with $V(G_1) = V(G_2)$. Show that
$$c(G_1) + c(G_2) \leq c(G_1 \cup G_2) + c(G_1 \cap G_2).$$

(b) This holds without assuming $V(G_1) = V(G_2)$.

3. Let $d_1 \leq \ldots \leq d_n$ be the degrees of the simple graph G and assume that $d_k \geq k$ for every $k \leq n - d_n - 1$. Then G is connected.

4. Show that $G_1 \times G_2$ is connected if and only if G_1, G_2 are connected and one of them contains an odd circuit.

5. A connected k-regular bipartite graph is 2-connected.

6. (a) Every connected graph G has a point whose removal does not disconnect the graph. What is the situation for strongly connected digraphs?

(b) Let G be a connected graph which does not contain a "cherry", i.e. two points of degree 1 with a common neighbor. Show that one can remove two adjacent points without disconnecting G.

(c) Let G be a connected graph which is neither a circuit nor complete. Show that one can remove two non-adjacent points without disconnecting G.

7. (a) Let T_1, T_2 be two spanning trees of a connected graph G. Prove that T_1 can be transformed into T_2 through a sequence of "intermediate" trees, each arising from the previous one by removing an edge and adding another.

(b) Suppose that G is 2-connected. Then to obtain T_1 from T_2 it suffices repeatedly to apply the following transformation: we remove the edge adjacent to some endpoint x of the tree and connect x to the rest by some other edge of G.

8. (a) Let G be a 2-connected graph on n points and $n_1 + n_2 = n$. Then $V(G)$ has a 2-coloration $\{A_1, A_2\}$ such that $|A_i| = n_i$ and A_1, A_2 induce connected subgraphs.

(b) Let G be a 2-connected non-bipartite graph on $2m$ points. Then $V(G)$ has a partition $\{A_1, A_2\}$ with $|A_i| = m$ such that the (A_1, A_2)-edges form a connected spanning subgraph.

9. A digraph G is strongly connected iff there is at least one edge leaving each set $X \subset V(G)$, $X \neq \emptyset$.

10. Let G be a digraph, $x, y \in V(G)$ and suppose that the edges of G are colored red, green and black. Then exactly one of the following two assertions holds:

(i) There is a black and red undirected (x,y)-path P in G such that all black edges of P are oriented from x to y,

(ii) There is a set $S \subset V(G)$, $x \in S$, $y \notin S$, such that no red edge connects S and $V(G) - S$ in any direction and no black edge goes from S to $V(G) - S$.

11. (a) Suppose that we can disconnect the strongly connected digraph G by removing $\leq k$ edges. Prove that we can also disconnect it by inverting $\leq k$ edges.

(b) Suppose that G is a digraph without isthmuses and we can get a strongly connected digraph from G by contracting at most k edges. Then we can also get a strongly connected graph by inverting at most k edges.

(c) If we can destroy all cycles of the digraph G having no loops by removing at most k edges, then we can also destroy them by inverting at most k edges.

12. A tournament is strongly connected if and only if it contains a Hamiltonian cycle.

13. A strongly connected tournament T on $n \geq 4$ points contains at least two points x such that $T - x$ is strongly connected.

*

14. Let G be a directed tree and $F \subseteq E(G)$. Prove that there is a point $x \in V(G)$ such that the edges of F incident with x have their head at x and the other edges incident with x have their tail at x.

15. Let G be a tree and let

$$\varphi : V(G) \rightarrow V(G)$$

be a mapping such that, whenever $(x,y) \in E(G)$, then

$$\varphi(x) = \varphi(y) \quad \text{or} \quad (\varphi(x), \varphi(y)) \in E(G).$$

Prove that φ has a fixed point or a fixed edge.

16. If the intersection of a set of subtrees of a tree G is non-empty, then it is a subtree.

17. (a) If G is a connected graph, then any two paths of maximum length have a point in common.

(b) If G is a tree, then all maximum paths of G have a point in common.

18. If $G_1,\ldots,G_k$ are subtrees of the tree G such that any two intersect, then they have a point in common. (This yields a new proof of part (b) of the preceding problem.)

19. Prove that, in a connected graph G, the distance $d(x,y)$ of the points x,y as well as the maximum length $D(x,y)$ of paths connecting x and y are metrics. That is

(a) $d(x,y) \geq 0$, with equality iff $x=y$,

(b) $d(x,y) = d(y,x)$,

(c) $d(x,y) + d(y,z) \geq d(x,z)$,

and similarly for $D(x,y)$.

20. If G is a tree and $p_1,\ldots,p_n,q_1,\ldots,q_{n+1}$ are points of it, then

$$\sum_{1\leq i<j\leq n} d(p_i,p_j) + \sum_{1\leq i<j\leq n+1} d(q_i,q_j) \leq \sum_{i=1}^{n}\sum_{j=1}^{n+1} d(p_i,q_j).$$

21. (a) Let $\tilde{d}(x) = \max_y d(x,y)$, where x, y are points of a tree. Show that the minimum of $\tilde{d}(x)$ is attained at one point (the *center* of G) or at two adjacent points (the *bicenter* of G).

(b) Show further that $\tilde{d}(x)$ is a convex function in the sense that, if y, z are neighbors of x, then

$$2\tilde{d}(x) \leq \tilde{d}(y) + \tilde{d}(z).$$

22. (a) Let $s(x) = \sum_y d(x,y)$, where x, y are points of a tree G. Show that $s(x)$ is *strictly convex* in the sense that, whenever x is a point and y, z are two neighbors of it, then

$$2s(x) < s(y) + s(z).$$

(b) Show that $s(x)$ assumes its minimum at one point (the *baricenter*) or at two adjacent points of G.

(c) Construct a tree which has a center and a baricenter whose distance apart is greater than 1000.

23. Determine those trees G on n points for which $\sum_{x,y} d(x,y)$ is maximal and, respectively, minimal.

24. In a tree on n points of diameter at least $2k-3$, there are at least $n-k$ paths of length k.

25. Given n cities, we want to build a connected telephone network between them. The expense $v(e)$ of a line e between any two cities is known, and we want to minimize the total expense. Thus, we want to find a tree on the given points such that the sum of the expenses of edges is minimal. Show that the following algorithm produces the desired result: At the ith step, choose a component G_i of the graph formed by the edges already selected, and select an edge connecting G_i to a point not in G_i with minimum expense. Stop when G_i contains every point.

26. (Cont'd) Show that if the edges have distinct expenses, then the optimal tree is unique.

*

27. Call two edges of a 2-edge-connected graph G *equivalent* if they are equal or their removal disconnects the graph. Show that

(a) this is an equivalence-relation,

(b) all edges of an equivalence-class lie on a circuit (which may contain other edges),

(c) removing the edges of an equivalence-class P, the components of the remaining graph are 2-edge-connected,

(d) contracting the components of $G-P$, we obtain a circuit.

28. Every 2-edge-connected graph G can be built up as follows: $G=G_1\cup\ldots\cup G_r$, where G_1 is a circuit and G_{i+1} is either a path which has exactly its endpoints in common with $G_1\cup\ldots\cup G_i$ or a circuit which has one point in common with $G_1\cup\ldots\cup G_i$ (Fig. 3). (Such a system of subgraphs is called an *ear-decomposition* of G.)

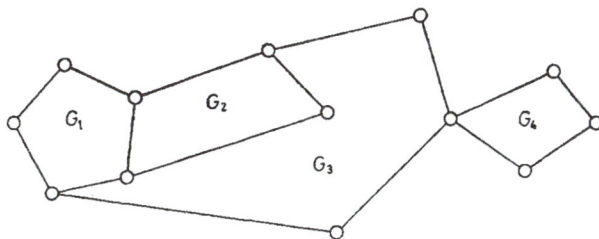

FIGURE 3

29. A graph G can be oriented so that the resulting digraph $\overrightarrow{G}$ is strongly connected iff G is 2-edge-connected.

30*. We say that an edge is *well fitted* with the circuit C if it lies on C or has no point in common with C.

Let $e_1,\ldots,e_k$ be independent edges of a 2-edge-connected graph G. Show that there is a circuit C into which all edges $e_1,\ldots,e_k$ are well fitted.

31. Show that "edges e, f are equal or lie on a circuit" is an equivalence relation.

32. The following properties of a simple graph G are equivalent:

(i) G is 2-connected,

(ii) any two points of G lie on a circuit,

(iii) any two edges of G lie on a circuit, G has no isolated points, and $|V(G)| \geq 3$.

33. (a) Let G be a 2-connected graph, which is not a circuit. Then there is a path P in G such that the inner points of the path P (if any) are of degree 2, and removing the edges and inner points of P the remaining graph is 2-connected.

(b) Formulate a characterization of 2-connected graphs in terms of an "ear-decomposition", analogous to 6.28.

34. Let p, q be two points of a 2-connected graph G. Show that G can be oriented in such a way that each edge is contained in a (p, q)-path.

35. Let G be a 2-connected graph. Then the following properties are equivalent:

(i) G is critically 2-connected,

(ii) no circuit of G has a chord.

36*. Let G be a critically 2-connected graph. Show that each circuit C of G has a point of degree 2.

37*. Let G be a critically 2-connected graph. Remove all points of degree 2 from G. Show that the resulting graph G' is a disconnected forest.

38. Construct a critically 2-connected graph G such that some $x \in V(G)$ is at distance at least 1000 from every point of degree 2.

*

39*. (Menger's Theorem). Let G be a digraph and a, $b \in V(G)$. Prove that

(a) there are k edge-disjoint (a, b)-paths iff G is k-edge-connected between a and b;

(b) there are k independent (a, b)-paths iff G is k-connected between a and b;

(c) analogous statements hold for undirected graphs.

40. Let A, B be disjoint subsets of $V(G)$ and assume that any set X which has a point from each (A, B)-path has at least k elements. Show that there are k vertex-disjoint (A, B)-paths.

41. Formulate and prove a common generalization of the undirected vertex-connectivity version of Menger's theorem 6.39b and 6.40.

42*. Let $B \subset V(G)$ and $a \in V(G) - B$; suppose that we are given k independent (a, B)-paths $P_1, \ldots, P_k$ and $k + r$ independent (a, B)-paths $Q_1, \ldots, Q_{k+r}$. Show that there are $1 \leq j_1 < \ldots < j_r \leq k + r$ and (a, B)-paths $R_1, \ldots, R_k$, R_i ending in the same points as P_i, such that $R_1, \ldots, R_k$, $Q_{j_1}, \ldots, Q_{j_r}$ are independent.

43. Let a, b, c be distinct points of G. Assume that there are k independent (a, b)-paths $P_1, \ldots, P_k$ and a (b, c)-path P_0 independent of them. Also, there are k (other) independent (a, b)-paths $Q_1, \ldots, Q_k$ and an (a, c)-path Q_0 independent of them. Show that there are $k + 1$ independent (c, b)-paths in G. (Try to avoid the use of Menger's theorem.)

44. Without using Menger's theorem, describe every graph G having two non-adjacent points a, b such that

(i) there are at most k independent (a,b)-paths in G, and

(ii) if any new edge is added to G, there will be at least $k+1$ independent (a,b)-paths in the resulting graph.

Deduce the undirected independent-path version of Menger's theorem from this.

45. Let G be a digraph and a, $b \in V(G)$. Decide whether the following statements are true:

(i) If any two paths which connect one of a and b to the other have a common edge, then there is an edge which is contained in all of these paths.

(ii) If G is k-edge-connected between a and b, then it is k-edge-connected between b and a.

(iii) If G is k-edge-connected between a and b and also between b and a, then there are k (a,b)-paths $P_1, \ldots, P_k$ and k (b,a)-paths $Q_1, \ldots, Q_k$ which are mutually edge-disjoint.

46. Assume that each point $x \neq a$, b of G has the same indegree and outdegree and that $d_G^+(a) - d_G^-(a) = k > 0$. Prove that there are k edge-disjoint (a,b)-paths in G.

47. If the outdegree of each point of G is equal to its indegree, then (i), (ii) and (iii) of 6.45 are true.

                              *

48. Let G be a graph and X, Y, $Z \subseteq V(G)$. Establish the inequalities:

(a) $\delta_G(X \cup Y) + \delta_G(X \cap Y) \leq \delta_G(X) + \delta_G(Y)$;

(b) $\delta_G(X - Y) + \delta_G(Y - X) \leq \delta_G(X) + \delta_G(Y)$;

(c) $\delta_G(X - Y - Z) + \delta_G(Y - Z - X) + \delta_G(Z - X - Y) + \delta_G(X \cap Y \cap Z) \leq \delta_G(X) + \delta_G(Y) + \delta_G(Z)$.

49*. Prove that a critically k-edge-connected graph has a point of degree k.

50. Let G be a k-edge connected graph and let $F_1, \ldots, F_m$ be k-element cutsets of edges. Prove that $G - (F_1 \cup \ldots \cup F_m)$ has at most $2m$ connected components.

51*. Let G be an Eulerian graph, $x \in V(G)$ and suppose that G is k-edge-connected between any two points u, $v \neq x$. Then we can find two neighbors[†] y, z of x such that, if we remove two edges (x,y) and (x,z) but join y to z by a new edge, the resulting graph is still k-edge-connected between any two points u, $v \neq x$.

52*. Let G be a $2k$-edge-connected, $2k$-regular graph. Prove that G can be obtained by the following construction:

———————

† If x has only one neighbor, y and z may be identical; but this case is trivial.

I. We start with two points joined by $2k$ edges.

II. Having constructed a graph, we select k edges, subdivide them by one point each, and identify the new points.

53*. Prove that the assertion of 6.51 holds for non-Eulerian graphs as well, provided $k \geq 2$ and x has even degree.

54. (a) Let G be a graph and $\vec{G}$ an orientation of G. If $\vec{G}$ is k-edge-connected, then G is $2k$-connected.

(b)* Conversely, if G is $2k$-edge-connected, then it has an orientation $\vec{G}$ which is k-connected (generalization of 6.29).

55. Let α, β be integers, G a graph and a, a', b, b' four points in G. We want to find α (a,a')-paths $P_1, \ldots, P_\alpha$ and β (b,b')-paths $Q_1, \ldots, Q_\beta$ such that $P_1, \ldots, P_\alpha$, $Q_1, \ldots, Q_\beta$ are edge-disjoint. Show that the following conditions are necessary:
(∗) We need to remove at least
   α edges to separate a from a',
   β edges to separate b from b', and
   $\alpha + \beta$ edges to separate $\{a,b\}$ from $\{a',b'\}$ or $\{a,b'\}$ from $\{a',b\}$.
Show by an example that these conditions are not always sufficient.

56*. Let a, a', b, b' be points of an Eulerian graph G, and α, β positive even integers. Assume that condition (∗) formulated in the previous problem is satisfied.

(a) Show that G has an orientation $\vec{G}$ such that

$$d_{\vec{G}}^+(a') - d_{\vec{G}}^-(a') = -\alpha, \quad d_{\vec{G}}^+(b') - d_{\vec{G}}^-(b') = -\beta,$$
$$d_{\vec{G}}^+(a) - d_{\vec{G}}^-(a) = \alpha, \quad d_{\vec{G}}^+(b) - d_{\vec{G}}^-(b) = \beta,$$
$$d_{\vec{G}}^+(x) = d_{\vec{G}}^-(x) \quad \text{for} \quad x \in V(G) - \{a, b, a', b'\}.$$

(b) Prove that G contains α (a,a')-paths and β (b,b')-paths, all edge-disjoint.

57. Construct a 5-connected graph G and 4 points a, b, c, d of it such that each (a,b)-path has a point in common with each (c,d)-path.

*

58. Let A, B be k-element sets of points of a k-connected graph G, separating a and b. Show that the set C of those points in $A \cup B$ which are the endpoints of $(a, A \cup B)$-paths is a k-element set separating a and b.

59. Let G be a critically k-connected graph and H a k-connected subgraph of it. Show that H is critically k-connected.

60*. (a) Let G be a k-connected graph, A a k-element separating set of it, G_1 a component of $G - A$ and assume that A is chosen so that $|V(G_1)|$ is minimal. Show that for any k-element separating set B, either $V(G_1) \subseteq B$ or $V(G_1) \cap B = \emptyset$. Moreover, $|V(G_1)| \leq \frac{k}{2}$ in the former case.

(b) Every critically k-connected graph G has a point of degree k.

61*. Let G be a simple 3-connected graph, and assume that the endpoints of the edge e are both of degree at least 4. Then one of G/e, $G-e$ is 3-connected.

62. Let G be a critically 3-connected graph, and e an edge of it which connects two points of degree at least 4. Then G/e is a critically 3-connected graph.

63. If G is a critically 3-connected graph, then every circuit of G contains at least two points of degree 3.

64. Suppose that a 3-connected graph G is such that for every edge e, $G-e$ and G/e are not 3-connected. Show that $G \cong K_4$.

65. Construct a critically 3-connected graph and a point x in it such that every point at distance at most 1000 from x has degree at least 1000.

*

66. In a k-connected graph G, any k points are on a circuit.

67*. Let e_1, e_2, e_3 be independent edges of a 3-connected graph G. Then there is a circuit containing e_1, e_2, e_3 except when e_1, e_2, e_3 form a separating system.

68. Let a, b, $x_1,\ldots,x_k$ be distinct points of a $(k+1)$-connected graph G. Prove that there is an (a,b)-path which contains $x_1,\ldots,x_k$.

69. Show that a circuit C in a simple planar 3-connected map has exactly one bridge iff it is the boundary of a face. Consequently, 3-connected simple planar graphs have an essentially unique embedding in the plane.

70*. Every 3-connected graph G has a circuit C with exactly one bridge.

*

For the rest of this chapter, let G be a digraph with a two specified vertices a and b. We also assume that a value ("capacity") $v(e) \geq 0$ is associated with each edge e.

71. (Bottleneck Theorem) Let P and C run over all (a,b)-paths and (a,b)-cuts, respectively. Prove that

$$\max_P \min_{e \in P} v(e) = \min_C \max_{e \in C} v(e).$$

72. (Min-Path-Max-Potential Theorem) A *potential* φ is a function on $V(G)$ satisfying $(*)$ $\varphi(a)=0$, $\varphi(y)-\varphi(x) \leq v(x,y)$ $((x,y) \in E(G))$. We define the "length" $u(P)$ of an (a,b)-path P to be $\sum_{e \in E(P)} v(e)$. Prove that

$$\min_P u(P) = \max_\varphi \varphi(b).$$

73. Let f be an (a,b)-flow and C an (a,b)-cut. Show that the value of f is given by

$$w(f) = \sum_{e \in C} f(e) - \sum_{e \in C^*} f(e).$$

74. (Max-Flow-Min-Cut Theorem) Show that the maximum value of an (a,b)-flow f satisfying $f(e) \le v(e)$ is given by

$$\min_C \sum_{e \in C} v(e),$$

where C runs over all (a,b)-cuts.

75*. The preceding solution suggests the following algorithm to construct a maximum (a,b)-flow.

Let f_k be an (a,b)-flow and P_k an (undirected) (a,b)-path. Let A, B be the sets of edges of P_k directed toward a and b, respectively. Suppose that P_k is such that $f(e) > 0$ for $e \in A$ and $f(e) < v(e)$ for $e \in B$ (if no such path P_k exists we know f_k is optimal). Set

$$\varepsilon = \min_{e \in A, e' \in B} (f_k(e), v(e') - f_k(e')).$$

Increase the value $f_k(e)$ by ε for $e \in B$ and decrease it by ε for $e \in A$, to get a new flow f_{k+1}. Prove the following assertions:

(a) Even repeating the above procedure infinitely many times, the values of the flows obtained do not necessarily converge to the values of a maximum flow.

(b) If P_k is always chosen to be a path with minimum possible length then, repeating the above procedure, the length of P_k does not decrease.

(c) If P_k is always a path with minimum length, we get an optimum flow in at most n^3 steps ($n = |V(G)|$).

76. Suppose that there is a value $u(x)$ prescribed for each point x, and a value $v(e)$ prescribed for each edge e. When does a function $f(e)$ exist such that

$$0 \le f(e) \le v(e); \qquad \sum_{e=(a,y)} f(e) - \sum_{e=(y,a)} f(e) = u(a)$$

for every a?

77. (a) Assume that the capacities $v(e)$ are integers. Show that there is a maximum value (a,b)-flow with integral entries.

(b) Show that the Max-Flow-Min-Cut theorem is a consequence of the directed edge version of Menger's theorem 6.39.

78. Let f be an integral (a,b)-flow of value $w(f) = w_1 + \ldots + w_k$ ($w_i = 1, 2, \ldots$). Show that f is the sum of k integral (a,b)-flows $f_1, \ldots, f_k$, of values $w_1, \ldots, w_k$, respectively.

§ 7. Factors of graphs

Here we consider the general question: given a graph, does there exist a subgraph whose degrees meet certain prescribed conditions? The characteristic example of such a question is the problem of the existence of 1-factors (perfect matchings), the solution of which (the König–Hall theorem for bipartite graphs and Tutte's theorem for the general case) is an outstanding result making this probably the most developed field of graph theory.

In addition to the question of their existence, the problem of describing the structure of all factors is also of importance. We touch only briefly on this problem (cf. also § 4).

One question which can be handled with the help of factorization results is whether a given sequence of integers can be the degree sequence of a graph (digraph, simple graph). This leads in fact to factorization problems for the complete graph and the answers are, correspondingly, much simpler.

Factorization problems constitute a class of integer programming problems for which there is a satisfactory solution. The basic idea of the correspondence between these two fields will be given in § 13, for the general case of hypergraphs. However, we shall not be able to go into the details in this volume.

1. Let G be a graph without isolated points. Then $\nu(G) + \varrho(G) = |V(G)|$.

2. (König's Theorem) Let G be a bipartite graph. Then $\nu(G) = \tau(G)$, and $\varrho(G) = \alpha(G)$.

3. (Hungarian Method) Let G be a bipartite graph with 2-coloration $\{A, B\}$ and let M be a matching of G. Let A_1 and B_1 be the sets of points in A and B not covered by M. Form a maximal forest $F \subseteq G$ with the following properties:

$(*)$ each point x of F in B has degree 2 and one of the edges adjacent to x belongs to M;

$(**)$ each component of F contains a point of A_1.

Prove that M is a maximum matching iff no point of B_1 is adjacent to any point of F. Deduce König's theorem 7.2 from this. Derive an algorithm to find a maximum matching in a bipartite graph from this result.

4. (a) Let G be a bipartite graph with bipartition $\{A, B\}$. Suppose that every $X \subseteq A$ is adjacent to at least $|X|$ points of B. Prove that G has a matching which matches all the points of A with (certain) points of B. Find different proofs which use and, respectively, do not use König's theorem.

(b) When does a bipartite graph G have a 1-factor?

5. Let G be a bipartite graph with bicoloration $\{A, B\}$ and

$$\delta = \max_{X \subseteq A} \{|X| - |\Gamma_G(X)|\}.$$

Prove that

$$\nu(G) = |A| - \delta.$$

6. (a) Suppose that G is a bipartite graph with bicoloration $\{A, B\}$ and let $k \geq 0$ be a given integer such that

$(*)$ $\qquad\qquad\qquad\qquad\qquad |\Gamma_G(X)| \geq |X| + k$

holds for each $X \subseteq A$, $X \neq \emptyset$. Let X_1, X_2 be two sets that satisfy $(*)$ with equality and suppose that $X_1 \cap X_2 \neq \emptyset$. Show that $X_1 \cap X_2$ also gives equality in $(*)$.

(b) Show that G as defined in part (a) has a subgraph G_1 containing A such that

(1) $d_{G_1}(x) = k + 1$ for all $x \in A$,

(2) $|\Gamma_{G_1}(X)| \geq |X| + k$ for all $X \subseteq A$, $X \neq \emptyset$.

7. Show that for any bipartite graph G with bicoloration $\{A, B\}$, the following three statements are equivalent:

(i) G is connected and each edge of G is contained in a 1-factor,

(ii) G is not $\overline{K}_2$ and for each $x \in A$ and $y \in B$, $G - x - y$ has a 1-factor,

(iii) G is not $\overline{K}_2$, $|A| = |B|$ and for each $\emptyset \neq X \subset A$, $|\Gamma(X)| > |X|$.

(Such a graph is called an *elementary bipartite graph*.)

8. Prove that a bipartite graph G is elementary if and only if it can be written in the form

$$G = G_0 \cup P_1 \cup \ldots \cup P_k,$$

where G_0 consists of two points and an edge joining them and P_i is an odd path which joins two points of $G_0 \cup P_1 \cup \ldots \cup P_{i-1}$ in different color classes and has no other point in common with $G_0 \cup P_1 \cup \ldots \cup P_{i-1}$.

9. Let G be an elementary bipartite graph different from K_2 and assume that G does not remain elementary if we remove any edge of it. Show that it has a point of degree 2. Does every edge of it have an endpoint of degree 2?

10. A bipartite graph G with a maximum degree $r = d(G)$ is the union of r matchings (i.e. has chromatic index r).[†]

11. Let G be any bipartite graph and suppose that $k \geq 1$. Then G is the union of k edge-disjoint spanning subgraphs $G_1, \ldots, G_k$ such that

$$\left\lfloor \frac{d(x)}{k} \right\rfloor \leq d_{G_i}(x) \leq \left\lceil \frac{d(x)}{k} \right\rceil \qquad \text{for each } x \in V(G).$$

12. Let G be a bipartite graph with minimum degree r. Then G is the union of r disjoint edge-covers.

[†] In general, the chromatic index of a graph G is bounded by $\left\lfloor \frac{3}{2} d(G) \right\rfloor$ [Shannon] and by $d(G) + 1$, if G is simple [Vizing], [see B].

13. Determine the least number $r = r(n)$ with the property that every r-regular bipartite graph G with $2n$ points has a 1-factor such that each edge of this 1-factor has a parallel in G.

14. Let G be an r-regular bipartite graph $(r \geq 2)$ with 2-coloration $\{A, B\}$, where $A = \{a_1, \ldots, a_n\}$ and $B = \{b_1, \ldots, b_n\}$. Suppose that not all edges incident with a_1 are parallel. Let a point x be moving in $V(G)$ and transforming the graph as follows:

(1) x starts at a_1 and moves along any (a_1, b_i) to b_i at which time this edge is removed.

(2) If x is at b_i, which it has just entered from a_μ, then it moves to the *first* point a_ν such that $a_\mu \neq a_\nu$ and a_ν is adjacent to b_i, and the edge (a_ν, b_i) is doubled. (Note that G may have multiple edges and thus, a_μ may still be adjacent to b_i though one (a_μ, b_i)-edge is removed.)

(3) If x is at a_ν which it entered from b_i, then it moves over to the *first* neighbor $b_j \neq b_i$ of a_ν and an (a_ν, b_j)-edge is removed.

Prove that, in a finite time, the procedure terminates with the situation that $x = a_1$ and a_1 is adjacent to some b_i by r parallel edges (and to no other point). Use this procedure to get a 1-factor.

15. (a) Suppose that the simple bipartite graph G with bipartition $\{A, B\}$ has a 1-factor and every $x \in A$ has degree at least k. Show that G has at least $k!$ 1-factors.

(b) An elementary bipartite graph with n points and m edges has at least $m - n + 2$ 1-factors.

16. Let G be a bipartite graph with bipartition $\{A, B\}$ and $f(x) \geq 0$ an integer valued function on $V(G)$. Show that G has an f-factor if and only if

(i) $\sum_{x \in A} f(x) = \sum_{y \in B} f(y)$

and

(ii) $\sum_{x \in X} f(x) \leq m(X, Y) + \sum_{y \in B - Y} f(y)$

for all $X \subseteq A$, and $Y \subseteq B$, where $m(X, Y)$ is the number of edges connecting X to Y.

17. Let G be a bipartite graph with n points and m edges and maximum degree d. Show that G can be embedded *as an induced subgraph* in a d-regular bipartite graph with $2n - 2\lfloor \frac{m}{d} \rfloor$ points, but never with fewer points.

18. Let G be a simple bipartite graph with bipartition $\{A, B\}$ such that $|A| = |B| = n$, and with maximum degree $d < \frac{n}{2}$. Show that G can be embedded in a *simple* regular bipartite graph on the same set $V(G)$ of points with degree $2d$.

19. Let $A = (a_{ij})$ be a non-negative $n \times n$ matrix such that all row- and column-sums of A are 1 (such a matrix is called *doubly stochastic*). Prove that $\operatorname{per} A > 0$, i.e. the determinant of A has a non-zero expansion term.

20. (a) Let G be a simple bipartite graph with 2-coloration $\{A, B\}$, and set $A = \{a_1, \ldots, a_n\}$ and $B = \{b_1, \ldots, b_n\}$. Let $a_{ij} \neq 0$ iff $(a_i, b_j) \in E(G)$ and assume that the non-zero a_{ij}'s are algebraically independent transcendentals. Prove that G has a 1-factor iff $\det(a_{ij}) \neq 0$.

(b) Prove the König–Hall criterion (7.4a) for the existence of a 1-factor in a bipartite graph from this observation.

21. Let G be a connected graph with n points and $E(G) = \{e_1, \ldots, e_m\}$. Consider all vectors $(y_1, \ldots, y_m)$ such that

(1)
$$\sum_{e_i \ni v} y_i = 0$$

for each point v. What is the dimension of the space spanned by them?

*

22. Let G be a simple graph on $2n$ points with all degrees at least n. Show that G has a 1-factor.

23. Let F_0 be any matching of G. Then G has a maximum matching which covers all points covered by F_0.

24. (a) Construct a graph with a unique 1-factor having all its degrees at least k.

(b) Prove that a graph with a unique 1-factor has a cutting edge.

(c) If a simple graph G on $2n$ points has a unique 1-factor, then

$$|E(G)| \leq n^2.$$

25. If G is a graph without isolated points and with maximum degree d, then

$$\nu(G) \geq \frac{|V(G)|}{d+1}.$$

26*. Let G be a connected graph such that, for each $x \in V(G)$, $\nu(G - x) = \nu(G)$. Show that G is a factor-critical.

27*. (a) (Berge's Formula) Show that for any graph G,

$$\delta(G) \overset{\text{def}}{=} |V(G)| - 2\nu(G) = \max_{X \subseteq V(G)} \{c_1(G - X) - |X|\}.$$

(b) (Tutte's Theorem) A graph has a 1-factor if and only if $c_1(G - X) \leq |X|$ for every $X \subseteq V(G)$.

28. (a) Let G be a simple graph such that

(i) G has no 1-factor, but

(ii) joining any two non-adjacent points of G by an edge results in a graph with a 1-factor.

Show (without using Tutte's theorem) that G has the following structure: it has a set V_1 of points joined to every other point and the remaining points span disjoint complete graphs.

(b) Give a new proof of Tutte's theorem 7.27b based on (a).

29. (a) (Petersen's Theorem) Every 2-connected 3-regular graph has a 1-factor.

(b) Construct a simple 3-regular graph having no 1-factor.

30. Let G be a $(k-1)$-edge-connected, k-regular graph with an even number of points. Remove $k-1$ edges of G. Show that the remaining graph G' has a 1-factor.

31. Let G be a factor-critical graph with more than one point. Show that it can be expressed in the form

$$G = P_0 \cup P_1 \cup \ldots \cup P_k,$$

where P_0 is an odd circuit and P_{i+1} is either an odd path with both endpoints in $P_0 \cup P_1 \cup \ldots \cup P_i$ but no inner point in $P_0 \cup P_1 \cup \ldots \cup P_i$ or an odd circuit having exactly one point in common with $P_0 \cup P_1 \cup \ldots \cup P_i$ $(i = 0, \ldots, k-1)$.

32*. (The Gallai–Edmonds Structure Theorem) Let G be a graph. Denote by D_G the set of those points which are left uncovered by at least one maximum matching. Let A_G be the set of neighbors of D_G and $C_G = V(G) - A_G - D_G$. Prove:

(a) Removing a point x of A_G, the sets D_G and C_G do not change.

(b) Every component of the graph induced by D_G is factor-critical. The graph induced by C_G has a 1-factor.

(c) If M is any maximum matching, it contains a maximum matching of each component of the graph induced by D_G and a 1-factor of the graph induced by C_G.

(d) $\nu(G) = \frac{1}{2}\{|V(G)| - c(D_G) + |A_G|\}$.

(e) Tutte's theorem follows from this.

(f) Determine the sets A, C, D for the graph shown in Fig. 4.

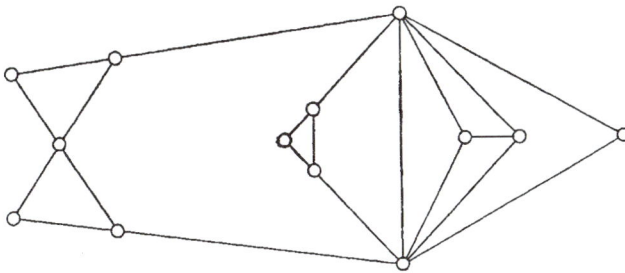

FIGURE 4

33. Let M be a matching of G and C a circuit of length $2k+1$ which contains k edges of M, but meets no other edge in M. Let G' be the graph obtained by contracting C and $M' = M - E(C)$. Then M is a maximum matching of G iff M' is a maximum matching of G'.

34*. (Edmonds' Matching Algorithm) (a) Let G be a graph, M be a matching of G and consider a maximal forest F with the following properties:

($*$) Each component of F contains exactly one point not covered by M, called a *root*.

($**$) Calling those points of F being at an odd distance from a root *inner points* and the other points, (including the roots) *outer points*, each inner point has degree 2 and one of the two edges incident with it belongs to M (Fig. 5).

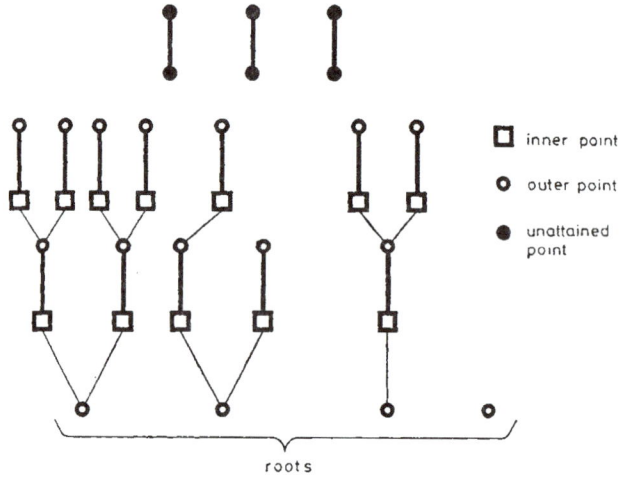

FIGURE 5

Then, if two outer points in distinct components are adjacent, M is not a maximum matching. Design an algorithm to find a maximum matchings based on this observation and the preceding problem.

(b) Test your algorithm on the graph in Fig. 4.

35. Assume that the k-regular graph G on $2m$ points has the property that any two odd circuits of it either intersect or are joined by an edge. Show that G has a 1-factor.

36. (a) If G has no 1-factor, then it has a point x such that each edge adjacent to x is contained in some maximum matching.

(b)* If G is a 2-connected graph with a 1-factor then it has a point such that each edge adjacent to this point occurs in some 1-factor. (Such a point is often said to be *totally covered*.)

37. The maximum size of a 2-matching of G is

$$|V(G)| - \max_{\substack{X \subseteq V(G) \\ X \text{ independent}}} \{|X| - |\Gamma_G(X)|\}.$$

38*. Every graph G has a maximum matching which is contained in a maximum 2-matching.

*

39. Let G be a connected $2d$-regular graph with an even number of edges. Prove that G has a d-factor.

40. Every $2d$-regular graph is the union of d 2-factors.

41. Which connected graphs G have a spanning subgraph F with

$$d_F(x) = \left\lfloor \frac{d(x)}{2} \right\rfloor \quad \text{or} \quad \left\lceil \frac{d(x)}{2} \right\rceil$$

for each x?

42. (a) Which connected graphs have a spanning subgraph with all degrees odd?

(b) Every 2-edge-connected graph G with all degrees at least 3 has a spanning subgraph in which each degree is positive and even.

43*. Let G be a graph and let us associate a non-negative integer $f(x)$ with each point of G. Construct a graph G' which has a 1-factor if and only if

(a) G has a perfect f-matching,

(b) G has an f-factor.

44. We associate a pair $(f(x), g(x))$ of non-negative integers with each point x of a digraph G. When does G have a spanning subgraph with indegree $f(x)$ and outdegree $g(x)$ at each point x?

45. Given a graph G, when can one orient its edges in such a way that the outdegree of every point x is equal to a prescribed number $f(x)$?

46. (a) Let G be a connected graph which contains a circuit. We associate a number $g(x)$ with each point x of G. Show that G has a spanning subgraph F whose degree differs from $g(x)$ at every point x.

(b) Let G be a tree, and let us associate an integer $g(x)$ with each point x. Show that exactly one of the following statements is true:

(i) G has an orientation with indegrees $g(x)$;

(ii) G has a spanning subgraph with all degrees different from $g(x)$.

*

47. Let $0 < d_1 \leq \ldots \leq d_n$ be integers. Show that there exists a tree with degrees $d_1, \ldots, d_n$ if and only if

$$d_1 + \ldots + d_n = 2n - 2.$$

48. Let $0 \leq d_1 \leq \ldots \leq d_n$ be integers. Show that they are the degrees of a graph without loops iff

(1) $d_1 + \ldots + d_n$ is even, and

(2) $d_n \leq d_1 + \ldots + d_{n-1}$.

49. Let $f_1, \ldots, f_n, g_1, \ldots, g_n \geq 0$ be integers. Prove that there exists a simple digraph on $\{v_1, \ldots, v_n\}$ without loops in which $d_H^+(v_i) = f_i$ and $d_H^-(v_i) = g_i$ iff

(a) $f_1 + \ldots + f_n = g_1 + \ldots + g_n$, and

(b) $\sum_{i \in I} f_i \leq \sum_{j \in J} g_j + |I|(n - |J|) - |I - J| \qquad (I, J \subseteq \{1, \ldots, n\})$.

Simplify the condition if $f_1 \leq \ldots \leq f_n$ and $g_1 \leq \ldots \leq g_n$.

50. Prove that there exists a simple graph with degrees $d_1 \leq d_2 \leq \ldots \leq d_n$ if and only if there exists one with degrees $d'_1, \ldots, d'_{n-1}$, where

$$d'_k = \begin{cases} d_k & \text{for } k = 1, \ldots, n - d_n - 1, \\ d_k - 1 & \text{for } k = n - d_n, \ldots, n - 1. \end{cases}$$

51. (a) Suppose that $d_i \geq 0$ $(i = 1, \ldots, n)$ and $d_1 + \ldots + d_n$ is even. Suppose that there is a simple digraph H without loops on $V = \{v_1, \ldots, v_n\}$ such that $d_H^+(v_i) = d_H^-(v_i) = d_i$ $(i = 1, \ldots, n)$. Then there exists a simple graph G on V with $d_G(v_i) = d_i$ $(i = 1, \ldots, n)$.

(b) Let $0 \leq d_1 \leq \ldots \leq d_n$. Prove that there exists a simple graph with degrees $d_1, \ldots, d_n$ if and only if

(1) $d_1 + \ldots + d_n$ is even, and

(2) $\sum_{i=n-k+1}^{n} d_i \leq k(k-1) + \sum_{i=1}^{n-k} \min(d_i, k)$

for each $1 \leq k \leq n$.

52. Let $0 \leq d_1 \leq d_2 \leq \ldots \leq d_n$ be given integers. Show that they are the degrees of a simple connected graph if and only if

(1) they satisfy the conditions of 7.51b, and moreover

(2) $d_1 > 0$, and

(3) $\sum_{i=1}^{n} d_i \geq 2(n-1)$.

53*. Let $d_1, \ldots, d_n$ be given integers. Prove that they are the degrees of a simple graph with a 1-factor if and only if $d_1, \ldots, d_n$ as well as $d_1 - 1, \ldots, d_n - 1$ are the degrees of some simple graph.

§ 8. Independent sets of points

The concept of an independent set of points is analogous to, and sounds simpler than the concept of an independent set of edges. However, it is more difficult to handle and much less is known. There is no "good" way to determine the maximum number of independent points; there are some fundamental reasons for this (see the work of Cook, Karp and Levin quoted in the introduction).

The problem of considering graphs critical with respect to independence number has proved easier to tackle. These graphs have many nice properties and, extending problem 8.25, it is even possible to find a classification of them. The quantity and depth of results concerning α-critical graphs is considerably larger than those concerning chromatic-critical graphs (see the next section).

A related problem is that of the problem of games on graphs, which is a very general setting for game theory. The reader is advised to translate the games of chess, tic-tac-toe etc. into games on directed graphs as in problem 8.8, and design games to which the surprising results of these problems can be applied (e.g. chess with "pass").

1. A graph with all degrees at most d satisfies

$$\alpha(G) \geq \frac{|V(G)|}{d+1}.$$

2. The points of a graph G can be covered by not more than $\alpha(G)$ vertex-disjoint paths.

3. The points of a graph G can be covered by not more than $\alpha(G)$ disjoint circuits, edges and points.

4*. Let G be a digraph and S a subset of $V(G)$ such that there are disjoint (directed) paths starting from the points of S and covering $V(G)$. Show that S contains a subset S_0 with the same property such that

$$|S_0| \leq \alpha(G).$$

5. (a) Every symmetric digraph has a kernel.

(b) An acyclic digraph G has a unique kernel.

(c) If every cycle of G has even length, then G has a kernel.

6. Every tournament T has a point from which every other point can be reached on a (directed) path of length at most 2.

7. Every digraph G contains an independent set $S \subseteq V(G)$ such that every point can be reached from S by a (directed) path of length at most 2.

8. Let G be a digraph and let players A, B play the following game. A occupies a point and then, taking turns, they each occupy a point which is accessible from the opponents last move and still unoccupied. A player loses whenever he has no possible move on his turn.

(a) Suppose that G has no cycles. Show that A has a winning strategy and determine the set of points he can start with.

(b) Now drop the assumption that G is acyclic but assume that it has a point x_0 of indegree 0. Show that the opening player A still has a winning strategy.

(c) Now let G be an arbitrary digraph. Show that the second player has a winning strategy if and only if he has a winning strategy on each strong component of G.

9. (Cont'd) Let G be a graph (or, equivalently, a symmetric digraph). Show that the opening player has a winning strategy if and only if G has no 1-factor.

*

10. Let $T_1, \ldots, T_k$ be maximum independent sets in a graph G. Show that

$$|T_1 \cup \ldots \cup T_k| + |T_1 \cap \ldots \cap T_k| \geq 2\alpha(G).$$

11. Let $T_1, \ldots, T_k$ be maximum independent sets of G and X any independent set. Let

$$S = X \cap T_1 \cap \ldots \cap T_k.$$

Then

$$|\Gamma(S)| - |S| \leq |\Gamma(X)| - |X|.$$

12. Let G be an α-critical graph without isolated points. Then every point x is contained in some maximum independent set, but not in all. If x, y are two points which do not constitute a component of G, then there is a maximum independent set containing exactly one of them. If x, y are adjacent then there is another maximum independent set missing both of them.

13. Substituting a complete graph for each point of an α-critical graph, we get an α-critical graph.

14. Every graph is an induced subgraph of an α-critical graph.

15. (a) Find infinitely many connected r-regular α-critical graphs $(r \geq 2)$.

(b) Which Platonic bodies are α-critical?

16. Which bipartite graphs are α-critical?

17*. Any two adjacent edges of an α-critical graph are contained in a chordless odd circuit.

18. An α-critical graph has no cutset spanning a complete graph (in particular, it has no cutpoint).

19*. (a) Let G_1, G_2 be connected α-critical graphs other than K_2. Split a point x of G_1 into two non-isolated points x_1 and x_2, remove an edge (y_1, y_2) of G_2, and identify x_i with y_i. (So two triangles yield a pentagon.) Show that the resulting graph is α-critical.

(b) Show further that every connected but not 3-connected α-critical graph arises in this way.

20. An α-critical graph G without isolated points has at least $2\alpha(G)$ points.

21. In an α-critical graph $|\Gamma(X)| \geq |X|$ for any independent set X.

22. Let X be an independent set of an α-critical graph G and $x \in X$. Then

$$d_G(x) \leq |\Gamma(X)| - |X| + 1.$$

23. (a) The maximum degree of an α-critical graph on n points, without isolated points, is at most $n - 2\alpha(G) + 1$.

(b) The number of edges in an α-critical graph on n points is at most $\binom{n-\alpha(G)+1}{2}$.

24*. Which connected α-critical graphs G are regular of degree $|V(G)| - 2\alpha(G) + 1$?

25. Characterize the connected α-critical graphs with $|V(G)| - 2\alpha(G) = 0, 1, 2$.

26. Assume that $\alpha(G - x - y) = \alpha(G)$ for every $x, y \in V(G)$ and $|V(G)| = 2\alpha(G) + 1$. Show that G is an odd circuit.

27. Let G be a graph such that $\alpha(G - \{x, y, z\}) = \alpha(G)$ for all $x, y, z \in V(G)$. Show that either $|V(G)| \geq 2\alpha(G) + 3$ or $G \cong K_4$.

§ 9. Chromatic number

The chromatic number is the most famous graphical invariant; its fame being mainly due to the Four Color Conjecture, which asserts that all planar graphs are 4-colorable. This has been the most challenging problem of combinatorics for over a century and has contributed more to the development of the field than any other single problem. A computer-assisted proof of this conjecture was finally found by Appel and Haken in 1977. Although today chromatic number attracts attention for several other reasons too, many of which arise from applied mathematical fields such as operations research, attempts to find a simpler proof of the Four Color Theorem is still an important motivation of its investigation.

Despite its popularity, we know rather little about chromatic number. It is difficult to compute it; there is no good characterization of k-chromatic graphs except for the trivial case $k = 2$ (cf. the non-trivial but "non-good" characterizations given in problems 9.6, 9.9, 9.11, 9.15, 9.16); and chromatic critical graphs have much weaker properties than α-critical graphs. The most difficult proofs often concern counterexamples.

The collection of problems presented here is an attempt to cover some important ideas developed so far in this area. None of these have lead to a satisfactory theory of chromatic number as yet (in particular, none of them has enabled us to give a computer-free proof of the Four Color Conjecture), but they may become elements of a more advanced theory. The main ideas dealt with in this section are: graphs with small degrees have small chromatic number; attempts to generalize the idea of "potential" used for bipartite graphs in § 5; the consideration of critical and saturated graphs; perfect graphs; chromatic polynomials, and other algebraic approaches; and Kempe chaining.

1. If every point of a graph G has degree at most k, then $\chi(G) \leq k+1$.

2. Prove that for any graph G, we can find a partition $V(G) = V_1 \cup V_2$ (V_1, $V_2 \neq \emptyset$) such that
$$\chi(G[V_1]) + \chi(G[V_2]) = \chi(G).$$
Further, if G is not complete, we can find a partition $V(G) = V_1 \cup V_2$ with
$$\chi(V[G_1]) + \chi(G[V_2]) > \chi(G).$$

3. Establish the inequality $\chi(G_1 \cup G_2) \leq \chi(G_1) \cdot \chi(G_2)$.

4. Suppose that G has a (good) coloration in which each color occurs at least twice. Show that G has such a coloration with $\chi(G)$ colors.

5. (a) Suppose that $|V(G)| = n$ and $V(G)$ has a partition $\{V_1, \ldots, V_k\}$ such that, for each $1 \leq i < j \leq k$, there exists an $x \in V_i$ and a $y \in V_j$ which are non-adjacent. Then
$$\chi(G) \leq n - k + 1.$$
(b) Let G be a simple graph on n points. Then
$$\chi(G) + \chi(\overline{G}) \leq n + 1, \qquad \chi(G) \cdot \chi(\overline{G}) \geq n.$$

6. Show that the chromatic number of a graph G is equal to the least number m such that
$$\alpha(G \oplus K_m) = |V(G)|.$$
$(G \oplus K_m)$ is the Cartesian product of G and the complete m-graph).

7. (a) $\chi(G_1 \times G_2) \leq \min(\chi(G_1), \chi(G_2))$.

(b) $\chi(G \times G) = \chi(G)$.

(c) $\chi(G \times K_n) = \min(\chi(G), n)$.

(d)* Show that if G is connected and $\chi(G) > n$ then $G \times K_n$ has a unique n-coloration (up to permuting colors).

(e) How many n-colorations can a graph have?

8. Let S be a set of points. If $S \subseteq V(G)$, then any coloration of G induces a partition of S. Embed S into some graph G in such a way that the partitions of S induced by k-colorations of G ($k \geq 3$) are

(a) the partition of S into 1-element classes only (this makes sense, of course, only if $|S| \leq k$);

(b) all partitions except the previous one;

(c) the partition $\{S\}$ only;

(d) all partitions except the previous one;

(e)* a given set $\{P_1, \ldots, P_N\}$ of partitions.

*

9. If a digraph G has no path of length m then $\chi(G) \leq m$.

10. When can we color the points of a digraph G by any number of colors 1, 2, ... in such a way that every edge joins a point of color i to a point of color $i+1$ $(1 \leq i)$?

11*. Prove that G is k-colorable if and only if it can be oriented in such a way that for any circuit C of G and any given direction C, at least $\frac{|E(C)|}{k}$ edges of C are oriented in this direction.

*

12. (a) Let G be a connected graph and suppose that there is a set $C(x)$ of colors associated with each point x of G. Assume that $|C(x)| \geq d_G(x)$ for every x and that strict inequality holds for at least one x_0. Then we can find a (good) coloration of G which uses one of the prescribed colors at each point.

(b) Suppose again that a set $C(x)$ of colors is associated with each point x of graph G, but now let $|C(x)| = d_G(x)$ for each point and $C(a) \neq C(b)$ for some a, b. Further assume that G is 2-connected. Show that G admits a (good) coloration which uses an element of $C(x)$ to color x for each point x. (Note the important special case when G is a circuit; cf. the solution of 9.8d).

13*. (Brook's Theorem) (a) Show that if every point of G has degree at most k (≥ 3) and G is 3-connected but is not a complete $(k+1)$-graph, then $\chi(G) \leq k$ (i.e. in this case the condition that $C(x)$ is not the same at every point can be dropped).

(b) Show that the condition of 3-connectivity can be replaced by connectivity.

*

14*. (a) Let G be an infinite[†] graph such that all finite subgraphs of G are k-colorable, but if we join any pair of non-adjacent points by a new edge in G, it will have a finite subgraph which is not k-colorable. Prove that $V(G)$ has a partition such that two points are adjacent iff they belong to distinct classes.

(b) (Erdős–de Bruijn Theorem) Suppose that all finite subgraphs of an infinite graph G are k-colorable. Prove that G itself is k-colorable.

(c) For each point $v \in V(G)$, let T_v be a copy of the discrete topological space on points $1, \ldots, k$. Then the colorations of $V(G)$ with k colors can be considered as points of the topological product space $\underset{v \in V(G)}{\times} T_v$. Prove that the legitimate colorations of G form a closed subset. Give another proof of the Erdős–de Bruijn theorem.

15. Let $\mathcal{K} \neq \emptyset$ be a class of simple graphs satisfying the following conditions:

(i) If $G \in \mathcal{K}$ and G has a homomorphism into G', then $G' \in \mathcal{K}$.

[†] This exercise shows that problems concerning infinite graphs with finite chromatic number can usually be reduced to the finite case. This is why we include it here, cf. 14.19.

(ii) If G is a graph, a, b, $c \in V(G)$ are such that a, c are adjacent but b is not adjacent to either of them, and if $G + (a,b) \in \mathcal{K}$ and $G + (b,c) \in \mathcal{K}$, then $G \in \mathcal{K}$.

Prove that $\mathcal{K}$ consists of all non-k-colorable graphs for some $k \geq 0$.

16*. (Hajós' Construction) Consider the following operations on simple graphs:

(α) Addition of edges and/or points to the graph,

(β) Identification of two non-adjacent points (and cancellation of the resulting multiplicities of edges),

(γ) For two graphs G_1, G_2 and $(x_i, y_i \in E(G_i))$, removal of (x_i, y_i) $(i = 1, 2)$, addition of a new edge (y_1, y_2), and identification of x_1 and x_2.

Prove that these operations produce non-k-colorable graphs from non-k-colorable ones and that every non-k-colorable graph arises by the repetition of these operations, from the initial graph K_{k+1} (see in Fig. 6 how the 5-wheel is obtained from K_4's).

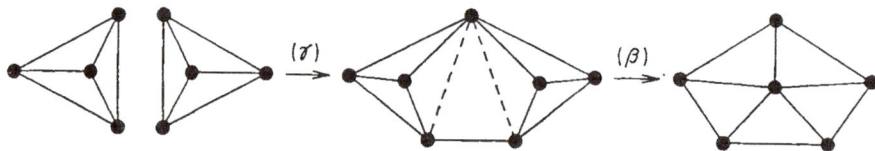

FIGURE 6

*

17. (a) Which graphs are critically 3-chromatic?

(b) Construct critically 4-chromatic graphs on $4n$ points in which the number of edges is at least n^2.

(c) Construct critically 6-chromatic graphs on $2n$ points in which each point has degree at least n.

18. We associate a new point x' with each point x of a χ-critical graph G, and join it to all neighbors of x in G. We take another new point y and join it to the points x' $(x \in V(G))$. Show that the resulting graph G' is χ-critical with $\chi(G') = \chi(G) + 1$.

19. (a) Is the graph in Fig. 7 an induced subgraph of some critically 4-chromatic graph?

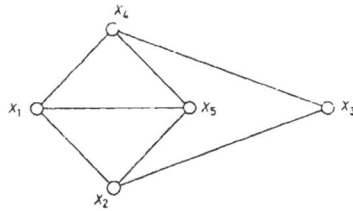

FIGURE 7

(b) A graph G_0 is a subgraph (or induced subgraph) of a critically $(k+1)$-chromatic graph G iff $\chi(G_0/e) \leq k$ for every edge e.

20. If we split a point of a critically $(k+1)$-chromatic graph G, the resulting graph G' is either k-chromatic or critically $(k+1)$-chromatic. For which values of k can the latter happen?

21. A critically $(k+1)$-chromatic graph is at least k-edge-connected.

22. Every χ-critical graph is 2-connected. Which ones are not 3-connected?

23. Show that if G is a χ-critical graph, $\chi(G) = k+1$ and S is a separating set of points of cardinality m, then the number of components of $G - S$ is not more than the number $B_{m,k}$ of partitions of m objects into at most k classes. Also show that this is best possible.

24. Let G be a critically $(k+1)$-chromatic graph. Prove that every pair of adjacent points can be connected by $k-1$ edge-disjoint chordless even paths.

$*$

25. Determine the chromatic number of

(a) the line-graph of K_n,

(b) its complement, and

(c) the line-graph of the symmetric digraph $\overrightarrow{K}_n$ obtained by replacing each edge of K_n by two, oppositely oriented edges. (You may use problem 13.21.)

26. (a) Let G be a digraph and $L(G)$ its line-graph. Prove that

$$\chi(L(G)) \geq \lceil \log_2 \chi(G) \rceil.$$

Moreover, the edges of G can be reoriented so that equality holds here.

(b)* If G is symmetric then

$$\chi(L(G)) \leq k \quad \text{if and only if} \quad \chi(G) \leq \binom{k}{\lfloor k/2 \rfloor}.$$

27. (a) Construct a k-chromatic graph without triangles.

(b)* Construct a k-chromatic graph without 3-, 4- and 5-circuits.

(c) Construct graphs having no odd circuits shorter than $2s+1$ and with chromatic number greater than k[†].

28. Let $I_1,\ldots,I_n$ be closed intervals on a line. Define a graph G on a set $\{x_1,\ldots,x_n\}$ by joining x_ν to x_μ iff $I_\nu \cap I_\mu \neq \emptyset$.

(a) Show that the resulting graph G has $\chi(G)=\omega(G)$.

(b) Show that the complement of this graph $\overline{G}$ also satisfies $\chi(\overline{G})=\omega(\overline{G})$.

(c) Show that every circuit C of G of length greater than 3 has a chord.

29. (a) Let G be a graph such that any circuit of G of length greater than 3 has a chord. Show that every (inclusionwise) minimal cutset S of G induces a complete graph.

(b) A graph G has the property that every circuit of length greater than 3 has a chord if and only if it can be represented as follows: Let $F_1,\ldots,F_n$ be subtrees of a tree T; set $V(G)=\{x_1,\ldots,x_n\}$ and join x_i to x_j if and only if $F_i \cap F_j \neq \emptyset$.

(c) If G has no chordless circuits of length greater than 3, then G has $\chi(G)=\omega(G)$ and the same holds for its complement.

30. Let G be a bipartite graph. Then $\chi(\overline{G})=\omega(\overline{G})$.

31. If G_1 and G_2 both satisfy $\chi(G_i)=\omega(G_i)$, then so does their strong product $G_1 \cdot G_2$.

32. (a) Let P be a partially ordered set and define a graph G on P by joining x, $y \in P$ if and only if $x \leq y$ or $y \leq x$.

(a) Show that $\chi(G)=\omega(G)$.

(b) Show the same for the complement of G.

33. Which of the previous exercises yield perfect graphs[‡]?

34. A graph G is perfect if and only if every induced subgraph G' of it contains an independent set which meets all maximum cliques of G'.

35*. Show that, substituting a perfect graph G_x for each point x of a perfect graph G produces a perfect graph G'.

*

36. Show that the chromatic polynomial of a graph G is a polynomial in the number λ of colors, and determine its degree.

[†] In fact, there exist graphs with arbitrarily large girth and chromatic number. [P. Erdős; see ES]

[‡] Problems 9.28–32 suggest that the complement of any perfect graph is perfect. This is in fact true, and a proof of it will follow from the theory of hypergraphs. See 13.55–57.

37. Show that
$$P_G(\lambda) = \sum_{T \subseteq E(G)} (-1)^{|T|} \lambda^{c(T)},$$
where $c(T) = c(V(G); T)$.

38. Show that
$$P_G(\lambda) = P_{G-e}(\lambda) - P_{G/e}(\lambda).$$

39. Determine the chromatic polynomials of complete graphs, trees, circuits and wheels.

40. Express the chromatic polynomial of G given the chromatic polynomials of
(a) its components, and
(b) its blocks.

41. Determine the chromatic polynomial of an interval graph.

42. If G has no loops then
$$P_G(x) = x^n - a_{n-1}x^{n-1} + a_{n-2}x^{n-2} - \ldots + (-1)^{n-1}a_1 x,$$
where $a_i \geq 0$. What is a_{n-1}? Find an interpretation for a_1. (Here and later n is the number of points.)

43. If G connected, then $a_i > 0$ $(i = 1, \ldots, n-1)$ and $a_{n-1} < a_{n-2} < \ldots a_{\lfloor n/2 \rfloor + 1}$.

44. Let x_0 be the largest real root of $P_G(\lambda)$. Show that $x_0 \leq n - 1$.

45. Show that $P_G(\lambda)$ has no root in the interval $(0, 1)$.

46. (a) Determine the meaning of the multiplicities of the roots 0 and 1.
(b) Prove that $|P_G(-1)|$ is the number of acyclic orientations of G.

47*. (a) Let G be a planar map with a quadrilateral face F. Let e_1, e_2 be the two diagonals of F, and $G_i = G + e_i$ $(i = 1, 2)$. Find a relation between $P_{G_1}(\lambda)$, $P_{G_2}(\lambda)$, $P_{G_1/e_1}(\lambda)$, $P_{G_2/e_2}(\lambda)$.
(b) Is there any linear relation
$$aP_G(\lambda) + bP_{G_1}(\lambda) + cP_{G_2}(\lambda) = 0$$
with some fixed integers a, b, c not all 0, which holds for every G and λ? Show that, except for at most five cases, there is no such relation even for a fixed value of λ.
(c) Set $\tau = \frac{\sqrt{5}+1}{2}$. Then
$$P_G(\tau + 1) = (\tau + 1)(P_{G_1}(\tau + 1) + P_{G_2}(\tau + 1)).$$

48. Show that $\tau + 1$ is not a root of any chromatic polynomial.

49*. (The Golden Ration Theorem) (a) If G is a planar triangulation with n points, then

$$P_G(\tau + 2) = (\tau + 2)\tau^{3n-10}P_G^2(\tau + 1).$$

(b)[†] For any planar graph G,

$$P_G(\tau + 2) > 0.$$

*

50. A planar graph G without loops has chromatic number at most 5.

51. Given a planar 3-regular 2-connected map, we can 4-color the faces (in such a way that adjacent faces get different colors) if and only if its chromatic index is 3.

52. Let G be a 3-regular 2-connected planar map. Then the faces of G are 4-colorable if and only if we can associate one of the numbers $+1$, -1 with each point in such a way that the sum of numbers associated with the points of any face is divisible by 3.

53. Assume that a planar map has a Hamiltonian circuit. Show that its faces are 4-colorable.

54*. (a) The connected planar graph G has all degrees at most 5, and at least one point of degree at most 4. Prove that G is 4-colorable.

(b) Prove that all 5-regular planar graphs are also 4-colorable.

55. We can two-color the points of every simple planar map G in such a way that each face receives both colors.

56. A planar triangulation is 3-colorable if and only if every point has even degree.

57. Let us draw some lines in the plane, such that no 3 are concurrent. Consider their points of intersection as the points of a graph and the segments between neighboring intersection points as edges. Prove that the resulting planar graph is 3-colorable.

§ 10. Extremal problems for graphs

Turán raised and solved the following problem in 1943: How many edges guarantee that a graph with this number of edges (and n given points) has a complete subgraph with k points? This type of question has been extensively investigated since then. Actually, any problem of the form "what is the extremal value of a given parameter for a given class of graph?" is an "extremal problem"; in this sense, almost everything in combinatorics could be included in this chapter.

[†] The Four Color theorem asserts $P_G(4) > 0$ for every planar graph. Since $\tau + 2 = 3,618\ldots$ is quite close to 4, $P_G(\tau + 2) > 0$ is interesting from this point of view.

However, we shall mainly restrict ourselves to the case when we are looking for the maximum number of edges among graphs on n points not containing certain subgraphs. Still, these subgraphs may only be prescribed up to isomorphism, up to homeomorphism (if, for example, circuits are excluded) or in some other way (if, for example, Hamiltonian cycles are excluded), and this leads to investigations with quite different methods.

One feature common to all of them, however, is that they can all be translated from questions involving the number of edges into ones involving instead the degrees of the points. There are several well-known tricks for interpreting information from one formulation in terms of the other.

Again we are able to select only a few characteristic problems from this large field. The reader will find other such problems in all graph theory books and a large variety of solved and unsolved problems in papers of P. Erdős.[†]

1. Suppose that the simple graph G on n points has more than $3(n-1)/2$ edges. Prove that it contains a Θ-graph, i.e. three independent paths connecting the same pair of points.

2. (a) If every point of a simple graph G has degree at least 3, then G contains a circuit with a chord.

(b) If G is a simple graph with $n \geq 4$ points and $2n-3$ edges, then G contains a circuit with a chord.

3. (a) Suppose that the simple graph G has all degrees at least 3. Prove that it contains a subdivision of K_4.

(b) Suppose that G is a simple graph with n points and $2n-2$ edges. Prove that G contains a subdivision of K_4.

4*. Let G be a simple graph with all degrees at least 3 and containing no two disjoint circuits. Prove that G is one of the following graphs: (i) K_5, (ii) a wheel, (iii) $K_{3,n-3}$ with any set of edges connecting points in the 3-element class added (Fig. 8).

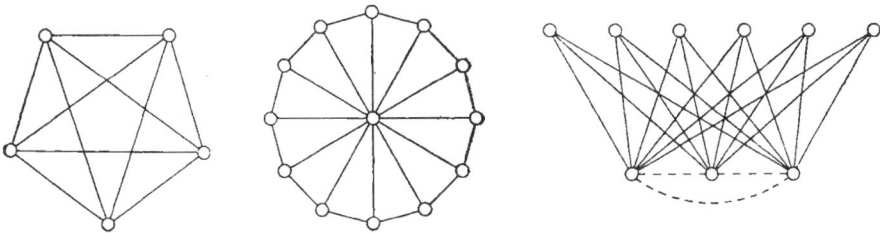

FIGURE 8

[†] See: P. Erdős, *The art of Counting, Selected Writings*, The MIT Press, Cambridge–London, 1973.

5. (a) Every simple 2-connected graph with $n \geq 5$ points and $2n-2$ edges contains a subdivision of $K_{2,3}$.

(b) Every simple 3-connected graph with $n \geq 6$ points and $3n-5$ edges contains a subdivision of $K_{3,3}$.

(c) How many edges guarantee the existence of a subdivision of $K_{2,3}$ and $K_{3,3}$ respectively, without the above connectivity assumptions?

6. (a) If a simple graph G can be contracted onto K_4, it also contains a subdivision of K_4.

(b) This is not true for K_5 instead of K_4. Find a 4-connected counterexample!

7. Let G be a simple graph with m edges and n points. Prove that it contains a connected (non-spanning) subgraph G_1 such that the neighbors of G_1 (i.e. those points of $V(G) - V(G_1)$ adjacent to some point of G_1) induce a graph with all degrees larger than $\frac{m}{n} - 1$.

8. Let G be a simple graph on n points, $m \geq 3$ and $|E(G)| \geq 2^{m-3}n$. Then G can be contracted onto K_m.

9. Let F be any graph with k edges and no loops or isolates. Prove that if G is a simple graph with

$$|E(G)| \geq 2^k |V(G)|,$$

then G contains a subdivision of F.

*

10. Determine the smallest 3-regular graphs with girths 4 and 5.

11. If G is an r-regular graph with girth g, then

$$|V(G)| \geq \begin{cases} 1 + r + r(r-1) + \ldots + r(r-1)^{\frac{g-3}{2}} & \text{if } g \text{ is odd,} \\ 2(1 + (r-1) + \ldots + (r-1)^{\frac{g}{2}-1}) & \text{if } g \text{ is even.} \end{cases}$$

12. Let $r \geq 2$ and $g \geq 2$ be given. Then there exists an r-regular graph with girth g.

13. Let G be an r-regular graph with girth at least g having the least number of points. Prove the following assertions:

(a) The diameter of G is at most g.

(b) The girth of G is g.

(c) $|V(G)| \leq \frac{r}{r-2}(r-1)^g$.

14. If G is an r-regular graph ($r \geq 2$) with girth at least g, and G has $2n$ points, where $n \geq 2r^g$, then G can be embedded in an $(r+1)$-regular graph with girth at least g on the same set of points.

15. (a) Construct a $(p+1)$-regular graph on $2(p^2+p+1)$ points with girth 6 (p prime).

(b)* Construct a $(p+1)$-regular graph on $2(p^3+p^2+p+1)$ points with girth 8 (cf. 10.11).

16. (a) Let G be a graph on n points such that $3 \le d_G(x) \le d$ for each point x. Then, to represent all circuits of G, we need at least $\frac{n+2}{d+1}$ points.

(b) If G has girth g and minimum degree at least 3 then to represent all circuits of G we need at least $\frac{3}{8}2^{g/2}$ points.

17*. Let G be a graph with all degrees at least 3 and girth $g \ge 3$. Prove that G contains $\frac{3}{8g}2^{g/2}$ disjoint circuits.

18. Let G be a graph, let ν denote the maximum number of disjoint circuits in G and τ the minimum number of points that represent all circuits of G. Prove that

(a) if $\nu = 1$, then $\tau \le 3$,

(b) $\tau \ge \nu \ge \frac{\tau}{4\log\tau}$,

(c) for infinitely many values of τ there are graphs G with

$$\nu \le \frac{4\tau}{\log\tau}.$$

*

19. Let G be a 2-connected graph, x, $y \in V(G)$ and assume that each point different from x, y has degree at least k. Prove that x, y can be connected by a path of length at least k.

20*. Let G be a simple graph such that, for each $X \subset V(G)$ with $|X| \le k$,

$$|\Gamma(X) - X| \ge 2|X| - 1.$$

Prove that G contains a path of length $3k-2$.

21. Let G be a simple graph on n points $x_1, \ldots, x_n$ with degrees $d_1 \le d_2 \le \ldots \le d_n$. Prove that G has a Hamiltonian circuit if it satisfies any one of the following:

(a) (Dirac's Condition) $d_1 \ge n/2$,

(b) (Pósa's Condition) $d_k \ge k+1$, for all $k < \frac{n}{2}$,

(c) (Bondy's Condition) $d_l \le l$, $d_k \le k$ implies that $d_l + d_k \ge n$ $(k \ne l)$,

(d) (Chvátal's Condition) $d_k \le k < \frac{n}{2}$ implies that $d_{n-k} \ge n-k$.

22. If G is a simple graph on n points with all degrees at least $\frac{n+q}{2}$, then any set F of q edges which form disjoint paths is contained in a Hamiltonian circuit.

23*. Let G be an r-regular simple graph on $2n+1$ points. Prove that G has a Hamiltonian circuit.

24. In the simple graph G on $n \ge 2$ points, every point has degree greater than $n/2$. Show that any two points of G can be joined by a Hamiltonian path.

25. Let G be a simple digraph on n points with all indegrees and outdegrees at least $n/2$. Prove that G contains a Hamiltonian cycle (cf. 10.21a).

26. Let G be a k-connected simple graph such that G contains no set of $k+1$ independent points ($k \geq 2$). Prove that G has a Hamiltonian circuit.

27. Let G be a simple graph on n points and with all degrees at least k. Then

(a) G contains a circuit of length at least $k+1$,

(b)* if G is 2-connected, it contains a circuit of length at least $2k$ or a Hamiltonian circuit.

28. Let G be a simple graph on n points with more than $\frac{k(n-1)}{2}$ edges ($k \geq 2$). Then G contains a circuit of length at least $k+1$.

29*. Let G be a 2-connected simple graph and let l be the maximum length of circuits in G. Then G contains no path of length $\left\lfloor \frac{l^2}{4} \right\rfloor + 1$.

*

30. The simple graph G on n points contains no triangles. Prove that

$$|E(G)| \leq \frac{n^2}{4}.$$

31. If G is a simple graph without triangles, then

$$|E(G)| \leq \alpha(G) \cdot \tau(G).$$

Prove the result in the preceding problem from this.

32. (a) If G is a simple k-regular graph on n points, then the total number of triangles in G and $\overline{G}$ is

$$\binom{n}{3} - \frac{n}{2}k(n - k - 1).$$

(b) Any simple graph G on n points and its complement $\overline{G}$ contain together at least

$$\frac{n(n - 1)(n - 5)}{24}$$

triangles.

33. The number of triangles in a graph with n points and m edges is at least

$$\frac{4m}{3n}\left(m - \frac{n^2}{4}\right).$$

34. (Turán's Theorem) Let G be a simple graph with mk points and more than $\binom{k}{2}m^2$ edges. Prove that G contains a complete $(k+1)$-graph. Generalize for the case $n = mk + r$, $1 \leq r \leq k - 1$.

35. Assume that G is a simple graph containing no complete $(k+1)$-graph. Prove that there is a simple k-chromatic graph H on $V(G)$ such that

$$d_H(x) \geq d_G(x) \qquad (x \in V(G)).$$

Deduce Turán's theorem from this.

36. (a) Every simple graph on n points with at least $\frac{n}{4}(1+\sqrt{4n-3})$ edges contains a quadrilateral (4-circuit).

(b) If $n = p^2 + p + 1$ (p prime), then there exists a 4-circuit-free simple graph on n points with $\frac{1}{2}p(p+1)^2 \sim \frac{1}{2}n^{3/2}$ edges.

37. Assume that the simple graph G with n points and m edges contains no $K_{r,r}$. Prove that

$$m < C \cdot n^{2-\frac{1}{r}},$$

where C depends only on r.

38*. (Erdős–Stone Theorem) (a) Let $\varepsilon > 0$ and k, $t \geq 1$ be given. Prove that if n is large enough, then every graph on n points and with all degrees at least $(1 - 1/k + \varepsilon)n$ contains $k+1$ disjoint t-sets such that any two points in different t-sets are adjacent.

(b) Let G be a simple graph with n points and $(1 - 1/k + \varepsilon)\frac{n^2}{2}$ edges. Then, if n is large enough, G contains $k+1$ disjoint t-sets such that any two points in different t-sets are adjacent.

(c) For each graph G_0 with at least one edge, denote by $M(n, G_0)$ the maximum number of edges in a simple graph on n points containing no subgraph isomorphic to G_0. Prove that

$$\lim_{n \to \infty} \frac{M(n, G_0)}{n^2} = \frac{1}{2}\left(1 - \frac{1}{\chi(G_0) - 1}\right).$$

39*. Let G_0 be a $(k+1)$-chromatic graph such that $G_0 - e$ is k-colorable for some edge e. Prove that the graph $H_{n,k}$ in the solution of 10.34 does not contain G_0 but if n is large enough, any other graph with this number of edges contains G_0 as a subgraph. So $H_{n,k}$ is the unique extremal graph.

40*. Let G be a simple graph on n points and denote by N_k the number of complete k-graphs in G. Prove that

(a) $\dfrac{N_{k+1}}{N_k} \geq \dfrac{1}{k^2 - 1}\left(k^2 \dfrac{N_k}{N_{k-1}} - n\right),$

(b) If $|E(G)| = \left(1 - \dfrac{1}{\vartheta}\right)\dfrac{n^2}{2}$, then $N_k \geq \dbinom{\vartheta}{k}\left(\dfrac{n}{\vartheta}\right)^k$ $(k \leq \vartheta + 1, \ \vartheta$ real$)$.

41. (a) A tournament T on n points contains at most $\frac{1}{4}\binom{n+1}{3}$ 3-cycles.

(b) A strongly connected tournament T on n points contains at least $n-2$ 3-cycles.

42. A strongly connected tournament T on n points contains at least $\binom{n-1}{2}$ cycles.

43. A tournament on n points contains at least one and at most $n!/2^{n/2}$ Hamiltonian paths.

44. (a) Every tournament T on n points contains a transitive subtournament on $\lfloor \log_2 n \rfloor + 1$ points.

(b) The number of k-element transitive subtournaments $(1 \leq k \leq \log_2 n)$ is at least

$$\prod_{j=0}^{k-1} \left(\frac{n+1}{2^j} - 1 \right).$$

(c) If T is strongly connected the number of k-element transitive subtournaments $(k \geq 3)$ is at most

$$\binom{n}{k} - \binom{n-2}{k-2}.$$

§ 11. Spectra of graphs and random walks

It is a classical approach to the study of a structure to describe it in terms of its numerical "invariants". With properly chosen invariants, the problems considered in the theory may be transformed into numerical or algebraic problems concerning the invariants and powerful methods of classical algebra may then be brought into play.

The invariants considered in the previous chapters (connectivity, chromatic number, chromatic polynomial) were defined combinatorially and their algebraic properties played only a minor role. On the other hand, introducing the spectrum of a graph as the spectrum of its adjacency matrix, we get an algebraically defined invariant system. The spectrum does not, in general, characterize the graph uniquely, though it does reflect many properties of it and more and more such connections are continually being discovered. Thus the introduction of spectra of graphs is not a universal method to solve all problems, but it does prove to be very powerful in some purely graph-theoretic situations (for example, in classifying strongly regular graphs).

The effective use of spectra in graph theoretic investigations depends on our ability to take two major steps. First we must be able to calculate the spectra of large classes of graphs (or, more generally, we must be able to translate graph theoretic information into spectral information) and, second, we must be able to deduce properties of graphs from their spectral properties. It is interesting that there is a large variety of methods for linking graphs with their spectra in both senses.

Some of the connections are sporadic; but certain eigenvalues are clearly linked to certain groups of combinatorial properties. Thus, the largest eigenvalue corresponds to density; the second largest, to "global connectivity" or conductance (expansion rate); some of the smallest eigenvalues, to chromatic number. The expansion rate, in turn, is closely related to various properties of the random walk on the graph, and we give a series of exercises about random walks.

*

Some properties of the spectra of graphs are special consequences of the Frobenius–Perron theory of non-negative matrices. We shall use without proof or reference the following facts (see, for example, F. R. Gantmacher, *Applications of the Theory of Matrices*, Interscience, 1959, or H. Minc, *Non-negative Matrices*, Wiley–Interscience, 1988). The maximum eigenvalue of any graph G is non-negative, and there is a non-negative eigenvector belonging to it. Moreover, if the graph is connected then the maximum eigenvalue has multiplicity 1 and there is a strictly positive eigenvector belonging to it. Any non-negative eigenvector of unit length belonging to the maximum eigenvalue maximizes the quadratic form $\mathbf{x}^T A_G \mathbf{x}$ on the unit sphere. The maximum eigenvalue has maximum absolute value among all eigenvalues.

It will be convenient to assume in this chapter that $V(G) = \{1, \ldots, n\}$, unless specified differently.

1. Determine the spectra and eigenvectors of a complete n-graph K_n, a star S_n, a complete bipartite graph $K_{n,m}$ and a circuit C_n on n points.

2. Suppose that we know the spectrum of a regular graph. Determine the spectrum of

(a) its complement $\overline{G}$.

(b) its line-graph $L(G)$.

(c) What is the spectrum of the Petersen-graph (Fig. 9)?

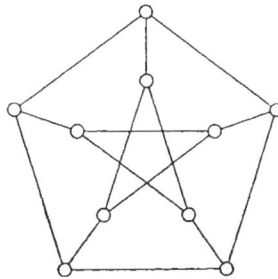

FIGURE 9

3. Let T be a forest and $e = (x, y) \in E(F)$. Then

$$p_T(\lambda) = p_{T-e}(\lambda) - p_{T-x-y}(\lambda).$$

4. Let T be a forest on n points and let a_k denote the number of k-element matchings in T. Then

$$p_T(\lambda) = \lambda^n - a_1\lambda^{n-2} + a_2\lambda^{n-4} - \ldots + (-1)^{\lfloor n/2 \rfloor} a_{\lfloor n/2 \rfloor} \lambda^{n-2\lfloor n/2 \rfloor}.$$

5. Let P_n denote a path with n points. Show that its characteristic polynomial can be written in either of the following forms:

(a) $p_{P_n}(\lambda) = \lambda^n - \binom{n-1}{1}\lambda^{n-2} + \binom{n-2}{2}\lambda^{n-4} - \binom{n-3}{3}\lambda^{n-6} \ldots,$

(b) $p_{P_n}(\lambda) = \dfrac{1}{\sqrt{\lambda^2 - 4}}\left(\left(\dfrac{\lambda + \sqrt{\lambda^2 - 4}}{2}\right)^{n+1} - \left(\dfrac{\lambda - \sqrt{\lambda^2 - 4}}{2}\right)^{n+1}\right).$

(c) Determine the eigenvalues of P_n.

(d) Determine the eigenvalues of a complete rooted D-ary tree of depth t.

6. Find infinitely many non-isomorphic pairs of trees having the same spectrum.

7. Suppose that the eigenvalues of G_1 and G_2 are given. Determine those of

(a) the Cartesian product of G_1 and G_2.

(b) the (strong) direct product of $G_1 \cdot G_2$ of G_1 and G_2.

8. Let G be a graph whose automorphism group contains a regular, commutative subgroup Γ. Let $\gamma_{x,y}$ be the element of Γ which moves x to y. Let χ be a multiplicative character of Γ. Prove that

$$\sum_{(1,i)\in E(G)} \chi(\gamma_{1,i})$$

is an eigenvalue of G.

9. Determine the eigenvalues of the n-cube Q_n.

10. Let G, H be two graphs on the point set $1, \varepsilon, \varepsilon^2, \ldots, \varepsilon^{p-1}$, where $\varepsilon = e^{2\pi i/p}$ and p is prime, both of which are invariant under the rotation by $2\pi/p$. Suppose that $G \cong H$. Then there is an integer t such that taking the tth power of each point maps G onto H isomorphically.

11. If the bipartite graph G has no 1-factor then 0 is an eigenvalue of it.

12. If the 3-regular graph G has a system of disjoint subgraphs, each isomorphic with the graph in Fig. 10, which covers all points, then 0 is an eigenvalue of G.

FIGURE 10

13. If λ_1 and λ_1' are the greatest eigenvalues of G and G', respectively, where G' is a subgraph of G, then $\lambda_1 \geq \lambda_1'$.

14. (a) Let λ_1 be the largest eigenvalue of the graph G and let d, D be the minimum and maximum degrees of G. Then

$$\max(d, \sqrt{D}) \leq \lambda_1 \leq D.$$

For which connected graphs is $\lambda_1 = D$?

(b) If G has n points and m edges, then

$$\frac{2m}{n} \leq \lambda_1 \leq \sqrt{\frac{2m(n-1)}{n}}.$$

(c) If G is a tree, then

$$\lambda_1 \leq 2\sqrt{D-1}.$$

15. Show that the line-graph of any graph G has all its eigenvalues at least -2, and that, if $|E(G)| > |V(G)|$, then its least eigenvalue is -2.

16*. Suppose that the eigenvalues of G are distinct. Show that every non-identity automorphism of G is of order 2.

17*. Suppose that the eigenvalues of the simple graph G are distinct and that its automorphism group is transitive. Show that $G \cong K_2$.

18. Let $\lambda_1 \geq \ldots \geq \lambda_n$ be the eigenvalues of the graph G. Show that

$$\lambda_{\tau(G)+1} \leq 0, \quad \lambda_{\alpha(G)} \geq 0, \quad \lambda_{\omega(G)} \geq -1, \quad \lambda_{n-\omega(G)+2} \leq -1.$$

19. (a) A connected graph G with maximum eigenvalue λ_1 is bipartite iff $-\lambda_1$ is an eigenvalue.

(b) A graph is bipartite iff its spectrum is symmetric with respect to the origin.

20. Let λ_1 be the maximum eigenvalue of G. Then

$$\chi(G) \le \lambda_1 + 1.$$

21*. (a) Let $\lambda_1 \ge \ldots \ge \lambda_n$ be the eigenvalues of G, and set $k = \chi(G)$. Then

$$\lambda_n + \ldots + \lambda_{n-k+2} \le -\lambda_1.$$

(b) Show that, if the graph G is k-colorable in such a way that two points are adjacent iff they have different colors, then we have equality in (a).

22. Let G be a d-regular graph on vertices $\{1, \ldots, n\}$, having eigenvalues $d = \lambda_1 \ge \lambda_2 \ge \ldots \ge \lambda_n$. Let $\mathbf{x}$ be any vector on its vertices with $\sum_i x_i = 0$. Prove that

$$(d - \lambda_2) \sum_{i=1}^{n} x_i^2 \le \sum_{(i,j) \in E} (x_i - x_j)^2 \le (d - \lambda_n) \sum_{i=1}^{n} x_i^2.$$

23. Prove that for every d-regular graph G with eigenvalues $d = \lambda_1 \ge \lambda_2 \ge \ldots \ge \lambda_n$, we have

$$\alpha(G) \le \frac{-n\lambda_n}{d - \lambda_n}.$$

24*. Suppose that G has k negative and l positive eigenvalues. Also suppose that G is the union of m edge-disjoint complete bipartite graphs. Then

$$m \ge \max(k, l).$$

25. The number of distinct eigenvalues of a connected graph is greater than its diameter.

26. Given a graph G, we can find a polynomial $f(x)$ such that

$$f(A_G) = J$$

if and only if G is regular and connected.

27. Given a projective plane, its *line-graph* L is formed as follows: Let U be the set of points of the plane and V the set of lines, then $V(L) = U \cup V$ and $u \in U$ is joined to $v \in V$ iff $u \in v$. Determine the spectrum of L.

28*. The simple graph G has the following properties:

    (i) each point has degree d,

    (ii) for each pair of points, there are $b(>0)$ other points adjacent to both.

(a) Determine the spectrum of G.

(b) Determine all such graphs with $b = 1$.

(c) Find infinitely many non-trivial examples of such graphs (a trivial example being a complete graph).

29. (a) Let G be a d-regular connected graph with diameter D and eigenvalues $d = \lambda_1 \geq \lambda_2 \geq \ldots \geq \lambda_n$. Prove that

$$d - \lambda_2 > \frac{1}{Dn}.$$

(b) If G is not bipartite then

$$d + \lambda_n > \frac{1}{2dDn}.$$

30. Let G be a d-regular graph with conductance Φ. Let **y** be any vector on $V(G) = \{1, \ldots, n\}$, and let $\bar{y}$ be a median of y, i.e., a value such that at most $n/2$ entries of y are smaller than $\bar{y}$ and at most $n/2$ entries are larger. (For odd n, $\bar{y}$ is the $((n+1)/2)^{\text{nd}}$ largest entry of y.) Then

$$\sum_{(i,j) \in E} |y_i - y_j| \geq \Phi \sum_i |y_i - \bar{y}|.$$

31. Let G be a d-regular graph with conductance Φ.
    (a) Prove that

$$\lambda_1 - \lambda_2 \leq 2\Phi.$$

    (b)* Prove that

$$\lambda_1 - \lambda_2 \geq \frac{\Phi^2}{4d}.$$

32. (a) Let G be a d-regular graph on $\{0, \ldots, n-1\}$ such that the mapping $i \mapsto i+1 \pmod{n}$ is an automorphism of G. Prove that the conductance of G is less than $10dn^{-1/d}$.

    (b) Let G be a connected graph with a vertex-transitive automorphism group, with n vertices and diameter D. Show that the conductance of G is at least $1/(2D)$.

33*. (a) Let G be the union of three random 1-factors on $\{1, \ldots, 2n\}$. Prove that if n is large enough, the probability that G is a 3-regular simple graph is between 0.1 and 0.9.

    (b) Prove that with probability at least 95%, the conductance of G is at least 1/1000. (Graphs with constant degree and conductance bounded from below by a positive constant are called *expanders*.)

*

For the rest of this chapter, let G be a connected graph with n vertices and m edges, and $(v_0, v_1, v_2, \ldots)$, a random walk on G.

34. Express the distribution of v_k in terms of the distribution of v_0, the degrees, and the adjacency matrix of G.

35. (a) Find a distribution for v_0 such that the distribution of v_k is the same for all k (*stationary distribution*).

(b) Prove that the stationary distribution is unique.

(c) If G is non-bipartite, then the distribution of v_k tends to a stationary distribution. This is not true for bipartite graphs if $n > 1$.

36. If G is non-bipartite, then the events $v_i = x$ and $v_j = y$ are "almost independent" if $j - i$ is large. More precisely, for every $\varepsilon > 0$ there exists a $t_0 > 0$ such that if $j - i > t_0$ then

$$|P(v_i = x, \ v_j = y) - P(v_i = x)P(v_j = y)| < \varepsilon.$$

37. Let $\nu_t(x)$ denote the number of times the vertex x occurs among $v_0, v_1, \ldots, v_{t-1}$. Show that

(a) $E(\nu_t(x)/t) \to \dfrac{d(x)}{2m}$ $(t \to \infty)$.

(b) $D^2(\nu_t(x)/t) \to 0$.

38. (a) The mean return time of a vertex u is $2m/d(u)$.

(b) The expected number of steps before a random walk starting at u returns to u through an edge vu is $2m$.

39. Find the mean access time between two vertices of a complete graph and between the endpoints of a path.

40. (a) Show by an example that the mean access time from u to v may be different from the mean access time from v to u, even in a regular graph.

(b) If u and v have the same degree, then the probability that a random walk starting at u visits v before returning to u is equal to the probability that a random walk starting at v visits u before returning to v. What can be said if the degrees of u and v are different?

41. The probability that a random walk starting at u visits v before returning to u is $2m/(d(u)\kappa(u,v))$, where $\kappa(u,v)$ is the mean commute time between u and v.

42. (a)* Prove that for any three vertices u, v and w,

$$a(u,v) + a(v,w) + a(w,u) = a(u,w) + a(w,v) + a(v,u).$$

(b) Prove that the vertices of any graph can be ordered so that if u precedes v then $a(u,v) \leq a(v,u)$.

43. The mean commute time between two nodes at distance r is at most $2mr$.

44. Let $a(i,j)$ denote the mean access time from i to j and let B denote the matrix $(a(i,j))_{i,j=1}^n$. As before, let D denote the diagonal matrix with $D_{ii} = 1/d(i)$, and let $M = DA_G$.

(a) Prove the identity $(I - M)B = J - 2mD$.

(b) Give a new proof of 11.38(a), based on this approach.

(c)* If G is d-regular, then

$$a(i,j) = nd \sum_{k=2}^{n} \frac{w_{kj}^2 - w_{ki}w_{kj}}{d - \lambda_k},$$

where $\lambda_1 = d > \lambda_2 \geq \ldots \geq \lambda_n$ are the eigenvalues of G and $\mathbf{w}_1, \ldots, \mathbf{w}_n$ are the corresponding orthonormal eigenvectors.

45. (a) Use 11.44(c) to give a new proof of 11.42(a).

(b) Find (asymptotically) the mean access time between two antipodal vertices of the k-cube Q_k.

46. (a) Find a formula for the mean commute time analogous to 11.44(c).

(b) Prove that the mean commute time between any pair of vertices of a d-regular graph is at least n and at most $2nd/(d - \lambda_2)$. Is there a lower bound on the mean access time that tends to ∞ with n?

47. (a) Find the mean cover time of K_n.

(b) The mean cover time of any graph (from any vertex) is at most $4nm$.

48. (a) Let $\mu(u, v)$ denote the expected number of vertices visited by a random walk starting at u, before hitting v. Prove that for every $u \in V(G)$ there exists a $v \in V(G) \setminus \{u\}$ such that $\mu(u, v) \geq n/2$.

(b) Let b be the expected number of steps before our random walk visits more than half of the vertices, and let a be the maximum mean access time. Prove that $b \leq 2a$.

49. The mean cover time (from any node) is at most $2\log_2 n$ times the maximum mean access time a between any two nodes.

50. Let G be a d-regular connected graph and let p_{ij}^t denote the probability that a random walk starting at vertex i will be at vertex j after t steps. Set $\lambda = \max\{|\lambda_2|, |\lambda_n|\}$. Prove that

$$\left| p_{ij}^t - \frac{1}{n} \right| \leq \left(\frac{\lambda}{d} \right)^t.$$

51. Consider a random walk on the k-cube Q_k. Prove that after $t = k^2 + 5k$ steps, the distribution of v_t is "almost uniform" on the appropriate color class, in the sense that for every vertex u that is at an even distance from v_0,

$$0.99 \cdot 2^{1-k} < \mathsf{P}(v_t = u) < 1.01 \cdot 2^{1-k}.$$

[Compare this time bound with the fact that the mean access time between antipodal vertices of Q_k is approximately 2^k.]

52. (a) Let $s \neq t$ be fixed vertices of G, and let $\phi(v)$ denote the probability that a random walk starting at vertex v hits s before it hits t (so $\phi(s) = 1$ and $\phi(t) = 0$). Prove that ϕ is a "harmonic function with poles s and t", $i.e.$,

$$\frac{1}{d(v)} \sum_{u \in \Gamma(v)} \phi(u) = \phi(v)$$

for every $v \neq s, t$.

(b) Consider the graph G as an electrical network, where each edge represents a unit resistance. Assume that an electric current is flowing through G, entering at s and leaving at t. Let $\phi(v)$ be the voltage of vertex v. Prove that ϕ is a harmonic function with poles s and t.

(c) Consider the edges of the graph G as ideal springs with unit Hooke constant ($i.e.$, it takes h units of force to stretch them to length h). Let us nail down vertices s and t to points 1 and 0 on the real line, and let the graph find its equilibrium. Prove that the positions of the vertices define a harmonic function with poles s and t.

(d) For every pair $s, t \in V(G)$, there is a unique harmonic function $\phi = \phi_{st}$ on G with poles s and t such that $\phi(s) = 1$ and $\phi(t) = 0$.

53. (a) If G is considered as an electrical network as in the previous problem and R_{st} denotes the resistance between vertices s and t, then the mean commute time between s and t is exactly $2mR_{st}$.

(b) Find a similar relation between the resistance R_{st} (or the mean commute time) and the forces and energies in the model with springs (11.52(c)).

(c) Let G' denote the graph obtained from G by identifying s and t, and let $T(G)$ denote the number of spanning trees of G. Prove that

$$R_{st} = \frac{T(G')}{T(G)}.$$

54. (Raleigh's Principle) Adding any edge to a graph G does not increase the resistance R_{st}.

55. Let G' be obtained from the graph G by adding a new edge (a, b), and let $s, t \in V(G)$.

(a) Prove that the mean commute time between s and t in G' is not larger than the mean commute time in G.

(b) Show by an example that similar assertion is not valid for the mean access time.

(c) If $a = t$ then the mean access time from s to t is not larger in G' than in G.

56. Let G be a regular graph and $t \in V(G)$.

(a) The average of $a(s, t)$ over all $s \in \Gamma(t)$ is exactly $n - 1$.

(b)* The average of $a(t, s)$ over all $s \in V(G) \setminus \{t\}$ is at least $n - 1$.

(c)* The average of $a(s,t)$ over all $s \in V(G) \setminus \{t\}$ is at least $n-1$.

57. Let $\pi(u,v)$ denote the probability that a random walk starting at u visits every vertex before hitting v.

(a) Prove that if G is a circuit of length n then $\pi(u,v) = 1/(n-1)$ for all $u \neq v$. Is there any other graph with this property?

(b) If u and v are two non-adjacent vertices of a connected graph G such that $\{u,v\}$ is not a cutset then there is a neighbor w of u such that $\pi(w,v) < \pi(u,v)$.

(c) Assume that for every pair $u \neq v \in V(G)$, $\pi(u,v) = 1/(n-1)$. Show that G is either a circuit or a complete graph.

58*. Consider a random walk on a graph G starting at vertex u, and mark, for each vertex different from u, the edge through which the vertex was first entered. The edges marked form a subtree T of G, which is spanning with probability 1. Prove that every spanning tree occurs with the same probability.

59*. Let G be a connected d-regular graph, $v_0 \in V(G)$, and assume that at each node, the ends of the edges incident with the node are labelled $1,2,\ldots,d$. A *traverse sequence* (for this graph, starting point, and labelling) is a sequence $(h_1, h_2, \ldots, h_t) \subseteq \{1,\ldots,d\}^t$ such that if we start a walk at v_0 and at the ith step, we leave the current vertex through the edge labelled h_i, then we visit every vertex. A *universal traverse sequence* is a sequence which is a traverse sequence for every d-regular graph on n vertices, every labelling of it, and every starting point.

Prove that for every $d \geq 2$ and $n \geq 2$, there exists a universal traverse sequence of length $O(d^2 n^4)$.

§ 12. Automorphisms of graphs

Although most graphs have no automorphisms other than the identity, many graphs arising from different combinatorial, geometric or algebraic structures possess, or are characterized by, a large automorphism group. This makes the study of relationships between the structure of graphs and their automorphism groups of some importance.

The early trend in these investigations was to deal with independence results; i.e. results which show that even the imposition of rather strong restrictions on the graphs may not restrict their automorphism groups. This approach involves the construction of graphs, for which fairly standard methods are now available. The other approach, in which properties of the graphs are found which do restrict the properties of their automorphism groups, needs more "ad hoc" methods.

If the automorphism group is not only considered as an abstract group but also as a permutation group, then, of course, we find stricter interrelationships with the graph. For example, the property that the automorphism group is transitive implies many nice properties for the graph.

Investigation on the endomorphism semigroup instead of the automorphism group mostly yields similar results with more complicated constructions. However, there are some surprising differences (see the last exercise).

1. What is the automorphism group of the Petersen graph (Fig. 9, problem 11.2)?

2. (a) Show that the automorphism group of the dodecahedron graph (Fig. 11) is isomorphic with $A_5 \times Z_2$.

(b) The automorphism group of the cube is isomorphic with $S_4 \times Z_2$. What are the automorphism groups of the other Platonic bodies?

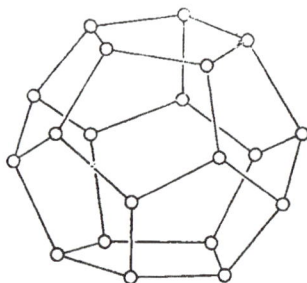

FIGURE 11

3. Construct a graph having automorphism group isomorphic with Z_k, a cyclic group of order k.

4. Let $\Gamma = \{g_1, \ldots, g_n\}$ be a group. Define a digraph G by joining g_i to g_j by an edge of color k if $g_i g_j^{-1} = g_k$. What is the automorphism group of the resulting colored digraph?

5. (Frucht's Theorem) Given a finite group Γ, there exists a simple graph G with automorphism group isomorphic to Γ.

6*. Let Γ be a group of order n $(n \geq 6)$. Construct a simple graph G on $2n$ points with $A(G) \cong \Gamma$.

7. (a) Any tournament has an odd number of automorphisms.

(b)* Every group Γ of odd order n is the automorphism group of a tournament on $2n$ points.

8*. Given a finite group Γ, construct a 3-regular graph with automorphism group Γ.

9. If G is a connected graph with one point x_1 of degree 2 and all other points of degree 3 then the automorphism group of G is of order 2^k for some k.

10. Let G be a graph such that $A(G)$ acts semiregularly on $E(G)$. Let Γ be any subgroup of $A(G)$. Then G has an orientation $\overrightarrow{G}$ such that $A(\overrightarrow{G}) = \Gamma$.

11*. Let Γ_1, Γ_2 be two finite groups. Construct a graph G which has an edge e such that $A(G) \cong \Gamma_1$ and $A(G-e) \cong \Gamma_2$.

12. (a) Let G be a connected graph and Γ a subgroup of $A(G)$ acting semiregularly. Prove that G can be contracted onto a graph G' such that $A(G')$ has a subgroup $\Gamma' \cong \Gamma$ which acts regularly.

(b) Let G be a connected graph and Γ a simple subgroup of $A(G)$. Then some connected subgraph of G can be contracted onto a graph G' such that $A(G')$ has a subgroup $\Gamma' \cong \Gamma$ which acts edge-transitively on G'. If, in addition, the elements of Γ have no fixed point in common, then Γ' has the same property.

*

13. (a) If the automorphism group of a graph G is commutative and transitive it is isomorphic to the direct product of cyclic groups of order 2.

(b) Construct a graph whose automorphism group is transitive and isomorphic with the direct product of n cyclic groups of order 2, for any given n.

(c)* Construct a simple graph with this property for large enough n.

14*. The r-regular connected graph G has a transitive automorphism group. Show that G is r-edge-connected.

15. (a) Let G be a graph with transitive automorphism group which is exactly 3-connected. Show that G is 3-regular. Is this true with 4 instead of 3?

(b) How large must the connectivity of an r-regular connected graph with transitive automorphism group be?

(c) A connected simple graph with an edge-transitive automorphism group and with all degrees at least r is r-connected.

16*. A connected graph G with an even number of points and transitive automorphism group has a 1-factor and, in fact, each edge of it belongs to a 1-factor.

17. The graph G is connected and $A(G)$ contains a commutative transitive subgroup Γ. Prove that G has a Hamiltonian circuit.

18. Which planar graphs have edge-transitive automorphism groups?

19*. Assume that the planar graph G has a non-cyclic simple automorphism group. Prove that $A(G) \cong A_5$.

20*. (a) Assume that G is a connected graph and $A(G)$ contains a subgroup $\Gamma \cong (Z_p)^3$ (p prime) which acts semiregularly. Then G can be contracted onto K_p.

(b) Assume that G is a connected graph and $A(G) > \Gamma' > \Gamma \cong (Z_p)^3$, where p is a prime and Γ' is a simple group without a common fixed point. Then G can be contracted onto K_m, where $m = \lfloor \log p / \log 8 \rfloor$.

(c) Let $\mathcal{K}$ be a class of graphs, not containing all graphs, such that if $G \in \mathcal{K}$ and $e \in E(G)$, then both $G-e$ and G/e belong to $\mathcal{K}$ (e.g. the class of graphs embeddable in a surface F, or the class of graphs having no subgraph contractible onto K_m). Then there exists a group Γ such that no graph of $\mathcal{K}$ has automorphism group isomorphic to Γ.

21. Let Γ be a permutation group on a set Ω. Construct a simple graph G such that $V(G) \supseteq \Omega$, Ω is invariant under $A(G)$ and the restriction of $A(G)$ to Ω yields Γ.

*

22. Determine the endomorphisms of the Petersen graph (Fig. 9, problem 11.2).

23. Construct a rigid graph.

24*. (a) Let Σ be a finite monoid (semigroup with identity). Prove that there exists a digraph G with colored edges such that the endomorphism semigroup $\text{End}(G) \cong \Sigma$.

(b) Prove that there exists such a simple graph as well.

25. (a) If $\text{End}(G)$ is isomorphic to the multiplicative semigroup $\{0, 1, -1\}$ then $G - e$ cannot be rigid for any edge e.

(b)* For any monoid Σ there exists a simple graph G with $\text{End}(G) \cong \Sigma$ such that $G + e$ is rigid for some new edge e.

§ 13. Hypergraphs

A hypergraph, being a set with a specified collection of subsets, is a very general structure to consider. Graphs (in which each specified subset has just two elements) and several other combinatorial structures (block designs, · matroids) are very special cases. Consequently the investigation of hypergraphs throws up a very large variety of problems and the "theory" is very dispersed. However, when various problems are translated to hypergraphs, some concepts such as matching number, covering number and chromatic number seem to occur very frequently. In my opinion these concepts are almost as general as those of logic: for example, given a combinatorial quantity defined as a minimum, it is usually easy to express it as the covering number of a suitable hypergraph. The duality between matching number and covering number is the background of many more graph theoretic and combinatorial results than would appear at first glance.

There are also some general methods in hypergraph theory. The averaging method (cf. problems 12, 21, 27, 28, 32, 41, 52, 53) seems to be trivial, but its direct applications have settled problems which were unsolved for several years (e.g. 32, 57). The linear algebra method (cf. problems 13, 15, 32, 45) is also very powerful, and leads to the fast developing field of algebraic combinatorics.

Finally we remark that certain special questions concerning hypergraphs give rise to theories which are distinguished from the rest by their more developed methods. We only touch upon transversal theory in problems 5–7. The theory of block designs (i.e. hypergraphs with a high degree of homogeneity) is not treated here.

1. A connected hypergraph H contains no circuits if and only if

(∗)
$$\sum_{E \in E(H)} (|E| - 1) = |V(H)| - 1.$$

2*. (a) Let H be a totally balanced hypergraph with $|E(H)| \geq 2$. Then H has two edges E, F such that each point of $E - F$ has degree 1.

(b) Let H be a totally balanced hypergraph and suppose that any two edges have at most p points in common. Then

(∗∗)
$$\sum_{E \in E(H)} (|E| - p) \leq |V(H)| - p.$$

3. Let P be a path and let $P_1, \ldots, P_m$ be subpaths (intervals) on P. Setting $V = V(P)$, $E_i = V(P_i)$, prove that the hypergraph $(V; \{E_1, \ldots, E_m\})$ is totally balanced.

4. The 3-uniform hypergraph H (without multiple edges) has $|V(H)| - 1$ edges. Prove that it contains a circuit of length at least 3.

$*$

5. (a) (Hall's Theorem) A hypergraph H has a system of distinct representatives iff $|V(H')| \geq |E(H')|$ for each partial hypergraph H' of H.

(b) Given a hypergraph H, we want to select a 2-element subset $f(E)$ of each edge E in such a way that the sets $f(E)$, considered as the edges of a graph, form a forest. Prove that this is possible iff $|V(H')| \geq |E(H')| + 1$ for each partial hypergraph H' of H.

6. Let $T \subseteq V(H)$. When does H have a system of distinct representatives containing T?

7*. Let H_1, H_2 be two hypergraphs with $V(H_1) = V(H_2)$ and $|E(H_1)| = |E(H_2)| = m$. Then H_1, H_2 have a common system of distinct representatives iff

$$|V(H') \cap V(H'')| \geq |E(H')| + |E(H'')| - m$$

for any two partial hypergraphs H', H'' of H_1 and H_2, respectively.

$*$

8. Let H be a hypergraph with m distinct edges ($m \geq 2$). Prove the following assertions:

(a) The number of sets of the form $E \triangle F$ ($E, F \in E(H)$), is at least m.

(b)* The same holds for the sets of the form $E - F$, ($E, F \in E(H)$).

(c)* The same holds for the sets of the form $E \cup F$ or $E \cap F$, ($E, F \in E(H)$, $E \neq F$).

9*. Let H be a hypergraph such that if $E \in E(H)$ then every $F \subseteq E$ belongs to $E(H)$. Prove that there exists a permutation σ of $E(H)$ such that $E \cap \sigma(E) = \emptyset$ for each $E \in E(H)$.

10. (a) The hypergraph H on n points has n distinct edges. Prove that it has a point x such that $H \setminus x$ has n distinct edges.

(b) The hypergraph H on n points has $m \leq \frac{3}{2}n$ distinct edges. Prove that it has a point x such that $H \setminus x$ has at least $m - 1$ distinct edges.

(c) The hypergraph H on n points has

$$m > \sum_{i=0}^{k-1} \binom{n}{i}$$

distinct edges. Prove that there is a set $X \subseteq V(H)$ with $|X| = k$, such that every subset of X occurs among the sets $X \cap E$, $E \in E(H)$.

11. Let H, K be two hypergraphs on the same n-element set V and suppose that

$$A' \subseteq A \in E(H) \Longrightarrow A' \in E(H),$$

and

$$V \supseteq B' \supseteq B \in E(K) \Longrightarrow B' \in E(K).$$

Prove that

$$|E(H) \cap E(K)| \leq \frac{1}{2^n} |E(H)| \cdot |E(K)|.$$

12. (a) Let H be any r-uniform hypergraph. Prove that $V(H)$ can be split into two classes such that the number of edges contained in one of the two classes is at most $\frac{|E(H)|}{2^{r-1}}$.

(b) Prove that $V(H)$ can be partitioned into r classes such that at least $\frac{r!}{r^r}|E(H)|$ edges have the property that they contain exactly one element of each class.

13*. An r-uniform hypergraph has m edges, any two of which meet in at most k points. Prove that it has at least

$$\frac{r^2 m}{r + (m-1)k}$$

points.

14*. Let H be a hypergraph with n points and m edges. Denote by $d(x)$ the degree of x and suppose that $d(x) < m$, $0 < |E| < n$ holds for all edges E and points x. Suppose further that, if $x \notin E$, then

$$d(x) \leq |E|.$$

Prove that $m \leq n$.

15. (a) Let H be a hypergraph, $V(H) \neq \emptyset$, in which any two edges have exactly one element in common, and no point occurs in all edges. Then $|E(H)| \leq |V(H)|$.

(b)* (Fisher's Inequality) Prove that the same holds if any two edges have exactly λ elements in common ($\lambda > 0$) and, moreover, the incidence vectors of edges are linearly independent over the reals.

16. Let H, K be two hypergraphs on the same set of points, and suppose that $|E(H)| = m$, $|E(K)| = m'$, where $m' = m$ or $m + 1$. Then

$$\sum_{A \in E(H)} \sum_{B \in E(K)} |A \triangle B| \geq \sum_{\{A,A'\} \subseteq E(H)} |A \triangle A'| + \sum_{\{B,B'\} \subseteq E(K)} |B \triangle B'|.$$

17*. (a) Let $F_1, \ldots, F_m$ be sets such that $|F_i \triangle F_j| = 2k$ ($1 \leq i < j \leq m$). Then the degree d of any point satisfies

$$d(m - d) \leq km.$$

(b) If there is a point x with degree $\neq 0$, 1, $m - 1$, m, then $m \leq k^2 + k + 2$.

18. (a) Suppose that any two edges of the simple hypergraph H on n points intersect. Prove that $|E(H)| \leq 2^{n-1}$.

(b) If, in addition, no two edges cover every point, then $|E(H)| \leq 2^{n-2}$.

19. Let H be a hypergraph on n points without multiple edges, such that, for any two edges E, $F \in E(H)$, either $E \subset F$ or $F \subset E$ or $E \cap F = \emptyset$. What is the maximum number of edges in H?

20. Divide the set of all subsets of an n-element set S into symmetric chains. (By a symmetric chain we mean a chain $E_r \subset E_{r+1} \subset \ldots \subset E_{n-r}$, where $|E_i| = i$ and $0 \leq r \leq n/2$.)

21. (a) (Sperner's Theorem) If H is a clutter on n points (i.e. no two edges of H contain each other), then $|E(H)| \leq \binom{n}{\lfloor n/2 \rfloor}$.

(b) For which hypergraphs does equality hold?

22*. In any partially ordered set $(S, \leq)$ there is a maximum set of pairwise incomparable elements (a maximum anti-chain) which is invariant under the automorphisms of S.

Give a new proof of Sperner's theorem 13.21 using the above observation.

23*. Let us call a hypergraph H *cross-cutting* if each subset of $V(H)$ is comparable with an edge of H.

(a) Does every cross-cutting hypergraph contain a cross-cutting clutter?

(b) A minimal cross-cutting hypergraph has at most $\binom{n}{\lfloor n/2 \rfloor}$ edges ($n = |V(H)|$).

24. If k is the maximum length of chains $E_1 \subset E_2 \subset \ldots \subset E_k$ of edges in a hypergraph H on n points without multiple edges, then $|E(H)|$ is not greater than the sum of the k largest binomial coefficients in the nth row of the Pascal triangle.

25. (a) Suppose that every set of $r+1$ edges of an r-uniform hypergraph H has a point in common. Then all edges of H have a point in common.

(b) Suppose that any k edges of an r-uniform hypergraph have a point in common. Then

$$\tau(H) \leq \frac{r-1}{k-1} + 1.$$

26. Let H be an r-uniform hypergraph with $\nu(H) = 1$ and $\tau(H) > 1$. Prove that there exists a set $S \subseteq V(H)$ such that $|E \cap S| \geq 2$ for every edge E and $|S| \leq 3r - 3$.

27*. Let H be an r-uniform hypergraph such that any two edges of H intersect. Prove that there is a set $W \subseteq V(H)$ such that

$$|W| \leq (2r - 1)\binom{2r - 3}{r - 1}$$

and any two edges in the subhypergraph H_W induced by W meet. This is not always true with

$$|W| \leq \binom{2r - 3}{r - 1}.$$

28. (a) Let $A_1, \ldots, A_t$ be distinct arcs of length k of a circuit C ($|V(C)| \geq 2k$) and suppose that any two have an edge in common. Then $t \leq k$.

(b) (Erdős–Ko–Rado Theorem) Let H be an r-uniform hypergraph on n points ($n \geq 2r$) without multiple edges such that any two edges of H intersect. Prove that

$$|E(H)| \leq \binom{n - 1}{r - 1}.$$

29*. (a) Let H be an r-uniform ν-critical hypergraph with n points and m distinct edges. Then (for r, ν fixed)

$$m = O(n^{r-2}).$$

(b) Let H be any r-uniform hypergraph with n points and m distinct edges. Then for $n \geq n_0(r,\nu)$ ($\nu = \nu(H)$) we have

$$m \leq \binom{n - 1}{r - 1} + \binom{n - 2}{r - 1} + \ldots + \binom{n - \nu}{r - 1}.$$

30*. Let H be a hypergraph and denote by d its maximum degree. Prove the following inequality between the sizes of optimal point-covers and fractional covers:

$$\tau(H) \leq (1 + \log d)\tau^*(H).$$

31*. (a) Let $0 < k \leq r$ be integers, u, v, w be reals with $u \geq v \geq r - 1$, $v \geq w \geq r - 2$, and assume that

$$\binom{u}{r} = \binom{v}{r} + \binom{w}{r-1}.$$

Prove that

$$\binom{u}{k} \leq \binom{v}{k} + \binom{w}{k-1}.$$

(b) Let H be an r-uniform hypergraph without multiple edges. Write $|E(H)|$ in the form $\binom{u}{r}$, where $u \geq r$ is real. Prove that the number of k-tuples contained in edges of H is at least $\binom{u}{k}$ $(1 \leq k \leq r)$.

(c) Deduce the Erdős–Ko–Rado theorem 13.28b from (b).

32*. (a) Let $|A_1| = \ldots = |A_m| = p$, $|B_1| = \ldots = |B_m| = q$ and $A_i \cap B_j = 0 \Longleftrightarrow i = j$. Then

$$m \leq \binom{p + q}{p}.$$

(b) An r-uniform τ-critical hypergraph H has at most $\binom{\tau(H) + r - 1}{r}$ edges.

*

33. If no two edges of a hypergraph H have exactly one point in common, then H is 2-colorable.

34. If every point of a connected hypergraph H has degree 2 and H is not a (2-uniform) odd circuit, then H is 2-chromatic.

35. Determine all hypergraphs in which any two edges have exactly one point in common and which are not 2-colorable.

36. Let H be a 3-uniform hypergraph on $n \geq 5$ points in which each pair of points occurs in the same (positive) number of edges. Prove that H is not 2-colorable.

37*. A hypergraph H has a totally unimodular incidence matrix if and only if each subhypergraph H_W has a bicoloration $\{A_0, A_1\}$ such that

$$\left\lfloor \frac{|E|}{2} \right\rfloor \leq |E \cap A_1| \leq \left\lceil \frac{|E|}{2} \right\rceil$$

for every $E \in E(H_W)$. (The *incidence matrix* of $H = (\{v_1, \ldots, v_n\}; \{E_1, \ldots, E_m\})$ is the matrix $A = (a_{ij})$ defined by

$$a_{ij} = \begin{cases} 1 & \text{if } v_i \in E_j \\ 0 & \text{otherwise.} \end{cases}$$

A matrix is *totally unimodular* if every square submatrix has determinant 0 or ± 1.)

38. (a) If every circuit of the hypergraph H has even length, then its points can be 2-colored so that the numbers of red and blue points in any edge differ by at most one.

(b) The incidence matrix of H is totally unimodular.

39. If the incidence matrix A of a hypergraph $H = (\{v_1, \ldots, v_n\}; \{E_1, \ldots, E_m\})$ is totally unimodular, then H is balanced.

40. A balanced hypergraph in which every edge has at least 2 elements is 2-colorable.

41. If an r-uniform hypergraph has at most 2^{r-1} edges, then it is 2-colorable.

42. Prove, using 13.30, that there exists a non-2-colorable r-uniform hypergraph on $2r^2$ points with $c \cdot r^2 2^r$ edges.

43*. If every edge of an r-uniform hypergraph H meets at most 2^{r-3} other edges, then H is 2-chromatic.

44*. Let H be an r-uniform hypergraph with n points and m edges such that two edges have at most one point in common.

(a) If $n \leq 2^{r-4}$, then H is 2-colorable.

(b) If H is not 2-colorable, then it contains at least $2^{r-4}/r$ points of degree at least $2^{r-4}/r$.

(c) If $m \leq 4^{r-4}/r^3$, then H is 2-colorable.
(For existence problems concerning such hypergraphs cf. 14.24).

45. (a) The hypergraph H has the property that, for all $k \geq 1$, the union of any k edges has at least $k+1$ points. Prove that H is 2-chromatic.

(b) Construct an r-uniform hypergraph H such that the union of any k edges $(k = 1, 2, \ldots)$ has at least k points but H is not 2-chromatic.

(c) Let H be a 3-chromatic hypergraph on n vertices, without isolated vertices, such that deleting any vertex of H, the remaining hypergraph is 2-colorable. Prove that the incidence vectors of edges of H generate the whole n-dimensional space. Give a new proof of (a), based on this.

46. Construct, for infinitely many values of r, r-uniform hypergraphs without multiple edges which are not 2-chromatic but any two edges of which intersect and have

(a) more than $r!$ edges,

(b) at most 3^{r-1} edges,

(c) more than 3^{r-1} points.

47. If a simple hypergraph H is r-uniform, 3-chromatic and any two edges of H intersect then

$$|E(H)| < r^r.$$

*

48. Prove the following relations between optimum matchings, covers, fractional matchings, and fractional covers [†]

$$\nu(H) \leq \nu^*(H) = \tau^*(H) \leq \tau(H).$$

49. (a) Let F be a graph. Then any k-cover is the sum of a 2-cover and of a $(k-2)$-cover.

(b) A similar assertion holds for k-matchings if k is even.

(c) There are optimal fractional covers and also optimal fractional matchings, where the weights are halves of integers. Consequently, $2\tau^*(G)$ is an integer.

(d) Deduce 7.37 from (c).

50. Let G be a bipartite graph. Prove that every k-cover is the sum of k 1-covers and that every k-matching is the sum of k 1-matchings. Consequently, $\tau(G) = \tau^*(G)$ and $\nu(G) = \nu^*(G)$.

51. (a) Prove that

$$\tau^*(G \otimes H) = \tau^*(G)\tau^*(H),$$
$$\tau(G)\tau(H) \geq \tau(G \otimes H) \geq \tau(G)\tau^*(H),$$
$$\nu(G)\nu(H) \leq \nu(G \otimes H) \leq \nu(G)\nu^*(H)$$

hold for any two hypergraphs G, H.

(b) Given a hypergraph H, $\tau(G \otimes H) = \tau(G)\tau(H)$ holds for every hypergraph G iff $\tau(H) = \tau^*(H)$.

(c) Setting $H^p = H \otimes H \otimes \ldots \otimes H$ (p factors), determine $\lim\limits_{p \to \infty} \sqrt[p]{\tau(H^p)}$.

52. If a hypergraph H is τ-critical, then either H consists of disjoint edges or

$$\tau(H) > \tau^*(H).$$

53. If a hypergraph H is ν-critical and has no empty edges, then

$$\nu(H) < \nu^*(H).$$

54. Every balanced hypergraph H has $\tau(H) = \nu(H)$.

55. If H is a hypergraph, then the following three assertions are equivalent.

(i) $\nu(H') = \nu^*(H')$, for every partial hypergraph H' of H.

(ii) $\tau(H') = \tau^*(H')$, for every partial hypergraph H' of H.

[†] One may use the duality theorem of linear programming: If $\max \mathbf{c}^T \mathbf{x}$ exists under the constraints $A\mathbf{x} \leq \mathbf{b}$, $\mathbf{x} \geq 0$ then also $\min \mathbf{b}^T \mathbf{y}$ exists under the constraints $A^T \mathbf{y} \geq \mathbf{c}$, $\mathbf{y} \geq 0$ and $\min \mathbf{b}^T \mathbf{y} = \max \mathbf{c}^T \mathbf{x}$ (here A is any $n \times m$ real matrix and $\mathbf{b}$ and $\mathbf{c}$ are n- and m-dimensional real vectors).

(iii) H is *normal*, i.e. $\nu(H') = \tau(H')$ for every partial hypergraph H' of H.

56*. (Cont'd) The following assertions are also equivalent to (i)–(iii):

(iv) $d(H') = q(H')$, for every partial hypergraph H' of H.

(v) H has the Helly-property (i.e. if edges $E_1, \ldots, E_m$ meet pairwise, then $E_1 \cap \ldots \cap E_m \neq \emptyset$) and $\overline{L(H)}$ is perfect.

57*. (Perfect Graph Theorem) A graph G is perfect iff $\overline{G}$ is perfect.

§ 14. Ramsey Theory

Every "irregular" structure, if it is large enough, contains a "regular" substructure of some given size. This phenomenon occurs in many situations. A typical example (the title of this chapter) is that, if we take a large complete graph whose edges are 2-colored (as "irregularly" as we like) there always exists a large monochromatic complete subgraph. However, one can color other objects (points in spaces, numbers, subsets, etc.) and also, one can look for other "regular" substructures (monochromatic subspaces, arithmetic progressions, configurations, etc.) Some of these questions turn out to be very difficult. However, a considerable number of them can be given a general treatment using categories (see problems 14.13–18 for the combinatorial background). The last sequence of problems deals with monotone subsequences of sequences and related problems in geometry and combinatorics. For example the famous problem of finding convex subsets of a set is treated here.

Ramsey theory in the narrower sense, when monochromatic subsets are looked for, is a special case of the theory of chromatic number of hypergraphs. Most of the results here can be formulated as assertions that a certain hypergraph "regular" and large enough, has high chromatic number.

We shall not deal with infinite Ramsey theory, although this is more closely related to the finite theory than, perhaps, are the finite and infinite counterparts of any other field in combinatorics. We hope that problems 9.14 and 14.19 will shed light on this connection.

1. (a) Every graph on $\binom{k+l}{k}$ points contains either a complete graph with $k+1$ points or an independent set of $l+1$ points.

(b) Let $a_1, \ldots, a_k \geq 1$ be given integers ($k \geq 2$). Prove that there exists a (least) natural number $n = R_k(a_1, \ldots, a_k)$ such that, if we k-color the edges of K_n in any way, there will be an $1 \leq i \leq k$ and a complete a_i-graph all of whose edges are colored with the ith color.

2. (a) Prove that

$$R_k(3) \stackrel{\text{def}}{=} R_k(3, \ldots, 3) \leq \lfloor e \cdot k! \rfloor + 1;$$

and that equality holds for $k = 2, 3$.

(b)* Prove that for $k \geq 2$,

$$2^{k/2} \leq R_2(k,k) < 2^{2k}.$$

3. (a) (Ramsey's Theorem) Let K_n^r denote the hypergraph consisting of all r-tuples of n points and let $a_1, \ldots, a_k \geq 1$. Prove that there exists a (least) number $n = R_k^r(a_1, \ldots, a_k)$ such that, if we k-color the edges of K_n^r, there will be an a_i-subset of $V(K_n^r)$ all of whose k-subsets have the ith color, for at least one $1 \leq i \leq k$.

(b) Define $R_k^r(a) = R_k^r(a, \ldots, a)$. Prove that

$$R_k^{r+1}(a) < k^{\lfloor R_k^r(a) \rfloor^r}.$$

4. 2-color the edges of a complete n-graph ($n \geq 3$). Prove that there will be a Hamiltonian circuit which either is monochromatic or consists of two monochromatic arcs.

5*. We 2-color the edges of a complete $\left\lfloor \frac{3k+1}{2} \right\rfloor$-graph. Prove that there is a monochromatic path of length k.

6*. Let us 2-color the edges of a complete n-graph. Prove the following assertions:

(a) If there is a monochromatic $(2k+1)$-circuit ($k \geq 3$), then there is a monochromatic $2k$-circuit.

(b) If there is a monochromatic $2k$-circuit ($k \geq 3$), then there is either a monochromatic $(2k-1)$-circuit or two monochromatic disjoint complete k-graphs.

(c) If $n \geq 2m-1$ and $n \geq 6$, then there will be a monochromatic m-circuit.

*

7. Let us 3-color the points of the plane. Prove that there will be two points at distance 1 with the same color.

8*. We 2-color the points of the plane. Suppose that there are three points of the same color forming a regular (equilateral) triangle with side 1 (a *monochromatic regular triangle with side 1*, for short). Then for each a, $b > 0$ such that $|a - b| < 1 < a + b$, we have a monochromatic triangle with sides 1, a, b.

9. (a) Show that if we 2-color the points of the plane, then we do not necessarily have a monochromatic regular triangle with side 1.

(b) Prove that, 2-coloring the points of the plane, we always have a monochromatic triangle with sides $\sqrt{2}$, $\sqrt{6}$, π.

10*. Does there exist a natural number $n = n_0(k, R)$ such that, if we k-color the points of the n-dimensional Euclidean space, one of the colors will contain a configuration congruent to R, where

(a) R is a rectangle,

(b) R is a non-rectangular parallelogram.

*

11. Let us split the first n natural numbers into k classes. Prove that if $n \geq k!e$, then one of the classes contains three integers x, y, z with $x + y = z$.

12. Let k, $r \geq 1$ be given. Then there is an $n_0(k,r)$ such that if $n \geq n_0(k,r)$ and $\{1, \ldots, n\}$ is k-colored, then we can find natural numbers $a, d_1, \ldots, d_r$ such that all sums

$$a + d_{i_1} + \ldots + d_{i_\nu} \qquad (1 \leq i_1 < \ldots < i_\nu \leq r. \quad 0 \leq \nu \leq r)$$

have the same color (and, of course, $a + d_1 + \ldots + d_r \leq n$).

13. Let us k-color all non-empty subsets of an n-element set. Prove that if n is large enough, there are two disjoint non-empty subsets X, Y such that X, Y, $X \cup Y$ have the same color.

14*. (a) Suppose that the set of all subsets of an n-element set S is k-colored, where $n \geq N(k,t)$. Find disjoint sets $A_1, B_1, \ldots, A_t, B_t$ such that for any fixed sequence $1 \leq i_1 < \ldots < i_\nu \leq t$, all unions of the form

$$C_{i_1} \cup \ldots \cup C_{i_\nu} \qquad (C_i = A_i \quad \text{or} \quad B_i \quad \text{or} \quad A_i \cup B_i)$$

have the same color.

(b) Prove that for any given k and r there is an $n = n(k,r)$ with the following property: whenever the set of all subsets of an n-element set S is k-colored, then there exist non-empty disjoint subsets $X_1, \ldots, X_r \subseteq S$ such that all non-empty unions of any of them have the same color.

15. Strengthen 14.12 as follows: For any k and r there exists a natural number n such that, if $\{1, \ldots, n\}$ is k-colored, then there always exist natural numbers $d_1, \ldots, d_r$ such that $d_1 + \ldots + d_r \leq n$ and all sums $d_{i_1} + \ldots + d_{i_\nu}$ $(1 \leq i_1 < \ldots < i_\nu \leq r, \quad \nu \geq 1)$ have the same color.

16*. Let us call a mapping $S \to \{0, \ldots, m-1\}$ and m-vector on S. Consider the set m^S of all m-vectors on S, where $m \geq 2$, and let α be a k-coloration of m^S.

(a) There is an $n_1(k,m)$ such that if $|S| \geq n_1(k,m)$, then there is a non-empty set $X \subseteq S$ and an m-vector $\mathbf{b}$ on $S - X$ such that $(m-1)X + \mathbf{b}$ and $(m-2)X + \mathbf{b}$ have the same color. (We abuse notation in two ways: we denote by X the all-1 vector on set X and we add up two m-vectors on different sets by extending them with 0's over the whole union.)

(b) There is an $n_r(k,m)$ such that if $|S| \geq n_r(k,m)$, then there are disjoint nonempty sets $X_1, \ldots, X_r \subseteq S$ and an m-vector $\mathbf{b}$ on $S - X_1 - \ldots - X_r$ such that each vector

$$\mathbf{a} = \sum_{i=1}^{r} a_i X_i + \mathbf{b} \qquad (0 \leq a_i \leq m-1)$$

has the same color as

$$\mathbf{a}' = \sum_{i=1}^{r} \min(a_i, m-2) X_i + \mathbf{b}.$$

17*. (Cont'd) Prove that there is an $N(k,r,m)$ such that if $|S| \geq N(k,r,m)$, then there are disjoint non-empty subsets $X_1, \ldots, X_r$ of S and an m-vector $\mathbf{b}$ of points of $S - X_1 - \ldots - X_r$ such that all m-vectors

$$\mathbf{a} = \sum_{i=1}^{r} a_i X_i + \mathbf{b}, \qquad (0 \leq a_i \leq m-1)$$

have the same color.

18. (Van der Waerden's Theorem) There exists a number $w = w(k,m)$ such that if the natural numbers $0, 1, \ldots, w$ are k-colored, then there is a monochromatic arithmetic progression of length m.

19. Let P be any property of k-colored finite sets of natural numbers (e.g. a set has property P if it is monochromatic and forms an arithmetic progression). Suppose that, k-coloring all natural numbers in any way, some finite subset will have property P. Then there exists a natural number N such that k-coloring the set $\{1, \ldots, N\}$ arbitrarily, some subset of it will have property P. †

20*. Decide whether the following assertion is true: If the set of all natural numbers is k-colored, one of the classes contains x, y, z with

(a) $\qquad\qquad\qquad\qquad x + y = 3z,$

(b) $\qquad\qquad\qquad\qquad x + 2y = z.$

21*. Let

$(*)$ $\qquad\qquad\qquad\qquad a_1 x_1 + \ldots + a_n x_n = 0$

be an equation with integral coefficients. Then the following two properties are equivalent:

(i) $(*)$ has a non-trivial $(0,1)$-solution.

(ii) If we color the natural numbers with finitely many colors, $(*)$ will have a monochromatic solution.

---

† This assertion is a very special case of a general principle called *compactness*. This principle would allow for example the deduction of Ramsey's theorem 14.2 from an infinite version of it. Since we cannot go into set-theory, this exercise is to serve as an illustration.

22*. Prove that there exists a function $f(k)$ with the property that, if S is any set of integers with $|S| \geq f(k)$, then the set of all integers can be k-colored so that $S+j$ meets all the colors for every integer j $(S+j = \{s+j : s \in S\})$. The k-coloration can even be required to be periodic.

23. Let q be a prime power and $k \geq 1$, $r \geq 0$.

(a) Let $n \geq N(k,r,q-1)$. Then, if we k-color the points of the the affine n-dimensional space over $GF(q)$, then some r-dimensional subspace will be monochromatic.

(b) Prove an analogous theorem for projective spaces.

24. Give two constructions, based on Ramsey's theorem and 14.23a, of an r-uniform hypergraph H which has chromatic number at least $k+1$ and in which any two edges have at most one point in common.

*

25. Let $(a_1,\ldots,a_{k^2+1})$ be any sequence of integers. Prove that it contains a monotone subsequence of length $k+1$.

26. (a) Let f be an integral-valued function defined on $\{1,\ldots,2^{n-1}\}$, such that $1 \leq f(i) \leq i$, for $i = 1,\ldots,2^{n-1}$. Prove that there is a sequence

$$1 = a_1 < \ldots < a_n \leq 2^{n-1}$$
with
$$f(a_1) \leq \ldots \leq f(a_n).$$

(b) This is not always true for $2^{n-1} - 1$ instead of 2^{n-1}.

27*. (a) Let T_n denote a transitive tournament on n points and let a number $f(x,y)$ be associated with each edge $(x,y) \in E(T_n)$. Prove that if $n > \binom{p+q-2}{p-1}$, then there is either a (directed) path $(x_0,\ldots,x_p)$ with

$$f(x_0,x_1) \leq f(x_1,x_2) \leq \ldots \leq f(x_{p-1},x_p)$$
or a path $(y_0,\ldots,y_q)$ with
$$f(y_0,y_1) > f(y_1,y_2) > \ldots > f(y_{q-1},y_q).$$

(b) Prove that the bound is sharp.

28*. Let us associate an element $f(X) \in X$ with each non-empty subset $X \subseteq S$ of the set S.

(a) If $|S| = 2m$, then there is a set $T \subseteq S$ with $|T| = m$, and an ordering $(x_1,\ldots,x_m)$ of $S-T$, such that

$$f(S - \{x_1,\ldots,x_i\}) \in T \qquad (i = 0,\ldots,m).$$

(b) If $|S| = 2^n$, then there is a chain $X_0 \subset X_1 \subset \ldots \subset X_n \subseteq S$ with $f(X_1) = \ldots = f(X_n)$.

(c) This is no longer true when $|S| = 2^n - 1$.

29. Given $N = k^n + 1$ points in the plane, we can always find an "almost straight" broken line, i.e. a sequence $a_0, a_1, \ldots, a_k$ such that every angle $a_{i-1} a_i a_{i+1}$ is at least $(1 - (1/n))\pi$.

30. We call a polygon convex from below (above) if it is convex and any semi-line starting from its vertices and parallel to the negative (positive) half of the y-axis has only its endpoint in common with the polygon (Fig. 12).

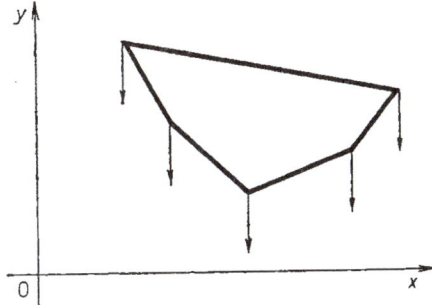

FIGURE 12

(a) Prove that any set S of $\binom{p+q}{p} + 1$ points in the plane, in general position (which includes here the property that no line determined by them is parallel to either axis), contains either a $(p+2)$-gon convex from below or a $(q+2)$-gon convex from above.

(b) This is not true with $\binom{p+q}{p}$ points.

31. (a) There exists a (least) natural number $K(m)$ such that, if we have $K(m)$ points in the plane in general position there always exists a convex m-gon of whose vertices are from the given set.

(b)* Establish the inequalities†

$$2^{m-2} + 1 \le K(m) \le \binom{2m-4}{m-2} + 1.$$

(c) Show that $K(4) = 5$ and $K(5) = 9$.

§ 15. Reconstruction

The problem of reconstruction is a version of the classical principle of invariants. If we associate some structure A' with each structure A (for example the line-graph of a graph, or the generating function of a sequence, or the Betti

---

† It is conjectured that the lower bound is the exact value of $K(m)$.

numbers of a manifold), we may be interested in the question as to whether A' uniquely determines A. For example, the fact that a power series determines its coefficients is a result of this type. Also we may be interested in characterizing those structures which arise in the form A'.

We restrict ourselves to reconstruction-type problems of combinatorial character; reconstructing graphs from their line-graphs, "matroids", direct products with themselves, etc. The best known problem in this area is the Reconstruction Conjecture, which asserts that a graph can be reconstructed from the collection of its maximal proper induced subgraphs. This is solved in special cases only.

We mention here for the sake of completeness one class of problems in this connection which we shall not be able to deal with in this volume. These are the problems concerning the construction of symmetric combinatorial structures (block designs, partial geometries, Latin squares) from finite fields. It is often the case for certain values of parameters that these are the only structures of the given type. This is so for example in the case of finite projective geometries of dimension at least three. However, the techniques needed usually involve an extensive use of theory of block designs and this is why we cannot include them here.

1. (a) Let G_1, G_2 be two simple graphs with all degrees at least 4 and suppose that their line-graphs $L(G_1)$ and $L(G_2)$ are isomorphic. Prove that $G_1 \cong G_2$.

(b) Let G_1, G_2 be two connected simple graphs and suppose that $L(G_1) \cong L(G_2)$, but $G_1 \not\cong G_2$. What are G_1 and G_2?

(c) Let G_1, G_2 be connected simple graphs and $\varphi: L(G_1) \to L(G_2)$ an isomorphism between $L(G_1)$ and $L(G_2)$. We say that φ is trivial if there is an isomorphism $\psi: G_1 \to G_2$ such that ψ induces φ on $E(G_1)$. For which graphs are there non-trivial isomorphisms between $L(G_1)$ and $L(G_2)$?

2. (a) Let α be an automorphism of the line-graph of the hypergraph K_n consisting of all r-tuples of n points ($n \geq 3r$). Then α is induced by an automorphism of K_n^r (i.e. a permutation of $V(K_n^r)$).

(b) The preceding result holds for $3r \geq n > 2r$ but fails for $n \leq 2r$.

3. Let $0 \leq i < r$ be given. Form the graph $L_i(K_n^r)$ whose points are the r-tuples in K_n^r and in which two r-tuples A, B are adjacent iff $|A \cap B| = i$. Show that for sufficiently large n, every automorphism of $L_i(K_n^r)$ is induced by a permutation of $V(K_n^r)$ in the natural way.

4. (a) Every simple graph is the line-graph of some hypergraph. Describe these hypergraphs.

(b) Show that every graph G on n points without isolated points is the line-graph of a hypergraph on $\lfloor n^2/4 \rfloor$ points.

5. (a) Let G be a simple graph with more than 9 points. Suppose that, for each point x, $G - x$ is the line-graph of some simple graph H_x. Prove that G is the line-graph of a simple graph.

(b) One can find a finite number of "sample graphs" $G_1,\ldots,G_k$ so that a simple graph G' is the line-graph of a simple graph if and only if it does not contain any of $G_1,\ldots,G_k$ as an induced subgraph.

6. (a) Call a triangle abc of a graph G *odd* if there is a point $x \neq a$, b, c which is adjacent either to one or to three of the points a, b, c. Show that if the simple graph G is a line-graph, then

(i) it does not contain the 3-star as an induced subgraph,

(ii) whenever abc, bcd are odd triangles $(a \neq d)$, then a is adjacent to d.

(b) If G is a simple graph such that (i) and (ii) hold, then G is the line-graph of some simple graph.

(c) Determine the "sample graphs" in 15.5b.

7*. Show that the statement of 15.5b is not valid for line-graphs of 3-uniform hypergraphs (without multiple edges).

8. (a) Construct two isomorphic 2-connected planar graphs whose duals are not isomorphic.

(b) If two 3-connected planar graphs are isomorphic so are their duals.

9. Let G_1, G_2 be two 3-connected graphs and $\varphi : E(G_1) \to E(G_2)$ a bijection between their edges such that edges forming a circuit in G_1 are mapped onto the edges of a circuit in G_2 and conversely.

(a) Prove that φ preserves adjacency of edges.

(b) Prove[†] that $G_1 \cong G_2$.

10. Let T, T' be two trees and assume that they have the same set $\{x_1,\ldots,x_r\}$ of endpoints. Suppose that

$$d_T(x_i, x_j) = d_{T'}(x_i, x_j) \qquad (1 \leq i < j \leq r).$$

Show that T, T' are isomorphic.

11. Let $x_1,\ldots,x_r$ be the endpoints of a tree T and set

$$d_{ij} = d(x_i, x_j).$$

Show that
(a) for any three indices i, j k,

$$d_{ij} + d_{jk} - d_{ik} \geq 0, \qquad d_{ij} + d_{jk} - d_{ik} \equiv 0 \pmod 2,$$

(b) for any four indices i, j, k, l, two of the numbers $d_{ij}+d_{kl}$, $d_{ik}+d_{jl}$, $d_{il}+d_{jk}$ are equal and the third one is not greater than these two.

[†] This problem says that 3-connected graphs can be reconstructed from their matroids.

(c) Suppose that $(d_{ij})_{i,j=1}^r$ is a symmetric matrix consisting of non-negative integers for which $d_{ij} = 0$ iff $i = j$, and which satisfies (a) and (b). Then there exists a tree T with r endpoints $x_1, \ldots, x_r$ such that

$$d(x_i, x_j) = d_{ij}.$$

12*. (a) Let $n = 2^k$. Construct two collections $\{a_1, \ldots, a_n\} \neq \{b_1, \ldots, b_n\}$ of integers (the same integer may occur more than once) such that the two collections $\{a_i + a_j : 1 \leq i < j \leq n\}$ and $\{b_i + b_j : 1 \leq i < j \leq n\}$ are equal.

(b) This is not possible when n is not a power of 2; i.e. in this case $\{a_1, \ldots, a_n\}$ is reconstructible from the collection $\{a_i + a_j : 1 \leq i < j \leq n\}$.

13. Let (a_{ij}), (b_{ij}) be complex matrices in each of which no two rows are equal and let

$$f(s_1, \ldots, s_k) = \sum_{i=1}^N \binom{a_{i1}}{s_1} \cdots \binom{a_{ik}}{s_k},$$

$$g(s_1, \ldots, s_k) = \sum_{i=1}^N \binom{b_{i1}}{s_1} \cdots \binom{b_{ik}}{s_k}.$$

Suppose further that $f(s_1, \ldots, s_k) = g(s_1, \ldots, s_k)$ for every choice of non-negative integers $s_1, \ldots, s_k$. Prove that $f \equiv g$, i.e. the matrices (a_{ij}) and (b_{ij}) arise from each other by permutation of the rows.

14. Let G_1, G_2 be two simple graphs with a common underlying set V, $|V| > 3$. Suppose that $G_1 - x - y \cong G_2 - x - y$ for all x, $y \in V$. Show that G_1 and G_2 are identical.

15. (a) Is the following statement true? If G_1 and G_2 are simple graphs with a common underlying set V (which is large enough) and $G_1 - x \cong G_2 - x$ for all $x \in V$, then G_1 and G_2 are identical.

(b) Let G_1, G_2 be two graphs on a common underlying set V and suppose that $G_1 - x \cong G_2 - x$ for all $x \in V$. Let H be any given graph with less than $|V|$ points. Show that G_1 and G_2 have the same number of subgraphs [induced subgraphs] isomorphic to H.

(c) If G_1 is disconnected, then so is G_2 and they are isomorphic.[†]

16*. Let T_1, T_2 be two trees on the same set V and suppose that for each $x \in V$,

$$T_1 - x \cong T_2 - x.$$

(a) Prove that T_1, T_2 have the same diameter d.

[†] Ulam's famous conjecture states that this assertion holds for *any* pair of graphs.

(b) Prove further that $T_1 \cong T_2$.

17*. (a) Let G_1, G_2 be two simple graphs with $|V(G_1)| = |V(G_2)| = n$, $E(G_1) = \{e_1, \ldots, e_m\}$, $E(G_2) = \{f_1, \ldots, f_m\}$. Assume that $m > \frac{1}{2}\binom{n}{2}$ and $G_1 - e_i \cong G_2 - f_i$ for $i = 1, \ldots, m$. Then $G_1 \cong G_2$.

(b) Prove the same result when $m > \frac{1}{2}\binom{n}{2}$ is replaced by $m > n \log n$.

18. Let H_1, H_2 be two r-uniform hypergraphs on the same set V, and let $r \leq k \leq |V| - r$. Suppose that

(*) $$|E(H_1 - X)| = |E(H_2 - X)|$$

for all $X \subseteq V$, with $|X| = k$. Then H_1 and H_2 are identical.

19. Let H_1, H_2 be hypergraphs on the same set V and suppose that, for each $x \in V$,

$$H_1 \setminus x = H_2 \setminus x.$$

Show that either $H_1 = H_2$ or H_1 contains all even-element subsets of V and H_2 contains all odd-element subsets of V (or conversely).

20*. (a) Let G_1, G_2 be two simple graphs and suppose that they have the same number of homomorphisms into any third graph H. Then $G_1 \cong G_2$.

(b) Suppose that G_1, G_2 are two simple graphs and every graph H has the same number of homomorphisms into G_1 as into G_2. Then $G_1 \cong G_2$.

21. If $G_1 \times G_1 \cong G_2 \times G_2$, then $G_1 \cong G_2$.

22. (a) Construct simple graphs G_1, G_2, F without isolates such that $G_1 \not\cong G_2$, but $G_1 \times F \cong G_2 \times F$.

(b) Suppose that $G_1 \times F \cong G_2 \times F$, where G_1, G_2, F are simple graphs. Then

$$G_1 \times F_0 \cong G_2 \times F_0$$

for all subgraphs F_0 of F.

(c) Prove that the cancellation law holds for strong direct products, i.e. if $G_1 \cdot H \cong G_2 \cdot H$, then $G_1 \cong G_2$.

(d) The same cancellation law holds for direct products of finite groups and rings.

II. Hints

§ 1

1. Decide about the persons one by one.

2. Decide now about the postcards. The answer to the second question is $n! \left\{ {k \atop n} \right\}$.

3. If we form all permutations of the letters, how often does the same word occur?

4. (a) Imagine k 1-forint coins in a row and suppose that the people come one by one and pick up forints as long as you allow them. (b) Reduce to the previous case by borrowing one forint from each person.

5. One can decide about the different kinds of postcards independently.

6. (a) Possible recurrence relations are

$$\left\{ {n+1 \atop k} \right\} = \left\{ {n \atop k-1} \right\} + k \left\{ {n \atop k} \right\},$$

$$\left[{n+1 \atop k} \right] = \left[{n \atop k-1} \right] + n \left[{n \atop k} \right].$$

(b) Observe that in a partition of n elements into $n-k$ classes, at least $n-2k$ classes must be singletons.

(c) For the Stirling partition numbers, write the recurrence in (a) as

$$\left\{ {n \atop k-1} \right\} = \left\{ {n+1 \atop k} \right\} - k \left\{ {n \atop k} \right\},$$

to get a recurrence for negative values of k.

(d) $\left\{ {n \atop k} \right\}$ and $\left[{-k \atop -n} \right]$ satisfy the same recurrences and boundary conditions.

7. (a) If x is an integer, both sides of the first identity count mappings of an n-element set into an x-element set. (b) Both sides count pairs (π, α) where π is a permutation of and n-elements set S, and α is a coloration of S with x colors, invariant under π. (c) Combine the identities in (a) and (b).

8. Substitute

$$j^n = \sum_{r=0}^{n} \left\{ {n \atop r} \right\} j(j-1)\ldots(j-r+1).$$

9. (a) Use

$$B_n = \sum_{k=0}^{\infty} \left\{ {n \atop k} \right\}.$$

(b) Prove that the distribution of terms in the sum in (a) is asymptotically normal; if we set

$$g_n(y) = \frac{1}{\sqrt{2\pi}} y^{n-y-1/2} e^{y-1} \qquad (\sim y^n/ey!),$$

then

$$\frac{g_n(\lambda(n) + y\lambda(n)/\sqrt{n})}{g_n(\lambda(n))} \to e^{-y^2/2} \qquad (n \to \infty).$$

10. Consider the partition class containing a given element.

11. Use the previous recurrence relation to derive a differential equation for $p(x)$.

12. (a) For a partition, consider all permutations in which the classes are arranged in order of non-decreasing lengths. (b) The result is $\approx (e-1)^n$.

13. Consider the Taylor expansion of e^{e^x}.

14. (a) The second derivation of the formula for $p(x)$ (1.12) can be modified to apply here. (b) The method of the first proof of the formula for $p(x)$ (1.11) easily extends to this case.

15. (a) Use the recurrence relation

$$S_n = \sum_{k=1}^{n} \binom{n}{k} S_{n-k}.$$

(b) Use

$$S_n = \frac{n!}{2\pi i} \oint_{|z|=\varepsilon} \frac{s(z)}{z^n} dz.$$

16. Represent the partition $n = a_1 + \ldots + a_s$, $a_1 \geq \ldots \geq a_s$ by a diagram in which the ith column consists of a_i dots.

17. Associate the partitions $n = a_1 + \ldots + a_m$ and $n - m = (a_1 - 1) + \ldots + (a_m - 1)$ with each other.

18. Let
$$n = a_1 + \ldots + a_m$$
be a partition of n into distinct terms. Write
$$a_i = 2^{\beta_i} b_i, \text{ where } b_i \text{ is odd.}$$

19. If
$$n = a_1 + \ldots + a_m, \quad a_1 > \ldots > a_m \geq 1$$
is a partition of n, then try to form a partition into distinct terms with one more term by subtracting 1 from a_1 and taking a new term 1. This will not work in general as we may have $a_1 - 1 = a_2$. In that case, also subtract 1 from a_2 and let the new term be 2, etc.

20. It is
$$S(x) = \frac{1}{(1-x)(1-x^2)(1-x^3)\ldots}.$$

21. Represent each of the generating functions as a product similarly as in the preceding solution.

22. The difference between the numbers of partitions of n into an odd and an even number of distinct terms is the coefficient of x^n in
$$(1-x)(1-x^2)(1-x^3)\ldots$$

23. Identity (a) expresses the following fact that the number of partitions into distinct odd terms is equal to the number of partitions with symmetric Ferrer's diagrams. (Symmetric means axially symmetric with respect to the line starting from the left lower corner and ascending at $45°$.)

24. Decide, starting with 1, which numbers with what multiplicity must occur in such a partition. If 1 occurs k_1 times, the next least number occurring is $k_1 + 1$.

25. If
$$n = x + y + z$$
is a partition of n, then $x+y$, $y+z$, $x+z$ are the sides of a triangle with circumference $2n$.

26. Find a 2-step recurrence relation on M_n; guess a 1-step recurrence relation and prove it by induction.

27. Find a recurrence relation for A_n.

28. Deduce a recurrence relation for C_n and prove that the upper determinant satisfies it as well as the initial values. As initial values, take $C_{-k+1} = \ldots = C_{-1} = 0$, $C_0 = 1$. (b) Use the recurrence relation in the same way as in 1.27.

29. Prove the recurrence relation

$$p_n(\lambda) = \lambda p_{n-1}(\lambda) - p_{n-2}(\lambda)$$

for the characteristic polynomial

$$p_n(\lambda) = \det(\lambda I - A).$$

30. Denote by x_n and y_n the number of such sequences starting with a or b and c or d, respectively, and find recurrence relations on these numbers.

31. Use ideas similar to those in the solution of 1.17.

32. The graph of such a function can be considered as a polygon, joining the point $(1,1)$ to the point (n,n), moving to the right or one upwards at each step.

33. It is more convenient to deal with the monotonic mappings f of $\{1,\dots,n\}$ into $\{0,\dots,n-1\}$ with $f(x) < x$. The mappings not having this property are in a one-to-one correspondence with step-functions connecting $(0,1)$ to $(n+1,n-1)$. The result is $\frac{1}{n+1}\binom{2n}{n}$.

34. Denote by $g(n,r)$ the number in question. Establish and then use the recurrence relation

$$g(n,r) = \sum_{k=0}^{r} \binom{r}{k} g(n-1,k).$$

35. Consider the procedure backwards.

36. Use induction as follows: remove an element of S, then the sequence of partitions of the remaining set is almost a sequence arising by the procedure of the problem.

37. Reduce (b) to the preceding problem.

38. Use 1.37 or 1.33.

39. Associate a bracketing of the product $x_1 \dots x_{n-1}$ with a triangulation.

40. Use the recurrence relation

$$D_n = \sum_{k=2}^{n-2} D_k D_{n-k+1}.$$

41. The triangles form a chain.

42. (a) Evaluate $\sum_{k=0}^{\lfloor(n-1)/2\rfloor}\binom{n}{2k+1}$ at the same time. (b) Consider the Pascal triangle. (c) Find a combinatorial interpretation. (d) This is the coefficient of x^m in $(1-x)^u(1+x)^u$. (e) $\binom{k}{m}\binom{n}{k}=\binom{n}{m}\binom{n-m}{n-k}$. (f) Set $\varepsilon=e^{\frac{2\pi i}{7}}$ and use

$$\sum_{j=0}^{6}\varepsilon^{kj}=\begin{cases}7 & \text{if }7|k,\\0 & \text{otherwise.}\end{cases}$$

(g) Find a recurrence relation for these numbers (z fixed, n varies). (h) Prove by induction that the result is $(-1)^m\binom{n-1}{m}$. (i) The result is $\binom{u+v+1}{m}$.

43. (a) Find a combinatorial interpretation. (b) Substitute $-n-1$ for n in (a). (c) Substitute

$$\binom{n+k}{p+q}=\sum_{j=0}^{k}\binom{n}{p+q-j}\binom{k}{j}.$$

(d) Substitute $-n-1$ for n in (c). (e) Expand $\binom{x+t}{n}-\binom{x}{n}$ by 1.42c and i.

44. To prove (a), differentiate both sides with respect to y and use induction on n. (b), (c) will follow from (a).

45. Use $f_n(x)=n!\binom{x}{n}$.

§ 2

1. The number of pupils who like either mathematics or physics is not $12+14$. By how much is 26 too large?

2. Determine the contribution of any atom of the Boolean algebra generated by $A_1,\ldots,A_n$ on each side.

3. Let $S=\{1,\ldots,n\}$ and let A_i be the set of integers divisible by p_i.

4. Count the number of mappings of a set of k given elements onto a set of n given elements.

5. $p-\sigma^1p+\ldots$ vanishes for $x_i=0$ $(i=1,\ldots,n)$.

6. Consider the "contribution" of an atom B of the Boolean algebra generated by $A_1,\ldots,A_n$.

7. (a) Apply the preceding result. (b) Use

$$\sigma_j = E\left(\binom{\eta}{j}\right).$$

8. Sum the results of the preceding problem for p, $p+1$, $\ldots$

9. This goes back to the partial sums of

$$\binom{k}{0} - \binom{k}{1} + \binom{k}{2} - \cdots$$

10. Show that

$$\sigma_r \leq \frac{m-r}{r}\sigma_{r-1} + \frac{1}{r}\binom{m}{r-1}\sigma_m.$$

11. For each $J \subseteq \{1,\ldots,n\}$,

$$\mathsf{P}\left(A_J \cdot \prod_{j \notin J} \overline{A}_j \right) \geq 0.$$

12. Denote by $\mathcal{E}_0$ and $\mathcal{O}_1$ the set of independent even subsets and the set of odd subsets spanning at most one edge. Prove the inequalities

$$|\mathcal{E}_1| \geq |\mathcal{O}_0| \quad \text{and} \quad |\mathcal{O}_1| \geq |\mathcal{E}_0|$$

simultaneously, by induction.

13. If $\mathsf{P}(A_1) = \ldots = \mathsf{P}(A_n) = 1$, then the terms $(-1)^{|I|}$ and $(-1)^{|J|}$ with $J = I \cup \{n\}$ cancel out; the remaining terms are positive.

14. (a) Use 2.6. (b) A lower estimate is the following:

$$\mathsf{P}(\overline{A}_1 \ldots \overline{A}_n) \geq 1 - \sum_{\substack{I,J \subseteq \{1,\ldots,n\} \\ \max I = \max J}} \lambda_I \lambda_J \mathsf{P}(A_{I \cup J}),$$

where $\lambda_\emptyset = 0$, $\lambda_{\{k\}} = 1$ for $1 \leq k \leq n$ and λ_I is an arbitrary real number for $|I| \geq 2$.

15. Let $q_i = 1 - p_i$ and $q_I = \prod_{i \in I} q_i$ and show that

$$p_{I \cup J} = p_I p_J \sum_{K \subseteq I \cap J} \frac{q_K}{p_K}.$$

16. Set $\mathsf{P}(A_K) = p_K + R_K$, $\mathcal{H} = \left\{ K : p_K \geq \frac{1}{M} \right\}$, minimize

$$\sum_{I,J \in \mathcal{H}} \lambda_I \lambda_J p_{I \cup J} \qquad (\lambda_\emptyset = 1)$$

and estimate

$$\sum_{I,J \in \mathcal{H}} \lambda_I \lambda_J R_{I \cup J}$$

at this place.

17. Let $M = \sqrt{x}/\log^2 x$ and let $P_1, \ldots, P_n$ be those primes $\leq M$ which do not divide k. "Sift out" those numbers from the given sequence which are divisible by one of $P_1, \ldots, P_n$, using Selberg's method.

18. Prove that ·

$$P(A_1 | \overline{A}_2 \ldots \overline{A}_n) \leq \frac{1}{2d}$$

by induction on n.

19. Let ζ be the number of A_i's that occur. Use Chebyshev's inequality

$$P(\overline{A}_1, \ldots, \overline{A}_n) = P(\zeta = 0) \leq \frac{1}{\sigma_1^2} E((\zeta - \sigma_1)^2).$$

20. To compare the result of 2.19 with the estimate of Selberg's method, expand the latter as a power series in p.

21. Only the statement concerning the inverse is non-trivial. Towards this, show that (a_{ij}) is invertible iff $a_{ii} \neq 0$ $(i = 1, \ldots, n)$.

22. What will the matrix $M = (m_{ij})$, where $m_{ij} = \mu(x_i, x_j)$, be equal to?

23. $\mu(a, b)$ depends only on the structure of the interval $\{z : a \leq z \leq b\}$.

24. Show that μ^* defined by $(*)$ satisfies the identities in the definition of μ.

25. Use the fact that $(Z - I)^n = 0$.

26. (For the second part) Choose V to be the set of all subsets of $\{1, \ldots, n\}$ and

$$f(K) = P\left(\prod_{i \notin K} A_i \prod_{j \in K} \overline{A}_j\right).$$

27. Use

$$\sum_{x \leq b} \mu(0, x) = \sum_{a \leq b_1 \leq b} \sum_{a \vee x = b_1} \mu(0, x).$$

28. Let a be an atom such that $a \leq x$ and consider

$$\sum_{y \vee a = x} \mu(0, y) = 0.$$

29. Decompose

$$\sum_x \sum_{y \geq x} \mu(x, y)$$

into terms according to whether x, y belong to A, B, C.

30. (a) Use 2.27 and induction on n. (b) Use induction on d and Euler's formula: if f_i denotes the number of i-dimensional faces of a d-dimensional convex polytope ($f_{-1} = f_d = 1$), then

$$\sum_{i=-1}^{d} (-1)^i f_i = 0.$$

31. Define

$$f_{ij} = \begin{cases} f(x_i) & \text{if } i = j, \\ 0 & \text{otherwise.} \end{cases}$$

What is the relationship between the matrices $F = (f_{ij})$ and $G = (g_{ij})$?

32. Use the set $\{1, \ldots, n\}$ ordered partially by divisibility and a slight modification of 2.31.

33. Find a similar representation $D = Z^T A Z$ of the matrix $(d_{ij}) = D$ as for the matrix G in the solution of 2.32.

34. Write M as a polynomial in $Z - I$, in a way similar to the one in 2.25.

35. Denote by $q_k(x)$ the number of those k-tuples in R whose union is x. Show that

$$q_0(x) - q_1(x) + q_2(x) - q_3(x) + \ldots = \mu(0, x).$$

36. Use the identity of 2.29 in an argument similar to the preceding one.

37. Use 2.27 and induction on r.

§ 3

1. Two permutations are conjugate iff their cycles have the same cardinalities.

2. (a) Use inclusion-exclusion; (b) build up such a permutation, starting from a given point.

3. Count those permutations in which the set of elements in the cycle containing 1 is given.

4. Count similarly as before.

5. Determine the expected number of points in k-cycles.

6. Use the coding of the second solution of 3.3.

7. Prove the recurrence relation

$$n p_n(x_1, \ldots, x_n) = \sum_{k=1}^{n} x_k p_{n-k}(x_1, \ldots, x_{n-k}).$$

8. For (b), represent these permutations by a permutation of the set of classes and by k permutations, permuting each of the classes.

9. If the corresponding elements of the two groups have the same order, then their regular representations have the same cycle index.

10. Observe the identity

$$q_n(x_1, \ldots, x_n) = p_n(x_1, \ldots, x_n) + p_n(x_1, -x_2, \ldots (-1)^{n-1} x_n).$$

11. Consider

$$\sum_{n=0}^{\infty} (p_n - p_{n-1}) y^n.$$

12. (a) Denoting the left hand side by $f_n(x)$, observe that $f_n(x) = n! p_n(x, \ldots, x)$, where $p_n(x_1, \ldots, x_n)$ is the cycle index of S_n. (b) Evaluate $f_n'(1)$.

13. Such a graph consists of disjoint circuits. If we orient these, we get the graph of the cycle decomposition of a permutation of n elements.

14. (a) Join the two numbers in a pair by an edge. (b) How many ways can one orient the preceding graph?

15. Express these numbers as certain values of the cycle index.

16. The sequence $\overline{\pi}(1), \ldots, \overline{\pi}(n)$ uniquely characterizes the permutation π.

17. If the jumper to jump at the ith time executes the $(\pi(i))^{\text{th}}$ longest jump, then π is a random permutation of $\{1, \ldots, n\}$. The ith jump is a record iff $\overline{\pi}(i) = i$.

18. Observe that the optimal strategy to follow is independent of the order of the first k jumps. Determine $p_{n-1}, p_{n-2}, \cdots$

19. This is equivalent to $\overline{\pi}(j+1) \geq \overline{\pi}(j)$.

20. The number of inversions of π is

$$(\overline{\pi}(1) - 1) + (\overline{\pi}(2) - 1) + \ldots + (\overline{\pi}(n) - 1).$$

21. Suppose that we start with a large reserve in our tank and use it, if running out of gas. Consider the point where the reserve needed is maximal.

22. (a) Call a sequence $(j_1, \ldots, j_s)$ of different integers j_i $(1 \leq j_i \leq n)$ *ascending*, if

(i) $x_{j_1} + \ldots + x_{j_s} > 0$, but

(ii) $x_{j_1} + \ldots + x_{j_\nu} \leq 0$ for $1 \leq \nu < s$.

(For $s = 1$, the one-element sequence (j_1) is ascending iff $x_{j_1} > 0$.) Call the sequence *descending* if

(i') $x_{j_1} + \ldots + x_{j_s} \leq 0$, but

(ii') $x_{j_\nu} + \ldots + x_{j_s} > 0$ for $1 < \nu \leq s$.

(Note that the two definitions are not completely analogous.) Consider partitions of $\{1, \ldots, n\}$ into ascending and descending sequences, and build up permutations π from them for which $a(\pi)$ and $b(\pi)$, respectively, can be determined easily, cf. also 3.21.

(b) Put $n = 2m$, $x_1 = \ldots = x_m = 1$ and $x_{m+1} = \ldots = x_{2m} = -1$ in (a).

23. (a) Count the number of k-gons invariant under a given rotation. (b) Use a similar method.

24. Count the number of elements of Γ fixing a given point.

25. This is only a slight generalization of 3.24.

26. We want to determine the number of orbits of the permutation group induced by Γ on the set Ω of mappings of D into R.

27. First find a formula involving F; then use

$$F(u_1, u_2, \ldots) = F\left(\frac{\partial}{\partial z_1}, \frac{\partial}{\partial z_2}, \ldots\right) e^{\sum_{i=1}^{\infty} u_i z_i}\Big|_{z_i = 0}$$

28. Use a method similar to the previous one; a one-to-one mapping which is a fixed point of (π, ϱ) must map a cycle of π onto a cycle of ϱ with the same length.

29. First determine the generating function of the numbers $q_n(\gamma)$ of mappings f satisfying $(*)$ and invariant under a given $\gamma \in \Gamma$.

30. If D is the set of forints, R the set of people and Γ the symmetric group on D, 3.26 gives the answer.

31. Put $|D| = n$, $|R| = N \geq n$, $\Gamma = \{1\}$ and $\Gamma_1 = S_N$ in 3.27; let $N \to \infty$.

§ 4

1. Supposing $d_n = 1$, remove the point v_n.

2. Use the preceding result and the binomial theorem.

3. Show that

$$p_n(x_1, \ldots, x_{n-1}, 0) = (x_1 + \ldots + x_{n-1}) p_{n-1}(x_1, \ldots, x_{n-1})$$

and use the identity 2.5.

4. Contract each T_i onto a single point v_i.

5. Try to reconstruct the sequence $b_1, \ldots, b_{n-1}$ of removed points; observe that b_i is the least number not occurring among $b_1, \ldots, b_{i-1}, a_i, \ldots, a_{n-1}$.

6. Remove one edge of the tree in every possible way.

7. (a) Prove that

$$\left(\frac{t(x)}{x}\right)' = t'(x)\frac{t(x)}{x}.$$

(b) Use the fact that

$$T_n = \frac{1}{n}t^{(n)}(0) = \frac{(n-1)!}{2\pi i n}\oint\frac{t'(z)}{z_n}\,dz,$$

where C is any simple closed curve around the origin.

8. Use 4.1. The result is $\frac{n!}{(n-l)!}S(n-2,l)$.

9. (a) Use the Binet–Cauchy formula:

$$\det A_0 A_0^T = \sum(\det B)^2,$$

where B ranges over all $(n-1)\times(n-1)$ submatrices of A_0. (b) Two rows of A_0 have at most one common non-zero entry. (c) The number of trees on n points is the number of spanning trees of K_n, the complete n-graph.

10. Write D in the form $A_0 A_0^T$ and use an argument similar to that in the solution of 4.9.

11. This is the number of spanning trees of the complete bipartite graph on $\{v_1,\ldots,v_n;w_1,\ldots,w_m\}$.

12. (a) Use the inclusion-exclusion principle and 4.4. (b)–(c) Use 4.9 or 4.12a.

13. (a) Consider the binary tree with $k+1$ endpoints as a diagram to calculate a product with k factors. (b) Imagine that the edges of such a tree are walls. Starting from the root, walk around the wall, keeping it on your left. Write a 1 when walking along an edge away from the root and a -1 otherwise. Which ± 1-sequences arise in this way?

14. Use 4.4. Alternatively prove the recurrence relation

$$E(n, k-1) = \left(1 - \frac{1}{k}\right)nE(n,k).$$

15. Trivial: the edges entering the points $\neq a$ can be chosen independently.

16. (a) Use induction, by splitting the edges entering v_1 into two classes C_1, C_2 and considering the graphs $G - C_1$, $G - C_2$. (b) Consider the a_{ij}'s as variables and substitute $a_{ij} = y_i$.

17. If there is a fixed point x, orient such a tree T to get an arborescence $\overrightarrow{T}$ rooted at x. Contract all orbits of π and investigate the image of $\overrightarrow{T}$.

18. To prove the upper bound consider rooted plane trees.

19. (a) Write the right-hand side in the form

$$x \prod_T \left(1 + x^{|V(T)|} + x^{2|V(T)|} + \ldots \right),$$

where T ranges over all isomorphism types of trees. (b) Denote by $W_n^{(d)}$ the number of non-isomorphic rooted trees in which the root has degree d. Express first the generating function of the numbers $W_n^{(d)}$ for a fixed d in terms of $w(x)$ by the Pólya–Redfield method.

20. (a) $w(x)$ satisfies the equation

$$w(x)e^{-w(x)} = \varphi(x),$$

where

$$\varphi(x) = x \exp \left(\frac{w(x^2)}{2} + \frac{w(x^3)}{3} + \ldots \right)$$

is analytic in a larger circle than w. (b) Show that

$$w''(x) = B_1(\tau - x)^{-3/2} + B_2 2(\tau - x)^{-1/2} + h(x),$$

where $h(x)$ is continuous on the (closed) disc $|x| \leq \tau$.

21. Each expansion term of per A counts the number of 1-factors parallel to a given one.

22. Find a recurrence relation.

23. Use inclusion-exclusion.

24. Those expansion terms of $\det B$ which correspond to a permutation with at least one odd cycle will cancel out.

25. (a) The non-zero terms in Pf B correspond to 1-factors of G. (b) When do the terms in Pf B corresponding to two 1-factors have the same sign?

26. Every term in $\det B$ that does not correspond to a 1-factor has 0 expectation.

27. Prove that there is an orientation such that, if we go around the boundary of any bounded face in the positive sense, we pass an odd number of edges agreeing with its orientation.

28. Orient the ladder as in Fig. 13 and apply 4.21 and 4.23.

FIGURE 13

29. (a) Write a_n^2 in the form

$$\det p_n(A),$$

where

$$p_n(\lambda) = \begin{vmatrix} -\lambda & 1 & & & 0 \\ 1 & -\lambda & & & \\ & & \ddots & & 1 \\ 0 & & & 1 & -\lambda \end{vmatrix}, \qquad A = \begin{pmatrix} 0 & 1 & & & 0 \\ -1 & 0 & & & \\ & & \ddots & & 1 \\ 0 & & & -1 & 0 \end{pmatrix}.$$

Use 1.29 to obtain the roots of $p_n(\lambda)$. (b) Observe that the formula (a) is the resultant of two polynomials and use its Sylvester determinant form. (c) Using formula (a), $\frac{\log a_n}{n^2}$ will be a sum approximating

$$\frac{1}{\pi^2} \int_0^\pi \int_0^\pi \log(4\cos^2 x + 4\cos^2 y) \mathrm{d}x\, \mathrm{d}y.$$

30. (a) Consider the $n \times n$ chessboard as consisting of every second square of the $(2n-1) \times (2n-1)$ chessboard. (b) Take the "union" of G and its dual.

31. Identify u_i and v_i.

32. Find a recurrence relation.

33. Consider a_n as the permanent of a matrix and expand it by its first row. Find simultaneous recurrence relations on a_n and some other analogous numbers.

34. This number is

$$\mathrm{per} \begin{pmatrix} \overbrace{1\ldots1}^{p} & \overbrace{00\ldots00}^{n-p} \\ 1\ldots1 & 10\ldots00 \\ \vdots & \ddots \quad \vdots \\ 1\ldots1 & 11\ldots10 \\ 1\ldots1 & 11\ldots11 \\ \vdots & \vdots \\ 1\ldots1 & 11\ldots11 \end{pmatrix} \begin{matrix} \left.\vphantom{\begin{matrix}1\\1\\1\\1\end{matrix}}\right\} n-p \\ \\ \left.\vphantom{\begin{matrix}1\\1\\1\end{matrix}}\right\} p \end{matrix}.$$

Expand by the first row.

35. Use inclusion-exclusion.

36. The number is the permanent of the matrix

$$
\begin{pmatrix}
\overbrace{11\ldots11}^{n} & \overbrace{00\ldots00}^{n} \\
11\ldots11 & 10\ldots00 \\
\vdots\ \vdots & \quad\ddots \\
11\ldots11 & 11\ldots10 \\
01\ldots11 & 11\ldots11 \\
\vdots\ \ \ddots & \quad\vdots \\
00\ldots01 & 11\ldots11 \\
00\ldots00 & 11\ldots11
\end{pmatrix}
\begin{matrix}
\left.\vphantom{\begin{matrix}1\\1\\1\\1\end{matrix}}\right\}n \\
\left.\vphantom{\begin{matrix}1\\1\\1\\1\end{matrix}}\right\}n
\end{matrix}
$$

Count the expansion terms having k entries from the upper left block.

§ 5

1. (a) What is the sum of the degrees of a graph? (b) The title of the chapter is misleading, use divisibility by 3. (c) How are the first and last two points joined to each other?

2. Use the solution of 5.1a.

3. If every circuit is even, then lengths of any two paths between a pair x, y of points have the same parity.

4. If the total work around any circuit is 0, the work needed to go from x to y is independent of the path used.

5. If there is an even cycle C, start defining a 2-coloration of G by first 2-coloring C and extending.

6. Take a maximum closed trail in G and show that it contains all the edges.

7. (a) Trivial by 6. (b) Note that $G_{k,n} = L(G_{k-1,n})$.

8. This means the same as "$G_{k,2}$ has a Hamiltonian cycle".

9. Look at the sections of an Euler trail between two consecutive occurrences of a point x of degree at least 3.

10. To show that we have used every edge consider an edge of T not used and nearest to x_0.

11. The Euler trails can be considered as starting from x_0 trough a specified edge. Show that every Euler trail arises by the algorithm given in 5.10, from some spanning arborescence.

12. Call a point "good" if all edges incident with it have been traversed in both directions. Look at the first "bad" point met during the walk.

13. Show and then use the fact that a graph with even degrees and without isolated points is the union of edge-disjoint circuits.

14. (a) Use 5.13 and 5.6. (b) Take a new point and join it to all points with odd degree.

15. Use induction on the number of edges.

16. Use induction on the number of edges; observe that if G_1 is a spanning subgraph with even degrees and K is a circuit, then $(V(G), E(G_1)\triangle E(K))$ is a spanning subgraph with even degrees (briefly; a "good" subgraph) as well.

17. (a) Use induction on n. Remove a point of odd degree and take the complement within the set of its neighbors. (b) Add a new point and join it to all points of $V(G)$. (c) Add a new point and join it to all points with even degree.

18. (a) Take all r-element matching of the complete graph on $V(G)$ and sieve out those having an edge from G. (b) By (a). (c) The number of 1-factors has the same parity as the determinant of the adjacency matrix.

19. Use an argument similar to the one above. For undirected graphs, the statement is true if $|V(G)| > 3$.

20. Show that reversing an edge, the parity of the number of Hamiltonian paths does not change.

21. Consider pairs (F_1, F_2) of 1-factors such that $F_1 \cap F_2 = \emptyset$, $e \in F_1$. Show that the number of such pairs is even, and make use of this.

22. Reduce the graph as shown in Fig. 14 and distinguish Hamiltonian circuits going through the edges in different directions.

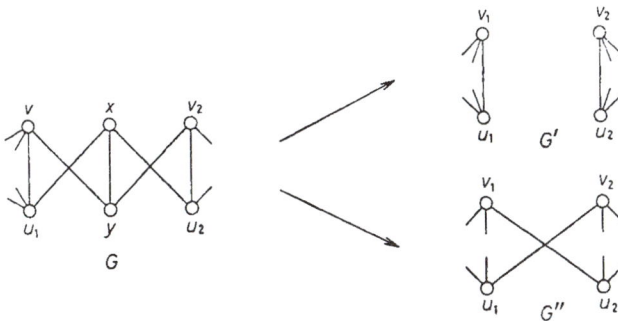

FIGURE 14

23. Let F be a spanning tree of G. Show that those edges of G^* which do not correspond to edges of F form a spanning tree of G^*.

24. Use the previous proof.

25. Each face has at least three edges on the boundary.

26. Remove the edges on the boundary of a face.

27. (a) Use the 2-coloration of the faces. (b) Count red-blue corners, using Euler's formula.

28. (a) Consider a 2-coloration of the faces. (b) Take a new point outside the pentagon and join it to the points of odd degree.

29. Count the number of edges joining a red point to a blue point.

30. Color a point x red, if $x \in V_1$ and V_1 contains an (x,a)-path; blue, if $x \in V_2$ and V_2 contains an (x,b)-path; green otherwise. Show that there is no triangle with all three colors.

31. Only the proof of (f) is somewhat difficult. Take a basis $\{\mathbf{v}_1, \ldots, \mathbf{v}_k, \mathbf{v}_{k+1}, \ldots, \mathbf{v}_n\}$ of V with $\langle \mathbf{v}_1, \ldots, \mathbf{v}_k \rangle = M$ and consider the transformation A defined by

$$A : \mathbf{e}_i \to \mathbf{v}_i.$$

32. (a) Show and then use the fact $\langle M \cup M^\perp \rangle = (M \cap M^\perp)^\perp$. (b) If $\mathbf{a}$ is the column vector formed by the diagonal of A, then $\mathbf{v}^T A \mathbf{v} = \mathbf{a}^T \mathbf{v}$ holds for every vector v over $GF(2)$.

33. (a) The stars generate the subspace of cuts, the orthogonal subspace consists of those sets of edges which constitute a subgraph with even degrees. (b) Use 4.9.

34. Show that each circuit C is the sum of those faces contained in C.

35. (a) Show that $\sum_{i \in I \cup J} C_i \subseteq K$. (b) Consider a set I such that $K = \sum_{i \in I} C_i$ is a circuit, $|I| \geq 2$ and $|I|$ is minimal. (c) Use (b) and induction on f. (d) By (c) and 5.34.

36. Use MacLane's criterion.

37. (a) Supposing $G = G_1 \cup G_2$, $V(G_1) \cap V(G_2) = \{x, y\}$, contract $G_1 - x$ and $G_2 - x$. (b) Consider a maximum path. (c) Remove the chord of the circuit, drow the rest in the plane and choose the circuit so that the number of regions inside be maximal. Prove that the graph has only chords outside the circuit. (d) Clear from (c) and 5.25b.

38. It suffices to consider triangulations. (Prove!) Find an edge e which is contained in exactly two triangles. Contract e and use induction.

§ 6

1. Addition of an edge decreases the number of components by at most one.

2. (a) Let $T_1, \ldots, T_{c(G_1)}$ be the components of $G_1, S_1, \ldots, S_{c(G_2)}$ the components of G_2. Construct a graph G^* with $V(G^*) = \{t_1, \ldots, t_{c(G_1)}, s_1, \ldots, s_{c(G_2)}\}$ joining t_i to s_j iff $T_i \cap S_j \neq \emptyset$ and look at the components of this graph.

(b) Add the points of $V(G_1) - V(G_2)$ to G_2 as isolated points.

3. Let x_i have degree d_i. Consider a component G_1 of G not containing x_n; this has points of degree less than $|V(G_1)|$ only!

4. If G_1 contains an odd circuit (say) and is connected, show that G_1 contains a walk of any large enough length connecting any two points, and make use of this fact.

5. Cf. 5.1b.

6. (a) Consider a longest path.
(b) Consider an endpoint of a longest path again.
(c) If no such pair exists, then the endpoints of a spanning tree must induce a complete subgraph.

7. (a) One can increase the number of common edges of T_1 and T_2 at each step.
(b) Use "backwards" induction on the number of edges in a common subtree of T_1 and T_2.

8. (a)–(b) Use 6.7b.

9. To show that there is an (a, b)-path, consider the set X of all points accessible from a.

10. Contract the red edges and remove the green ones.

11. (a)–(b)–(c). If F is a minimal set whose removal (contraction) yields a digraph with the desired property, then the inversion of F also results in such a graph.
(b) Consider the case $|F| = 1$ first.
(c) Show first that the points of an acyclic graph G can be ordered in such a way that every edge has larger head than tail.

12. To show that a strongly connected tournament contains a Hamiltonian cycle, consider a maximal cycle.

13. Consider a longest cycle which is not a Hamiltonian cycle.

14. Invert the edges of F.

15. Find a proper subtree mapped into itself by φ.

16. Consider two points of the intersection and the path connecting them.

17. (a) Assuming P_1, P_2 are disjoint maximum paths, consider a (P_1, P_2)-path Q.
(b) The middle point of a fixed maximum path is contained in every other maximum path.

18. Use induction on k or n. In the first case find an edge separating G_k from $G_1 \cap \ldots \cap G_{k-1}$.

19. Only (c) is non-trivial. In the case of $d(x, y)$ consider a minimum (x, y)-path and a minimum (y, z)-path; in the case of $D(x, y)$, consider a maximal (x, z)-path.

20. Take an edge and count how many paths of type (p_i, p_j), (p_i, q_j), (q_i, q_j) contain it.

21. (a) Show that removing all points of degree 1, $\tilde{d}(x)$ decreases by 1 for all of the remaining points.
 (b) Consider a longest path with endpoint x.

22. (a) Moving from x to y, for which points z does $d(x, z)$ increase and decrease?
 (b) Easy from (a).
 (c) Take a long path and a large "fan" at one of its ends.

23. To determine the minimum of $\sum_{x,y} d(x, y)$ observe that exactly $n - 1$ terms are equal to 1; can all the others be equal to 2? On the other hand, show that $s(x)$ is maximal when G is a path and x an endpoint of it.

24. Associate, with each of all but k of the points, a path of length k starting from it, such that different points are associated with different paths.

25. Let G be the tree constructed by the algorithm, and H a tree of minimum expense having the largest possible number of edges in common with G. Assuming $G \neq H$, consider an edge of G not in H.

26. Use an argument similar to the one in 6.25.

27. Two edges are equivalent iff they lie on the same circuits!

28. Choose subgraphs $G_1, \ldots, G_i$ satisfying the condition and show that if $G_1 \cup \ldots \cup G_i \neq G$ you can choose a G_{i+1}.

29. The non-trivial part is that if G is 2-edge-connected it has a desired orientation. Use 6.27 or 6.28.

30. Find a circuit with which all but one of the edges are well-fitted and contract it, and proceed by induction.

31. To show that if e_1, e_2 lie on a circuit C_1 and e_2, e_3 lie on a circuit C_2, then e_1, e_3 also lie on a circuit, consider the common points of C_1 and C_2 next to e_3 on C_2 in both directions.

32. In the cycle (i)$\Rightarrow$(iii)$\Rightarrow$(ii)$\Rightarrow$(i) only the first step is non-trivial. Prove (iii) for two adjacent edges first!

33. (a) Select 3 paths P_1, P_2, P_3 connecting the same pair of points, such that the length of P_1 is minimal, and show that P_1 satisfies the requirements. (b) Use either (a) or an argument analogous to the solution of 6.28.

34. Arrange the points of G in a sequence such that p is the first point, q is the last point and from any other point there are edges connecting it both to earlier and to later points.

35. A chord of circuit could be removed without violating 2-connectivity.

36. Select a (C, C)-path P with endpoints x, y for which the arc R connecting x and y on C is minimal. Then the neighbor z of x on R has degree 2.

37. If G' were connected, G would contain a circuit whose points of degree 2 would constitute an arc.

38. Identify the corresponding endpoints of two isomorphic trees.

39. (a) Use induction on $|E(G)|$. If there is an $S \subseteq V(G)$ which defines a k-element (a, b)-cut and $|S| \geq 2$, $|V(G) - S| \geq 2$, then apply the induction hypothesis on the graphs obtained by contracting S and $V(G) - S$, respectively.

(b) Split each point $x \neq a$, b into two points x_1, x_2, where x_1 is joined to x_2, and x_2 is joined to y_1 iff x is joined to y.

(c) Replace each (undirected) edge by two oppositely oriented edges.

40. Take two new points a and b; join a to all points of A, b to all points of B.

41. A common generalization is the following. If A, $B \subseteq V(G)$ are disjoint sets and there is a positive integer "capacity" $w(x)$ associated with each point x, and for every set S meeting all (A, B)-paths we have

$$\sum_{x \in S} w(x) \geq k,$$

then there are k (A, B)-paths such that each point x is contained in at most $w(x)$ of them.

42. Choose k independent (a, B)-paths $R_1, \ldots, R_i$ such that R_t ends at the some point as P_t and

$$|E(R_1 \cup \ldots \cup R_k) - E(Q_1 \cup \ldots \cup Q_{k+r})|$$

is minimal.

43. Choose $B = V(P_0)$ and apply 6.42.

44. Knowing Menger's theorem it is easy to see that G will consist of two complete graphs G_1, G_2 with k common points, $a \in V(G_1)$, $b \in V(G_2)$, a, $b \notin V(G_1 \cap G_2)$. To prove this without Menger's theorem show that every point is connected to a or to b and the points connected to both separate a and b.

45. None of them is true.

46. Use Menger's theorem.

47. For (i): Remove the edges of an (a, b)-path (if any) and apply 6.46.

48. By straightforward counting of the edges on both sides.

49. Consider a set X with $\delta_G(X) = k$ and $|X|$ minimal.

50. Contract those edges not in $F_1, \ldots, F_m$ and consider the resulting graph.

51. Let $U \subset V(G) - \{x\}$ be a maximal set such that $\delta_G(U) = k$ and a certain neighbor y of x belongs to U. Show that there exists a neighbor z of x not in U.

52. Apply the previous "point-splitting" repeatedly until x disappears.

53. Use 6.48c instead of 6.48a.

54. (a) is trivial. For (b), find a point of degree $2k$ and "split" it as in the solution of 6.52.

55. Consider a quadrilateral.

56. (a) Show that there are $\alpha + \beta$ edge-disjoint paths connecting $\{a,b\}$ to $\{a',b'\}$ such that exactly α of them start at a and exactly α of them (some other α, possibly) ends at a'.
  (b) Use a similar argument to the one as in 6.46 and 6.47.

57. Look for a planar example.

58. C obviously separates a and b. To show $|C|=k$, consider the set D of points accessible from b on a $(b, A \cup B)$-path and show that

$$C \cap D \subseteq A \cap B.$$

As another possibility, use Menger's theorem.

59. If (x,y) is an edge of H and T a $(k-1)$-element separating set of $G-(x,y)$, then $T \cap V(H)$ separates $H-(x,y)$.

60. (a) Assume the statement is false. Let $G - A = G_1 \cup G_2$, $G - B = G_3 \cup G_4$, $(G_1 \cap G_2 = G_3 \cap G_4 = \emptyset)$ and assume, e.g., that $G_1 \cap G_3 \neq \emptyset$, $V(G_1) \cap B \neq \emptyset$. Let $a \in G_1 \cap G_3$, and let C be the set of points of $A \cup B$ accesible from a on an $(a, A \cup B)$-path. Show that

$$1° \ |C| > k,$$
$$2° \ C \subseteq (A \cap V(G_3)) \cup (A \cap B) \cup (B \cap V(G_1)),$$
$$3° \ G_2 \cap G_4 = \emptyset,$$
$$4° \ |V(G_4)| < |V(G_1)|.$$

  (b) Show that G_1 defined in (a) is a one-element set.

61. Set $e = (a,b)$ and assume both G/e and $G-e$ are only 2-connected. Show
  (a) there are two points x,y in $G-e$ separating a and b,
  (b) there is a point u in G such that $\{a,b,u\}$ separates x and y,
  (c) if $\{x,y\}$ separates a from u, then x,y and b are the only neighbors of a.

62. Show that if G is a simple graph and e is an edge of G which connects points with degree at least k and, moreover, G/e is k-connected, then so is G.

63. Use induction on the length of the circuit.

64. Let a be a point of degree 3, joined to b_1, b_2, b_3. Show that, if $G \neq K_4$, there is a point u_i separating b_{i+1} and b_{i+2} in $G - a - b_i$ ($i = 1,2,3$; $b_{j+3} = b_j$), and then use this fact.

65. Draw a tree T without points of degree 2 in the plane so that its endpoints are on its convex hull; then add the edges of the convex hull C. Show that this graph G is a line-critically 3-connected graph.

66. Use induction on k. If $x_1, \ldots, x_{k-1}$ are on C, but x_k is not, consider k independent (x_k, C)-paths.

67. Assume no circuit contains all the three given edges. Find a circuit which contains two of them and does not touch the third one, and, as in 6.66, investigate the paths joining the third edge to this circuit.

68. Use an argument similar to the one in 6.66.

69. Show that every bridge of the boundary of a face contains all points of the circuit.

70. Let $f \in E(G)$ and select a circuit C of $G - f$ such that the bridge B of C containing f is maximal.

71. Assume $v(e_1) \geq \ldots \geq v(e_m)$, where $E(G) = \{e_1, \ldots, e_m\}$, and consider the first index k such that $\{e_1, \ldots, e_k\}$ contains an (a, b)-path.

72. Determine a potential function φ such that, for each point x, there is an (a, x)-path with value $\varphi(x)$.

73. Let C be given by $S \subseteq V(G)$. Consider

$$\sum_{x \in S} \left(\sum_{e=(x,y)} f(e) - \sum_{e=(y,x)} f(e) \right).$$

74. The non-trivial part of the proof is to find an (a, b)-flow f and an (a, b)-cut C such that

$$w(f) = \sum_{e \in C} v(e).$$

Consider a flow f of maximum value, satisfying $f \leq v$. For each $e \in E(G)$, we introduce a new edge e' having the same endpoints but converse orientation, and let

$$v_0(e) = \begin{cases} v(e) - f(e) & \text{if } e \in E(G), \\ f(e_1) & \text{if } e = e_1', \ e_1 \in E(G). \end{cases}$$

We then consider the digraph G_1 determined by those edges e, e' for which $v_0(e) > 0$ and $v_0(e') > 0$; respectively. Show that G_1 is not connected between a and b.

75. (a) Let ω_n be the increase of flow value at the nth step; then $\sum_{n=1}^{\infty} \omega_n$ must be convergent. Try to achieve $\omega_n = \alpha^n$, $\alpha < 1$. Also, there must be edges with $f_k(e) = 0$ or $v(e)$ at each step. How can the value of $f_k(e)$ change between two successive equal states?

(b) Show that if we omit those edges which the paths P_k and P_{k+1} use in different directions, $P_k \cup P_{k+1}$ still contains two (a,b)-paths Q_1, Q_2 with $Q_1 \cap Q_2 \subseteq P_k \cap P_{k+1}$.

(c) When an edge is used in opposite directions by P_k and P_{k+l}, then P_{k+l} is longer than P_k.

76. Introduce two new points a_0, b_0; join a_0 to each point x with $u(x) \geq 0$ and give the new edge e the capacity $v(e) = u(x)$; join each point x with $u(x) < 0$ to b_0 and give the new edge e the capacity $v(e) = -u(x)$. The problem is equivalent to finding a suitable (a_0, b_0)-flow in the resulting digraph G_1.

77. (a) Substitute $v(e)$ parallel edges for every edge e and apply Menger's theorem.

(b) Demonstrate the existence of an (a,b)-flow $f(e) \leq v(e)$ and an (a,b)-cut C such that

$$\sum_{e \in C} v(e) = w(f)$$

first for the case of rational capacities $v(e)$.

78. One may assume $k = 2$. Consider an integral (a,b)-flow f' such that $w(f') \geq w_1$; $f'(e) \leq f(e)$ and $\sum_e f'(e)$ is minimal.

§ 7

1. Show that a minimum system of edges which cover all points consists of disjoint stars.

2. One can use Menger's theorem (6.39). In order to get a direct proof, consider a minimal subgraph G' of G such that $\tau(G') = \tau(G)$ and show that G' consists of independent edges.

3. From each point y of F in A, there is a path alternating relative to M connecting y to A_1.

4. To get a direct proof, use induction on $|A|$; try to find a subset $\emptyset \neq A_1 \subset A$ such that $|\Gamma(A_1)| = |A_1|$.

5. Use an argument similar to that in the first solution of 7.4.

6. (a) Observe that

$$\Gamma(X_1 \cup X_2) = \Gamma(X_1) \cup \Gamma(X_2), \quad \Gamma(X_1 \cap X_2) \subseteq \Gamma(X_1) \cap \Gamma(X_2).$$

(b) Consider a minimal subgraph G_1 with $V(G_1) = V(G)$ and property (2). Observe that (1) means that the one-element sets satisfy (2) with equality.

7. The implications (i)$\Rightarrow$(iii)$\Rightarrow$(ii)$\Rightarrow$(i) are easily verified.

8. "If" is easy by the preceding result. Conversely, fix a 1-factor F, an edge of F as G_0 and choose the paths $P_1, \ldots, P_k$ alternating with respect to F (cf. the solution of 6.28).

9. P_k in the previous problem cannot be a single edge!

10. Prove it for r-regular graphs first. Show that G has a 1-factor, and remove the edges of this 1-factor.

11. Split each point x into $\left\lfloor \frac{d(x)}{k} \right\rfloor$ points of degree k and, if necessary, one point of degree $d(x) - k \left\lfloor \frac{d(x)}{k} \right\rfloor$.

12. Use 7.11 with $k=r$.

13. Define a graph G_1 on $V(G)$ by joining two points iff they are joined in G by more than one edge. For which values of r can G_1 fail to satisfy the condition given in 7.4b?

14. To show we cannot go into an infinite cycle, consider all edges which x passes infinitely many times and show that every point is adjacent to at most two of them.

15. Distinguish between elementary and non-elementary graphs.

16. Reduce to a network-flow problem, similar to that in the first solution of 7.2.

17. Embedding into a G' of this size is straightforward. To show that smaller G' would not do, count the edges leaving a color-class of G by two different methods.

18. It should be shown that, if $\tilde{G}$ is the bipartite graph on $V(G)$, where $x \in A$ is joined to $y \in B$ iff they are not adjacent in G, then $\tilde{G}$ has an $(n-2d)$-factor.

19. Construct a bipartite graph G on $\{u_1,\ldots,u_n, v_1,\ldots,v_n\}$, where u_i is adjacent to v_j iff $a_{ij} > 0$. One has to show that this bipartite graph has a 1-factor.

20. (a) If G has a 1-factor the expansion term corresponding to this cannot cancel out as the entries are algebraically independent. (b) If G has no 1-factor the columns of (a_{ij}) are linearly dependent. Consider a minimum set of columns which are linearly dependent.

21. If G is bipartite, the conditions (1) are not linearly independent.

22. Suppose not. Let F be a maximum matching and u, v two points not in F. How many edge can join $\{u,v\}$ to any edge of F?

23. Consider a maximum matching F which has as many edges in common with F_0 as possible.

24. (a) Use induction on k. (b) Consider the decomposition given in 6.27. You may use 6.30 as well. (c) Use this sharpening of (b): If G has a unique 1-factor F_1 then it has a cut-edge which belongs to F.

25. Consider a maximum matching F and select an edge from each point not covered by F.

26. Show by induction on the distance between x and y that $\nu(G-x-y) < \nu(G)$ for each $x \neq y \in V(G)$.

27. (a) $\delta(G) \geq \max \ldots$ is easy. To establish $\leq$, use induction and the previous result. (b) Trivial from (a).

28. (a) One has to show that "adjacency" is an equivalence-relation in $G-V_1$. (b) Adding edges does not disturb the inequalities of the theorem.

29. (a) Apply Tutte's theorem. (b) A cut-edge of a 3-regular graph must be contained in every 1-factor.

30. Tutte's condition is satisfied. This follows from counting the edges and removed edges in a way similar to that in the solution of 7.29a.

31. Use an argument similar to that in 7.8, selecting paths $P_0, P_1, \ldots, P_k$ one by one so that they alternate relative to a given maximum matching F.

32. (a) Let $x \in A_G$ and let $z \in D_G$ be a neighbor of x. Supposing indirectly that $y \notin D_G$, but some maximum matching M' of $G-x$ avoids y, consider a maximum matching M of G avoiding z and form $M \cup M'$. (b) Remove all points of A_G, using (a).

    For (c)–(e), use (a) and (b).

33. From any matching M_0 of G' we get a matching with $|M_0| + k$ edges in G. For the proof of the converse, consider the Gallai–Edmonds structure described in the preceding problem.

34. If two outer points in the same component are adjacent we have an odd circuit as in 7.33. Otherwise, remove the inner points.

35. Consider the components of the set D_G in 7.32 and use an argument similar to that in 7.29.

36. (a) Let F be a maximum matching and x a point not covered by F. (b) Find an edge e such that e is in some, but not all, 1-factors and $G - e$ is 2-connected (cf. 6.36).

37. It is easy to see that no 2-matching can be larger than the value given. To find a 2-matching of this size, take two copies G, G' of G, where $V(G) = \{v_1, \ldots, v_n\}$ and $V(G') = \{v'_1, \ldots, v'_n\}$ and define a bipartite graph G_0 by joining v_i to v'_j iff $(v_i, v_j) \in E(G)$.

38. First consider factor-critical graphs and then use the Gallai–Edmonds structure theorem 7.32.

39. Consider an Euler trail of G.

40. Consider a pseudosymmetric orientation (5.13) and split each point into two, separating the outgoing and incoming edges. Apply 7.10 to this bipartite graph.

41. Add a new point x and join it to all points of odd degree adding, if necessary, a loop at x.

42. (a) The number of points must be even! (b) G is a contraction of a 2-connected 3-regular graph.

43. (a) Replace each point by $f(x)$ independent points. (b) If $f(x) = 1$ for at least one endpoint of any edge, then an f-matching is an f-factor.

44. Split each point x into two points x', x'' incident with the incoming and outgoing edges, respectively.

45. Subdivide each edge and define f to be 1 for the new points. Look for an f-factor of the resulting graph.

46. Use induction.

47. Use induction.

48. Show that the sequence $d_1, \ldots, d_{n-2}, d_{n-1} - 1, d_n - 1$ satisfies the same conditions.

49. Use 7.44.

50. Transform G with degrees $d_1, \ldots, d_n$ to a graph with the same degrees in which v_n is adjacent to $v_{n-d_n}, \ldots, v_{n-1}$.

51. (a) Choose H with as many pairs of oppositely directed edges as possible. Show that those edges of H whose converse is not in H can form neither an even cycle nor two disjoint odd cycles. (b) Trivial by (a) and 7.49.

52. To show the sufficiency of the condition, consider a graph having these degrees and reduce the number of components one by one.

53. Let G be one graph on $V = \{v_1, \ldots, v_n\}$ with $d_G(v_i) = d_i$ and G' another with $d_{G'}(v_i) = d_i - 1$. Choose G' and G as "close" as possible.

§ 8

1. Consider a maximal independent set.

2. Let S_1 be a maximum independent set of G and let S_{i+1} be a maximum independent set in $G - S_1 - \ldots - S_i$. Show that S_{i+1} has a matching into S_i.

3. Find a circuit or an edge which contains a point x and all neighbors of x.

4. If $(x, y) \in E(G)$, where $x, y \in S$, consider $G - x$.

5. (a) Take any maximal independent set. (b) Use induction, removing a point with indegree 0 and its neighbors. (c) Use 5.3.

6. Consider the point of highest outdegree.

7. Remove a point and all points which can be reached from it on an edge, and use induction.

8. (a) Observe that whether or not the player to move can win depends only on his opponent's last move; call this last move a "winning move" (for the opponent) if he cannot win. What are the characteristic properties of the set of winning moves? (b) The point x_0 is only involved if A starts with it. (c) Observe that when they have once left a component they will make no more moves there.

9. If G has no 1-factor, A should select a maximum matching and start with a point not on this.

10. Use induction on k.

11. Use an argument similar to the one in the previous problem.

12. Remove an edge incident with the point x.

13. It is enough to show this if a single point x is replaced by two adjacent points x_1, x_2 both joined to the same "old" points as x.

14. It is enough to embed the given graph G_0 into a G such that $\alpha(G-e) > \alpha(G)$ holds for the edges of G_0.

15. (a) For $r = 2$, odd circuits do. For $r > 2$, increase the degree by substituting complete graphs for points. (b) The tetrahedron and icosahedron.

16. An α-critical bipartite graph consists of independent edges. Supposing there was an $x \in V(G)$ joined to two points y_1, y_2, consider a maximum independent set S_i of $G-(x,y_i)$ and the subgraph induced by $S_1 \triangle S_2 \cup \{x\}$ (cf. the solution of 7.2).

17. Follow the previous solution.

18. If S were a minimal cutset spanning a complete graph and $x \in S$, consider two edges joining x to different components of $G - S$.

19. (a) Show that $\alpha(G) = \alpha(G_1) + \alpha(G_2)$. (b) Suppose that $G = G_1 \cup G_2$ with $V(G_1) \cap V(G_2) = \{x,y\}$, $|V(G_i)| \geq 3$. Determine, for each $X \subseteq \{x,y\}$ and $i = 1$, 2, the maximum size of an independent set in G_i which meets $\{x,y\}$ in X, by writing out several inequalities involving these numbers.

20. Apply 8.10 with the set of all maximum independent sets.

21. Apply 8.11.

22. Let $T_1, \ldots, T_k$ be the set of all maximum independent sets containing x; apply 8.11 again.

23. (a) Use the previous result. (b) Use part (a).

24. These are the complete graphs and odd cycles only; show and use the fact that, if T, S are maximum independent sets with $T \cap S \neq \emptyset$, then $\Gamma(T \cup S) = V(G) - T - S$.

25. Use 8.19 and the previous problem.

26. Consider an α-critical spanning subgraph.

27. Again consider an α-critical spanning subgraph of G.

§ 9

1. Use induction on $|V(G)|$ removing a point.

2. To show the second statement, let V_1 be the set of points of a maximum clique.

3. Use pairs of colors to color $G_1 \cup G_2$.

4. Consider a coloration α such that each color occurs at least twice and a coloration β with $\chi(G)$ colors, so that they have as many color classes in common as possible.

5. Use induction on k, and a re-coloration similar to the one in the solution of 9.2.

6. It helps to answer the question: can $\alpha(G \oplus K_m)$ be larger than $|V(G)|$?

7. (a) If a coloration of G_1 is given, color a point of $G_1 \times G_2$ according to its first coordinate. (b) Observe that $G \times G \supseteq G$.
  (c) If α is a k-coloration of $G \times K_n$, $n > k$, use the fact that some color occurs twice in the set $\{(v,x) : x \in V(K_n)\}$ for each $v \in V(G)$.
  (d) For an n-coloration α of $G \times K_n$, show that if $\{\alpha(v,x) : x \in V(K_n)\}$ are different, the so are $\{\alpha(u,x) : x \in V(K_n)\}$ for any neighbor u of v.
  (e) What about $K_n^m = \underbrace{K_n \times \ldots \times K_n}_{m}$?

8. For (d), observe that, if we want to 3-color an odd circuit but one of the colors is excluded at each point, then it is possible iff the excluded color is not the same at every point. For (e), consider first the case when all but one partition are allowed.

9. First prove this for acyclic graphs; for the general case, see the solution of 6.11.

10. If and only if every circuit has the same number of edges oriented in each direction around the circuit.

11. If $\overrightarrow{G}$ is an orientation of G with the property mentioned then, for a walk of G, define the expense of it by saying that we gain one if we go along an edge in the right way (from tail to head) but lose $k - 1$ otherwise. Verify and then use the following:
  (a) the expenses of walks from a to b are bounded from below,
  (b) if a, b are adjacent, then the minimum expenses of (x_0, a)-walks and (x_0, b)-walks are different but differ by at most $k - 1$.

12. (a) Find an arrangement $(x_0, \ldots, x_{n-1})$ of the points of G such that x_i is adjacent to some x_j $(j < i)$; start by coloring x_{n-1}. (b) Show that the sequence used in the previous solution can be chosen so that x_0 is adjacent to x_{n-1} and $C(x_0) \neq C(x_{n-1})$.

13. (a) Consider x_0, x_{n-2}, x_{n-1} such that $(x_{n-2}, x_{n-1}) \notin E(G)$, $(x_0, x_{n-1}) \in E(G)$, $(x_0, x_{n-2}) \in E(G)$. (b) If $G = G_1 \cup G_2$ with $V(G_1) \cap V(G_2) = \{x_1, x_2\}$, $|V(G_i)| \geq 3$, consider $G_1 + (x_1, x_2)$ and $G_2 + (x_1, x_2)$.

14. (a) One has to show that "to be non-adjacent" is an equivalence relation. (b) Embed G in a maximal graph on the same set with the same property. (c) Consider the set of those colorations coloring the endpoints of a given edge e differently.

15. Let K_{k+1} be the smallest complete graph in $\mathcal{K}$; then $\mathcal{K}$ will be the class of all non-k-colorable graphs. To show that these are, in fact, in $\mathcal{K}$, consider a counterexample with a fixed number of points but maximum number of edges and prove that "to be non-adjacent" is an equivalence relation on the points.

16. Use the preceding result.

17. (a) Odd circuits. (b) In every 3-coloration of $K_{n,n'}$ one of the two classes is monochromatic. (c) In every 5-coloration of $K_{n,n'}$ one of the two classes receives only two colors.

18. To show $\chi(G') > \chi(G)$, let α be a coloration of G' with colors $1, \ldots, \chi(G)$ and $\alpha(y) = 1$. "Project" α to get a coloration of G not using color 1.

19. (a) What would a 3-coloration of $G - (x_3, x_4)$ look like? (b) Use 9.8 to show the sufficiency.

20. It can happen for all $k \geq 3$.

21. If $k - 1$ edges separate G into two pieces G_1, G_2, consider k-colorations of these and permute the colors to fit the two colorations together.

22. If G is critically $(k+1)$-chromatic and $G = G_1 \cup G_2$ with $V(G_1) \cap V(G_2) = \{x, y\}$, then investigate the k-colorability of the graphs obtained from G_1 and G_2 by connecting or identifying the points x and y.

23. Let $G_1, \ldots, G_N$ be the components of $G - S$. Consider the partitions of S induced by k-colorations of $G - V(G_i)$ $(i = 1, \ldots, N)$.

24. In every k-coloration of $G - (x, y)$ the points x, y have the same color. If this color is i, then for any other color j, the set of points with color i and j must contain an (x, y)-path.

25. (a) Consider K_n as the graph formed by the edges and diagonals of a regular n-gon. (b) The edges corresponding to a color class in a coloration of $\overline{L(K_n)}$ form either a star or a triangle. (c) Considering a coloration of $L(\overrightarrow{K_n})$, for any two vertices u and v of $\overrightarrow{K_n}$ there must be a color that occurs among the edges leaving u but not among the edges leaving v.

26. (a) Assuming that a coloration of $L(G)$ is given, color each vertex v with the set of colors occurring on edges leaving v. (b) To improve the bound given in (a), use 13.20.

27. (a) Use e.g. 9.18. (b) If we have G_k with $\chi(G_k) = k$ and join each point x of it to a new point x' by an edge, then, n-coloring this piece, the set of new points cannot be monochromatic. (c) Take iterated line-graphs of a transitive tournament.

28. (a) Remove an interval whose right endpoint is as far to the left as possible. (b) Consider I_1 and a as before, but now remove all intervals containing a from the system. (c) We may assume that $V(G) = V(C)$. Consider I_1 and a as before.

29. (a) Consider circuits formed by two arcs joining $v \in S$ to $u \in S$ through different components of $G - S$. (b) The "if" part follows like 9.28c. For the "only if" part use (a) and induction. (c) Use induction and (a).

30. This is equivalent to $\alpha(G) = \varrho(G)$.

31. Show that
$$\chi(G_1 \cdot G_2) \leq \chi(G_1) \cdot \chi(G_2),$$
$$\omega(G_1 \cdot G_2) \geq \omega(G_1) \cdot \omega(G_2).$$

32. Heuristically, you have to partition P into "levels". Take the points on the top as "first level", the next largest points as "second level", etc.
   b) Use 8.4.

33. 9.31 does not yield a perfect graph, even if G_1 and G_2 are perfect.

34. A color-class of an $\omega(G)$-coloration does so.

35. It suffices to prove the previous condition for G', in the case when only one G_x has more than one point.

36. Consider the partitions of $V(G)$ induced by λ-colorations.

37. Count the number of λ-colorations using inclusion-exclusion.

38. Which λ-colorations of $G - e$ are not λ-colorations of G?

39. For circuits, use the previous formula.

40. If G has several components, we can color them independently. If it has several blocks, we can color them almost independently.

41. Remove the interval whose right endpoint is the farthest to the left.

42. Use 9.38.

43. Use induction on $|E(G)|$, but now starting with a tree.

44. Use the formula given in the solution of 9.36.

45. In fact, $(-1)^{n-1} P_G(\lambda) > 0$ in $(0,1)$ for connected graphs. Show this by an induction similar to the one in the solution of 9.43.

46. (a) Show that 0 is a simple root iff G is connected, 1 is a simple root iff G is 2-connected or $\cong K_2$. (b) Use induction and 9.38.

47. (a) Trivial by 9.38. (b) Any linear relation like this ought to be satisfied in the three cases when $V(G) = V(F)$. (c) Show that this (in fact, any) linear relation holds generally if it holds for the three cases when $V(G) = V(F)$.

48. What about $\frac{3-\sqrt{5}}{2}$?

49. (a) Show that the relation holds when G is some very simple triangulation and reduce the problem to this case by "switching" the diagonals of quadrilaterals; use the identity

$$P_{G_1}^2(\tau + 1) - P_{G_2}^2(\tau + 1) = \tau^{-3}(P_{G_1|e_1}^2(\tau + 1) - P_{G_2|e_2}^2(\tau + 1)).$$

(b) Use "backward" induction, removing edges from a triangulation.

50. Show that G has a point of degree at most 5.

51. If the faces are colored with colors 1, 2, 3, 4, give color α to edges adjacent to faces of color 1 and 2 or 3 and 4; color β to edges adjacent to faces of color 1 and 3 or 2 and 4; color γ to the rest.

For the converse, first color the regions of the maps formed by red-green and red-blue edges in a 3-coloration of the edges.

52. If the edges are 3-colored with 1, 2, 3, associate the number +1 with those points where 1, 2, 3 is the clockwise ordering of the edges incident with the point and associate −1 with the rest. For the converse, consider $L(G)$ and 5.4.

53. Use two colors inside and two colors outside the Hamiltonian circuit.

54. (a) Use induction, removing a point of degree at most 4, coloring the rest with 4 colors and modifying this coloration to get only three colors in the neighborhood of the removed point. (b) Identify two non-adjacent points on the boundary of a face and apply an argument similar to the one in part (a).

55. We may assume that G is a triangulation. Then consider a 1-factor in its dual map G^*.

56. For part "if", orient the edges in such a way that, if 1 unit of work is needed to pass a given edge in the given direction, then 0 (mod 3) units of work are needed to go round each face.

57. Suppose that no line is parallel to the x-axis. Then color the points one by one, in the order of their ordinates.

<h2 style="text-align:center">§ 10</h2>

1. Otherwise, each block of G would consist of a single circuit.

2. (a) Consider a longest path. (b) Use induction based on (a).

3. (a) Instead of (a), prove the slightly stronger statement: If a simple graph G has at most one point with degree at most 2, then G contains a subdivision of K_4. (b) Reduce to (a) as before.

4. Show and use the fact that G cannot contain a subdivision of K_4 and an edge disjoint from it.

5. (a) The graph contains a subdivision of K_4. (b) Use Kuratowski's theorem 5.39d.

   (c) The only exceptions to (a) when dropping the connectivity assumption are connected graphs whose blocks are complete 4-graphs.

6. The point is that K_4 has degrees at most 3.

7. Contract edges forming a connected subgraph as long as

$$\frac{|E(G')|}{|V(G')|} \geq \frac{m}{n}$$

holds for the resulting graph G'.

8. Use induction on m, and the preceding problem.

9. Use a method similar to the one in 10.8.

10. If one builds them up starting with a given point and a given circuit, respectively, when can the branches "grow together"?

11. The points which can be reached from a point x_0 on paths of length at most $\frac{g-1}{2}$ are distinct.

12. Supposing $G(r,g)$ and $G(r',g-1)$ are available, where $r' = |V(G(r,g))|$, construct $G(r+1,g)$.

13. (a) If two points a, b were at a distance greater than g, remove a, b and match up their neighbors. (b) Treat the cases $r = 2l$, $r = 2l+1$ separately. In the first case, there must be a circuit of length g through any given point. In the second case, one of any two adjacent points has this property.

   (c) Use a counting similar to that in 10.11.

14. If G' has girth at least g, $(u,v) \in E(G')$ and both x, y are at distance at least $g-1$ from $\{u,v\}$, then $G' - (u,v) + (x,u) + (y,v)$ has girth at least g.

15. (a) Use a finite projective plane. (b) Let the points of G be the lines and points on the hypersurface $x_1^2 + x_2^2 + x_3^2 + x_4^2 + x_5^2 = 0$ in the 4-dimensional projective plane over $GF(p)$ and connect a point to a line iff they are incident.

16. (a) Let Z be a minimum set of points representing all circuits and count the number of edges between Z and $V(G) - Z$. (b) Use (a) and an argument similar to that in 10.11.

17. Let $G_1 = G$; if $G_1, \ldots, G_i$ are already defined, then let C_i be a minimum circuit in G_i, and let G_{i+1} be obtained from G_i by removing the points of C_i and also all points not in circuits of $G_i - V(C_i)$ and by "smoothing out" points of degree 2. Consider an i for which G_i has largest girth.

18. (a) Use 10.4. (b) By a similar argument to the one in the previous proof. (c) A minimal 3-regular graph with girth $\log_2 k$ has this property.

19. Use induction removing x.

20. Consider a maximum path $P = (x_0, x_1, \ldots, x_m)$. We say that the path $(x_0, \ldots, x_i, x_m, x_{m-1}, \ldots, x_{i+1})$ arises *by deformation* from P (provided $(x_i, x_m) \in E(G)$). Let X be the set of all endpoints of paths arising by repeated deformation from P.

21. (a) One may suppose that adding any edge (x, y) to G a Hamiltonian circuit arises. Consider the neighbors of x and y on this Hamiltonian circuit. (b) (c) — prove (d) first. (d) The argument in the solution of (a) proves that if $d_k + d_l \geq n$, then x_k and x_l are adjacent. Consider a missing edge (x_k, x_l) with $k+l$ maximal.

22. Show that if G is saturated and x_i, x_j are non-adjacent points, then

$$d_G(x_i) + d_G(x_j) \leq n + q - 1.$$

23. Find a circuit of length $2n$.

24. Join the two points x, y if necessary and subdivide the edge (x, y).

25. Consider a maximum cycle C. First prove that $|V(G)| > \frac{n}{2}$, then consider a maximum path P of $G - V(C)$ and try to replace an arc of C by a path going through P.

26. Consider a maximum circuit C, a component G_1 of $G - V(C)$ and the neighbors of those points of C adjacent to G_1.

27. (a) Consider an endpoint of a longest path in G. (b) Let $(x_0, \ldots, x_m)$ be a longest path. Consider first the case when there are edges (x_0, x_j) and (x_m, x_i) such that $i < j$.

28. Use induction on n and 7.27b.

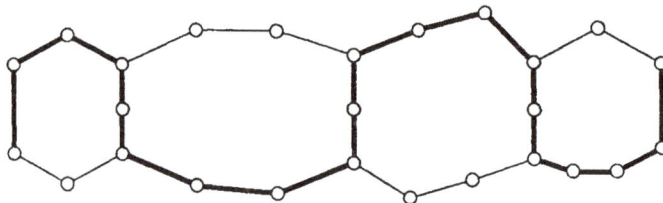

FIGURE 15

29. Any path P in G is contained in a subgraph as in Figure 15.

30. Use the fact that if x, y are adjacent points, then

$$d_G(x) + d_G(y) \le n.$$

31. Observe that

$$d_G(x) \le \alpha(G)$$

for each point x.

32. If a triple $\{x, y, z\} \subseteq V(G)$ does not span a triangle of G or $\overline{G}$, then it contains exactly two points such that the two edges in the triple adjacent to any one of these points do not belong to the same G or $\overline{G}$.

33. The edge (x, y) is contained in at least $d(x) + d(y) - n$ triangles.

34. Use induction on m, starting with a complete k-graph.

35. Consider the neighborhood of a point with maximum degree and use induction on k.

36. (a) Count the number of "cherries" (3-point trees) in the graph in two different ways. (b) Use a method similar to the one in 10.15.

37. Now count the number of $K_{1,r}$'s in G.

38. (a) By induction on k, choose k disjoint s-sets $A_1, \ldots, A_k$ such that any two points in distinct s-sets are adjacent (s is an appropriately large number) and show that many points outside are adjacent to at least t points of every A_i. (b) Remove points with degree $\left(1 - \frac{1}{k} + \frac{\varepsilon}{2}\right) \cdot |V(G)|$ repeatedly; prove that when one gets stuck the number of points is still large. (c) By (b).

39. Prove that if $G \ne H_{n,k}$ does not contain G_0, then $2k \cdot |V(G_0)|$ points can be removed from G so that the remaining graph G_1 satisfies

$$|E(G_1)| - |E(H_{n_1,k})| > |E(G)| - |E(H_{n,k})| \qquad (n_1 = |V(G_1)|).$$

40. (a) If a given complete k-graph is contained in "few" complete $(k+1)$-graphs, then we find "many" pairs (A, U) of k-sets of points such that A is a complete graph, U is not, and $|A \cap U| = k - 1$. (b) First prove that

$$(*) \qquad \frac{N_k}{N_{k-1}} \ge \frac{\vartheta - k + 1}{\vartheta} \cdot \frac{n}{k}.$$

41. (a) Let $d_1, \ldots, d_n$ be the outdegrees of points in T. Then T contains

$$\binom{n}{3} - \binom{d_1}{2} - \cdots - \binom{d_n}{2}$$

3-cycles. (b) Observe that each point is adjacent to at least one 3-cycle.

42. Prove that it contains at least $n-k+1$ k-cycles for each $3 \le k \le n$.

43. If we specify $\left[\frac{n}{2}\right]$ independent edges

$$e_1, \ldots, e_{\left[\frac{n}{2}\right]},$$

there is at most one Hamiltonian path whose 1st, 3rd, ... edges are these.

44. (a) The assertion says that, if $n = 2^{k-1}$, then there is a transitive subtournament with k points. Prove this by induction on k. (b) Using induction on k, observe that the set of points accessible from x contains at least

$$\prod_{j=0}^{k-2} \left(\frac{d+1}{2^j} - 1 \right)$$

transitive subtournaments with $k-1$ points, where d is the outdegree of x.

   (c) Prove by induction on n that there are at least $\binom{n-2}{k-2}$ k-tuples that contain a 3-cycle.

§ 11

1. What does an eigenvector mean combinatorially?

2. (a) Use the fact that $(1,\ldots,1)$ is an eigenvector and the eigenvectors are pairwise orthogonal. (b) Establish and use the identities

$$A_G = B_G B_G^T - dI, \quad A_{L(G)} = B_G^T B_G - 2I.$$

(c) The Petersen graph is $\overline{L(K_5)}$.

3. We may assume that $V(T) = \{x_1, \ldots, x_n\}$, $e = (x_k, x_{k+1})$ $(1 \le k \le n-1)$ and that $T - e = T_1 \cup T_2$, where $V(T_1) = \{x_1, \ldots, x_k\}$ and $V(T_2) = \{x_{k+1}, \ldots, x_n\}$. Consider the expansion of $\det(\lambda I - A)$ by its first k rows.

4. Use induction and the previous excercise.

5. (a) follows by 11.4 and 1.31. Both (a) and (b) follow by induction. Both (b) and (c) are essentially solved in 1.29. For (d), consider an eigenvector $\mathbf{y}$, belonging to an eigenvector λ, invariant under the automorphisms of the tree, and derive the recurrence

$$D y_{k+1} + y_{k-1} = \lambda y_k,$$

where y_k is any entry of $\mathbf{y}$ at a distance k from the root.

6. By 11.4, we only have to find two trees with the same number of k-element matchings for every k. Look for them among those trees with no 3 independent edges.

7. (a) Prove that
$$p_G(\lambda) = \det p_{G_1}(\lambda I - A_{G_2}).$$
(b) How does the adjacency matrix $A_{G_1 \cdot G_2}$ arise?

8. The vector $\mathbf{v} = (\chi(\gamma_{1,1}), \dots, \chi(\gamma_{1,n}))^T$ is an eigenvector.

9. The automorphism group of Q_n contains the commutative regular subgroup $(Z_2)^n$; one can also use 11.7a.

10. If $G \cong H$, their eigenvalues are the same.

11. Note that 0 is not an eigenvalue iff $\det A \neq 0$.

12. Find an eigenvector which associates a with those points of degree 3 in the covering subgraphs and b with the other points.

13. Use the fact that $\lambda_1 = \max\limits_{|\mathbf{v}|=1} \mathbf{v}^T A_G \mathbf{v}$.

14. (a) Use the previous result; to determine the case of equality use the fact that the adjacency matrix of a connected graph has a (strictly) positive eigenvector. (b) Use the fact that the sum of all eigenvalues is 0 and the sum of their squares is 2m. (c) Embed G in a complete rooted $(D-1)$-ary tree.

15. Use the formula $A_{L(G)} = B^T B - 2I$, where B is the point-edge incidence matrix:
$$B = (b_{ij}), \quad b_{ij} = \begin{cases} 1 & \text{if } i \text{ belongs to the } j^{\text{th}} \text{ edge,} \\ 0 & \text{otherwise.} \end{cases}$$

16. To any authomorphism α of G there corresponds a permutation matrix P such that $A_G P = P A_G$. Also use the fact that $A = T^{-1} L T$, where T is an orthogonal matrix and L is a diagonal matrix with the eigenvalues in the diagonal.

17. Use the previous result to show that Γ is commutative and regular and apply 11.8 to calculate the eigenvalues.

18. Use the Interlacing Eigenvalue Theorem: if A is a symmetric matrix with eigenvalues $\lambda_1 \geq \lambda_2 \geq \dots \geq \lambda_n$ and B is a symmetric submatrix of it with eigenvalues $\mu_1 \geq \dots \geq \mu_m$, then $\lambda_i \geq \mu_i \geq \lambda_{i+n-m}$.

19. (a) Suppose that $A_G \mathbf{w} = -\lambda_1 \mathbf{w}$, $\mathbf{w} = (w_1, \dots, w_n)^T$. Prove that $\mathbf{w}' = (|w_1|, \dots, |w_n|)^T$ satisfies $-\mathbf{w}^T A_G \mathbf{w} = \mathbf{w}'^T A_G \mathbf{w}'$. (b) Consider a component whose least eigenvalue is minimal.

20. Use 11.13, 11.14 and induction.

21. (a) Show that, with respect to a suitable basis, A has a $k \times k$ symmetric submatrix D with 0's in the main diagonal and such that λ_1 is an eigenvalue of D. (b) 0 is an eigenvalue with multiplicity $n - k$.

22. Show that for every vector **x**,

$$\mathbf{x}^T(dI - A_G)\mathbf{x} = \sum_{(i,j)\in E} (x_i - x_j)^2.$$

23. Consider a vector which is constant on a maximum independent set and on its complement.

24. Write $\mathbf{x}^T A_G \mathbf{x}$ as a sum of m positive and m negative squares.

25. If $f(\lambda)$ is the polynomial whose roots are the distinct eigenvalues of A_G, then $f(A_G)=0$.

26. Show that both statements are equivalent to "every eigenvector of A_G is an eigenvector of J".

27. Calculate A_L^2.

28. (a) A_G satisfies the equation

$$A_G^2 = bJ + (d - b)I.$$

(b) $G = K_3$. (c) Generalize the example of $K_4 \oplus K_4$.

29. (a) Consider an eigenvector **x** belonging to λ_2 and those terms in 11.22 that correspond to the edges of a path connecting the smallest and largest entries of **x**. (b) Consider $A_G^2 - dI$.

30. Order the points so that $y_1 \le y_2 \le \ldots \le y_n$ and use that for $i < j$, $y_j - y_i = (y_{i+1} - y_i) + \cdots + (y_j - y_{j-1})$.

31. (a) Use 11.22. (b) Apply 11.30 with $y_i = (\max\{x_i - \bar{x}, 0\})^2$, where **x** is an eigenvector belonging to λ_2, and $\bar{x}$ is the median of x.

32. (a) Use 11.8 to express the eigenvalues of G. (b) Consider shortest paths between all pairs of points, and all their images under automorphisms.

33. (a) Express the probability in question by inclusion-exclusion and find its limit as $n \to \infty$ (cf. the solution of 4.23). (b) What is the probability that M_1 contains p edges between S and $V(G) \setminus S$ for a given partition?

34. Let $p_j^{(k)}$ denote the probability of $v_k = j$ $(j \in V(G))$. Show and use that

$$p_j^{(k+1)} = \sum_{i\in\Gamma(j)} p_i^{(k)} \frac{1}{d(i)}.$$

35. (a) Try $p_i^{(0)} = d(i)/(2m)$. (b) A stationary distribution is an eigenvector of M. (c) Use 11.34 and a version of 11.19.

36. Note that $P(v_j = y | v_i = x)$ is just the probability that a random walk starting at x hits y in the $(j-i)^{\text{th}}$ step.

37. (a) Use 11.35(c) and the equality

$$E(\nu_t(x)) = \sum_{i=1}^{t-1} P(v_i = x).$$

(b) Use 11.36 and the equality

$$D^2(\nu_t(x)) = \sum_{i=0}^{t-1} \sum_{j=1}^{t-1} [P(v_i = x, \ v_j = x) - P(v_i = x)P(v_j = x)].$$

38. (a) Use 11.37. (b) Find and prove an analogue of 11.37 for the count of passages through an edge.

39. For the case of a path with n vertices, find the mean access time from the $(n-1)^{\text{st}}$ vertex to the nth first.

40. (a) Consider a path of length 2. To get a regular example, choose u a cutpoint. (b) There is a natural bijection between walks from u to v and walks from v to u.

41. On a walk from u to v and back, consider the first return to u.

42. (a) Find a bijection between roundtrips starting at u, visiting v, and then w, and roundtrips starting at u, visiting w, and then v. (b) For a fixed vertex t, order the vertices according to the value of $a(u,t) - a(t,u)$.

43. For $r=1$, use 11.38(b).

44. (a) Use that for $i \neq j$,

$$a(i,j) = 1 + \frac{1}{d(i)} \sum_{k \in \Gamma(i)} a(k,j).$$

(b) What is the right hand side of this equation for $i=j$? (c) Let $p_{ij}^{(t)}$ denote the probability that a random walk starting at i is at j after t steps; let $q_{ij}^{(t)}$ denote the probability that the random walk starting at i hits vertex j the first time in the tth step. Show that

$$p_{ij}^{(t)} = \sum_{s=0}^{t} q_{ij}^{(s)} p_{jj}^{(t-s)},$$

and translate this relation in terms of generating functions.

45. (a) is trivial by adding up the result of 11.44(c). For (b), use the first solution of 11.9.

46. For (a), use 11.44(c). For (b), notice that the vectors $\mathbf{w}_i = (w_{ki})_{k=1}^n$ are mutually orthogonal unit vectors.

47. (a) Find the expected number of steps while exactly i vertices are unvisited. (b) Cf. 11.43.

48. (a) Prove that the average of $\mu(u,v)$, over all $v \in V(G) \setminus \{u\}$, is $n/2$. (b) If α_v is the number of steps before visiting v, then b is the expectation of the median of the α_v.

49. By 11.48(b), in $2a$ steps we have seen more than half of all vertices; in another $2a$ steps we have seen more than half of the rest, etc.

50. Use the solution method of 11.44(c).

51. Use 11.44A.

52. (a) is trivial by elementary probability theory. (b) follows from Kirchhoff's and Ohm's Laws, (c) follows by considering the forces acting on the vertices. To prove (d), consider the difference of two harmonic functions with the given properties, and the vertex where it assumes its maximum.

53. (a) Show and use that the resistance between s and t is $\left(\sum_{u \in \Gamma(t)} \phi_{st}(u) \right)^{-1}$. (b) First, relate R_{st} to the force acting on the nail at 0. (c) Compute the voltages of points from Kirchhoff's Laws, and use 4.9.

54. Use 11.53(b).

55. For (b), consider a path with endpoints a and b. For (c), note that if (a,b) is used then we have reached t.

56. For (b) and (c), use 11.44(c) and the inequality between arithmetic and harmonic means.

57. (a) Show that $\pi(u,v)$ is independent of v, by considering the first step when a neighbor of v is visited. (b) Use that there is a walk starting at any neighbor of u such that the next-to-last vertex visited is v and the last is u. (c) Use 6.6(c).

58. Let $P(S,u)$ be the probability that a random walk starting at u generates spanning tree S as described in the problem. For each edge $(u,v) \in E(G)$, let (u,v^S) denote the first edge on the (u,v)-path in S. Prove that

$$P(S,u) = \frac{1}{d_G(u)} \sum_{v \in \Gamma(u)} P(S - (u,v^S) \cup (u,v), v^S).$$

59. Consider a random sequence.

§ 12

1. The Petersen graph is the complement of the line-graph of K_5.

2. (a) Consider cubes of the type shown in Fig. 16. (b) For the case of the cube, consider the 4 main diagonals.

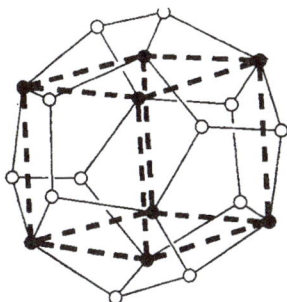

FIGURE 16

3. A k-circuit has some superfluous automorphisms; how can we prevent it from changing the direction?

4. Multiplication by any fixed element from the right is an automorphism of G.

5. Get rid of the directions and colorations by a method similar to the one in 12.3.

6. Let $\{h_1,\ldots,h_m\}$ be a minimal set of generators of Γ. Take the elements of $\Gamma \times \{1,2\}$ as points and connect $(g,1)$ to $(g',2)$ iff $g'g^{-1} = h_i$ for some i. Add edges within $\Gamma \times \{1\}$ and $\Gamma \times \{2\}$ to guarantee that there are no automorphisms other than those induced by the elements of Γ in the natural way.

7. (a) A group of even order has an element of order 2. (b) Find all tournaments which admit a regular group of automorphisms $\cong \Gamma$.

8. Trying the same kind of construction as in 12.5, it is easy to get all points not in $V(G_0)$ to have degree 3. Split each point of G_0 into points of degree 1 and connect them by a circuit.

9. Let $x_1, x_2,\ldots,x_n$ be an arrangement of the points of G such that for $i > 1$ x_i is adjacent to some x_j with $j < i$. Consider the subgroups of automorphisms fixing $x_1,\ldots,x_i$.

10. It suffices to consider one orbit of $A(G)$ on $E(G)$.

11. First solve the case $\Gamma_2 = \{1\}$. Then consider a certain "product" of examples.

12. (a) Consider a maximal tree T such that $T \cap \gamma(T) = \emptyset$ for every $\gamma \in \Gamma$, $\gamma \neq 1$, and contract each $\gamma(T)$. (b) Let $e \in E(R)$, G_1 the subgraph formed by the edges which are images of e under Γ and try to contract the components of G_1.

13. (a) A commutative and transitive permutation group is regular. Fix an $x_0 \in V(G)$ and consider the mapping $\alpha(x_0) \mapsto \alpha^{-1}(x_0)$ $(\alpha \in A(G))$. (b) The n-cube almost does it.

(c) Consider the n-cube Q_n and a vertex x of it. Add some edges between the neighbors of x and all edges which are forced by those automorphisms of Q_n preserving the directions of edges. What will the subgraph induced by the neighbors of x be?

14. Use an argument similar to the one in the solution of 6.49.

15. (a)–(b) Consider a minimal set X which is a component of $G - T$, for some minimum cutset T. Use 6.60a. The connectivity must be at least $\frac{2}{3}(r+1)$.

(c) Use 6.60a again.

16. Use Tutte's theorem 7.27 and 12.14.

17. The Cartesian product of a circuit and a path contains a Hamiltonian circuit.

18. The connected ones are the following: circuits, the platonic bodies, the line-graphs of the cube and the dodecahedron, graphs obtained from these by multiplying each edge by the same number, subdivisions of these examples by one point on each edge, the duals of the line-graphs of the cube and the dodecahedron, and the stars.

19. Use 12.12b and the previous solution.

20. (a) Show that the Cartesian product of two p–1-paths and a K_2 is contractible onto K_p. (b) One may assume that Γ' acts edge-transitively and Γ does not act semiregularly. Use 10.8.

(c) Consider a simple group.

21. Construct a simple graph G with $A(G) \cong \Gamma$ and link it to Ω in a suitable manner.

22. Every endomorphism of the Petersen graph is an automorphism.

23. Use the same idea as in the previous solution: any endomorphism is one-to-one on a minimal odd circuit.

24. (a) goes as the solution of 12.4. (b) When replacing colored edges by paths with attached pieces use many rigid graphs which have no homomorphisms into each other.

25. (a) Either 0 or -1 remains an endomorphism. (b) Modify the previous construction.

§ 13

1. Associate a bipartite graph G with H in the following way: Let $U = V(H)$, $W = E(H)$, $V(G) = U \cup W$, and join $u \in U$ to $w \in W$ iff $u \in w$.

2. (a) If E is an edge such that $H' = (H - \{E\}) \backslash E$ is connected (hence $\emptyset \notin E(H)$), then we can find an F with the desired property. (b) Use induction, removing the edge E in the preceding problem.

3. Consider any circuit of length > 3 and choose two points which are consecutive on the circuit but there is a further point of the circuit which lies between them in the ordering defined by the path P.

4. Otherwise it is totally balanced!

5. Apply 7.4 and 7.6.

6. Such a system of distinct representatives exists iff H has a system of distinct representatives and so does the hypergraph

$$V(H_T^*) = E(H),$$
$$E(H_T^*) = \{U_x : x \in T\},$$

where U_x is the set of edges of H containing x.

7. Represent the situation by a flow problem rather than a matching problem.

8. (a) The sets $E_0 \triangle F$ $(F \in E(H))$ are distinct! (b) Let $x \in V(H)$, and consider the hypergraphs $H \backslash x$ and H', where $V(H') = V(H) - \{x\}$,

$$E(H') = \{E \in E(H) : E \cup \{x\} \in E(H), x \notin E\}$$

(c) Consider the hypergraphs $H_1 = H - x$, $H_2 = H - E(H_1)$.

9. Consider the hypergraphs $H_1 = H - x$, $\overline{H}_2 = H - E(H_1)$, $H_2 = \overline{H}_2 \backslash x$, for some $x \in V(H)$.

10. (a) Prove, by induction on k, that there is a set $X \subseteq V(H)$ with $|X| = k$ such that H_X has at least $k+1$ distinct edges $(1 \leq k \leq n-1)$. (b) Form a graph similar to the one in the previous second solution. Prove that it is bipartite and contains no Θ-graph. (c) Set $H_1 = H \backslash x$, $V(H_2) = V(H) - x$, and $E(H_2) = \{E \in E(H) : x \notin E, E \cup \{x\} \in E(H)\}$. Prove that either H_1 satisfies the condition of the problem with k, or H_2 does so with $k-1$.

11. Consider the edges containing a given $x \in V$ and use induction.

12. (a) Average the number of edges contained in the classes over all partitions of $V(H)$ into two classes. (b) Average now over all r-colorations of $V(H)$.

13. Establish and use the inequality

$$\sum_{x \in E} d(x) \leq r + (m-1)k \qquad (E \in E(H)),$$

where $d(x)$ is the degree of the point x.

14. Suppose indirectly that $m > n$. Sum the inequalities

$$\frac{d(x)}{m - d(x)} < \frac{|E|}{n - |E|} \qquad (x \in E).$$

15. (a) H satisfies the conditions of the previous problem. (b) If $\mathbf{a}_1, \ldots, \mathbf{a}_m$ are the characteristic vectors of edges, then $\mathbf{a}_i \mathbf{a}_j = \lambda$ for $i \neq j$.

16. How many times is a given elements counted on each of the two sides?

17. (a) Suppose that $x \in F_1, \ldots, F_d$, $x \notin F_{d+1}, \ldots, F_m$. Take $F_1 - \{x\}, \ldots, F_d - \{x\}$ $m - d$ times, $F_{d+1}, \ldots, F_m$ d times and apply 13.16. (b) Suppose that $|F_1| = \ldots = |F_{m-1}| = 2k$, $F_m = \emptyset$, and (indirectly) $m = k^2 + k + 3$. Verify that there are not more than k points of degree at least $m - k - 1$, but every F_j $(1 \leq j \leq m-1)$ must contain at least $k + 1$ such points.

18. (a) At most one of X and $E(H) - X$ can be an edge. (b) Use 13.8b, 13.9, or 13.11.

19. Count the pairs (E_1, E_2) of edges such that $E_2 \subset E_1$ and there is no $F \in E(H)$ with $E_2 \subset F \subset E_1$.

20. Use induction on n.

21. (a) Consider the chains constructed in the previous problem. For another proof, count in two different ways those permutations of $V(H)$ which have an edge of H as beginning section. (b) The even case is easy. For $n = 2k + 1$, show that each edge must be of size k or $k + 1$ and that, if $X \subset Y$, $|X| = k$, $|Y| = k + 1$, then one of X, Y must belong to $E(H)$.

22. If A and B are maximum antichains, then $A \vee B = \{x \in A \cup B : x \not< y \text{ for all } y \in A \cup B\}$ is a maximum antichain as well ($x < y$ means $x \leq y$ and $x \neq y$).

23. To prove (b) consider the subsets S_E $(E \in E(H))$ which are comparable with the edge E only, and show that, if chosen properly, they form a clutter.

24. Let

$$\binom{n}{p}, \ldots, \binom{n}{p + k - 1}$$

be the k largest binomial coefficients and let M be the hypergraph on $V(H)$ whose edges are all the $p, \ldots, (p+k-1)$-tuples. Compare H with M in each of the symmetric chains of 13.20.

25. (a) Fix an edge E. What does it mean if no point of E occurs in all the edges? (b) Find j edges whose intersection has at most $r - (j-1)(\tau(H) - 1)$ points, for $j = 1, 2, \ldots, k$.

26. Take two edges with one point in common if possible and a third edge not containing this point.

27. Take a minimal W such that any two edges of H_W meet. For every ordering of W, count the number of those points $x \in W$ which are both first points and last points of edges of H_W.

28. (a) From any given point, there starts at most one A_i. (b) Count the edges which consist of consecutive points in cyclic permutations of $V(H)$ in two different ways (cf. the second solution of Sperner's theorem 13.21a).

29. (a) Select a set S such that $|S \cap E| \geq 2$ for each edge E and $|S|$ is bounded. (b) If H is not ν-critical, use induction on ν.

30. Choose points $x_1, \ldots, x_t$ such that x_{i+1} is a point of maximum degree δ_{i+1} of $H - x_1 - \ldots - x_i$. Show that $|E(H - x_1 - \ldots - x_i)| \leq \tau^* \delta_{i+1}$.

31. (a) Write

$$\binom{u}{r} = \binom{w}{r} + \binom{w-1}{r-1}\binom{u-w+1}{1} + \binom{w-2}{r-2}\binom{u-w+2}{2} + \ldots,$$

and similarly for $\binom{v}{r}$, $\binom{u}{r-1}$, $\binom{v}{r-1}$, and compare the corresponding terms in $\binom{u}{r} - \binom{v}{r} - \binom{w}{r-1}$ and $\binom{u}{r-1} - \binom{v}{r-1} - \binom{w}{r-2}$. (b) Consider a point x with minimum degree and consider the hypergraphs $H - x$, $H - E(H-x)$. Use induction on r and $|E(H)|$ on the basis of (a). (c) Suppose that any two edges of H meet, where H is r-uniform. How many r-tuples can be contained in the sets $V(H) - E$, $E \in E(H)$?

32. (a) Use an argument similar to the one in 13.21 and 13.27; for a given permutation of the points, there is at most one index i such that every element of A_i has smaller index than every element of B_i. (b) This is a special case of (a).

33. Color the points one-by-one; when would one be forced to complete a monochromatic edge?

34. Show first that if one point has degree 1 and the others have degree 2, then H is 2-chromatic.

35. Suppose that H has no 1-element edges and no isolated points. Show that
(a) every point has degree at least 2,
(b) a point of degree at least 4 belongs to 2-element edges only,
(c) any two points belong to an edge.

36. Supposing there was a 2-coloration, count the edges in two different ways; by the monochromatic pairs contained in them and also by the bichromatic pairs contained in them.

37. To show the "if" part find a set $W \subseteq V(H)$ such that either all of $|W \cap E_1|$, ..., $|W \cap E_m|$ are even or one of them is odd.

38. (a) Show that replacing an edge E by disjoint subsets $E_1, \ldots, E_k \subset E$ does not influence the assumption. (b) Use the preceding characterization of hypergraphs with totally unimodular incidence matrix.

39. The incidence matrix of an odd circuit is not unimodular.

40. Show that a minimum counterexample would contain a connected spanning 2-uniform (hyper)graph.

41. Show that, coloring the points at random, the probability of getting a monochromatic edge is less than 1 [cf. 13.12a].

42. Let $|S| = 2r^2$ and let the $\binom{2r^2}{r}$ r-tuples in S constitute the points of another hypergraph H; for each partition $P = \{X, Y\}$ of S let those r-tuples contained in X or in Y form an edge E_p of H. What are $\tau(H)$ and $\tau^*(H)$?

43. Consider a random coloration again but estimate by 2.18.

44. (a) Consider a point with maximum degree. (b) Remove the point with largest degree from every edge. (c) Count the edges adjacent to the point of high degree.

45. (a) Select a 2-element subset of each edge such that these subsets form a forest. (b) For $r = 2$, the triangle is such a hypergraph. Use induction on r. (c) Assume that there is a non-zero vector orthogonal to the incidence vector of every edge, and consider its negative and positive entries.

46. For (a), use a modification of the construction in 13.45; for (b), try to "multiply" two smaller constructions; for (c), look for a simple direct example.

47. Select a k-element set X for each $1 \le k \le r$ such that the number of edges containing X is more than $|E(H)|/r^k$.

48. Setting $V(H) = \{v_1, \ldots, v_n\}$, $E(H) = \{E_1, \ldots, E_m\}$,

$$a_{ij} = \begin{cases} 1 & \text{if } v_i \in E_j, \\ 0 & \text{otherwise,} \end{cases}$$

and $A = (a_{ij})_{i=1,j=1}^{n\ \ m}$, the value $\nu^*(H)$ is the maximum of $\mathbf{1}^T\mathbf{x}$ under the constraints $A\mathbf{x} \le 1$, $\mathbf{x} \ge 0$.

49. (a) Decompose directly, but take into consideration the multiplicities of points. (b) Use 7.40. (c) There are optimal fractional covers (matchings) with rational weights. Reduce the least common denominator of the weights using (a) and (b).
   (d) $2\nu^*(G)$ is the maximum size of 2-matchings. $2\tau^*(G)$ is the formula given in 7.37 for $\nu_2(G)$.

50. Use a construction similar to the one in the previous solution.

51. (a) To prove the second inequality observe that if $S \subseteq V(G \otimes H)$ covers all edges of $G \otimes H$, then $S \cap (V(G) \otimes F)$ has at least $\tau(G)$ elements for every edge F of H. (b) "If" is easy by (a); to prove "only if", assume that $\tau_n(H) < n\tau(H)$ and let G consist of all $(N - n)$-element subsets of an N-set (N large). (c) Use 13.30 to estimate $\tau(H^p)$ from above.

52. If a point X is adjacent to two distinct edges E_1, E_2, take the "mean" of $(\tau - 1)$-element covers of $H - \{E_1\}$ and $H - \{E_2\}$.

53. Take the "mean" of maximum matchings of $H - v_i$ $v_i \in E_0 \in E(H)$.

54. One has to show that a τ-critical balanced hypergraph consists of disjoint edges. Copy the 2nd solution of 7.2.

55. To show (i)$\Rightarrow$(iii), note that a minimum counterexample would be ν-critical.

56. Show and use the following fact: If a hypergraph H satisfies (iv), then every hypergraph H_0 arising by multiplication of edges also satisfies (iv). (Multiplication includes multiplication by 0, i.e., removal.)

57. Suppose that G is perfect; associate a hypergraph H with G such that $L(H) \cong G$.

§ 14

1. (a) A point x is either adjacent to at least $\binom{k+l-1}{k-1}$ points or non-adjacent to at least $\binom{k+l-1}{k}$ points. (b) The existence of $R_2(a_1, a_2)$ follows from the preceding result. Use induction on k.

2. (a) $[ek!] = k[e(k-1)!] + 1$. (b) The upper bound is trivial from 14.1a. To prove the lower bound consider the hypergraph H with

$$V(H) = E(K_n) \quad \text{and} \quad E(H) = \{E(A) : A \text{ is a complete } k\text{-subgraph of } K_n\}.$$

3. (a) For $r = 2$ we already know this. Use induction on $a_1 + \ldots + a_k$ and r. (b) Select an ordered subset $(x_1, \ldots, x_m)$ of elements such that for each r-tuple $1 \leq \nu_1 < \ldots < \nu_r < m$ of indices, all $(r+1)$-sets of the form $(x_{\nu_1}, \ldots, x_{\nu_r}, x_\mu)$ $(\nu_r < \mu \leq m)$ have the same color.

4. Easy by induction on n.

5. Consider a maximum red path P and two disjoint blue paths Q_1, Q_2 such that Q_1, Q_2 consist of $(V(P), V(G) - V(P))$-edges, their endpoints are in $V(G) - V(P)$, they do not contain the endpoints of P and their length is maximal. Prove that $Q_1 \cup Q_2 \cup P$ covers all points.

6. (a) Let $(x_0, \ldots, x_{2k})$ be a red circuit. Prove that (x_i, x_{i+2}) is blue, (x_i, x_{i+4}) is red and (x_i, x_{i+3}) is blue $(i = 0, \ldots, 2k; x_{2k+1+j} = x_j)$. (b) Let $(x_1, \ldots, x_{2k})$ be a red circuit. Prove that $\{x_1, x_3, \ldots, x_{2k-1}\}$ and $\{x_2, x_4, \ldots, x_{2k}\}$ induce complete subgraphs. (c) Show that there is a monochromatic M-circuit for some $M \geq m$, then use (a), (b).

7. Consider the configuration in Fig. 17, in which all lines have unit length.

8. Consider the arrangement of six triangles with sides 1, a, b shown in Fig. 18.

9. (a) Take a stripwise coloration. (b) See Fig. 19.

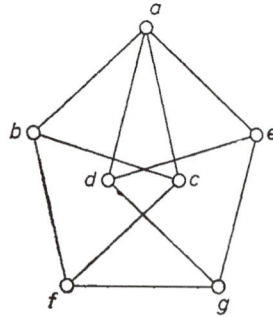

FIGURE 17

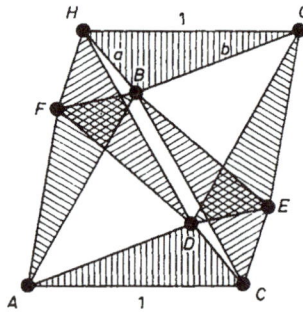

FIGURE 18

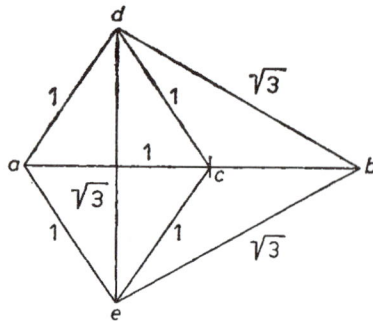

FIGURE 19

10. (a) Consider the Cartesian product of two large simplices. (b) One may assume that $\mathbf{ab}=1$, where $\mathbf{a}$, $\mathbf{b}$ are the two side vectors of R. Color the point $\mathbf{w}$ with $[\mathbf{w}^2]$ modulo 4.

11. Let $\{A_1,\ldots,A_k\}$ be the partition of $\{1,\ldots,n\}$ in question. Form a complete graph on $\{1,\ldots,n+1\}$ and color the edge (i,j) with color ν iff $|i-j|\in A_\nu$.

12. Use induction on r: Find a sequence $I_1,\ldots,I_r$ of "long" intervals whose first terms constitute an arithmetic progression and which are similarly colored.

13. Use 14.2a in a similar way to that in 14.11.

14. (a) First select a large set T and A_1, $B_1 \subseteq S - T$ with $A_1 \cap B_1 = \emptyset$ such that, for every $X \subseteq T$, the colors of $A_1 \cup X$, $B_1 \cup X$, $A_1 \cup B_1 \cup X$ are the same. (b) Let t be large in (a) and color the subset $\{i_1, \ldots, i_\nu\} \subseteq \{1, \ldots, t\}$ by the color of $A_{i_1} \cup \ldots \cup A_{i_\nu}$.

15. Color the subsets of an n-element set corresponding to the colors of their cardinalites.

16. (a) Trivial for $n_1(k,m) = k$. (b) Let $|T| = k^{m^{r-1}}$, $T \subseteq S$. Color vectors $\mathbf{a}$ of points of $S - T$ with the $m^{|T|}$-tuples $(\alpha(\mathbf{a}+\mathbf{d}); \mathbf{d} \in m^T)$.

17. By repeated application of 14.16b, we can find disjoint non-empty sets $X'_1, \ldots,$ $X'_t \subseteq S$ and an m-vector $\mathbf{b}$ of points of $S - X'_1 - \ldots - X'_t$ such that the color of every m-vector

$$\mathbf{a} = \sum_{i=1}^{t} c_i X'_i + \mathbf{b}'$$

is the same as the color of the m-vector
$$\bar{\mathbf{a}} = \bigcup_{c_i \neq \emptyset} X'_i + \mathbf{b}'.$$

Now apply 14.14b.

18. Use a method similar to the one in 14.15.

19. Suppose that there were a k-coloration of $\{1, \ldots, N\}$ for each N with no subset having property P. Find such k-colorations which could be combined.

20. (a) Use the following observation: if x, y, z satisfy (a) and x, y, $z \equiv 0$ or i (mod 5), then x, y, $z \equiv 0$ (mod 5) (for each fixed $1 \leq i \leq 4$). (b) Use induction on k and Van der Waerden's theorem.

21. Translate parts (a) and (b) of the previous solution to the general case.

22. Color the points at random and use 2.18 to ensure that every edge meets every color.

23. (a) This is a special case of 14.17. (b) Given $r_1, \ldots, r_k \geq 0$, prove that a number $\overline{N}(k; r_1, \ldots, r_k; q-1)$ exists such that if $n \geq \overline{N}$ and all points of the n-dimensional projective space over $GF(q)$ are k-colored, there always exists an r_i-dimensional subspace all of whose points have color i for some $1 \leq i \leq k$. Try to reduce this to the affine case by removing the "infinitely distant" hyperplane.

24. For $r = q$, a prime power, the lines of an affine space over $GF(q)$ of high enough dimension will do it. Otherwise take a prime power $q \geq r$ and reduce the edges of a q-uniform example.

25. For each a_i, consider the largest monotonically increasing subsequence starting with a_i.

26. (a) Let $t(x)$ be the maximum length of a sequence $x = a_1 < a_2 < \ldots$ with $f(a_1) \leq f(a_2) \leq \ldots$ Prove that, if $t(x) \geq t(1) - k$, then $f(x) \leq 2^k$ $(0 \leq k \leq t(1) - 1)$. (b) Let f consist of $(n-1)$ monotone decreasing parts.

27. (a) Denote by $\eta(x,y)$ and $\varphi(x,y)$ the lengths of the longest monotone increasing and decreasing paths starting with the edge $(x,y) \in E(T_n)$. For any two points a, b,

$$\{(\eta(a,x), \varphi(a,x)); \ (a,x) \in E(T_n)\} \neq \{(\eta(b,x), \varphi(b,x)); \ (b,x) \in E(T_n)\}.$$

(b) To prove that the bound is sharp use induction on p and q.

28. (a) Let x_{i+1} be a point different from

$$f(S - \{x_1, \ldots, x_\nu\}), \ \nu = 0, \ldots, i \quad (0 \leq i \leq m - 1).$$

(b) Prove a common generalization of (a) and (b): There is an ordering $(x_1, \ldots, x_{2^n})$ of S such that, for each $1 \leq j \leq n$,

$$f(\{x_1, \ldots, x_\nu\}) \in \{x_1, \ldots, x_{2j}\},$$
$$2^j < \nu \leq 2^n$$

holds for at least $(n-j)2^j$ values of ν. (c) Look for an f in the following form. Given a function φ with $1 \leq \varphi(i) \leq i$, let $f(X)$ be the $\varphi(|X|)^{\text{th}}$ element of X $(X \subseteq S = \{1, \ldots, 2^{n-1} - 1\})$.

29. Direct each segment (p,q) formed by two given points upwards and color it with color i if its angle formed with the positive half of the x-axis is between $\frac{i}{n}\pi$ and $\frac{i+1}{n}\pi$. What can the chromatic number of the graph G_i formed by the segments of color i be, if it contains no "almost straight" broken line?

30. (a) Associate the value $\frac{y_1 - y_2}{x_1 - x_2}$ with the edge $((x_1, y_1), (x_2, y_2))$ of the complete graph on S; orient this from (x_1, y_1) toward (x_2, y_2) if $x_1 < x_2$ and apply 14.27. (b) Carry out the construction in 14.27 with tournaments of the special type considered here.

31. (a) Color a quadruple of points red if it forms a convex quadrilateral and blue otherwise; apply Ramsey's theorem. (b) The upper bound is immediate from the preceding problem. The lower bound follows by putting together sets with cardinalities $\binom{m-2}{i}$ $(i = 0, \ldots, m-2)$ containing no $(i+2)$-gons convex from below and no $(m-i)$-gons convex from above.

§ 15

1. (a) How can one realize the points of G in $L(G)$? (b) There are three edges e_1, e_2, $e_3 \in E(G_1)$ starting from some point such that the corresponding edges e'_1, e'_2, $e'_3 \in E(G_2)$ form a triangle. (c) The solution of (b) contains the answer.

2. (a) Show that $|E \cap F| = |\alpha(E) \cap \alpha(F)|$, for any two E, $F \in E(K_n^r)$. (b) Show that it can be "recognized" in $L(K_n^r)$ when two r-tuples have k points in common for $k = r - 1$; then show this for $k = 1, 2, \ldots$.

3. Again, "recover" $|E \cap F|$ from $L_i(K_n^r)$ by counting the r-tuples adjacent to both E and F.

4. (a) Consider hypergraphs of stars. (b) What we need is a covering of the edges by $\lfloor n^2/4 \rfloor$ cliques. This can be done by a method similar to the one in Turán's theorem 10.28.

5. (a) Use 15.1b to show that the graphs H_x "fit together". (b) Obvious from the preceding problem.

6. (a) If $G = L(H)$, then those triples of edges of H which form an odd triangle in G belong to a star and conversely. (b) Define two edges (a, b) and (c, d) of G to be equivalent if they are identical or they belong to an odd triangle or they span a complete quadrilateral. Show that this is an equivalence relation and that the equivalence classes are complete covering subgraphs as in 15.4a and they will give a desired graph H with $L(H) = G$ in all but a few cases. (c) How can (ii) in (a) be violated?

7. Verify that the graph shown in Fig. 20 is not the line-graph of any 3-uniform hypergraph without multiple edges but every proper subgraph of it is such a line-graph.

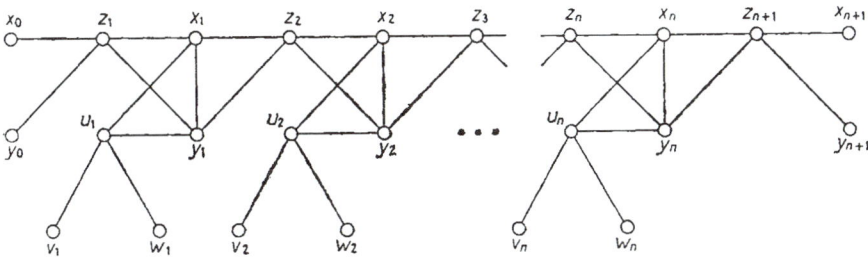

FIGURE 20

8. (a) This means: find two essentially different embeddings in the plane of the same planar graph. (b) Use 6.69.

9. (a) Consider two edges (x, y), (x, z) and a circuit C through y, z in $G_1 - x$. (b) The solution of 15.1b simplifies since, in this case, no star can correspond to a triangle.

10. Use induction on r, showing that T and T' are isomorphic, where x_i' corresponds to x_i.

11. (a) follows from an identity used in the preceding solution. (b) follows similarly. (c) Construct T by induction. Consider two points x_i, x_j such that

$$d_{ri} + d_{rj} - d_{ij}$$

is minimal.

12. (a) Let $\{a_1, \ldots, a_n\}$ consist of $\binom{k+1}{2l}$ copies of $2l$, for $0 \le l \le \frac{k+1}{2}$ and $\{b_1, \ldots, b_n\}$ consist of $\binom{k+1}{2l+1}$ copies of $2l+1$, $0 \le l \le \frac{k}{2}$. (b) Consider the polynomials

$$f(x) = \sum_{i=1}^{n} x^{a_i}, \qquad g(x) = \sum_{i=1}^{n} x^{b_i}.$$

13. By considering $f - g$, it suffices to show that

$$\sum_{i=1}^{M} \alpha_i \binom{c_{i1}}{s_1} \cdots \binom{c_{ik}}{s_k}$$

cannot vanish for all integers $s_1, \ldots, s_k \ge 0$ unless it is identically 0.

14. Prove first that G_1, G_2 have the same number of edges, then that they have the same degrees.

15. (a) Consider two Hamiltonian circuits on V. (b) Count H in all subgraphs $G - x$. (c) Prove that each k-point-connected graph occurs among the components of G_1 and G_2 the same number of times, for $k = |V| - 1, \ldots$.

16. (a) The diameter of T_i is the maximum of diameters of $T_i - x$, except when T_i is a path. (b) If $T_i - x$ has diameter d for each endpoint x, then the branches of T_i relative to its center can be reconstructed using a method similar to the one in 15.9c. If an endpoint x_0 is contained in all paths of length d, then examine whether all $(d-1)$-paths of $T_i - x_0$ can be covered by one point.

17. (a) Express the numbers of isomorphisms of G_1 into G_2 and onto itself by a formula similar to that in 5.18, and observe that the right-hand sides are identical. (b) Use the identity in 2.7b instead of the sieve formula.

18. Show first that if $(*)$ holds for every $|X| = k$, then it also holds for every $|X| \le k$ (provided $k \le |V| - r$).

19. Suppose that $H_1 \ne H_2$. Show that $\emptyset \in E(H_1)$ or $\emptyset \in E(H_2)$; then use induction on k to show that each k-element subset belongs to H_1 or to H_2, depending on the parity of k.

20. (a) With epimorphisms (homomorphisms onto H) the assertion would be trivial. Express the number of epimorphisms in terms of the number of homomorphisms. (d) "Dualize" the solution of (a).

21. Show that the number of homomorphisms of H into $G \times G$ is the square of the number of homomorphisms of H into G.

22. (a) Try $F = K_2$; observe that G_1, G_2 must have the same degree sequence. (b) Use an argument similar to the one in the preceding solution. (c) Let G° denote the graph obtained by adding a loop at each point of G. The $(G \cdot H)^\circ = G^\circ \times H^\circ$. (d) Again, if two groups have the property that every group has the same number of homomorphisms into each of them, then they are isomorphic.

III. Solutions

§ 1. Basic enumeration

1. I. The first person may be sent any of the k kinds of postcards. No matter which one he is sent, we may still send the second one any of the k kinds, so there are $k \cdot k = k^2$ ways to send cards to the first two friends. Again, whatever they are sent, the third friend can still be sent k kinds, etc. So there are k^n ways to send out the cards.

II. If they have to be sent different cards, the first person can still be sent any of the k cards. But for any choice of this card, there are only $k-1$ kinds of cards left for the second person; whatever the first and second friends receive the third one can get one of $k-2$ postcards, etc.... Thus the number of ways to send them postcards is $k(k-1)\dots(k-n+1)$ (which is, of course, 0 if $n > k$).

III. This is the same as the first question but we have $\binom{k}{2}$ pairs of postcards instead of k postcards. Thus the result is $\binom{k}{2}^n$.

2. I. We have to decide about the postcards independently. Any postcard can be sent to any of the n friends. Hence, the result is n^k.

II. Let $C_1,\dots,C_k$ be the cards. The set $S = \{C_1,\dots,C_k\}$ must be split into n disjoint non-empty sets $S_1,\dots,S_n$. Thus $\{S_1,\dots,S_n\}$ is a partition of S. From any partition of S into n (non-empty) classes we get $n!$ possibilities to send out the postcards. Hence the answer is $n! \cdot \left\{ {k \atop n} \right\}$.

3. There are 16! permutations of the letters of CHARACTERIZATION. However, not all of these give new words; in fact, in any permutation, if we exchange the three A's, the two C's, the two R's, the two I's or the two T's we get the same word. Thus for any permutation, there are $3! \cdot 2 \cdot 2 \cdot 2 \cdot 2 = 96$ permutations which give the same word, so the result is

$$\frac{16!}{96}$$

In general, if there are k_A A's, k_B B's, etc. then the result is

$$\frac{(k_A + k_B + \dots)!}{k_A! k_B! \dots}$$

4. (a) If we distribute the forints by the procedure described in the hint we have to say "Next please" $n-1$ times. If we determine at which points (after which coins) we say this we uniquely determine the distribution. There are $k-1$ possible points to switch and we have to choose $n-1$ out of these. Hence the result is

$$\binom{k-1}{n-1}.$$

(b) Borrow one forint from each person. If we distribute the $n+k$ forints we then have in such a way that each person gets at least one, we would then have done the same as if we had distributed the k forints without this requirement. More precisely, distributions of $n+k$ forints among persons so that each one gets at least one are in a one-to-one correspondence with all distributions of n forints among k persons. Hence the answer is

$$\binom{n+k-1}{n-1}.$$

5. We can distribute the first kind of postcard in

$$\binom{a_1+n-1}{n-1}$$

ways by the preceding solution. Whatever the choice here, there are

$$\binom{a_2+n-1}{n-1}$$

ways to send out the second one, etc. Thus the result is

$$\binom{a_1+n-1}{n-1}\binom{a_2+n-1}{n-1}\cdots\binom{a_k+n-1}{n-1}.$$

6. (a) We interpret $\left\{\begin{smallmatrix} n+1 \\ k \end{smallmatrix}\right\}$ as the number of partitions of $\{1,\ldots,n+1\}$ into exactly k classes. Consider the restrictions of partitions of $\{1,\ldots,n+1\}$ onto $\{1,\ldots,n\}$. This way we obtain each partition of $\{1,\ldots,n\}$ into k classes exactly k times, and each partition into $k-1$ classes exactly once. Hence

$$\left\{\begin{matrix} n+1 \\ k \end{matrix}\right\} = \left\{\begin{matrix} n \\ k-1 \end{matrix}\right\} + k\left\{\begin{matrix} n \\ k \end{matrix}\right\}.$$

We can get another recurrence relation if we remove the class containing $n+1$ to get a partition of $n+1-r$ elements into $k-1$ classes, where r is the number of elements in this class. We have $\binom{n}{r-1}$ possible ways of choosing this class. Hence

$$\left\{\begin{matrix} n+1 \\ k \end{matrix}\right\} = \sum_{r=1}^{n+1}\binom{n}{r-1}\left\{\begin{matrix} n+1-r \\ k-1 \end{matrix}\right\} = \sum_{m=0}^{n}\binom{n}{m}\left\{\begin{matrix} m \\ k-1 \end{matrix}\right\}.$$

For the Stirling cycle numbers, we can restrict each permutation of $\{1,\dots,n+1\}$ with k cycles to the set $\{1,\dots,n\}$: if $n+1$ is a fixed point, we simply delete it; else, we skip it in the cycle containing it. This way every permutation of $\{1,\dots,n\}$ with $k-1$ cycles is obtained exactly once, and each permutation of $\{1,\dots,n\}$ with k cycles is obtained exactly n times (since $n+1$ may be inserted after any element). Hence we obtain

$$\begin{bmatrix} n+1 \\ k \end{bmatrix} = \begin{bmatrix} n \\ k-1 \end{bmatrix} + n \begin{bmatrix} n \\ k \end{bmatrix}.$$

On the basis of these recurrences, it is easy to tabulate both kinds of Stirling numbers:

n \ k	0	1	2	3	4	5	6	7
0	1	0	0	0	0	0	0	0
1	0	1	0	0	0	0	0	0
2	0	1	1	0	0	0	0	0
3	0	1	3	1	0	0	0	0
4	0	1	7	6	1	0	0	0
5	0	1	15	25	10	1	0	0
6	0	1	31	90	65	15	1	0
7	0	1	63	301	350	141	21	1

Table 1a

The Stirling partition numbers

n \ k	0	1	2	3	4	5	6	7
0	1	0	0	0	0	0	0	0
1	0	1	0	0	0	0	0	0
2	0	1	1	0	0	0	0	0
3	0	2	3	1	0	0	0	0
4	0	6	11	6	1	0	0	0
5	0	24	50	35	10	1	0	0
6	0	120	274	225	85	15	1	0
7	0	720	1764	1624	735	175	21	1

Table 1b

The Stirling cycle numbers

(b) Classify partitions of $\{1,\dots,n\}$ into $n-k$ classes by considering the union S of those classes with more than one element. If $|S|=j$, then by the remark in the hint, we have $j \leq 2k$. Let $a_{j,k}$ be the number of partitions of a j-element set into $j - k$ classes of size at least 2. For a given j, we have $\binom{n}{j}$ ways to select S, and thus

$$\left\{ \begin{matrix} n \\ n-k \end{matrix} \right\} = \sum_{j=0}^{2k} a_{j,k} \binom{n}{j},$$

which is clearly a polynomial in n (of degree $2k$). The assertion concerning Stirling cycle numbers can be proved along the same lines.

(c) The boundary conditions determine $\left\{ \begin{matrix} n \\ k \end{matrix} \right\}$ if either n or k is 0. The values for $n = 0$ uniquely determine the values for all $n > 0$ by the recurrence (it is easy to check that the values for $k = 0$, $n > 0$ are 0, as they should be). Using the recurrence in the form given in the hint, we see that the values for $k = 0$ determine the values for all $k < 0$ (again, the values with $n = 0$, $k < 0$ come out right). Finally, writing the recurrence as

$$k \left\{ \begin{matrix} n \\ k \end{matrix} \right\} = \left\{ \begin{matrix} n+1 \\ k \end{matrix} \right\} - \left\{ \begin{matrix} n \\ k-1 \end{matrix} \right\},$$

we obtain the values for all $n < 0$ and $k > 0$, by recurrence on $k - n$. Again, the assertion for the Stirling cycle numbers follows similarly.

(d) Write up the recurrence relation of (a) for $\left[\begin{matrix} -k \\ -n \end{matrix} \right]$:

$$\left[\begin{matrix} -k+1 \\ -n \end{matrix} \right] = \left[\begin{matrix} -k \\ -n-1 \end{matrix} \right] - k \left[\begin{matrix} -k \\ -n \end{matrix} \right],$$

or

$$\left[\begin{matrix} -k \\ -n+1 \end{matrix} \right] = \left[\begin{matrix} -k+1 \\ -n \end{matrix} \right] + k \left[\begin{matrix} -k \\ -n \end{matrix} \right].$$

So $\left[\begin{matrix} -k \\ -n \end{matrix} \right]$ satisfies the same recurrence relation as $\left\{ \begin{matrix} n \\ k \end{matrix} \right\}$. Since they also satisfy the same boundary conditions, they are identical by (c). [cf. D. E. Knuth, Two notes on notation, *Amer. Math. Monthly*, 1992.]

7. (a) Suppose first that x is a positive integer. Let $|X| = x$, $|N| = n$. The number of mappings of N into X is x^n (cf. 1.1). On the other hand, let k denote the cardinality of the range of a mapping of N into X. For k fixed, we can specify in $\left\{ \begin{matrix} n \\ k \end{matrix} \right\}$ ways which elements of N are mapped onto the same element of X. Once this partition of N is specified, we have to find an image for each class of it, distinct images for distinct classes. This can be done in $x(x-1)\ldots(x-k+1)$ ways. Thus

$$\left\{ \begin{matrix} n \\ k \end{matrix} \right\} x(x-1)\ldots(x-k+1)$$

is the number of mappings of N into X with range of cardinality k. This proves the identity when x is a positive integer. But this means that if we consider x as a variable the polynomials on the two sides have infinitely many values in common. Therefore, they must be identical.

(b) Again, we may assume that x is a positive integer. If a permutation π of a set S has exactly k cycles, then x^k is the number of x-colorations of S invariant under π. The left hand side of the identity sums these numbers for all permutations π of $S = \{1,\dots,n\}$. A given x-coloration of S is counted $k_1!\dots k_x!$ times, where k_i is the number of elements with color i. The number of occurrences of a given sequence $k_1,\dots,k_x$ is $n!/k_1!\dots k_x!$ (see 1.3) and so by 1.4b, this sum is

$$\sum_{\substack{k_1,\dots,k_x \geq 0 \\ k_1+\dots+k_x=n}} \frac{n!}{k_1!\dots k_x!}k_1!\dots k_x! = n!\binom{x+n-1}{n} = x(x+1)\dots(x+n-1).$$

(c) We have by (a) and (b) (the latter applied with $-x$ substituted for x)

$$x^n = \sum_{k=0}^{n} \left\{{n \atop k}\right\} x(x-1)\dots(x-k+1) =$$

$$\sum_{k=0}^{n} \left\{{n \atop k}\right\} \sum_{j=0}^{k}(-1)^{k-j}\left[{k \atop j}\right] x^j = \sum_{j=0}^{k}(-1)^j x^j \sum_{k=0}^{n}(-1)^k \left\{{n \atop k}\right\}\left[{k \atop j}\right],$$

whence the assertion follows by comparing coefficients.

8. $$\frac{1}{k!}\sum_{j=0}^{k}(-1)^{k-j}\binom{k}{j}j^n = \sum_{j=0}^{k}\frac{(-1)^{k-j}}{j!(k-j)!}\sum_{r=0}^{n}\left\{{n \atop r}\right\}j(j-1)\dots(j-r+1) =$$

$$= \sum_{j=0}^{k}\sum_{r=0}^{j}(-1)^{k-j}\left\{{n \atop r}\right\}\frac{1}{(k-j)!(j-r)!} =$$

$$= \sum_{r=0}^{k}\frac{\left\{{n \atop r}\right\}}{(k-r)!}\sum_{j=r}^{k}(-1)^{k-j}\binom{k-r}{k-j} = \sum_{r=0}^{k}\frac{\left\{{n \atop r}\right\}}{(k-r)!}(1-1)^{k-r} = \left\{{n \atop k}\right\}.$$

9. (a) Since $\left\{{n \atop k}\right\} = 0$ for $k > n$, we have

$$B_n = \sum_{k=0}^{\infty}\left\{{n \atop k}\right\} = \sum_{k=0}^{\infty}\frac{1}{k!}\sum_{j=0}^{k}(-1)^{k-j}\binom{k}{j}j^n =$$

$$= \sum_{j=0}^{\infty}\frac{j^n}{j!}\sum_{k=j}^{\infty}\frac{(-1)^{k-j}}{(k-j)!} = \sum_{j=0}^{\infty}\frac{j^n}{j!}\cdot\frac{1}{e}.$$

(b) First we remark that

$$(1) \qquad B_n \sim \frac{1}{\sqrt{2\pi}} \sum_{k=1}^{\infty} k^{n-k-\frac{1}{2}} e^{k-1} \qquad (n \to \infty).$$

In fact, let k_0 be chosen such that for $k \geq k_0$,

$$\left| \frac{k!}{\sqrt{2\pi k}\,(k/e)^k} - 1 \right| < \varepsilon,$$

where ε is any given positive number. Since

$$\sum_{k=0}^{k_0-1} \frac{k^n}{k!} = o\left(\frac{k_0^n}{k_0!}\right) = o(B_n),$$

we have

$$B_n \sim \frac{1}{e} \sum_{k=k_0}^{\infty} \frac{k^n}{k!}.$$

Similarly,

$$\frac{1}{\sqrt{2\pi}} \sum_{k=1}^{\infty} k^{n-k-1/2} e^{k-1} \sim \frac{1}{\sqrt{2\pi}} \sum_{k=k_0}^{\infty} k^{n-k-1/2} e^{k-1}.$$

By the choice of k_0, we have

$$1 - \varepsilon < \frac{\dfrac{1}{\sqrt{2\pi}} \displaystyle\sum_{k=k_0}^{\infty} k^{n-k-1/2} e^{k-1}}{\dfrac{1}{e} \displaystyle\sum_{k=k_0}^{\infty} k^n/k!} < 1 + \varepsilon.$$

This proves (1).

Now put

$$g_n(x) = \begin{cases} \dfrac{1}{\sqrt{2\pi}} x^{n-x-\frac{1}{2}} e^{x-1} & \text{for } x \geq 0, \\ 0 & \text{for } x \leq 0. \end{cases}$$

Then $g_n(x)$ has a unique maximum at the point $\lambda(n)$. Put

$$h_n(y) = \frac{g_n(\lambda(n)(1 + y/\sqrt{n}))}{g_n(\lambda(n))}.$$

Then $h_n(y)$ has a unique maximum at $y = 0$, this maximal value is 1 and we have, by the above,

$$(2) \qquad B_n \sim \left\{ \sum_{k=-\infty}^{\infty} \frac{\sqrt{n}}{\lambda(n)} h_n(y_k) \right\} \cdot \frac{\lambda(n)}{\sqrt{n}} g_n(\lambda(n)),$$

where

$$y_k = k\frac{\sqrt{n}}{\lambda(n)} - \sqrt{n}.$$

The sum in $\{\}$ is approximating the integral $\int\limits_{-\infty}^{\infty} h_n(y)\,dy$ and the difference is less than $(n/\lambda(n))\max h_n(y) = o(1)$. We shall show that

(3) $$h_n(y) \to e^{-y^2/2} \qquad (n \to \infty)$$

and that the functions $h_n(y)$ have a common integrable majorant. It will then follow by Lebesgue's theorem that

$$B_n \sim \frac{\lambda(n)}{\sqrt{n}}g_n(\lambda(n)) \int\limits_{-\infty}^{\infty} e^{-y^2/2}dy = \frac{\lambda(n)}{\sqrt{n}}g_n(\lambda(n))\sqrt{2\pi} = \frac{1}{\sqrt{n}}\lambda(n)^{n+\frac{1}{2}}e^{\lambda(n)-n-1}.$$

Now (3) follows from the transformation

(4) $$\log h_n(y) = -(n - \lambda(n))\left\{\frac{y}{\sqrt{n}} - \log\left(1 + \frac{y}{\sqrt{n}}\right)\right\} -$$
$$- \left(\lambda(n)\frac{y}{\sqrt{n}} + \frac{1}{2}\right)\log\left(1 + \frac{y}{\sqrt{n}}\right).$$

The second term is $\sim \frac{y}{\sqrt{n}}\left(\lambda(n)\frac{y}{\sqrt{n}} + \frac{1}{2}\right) = o(1)$ (as $n \to \infty$), while the first one is equal to

$$(n - \lambda(n))\left(-\frac{y^2}{2n} + \frac{y^3}{3n\sqrt{n}} - \dots\right) = -\frac{y^2}{2} + o(1)$$

for $n > y^2$. This proves (3).

To establish an integrable majorant for $h_n(y)$, first let $y > 0$. Then by (4),

$$\log h_n(y) < -\frac{n}{2}\left(\frac{y}{\sqrt{n}} - \log\left(1 + \frac{y}{\sqrt{n}}\right)\right)$$

(for $n \geq 16$). This upper bound is monotone decreasing and so

$$h_n(y) < \exp\left(-2y + 8\log\left(1 + \frac{y}{4}\right)\right) \qquad (n \geq 16).$$

For $y < 0$ we have similarly

$$\log h_n(y) < -\frac{n}{2}\left(\frac{y}{\sqrt{n}} - \log\left(1 + \frac{y}{\sqrt{n}}\right)\right) - \frac{1}{2}\log\left(1 + \frac{y}{\sqrt{n}}\right)$$

(provided $n > y^2$) and then it is easy to see that, for $y < -3$ the right-hand side is monotone increasing and hence that

$$h_n(y) < e^{-y^2}.$$

This holds for $n \le y^2$ as well since then $h_n(y) = 0$. Estimating $h_n(y)$ by 1 in the interval $-3 \le y \le 0$, we obtain the desired integrable majorant. [L. Moser–M. Wyman, *Trans. Royal Soc. Can.* **49** (1955) 49–54.]

10. Let S be the set to be partitioned and $x \in S$. If the class containing x has k elements, it can be chosen in $\binom{n-1}{k-1}$ ways and the remaining $n - k$ elements can be partitioned in B_{n-k} ways. So the number of partitions in which the class containing x has k elements is

$$\binom{n-1}{k-1} B_{n-k}.$$

This remains true for $k = n$ if we $B_0 = 1$. Thus

$$B_n = \sum_{k=1}^{n} \binom{n-1}{k-1} B_{n-k} = \sum_{k=0}^{n-1} \binom{n-1}{k} B_k.$$

11. From the recurrence relation in the previous solution, we have

$$p(x) = \sum_{n=0}^{\infty} \frac{B_n}{n!} x^n = 1 + \sum_{n=1}^{\infty} \frac{x^n}{n!} \sum_{k=0}^{n-1} \binom{n-1}{k} B_k =$$

$$= 1 + \sum_{k=0}^{\infty} \frac{B_k}{k!} \sum_{n=k+1}^{\infty} \frac{x^n}{n} \frac{1}{(n-k-1)!}$$

and so,

$$p'(x) = \sum_{k=0}^{\infty} \frac{B_k}{k!} \sum_{n=k+1}^{\infty} \frac{x^{n-1}}{(n-k-1)!} = \sum_{k=0}^{\infty} \frac{B_k x^k}{k!} \sum_{r=0}^{\infty} \frac{x^r}{r!} = p(x)e^x$$

or

$$(\log p(x))' = \frac{p'(x)}{p(x)} = e^x, \qquad p(x) = e^{e^x + c}$$

for some constant c, which can be determined by setting $x = 0$:

$$1 = p(0) = e^{e^0 + c}, \qquad c = -1.$$

Thus we have

$$p(x) = e^{e^x - 1}.$$

12. (a) Let k_i denote the number of i-element classes in a partition. Then

$$k_1 + 2k_2 + \ldots + nk_n = n.$$

Let us count partitions with $k_1, \ldots, k_n$ fixed. If we consider any arrangement of the n objects we can get such a partition by taking the first k_1 elements as 1-element classes, the next $2k_2$ elements as 2-element classes, etc. We get any given partition exactly

$$k_1!(1!)^{k_1} \cdot k_2!(2!)^{k_2} \ldots k_n!(n!)^{k_n}$$

times; for we can construct the arrangement of the objects by putting the one-element classes first, then the two-element ones, etc. However there are $k_i!$ possible ways to order the i-element classes and $(i!)^{k_i}$ possible ways to order the elements in the i-element classes. Thus the number of partitions with k_i i-element classes $(i=1,\ldots,n;\ k_1+2k_2+\ldots+nk_n=n)$ is

$$\frac{n!}{k_1!(1!)^{k_1} k_2!(2!)^{k_2} \ldots k_n!(n!)^{k_n}}.$$

This proves the assertion. Now

$$p(x) = \sum_{n=0}^{\infty} \frac{B_n}{n!} x^n = \sum_{n=0}^{\infty} \sum_{\substack{k_1,\ldots,k_n \geq 0 \\ k_1+2k_2+\ldots nk_n=n}} \frac{1}{k_1!(1!)^{k_1} k_2!(2!)^{k_2} \ldots k_n!(n!)^{k_n}} x^n.$$

We may actually allow infinitely many k_i's in each term with the restriction $k_1 + 2k_2 + \ldots = n$, since clearly $k_r = 0$ for $r > n$. Thus

$$p(x) = \sum_{n=0}^{\infty} \sum_{\substack{k_i \geq 0 \\ k_1+2k_2+\ldots=n}} \frac{x^n}{\prod_{i=1}^{\infty} k_i!(i!)^{k_i}} = \sum_{k_i \geq 0} \frac{x^{\sum_{i=1}^{\infty} ik_i}}{\prod_{i=1}^{\infty} k_i!(i!)^{k_i}},$$

where the summation extends over all sequences $k_1, k_2, \ldots$ of non-negative integers containing finitely many non-zero entries. Consequently,

$$p(x) = \prod_{i=1}^{\infty} \left(\sum_{k_1=0}^{\infty} \frac{x^{ik_i}}{k_i!(i!)^{k_i}} \right) = \prod_{i=1}^{\infty} \frac{x^i}{e^{i!}} = e^{e^x-1}.$$

(b) If we consider

$$\sum_{\substack{k_1,\ldots,k_n \geq 0 \\ k_1+\ldots+k_n=n}} \frac{n!}{k_1!(1!)^{k_1} k_2!(2!)^{k_2} \ldots k_n!(n!)^{k_n}},$$

then we can observe that this is the polynomial expansion of

$$\left(\sum_{i=1}^{n}\frac{1}{i!}\right)^{n} \approx (e-1)^{n}.$$

In fact the difference is

$$(e-1)^{n} - \left(\sum_{i=1}^{n}\frac{1}{i!}\right)^{n} = \left(e - 1 - \sum_{i=1}^{n}\frac{1}{i!}\right)\left(\sum_{k=0}^{n-1}(e-1)^{k}\left(\sum_{i=1}^{n}\frac{1}{i!}\right)^{n-k}\right) <$$

$$< \left(\frac{1}{(n+1)!} + \frac{1}{(n+2)!} + \ldots\right)n(e-1)^{n} \to 0.$$

13. By 1.12,

$$e^{e^{x}} = \sum_{n=0}^{\infty}\frac{eB_{n}}{n!}x^{n}.$$

On the other hand,

$$e^{e^{x}} = \sum_{k=0}^{\infty}\frac{e^{kx}}{k!} = \sum_{k=0}^{\infty}\frac{1}{k!}\sum_{n=0}^{\infty}\frac{(kx)^{n}}{n!} = \sum_{n=0}^{\infty}\frac{1}{n!}\sum_{n=0}^{\infty}\frac{k^{n}}{k!}x^{n}.$$

Hence

$$\frac{eB_{n}}{n!} = \frac{1}{n!}\sum_{k=0}^{\infty}\frac{k^{n}}{k!}$$

which proves the assertion.

We remark that, conversely, the formula in this problem implies 1.11.

14. (a) We have

$$Q_{n} = \sum_{\substack{k_{1},\ldots,k_{n}\geq 0 \\ k_{1}+2k_{2}+\ldots+nk_{n}=n \\ k_{1}+k_{2}+\ldots+k_{n}\text{ even}}}\frac{n!}{k_{1}!(1!)^{k_{1}}\ldots k_{n}!(n!)^{k_{n}}}$$

and so

$$q(x) = \sum_{n=0}^{\infty}\sum_{\substack{k_{1},\ldots,k_{n}\geq 0 \\ k_{1}+2k_{2}+\ldots+nk_{n}=n \\ k_{1}+k_{2}+\ldots+k_{n}\text{ even}}}\frac{x^{n}}{k_{1}!(1!)^{k_{1}}\ldots k_{n}!(n!)^{k_{n}}} =$$

$$= \sum_{\substack{k_{1},k_{2},\ldots\geq 0 \\ k_{1}+k_{2}+\ldots\text{ finite and even}}}\prod_{i=1}^{\infty}\frac{x^{ik_{i}}}{k_{i}!(i!)^{k_{i}}} =$$

$$= \sum_{r=0}^{\infty} \frac{1}{(2r)!} \sum_{\substack{k_1,k_2,\ldots \geq 0 \\ k_1+k_2+\ldots=2r}} \frac{(2r)!}{k_1!k_2!\ldots} \prod_{i=1}^{\infty} \left(\frac{x^i}{i!}\right)^{k_i}.$$

The second summation is just the polynomial expansion of

$$\left(\sum_{i=1}^{\infty} \frac{x^i}{i!}\right)^{2r} = (e^x - 1)^{2r}$$

and so

$$q(x) = \sum_{r=0}^{\infty} \frac{1}{(2r)!}(e^x - 1)^{2r} = \mathrm{ch}\,(e^x - 1).$$

To get an expression for the coefficients we write

$$\mathrm{ch}\,(e^x - 1) = \frac{e^{e^x-1} + e^{1-e^x}}{2} = \sum_{n=0}^{\infty} \frac{B_n + D_n}{2} x^n,$$

where

$$B_n = \frac{1}{e} \sum_{k=0}^{\infty} \frac{k^n}{k!}$$

is the number of all partitions and

$$D_n = e \sum_{k=0}^{\infty} \frac{(-1)^k k^n}{k!}$$

is the difference between the number of partitions into an even and an odd number of classes.

(b) If the class containing x has $2k$ elements, then we get the recurrence relation

(1)
$$R_n = \sum_{k=1}^{\lfloor n/2 \rfloor} \binom{n-1}{2k-1} R_{n-2k}.$$

We remark that, obviously, $R_{2m+1}=0$.

$$r'(x) = \sum_{n=1}^{\infty} \frac{R_n}{(n-1)!} x^{n-1} = \sum_{n=1}^{\infty} \frac{x^{n-1}}{(n-1)!} \sum_{k=1}^{\lfloor \frac{n}{2} \rfloor} \binom{n-1}{2k-1} R_{n-2k} =$$

$$= \sum_{k=1}^{\infty} \frac{1}{(2k-1)!} \sum_{n=2k}^{\infty} \frac{R_{n-2k}}{(n-2k)!} x^{n-1} =$$

$$= \sum_{k=1}^{\infty} \frac{x^{2k-1}}{(2k-1)!} \sum_{n=2k}^{\infty} \frac{R_{n-2k}x^{n-2k}}{(n-2k)!} = r(x) \sum_{k=1}^{\infty} \frac{x^{2k-1}}{(2k-1)!} = r(x)\mathrm{sh}x,$$

whence

$$r(x) = e^{\mathrm{ch}x - 1}.$$

(The constant -1 in the exponent follows, as before, by substituting $x = 0$.)
[A. Rényi, *MTA III. Oszt. Közl.* **16** (1966) 77–105.]

15. (a) If there are exactly $k \geq 1$ elements mapped onto 1 by f, then these elements
can be chosen in $\binom{n}{k}$ ways and the rest can be mapped into $2, \ldots, n$ in exactly
S_{n-k} ways. Hence

$$S_n = \sum_{k=1}^{n} \binom{n}{k} S_{n-k}$$

or

$$\frac{2S_n}{n!} = \sum_{k=0}^{n} \frac{1}{k!} \frac{S_{n-k}}{(n-k)!} \qquad (n \geq 1).$$

Thus

$$2s(x) = \sum_{n=0}^{\infty} \frac{2S_n}{n!} x^n = 1 + \sum_{n=0}^{\infty} \left(\sum_{k=0}^{n} \frac{1}{k!} \frac{S_{n-k}}{(n-k)!} \right) x^n =$$

$$= 1 + \left(\sum_{m=0}^{\infty} \frac{S_m}{m!} x^m \right) \left(\sum_{k=0}^{\infty} \frac{x^k}{k!} \right) = 1 + s(x)e^x,$$

whence

$$s(x) = \frac{1}{2 - e^x}.$$

Let us expand this:

$$s(x) = \frac{1}{2} \frac{1}{1 - e^x/2} = \frac{1}{2} \sum_{k=0}^{\infty} \left(\frac{e^x}{2} \right)^k = \sum_{k=0}^{\infty} \frac{1}{2^{k+1}} \sum_{n=0}^{\infty} \frac{k^n x^n}{n!},$$

whence the formula for S_n follows by comparing the coefficients of x^n on both
sides.

(b) By the Cauchy formula,

$$\frac{S_n}{n!} = \frac{1}{2\pi i} \oint_{|z|=\varepsilon} \frac{dz}{(2 - e^z)z^{n+1}}$$

for any ε small enough. This can be re-written as

$$\frac{1}{2\pi i} \oint_{|z|=2} - \frac{1}{2\pi i} \oint_{|z-\log 2|=\varepsilon} ,$$

because the only singularities of the integrand are at $z=0$ and $z=\log 2+2k\pi i$, and hence only $z=\log 2$ lies in the region between the circles $|z|=2$ and $|z|=\varepsilon$. The first term tends to 0 as $n\to\infty$. The second term is known to be equal to

$$\left.\frac{-1}{z^{n+1}\frac{d}{dz}(2-e^z)}\right|_{z=\log 2} = \frac{1}{2(\log 2)^{n+1}}(\to\infty).$$

This proves the assertion.

16. If $n=a_1+\ldots+a_s$ is a partition of n, $a_1\geq\ldots\geq a_s$, then we construct a diagram, the so-called *Ferrer's diagram* of the partition as follows: we put a_i dots in the ith column, starting from the bottom (Fig. 21).

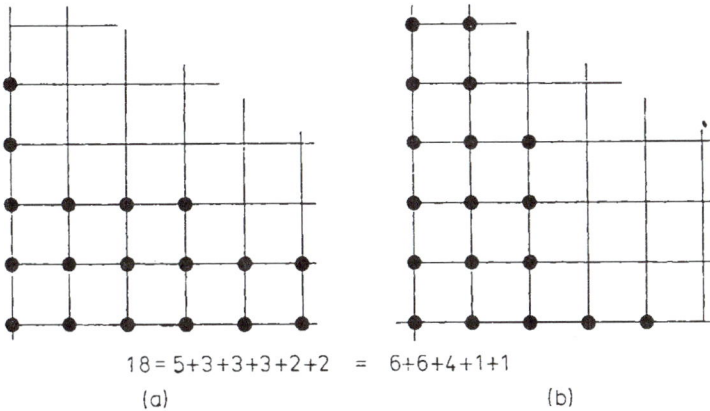

$$18 = 5+3+3+3+2+2 \quad = \quad 6+6+4+1+1$$

(a)   (b)

FIGURE 21

This construction gives a one-to-one correspondence between partitions of the number n and arrays composed of n points, arranged in columns such that the number of points in the successive columns decreases (not strictly). We may also consider these points as lattice points. In that case we will have a collection S of lattice points such that if $(i,j)\in S$, then $i\geq 0$, $j\geq 0$ and if $(i,j)\in S$, $0\leq i'\leq i$, $0\leq j'\leq j$ then $(i',j')\in S$. Hence it follows that reflecting a Ferrer's diagram on the line $x=y$ we get another Ferrer's diagram. Now, if a Ferrer's diagram has at most r columns, then this reflected Ferrer's diagram has at most r points in any column, i.e. it represents a partition into terms, each at most r. This gives a one-to-one correspondence between partitions into at most r terms and partitions into terms, each at most r.

17. Let $a_1 + \ldots + a_m = n$ be a partition of n into exactly m terms; $a_1 \geq \ldots \geq a_m$. Then

(1) $$n - m = (a_1 - 1) + \ldots + (a_m - 1)$$

and here $a_1 - 1 \geq \ldots \geq a_m - 1$. Some terms at the end of (1) may be 0's; discarding these, we obtain a partition of $n - m$ into not more than m terms. Conversely, if we have a partition of $n - m$ into $r \leq m$ terms, then we can get a partition of n by adding 1 to each term and taking $m - r$ additional 1's. This one-to-one correspondence proves the assertion.

Let $a_1 + \ldots + a_m = n$ be a partition of n into exactly m distinct terms. Then $a_1 > a_2 \ldots > a_m \geq 1$, hence

$$a_1 - (m - 1) \geq a_2 - (m - 2) \geq \ldots \geq a_{m-1} - 1 \geq a_m \geq 1$$

and so, the decomposition

$$n - \binom{m}{2} = (a_1 - (m - 1)) + (a_2 - (m - 2)) + \ldots + a_m$$

is a partition of $n - \binom{m}{2}$ into m terms.

Conversely, if we are given a partition of $n - \binom{m}{2}$ into exactly m terms, then adding $m - 1, \ldots, 1, 0$ to the terms, we get a partition of n into m distinct terms. Thus:

The number of partitions of n into exactly m distinct terms is equal to the number of partitions of $n - \binom{m}{2}$ into exactly m terms.

18. If

(1) $$n = a_1 + \ldots + a_m, \qquad a_1 > \ldots > a_m \geq 1,$$

then set

$$a_i = 2^{\beta_i} b_i, \text{ where } b_i \text{ is odd.}$$

Replacing each a_i by $2^{\beta_i} b_i$'s in (1) we get (after rearranging the terms if necessary) a partition of n into odd terms. The number of occurrences of the odd number d is

(2) $$\sum_{b_i = d} 2^{\beta_i}.$$

We show that the above correspondence is a bijection; i.e. any partition of n into odd terms arises exactly once. In fact, consider a partition of n into odd terms; let there be γ_1 1's, γ_3 3's .... If this arises as above, then

$$\gamma_d = \sum_{b_i = d} 2^{\beta_i}$$

for any odd number d. Since here the numbers β_i must be distinct (because $\beta_i = \beta_j$, $b_i = b_j = d$ would imply $a_i = a_j = 2^{\beta_i} d$); it follows that they are uniquely determined by γ_d. Thus a partition of n into odd numbers arises in at most one way. On the other hand, γ_d can be written as a sum of distinct powers of 2:

$$\gamma = \sum 2^{\beta_i(d)},$$

and then the numbers $d2^{\beta_i(d)}$ form a partition of n into distinct terms such that the corresponding partition into odd terms is the same as the one we started with.

19. Consider a partition

$$(1) \qquad\qquad n = a_1 + \ldots + a_m, \qquad a_1 > \ldots > a_m$$

and its Ferrer's diagram. We draw lines through the first and the last column, through the lowest row and also a line decreasing at 45° through the dot at the top of the first column (Fig. 22).

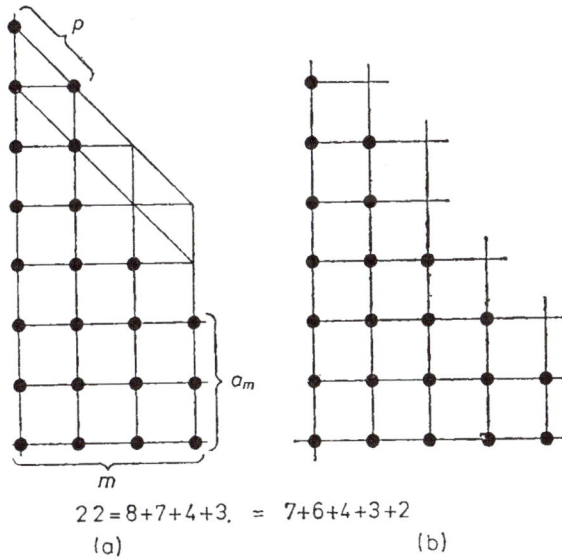

$$22 = 8+7+4+3. = 7+6+4+3+2$$

(a) \qquad\qquad\qquad\qquad\qquad (b)

FIGURE 22

The diagram is contained in the trapezoid formed by these four lines because $a_1 > \ldots > a_m$. We denote by p the number of dots on the upper (skew) line. Now we transform the diagram as follows. If $p < a_m$, then we take off the last dots of the first p columns and form a new column of p dots at the end (Fig. 22a→b). The resulting diagram will have distinct rows; this is trivial if a_m is not decreased, i.e. $p < m$; and it is also clear in case $p = m$ and $a_m > p+1$. Now $a_m = p+1$, $m = p$ is excluded, because then

$$n = (p+1) + (p+2) + \ldots + (2p) = \frac{3p^2 + p}{2}.$$

If $a_m \leq p$, then we remove the last column and add one point to the first a_m columns (Fig. 22b→a). This is possible except when there are no a_m columns, i.e. $m-1 < a_m$. But $m \geq p \geq a_m$ and hence, this could only arise if $m = p = a_m$; but then

$$n = p + (p+1) + \ldots + (2p-1) = \frac{3p^2 - p}{2},$$

which was excluded. Thus we have defined a transformation which associates Ferrer's diagrams of partitions into distinct terms with each other. It is easy to check (that is why Fig. 22 could be used to illustrate both cases we considered) that carrying out this transformation twice we get the original diagram, i.e. the transformation pairs off the partitions into distinct terms. Since each pair contains one partition into an odd number of terms and one into an even number of terms, this pairing proves the assertion of the problem. [Euler; cf. also 1.22]

20. If we consider

$$(1 + x + x^2 + \ldots)(1 + x^2 + x^4 + \ldots)(1 + x^3 + x^6 + \ldots),$$

then the coefficient of x^n is the number of ways x^n can be represented as $x^{k_1} x^{2k_2} \ldots x^{nk_n}$, i.e. the number of partitions of n. Hence

$$S(x) = \sum_{n=0}^{\infty} \pi_n x^n = (1 + x + x^2 + \ldots)(1 + x^2 + x^4 + \ldots)\ldots =$$

$$= \frac{1}{1-x} \cdot \frac{1}{1-x^2} \cdot \frac{1}{1-x^3} \ldots = \frac{1}{(1-x)(1-x^2)\ldots}.$$

(We have skipped the question of convergence of these infinite products; this could easily be settled but they could also be viewed as formal series.)

21. The coefficient of x^n in the product

$$(1+x)(1+x^2)(1+x^3)\ldots$$

is the number of partition of n into distinct terms; the coefficient of x^n in

$$(1 + x + x^2 + \ldots)(1 + x^3 + x^6 + \ldots)(1 + x^5 + x^{10} + \ldots)\ldots$$

is the number of partitions of n into odd terms. Thus we have to show that

$$(1+x)(1+x^2)(1+x^3)\ldots =$$
$$= (1 + x + x^2 + \ldots)(1 + x^3 + x^6 + \ldots)(1 + x^5 + x^{10} + \ldots)\ldots$$

or, equivalently,

(1) $$(1+x)(1+x^2)(1+x^3)\ldots = \frac{1}{(1-x)(1-x^3)(1-x^5)\ldots}.$$

Now

$$(1-x)(1+x)(1+x^2)(1+x^4)\ldots(1+x^{2^k})\ldots = 1,$$

and, substituting x^{2k+1} for x, we get

$$(1-x^{2k+1})(1+x^{2k+1})(1+x^{2(2k+1)})(1+x^{4(2k+1)})\ldots = 1.$$

Multiplying these identities together for $k = 0, 1, 2, \ldots$, we get all binomials $1 - x^{2k+1}$, also all binomials $1 + x^n$; thus

$$(1-x)(1-x^3)(1-x^5)\ldots(1+x)(1+x^2)(1+x^3)\ldots = 1,$$

which proves (1).

22. If we expand

$$(1-x)(1-x^2)(1-x^3)\ldots$$

we get $-x^n$ from each partition of n into an odd number of terms and $+x^n$ from each partition of n into an even number of terms. So, by 1.19, x^n cancels out unless $n = (3k^2 \pm k)/2$.

Let us determine what happens if $n = (3k^2 + k)/2$. The correspondence between partitions of n constructed in the solution of 1.19, pairs off partitions into an even number of terms with partitions into an odd number of terms except for the partition $n = (k+1) + (k+2) + \ldots + (2k)$, which will give a coefficient $(-1)^k$.

Similarly, if $n = (3k^2 - k)/2$, we get $(-1)^k x^k$. Hence

$$(1-x)(1-x^2)(1-x^3)\ldots = 1 + \sum_{k=1}^{\infty}(-1)^k\left\{x^{\frac{3k^2-k}{2}} + x^{\frac{3k^2+k}{2}}\right\}.$$

[For a tricky analytic proof, which of course also proves 1.19, see e.g. Rademacher, *Lectures on Elementary Number Theory*, Blaisdell, 1964.]

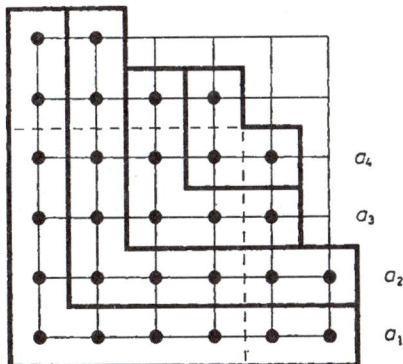

FIGURE 23

23. (a) Let us prove the combinatorial assertion formulated in the hint. If we are given a symmetric Ferrer's diagram with k elements in the "diagonal", then let a_1 denote the total number of dots in the first row and first column; let a_2 be the number of dots in the second row and column which are not in the first row or column, etc. (Fig. 23). Then $a_1 > a_2 > \ldots > a_k$ are odd numbers which partitions

n. Conversely, we can construct from any partition into k distinct odd numbers a symmetric Ferrer's diagram with k dots in the "diagonal". This proves the assertion of the hint.

Now the number of partitions of n into distinct odd numbers is the coefficient of x^n in

$$(1 + x)(1 + x^3)(1 + x^5) \ldots .$$

The number of symmetric Ferrer's diagrams with k elements in the diagonal can be counted as follows. We remove the $k \times k$ square in the corner and consider the ith row above the square plus the ith column to the right of the square as a summand. This way we get a partition of n into k^2 and some even numbers $\leq 2k$. The number of such partitions is the coefficient of x^n in

(1)
$$x^{k^2}(1 + x^2 + x^4 + \ldots)(1 + x^4 + x^8 + \ldots) \ldots (1 + x^{2k} + x^{4k} + \ldots)$$
$$= \frac{x^{k^2}}{(1 - x^2)(1 - x^4) \ldots (1 - x^{2k})}.$$

Hence

$$\sum_{k=0}^{\infty} \frac{x^{k^2}}{(1 - x^2)(1 - x^4) \ldots (1 - x^{2k})}$$

in the generating function of the numbers of symmetric Ferrer's diagrams with n dots. This proves (a).

(b) Now (b) could be proved similarly but it also follows from another observation. The number of partitions of n into k distinct odd terms is the coefficient of $x^n y^k$ in

$$(1 + xy)(1 + x^3 y)(1 + x^5 y) \ldots .$$

On the other hand, we know that this number is equal to the number of symmetric Ferrer's diagrams with n dots and k dots in the "diagonal", i.e. the coefficient of x^n in (1). This gives the identity

$$(1 + xy)(1 + x^3 y) \ldots = \sum_{k=0}^{\infty} \frac{x^{k^2} y^k}{(1 - x^2)(1 - x^4) \ldots (1 - x^{2k})}.$$

Setting $y = x$ we get (b) [Euler].

24. Suppose that there are k_1 1's in the partition. The partial sums formed by these 1's represent uniquely each number up to k_1. So none of $2, \ldots, k_1$ can occur in the partition. Since $k_1 + 1$ must be representable as a partial sum, it must occur in the partition with some multiplicity $k_2 > 0$. Then all numbers are taken care of till $k_2(k_1 + 1) + k_1$ (i.e. they all have a unique representation). Going on similarly, we get that the numbers occurring in the partition are $1, k_1, (k_1 + 1)(k_2 + 1), \ldots,$

$(k_1 + 1)(k_2 + 1) \ldots (k_m + 1)$, where $(k_1 + 1) \ldots (k_i + 1)$ has multiplicity k_{i+1} $(i = 1, \ldots, m)$. Since the sum of all numbers must be n, we have

$$n = k_1 \cdot 1 + k_2(k_1 + 1) + \ldots + k_{m+1}(k_1 + 1) \ldots (k_m + 1)$$

or

$$n + 1 = (k_1 + 1) \ldots (k_{m+1} + 1).$$

If $m = 0$ we have the trivial partition consisting of 1's; if $n+1$ is a prime then this must be the case. On the other hand, if $n+1$ is composite then

$$n + 1 = (k_1 + 1)(k_2 + 1)$$

with $k_1, k_2 \geq 1$ and then the partition

$$n = \underbrace{1 + \ldots + 1}_{k_1} + \underbrace{(k_1 + 1) + \ldots + (k_1 + 1)}_{k_2}$$

is a suitable one [see R].

25. The first assertion is easy: if integers a, b, c are the sides of a traingle with circumference $2n-3$, then $a+1, b+1, c+1$ trivially also satisfy the triangle inequality, and so they are the sides of a triangle with circumference $2n$. Conversely, if a, b, c are the sides of a triangle with circumference $2n$, then

$$(a - 1) + (b - 1) - (c - 1) = a + b - c - 1 \geq 0,$$

and since $a + b - c - 1 = 2n - 2c - 1$ is odd, we have

$$(a - 1) + (b - 1) - (c - 1) > 0.$$

Hence $a - 1$, $b - 1$, $c - 1$ are the sides of a triangle with circumference $2n - 3$.

To prove the second assertion, observe that if

$$n = x + y + z,$$

where x, y and z are positive integers, then

$$(x + y) + (x + z) + (y + z) = 2n,$$
$$(x + y) + (x + z) = y + z + 2x > y + z,$$

i.e. $x+y$, $y+z$, $x+z$ are the sides of a triangle with circumference $2n$. Conversely, if integers a, b, c are the sides of such a triangle then, setting

$$x = n - a, \quad y = n - b, \quad z = n - c,$$

we have

$$x = \frac{b + c - a}{2} > 0, \quad y > 0, \quad z > 0,$$

and also

$$x + y = c, \quad x + z = b, \quad y + z = a.$$

Thus, we have a one-to-one correspondence between partitions of n and triangles with circumference $2n$ (and integer sides).

To determine the number of partitions of n into exactly three terms we observe first that the number of partitions of m into exactly two terms not exceeding a is

$$\begin{cases} 0 & \text{if } a < \dfrac{m}{2}, \\ a - \left\lfloor \dfrac{m-1}{2} \right\rfloor & \text{if } \dfrac{m}{2} \leq a < m, \\ \left\lfloor \dfrac{m}{2} \right\rfloor & \text{if } a \geq m. \end{cases}$$

Hence, the number of partitions

$$n = a + b + c, \quad a \geq b \geq c$$

is

$$\sum_{\frac{n-a}{2} \leq a < n-a} \left(a - \left\lfloor \frac{n-a-1}{2} \right\rfloor \right) + \sum_{n-a \leq a} \left\lfloor \frac{n-a}{2} \right\rfloor$$

or, equivalently

$$\sum_{\frac{n}{3} \leq a < \frac{n}{2}} \left(a - \left\lfloor \frac{n-a-1}{2} \right\rfloor \right) + \sum_{\frac{n}{2} \leq a \leq n} \left\lfloor \frac{n-a}{2} \right\rfloor.$$

It is seen that the result depends on the residue of $n \bmod 6$, e.g. if $n = 6k$, then the result is $3k^2$.

26. On the first day, we have three choices: we may buy a pretzel, in which case we have M_{n-1} further possible ways to spend the remaining $n-1$ forints; or we may buy candies for 2 forints, and then we can spend the rest in M_{n-2} ways; similarly we have M_{n-2} possibilities if we first buy an ice-cream. Thus we have

$$(1) \qquad M_n = M_{n-1} + 2M_{n-2}.$$

We need some values at the beginning and obviously,

$$M_0 = M_1 = 1 \quad M_2 = 3.$$

Hence by (1)

$$M_3 = 5, \quad M_4 = 11, \quad M_5 = 21, \quad M_6 = 43.$$

We see here and conjecture that

$$(2) \qquad M_n = 2M_{n-1} + (-1)^n.$$

In fact, (2) follows easily from (1) by induction:

if

$$M_{n-1} = 2M_{n-2} + (-1)^{n-1}, \quad 2M_{n-2} = M_{n-1} + (-1)^n,$$

then

$$M_n = M_{n-1} + 2M_{n-2} = M_{n-1} + (M_{n-1} + (-1)^n) = 2M_{n-1} + (-1)^n.$$

Now we have

$$M_n = 2M_{n-1} + (-1)^n = 2(2M_{n-2} + (-1)^{n-1}) + (-1)^n =$$
$$4M_{n-2} + 2(-1)^{n-1} + (-1)^n = 8M_{n-3} + 4(-1)^{n-2} + 2(-1)^{n-1} + (-1)^n =$$
$$= \ldots = 2^n - 2^{n-1} + 2^{n-2} - \ldots + (-1)^n =$$
$$= \frac{2^{n+1} - (-1)^n}{2 - (-1)} = \frac{1}{3}(2^{n+1} + (-1)^n).$$

Of course, if one guesses this result earlier one can prove it easily by induction from (1), without bringing (2) into consideration.

27. We may begin by taking one or two steps. In the first case, we have A_{n-1} possibilities to continue; in the second, we have A_{n-2} possibilities. Thus

(1) $$A_n = A_{n-1} + A_{n-2}.$$

Since we have

$$A_1 = 1, \quad A_2 = 2,$$

the sequence is essentially the sequence of the Fibonacci numbers. We set $A_0 = 1$ and

$$f(x) = \sum_{n=0}^{\infty} A_n x^n.$$

Then

$$xf(x) = \sum_{n=1}^{\infty} A_{n-1} x^n,$$
$$x^2 f(x) = \sum_{n=2}^{\infty} A_{n-2} x^n,$$

and by (1)

$$f(x) - xf(x) - x^2 f(x) = A_0 + (A_1 - A_0)x + \sum_{n=2}^{\infty} (A_n - A_{n-1} - A_{n-2})x^n = A_0 = 1$$

and hence

$$f(x) = \frac{1}{1 - x - x^2}.$$

To get an explicit formula for A_n we write this as

$$f(x) = \frac{1/\sqrt{5}x}{1 - \dfrac{1+\sqrt{5}}{2}x} - \frac{1/\sqrt{5}x}{1 - \dfrac{1-\sqrt{5}}{2}x} =$$

$$= \frac{1}{\sqrt{5}x} \sum_{k=0}^{\infty} \left(\left(\frac{1+\sqrt{5}}{2}x \right)^k - \left(\frac{1-\sqrt{5}}{2}x \right)^k \right) =$$

$$= \frac{1}{\sqrt{5}} \sum_{n=0}^{\infty} \left\{ \left(\frac{1 + \sqrt{5}}{2} \right)^{n+1} - \left(\frac{1 - \sqrt{5}}{2} \right)^{n+1} \right\} x^n.$$

Hence

$$A_n = \frac{1}{\sqrt{5}} \left\{ \left(\frac{1 + \sqrt{5}}{2} \right)^{n+1} - \left(\frac{1 - \sqrt{5}}{2} \right)^{n+1} \right\}.$$

28. (a) If on the first day, we buy a product costing i forints, we have C_{n-i} possible ways to spend the rest of the money. Hence

$$(1) \qquad\qquad C_n = \sum_{i=1}^{k} a_i C_{n-i} \qquad (n > k).$$

It will be convenient to set $C_0 = 1$, $C_{-1} = \ldots = C_{-k+1} = 0$. These new values will satisfy (1) for $n \geq 1$.

Observe that the sequence $M \vartheta_\nu^n$ satisfies (1) for any fixed ν, for

$$M \vartheta_\nu^n = \sum_{i=1}^{k} a_i M \vartheta_\nu^{n-i}$$

is equivalent to

$$M \vartheta_\nu^{n-k} (\vartheta_n^k - a_1 \vartheta_\nu^{k-1} - \ldots - a_{k-1} \vartheta_\nu - a_k) = 0,$$

which is clearly satisfied. Hence any sequence

$$(2) \qquad\qquad x_n = \sum_{\nu=1}^{k} M_\nu \vartheta_\nu^n$$

satisfies (1). Let us try to look for our sequence C_n in the form (2). It will be convenient to set

$$x_n = C_{n-k+1}.$$

Then $x_0 = x_1 = \ldots = x_{k-2} = 0$, $x_{k-1} = 1$, or

$$M_1 + M_2 + \ldots + M_k = 0$$
$$\vartheta_1 M_1 + \vartheta_2 M_2 + \ldots + \vartheta_k M_k = 0$$

$$(3) \qquad\qquad\qquad\qquad \vdots$$

$$\vartheta_1^{k-2} M_1 + \vartheta_2^{k-2} M_2 + \ldots + \vartheta_k^{k-2} M_k = 0$$
$$\vartheta_1^{k-1} M_1 + \vartheta_2^{k-1} M_2 + \ldots + \vartheta_k^{k-1} M_k = 1.$$

Denote the Vandermonde determinant

$$
\begin{vmatrix}
1 & \vartheta_1 & \cdots & \vartheta_1^{k-1} \\
\vdots & \vdots & & \\
1 & \vartheta_k & \cdots & \vartheta_k^{k-1}
\end{vmatrix}
$$

by D, then, by Cramer's rule, (3) has a solution for $M_1,\ldots,M_k$; in fact

$$
M_\nu = \frac{1}{D}
\begin{vmatrix}
1 & \vartheta_1 & \cdots & \vartheta_1^{k-1} \\
\vdots & & & \\
0 & 0 & \cdots & 1 \\
\vdots & \vdots & & \\
1 & \vartheta_k & \cdots & \vartheta_k^{k-1}
\end{vmatrix}
= \frac{1}{D}
\begin{vmatrix}
1 & \vartheta_1 & \cdots & \vartheta_1^{k-2} & 0 \\
\vdots & \vdots & & & \\
1 & \vartheta_\nu & \cdots & \vartheta_\nu^{k-2} & 1 \\
1 & \vartheta_k & \cdots & \vartheta_k^{k-1} & 0
\end{vmatrix}
$$

(the second form arises by subtracting ϑ_i^{k-1} times the νth row from the ith ($i \neq \nu$) and then adding ϑ_ν^{j-1} times the last column to the jth ($j=1,\ldots,k-1$). Hence

$$
x_n = \sum_{\nu=1}^{k} M_\nu \vartheta_\nu^n = \frac{1}{D}
\begin{vmatrix}
1 & \vartheta_1 & \cdots & \vartheta_1^{k-2} & \vartheta_1^n \\
\vdots & & & & \\
1 & \vartheta_k & \cdots & \vartheta_k^{k-2} & \vartheta_k^n
\end{vmatrix}
$$

and $C_n = x_{n+k-1}$ is the value given in the problem.

(b) Set

$$
f(x) = \sum_{n=0}^{\infty} C_n x^n.
$$

Then

$$
f(x) - \sum_{i=1}^{k} a_i x^i f(x) = C_0 + \sum_{n=1}^{\infty} x^n (C_n - a_1 C_{n-1} - \ldots - a_k C_{n-k})
$$

(where again $C_{-1}=\ldots=C_{-k+1}=0$, $C_0=1$). Hence the last sum is 0, so

$$
f(x)(1 - a_1 x - \ldots - a_k x^k) = 1,
$$

and $f(x) = \dfrac{1}{1 - a_1 x - \ldots - a_k x^k}.$

To get an explicit formula for the coefficients we observe that the roots of the denominator are $1/\vartheta_1,\ldots,1/\vartheta_k$ and hence, $f(x)$ can be written as a sum of partial functions:

$$f(x) = \frac{1}{(1-\vartheta_1 x)\ldots(1-\vartheta_k x)} = \sum_{\nu=1}^{k} \frac{A_\nu}{1-\vartheta_\nu x}.$$

Multiplying by $1-\vartheta_\nu x$ and substituting $x=1/\vartheta_\nu$, we get

$$A_\nu = \prod_{j\neq\nu}\left(1-\frac{\vartheta_j}{\vartheta_\nu}\right)$$

and so

$$f(x) = \sum_{\nu=1}^{k} A_\nu \frac{1}{1-\vartheta_\nu x} = \sum_{\nu=1}^{k} A_\nu \sum_{n=0}^{\infty}\vartheta_\nu^n x^n = \sum_{n=1}^{\infty}\left(\sum_{\nu=1}^{k}\prod_{j\neq\nu}\left(1-\frac{\vartheta_j}{\vartheta_\nu}\right)\vartheta_\nu^n\right)x^n$$

and hence

$$C_n = \sum_{\nu=1}^{k}\prod_{j\neq\nu}\left(1-\frac{\vartheta_j}{\vartheta_\nu}\right)\vartheta_\nu^n = \sum_{\nu=1}^{k}\prod_{j\neq\nu}(\vartheta_\nu-\vartheta_j)\vartheta_\nu^{n-k+1}.$$

It is left to the reader to verify that this formula is equal to the one given in part (a).

29. We have $p_n(\lambda)=$

$$
= \underbrace{\begin{vmatrix} \lambda & -1 & & & 0 \\ -1 & \lambda & & & \\ & & \ddots & & -1 \\ 0 & & & -1 & \lambda \end{vmatrix}}_{n} = \underbrace{\begin{vmatrix} \lambda & -1 & & & 0 \\ -1 & \lambda & & & \\ & & \ddots & & -1 \\ 0 & & & -1 & \lambda \end{vmatrix}}_{n-1} + \underbrace{\begin{vmatrix} -1 & 0 & \cdots & & 0 \\ -1 & \lambda & -1 & & \\ & -1 & \lambda & & \\ & & & \ddots & -1 \\ 0 & & & -1 & \lambda \end{vmatrix}}_{n-1} =
$$

$$= \lambda p_{n-1}(\lambda) - p_{n-2}(\lambda) \qquad (n\geq 3).$$

This recurrence relation remains valid for $n\geq 1$ if we set $p_0(\lambda)=1$ and $p_{-1}(\lambda)= 0$. As in the solution of 1.27, we denote by ϑ_1, ϑ_2 the roots of

$$x^2 - \lambda x + 1 = 0$$

$\Bigg(\vartheta_1$, ϑ_2 are of course, functions of λ :

$$\vartheta_1 = \frac{\lambda + \sqrt{\lambda^2 - 4}}{2}, \quad \vartheta_2 = \frac{\lambda - \sqrt{\lambda^2 - 4}}{2}\Bigg).$$

Then

$$p_n(\lambda) = c_1 \vartheta_1^{n+1} + c_2 \vartheta_2^{n+1}.$$

Taking $n = 0$ and -1 we get

$$c_1 + c_2 = 0,$$
$$c_1 \vartheta_1 + c_2 \vartheta_2 = 1.$$

Hence

$$c_1 = \frac{1}{\sqrt{\lambda^2 - 4}}, \qquad c_2 = \frac{-1}{\sqrt{\lambda^2 - 4}}$$

and

$$p_n(\lambda) = \frac{1}{\sqrt{\lambda^2 - 4}} (\vartheta_1^{n+1} - \vartheta_2^{n+1}).$$

Thus if

$$p_n(\lambda) = 0,$$

then

$$\vartheta_1^{n+1} = \vartheta_2^{n+1}$$

or, equivalently,

$$\vartheta_1 = \varepsilon^2 \vartheta_2,$$

where

$$\varepsilon = e^{\frac{k\pi i}{n+1}}, \qquad 0 \le k \le n.$$

Solving $\vartheta_1 = \varepsilon^2 \vartheta_2$ for λ we obtain

$$\lambda = \pm \left(\varepsilon + \frac{1}{\varepsilon} \right) = \pm 2 \cos \frac{k\pi}{n+1}.$$

Here we may omit $\pm$ since $-\cos \frac{k}{n+1} = \cos \frac{n+1-k}{n+1}$. It is easily seen by substitution that these numbers are roots of $p_n(\lambda)$ for $k = 1, \ldots, n$; therefore, it is not a root for $k = 0$. Thus the eigenvalues of A are

$$2 \cos \frac{k\pi}{n+1}, \qquad k = 1, \ldots, n$$

[cf. also problems 4.28 and 11.5].

30. If we have a sequence of length $n-1$ in which a, b are not neighbors we can put in front of it
     (1) c or d or a [b] if it starts with b [a];
     (2) a, b, c or d if it start with c or d.
Hence if x_n and y_n denote the numbers of sequences of length n starting with a or b and c or d, respectively, we have the recurrence relations

$$n_x = x_{n-1} + 2y_{n-1},$$
$$y_n = 2x_{n-1} + 2y_{n-1}.$$

To solve these recurrence relations, put

$$v_n = \begin{bmatrix} x_n \\ y_n \end{bmatrix}, \qquad A = \begin{pmatrix} 1 & 2 \\ 2 & 2 \end{pmatrix}.$$

Then we have

$$v_n = Av_{n-1} = \ldots = A^{n-1}\begin{bmatrix} 2 \\ 2 \end{bmatrix} = A^n \begin{bmatrix} 0 \\ 1 \end{bmatrix}.$$

We can transform A into a diagonal form by setting

$$L = \begin{pmatrix} \frac{3+\sqrt{17}}{2} & 0 \\ 0 & \frac{3-\sqrt{17}}{2} \end{pmatrix},$$

$$T = \begin{pmatrix} \sqrt{\frac{\sqrt{17}-1}{2\sqrt{17}}} & -\sqrt{\frac{\sqrt{17}+1}{2\sqrt{17}}} \\ \sqrt{\frac{\sqrt{17}+1}{2\sqrt{17}}} & \sqrt{\frac{\sqrt{17}-1}{2\sqrt{17}}} \end{pmatrix};$$

then $A = TLT^{-1}$ (L has the eigenvalues of A in the diagonal and T is composed of the eigenvectors of A). We want to determine

$$x_n + y_n = [1,1]A^n \begin{bmatrix} 0 \\ 1 \end{bmatrix} = \frac{1}{2}[0,1]A^{n+1}\begin{bmatrix} 0 \\ 1 \end{bmatrix} = \frac{1}{2}[0,1]TL^{n+1}T^{-1}\begin{bmatrix} 0 \\ 1 \end{bmatrix}.$$

Here

$$[0,1]T = \left[\sqrt{\frac{\sqrt{17}+1}{2\sqrt{17}}}, \sqrt{\frac{\sqrt{17}-1}{2\sqrt{17}}} \right]$$

whence

$$x_n + y_n = \frac{\sqrt{17}+1}{4\sqrt{17}}\left(\frac{3+\sqrt{17}}{2}\right)^{n+1} + \frac{\sqrt{17}-1}{4\sqrt{17}}\left(\frac{3-\sqrt{17}}{2}\right)^{n+1}.$$

31. Let $\{a_1,\ldots,a_k\} \subseteq \{1,2,\ldots,n\}$ be a selection such that $a_{\nu+1} \geq a_\nu + 2$. Then

$$a_1 < a_2 - 1 < \ldots < a_k - (k-1),$$

i.e. $\{a_1, a_2-1, \ldots, a_k-(k-1)\}$ is a k-element subset of $\{1,\ldots,n-k+1\}$. Conversely, for any k-element subset $\{a_1', a_2', \ldots, a_k'\} \subseteq \{1,\ldots,n-k+1\}$ the set $\{a_1', a_2'+1, \ldots, a_k'+(k-1)\}$ is a subset of $\{1,\ldots,n\}$ containing no two consecutive integers. Hence the answer is

$$\binom{n-k+1}{k}.$$

32. For any such function f, consider the following polygon: connect $(x, f(x))$ to $(x+1, f(x))$ and also connect $(x, f(x))$, to $(x, f(x)-1)$, to $(x, f(x)-2)$, $\ldots$, to

$(x, f(x-1))$ (here $f(0)=1$). Also join $(n+1, f(n))$ to $(n+1, f(n)+1)$, ..., to $(n+1, n)$ (Fig. 24). Then we get a polygon connecting $(1,1)$ to $(n+1, n)$ whose edges join neighboring lattice points, and if we move along it from $(1,1)$ to $(n+1, n)$, then we always step either to the right or upwards. Call such a polygon a *step-polygon*. Conversely, a step-polygon connecting $(1,1)$ to $(n+1, n)$ represents a monotonic function of $\{1, \ldots, n\}$ into itself. Thus it suffices to enumerate such polygons.

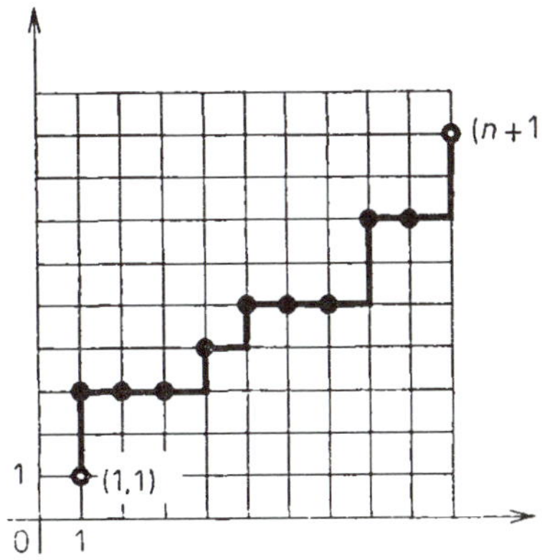

FIGURE 24

Let $e_1, \ldots, e_{2n-1}$ be the edges of a step-polygon. There are n edges which are "horizontal" and $n-1$ edges which are "vertical". Conversely, if we specify which n edges should be horizontal, then we uniquely determine a step-polygon. Thus the number of step-polygons connecting $(1,1)$ to $(n+1, n)$ is

$$\binom{2n-1}{n}.$$

33. (a) We may as well enumerate all monotone mappings of $\{1, \ldots, n\}$ into $\{0, \ldots, n-1\}$ with the property that $f(x) < x$. Constructing the step-polygon corresponding to such an f, it will connect the points $(1,0)$ and $(n+1, n-1)$ and will not intersect the line $x=y$.

We know the number of all step-polygons; it is $\binom{2n-1}{n}$ by the preceding solution. So, if we enumerate those meeting the line $x=y$ we will know the number of those not meeting this line as well.

Let P be such a polygon which meets the line $x=y$ and let (a, a) be their first point of intersection. Reflect the piece between $(1,0)$ and (a, a) in the line $x=y$. Then we get a step-polygon P' connecting $(0,1)$ and $(n+1, n-1)$. Conversely, if P' is a step-polygon connecting $(0,1)$ to $(n+1, n-1)$ then, obviously, it must intersect the line $x=y$ and if we again denote by (a, a) their first point of intersection, we get a step-polygon connecting $(1,0)$ to $(n+1, n-1)$ which meets the line $x=y$. Hence, the number of all step-polygons connecting $(0,1)$ to $(n+1, n-1)$ is equal to the number of step-polygons connecting $(1,0)$ to $(n+1, n-1)$ which intersect $x=y$ (Fig. 25). Now the number of step-polygons between $(0,1)$ and $(n+1, n-1)$

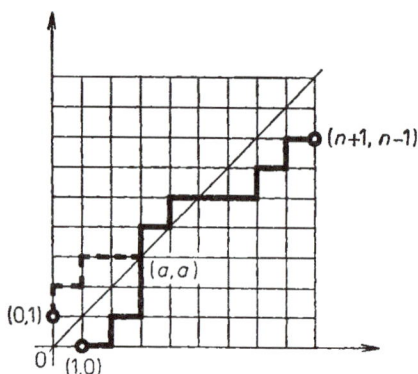

FIGURE 25

can be determined as in the previous solution. Such a polygon has $2n-1$ edges, $n-2$ going upwards. Hence their number is

$$\binom{2n-1}{n-2}$$

and so, the answer to our problem is

$$\binom{2n-1}{n-1} - \binom{2n-1}{n-2} = \frac{1}{n+1}\binom{2n}{n}.$$

(b) Denoting by $f(k)$ the sum of the first k digits, the problem becomes equivalent to (a). (The numbers $\frac{1}{n+1}\binom{2n}{n}$, called *Catalan numbers*, occur as answers to many enumeration problems; see 1.37–40.)

34. Consider such a sequence and cancel all entries n. The resulting sequence has some length k, say $0 \leq k \leq r$ and also satisfies the assumption. Conversely, if a sequence composed of $1,\ldots,n-1$ of length k has the property that less than i entries of it are $\leq i$ $(i=1,\ldots)$, then we may insert $r-k$ new entries n (which can be done in $\binom{r}{k}$ ways), and get a $(1,\ldots,n)$-sequence of length r with the same property. Thus,

(1) $$g(n,r) = \sum_{k=0}^{r} \binom{r}{k} g(n-1,k) \qquad (1 \leq r < n),$$

where $g(n,r)$ denotes the number of $(1,\ldots,n)$-sequences of length r with the property than less than i entries are $\leq i$, $i=1,2,\ldots$.
    Now

(2) $$g(n,r) = (n-r)n^{r-1}$$

follows easily by induction on n. For $n=r$ (2) is valid since then, each $(1,\ldots,n)$-sequence of length n contains n entries $\leq n$ and thus, $g(n,r)=0$. Let $n>r$. Then, using (1),

$$g(n,r) = \sum_{k=0}^{r}\binom{r}{k}g(n-1,k) = \sum_{k=0}^{r}\binom{r}{k}(n-1-k)(n-1)^{k-1} =$$

$$= \sum_{k=0}^{r}\binom{r}{k}(n-1)^k - \sum_{k=0}^{r}\binom{r}{k}k(n-1)^{k-1} = n^r - r\sum_{k=1}^{r}\binom{r-1}{k-1}(n-1)^{k-1} =$$

$$= n^r - r\cdot n^{r-1}$$

which proves (2). [H. E. Daniels, *Proc. Roy. Soc.* A **183**, (1945) 405–435.]

35. If we write down the partitions in converse order, starting with the identity partition (the partition into one-element sets) and ending with the zero partition $\{S\}$, we can observe that each row arises from the previous one by "sticking two classes together". In the ith row we will always have $n-i+1$ classes, thus the number of ways to get the $(i+1)^{\text{st}}$ row is $\binom{n-i+1}{2}$. Hence the number of all such procedures is

$$\binom{n}{2}\binom{n-1}{2}\cdots\binom{2}{2} = \frac{n!(n-1)!}{2^{n-1}}.$$

36. Let P_i be the partition arising at the ith step, $P_0=\{S\}$; we call the sequence $P_0,P_1,\ldots$ a *splitting procedure*. Denote by P_i' the partition of $S-\{x\}$ induced by P_i ($x\in S$). Then P_{i+1}' arises from P_i' by splitting all classes of P_i with more than one element into two classes, except possibly for a particular value i_0 of i in which case one of the classes pf P_{i_0}', though having more than one element, may not split: in the corresponding step the class C of P_{i_0} containing x is split into $\{x\}$ and $C-\{x\}$. Now, let P_i^* and P_i^{**} denote the partitions of $C-\{x\}$ and $S-C$ induced by P_i' and set

$$Q_i = \begin{cases} P_i' & \text{if } i \leq i_0 \\ P_{i+1}^* \cup P_i^{**} & \text{if } i > i_0. \end{cases}$$

Then $Q_0=\{S\}$, $Q_1,\ldots$ is a *splitting procedure* of $S-\{x\}$.

Conversely, let $Q_0,Q_1,\ldots$ be such a *splitting procedure* of $S-\{x\}$. We determine how many ways $Q_0,Q_1,\ldots$ arises from a splitting procedure $P_0,P_1,\ldots$ of S. We specify the class $C-\{x\}\in Q_0\cup Q_1\cup\ldots$. Once we have done so we have a unique $P_0,P_1,\ldots$. In fact, the index i_0 is the first for which $C-\{x\}\in Q_{i_0-1}$. Hence, we get P_i ($1\leq i\leq i_0$) from Q_i by adding x to the class containing $C-x$, and we can construct P_i ($i_0<i$) from Q_i by replacing the partition of $C-\{x\}$ induced by Q_i by the partition of $C-\{x\}$ induced by Q_{i-1} and adding x as a separate class. It is easy to check that, this way, we get a unique splitting procedure $P_0,P_1,\ldots$ from any splitting procedure $Q_0,Q_1,\ldots$ of $S-\{x\}$ together with a specified number of

$Q_0 \cup Q_1 \cup \dots$. Let us calculate $|Q_0 \cup Q_1 \cup \dots|$. Luckily, this will depend only on n. Obviously, any one-element subset of $S - \{x\}$ belongs to $Q_0 \cup Q_1 \cup \dots$, this makes $n - 1$. To each $C \in Q_0 \cup Q_1 \cup \dots$, $|C| \geq 2$, let us observe that there corresponds a splitting and conversely. There are $n - 2$ splittings (since each splitting increases the number of classes by one) and hence, we have

$$|Q_0 \cup Q_1 \cup \dots| = (n - 1) + (n - 2) = 2n - 3.$$

Thus, the number of splitting procedures of S in $2n - 3$ times the number of splitting procedures of $S - \{x\}$. Hence, by induction, the looked-for number is

$$(2n - 3)(2n - 5) \dots 1 = (2n - 3)!!$$

37. (a) We have to break at each of $n - 1$ points; the procedure is determined if we specify the order in which we break at the different places. Hence, the number is $(n - 1)!$

(b) Let us call such a procedure a *breaking procedure*. Consider any splitting procedure $P_0, P_1, \dots$ of an n-element set S (see the previous solution). We may interpret this as follows: whenever we split a class, we specify one half of it and put this in front of the other half. So, the classes of the partition arising at each step will be ordered and the final partition gives an ordering of S; moreover, the splitting procedure corresponds to a breaking procedure of the "stick" consisting of the elements of S in the given order.

Conversely, if a breaking procedure and an ordering of S is given then applying the breaking procedure with the ordered set S we get a splitting procedure. Let b_n and s_n denote the number of breaking and splitting procedures, respectively, then the above correspondence between splitting procedures with orderings of the pairs the classes are split into and pairs consisting of an ordering and a breaking procedure implies

$$2^{n-1} s_n = n! b_n.$$
$$s_n = (2n - 3)(2n - 5) \dots 1, \text{ and therefore}$$
$$b_n = \frac{2^{n-1}(2n - 3)(2n - 5) \dots 1}{n!} = \frac{(2n - 2)!}{n!(n - 1)!} = \frac{1}{n}\binom{2n - 2}{n - 1}.$$

38. *First solution.* Consider a bracketed product. The multiplication carried out at the last time means breaking the product into two pieces, where inside each piece there is some bracketing:

$$(x_1 \dots x_i)(x_{i+1} \dots x_n).$$

The two multiplications before this last one mean breaking each half of this into two pieces

$$((x_1 \dots)(\dots x_i))((x_{i+1} \dots)(\dots x_n))$$

(some of these brackets may contain a single variable only, in which case it does not occur). This way we see that the number of bracketings of a product with n

factors is equal to the number of breaking procedures of a stick of length n, i.e. by the previous result,

$$\frac{1}{n}\binom{2n-2}{n-1}.$$

Second solution. Let $f(x)$ denote the number of closing brackets between the first $x+1$ symbols ($x = 1, 2, \ldots, n-1$). Then

(a) $0 \le f(x) \le n-2$,

(b) $f(x)$ is monotonic and

(c) $f(x) < x$ as there are at most $x-1$ pairs of brackets between $x+1$ symbols.

Conversely, let $f(x)$ be a function defined on $\{1, \ldots, n-1\}$ with properties (a), (b), (c), then $f(x)$ uniquely determines a bracketing of the product. It will be more convenient to describe a bracketing as a way to multiply the n elements together (in a non-associative non-commutative structure). Obviously, there must be $f(i) - f(i-1)$ closing brackets between x_i and x_{i+1}, and $n-2-f(n-1)$ closing brackets at the end.

Now consider the first closing bracket (from the left), multiply the two elements in front of it and remove this bracket. Again, find the first remaining closing bracket, multiply the two elements in front of it (one of these two may be the result of the previous multiplication), etc. This must be the same order of multiplication as the one prescribed by the bracketing which determines $f(x)$. It is also easy to check that we will be able to carry this out, i.e. we never get a closing bracket with just one element in front of it, by (c). So we have a bijection between functions f with properties (a), (b), (c) and bracketings of products. The functions with properties (a), (b), (c) have been enumerated in 1.33, so the result is

$$\frac{1}{n}\binom{2n-2}{n-1}.$$

39. Let $v_1, \ldots, v_n$ be the vertices of our n-gon. If v_i is joined to v_j by a diagonal in a given triangulation, put a pair of brackets into the product:

$$x_1 \ldots (x_i \ldots x_{j-1}) x_j \ldots x_{n-1}.$$

Then it is easy to see that this bracketing is a "correct" one and conversely, from each correct bracketing we get a triangulation. Thus, the answer is

$$D_n = \frac{1}{n-1}\binom{2n-4}{n-2},$$

by 1.38.

40. Let $v_1, \ldots, v_n$ be the vertices of our n-gon (in this cyclic order). In any triangulation, the edge (v_1, v_n) is contained in a unique triangle (v_1, v_n, v_k). If k is fixed, we have D_k possible ways to triangulate the k-gon $(v_1, \ldots, v_k)$ and D_{n-k+1}

possible ways to triangulate the $(n-k+1)$-gon $(v_k,\ldots,v_n)$. If we set $D_2=1$, this will also be true for $k=2$ and $k=n-1$. Thus

$$D_n = \sum_{k=2}^{n-1} D_k D_{n-k+1} \qquad (n \geq 3),$$
$$D_2 = 1.$$

Now let

$$D(x) = \sum_{n=2}^{\infty} D_n x^n.$$

Then (1) implies that

$$D(x)^2 = \sum_{k=2}^{\infty} D_k x^k \sum_{l=2}^{\infty} D_l x^l = \sum_{k=2}^{\infty}\sum_{l=2}^{\infty} D_k D_l x^{k+l} = \sum_{n=4}^{\infty} x^n \sum_{k=2}^{n-2} D_k D_{n-k} =$$
$$= \sum_{n=4}^{\infty} D_{n-1} x^n = x\left(\sum_{n=3}^{\infty} D_n x^n\right) = x(D(x) - x^2).$$

Hence

$$D(x)^2 - xD(x) + x^3 = 0,$$
$$D(x) = \frac{x \pm \sqrt{x^2 - 4x^3}}{2}.$$

Since $D'(0)=0$, we must take the negative sign for $x=0$; since $D(x)$ is continuous, we must take the negative root for $x<\frac{1}{4}$. Thus

$$D(x) = \frac{1 - \sqrt{1-4x}}{2} \cdot x .$$

Expanding this by Newton's formula we get

$$D(x) = \frac{x}{2}\left(1 - \sum_{k=0}^{\infty} \binom{1/2}{k}(-1)^k 4^k x^k\right) = \sum_{k=1}^{\infty}(-1)^{k-1}\binom{1/2}{k}2^{2k-1}x^{k+1}.$$

Here

$$\binom{1/2}{k} = \frac{(1/2)(-1/2)\ldots(1/2-k+1)}{k!} =$$
$$= (-1)^{k-1}\frac{1}{2^k}\frac{1 \cdot 3 \cdot \ldots \cdot (2k-3)}{k!} = (-1)^{k-1}\frac{1}{2^{2k-1}}\frac{1}{k}\binom{2k-2}{k-1}$$

and hence

$$D(x) = \sum_{k=1}^{\infty} \frac{1}{k} \binom{2k-2}{k-1} \cdot x^{k+1} = \sum_{n=0}^{\infty} \frac{1}{n-1} \binom{2n-4}{n-2} x^n.$$

Hence the result follows.

41. We may assume that $n \geq 4$. Let us consider a vertex v_1; there are n possible ways to choose this. Connect the two neighbors of v_1 by a diagonal d_1. Now the triangle incident with this diagonal from the other side must have an edge in common with the boundary of the n-gon; so its third edge must be one of the two diagonals connecting an endpoint of d_1 to the neighbor of the other endpoint. So we have two possible ways of choosing this second diagonal d_2. Similarly, if $d_1, \ldots, d_i$ $(i < n-3)$ have been chosen, then there are two possibilities for d_{i+1}. So we have found $n2^{n-4}$ ways to choose the point x_1 and a sequence $d_1, d_2, \ldots, d_{n-3}$ of diagonals dividing the convex n-gon into triangles having an edge in common with the n-gon. However, every triangulation (with the required property) has two triangles which contain two (adjacent) edges of the n-gon; thus, each is counted twice. So the result is $n2^{n-5}$.

42. (a) We have

$$\sum_{k=0}^{\lfloor \frac{n}{2} \rfloor} \binom{n}{2k} + \sum_{k=0}^{\lfloor \frac{n-1}{2} \rfloor} \binom{n}{2k+1} = \sum_{k=0}^{n} \binom{n}{k} = 2^n,$$

$$\sum_{k=0}^{\lfloor \frac{n}{2} \rfloor} \binom{n}{2k} - \sum_{k=0}^{\lfloor \frac{n-1}{2} \rfloor} \binom{n}{2k+1} = \sum_{k=0}^{n} (-1)^k \binom{n}{k} = (1-1)^n = 0,$$

thus

$$\sum_{k=0}^{\lfloor \frac{n}{2} \rfloor} \binom{n}{2k} = \frac{2^n + 0}{2} = 2^{n-1} \quad (n \geq 1).$$

(b) We can replace the last term $\binom{n-m}{0}$ by $\binom{n-m+1}{0}$; then the sum of the last two terms is $\binom{n-m+1}{0} + \binom{n-m+1}{1} = \binom{m-m+2}{1}$; now the sum of the last two terms is $\binom{n-m+2}{1} + \binom{n-m+2}{2} = \binom{n-m+3}{2}$, etc. Finally we get $\binom{n+1}{m}$ (Table 2).

(c) *First solution.* Let $|U| = u$, $|V| = v$, $U \cap V = \emptyset$. Then, $\binom{u}{k}$ is the number of k-tuples from U, $\binom{v}{m-k}$ is the number of $(m-k)$-tuples from V; $\sum_{k=0}^{m} \binom{u}{k} \binom{v}{m-k}$ is the number of m-tuples from $U \cup V$, i.e.

(1)
$$\sum_{k=0}^{m} \binom{u}{k} \binom{v}{m-k} = \binom{u+v}{m}.$$

$$1$$

$$1 \quad 1$$

$$1 \quad 2 \quad 1$$

$$1 \quad 3 \quad 3 \quad 1$$

$$1 \quad 4 \quad 6 \quad 4 \quad 1$$

$$1 \quad 5 \quad 10 \quad 10 \quad 5 \quad 1$$

$$1 \quad 6 \quad 15 \quad 20 \quad 15 \quad 6 \quad 1$$

$$1 \quad 7 \quad 21 \quad 35 \quad 35 \quad 21 \quad 7 \quad 1$$

$$1 \quad 8 \quad 28 \quad 56 \quad 70 \quad 56 \quad 28 \quad 8 \quad 1$$

Table 2

Second solution. The coefficient of x^m on each side of the identity

$$(1+x)^u (1+x)^v = (1+x)^{u+v}$$

is the left- and right-hand side of (1), respectively.

(d) The coefficient of x^m in $(1-x)^u(1+x)^u = (1-x^2)^u$ is 0, if m is odd and $(-1)^{m/2}\binom{u}{m/2}$, if m is even.

(e) We have

$$\binom{k}{m}\binom{n}{k} = \frac{k!}{m!(k-m)!}\frac{n!}{k!(n-k)!} = \frac{n!}{m!}\frac{1}{(k-m)!(n-k)!} =$$

$$= \frac{n!}{m!(n-m)!}\frac{(n-m)!}{(k-m)!(n-k)!} = \binom{n}{m}\binom{n-m}{n-k}.$$

Thus

$$\sum_{k=m}^{n}\binom{k}{m}\binom{n}{k} = \sum_{k=m}^{n}\binom{n}{m}\binom{n-m}{n-k} = \binom{n}{m}\sum_{k=m}^{n}\binom{n-m}{n-k} = \binom{n}{m}2^{n-m}.$$

(f)

$$\sum_{k=0}^{\lfloor \frac{n}{7}\rfloor}\binom{n}{7k} = \sum_{k=0}^{n}\left(\frac{1}{7}\sum_{j=0}^{6}\varepsilon^{kj}\right)\binom{n}{k} = \frac{1}{7}\sum_{j=0}^{6}\sum_{k=0}^{n}\binom{n}{k}\varepsilon^{kj} = \frac{1}{7}\sum_{j=0}^{6}(1+\varepsilon^{j})^n =$$

$$= \frac{2^n}{7} \left[1 + 2 \sum_{j=1}^{3} \cos \frac{\pi j n}{7} \cos^n \frac{\pi j}{j} \right].$$

(g) Let

$$a_n = \sum_{k=0}^{\lfloor \frac{n}{2} \rfloor} \binom{n-k}{k} z^k.$$

Then

$$a_n = 1 + \sum_{k=1}^{\lfloor \frac{n}{2} \rfloor} \left(\binom{n-1-k}{k} + \binom{n-k-1}{k-1} \right) z^k =$$

$$= \sum_{k=0}^{\lfloor \frac{n}{2} \rfloor} \binom{n-1-k}{k} z^k + \sum_{k=1}^{\lfloor \frac{n}{2} \rfloor} \binom{n-k-1}{k-1} z^k.$$

The bound $\lfloor \frac{n}{2} \rfloor$ can be replaced by $\lfloor \frac{n-1}{2} \rfloor$ in the first term since $k = \frac{n}{2}$ (if this is an integer) gives a 0 term. Hence the first term is

$$\sum_{k=0}^{\lfloor \frac{n-1}{2} \rfloor} \binom{n-1-k}{k} z^k = a_{n-1}.$$

Similarly, the second term is

$$\sum_{k=1}^{\lfloor \frac{n}{2} \rfloor} \binom{n-1-k}{k-1} z^k = z \sum_{k=0}^{\lfloor \frac{n-2}{2} \rfloor} \binom{n-2-k}{k} z^k = z a_{n-2}.$$

So we have the recurrence relation

(1) $$a_n = a_{n-1} + z a_{n-2}, \quad a_0 = a_1 = 1.$$

As in 1.28, (1) remains valid if we put $a_{-1} = 0$. Then by 1.28

$$a_n = \frac{\begin{vmatrix} 1 & \alpha^{n+1} \\ 1 & \beta^{n+1} \end{vmatrix}}{\begin{vmatrix} 1 & \alpha \\ 1 & \beta \end{vmatrix}} = \frac{\beta^{n+1} - \alpha^{n+1}}{\beta - \alpha},$$

where α, β are two roots of the equation

$$x^2 - x - z = 0,$$

i.e.

$$\alpha = \frac{1 - \sqrt{1 + 4z}}{2}, \quad \beta = \frac{1 + \sqrt{1 + 4z}}{2},$$

and thus

$$a_n = \frac{1}{\sqrt{1 + 4z}} \left(\left(\frac{1 + \sqrt{1 + 4z}}{2} \right)^{n+1} - \left(\frac{1 - \sqrt{1 + 4z}}{2} \right)^{n+1} \right).$$

This holds if $z \neq -1/4$, i.e. $\alpha \neq \beta$. In case $z = -1/4$ we can rewrite (1) as

$$a_n - \frac{1}{2} a_{n-1} = \frac{1}{2} \left(a_{n-1} - \frac{1}{2} a_{n-2} \right).$$

Hence

$$a_n - \frac{1}{2} a_{n-1} = \frac{c}{2^n},$$

where from $a_1 - \frac{1}{2} a_0 = 1 - \frac{1}{2} = \frac{1}{2}$ it follows that $c = 1$, i.e.

$$a_n - \frac{1}{2} a_{n-1} = \frac{1}{2^n}.$$

Thus

$$a_n = \frac{1}{2^n} + \frac{1}{2} a_{n-1} = \frac{1}{2^n} + \frac{1}{2} \left(\frac{1}{2^{n-1}} + \frac{1}{2} a_{n-2} \right) =$$

$$= \frac{1}{2^n} + \frac{1}{2^n} + \frac{1}{4} a_{n-2} = \ldots = \frac{n}{2^n} + \frac{1}{2^n} a_0 = \frac{n+1}{2^n}.$$

(h) The result is $(-1)^m \binom{n-1}{m}$. This is clear for $m = 0$. We prove it by induction:

$$\sum_{k=0}^{m} (-1)^k \binom{n}{k} = \sum_{k=0}^{m-1} (-1)^k \binom{n}{k} + (-1)^m \binom{n}{m} =$$

$$= (-1)^{m-1} \binom{n-1}{m-1} + (-1)^m \binom{n}{m} = (-1)^m \binom{n-1}{m}.$$

(i) The number of those $(u+v+1-m)$-tuples of $\{1, \ldots, u+v+1\}$ whose $(u+1)^{\text{st}}$ element is $u + k + 1$ is $\binom{u+k}{k} \binom{v-k}{m-k}$. Summing for $k = 0, \ldots, m$ we get the result stated in the hint.

43. (a) The left-hand side can be interpreted as follows: we select k points of an m-element set M; then we select m points out of an n-element set N and the k points chosen before. The result is, therefore, a pair (X, Y), where $X \subseteq M$, $Y \subseteq N \cup X$ and $|Y| = m$. The number of possible ways of choosing this is

$$\sum_{k=0}^{m} \binom{m}{k} \binom{n+k}{m}.$$

The same enumeration problem can be solved by selecting $Y \cap M$ and $Y \cap N$ first; if $|Y \cap N| = j$, then this can be done in $\binom{n}{j}\binom{m}{m-j}$ ways; then we have to select on X with $Y \cap M \subseteq X \subseteq M$, which can be done in 2^j ways; so the total number of choices is

$$\sum_{j=0}^{m} \binom{n}{j}\binom{m}{j} 2^j.$$

(b) Since identity (a) means the equality of two polynomials in n for infinitely many values of n, it follows that the two polynomials are identical. Thus we may substitute $-n-1$ for n and get, on the left hand side,

$$\sum_{k=0}^{m} \binom{m}{k}\binom{-n-1+k}{m} = \sum_{k=0}^{m} \binom{m}{k}(-1)^m \binom{n+m-k}{m}$$

$$= (-1)^m \sum_{k=0}^{m} \binom{m}{k}\binom{n+k}{m}.$$

The right-hand side becomes

$$\sum_{k=0}^{m} \binom{n}{k}\binom{-n-1}{k} 2^k = \sum_{k=0}^{m} \binom{m}{k}(-1)^k \binom{n+k}{k} 2^k = \sum_{k=0}^{m} \binom{m}{k}\binom{n+k}{k}(-2)^k.$$

This proves the assertion.

(c)

$$\sum_{k=0}^{\infty} \binom{p}{k}\binom{q}{k}\binom{n+k}{p+q} = \sum_{k=0}^{\infty} \binom{p}{k}\binom{q}{k} \sum_{j=0}^{k} \binom{k}{j}\binom{n}{p+q-j} =$$

$$= \sum_{j=0}^{\infty} \binom{n}{p+q-j} \sum_{k=0}^{\infty} \binom{p}{k}\binom{q}{k}\binom{k}{j}.$$

Here

$$\sum_{k=0}^{\infty} \binom{p}{k}\binom{q}{k}\binom{k}{j} = \sum_{k=0}^{\infty} \binom{p}{k} \frac{q!}{(q-k)!j!(k-j)!} =$$

$$= \sum_{k=0}^{\infty} \binom{p}{k}\binom{q}{j}\binom{q-j}{q-k} = \binom{q}{j}\binom{p+q-j}{q}$$

by 1.42c. Thus the sum on the left-hand side is

$$\sum_{j=0}^{\infty} \binom{n}{p+q-j}\binom{q}{j}\binom{p+q-j}{q} = \sum_{j=0}^{\infty} \frac{n!}{(n-p-q+j)!(q-j)!(p-j)!j!} =$$

$$= \sum_{j=0}^{\infty} \binom{n}{p}\binom{p}{j}\binom{n-p}{q-j} = \binom{n}{p} \sum_{j=0}^{\infty} \binom{p}{j}\binom{n-p}{q-j} = \binom{n}{p}\binom{n}{q}$$

by 1.42c again.

(d) This follows in exactly the same way as (b).

(e) We have

$$\binom{x+t}{n} - \binom{x}{n} = \sum_{k=1}^{n} \binom{t}{k}\binom{x}{n-k}$$

by 1.42c. Dividing by t and letting $t \to 0$ the first expression follows. The second one is obtained analogously by 1.42i and

$$\binom{x+t}{n} - \binom{x}{n} = \sum_{k=1}^{n} \binom{t+k-1}{k}\binom{x-k}{n-k}.$$

44. We prove (a) by induction on n. For $n=0$ it is trivial. Let $n>0$. We have

$$\frac{\partial}{\partial y}(x+y+n)^n = n(x+y+n)^{n-1} = n(x+(y+1)+(n-1))^{n-1},$$

$$\frac{\partial}{\partial y} \sum_{k=0}^{n} \binom{n}{k} x(x+k)^{k-1}(y+n-k)^{n-k} =$$

$$= \sum_{k=0}^{n} \binom{n}{k} x(x+k)^{k-1}(n-k)(y+n-k)^{n-k-1} =$$

$$= n \sum_{k=0}^{n-1} \binom{n-1}{k} x(x+k)^{k-1}((y+1)+(n-1)-k)^{(n-1)-k}.$$

The two right-hand sides here are equal by the induction hypothesis. To prove the identity it suffices to show that (a) holds for any one value of y. Choose $y = -x-n$. Then the right-hand side vanishes while the left-hand side is

$$\sum_{k=0}^{n} \binom{n}{k} x(x+k)^{k-1}(-x-k)^{n-k} = x \sum_{k=0}^{n} \binom{n}{k}(-1)^{n-k}(x+k)^{n-1} =$$

$$= \sum_{k=0}^{n} \binom{n}{k}(-1)^{n-k} \sum_{j=0}^{n-1} \binom{n-1}{j} k^j x^{n-j} =$$

$$= \sum_{j=0}^{n-1} \binom{n-1}{j} x^{n-j} \sum_{k=0}^{n} \binom{n}{k} k^j (-1)^{n-k} =$$

$$= \sum_{j=0}^{n-1} \binom{n-1}{j} x^{n-j} S(j,n) n! = 0 \qquad\qquad \text{as } j < n \text{ (by 1.8).}$$

(b) The left-hand side of (a) can be rewritten as

$$\sum_{k=0}^{n} \binom{n}{k} x(x+k)^{k-1}(y+n-k)^{n-k-1}(y+n-k) =$$

$$= \sum_{k=0}^{n} \binom{n}{k} x(x+k)^{k-1} y(y+n-k)^{n-k-1} +$$

$$+ \sum_{k=0}^{n} \binom{n}{k} x(x+k)^{k-1}(y+n-k)^{n-k-1}(n-k).$$

Here the second term is

$$\sum_{k=0}^{n-1} \binom{n-1}{k} x(x+k)^{k-1}((y+1)+(n-1)-k)^{(n-1)-k} =$$

$$= n(x+(y+1)+(n-1))^{n-1} = n(x+y+n)^{n-1}$$

by (a). Hence

$$\sum_{k=0}^{n} \binom{n}{k} x(x+k)^{k-1} y(y+n-k)^{n-k-1} = (x+y+n)^n - n(x+y+n)^{n-1} =$$

$$= (x+y)(x+y+n)^{n-1}.$$

Dividing by xy identity (b) follows.

(c) Subtract $\frac{1}{x}(y+n)^{n-1} + \frac{1}{y}(x+n)^{n-1}$ from both sides of (b) then we get

$$\sum_{k=1}^{n-1} \binom{n}{k} (x+k)^{k-1}(y+n-k)^{n-k-1} = \frac{1}{x}\{(x+y+n)^{n-1} - (y+n)^{n-1}\} +$$

$$+ \frac{1}{y}\{(x+y+n)^{n-1} - (x+n)^{n-1}\}.$$

Letting $x, y \to 0$ we obtain (c).

45.

$$\sum_{k=0}^{n} \binom{n}{k} f_k(x) f_{n-k}(y) = \sum_{k=0}^{n} \binom{n}{k} k! \binom{x}{k} (n-k)! \binom{y}{n-k} =$$

$$= \sum_{k=1}^{n} n! \binom{x}{k} \binom{y}{n-k} = n! \binom{x+y}{n} =$$

by 1.42c (strictly speaking, the identity

$$\sum_{k=0}^{n} \binom{x}{k} \binom{y}{n-k} = \binom{x+y}{n}$$

was proved for the case when, x, y are natural numbers; but if the polynomials on the two sides coincide for every such choice of the variables they are identical).

Other instances of such polynomials: $f_n(x) = x^n$ (then the formula in the problem is the binomial theorem), $f_n(x) = x(x+1)\ldots(x+n-1)$ (this follows in the same way as $f_n(x) = x(x-1)\ldots(x-n+1)$), $f_n(x) = x(x+n)^{n-1}$ (this follows from identity (b) in the preceding problem). [See also G. C. Rota–R. Mullin, in *Graph Theory and its Applications*, Academic Press, New York 1970, 167–213.]

§ 2. The sieve

1. Let us subtract from 30 the number of pupils who like mathematics, physics, chemistry, respectively:

$$30 - 12 - 14 - 13.$$

This way, however, a student who likes both mathematics and physics is subtracted twice; so we have to add them back, and also for the two other pairs of subjects:

$$30 - 12 - 14 - 13 + 5 + 7 + 4.$$

There is still trouble with those who like all three subjects. They were subtracted 3 times, but put back 3 times, so we have to subtract them once more to get the result

$$30 - 12 - 14 - 13 + 5 + 7 + 4 - 3 = 4.$$

2. (a) Let

$$B = A_1 A_2 \ldots A_k \overline{A}_{k+1} \ldots \overline{A}_n$$

be any atom of the Boolean algebra generated by $A_1, \ldots, A_n$ (with an appropriate choice of indices, every atom has such a form.) Every event in the formula is the union of certain (disjoint) atoms; let us express each $\mathsf{P}(A_I)$ and $\mathsf{P}(A_1 + \ldots + A_n)$ as the sum of the probabilities of the corresponding atoms. We show that the probability of any given atom cancels out.

The coefficient of $P(B)$ on the left-hand side is

$$1 \quad \text{if} \quad k \neq 0$$
$$0 \quad \text{if} \quad k = 0$$

B occurs in A_I iff $I \subseteq \{1, \ldots, k\}$, so its coefficient on the right-hand side is

$$\sum_{j=1}^{k} \binom{k}{j} (-1)^{j-1} = 1 - \sum_{j=0}^{k} \binom{k}{j} (-1)^j = 1 - (1-1)^k = \begin{cases} 1 & \text{if } k \neq 0, \\ 0 & \text{if } k = 0. \end{cases}$$

Thus $P(B)$ has the same coefficient on both sides, which proves (a).

(b) Choose an element x of S by a uniform distribution. Then A_i can be identified with the event that $x \in A_i$, and we have

$$P(A_i) = \frac{|A_i|}{|S|}.$$

So we have, by the above

$$P(A_1 + \ldots + A_n) = \sum_{j=1}^{n} (-1)^{j-1} \sum_{|I|=j} \frac{|A_I|}{|S|} = \sum_{\emptyset \neq I \subseteq \{1,\ldots,n\}} (-1)^{|I|-1} \frac{|A_I|}{|S|},$$

or, equivalently,

$$P(\overline{A}_1 \ldots \overline{A}_n) = 1 - P(A_1 + \ldots + A_n) = 1 - \sum_{\emptyset \neq I \subseteq \{1,\ldots,n\}} (-1)^{|I|-1} \frac{|A|}{|S|}.$$

The assertion (b) follows on multiplying by $|S|$.

3. Let A_i be the set of integers between 1 and n divisible by p_i. Then, for $I \subseteq \{1, \ldots, r\}$,

$$A_I = \bigcap_{i \in I} A_i = \left\{ k : 1 \leq k \leq n, \ \prod_{i \in I} p_i | k \right\}$$

and so,

$$|A_I| = \frac{n}{\displaystyle\prod_{i \in I} p_i}.$$

Hence,

(1) $$\varphi(n) = \sum_{I \subseteq \{1,\ldots,r\}} (-1)^{|I|} \frac{n}{\displaystyle\prod_{i \in I} p_i}.$$

Now $\prod_{i \in I} p_i$ ranges over divisors of n; over those divisors of n, in fact, which are not divisible by the square of any prime. To write (1) in a neater form, define

$$\mu(k) = \begin{cases} 0 & \text{if } k \text{ is divisible by the square of some prime,} \\ (-1)^i & \text{if } k \text{ has } i \text{ distinct prime factors.} \end{cases}$$

Then

$$\varphi(n) = \sum_{k|n} \frac{\mu(k)}{k} n.$$

Also, one can observe that (1) is the expansion of the product

$$\varphi(n) = n \prod_{i=1}^{r} \left(1 - \frac{1}{p_i}\right).$$

4. Let $|X| = k$, $Y = \{y_1, \dots, y_n\}$. Then the number of those mappings of X into Y which are onto Y is, obviously,

$$0 \quad \text{if} \quad k < n,$$
$$n! \quad \text{if} \quad k = n.$$

On the other hand, let A_i denote the set of mappings of X into $Y - y_i$ and let S be the set of all mappings of X into Y. Then we are interested in $S - (A_1 \cup \dots \cup A_n)$. By the inclusion-exclusion formula.

$$|S - (A_1 \cup \dots \cup A_n)| = \sum_{I \subseteq \{1,\dots,n\}} (-1)^I |A_I|.$$

Here A_I is the set of mappings of X into $Y - \{y_i : i \in I\}$, hence

$$|A_I| = (n - |I|)^k.$$

So

$$|S - (A_1 \cup \dots \cup A_n)| = \sum_{j=0}^{n} (-1)^j \binom{n}{j} (n-j)^k = \sum_{j=0}^{n} (-1)^{n-j} \binom{n}{j} j^k,$$

which proves the assertion.

If $k > n$, the result of the same inclusion-exclusion procedure is the number of mappings of X onto Y, which is $n! \left\{ \begin{matrix} k \\ n \end{matrix} \right\}$. Thus the result is a new proof of 1.8.

5. Put $x_i = 0$ into $\sigma^0 p - \sigma^1 p + \dots$. Then, if we set

$$\tilde{p}(x_1, \dots, x_{i-1}, x_{i+1}, \dots, x_n) =$$
$$= p(x_1, \dots, x_{i-1}, 0, x_{i+1}, \dots, x_n),$$

we have $\sigma^k p|_{x_i=0} = \sigma^{k-1}\tilde{p} + \sigma^k \tilde{p}$ (a k-tuple of variables of p either contains or does not contain x_i),

$$\sigma^0 p|_{x_i=0} = p|_{x_i=0} = \tilde{p} = \sigma^0 \tilde{p}, \text{ so } (\sigma^0 p - \sigma^1 p + \sigma^2 p - \ldots)|_{x_i=0} =$$
$$= \sigma^0 \tilde{p} - (\sigma^0 \tilde{p} + \sigma^1 \tilde{p}) + (\sigma^1 \tilde{p} + \sigma^2 \tilde{p}) - \ldots = 0.$$

Therefore, $x_1 \ldots x_n$ divides $\sigma^0 p - \sigma^1 p + \ldots$. Since $\sigma^0 p - \sigma^1 p + \ldots$ has degree at most m, it follows that

$$\sigma^0 p - \sigma^1 p + \ldots = \begin{cases} 0 & \text{if } m < n, \\ cx_1 \ldots x_n & \text{if } m = n. \end{cases}$$

Also note that $\sigma^k p$ $(k \geq 1)$ certainly does not contain any term of the form $x_1 \ldots x_n$; hence c is the coefficient of $x_1 \ldots x_n$ in $\sigma^0 p = p$.

We remark that for $p(x_1, \ldots, x_n) = (x_1 + \ldots + x_n)^k$ and $x_1 = \ldots = x_n = 1$ we obtain the preceding result (cf. problem 4.3).

6. Let us express $\mathsf{P}(B)$ as the sum of probabilities of atoms; we show that each atom will have a non-negative coefficient. Let

$$B = A_1 \ldots A_k \overline{A}_{k+1} \ldots \overline{A}_n$$

be any atom of the Boolean algebra generated by $A_1, \ldots, A_n$. Define $A'_1 = \ldots = A'_k = 1$, $A'_{k+1} = \ldots = A'_n = 0$, and let $B'_i = f_i(A'_1, \ldots, A'_n)$. Then $B \subseteq B_i$ iff $B'_i \neq 0$. Therefore, the coefficient of $\mathsf{P}(B)$ in $\sum_{i=1}^{k} c_i \mathsf{P}(B_i)$ is

$$\sum_{B_i \supseteq B} c_i = \sum_{i=1}^{k} c_i \mathsf{P}(B'_i) \geq 0,$$

by the assumption [Ré].

7. (a) By the preceding result, we may assume that

$$\mathsf{P}(A_1) = \ldots = \mathsf{P}(A_k) = 1, \quad \mathsf{P}(A_{k+1}) = \ldots = \mathsf{P}(A_n) = 0.$$

Then the left-hand side is 1 if $k = q$ and is 0 if $k \neq q$; the right-hand side is

$$\sum_{j=q}^{k} (-1)^{j+q} \binom{k}{j} \binom{j}{q}.$$

If $q > k$, this sum is, obviously, 0. If $q = k$, the sum is 1. Finally, if $q < k$, then

$$\binom{k}{j}\binom{j}{q} = \frac{k!}{q!(j-q)!(k-j)!} = \binom{k}{q}\binom{k-q}{j-q}.$$

So we have

$$\sum_{j=q}^{k}(-1)^{j+q}\binom{k}{q}\binom{k-q}{j-q} = \binom{k}{q}\sum_{j=q}^{k}(-1)^{j-q}\binom{k-q}{j-q} =$$

$$= \binom{k}{q}(1-1)^{k-q} = 0 \quad \text{[K. Jordán; Ré].}$$

(b) $E\left(\binom{\eta}{j}\right)$ is the expected number of j-tuples of the events $A_1,\ldots,A_n$ such that all j occur. Since the probability of simultaneous occurrence of a given j-tuple $\{A_{i_1},\ldots,A_{i_j}\}$ is $\mathsf{P}(A_{i_1}\ldots A_{i_j})$, we have

$$E\left(\binom{\eta}{j}\right) = \sum_{1\leq i_1 < \ldots < i_j \leq n} \mathsf{P}(A_{i_1}\ldots A_{i_j}) = \sigma_j.$$

Thus

$$\mathsf{E}(x^\eta) = E\left(\sum_{j=0}^{\eta}\binom{\eta}{j}(x-1)^j\right) = \sum_{j=0}^{n}\sigma_j(x-1)^j.$$

8. By the preceding result, the probability to be determined is

$$\sum_{q=p}^{n}\sum_{j=q}^{n}(-1)^{j+q}\binom{j}{q}\sigma_j = \sum_{j=p}^{n}\sigma_j\sum_{q=p}^{j}(-1)^{j+q}\binom{j}{q} = \sum_{j=p}^{n}\sigma_j\sum_{\nu=0}^{j-p}(-1)^\nu\binom{j}{\nu} =$$

$$= \sum_{j=p}^{n}\sigma_j(-1)^{j-p}\binom{j-1}{j-p} = \sum_{j=p}^{n}(-1)^{j+p}\binom{j-1}{p-1}\sigma_j$$

by 1.42h [see Ré].

9. Let $j > 0$, be odd. Then we want to show that

$$\mathsf{P}(\overline{A}_1\ldots\overline{A}_n) - \sigma_0 + \sigma_1 - \ldots + \sigma_j \geq 0.$$

By 2.6, we may assume that

$$\mathsf{P}(A_1) = \ldots = \mathsf{P}(A_k) = 1, \quad \mathsf{P}(A_{k+1}) = \ldots = \mathsf{P}(A_n) = 0.$$

The case $k=0$ is trivial, so let $k \geq 1$. Then we have to show that

$$0 - 1 + \binom{k}{1} - \binom{k}{2} + \ldots + \binom{k}{j} \geq 0,$$

if j is odd. This is clear as by 1.42h,

$$1 - \binom{k}{1} + \binom{k}{2} - \ldots - \binom{k}{j} = (-1)^j\binom{k-1}{j}.$$

The case of even j follows similarly.

10. I. By 2.6 we may assume again that

$$P(A_1) = \ldots = P(A_k) = 1, \quad P(A_{k+1}) = \ldots = P(A_n) = 0.$$

Then

$$\sigma_r = \binom{k}{r}, \quad \sigma_{r-1} = \binom{k}{r-1},$$

so what we have to prove is

$$\binom{k}{r} \le \frac{n-r+1}{r}\binom{k}{r-1}$$

or, dividing by $\binom{k}{r}$,

$$1 \le \frac{n-r+1}{r} \cdot \frac{r}{k-r+1} = \frac{n-r+1}{k-r+1}$$

which is clear [M. Fréchet; Ré].

II. We prove the inequality

$$\sigma_r \le \frac{m-r}{r}\sigma_{r-1} + \frac{1}{r}\binom{m}{r-1}\sigma_m$$

by the usual method. It suffices to show that

$$\binom{k}{r} \le \frac{m-r}{r}\binom{k}{r-1} + \frac{1}{r}\binom{m}{r-1}\binom{k}{m}.$$

Rearranging, we get

$$\frac{k-m+1}{r}\binom{k}{r-1} \le \frac{1}{r}\binom{m}{r-1}\binom{k}{m}.$$

This is clear if $k < m$, so suppose that $k \ge m$. Dividing by the left hand side, we obtain

$$1 \le \frac{1}{k-m+1} \cdot \binom{m}{r-1}\binom{k}{m}\bigg/\binom{k}{r-1} = \frac{1}{m-r+1}\binom{k-r+1}{m-r}$$

which is obviously true.

In particular, if $\sigma_{r+1} = 0$, then we have $\sigma_r \le \frac{1}{r}\sigma_{r-1}$.

11. I. Suppose that we have such events $A_1, \ldots, A_n$ and let $J \subseteq \{1, \ldots, n\}$. By the sieve formula 2.2a

$$P(\prod_{i \notin J} \overline{A}_i \cdot A_J) = \sum_{I \subseteq \{1,\ldots,n\}-J} (-1)^{|I|}P(A_I A_J) = \sum_{J \subseteq K \subseteq \{1,\ldots,n\}} (-1)^{|K-J|}p_K.$$

Since a probability is non-negative, we have

(1) $$\sum_{J \subseteq K \subseteq \{1,\ldots,n\}} (-1)^{|K-J|}p_K \ge 0,$$

for every $J \subseteq \{1, \ldots, n\}$. Also, obviously,

(2)
$$p_\emptyset = 1.$$

II. We show that (1) and (2) are sufficient. Take 2^n atoms v_J ($J \subseteq \{1, \ldots, n\}$) and define

$$P(v_J) = \sum_{J \subseteq K \subseteq \{1,\ldots,n\}} (-1)^{|K-J|} p_K,$$

$$A_\emptyset = \{v_J : J \subseteq \{1, \ldots, n\}\},$$

$$A_i = \{v_J : i \in J\} \quad (i = 1, \ldots, n).$$

Then

$$A_I = \prod_{i \in I} A_i \{v_J : I \subseteq J \subseteq \{1, \ldots, n\}\}.$$

We claim

$$P(X) = \sum_{v_J \in X} P(v_J)$$

defines a probability measure on $A_\emptyset$ and $P(A_I) = p_I$. Let us show the second statement first:

$$P(A_I) = \sum_{I \subseteq J \subseteq \{1,\ldots,n\}} P(v_J) = \sum_{I \subseteq J \subseteq \{1,\ldots,n\}} \sum_{J \subseteq K \subseteq \{1,\ldots,n\}} (-1)^{|K-J|} p_K =$$

$$= \sum_{I \subseteq K \subseteq \{1,\ldots,n\}} p_K \sum_{I \subseteq J \subseteq K} (-1)^{|K-J|}.$$

Here

$$\sum_{I \subseteq J \subseteq K} (-1)^{|K-J|} = \begin{cases} 1 & \text{if } I = K, \\ 0 & \text{otherwise,} \end{cases}$$

so we have

$$P(A_I) = p_I.$$

To show P is a probability measure, we merely note that (1) implies $P \geq 0$ while (2) implies $P(A_\emptyset) = 1$ [Ré].

12. It suffices to prove the inequality if there is a set $K \subseteq \{1, \ldots, n\}$ such that

$$P(A_j) = \begin{cases} 1 & \text{if } j \in K, \\ 0 & \text{otherwise,} \end{cases}$$

by 2.6. If $K \neq \emptyset$ both sides are 1. So let $K \neq \emptyset$. Then we may assume without loss of generality that $K = \{1, \ldots, n\}$; otherwise we could consider the subgraph induced by K. Then the inequality has the form

(1)
$$|\mathcal{E}_1| \geq |\mathcal{O}_0|.$$

We prove this together with

(2) $$|\mathcal{O}_1| \geq |\mathcal{E}_0|.$$

by induction on n. If G is a complete graph, then

$$|\mathcal{O}_1| = n, \quad |\mathcal{E}_1| = \binom{n}{2} + 1, \quad |\mathcal{E}_0| = 1, \quad |\mathcal{O}_0| = n,$$

thus (1) and (2) are satisfied. Suppose that G is not complete and let, say, $n \in V(G)$ be adjacent to $1, \ldots, h$ but not adjacent to $h+1, \ldots, n-1$. Form the subgraphs G^*, G^{**} of G induced by $\{1, \ldots, n-1\}$ and $\{h+1, \ldots, n-1\}$, respectively. Denote by $\mathcal{E}_0^*$, $\mathcal{E}_0^{**}$, etc. the set $\mathcal{E}_0$, etc. for G^* and G^{**}. Then an independent set of G is either an independent set of G^* or consists of n and an independent set of G^{**}. Hence

(3)
$$\begin{aligned}|\mathcal{E}_0| &= |\mathcal{E}_0^*| + |\mathcal{O}_0^{**}|,\\ |\mathcal{O}_0| &= |\mathcal{O}_0^*| + |\mathcal{E}_0^{**}|.\end{aligned}$$

Also, each subset of G^* which spans at most one edge is such a subset of G; and if X is such a subset of G^{**}, then $X \cup \{n\}$ is such a subset of G. But now G may have other subsets inducing at most one edge, e.g. $\{1, n\}$. Hence, we can only claim

(4)
$$\begin{aligned}|\mathcal{E}_1| &\geq |\mathcal{E}_1^*| + |\mathcal{O}_1^{**}|,\\ |\mathcal{O}_1| &\geq |\mathcal{O}_1^*| + |\mathcal{E}_1^{**}|.\end{aligned}$$

Now by the induction hypothesis,

$$|\mathcal{E}_1^*| \geq |\mathcal{O}_0^*|, \quad |\mathcal{O}_1^*| \geq |\mathcal{E}_0^*|,$$

and similar inequalities hold for G^{**}. Thus (3) and (4) imply (1) and (2).

The lower estimate follows from (2) exactly as the upper one follows from (1) and reads

$$\mathsf{P}(\overline{A}_1 \ldots \overline{A}_n) \geq \sum_{I \in \mathcal{E}_0} \mathsf{P}(A_I) - \sum_{I \in \mathcal{O}_1} \mathsf{P}(A_I).$$

[A. Rényi, *J. Math. Pures Appl.* **37** (1958) 393–398.]

13. Again we may assume that

$$\mathsf{P}(A_i) = \begin{cases} 1 & \text{if } i \in K, \\ 0 & \text{otherwise} \end{cases}$$

for some $K \subseteq \{1, \ldots, n\}$. If $K = \emptyset$, both sides are 1, so suppose that $K \neq \emptyset$. Then, we have to show

(1)
$$\sum_{\substack{I \in \mathbf{I} \\ I \subseteq K}} (-1)^{|I|} \geq 0.$$

Let x be the largest element of K. In (1), those pairs of terms $(-1)^{|I|}$, $(-1)^{|J|}$, where $J = I \cup \{x\}$ ($x \notin I$) obviously cancel each other.

If $J \in \mathscr{I}$ and $x \in J$, then, obviously, $J - \{x\} \in \mathscr{I}$. Thus, in (1), only those terms $(-1)^{|I|}$ are left, where $x \notin I$ and $I \cup \{x\} \notin \mathscr{I}$. We show all such terms are positive. In fact, $I \cup \{x\} \notin \mathscr{I}$ means

$$|(I \cup \{x\}) \cap \{1, \ldots, k\}| > 2f(k)$$

for some k; since $I \in \mathscr{I}$ we must have

$$|I \cap \{1, \ldots, k\}| \leq 2f(k),$$

whence $k \geq x$ and

$$|I \cap \{1, \ldots, k\}| = 2f(k).$$

But $I \subseteq K \subseteq \{1, \ldots, x\} \subseteq \{1, \ldots, k\}$, and thus

$$|I \cap \{1, \ldots, k\}| = |I| = 2f(k),$$

i.e. $(-1)^{|I|} = (-1)^{2f(k)} = 1$ as stated.

The lower estimate is clearly the following: Let

$$\mathscr{I}' = \{I \subseteq \{1, \ldots, n\} : |I \cap \{1, \ldots, k\}| \leq 2f(k) + 1 \quad (k = 1, \ldots, n)\}$$

then

$$\mathsf{P}(\overline{A}_1 \ldots \overline{A}_n) \geq \sum_{I \in \mathscr{I}'} (-1)^{|I|} \mathsf{P}(A_I).$$

[For an account of sieve methods, from the number theoretic point of view, see H. Halberstam–K. F. Roth, *Sequences*, Clarendon Press, Oxford, 1966.]

14. By 2.6, we may assume that

(1) $\mathsf{P}(A_1) = \ldots = \mathsf{P}(A_k) = 1, \quad \mathsf{P}(A_{k+1}) = \ldots = \mathsf{P}(A_n) = 0.$

If $k = 0$, then $\mathsf{P}(A_{I \cup J}) = 0$ except when $I = J = \emptyset$, and then we have to show that

$$1 \leq \lambda_\emptyset^2,$$

which is clear. If $k > 0$, then we have to verify that

$$0 \leq \sum_{I,J \subseteq \{1, \ldots, k\}} \lambda_I \lambda_J = \left(\sum_{I \subseteq \{1, \ldots, k\}} \lambda_I \right)^2,$$

which is again clear.

For $\lambda_I = (-1)^{|I|}$ the right hand side is just the Sieve Formula, so equality holds.

(b) To prove the lower estimate given in the hint, we may again assume that

$$\mathsf{P}(A_i) = \begin{cases} 1 & \text{if } i \in K, \\ 0 & \text{otherwise} \end{cases}$$

(since the ordering of the indices is involved in the formula we may not assume this time that K is a section $\{1,\dots,k\}$). The relation is trivial if $K=\emptyset$ $(1\geq 1)$, so suppose that $K\neq\emptyset$. Then we have to show

$$0 \geq 1 - \sum_{\substack{I,J\subseteq K \\ \max I=\max J}} \lambda_I\lambda_J.$$

Let

$$K = \{k_1,\dots,k_\nu\}, \quad k_1 < \dots < k_\nu,$$

then this inequality can be written as

$$1 \leq \sum_{j=1}^{\nu} \sum_{\substack{I\subseteq K \\ \max I=k_j}} \sum_{\substack{J\subseteq K \\ \max J=k_i}} \lambda_I\lambda_J = \sum_{j=1}^{\nu}\left(\sum_{\substack{I\subseteq K \\ \max I=k_j}} \lambda_I\right)^2 = \lambda_{\{k_1\}}^2 + \sum_{j=2}^{\nu}\left(\sum_{\substack{I\subseteq K \\ \max I=k_j}} \lambda_I\right)^2,$$

which is clearly true.

15. We have

$$\sum_{I,J\subseteq\{1,\dots,n\}} \lambda_I\lambda_J p_{I\cup J} = \sum_{\substack{|I|\leq k \\ |J|\leq k}} \lambda_I\lambda_J p_I p_J \sum_{K\subseteq I\cap J} \frac{q_K}{p_K}$$

because

$$p_{I\cup J} = p_I p_J / p_{I\cap J}$$

and

$$\sum_{K\subseteq I\cup J} \frac{q_K}{p_K} = \prod_{i\in I\cap J}\left(1+\frac{q_i}{p_i}\right) = \frac{1}{p_{I\cap J}}.$$

Thus

$$\sum_{\substack{|I|\leq k \\ |J|\leq k}} \lambda_I\lambda_J p_{I\cup J} = \sum_{|K|\leq k} \frac{q_K}{p_K} \sum_{\substack{K\subseteq I \\ |I|\leq k}} \sum_{\substack{K\subseteq J \\ |J|\leq k}} \lambda_I\lambda_J p_I p_J =$$

$$= \sum_K \frac{q_K}{p_K}\left(\sum_{\substack{K\subseteq I \\ |I|\leq k}} p_I\lambda_I\right)^2 = \sum_K \frac{q_K}{p_K}\mu_K^2,$$

where

(1)
$$\mu_K = \sum_{\substack{K\subseteq I \\ |I|\leq k}} p_I\lambda_I.$$

Here we can express λ_I as

(2)
$$\lambda_I = \frac{1}{p_I} \sum_{\substack{I \subseteq K \\ |K| \le k}} (-1)^{|K-I|} \mu_K,$$

because the μ_K's obviously determine the λ_I's and the λ_I's defined by (2) satisfy (1):

$$\sum_{\substack{K \subseteq I \\ |I| \le k}} p_I \lambda_I = \sum_{\substack{K \subseteq I \\ |I| \le k}} \sum_{\substack{I \subseteq J \\ |J| \le k}} (-1)^{|J-I|} \mu_J = \sum_{\substack{K \subseteq J \\ |J| \le k}} \mu_J \sum_{K \subseteq I \subseteq J} (-1)^{|J-I|} = \mu_K.$$

Thus we have to minimize

(3)
$$\sum_{|K| \le k} \frac{q_K}{p_K} \mu_K^2$$

subject to

(4)
$$\sum_{|K| \le k} (-1)^{|K|} \mu_K = 1.$$

This can be done either by the Lagrange method or simply by the transformation:

(5)
$$\sum_{|K| \le k} \frac{q_K}{p_K} \mu_K^2 = \sum_{|K| \le k} \frac{p_K}{q_K} \left(\frac{q_K}{p_K} \mu_K - \frac{(-1)^{|K|}}{Q} \right)^2 + \frac{1}{Q},$$

where

$$Q = \sum_{|K| \le k} \frac{p_K}{q_K},$$

which holds under the assumption (4). Since

(6)
$$\mu_K = \frac{1}{Q} \cdot \frac{p_K}{q_K} \cdot (-1)^{|K|}$$

is a choice of the variables which minimizes the right-hand side of (5) and also satisfies the constraint (4), we get that the minimum of (3) is $1/Q$ and (6) gives the extremum. Hence

$$\lambda_I = \frac{1}{p_I} \sum_{\substack{I \subseteq K \\ |K| \le k}} (-1)^{|K-I|} \frac{1}{Q} \frac{p_K}{q_K} (-1)^{|K|} = \frac{(-1)^{|I|}}{Q q_I} \sum_{\substack{L \subseteq \{1,\dots,n\}-I \\ |L| \le k-|I|}} \frac{p_L}{q_L}.$$

16. By 2.14,

$$\mathsf{P}(\overline{A}_1 \dots \overline{A}_n) \le \sum_{I, J \subseteq \{1, \dots, n\}} \lambda_I \lambda_J \mathsf{P}(A_{I \cup J})$$

for any choice of λ_I ($I \subseteq \{1,\ldots,n\}$ with the only restriction that $\lambda_\emptyset = 1$. We will only use λ_I's that satisfy the requirement $\lambda_I = 0$ for $I \notin \mathcal{H}$. Setting

$$P(A_I) = p_I + R_I,$$

we have

(1) $$P(\overline{A}_1 \ldots \overline{A}_n) \leq \sum_{I,J \in \mathcal{H}} \lambda_I \lambda_J p_{I \cup J} + \sum_{I,J \in \mathcal{H}} \lambda_I \lambda_J R_{I \cup J}.$$

As in the preceding solution, the choice

(2) $$\lambda_I = \frac{(-1)^{|I|}}{Q q_I} \sum_{\substack{L \subseteq \{1,\ldots,n\} - I \\ L \cup I \in \mathbf{H}}} \frac{p_L}{q_L}$$

minimizes the first term and then the value of it is $1/Q$, where

$$Q = \sum_{K \in \mathcal{H}} \frac{p_K}{q_K}.$$

We now estimate the second term in (1) when λ_I is given by (2). We have

$$|\lambda_I| \leq \frac{1}{Q q_I} \sum_{\substack{L \subseteq \{1,\ldots,n\} - I \\ L \cup I \in \mathbf{H}}} \frac{p_L}{q_L} \leq \frac{1}{q_I},$$

because $L \cup I \in \mathcal{H}$ implies $L \in \mathcal{H}$. Thus

$$\left| \sum_{I,J \in \mathcal{H}} \lambda_I \lambda_J R_{I \cup J} \right| \leq \varepsilon \sum_{I,J \in \mathcal{H}} \frac{1}{q_I} \frac{1}{q_J} = \varepsilon \left(\sum_{I \in \mathcal{H}} \frac{1}{q_I} \right)^2 \leq \varepsilon \left(\sum_{I \in \mathcal{H}} \frac{M p_I}{q_I} \right)^2 \leq$$

$$\leq \varepsilon M^2 \left(\sum_{I \subseteq \{1,\ldots,n\}} \frac{p_I}{q_I} \right)^2 = \varepsilon M^2 \left(\prod_{i=1}^n \left(1 + \frac{p_i}{q_i}\right) \right)^2 = \frac{\varepsilon M^2}{q_1^2 \ldots q_n^2}.$$

The first term can be estimated as follows:

$$Q = \sum_{K \in \mathcal{H}} \frac{p_K}{q_K} = \sum_{K \in \mathcal{H}} \prod_{k \in K} \frac{p_k}{1 - p_k} = \sum_{K \in \mathcal{H}} \prod_{k \in K} (p_k + p_k^2 + \ldots).$$

If we multiply out we get exactly those terms $p_1^{l_1} \ldots p_n^{l_n}$ for which $\prod_{l_i > 0} p_i \geq \frac{1}{M}$; so we certainly get all terms of this form which are themselves at least $\frac{1}{M}$. Thus

$$Q \geq S$$

and the assertion follows.

17. Let $P_1,\ldots,P_n,M$ be defined as in the hint. Let us choose one of the numbers $l,k+l,\ldots,(x-1)k+l$ at random and let A_i be the event that it is divisible by P_i. Set

$$P_I = \prod_{i\in I} P_i \quad (I \subseteq \{1,\ldots,n\}).$$

$x\cdot \mathsf{P}(A_I)$ is the number of multiples of P_I among $l,k+l,\ldots,(x-1)k+l$. Since $(P_I,k)=1$, P_I has one multiple among $\mu P_I\cdot k+l, (\mu P_I+1)k+l,\ldots,(\mu P_I+P_I-1)k+l$ for every $0\le \mu \le \frac{x}{P_I}-1$. Hence

$$\left[\frac{x}{P_I}\right] \le x\mathsf{P}(A_I) \le \left\lfloor\frac{x}{P_I}\right\rfloor + 1$$

and thus

(1) $$\left|\mathsf{P}(A_I) - \frac{1}{P_I}\right| \le \frac{1}{x}.$$

Thus if we set $p_i = \frac{1}{P_i}$, $\varepsilon = \frac{1}{x}$, the conditions of 2.16 are satisfied and hence,

$$\mathsf{P}(\overline{A}_1\ldots\overline{A}_n) \le \frac{1}{S} + \frac{\varepsilon M^2}{(1-p_1)^2\ldots(1-p_n)^2}.$$

Now here

$$S = \sum \frac{1}{p_1^{l_1}\ldots p_n^{l_n}},$$

where the summation extends over all choices of $l_1,\ldots,l_n \ge 0$ with $P_1^{l_1}\ldots P_n^{l_n} \le M$. This sum is exactly

$$S = \sum_{\substack{m\le M \\ (m,k)=1}} \frac{1}{m}.$$

To get a lower bound on this, observe that

$$\left(\sum_{\substack{m\le M \\ (m,k)=1}} \frac{1}{m}\right) \prod_{P|k}\left(1 + \frac{1}{P} + \frac{1}{P^2} + \ldots\right) \ge \sum_{m\le M} \frac{1}{m} > \log M,$$

whence

$$S \ge \log M \frac{1}{\displaystyle\prod_{P|k}\left(1 + \frac{1}{P} + \ldots\right)} = \log M \prod_{P|k}\left(1 - \frac{1}{P}\right) = \log M \frac{\varphi(k)}{k}.$$

On the other hand, it is known that

$$(1-p_1)^2\ldots(1-p_n)^2 \ge \prod_{P\le M}\left(1 - \frac{1}{P}\right)^2 \ge \frac{c}{\log^2 M}$$

and thus

$$P(\overline{A}_1 \ldots \overline{A}_n) \le \frac{k}{\varphi(k) \cdot \log M} + \frac{M^2 \log^2 M}{cx} \le \frac{k}{\varphi(k)} \frac{2,5}{\log x}$$

if x is large enough. Thus the number of primes in the sequence $l, k+l, \ldots,$ $(x-1)k+l$ is

$$\le n + xP(\overline{A}_1 \ldots \overline{A}_n) \le \sqrt{x} + 2,5 \frac{k}{\varphi(k)} \frac{x}{\log x} \le 3 \frac{k}{\varphi(k)} \frac{x}{\log x},$$

if x is large enough.

18. We prove that

(1)
$$P(A_1 | \overline{A}_2 \ldots \overline{A}_n) \le \frac{1}{2d}$$

by induction on n. This, in fact, will imply that

$$P(\overline{A}_1 \ldots \overline{A}_n) = P(\overline{A}_2 \ldots \overline{A}_n) P(\overline{A}_1 | \overline{A}_2 \ldots \overline{A}_n) \ge P(\overline{A}_2 \ldots \overline{A}_n) \left(1 - \frac{1}{2d}\right) > 0$$

because by the induction hypothesis, $P(\overline{A}_2 \ldots \overline{A}_n) > 0$ (this guarantees that the conditional probability $P(A_1 | \overline{A}_2 \ldots \overline{A}_n)$ is meaningful).

Let, say $2, \ldots, h$ be the neighbors of 1. We have

$$P(A_1 | \overline{A}_2 \ldots \overline{A}_n) = \frac{P(A_1 \overline{A}_2 \ldots \overline{A}_h | \overline{A}_{h+1} \ldots \overline{A}_n)}{P(\overline{A}_2 \ldots \overline{A}_h | \overline{A}_{h+1} \ldots \overline{A}_n)}.$$

Here the numerator is

$$P(A_1 \overline{A}_2 \ldots \overline{A}_h | \overline{A}_{h+1} \ldots \overline{A}_n) \le P(A_1 | \overline{A}_{h+1} \ldots \overline{A}_n) = P(A_1) \le \frac{1}{4d}$$

(as A_1 is independent of $\overline{A}_{h+1}, \ldots \overline{A}_n$) while for the denominator we have

$$P(\overline{A}_2 \ldots \overline{A}_h | \overline{A}_{h+1} \ldots \overline{A}_n) = 1 - P(A_2 + \ldots + A_h | \overline{A}_{h+1} \ldots \overline{A}_n) \ge$$

$$\ge 1 - \sum_{i=2}^{h} P(A_i | \overline{A}_{h+1} \ldots \overline{A}_n).$$

Now applying our induction hypothesis with the graph induced by $\{i, h+1, \ldots, n\}$ we get

$$P(A_i | \overline{A}_{h+1} \ldots \overline{A}_n) \le \frac{1}{2d}$$

and so

$$P(\overline{A}_2 \ldots \overline{A}_h | \overline{A}_{h+1} \ldots \overline{A}_n) \geq 1 - \frac{h-1}{2d} \geq \frac{1}{2}$$

as $h - 1$ is the degree of 1 and so it is at most d. Thus

$$P(A_1 | \overline{A}_2 \ldots \overline{A}_n) \leq \frac{\frac{1}{4d}}{\frac{1}{2}} = \frac{1}{2d}.$$

[P. Erdős–L. Lovász, in: *Infinite and Finite Sets*, Coll. Math. Soc. J. Bolyai **10** (1974) Bolyai–North Holland, 609–627.]

19. We have

$$P(\overline{A}_1 \ldots \overline{A}_n) = P(\zeta = 0) \leq P((\xi - \sigma_1)^2 \geq \sigma_1^2) \leq$$

$$\leq \frac{1}{\sigma_1^2} E((\zeta - \sigma_1)^2) = \frac{1}{\sigma_1^2}(E(\zeta^2) - \sigma_1^2).$$

Now let

$$\zeta_i = \begin{cases} 1 & \text{if } A_i \text{ occurs,} \\ 0 & \text{otherwise,} \end{cases}$$

then

$$\zeta = \sum_{i=1}^{n} \zeta_i$$

and thus,

$$E(\zeta^2) = E\left(\sum_{i=1}^{n} \zeta_i^2 + 2 \sum_{1 \leq i < j \leq n} \zeta_i \zeta_j \right) = \sigma_1 + 2\sigma_2,$$

whence

$$P(\overline{A}_1 \ldots \overline{A}_n) \leq \frac{1}{\sigma_1^2}(\sigma_1 + 2\sigma_2 - \sigma_1^2).$$

The first step is actually Chebyshev's inequality;

$$P(\zeta = 0) \leq P\left((|\zeta - E(\zeta)| \geq \frac{E(\zeta)}{D(\zeta)} D(\zeta) \right) \leq \frac{D^2(\zeta)}{E^2(\zeta)} = \frac{E((\zeta - \sigma_1)^2)}{\sigma_1^2}.$$

20. I. By 2.9,

$$P(\overline{A}_1 \ldots \overline{A}_n) \leq 1 - np + \binom{n}{2} p^2 = \alpha.$$

II. By Selberg's method,

$$P(\overline{A}_1 \ldots \overline{A}_n) \leq \sum_{|I|,|J| \leq 1} \lambda_I \lambda_J p^{|I \cup J|},$$

where $\lambda_\emptyset = 1$, $\lambda_{\{i\}}$ arbitrary. By 2.15, the λ_I's can be chosen so that the right-hand side becomes

$$\frac{1}{\sum\limits_{|I| \leq 1} \frac{p^{|I|}}{(1-p)^{|I|}}} = \frac{1}{1 + \frac{np}{1-p}} = \beta.$$

III. The preceding formula gives

$$\mathsf{P}(\overline{A}_1 \ldots \overline{A}_n) \leq \frac{1-p}{np} = \gamma$$

by direct substitution.

Obviously, $\gamma \geq \beta$, thus II is better than III. However, I and II are incomparable. For let p be fixed and $n \to \infty$, then $\beta \to 0$ but $\alpha \to \infty$, thus II even III is better if n is large. On the other hand, fix n and let $p \to 0$. Then the power series of β is

$$1 - np + n(n-1)p^2 + \ldots .$$

which is larger than α if p is very small.

21. It is trivial that sum of compatible matrices is compatible. Let $A = (a_{ij})$, $B = (b_{ij})$ be compatible matrices, $AB = C = (c_{ij})$ and suppose that $c_{ij} \neq 0$. Since

$$c_{ij} = \sum_k a_{ik} b_{kj},$$

there exists a k such that $a_{ik} \neq 0$ and $b_{kj} \neq 0$. Hence $x_i \leq x_k$ and $x_k \leq x_j$, and so, by transitivity, $x_i \leq x_j$. Thus C is compatible.

Let A be an invertible compatible matrix. We may assume without loss of generality that $x_i \leq x_j \Rightarrow i \leq j$ (this can be achieved by indexing appropriately). Then, it is seen that all the non-zero entries of A are above the diagonal, so $\det A = \prod\limits_{i=1}^n a_{ii}$. So $a_{ii} \neq 0$. Now let $B = (b_{ij})$ be the inverse of A, and suppose indirectly that it is not compatible, i.e. there are $x_i \not\leq x_j$ with $b_{ij} \neq 0$. Choose i here to be maximal. We have

$$\sum_{k=1}^n a_{ik} b_{kj} = 0,$$

but

$$a_{ii} b_{ij} \neq 0,$$

so there must be a $k \neq i$ with

$$a_{ik} b_{kj} \neq 0.$$

Since (a_{ij}) is compatible, this implies $x_i \leq x_k$ and so, $i < k$. By the choice of i, $b_{kj} \neq 0$ implies that $x_k \leq x_j$ and so, $x_i \leq x_j$, a contradiction.

[Most problems in the rest of this section are based on G. C. Rota's paper in: *Zeitschr. f. Wahrscheinlichkeitstheorie*, **2** (1964) 340–368.]

22. Define

$$z_{ij} = \begin{cases} 1 & \text{if } x_i \leq x_j, \\ 0 & \text{otherwise}, \end{cases}$$

$Z = (z_{ij})$. Then the last two requirements on μ can be re-written as

$$MZ = I,$$

where $M = (m_{ij})$ with $m_{ij} = \mu(x_i, x_j)$ and I is the identity matrix. Since $z_{ii} \neq 0$, Z is invertible (see the preceding solution) and so, by the preceding result, $M = Z^{-1}$ is a compatible matrix. This defines the desired function $\mu(x, y)$.

23. Note that if we know $\mu(a, y)$ for every y such that $a \leq y < b$, then the equation

$$\sum_{a \leq y \leq b} \mu(a, y) = 0$$

uniquely determines $\mu(a, b)$. Therefore, we can calculate $\mu(a, x)$ for every $a \leq x \leq b$ inductively using only elements between a and b. Hence, if $a \leq b \in V$, $a' \leq b' \in V'$ and the partially ordered sets $\{z \in V : a \leq z \leq b\}$, $\{z' \in V' : a' \leq z' \leq b'\}$ are isomorphic as partially ordered sets, then

$$\mu(a, b) = \mu(a', b').$$

(a) We claim that $\mu(X, Y) = (-1)^{|Y - X|}$ $(X \subseteq Y \subseteq S)$. In fact,

$$\sum_{A \subseteq Y \subseteq B} (-1)^{|Y - A|} = \sum_{k=0}^{|B - A|} \binom{|B - A|}{k} (-1)^k = \begin{cases} 1 & \text{if } B = A, \\ 0 & \text{if } B \supset A. \end{cases}$$

(b) Here, any interval $a \leq z \leq b$ is a chain. So it suffices to calculate $\mu(a_1, a_n)$ for a chain

$$a_1 < a_2 < \ldots < a_n.$$

We have by definition,

$$\mu(a_1, a_1) = 1.$$

Also,

$$\mu(a_1, a_1) + \mu(a_1, a_2) = 0,$$

whence

$$\mu(a_1, a_2) = -1.$$

Now, for $i \geq 3$,

$$\mu(a_1, a_1) + \mu(a_1, a_2) + \ldots + \mu(a_1, a_i) = 0,$$

whence, obviously,

$$\mu(a_1, a_3) = \ldots = \mu(a_1, a_n) = 0.$$

Thus

$$\mu(x, y) = \begin{cases} 1 & \text{if } x = y, \\ -1 & \text{if } (x, y) \text{ is an edge,} \\ 0 & \text{otherwise.} \end{cases}$$

(c) We claim that

(1) $$\mu(x, y) = \mu\left(\frac{y}{x}\right),$$

where $\mu(k)$ is the (number theoretical) Möbius function: if $k = p_1^{\alpha_1} \ldots p_r^{\alpha_r}$ ($p_1, \ldots, p_r$ distinct primes), then

$$\mu(k) = \begin{cases} (-1)^r & \text{if } \alpha_1 = \ldots = \alpha_r = 1, \\ 0 & \text{otherwise.} \end{cases}$$

Obviously, $\mu(x, y)$ defined by (1) satisfies $\mu(a, a) = 1$. Also, if $a \neq b$, $a | b$, then

$$\sum_{a | y | b} \mu(a, y) = \sum_{a | y | b} \mu\left(\frac{y}{a}\right) = \sum_{d | \frac{b}{a}} \mu(d).$$

Set

$$\frac{b}{a} = k = p_1^{\alpha_1} \ldots p_r^{\alpha_r},$$

then

$$\sum_{d | k} \mu(d) = \sum_{1 \leq i_1 < \ldots < i_\nu \leq r} \mu(p_{i_1} \ldots p_{i_\nu}) = \sum_{\nu=0}^{r} \binom{r}{\nu}(-1)^\nu = 0.$$

24. We have to show

$$\mu^*(a, a) = 1,$$

$$\sum_{a \leq^* x \leq^* b} \mu^*(a, x) = 0 \qquad (a \neq b),$$

i.e.

$$\mu(a, a) = 1,$$

$$\sum_{b \leq x \leq a} \mu(x, a) = 0.$$

The first assertion is trivial. To show the second one, note that, in terms of M and Z, it can be expressed as

$$Z \cdot M = I,$$

which is obvious, because $M = Z^{-1}$.

25. Let $Z - I = (u_{ij})$, then

$$u_{ij} = \begin{cases} 1 & \text{if } x_i < x_j, \\ 0 & \text{otherwise.} \end{cases}$$

Then the $(i,j)^{\text{th}}$ entry of $(Z - I)^n$ is

$$\sum_{k_1,\ldots,k_{n-1}} u_{ik_1} u_{k_1 k_2} \cdots u_{k_{n-1} j} = 0,$$

because a non-zero term would correspond to a chain $x_i < x_{k_1} < x_{k_2} < \ldots < x_{k_{n-1}} < x_j$, which clearly does not exist. Hence

$$(I - Z)^n = 0$$

or, equivalently,

$$\sum_{i=1}^{n} \binom{n}{i} (-1)^{i-1} Z^i = I.$$

Multiplying by M, we get

$$\sum_{i=1}^{n} \binom{n}{i} (-1)^{i-1} Z^{i-1} = M.$$

Note that instead of n, we could consider the maximum length of chains in V.

26. $\sum\limits_{z \leq x} g(z)\mu(z,x) = \sum\limits_{z \leq x} \sum\limits_{y \leq z} f(y)\mu(z,x) = \sum\limits_{y \leq x} f(y) \sum\limits_{y \leq z \leq x} \mu(z,x) = f(x).$
(Note: setting $f = (f(x_1),\ldots,f(x_n))$, $g = (g(x_1),\ldots,g(x_n))$, the definition of g means

$$g = f \cdot Z$$

and the assertion is

$$f = gM = gZ^{-1}.)$$

To get the sieve formula let V consist of all subsets of $\{1,\ldots,n\}$, and let $\leq$ mean inclusion. By 2.23, $\mu(K,L) = (-1)^{|L-K|}$ $(K \subseteq L)$. Set

$$f(K) = \mathsf{P}\left(\prod_{i \notin K} A_i \prod_{j \in K} \overline{A}_j \right),$$

then

$$g(L) = \sum_{K \subseteq L} f(K) = \mathsf{P}\left(\prod_{i \notin L} A_i \right) = \mathsf{P}(A_{\overline{L}})$$

$(\overline{K} = \{1,\ldots,n\} - K)$, and so,

$$P\left(\prod_{i=1}^{n} \overline{A}_i\right) = f(\{1, \ldots, n\}) = \sum_{K \subseteq \{1,\ldots,n\}} (-1)^{n-|K|} P(A_{\overline{K}}) =$$

$$= \sum_{K \subseteq \{1,\ldots,n\}} (-1)^{|K|} P(A_K).$$

27. (a) We proceed by induction on the number of elements between a and b. If $a = b$, then the sum looks like

$$\sum_{x \leq b} \mu(0, x),$$

which is 0 by the definition of μ. Let $b > a$. Then

$$\sum_{x \leq b} \mu(0, x) = \sum_{a \leq b_1 \leq b} \left(\sum_{a \vee x = b_1} \mu(0, x)\right).$$

The left-hand side is 0 by definition of μ. All terms on the right-hand side with $b_1 < b$ are 0's by the induction hypothesis. Thus, the last one is 0 as well.

(b) follows immediately from (a) and 2.24 by inverting the order.

28. We use induction on the number of elements below x. If $x = 0$ the assertion is true (0 is the union of the empty set of atoms). Let $x > 0$. If $a \leq x$ is any atom, then, by the preceding result,

$$\sum_{y \vee a = x} \mu(0, y) = 0.$$

Here, obviously, no y is a union of atoms. So each term with $y < x$ is 0 by the induction hypothesis. It follows that so is $\mu(0, x)$.

29. We have

$$\sum_{y \geq a} \mu(a, y) = \begin{cases} 1 & \text{if } a = 1, \\ 0 & \text{otherwise}, \end{cases}$$

so

$$\sum_{x,y} \mu(x, y) = 1.$$

On the other hand,

(1) $$\sum_{x,y} = \sum_{x \in A} \sum_{y \in C} + \sum_{x,y \in A \cup B} + \sum_{x,y \in B \cup C} - \sum_{x,y \in B}.$$

Here

$$\sum_{x,y \in A \cup B} \mu(x, y) = \sum_{y \in A \cup B} \sum_{x \leq y} \mu(x, y) = 1,$$

and similarly,

$$\sum_{x,y \in B \cup C} \mu(x,y) = \sum_{x \in B \cup C} \sum_{y \geq x} \mu(x,y) = 1.$$

Hence (1) has the form

$$1 = \sum_{x \in A} \sum_{y \in C} \mu(x,y) + 1 + 1 - \sum_{x,y \in B} \mu(x,y),$$

which proves the assertion.

30. (a) Let $b = \{\{1\},\{2,\ldots,n\}\}$ and apply 2.27:

$$\sum_{x \wedge b = 0} \mu(x,1) = 0$$

or, equivalently,

(1) $$\mu(0,1) = - \sum_{\substack{x \wedge b = 0 \\ x \neq 0}} \mu(x,1).$$

Determine the partitions x with $x \wedge b = 0$, $x \neq 0$. Let $x = \{A_1,\ldots,A_p\}$ be such a partition. Let $1 \in A_1$, say. Then for $i > 1$, $A_i = A_i \cap \{2,\ldots,n\}$ is a class of $x \wedge b = 0$, hence $|A_i| = 1$. As $x \neq 0$, $|A_1| \geq 2$. But $A_1 \cap \{2,\ldots,n\}$ is also a class of $x \wedge b = 0$, so $|A_1| = 2$. So x is of the form

$$x = \{\{1,i\},\{2\},\ldots,\{i-1\},\{i+1\},\ldots,\{n\}\}.$$

The interval $\{z : x \leq z \leq 1\}$ consists of all partitions which do not separate 1 and i. This interval is, obviously, isomorphic to the lattice of all partitions of $n-1$ elements (1 and i "stick together"). Hence

$$\mu(x,1) = (-1)^{n-2}(n-2)!$$

by the induction hypothesis. Since there are $n-1$ partitions x to consider, (1) implies

$$\mu(0,1) = -(n-1)(-1)^{n-2}(n-2)! = (-1)^{n-1}(n-1)! .$$

[M. P. Schützenberger].

(b) For every face x, the interval $[0,x]$ in V is also the lattice of faces of a convex polytope (namely x). Hence we may assume, by induction, that

$$\mu(0,x) = (-1)^{\dim(x)+1}$$

for every face x except 1. Thus by the definition of μ,

$$\mu(0,1) = - \sum_{x \neq 1} \mu(0,x) = \sum_{x \neq 1} (-1)^{\dim(x)} = \sum_{i=-1}^{d-1} f_i(-1)^i.$$

By Euler's Formula, this is just $(-1)^{d+1}$.

31. If F is defined as in the hint, we have

$$Z^T F Z = G.$$

In fact,

$$\sum_{k=1}^{n}\sum_{l=1}^{n} z_{ki} f_{kl} z_{lj} = \sum_{k=1}^{n} z_{ki} f_{kk} z_{kj} = \sum_{\substack{k \\ x_k \leq x_i \\ x_k \leq x_j}} f(x_k) =$$

$$= \sum_{\substack{k \\ x_k \leq x_i \wedge x_j}} f(x_k) = g(x_i \wedge x_j) = g_{ij}.$$

Hence

$$\det G = \det(Z^T F Z) = (\det Z^T)(\det F)(\det Z) = \det(F) = f(x_1)\ldots f(x_k).$$

If V is only a partially ordered set, then we define F as before and consider

$$G = Z^T F Z = (g_{ij}).$$

We have, as before,

$$g_{ij} = \sum_{k=1}^{n}\sum_{l=1}^{n} z_{ki} f_{kl} z_{ij} = \sum_{\substack{x_k \leq x_i \\ x_k \leq x_j}} f(x_k).$$

We cannot rewrite this in terms of $x_i \wedge x_j$ (as this is not defined) so we have to state the formula as follows: let $f(x)$ be any function on V and set

$$g_{ij} = \sum_{\substack{x \leq x_i \\ x \leq x_j}} f(x).$$

Then

$$\det(g_{ij}) = f(x_1)\ldots f(x_n).$$

Note, that if any two elements x_i, x_j of V have a greatest lower bound $x_i \wedge x_j$, then

$$g_{ij} = \sum_{x \leq x_i \wedge x_j} f(x) = g(x_i \wedge x_j),$$

where

$$g(y) = \sum_{x \leq y} f(x)$$

[H. S. Wilf, *Bull. Amer. Math. Soc.* **74** (1968) 960–964].

32. I. Let $V = \{1,\ldots,n\}$, let the partial order be defined divisibility and $g(x) = x$. We try to apply the preceding formula (V is not a lattice but it is closed under the operation $x \wedge y = (x, y)$ and so, the same formula holds by the remark at the end of the previous solution). To get a suitable $f(x)$ we use 2.26:

$$f(x) = \sum_{y \le x} \mu(y, x) g(y) = \sum_{y | x} y\mu \left(\frac{x}{y} \right) = \varphi(x)$$

by the solution of 2.3. So the value of the determinant is $\varphi(1) \cdot \varphi(2) \ldots \varphi(n)$.

II. This determinant is just $\varphi(1) \ldots \varphi(n) \times \{$the upper left entry in $G^{-1}\}$ where, as in the preceding solution

$$G = ((i,j))_{i,j=1}^{n} = Z^T F Z,$$

$$F = (f_{ij}) = \begin{pmatrix} \varphi(1) & & 0 \\ & \ddots & \\ 0 & & \varphi(n) \end{pmatrix},$$

$$Z = (z_{ij}), \quad z_{ij} = \begin{cases} 1 & \text{if } i | j, \\ 0 & \text{otherwise.} \end{cases}$$

Thus

$$G^{-1} = (Z^T F Z)^{-1} = M F^{-1} M^T,$$

where $M = (m_{ij})$,

$$m_{ij} = \begin{cases} \mu \left(\frac{j}{i} \right) & \text{if } i | j, \\ 0 & \text{otherwise.} \end{cases}$$

Thus the upper left corner entry of G^{-1} is

$$\sum_{k=1}^{n} \sum_{l=1}^{n} \mu(1,k)(F^{-1})_{kl} \mu(1,l) = \sum_{k=1}^{n} \frac{\mu^2(k)}{\varphi(k)}.$$

Hence

$$\begin{vmatrix} (2,2) & \cdots & (2,n) \\ \vdots & & \\ (n,2) & \cdots & (n,n) \end{vmatrix} = \varphi(1) \ldots \varphi(n) \sum_{k=1}^{n} \frac{\mu^2(k)}{\varphi(k)}.$$

33. Let

$$A = \begin{pmatrix} 0 & 1 & \cdot & \cdot & \cdot & 1 \\ 1 & -2 & & & & \\ \cdot & & \cdot & & & 0 \\ \cdot & & & \cdot & & \\ \cdot & & 0 & & \cdot & \\ 1 & & & & & -2 \end{pmatrix}.$$

Consider x_1 as a root, then T can be considered as an arborescence and this defines an ordering of $V = V(T)$: $x_i \le x_j$, if the (unique) (x_1, x_j)-path contains x_i.

Then let Z be as in 2.22; we claim that

$$Z^T A Z = D.$$

In fact, the $(i,j)^{\text{th}}$ entry of $Z^T A Z$ is

$$d'_{ij} = \sum_{k=1}^{n}\sum_{l=1}^{n} z_{ki} a_{kl} z_{lj} = \sum_{x_k \leq x_i}\sum_{x_l \leq x_j} a_{kl}.$$

Now a_{kl} is non-zero only if $k=l$ or $k=1$ or $l=1$ so we have

$$d'_{ij} = \sum_{x_k \leq x_i, x_j}(-2) + \sum_{x_k \leq x_i} 1 + \sum_{x_l \leq x_j} 1 =$$

$$= -2[d(x_i \wedge x_j, x_1)+1] + [d(x_i, x_1)+1] + [d(x_j, x_1)+1],$$

where $x_i \wedge x_j$ is the last common point of the (x_1, x_i) and (x_1, x_j)-paths.

Hence

$$d'_{ij} = d(x_i, x_1) + d(x_j, x_1) - 2d(x_i \wedge x_j, x_1) = d(x_i, x_j)$$

as stated. Thus

$$\det D = \det A = -(n-1)(-2)^{n-2}$$

[R. L. Graham–H. O. Pollak, *Bell Sys. Tech. J.* **50** (1971) 2495–2519].

34. Set $U = Z - I = (u_{ij})$. Then $U^n = 0$. Further we have

$$(I+U)(I-U+U^2-U^3+\ldots+(-1)^{n-1}U^{n-1}) = I + (-1)^{n-1}U^n = I.$$

So

$$I - U + U^2 - U^3 + \ldots = (I+U)^{-1} = Z^{-1} = M.$$

Hence

$$\mu(x,y) = \sum_{v=1}^{n}(-1)^v p'_v,$$

where p'_v is the $(x,y)^{\text{th}}$ entry in U^v; but this entry is

$$p'_v = \sum_{i,\ldots,i_v} u_{0i_1} u_{i_1 i_2} \ldots u_{i_{v-1} 1} = \sum_{x < x_{i_1} < \ldots < x_{i_{v-1}} < y} 1 = p_v.$$

35. Set

$$q(x) = q_0(x) - q_1(x) = q_2(x) - q_3(x) + \ldots$$

Then

$$\sum_{x \leq y} q(x) = \sum_{k=0}^{n}(-1)^k \sum_{x \leq y} q_k(x).$$

Now let $a_1, \dots, a_m$ be all the atoms $\leq y$. Then $\sum_{x \leq y} q_k(x)$ counts all k-tuples from $\{a_1, \dots, a_m\}$. Therefore,

$$\sum_{x \leq y} q_k(x) = \binom{m}{k}$$

and so

$$\sum_{x \leq y} q(x) = \sum_{k=0}^{n} (-1)^k \binom{m}{k} = \begin{cases} 1 & \text{if } m = 0, \text{ i.e., } y = 0, \\ 0 & \text{if } m > 0, \text{ i.e., } y > 0. \end{cases}$$

By the Möbius inversion formula 2.26, we have

$$q(x) = \sum_{y \leq x} \mu(y, x) \sum_{z \leq y} q(z) = \mu(0, x).$$

36. There are three types of elements in L:

(a) elements of C,

(b) elements such that there is a $y \in C$, with $x < y$,

(c) elements such that there is a $y \in C$, with $x > y$.

(a), (b) and (c) partition the elements; in fact, there cannot be any x with two of these properties since this would give two comparable elements of C. Also, if an x did not have any of the properties (a), (b), (c), then no maximal chain containing x would meet C.

Let us write $x < C$ and $x > C$ if (b) and (c) holds, respectively. Consider the number $q_k(a, b)$ of k-tuples in C whose union is b and meet is a, and let $a < C < b$. Then we prove that

$$q(a, b) = \sum_{k=2}^{n} (-1)^k q_k(a, b) = \mu(a, b)$$

by induction on the number of elements between a and b. Let $a_1, \dots, a_m$ be those elements of C between a and b. We may assume that $a = 0$ and $b = 1$, because $\{a_1, \dots, a_m\}$ forms a set in the sublattice $\{z : a \leq z \leq b\}$ with the same properties as C. We have

$$\sum_{x < C < y} q_k(x, y) = \binom{m}{k}$$

and so

(1) $$\sum_{x < C < y} q(x, y) = \binom{m}{2} - \binom{m}{3} + \dots = m - 1.$$

By 2.29, μ satisfies the same identity:

(2) $$\sum_{x<C<y} \mu(x,y) = \sum_{x,y\in C} \mu(x,y) - 1 = \sum_{x\in C} \mu(x,x) - 1 = m - 1.$$

Now in (1) and (2) $q(x,y)=\mu(x,y)$ if $x\neq 0$ or $y\neq 1$ by the induction hypothesis. So we must have

$$q(0,1) = \mu(0,1).$$

37. Let a be any atom, then

$$\sum_{a\vee x=1} \mu(0,x) = 0$$

by 2.27. Since a covers 0, either $x=1$ or $a\vee x=1$ covers x, so x is a coatom.

If we use induction on r we may assume that $\mu(0,x)$ has sign $(-1)^{r-1}$ (the rank of the interval $0\leq z\leq x$ is $r-1$, since any maximal $(0,x)$-chain extended by 1, forms a maximal $(0,1)$-chain). So

$$\mu(0,1) = -\sum_{\substack{x\neq 1 \\ x\vee a=1}} \mu(0,x)$$

implies the assertion.

§ 3. Permutations

1. If π, ϱ are permutations of $\{1,\ldots,n\}$, then

$$\pi' = \varrho^{-1}\pi\varrho$$

is the permutation arising as follows: we draw the "digraph" of π, i.e. we connect i to $\pi(i)$ for $i=1,\ldots,n$; then consider the image of this graph by ϱ as the graph of a permutation π'. Hence it follows that two permutations are conjugate iff their graphs are isomorphic. These graphs consist of disjoint cycles (corresponding to the cycle decomposition of the permutation). Therefore, the graphs of two permutations are isomorphic if and only if the cardinalities of their cycles are the same.

Hence, a conjugacy class can be described by the cardinalities of cycles. These cardinalities form a partition of the number n and conversely, each partition describes a conjugacy class. Hence, the number of conjugacy classes of S_n is π_n, the number of partitions of the number n (see problem 1.20).

2. (a) By inclusion-exclusion, the number in question is

$$\sum_{X \subseteq \{1,\dots,n\}} (-1)^{|X|} S_X,$$

where S_X denotes the number of those partitions which fix all points in X. This number is, obviously $(n-|X|)!$ So we have

$$\sum_{X \subseteq \{1,\dots,n\}} (-1)^{|X|}(n-|X|)! = \sum_{k=0}^{n}(-1)^k \binom{n}{k}(n-k)! = n! \sum_{k=0}^{n} \frac{(-1)^k}{k!} \approx \frac{n!}{e},$$

where the difference at the $\approx$ sign is less than $\frac{1}{n+1}$.

(b) If we want to construct a permutation π consisting of a single cycle, we can map 1 onto any of $2,\dots,n$. This gives $n-1$ possibilities. We can map $\pi(1)$ onto any of $n-2$ elements, etc, if $\pi(1), \pi^2(1), \dots, \pi^k(1)$ are specified, we can choose any of the remaining $n-k-1$ elements for $\pi\left(\pi^k(1)\right)$. Thus the number of such permutations is $(n-1)(n-2)\dots 1 = (n-1)!$.

3. Let us count the permutations in which 1 is contained in a cycle of length k. There are $\binom{n-1}{k-1}$ possible ways to choose the elements of this cycle; there are $(k-1)!$ ways to order them in a cycle and $(n-k)!$ ways to permute the rest. Thus we get

$$\binom{n-1}{k-1}(k-1)!(n-k)! = (n-1)!$$

and the probability is

$$\frac{(n-1)!}{n!} = \frac{1}{n},$$

independently of k, a remarkable fact.

Second solution. Given a permutation π, let us write it as a product of cycles (we also write out one-element cycles i.e. fixed points) such that each cycle should end with its least element and these last elements of cycles should be in increasing order. Thus the first cycle ends with 1. The resulting string of numbers can be viewed as another permutation $\hat{\pi}$. It is easy to see that π can be recovered from $\hat{\pi}$ uniquely: the first cycle of π ends with 1; the second cycle ends with the last number not occurring in the first cycle, etc.

The length of the cycle of π containing 1 is determined by the position of 1 in $\hat{\pi}$. This clearly can be any one of the n positions with the same probability $1/n$.

4. We count those permutations in which 1 and 2 belong to distinct cycles. If the cycle containing 1 has length k we have $\binom{n-2}{k-1}$ possibilities to select its elements and $(k-1)!$ possibilities to order them into a cycle; we have $(n-k)!$ possibilities to order the rest. Hence the number of such permutations is

$$\sum_{k=1}^{n-1} \binom{n-2}{k-1}(k-1)!(n-k)! = (n-2)! \sum_{k=1}^{n-1}(n-k) =$$

$$= (n-2)!\frac{n(n-1)}{2} = \frac{n!}{2},$$

whence the probability in question is $1/2$.

Second solution. Consider the coding of permutations introduced in the second solution of 3.3. 1 and 2 belong to the same cycle of π iff 2 occurs in $\hat{\pi}$ before 1. Obviously this happens with probability $1/2$.

5. Let $1 \leq k \leq n$, and

$$\zeta_i = \begin{cases} 1 & \text{if } i \text{ is contained in a } k\text{-cycle}, \\ 0 & \text{otherwise}. \end{cases}$$

Then $\zeta_1 + \ldots + \zeta_n$ is the number of points which are contained in k-cycles and $\frac{1}{k}(\zeta_1 + \ldots + \zeta_n)$ is the number of k-cycles. Now

$$\mathsf{E}\left(\frac{1}{k}(\zeta_1 + \ldots + \zeta_n)\right) = \frac{1}{k}\sum_{i=1}^{n}\mathsf{E}(\zeta_i).$$

Here, by 3.3,

$$\mathsf{E}(\zeta_i) = \frac{1}{n}$$

and thus, the expected number of k-cycles is

$$\frac{1}{k}\sum_{i=1}^{n}\mathsf{E}(\zeta_i) = \frac{1}{k}.$$

The expected number of cycles is, therefore,

$$\frac{1}{1} + \frac{1}{2} + \ldots + \frac{1}{n} \sim \log n.$$

6. Let π denote the permutation defined by

$$\pi(j) = \text{the index of the box containing key } j.$$

Suppose that we have broken open boxes $1, \ldots, k$. Clearly we can open the rest iff each cycle of π contains one of $1, \ldots, k$.

Consider the permutation $\hat{\pi}$ introduced in the second solution of 3.3. Since the cycles of π are ordered in the definition of $\hat{\pi}$ such that their least elements are in increasing order, it suffices to require that the least element of the last cycle should be at most k; or, by the definition of $\hat{\pi}$, that $\hat{\pi}(n) \leq k$. The probability that this occurs is clearly k/n. [J. Bognár–J. Mogyoródi–A. Prékopa–A. Rényi–D. Szász,

Problem Book on Probability (in Hungarian), Tankönyvkiadó, Budapest, 1970, p. 56.]

7. We have

$$n!p_n(x_1,\ldots,x_n) = \sum_{\pi \in S_n} x_1^{k_1(\pi)} \ldots x_n^{k_n(\pi)},$$

where $k_i(\pi)$ denotes the number of i-cycles of π. Let us consider a point v of the underlying set and let C_π be the cycles of π which contains v; let it have length l_π. Also, let ϱ_π be the permutation of the elements not in C_π, induced by π. Thus

$$n!p_n(x_1,\ldots,x_n) = \sum_{\pi \in S_n} x_{l_\pi} x_1^{k_1(\varrho_\pi)} \ldots x_{n-l_\pi}^{k_{n-l_\pi}(\varrho_\pi)}.$$

The cycle C_π with length l_π can be chosen $(n-1)(n-2)\ldots(n-l_\pi+1)$ ways (see 3.3); with each of these cycles, we can pair any permutation ϱ_π of the remaining $n - l_\pi$ points. Thus

$$n!p_n(x_1,\ldots,x_n) = \sum_{l=1}^{n}(n-1)\ldots(n-l+1)x_l(n-l)!p_{n-l}(x_1,\ldots,x_{n-l}) =$$

$$= \sum_{l=1}^{n}(n-1)!x_l p_{n-l}(x_1,\ldots,x_{n-l}).$$

or, equivalently,

$$n \cdot p_n(x_1,\ldots,x_n) = \sum_{l=1}^{n} x_l p_{n-l}(x_1,\ldots,x_{n-l}).$$

Set

$$f(y) = \sum_{n=0}^{\infty} p_n(x_1,\ldots,x_n)y^n,$$

then

$$f'(y) = \sum_{n=0}^{\infty} n \cdot p_n(x_1,\ldots,x_n)y^{n-1} = \sum_{n=0}^{\infty}\sum_{l=1}^{n} x_l p_{n-l}(x_1,\ldots,x_{n-l})y^{n-1} =$$

$$= \sum_{l=1}^{\infty} x_l y^{l-1}\sum_{n=l}^{\infty} p_{n-l}(x_1,\ldots,x_{n-l})y^{n-l} = \left(\sum_{l=1}^{\infty} x_l y^{l-1}\right)f(y).$$

Hence

$$(\ln f(y))' = x_1 + x_2 y + \ldots + x_k y^{k-1} + \ldots,$$

$$\ln f(y) = C + x_1 y + x_2\frac{y^2}{2} + \ldots + x_k\frac{y^k}{k} + \ldots,$$

and substitution of $y = 0$ yields $C = 0$. Thus

$$f(y) = \exp\left(x_1 y + x_2 \frac{y^2}{2} + \ldots + x_k \frac{y^k}{k} + \ldots\right)$$

as stated. [G. Pólya, *Acta Math.* **68** (1937) 145–254; this reference is relevant for all problems on cycle index and the "Pólya–Redfield method"; also see N. G. de Bruijn in: *Applied Combinatorial Mathematics*, (E. E. Beckenbach, ed.) Wiley, 1964.]

8. (a) The rotation by m decomposes into (n,m) cycles of length $n/(n,m)$. Hence the cycle index is

$$\frac{1}{n}\sum_{m=1}^{n} x_{n/(n,m)}^{(n,m)} = \frac{1}{n}\sum_{d|n} x_d^{n/d} = \frac{1}{n}\sum_{\substack{m \le n \\ (m,n)=\frac{n}{d}}} 1 = \frac{1}{n}\sum_{d|n}\varphi(d)x_d^{n/d}.$$

More generally, let G be any group and consider its regular representation as a permutation group, i.e. define, for $a \in G$ and $x \in G$,

$$a(x) = xa.$$

Then a is a permutation of the set G which decomposes into $|G|/k$ cycles of length k, where k is the order of a; because

$$xa^m = x \Leftrightarrow a^m = 1.$$

Hence the cycle index is

$$\frac{1}{|G|}\sum_{d|\,|G|} x_d^{|G|/d}\varphi(G,d),$$

where $\varphi(G,d)$ is the number of elements of order d in the group G.

(b) Let $P = \{C_1, C_2, \ldots, C_k\}$, $C_i = \{c_{i1}, \ldots, c_{in}\}$. A permutation which keeps P invariant can be described as follows. We take a permutation π of $\{1, \ldots, k\}$ and k permutations $\varrho_1, \ldots, \varrho_k$ of $\{1, \ldots, n\}$. We denote by $\langle \pi; \varrho_1, \ldots, \varrho_k \rangle$ the permutation defined by

$$\langle \pi; \varrho_1, \ldots, \varrho_k \rangle (c_{ij}) = c_{\pi(i), \varrho_i(j)}$$
$$(1 \le i \le k, \quad 1 \le j \le n).$$

Let us determine the cycles of $\langle \pi; \varrho_1, \ldots, \varrho_k \rangle$. If, say, $(1, \ldots, l)$ is a cycle of π, then the iterated images of c_{1j} by $\langle \pi; \varrho_1, \ldots, \varrho_k \rangle$ are in $C_1, \ldots, C_l, C_1, \ldots$. Thus, the first image which is equal to c_{1j} is the $(lm)^{\text{th}}$ image for some m. To determine this m we observe that the lth image of c_{1j} is $c_{1j'}$ with $j' = \varrho_l(\ldots \varrho_1(j)\ldots)$. Hence, m is the length of the cycle of $\varrho_1 \ldots \varrho_l$ containing j. Thus, $C_1 \cup \ldots \cup C_l$ contains

$k_i(\varrho_1 \ldots \varrho_l)$ cycles with length li. Similar assertions hold for the other cycles of π. The term corresponding to this permutation is

$$\prod x_l^{k_1(\varrho_{i_1} \cdots \varrho_{i_l})} x_{2l}^{k_2(\varrho_{i_1} \cdots \varrho_{i_l})} \ldots x_{nl}^{k_n(\varrho_{i_1} \cdots \varrho_{i_l})},$$

where the product extends over all cycles $(i_1 \ldots i_l)$ of π. Summing this over all choices of $\varrho_1, \ldots, \varrho_l$ we get (since for different cycles $\varrho_{i_1}, \ldots, \varrho_{i_l}$ can be chosen independently),

$$\prod \left(\sum_{\varrho} x_l^{k_1(\varrho_{i_1} \cdots \varrho_{i_l})} \ldots x_{nl}^{k_n(\varrho_{i_1} \cdots \varrho_{i_l})} \right),$$

where again, the product extends over all cycles $(i_1 \ldots i_l)$ of π. Since $\varrho_{i_1} \ldots \varrho_{i_l}$ represents each permutation $(n!)^{l-1}$ times,

$$\sum_{\varrho_{i_1}, \ldots, \varrho_{i_l}} x_l^{k_1(\varrho_{i_1} \cdots \varrho_{i_l})} \ldots x_{nl}^{k_n(\varrho_{i_1}, \ldots, \varrho_{i_l})} = (n!)^{l-1} \sum_{\pi \in S_n} x_l^{k_1(\pi)} \ldots x_{nl}^{k_n(\pi)} =$$

$$= (n!)^l p_n(x_l, \ldots, x_{nl}),$$

where $p_n(x_1, \ldots, x_n)$ is the cycle index of S_n. Thus

$$\prod_{(i_1, \ldots, i_l)} \left(\sum_{\varrho_{i_1}, \ldots, \varrho_{i_l}} x_l^{k_1(\varrho_{i_1}, \ldots, \varrho_{i_l})} \ldots x_{nl}^{k_n(\varrho_{i_1}, \ldots, \varrho_{i_l})} \right) = \prod_{(i_1, \ldots, i_l)} (n!)^l p_n(x_l, \ldots, x_{nl}) =$$

$$= \prod_{l=1}^{k} (n!)^{lk_l(\pi)} p_n(x_l, \ldots, x_{nl})^{k_l(\pi)} = (n!)^k \prod_{l=1}^{k} p_n(x_l, \ldots, x_{nl})^{k_l(\pi)}.$$

If we sum over π, and divide by $k!(n!)^k$, the number of permutations in the group, we get

$$p_k\left(p_n(x_1, \ldots, x_n), p_n(x_2, \ldots, x_{2n}), \ldots, p_n(x_k, \ldots, x_{kn}) \right).$$

In exactly the same way we prove, that if Γ, Γ_1 are permutation groups with cycle indices $F(x_1, \ldots, x_k)$, $G(x_1, \ldots, x_k)$ respectively, then the cycle index of their wreath product (the group of permutations $\langle \pi, \varrho_1, \ldots, \varrho_k \rangle$, where $\pi \in \Gamma$, $\varrho_i \in \Gamma_1$) is

$$F\left(G(x_1, \ldots, x_n), G(x_2, \ldots, x_{2n}), \ldots, G(x_k, \ldots, x_{kn}) \right).$$

[G. Pólya, *Acta Math.* **68** (1937) 145–254.]

9. By (a) of the previous problem, if we can find two groups which have the same number of elements of order k for $k = 1, 2, \ldots$, then the regular permutation group representations of these groups yield a counterexample.

We will find two non-isomorphic groups of order p^3 $(p > 2, \text{prime})$ in which each element $\neq 1$ has order p. One of these is, plausibly, $G_1 = Z_p \times Z_p \times Z_p$. Let G_2 consist of all matrices of form

$$\begin{pmatrix} 1 & a & b \\ 0 & 1 & c \\ 0 & 0 & 1 \end{pmatrix},$$

where $a, b, c \in GF(p)$. Since

(1)
$$\begin{pmatrix} 1 & a & b \\ 0 & 1 & c \\ 0 & 0 & 1 \end{pmatrix} \begin{pmatrix} 1 & x & y \\ 0 & 1 & z \\ 0 & 0 & 1 \end{pmatrix} = \begin{pmatrix} 1 & a+x & b+y+az \\ 0 & 1 & c+z \\ 0 & 0 & 1 \end{pmatrix},$$

these matrices form a group with p^3 elements. Moreover,

$$\begin{pmatrix} 1 & a & b \\ 0 & 1 & c \\ 0 & 0 & 1 \end{pmatrix}^k = \begin{pmatrix} 1 & ka & kb + \binom{k}{2}ac \\ 0 & 1 & kc \\ 0 & 0 & 1 \end{pmatrix},$$

which follows easily by induction on k. Thus

$$\begin{pmatrix} 1 & a & b \\ 0 & 1 & c \\ 0 & 0 & 1 \end{pmatrix}^p = \begin{pmatrix} 1 & pa & p\left(b + \frac{p-1}{2}ac\right) \\ 0 & 1 & pc \\ 0 & 0 & 1 \end{pmatrix} = \begin{pmatrix} 1 & 0 & 0 \\ 0 & 1 & 0 \\ 0 & 0 & 1 \end{pmatrix},$$

i.e. each element other than 1 has order p in this group G_2. Finally, $G_1 \neq G_2$ as G_2 is non-commutative: the matrices in (1) commute iff $az = cx$. [G. Pólya, *Acta Math.* **68** (1937) 145–254.]

10. We have

$$p_n(x_1, -x_2, \ldots, (-1)^n x_n) = \frac{1}{n!} \sum_{\pi \in S_n} (-1)^{k_2(\pi)+k_4(\pi)+\cdots} x_1^{k_1(\pi)} \ldots x_n^{k_n(\pi)} =$$

$$= \frac{1}{n!} \left(\sum_{\pi \in A_n} x_1^{k_1(\pi)} \ldots x_n^{k_n(\pi)} - \sum_{\pi \in S_n - A_n} x_1^{k_1(\pi)} \ldots x_n^{k_n(\pi)} \right).$$

Hence

$$p_n(x_1, x_2, \ldots, x_n) + p_n(x_1, -x_2, \ldots, (-1)^n x_n) =$$

$$= \frac{2}{n!} \sum_{\pi \in A_n} x_1^{k_1(\pi)} \ldots x_n^{k_n(\pi)} = q_n(x_1, \ldots, x_n).$$

Thus

$$\sum_{n=0}^{\infty} q_n(x_1, \ldots, x_n) y^n =$$

$$= \exp\left(x_1 y + x_2 \frac{y^2}{2} + \ldots + x_k \cdot \frac{y^k}{k} + \ldots\right) + \exp\left(x_1 y - x_2 \cdot \frac{y^2}{2} + \ldots + (-1)^k x_k \cdot \frac{y^k}{k} + \ldots\right).$$

11. Let $p_{-1} = 0$, then

$$\sum_{n=0}^{\infty} (p_n - p_{n-1}) y^n = (1 - y) \sum_{n=0}^{\infty} p_n y^n = (1 - y) \exp \left(\sum_{k=1}^{\infty} \frac{x_k}{k} y^k \right)$$

$$= \exp \left(\sum_{k=0}^{\infty} \frac{x_k}{k} y^k + \ln(1 - y) \right) = \exp \left(\sum_{k=0}^{\infty} \frac{x_k - 1}{k} y^k \right).$$

By our assumption on the x_i's, there is a polynomial in the exponent and hence this function is an entire function. Therefore, its Taylor series converges at $y = 1$:

$$\lim_{N \to \infty} \sum_{n=0}^{N} (p_n - p_{n-1}) = \lim_{N \to \infty} p_N = \exp \left(\sum_{k=0}^{\infty} \frac{x_k - 1}{k} \right).$$

12. (a) Obviously,

$$n! p_n(x, \ldots, x) = \sum_{\pi \in S_n} x^{k_1(\pi) + \ldots + k_n(\pi)} = \sum_{k=0}^{n} \begin{bmatrix} n \\ k \end{bmatrix} x^k.$$

Hence

$$\sum_{n=0}^{\infty} \frac{f_n(x)}{n!} y^n = e^{xy + x \frac{y^2}{2} + \ldots + x \frac{y^k}{k} + \ldots} = e^{x(-\ln(1-y))} = (1 - y)^{-x} =$$

$$= \sum_{n=0}^{\infty} \binom{-x}{n} (-1)^k y^n = \sum_{n=0}^{\infty} \binom{x + n - 1}{n} y^n.$$

Thus

$$f_n(x) = n! \binom{x + n - 1}{n} = x(x + 1) \ldots (x + n - 1).$$

(b) Since

$$f_n'(1) = \sum_{k=0}^{n} k \begin{bmatrix} n \\ k \end{bmatrix}$$

is the sum of numbers of cycles in permutations, the expected number of cycles in a random permutation is $\frac{1}{n!} f_n'(1)$. Evaluating this, we get

$$\frac{1}{n!} f_n'(x) = \left(\frac{x(x + 1) \ldots (x + n - 1)}{n!} \right)' =$$

$$= \frac{x(x + 1) \ldots (x + n - 1)}{n!} \left(\frac{1}{x} + \ldots + \frac{1}{x + n - 1} \right)$$

and thus

$$\frac{1}{n!}f_n'(1) = \frac{1}{1} + \ldots + \frac{1}{n} \sim \log n.$$

13. Orient each circuit of such a graph. Let π denote the permutation we get this way. Then $k_1(\pi) = k_2(\pi) = 0$. Moreover, we get from the same graph $2^{k_3(\pi)+\ldots+k_n(\pi)}$ permutations. Thus, if we form the sum

$$\sum \frac{1}{2^{k_3(\pi)+\ldots+k_n(\pi)}}$$

over the permutations associated with a given graph we get 1; and if we sum over all permutations such that $k_1(\pi) = k_2(\pi) = 0$ we get the number of 2-regular simple graphs. Thus

$$g_n = \sum_{\substack{\pi \\ k_1(\pi)=k_2(\pi)=0}} \frac{1}{2^{k_3(\pi)+\ldots+k_n(\pi)}} = n!p_n\left(0,0,\frac{1}{2},\ldots,\frac{1}{2}\right).$$

Hence

$$\sum_{n=0}^{\infty} \frac{g_n}{n!}y^n = \sum_{n=0}^{\infty} p_n\left(0,0,\frac{1}{2},\ldots,\frac{1}{2}\right)y^n = e^{\frac{1}{2}\left(\frac{y^3}{3}+\frac{y^4}{4}+\ldots\right)} =$$

$$= e^{\frac{1}{2}(-\ln(1-y))-\frac{1}{2}y-\frac{1}{4}y^2} = \frac{1}{\sqrt{1-y}e^{\frac{1}{2}y+\frac{1}{4}y^2}}.$$

14. (a) If we join i and j ($1 \le i, j \le n$) by an edge whenever (i,j) is a pair in our partition we get a 2-regular graph; this time double edges and/or loops may occur. Similarly as before, we get that our number is

$$n!p_n\left(1,1,\frac{1}{2},\ldots,\frac{1}{2}\right)$$

and hence, the exponential generating function is

$$\sum_{n=0}^{\infty} p_n\left(1,1,\frac{1}{2},\ldots,\frac{1}{2}\right)y^n = e^{-\frac{1}{2}\ln(1-y)+\frac{y}{2}+\frac{y^2}{4}} = \frac{e^{\frac{y}{2}+\frac{y^2}{4}}}{\sqrt{1-y}}.$$

(b) Now join i to j by a directed edge whenever (i,j) belongs to the partition. By changing the directions in such a way as to get cycles, we associate a permutation with the given partition. As before, we get

$$2^{k_3(\pi)+\ldots+k_n(\pi)}$$

permutations from a given partition; but we obtain

$$3^{k_2(\pi)}2^{3k_3(\pi)+4k_4(\pi)+\ldots+nk_n(\pi)}$$

digraphs (i.e. partitions) from a given permutation. Thus the answer is

$$\sum_{\pi \in S_n} \frac{3^{k_2(\pi)} 2^{3k_3(\pi)+\dots+nk_n(\pi)}}{2^{k_3(\pi)+\dots+k_n(\pi)}} = n! p_n(1,3,2^2,2^3,\dots)$$

and the exponential generating function is

$$p_n(1,3,2^2,2^3,\dots) y^n = e^{y+\frac{3}{2}\cdot y^2 + \frac{2^2 y^2}{3} + \dots + \frac{2^{k-1}y^k}{k}} = e^{\frac{y^2}{2}+\frac{1}{2}(-\log(1-2y))} =$$

$$= \frac{e^{\frac{y^2}{2}}}{\sqrt{1-2y}} \qquad \text{[G. Baróti].}$$

15. (a) Obviously, the number of permutations having no fixed point is

$$n! p_n(0,1,\dots,1).$$

The exponential generating function of these numbers is

$$\sum_{n=0}^{\infty} p_n(0,1,\dots,1) y^n = e^{\frac{y^2}{2}+\frac{y^3}{3}+\dots+\frac{y^k}{k}+\dots} = e^{-y-\ln(1-y)} =$$

$$= \frac{e^{-y}}{1-y} = (1+y+y^2+\dots)\left(1 - \frac{y}{1!} + \frac{y^2}{2!} - \dots\right),$$

whence

$$p_n(0,1,\dots,1) = \sum_{k=0}^{n} \frac{(-1)^k}{k!} \sim \frac{1}{e}.$$

(b) The number of permutations consisting of exactly one cycle is, by 3.10, the coefficient of x in the polynomial

$$x(x+1)\dots(x+n-1),$$

i.e. it is equal to $(n-1)!$.

16. Let $k_1,\dots,k_n$ be integers, $1 \le k_i \le i$. We claim that there exists a unique permutation π with $\overline{\pi}(k) = k_i$. Since $\overline{\pi}(n)$ denotes the number of integers $1 \le j \le n$ with $\pi(j) \ge \pi(n)$, it follows that $\pi(n) = n - k_n + 1$.

Suppose that we have determined what the values of $\pi(n)$, $\pi(n-1)$, ..., $\pi(n-k+1)$ are. Then $\pi(n-k)$ must be the $(k_{n-k})^{\text{th}}$ number in the sequence obtained from $(n, n-1, \dots, 1)$ by deleting $\pi(n),\dots,\pi(n-k+1)$. It is immediately seen that in this way $\pi(n),\dots,\pi(1)$ are uniquely determined by $k_n,\dots,k_1$ and they form a permutation of $\{1,\dots,n\}$. Moreover, this permutation π will satisfy $\pi(i) = k_i$ $(i = 1,\dots,n)$ by the construction.

Let us determine

$$n! P\left(\overline{\pi}(i_1) = k_{i_1}, \dots, \overline{\pi}(i_r) = k_{i_r}\right),$$

where $1 \leq k_i \leq i$ are given integers. The values $\overline{\pi}(i)$, $i \neq i_1, \ldots, i_r$ can be chosen arbitrarily; hence

$$n! P\left(\overline{\pi}(i_1) = k_{i_1}, \ldots, \overline{\pi}(i_r) = k_{i_r}\right) = \prod_{\substack{i \neq i_1, \ldots, i_r \\ 1 \leq i \leq n}} i$$

and thus

$$P\left(\overline{\pi}(i_1) = k_{i_1}, \ldots, \overline{\pi}(i_r) = k_{i_r}\right) = \frac{1}{i_1 \ldots i_r}.$$

Thus

$$P\left(\overline{\pi}(i_1) = k_{i_1}, \ldots, \overline{\pi}(i_r) = k_{i_r}\right) = \prod_{\nu=i}^{r} P\left(\overline{\pi}(i_\nu) = k_{i_\nu}\right),$$

which shows that $\overline{\pi}(1), \ldots, \overline{\pi}(n)$ are independent. [C. Rényi and A. Rényi, in: *Combinatorial Theory Appl.* Coll. Math. Soc. J. Bolyai **4**, Bolyai-North-Holland (1970) 945–971.]

17. (a) By the remark in the hint, we have to determine the expected number of integers i with $\overline{\pi}(i) = i$. Let

$$\zeta_i = \begin{cases} 1 & \text{if } \overline{\pi}(i) = i, \\ 0 & \text{otherwise.} \end{cases}$$

Then we are interested in

$$\mathsf{E}(\zeta_1 + \ldots + \zeta_n) = \sum_{i=1}^{n} \mathsf{E}(\zeta_i) = \sum_{i=1}^{n} P\left(\overline{\pi}(i) = i\right) = \sum_{i=1}^{n} \frac{1}{i} \sim \log n.$$

(b) The probability that $i_1, \ldots, i_k$ (and only these) are records is, by the preceding solution,

$$\prod_{\nu=1}^{k} \frac{1}{i_\nu} \cdot \prod_{i \neq i_1, \ldots, i_k} \left(1 - \frac{1}{i}\right) = \frac{1}{n!} \prod_{i \neq i_1, \ldots, i_k} (i - 1)$$

and thus, the probability that there are exactly k records is

$$\frac{1}{n!} \sum_{1 \leq j_1 < \ldots < j_{n-k} \leq n} (j_1 - 1) \ldots (j_{n-k} - 1) = \frac{\left[\begin{smallmatrix} n \\ k \end{smallmatrix}\right]}{n!},$$

where the last step follows by identity 1.7(b).

Second solution. Observe that in $\hat{\pi}$ (see the second solution of 3.3) the last elements of cycles of π are exactly the "negative records", i.e. those elements which are not followed by any smaller element. Thus the number of "negative records" in $\hat{\pi}$ is equal to the number of cycles in π. Therefore the number of permutations with k "negative records" is the same as the number of those with k cycles, which is $\left[\begin{smallmatrix} n \\ k \end{smallmatrix}\right]$ by 3.12a. The number of permutations with k records is clearly the same. [C. Rényi and A. Rényi, ibid.]

18. (a) If we consider the same random permutation of $\{1,\ldots,n\}$ as in 3.16, then the winner is the last i with $\overline{\pi}(i)=i$.

A *strategy* is a function which associates with each sequence $(\overline{\pi}(1),\ldots,\overline{\pi}(k))$ $(1\leq k\leq n)$ one of "yes" or "no". The *value* of a strategy is the probability that the first "yes" is after the winner. Since there are finitely many strategies, there is a best one. Furthermore, if we are late, i.e. come after the kth jump, then we have to follow a strategy which says "no" for the first k jumps. Obviously, there are optimal strategies among these, too.

Let us denote by A_k the event that a (fixed) optimal strategy (with k "no"s first) wins, and put $\mathsf{P}(A_k)=p_k$. Observe that A_k is independent of the order of the first k jumpers. This is heuristically clear and can be shown as follows: Let

$$\mathsf{P}(A_k|\overline{\pi}(1) = l_1, \ldots, \overline{\pi}(k) = l_k) = q_{l_1,\ldots,l_k}$$

and let $q_{m_1,\ldots,m_k}$ be maximal among the numbers $q_{l_1,\ldots,l_k}$ $(1 \leq l_i \leq i)$. Then define a strategy which says "yes" for a sequence $(\overline{\pi}(1),\ldots,\overline{\pi}(\nu))$ iff our original strategy says "yes" for $(m_1,\ldots,m_k,\overline{\pi}(k+1),\ldots,\overline{\pi}(\nu))$. The probability that this new strategy wins is $q_{m_1,\ldots,m_k}$, whence

$$q_{m_1,\ldots,m_k} \leq \frac{1}{k!} \sum_{l_1,\ldots,l_k} q_{l_1,\ldots,l_k} = \mathsf{P}(A_k) = p_k.$$

We must have equality here, whence $q_{l_1,\ldots,l_k}=p_k$.

Now let $k<n$. Then

(1) $\qquad \mathsf{P}(A_k) = \mathsf{P}(\overline{\pi}(k + 1) = k + 1)\mathsf{P}(A_k|\overline{\pi}(k + 1) = k + 1)+$
$\qquad\qquad + \mathsf{P}(\overline{\pi}(k + 1) < k + 1) \cdot \mathsf{P}(A_k|\overline{\pi}(k + 1) < k + 1).$

Here the second term corresponds to the case when the $(k+1)^{\text{st}}$ jump is not a record. In this case we obviously will not bet on this jump; hence we say "no" and follow the optimal strategy after $k+1$ jumps, i.e.

$$\mathsf{P}(A_k|\overline{\pi}(k + 1) < k + 1) = \mathsf{P}(A_{k+1}|\overline{\pi}(k + 1) < k + 1) = p_{k+1}.$$

By 3.16,

$$\mathsf{P}(\overline{\pi}(k + 1) < k + 1) = \frac{k}{k + 1}.$$

If the $(k+1)^{\text{st}}$ jump is a record we have two choices. Either we say "yes", then we win with probability

$$\mathsf{P}(\overline{\pi}(k + 2) < k + 2,\ldots,\overline{\pi}(n) < n) = \frac{k + 1}{k + 2}\cdots\frac{n - 1}{n} = \frac{k + 1}{n},$$

or we say "no", in which case we follow the strategy after $k+1$ "no"s and have a chance of

$$\mathsf{P}(A_{k+1}|\overline{\pi}(k + 1) = k + 1) = \mathsf{P}(A_{k+1}) = p_{k+1}.$$

Thus

$$P_k(A_k|\bar{\pi}(k+1) = k+1) = \max\left(\frac{k+1}{n}, p_{k+1}\right).$$

Since

$$\mathsf{P}(\bar{\pi}(k+1) = k+1) = \frac{1}{k+1},$$

it follows from (1) that

(2) $$p_k = \max\left(\frac{p_{k+1}}{k+1}, \frac{1}{n}\right) + \frac{k}{k+1}p_{k+1} = \max\left(p_{k+1}, \frac{1}{n} + \frac{k}{k+1}p_{k+1}\right).$$

This makes it possible to determine $p_{n-1}, p_{n-2}, \ldots$ recursively. If we arrive after the $(n-1)^{\text{st}}$ jump we have to bet on the last jump and win with probability

$$p_{n-1} = \mathsf{P}(\bar{\pi}(n) = n) = \frac{1}{n}.$$

Now define $\tilde{p}_k$ for $k = n-1, \ldots, 1$ recursively as follows:

$$\tilde{p}_{n-1} = \frac{1}{n}, \quad \tilde{p}_k = \frac{1}{n} + \frac{k}{k+1}\tilde{p}_{k+1}.$$

Since

$$\frac{\tilde{p}_k}{k} = \frac{1}{nk} + \frac{\tilde{p}_{k+1}}{k+1},$$

it follows that

(3)
$$\frac{\tilde{p}_k}{k} = \frac{1}{nk} + \frac{1}{n(k+1)} + \ldots + \frac{1}{n(n-2)} + \frac{1}{n(n-1)},$$

$$\tilde{p}_k = \frac{k}{n}\left(\frac{1}{k} + \ldots + \frac{1}{n-1}\right) \sim \frac{k}{n}\log\frac{n}{k}.$$

Now we have $p_{n-1} = \tilde{p}_{n-1}$; let k be the largest index with $p_k \neq \tilde{p}_k$. Then

$$p_k \neq \frac{1}{n} + \frac{k}{k+1}p_{k+1} = \frac{1}{n} + \frac{k}{k+1}\tilde{p}_{k+1}$$

and so,

$$p_k = p_{k+1}.$$

(2) then implies that

$$p_k = p_{k-1} = \ldots = p_1.$$

This k is the largest values for which

$$\tilde{p}_{k+1} > \frac{1}{n} + \frac{k}{k+1}\tilde{p}_{k+1} = \tilde{p}_k,$$

i.e. by (3),

(4) $$\frac{1}{k+1} + \ldots + \frac{1}{n-1} > 1.$$

This value of k satisfies

$$\log \frac{n}{k+1} \approx 1, \qquad k+1 \approx \frac{n}{e}.$$

So $\quad p_m = \dfrac{m}{n}\left(\dfrac{1}{m} + \ldots + \dfrac{1}{n}\right) \approx \dfrac{m}{n}\log\dfrac{n}{m}$ for $m > k$, and $p_1 = \ldots = p_{k+1} =$

$$= \frac{k+1}{n}\left(\frac{1}{k+1} + \ldots + \frac{1}{n}\right) \approx \frac{1}{e}.$$

(b) The solution above already contains the optimum strategy. If k is such that $\tilde{p}_k \leq \tilde{p}_{k+1}$, then we get equality in (2), by the argument before it, if we say "no"; but if $p_k > p_{k+1}$ and we have a record we have to say "yes". Thus we have to say "no" for the first k cases, where k is the largest number such that (4) holds. After this, we say "yes" to the first record. [J. Bognár–J. Mogyoródi–A. Prékopa–A. Rényi–D. Szász, *Problem Book on Probability* (in Hungarian), Tankönyvkiadó, Budapest, 1970, p. 56.]

19. If there is no triple $i < j < k$ with $\pi(j) < \pi(i) < \pi(k)$, then each $i < j$ such that $\pi(i) > \pi(j)$ also satisfies $\pi(i) > \pi(j+1)$; thus $\overline{\pi}(j) \leq \overline{\pi}(j+1)$, i.e. the mapping $\overline{\pi}$ is monotone. Conversely, if $\overline{\pi}$ is monotone, there exists no $i < j < k$ with $\pi(j) < \pi(i) < \pi(k)$. Suppose indirectly that there is such a triple. Consider $\pi(j)$, $\pi(j+1)$, $\ldots$, $\pi(k)$. In this sequence there must be two consecutive terms $\pi(l)$, $\pi(l+1)$ such that $\pi(l) < \pi(i) < \pi(l+1)$. Then any $\nu < l+1$ with $\pi(\nu) > \pi(l+1)$ also satisfies $\nu < l$ and $\pi(\nu) > \pi(l)$; moreover, i satisfies $i < l$ and $\pi(i) > \pi(l)$ but $\pi(i) < \pi(l+1)$. Thus $\overline{\pi}(l) > \overline{\pi}(l+1)$, a contradiction.

Thus, the number we seek is equal to the number of monotone mappings φ of $\{1, \ldots, n\}$ into itself such that $1 \leq \varphi(i) \leq i$. By 1.33, this number is

$$\frac{1}{n+1}\binom{2n}{n}.$$

20. If π is any permutation, then the number of pairs $i < j$ with $\pi(i) > \pi(j)$ is, for a fixed j, equal to $\overline{\pi}(j) - 1$. Hence the total number of inversions is

$$(\overline{\pi}(1) - 1) + (\overline{\pi}(2) - 1) + \ldots + (\overline{\pi}(n) - 1).$$

Thus, the number of permutations with exactly k inversions is the number of solutions of the equation.

$$x_1 + \ldots + x_n = k$$

in integers satisfying $0 \leq x_i \leq i - 1$ (x_i will be $\overline{\pi}(i) - 1$). This number is, similarly as in 1.20, the coefficient of x^k in the product

$$1(1 + x) \ldots (1 + x + x^2 + \ldots + x^{n-1}).$$

21. Suppose that we have a very large tank with sufficient gasoline in it to go around the track. Let us start at an arbitrary point and go around. Stop at each station and buy all the gasoline in it. Then returning to the starting point we end up with the same amount of gasoline we started with, by the hypothesis.

Let A be the point, where the amount of gasoline in our tank was minimal. Then clearly A is at a gas station. Now if we start at A we can clearly go around the whole track.

22. Define ascending and descending sequences as in the hint. Let π be an arbitrary permutation and $C_1,\ldots,C_k$ its cycles. Assume that $\sum_{j\in C_i} x_j > 0$ for $i = 1,\ldots,p$, but ≤ 0 for $i=p+1,\ldots,k$.

We claim each of the cycles $C_1,\ldots,C_p$ has a unique decomposition into arcs which are ascending sequences. Let, e.g. $C_1 = (1,\ldots,r)$. Let us imagine C_1 as a speed track (cf. 3.21), in which each positive x_i corresponds to a gas station containing x_i liters of gasoline, while each negative x_i corresponds to a piece of the track which takes $-x_i$ liters of gasoline to travel (there may be two gas station immediately after each other or two consecutive pieces of the route with no gas station between them).

Going around the track we gain some gasoline, since $x_1 + \ldots + x_r > 0$. Let us keep on going around. Take those gas-stations after which the amount of gasoline in the tank is a record. There is at least one such station, since the amount of gasoline increases at each turn, and trivially if we return to a record point the amount is a record again. The pieces of the track between two consecutive record points yield the desired decomposition of C into arcs which are ascending sequences. It is also easy to verify that there is no other such decomposition.

Similarly we can decompose each C_i, $p+1 \leq i \leq k$ into descending arcs. So each permutation π yields a decomposition of $\{1,\ldots,n\}$ into disjoint ascending and descending sequences.

Now suppose that we are given a partition of $\{1,\ldots,n\}$ into u ascending and v descending sequences. Let, e.g. $1,\ldots,m$ be those numbers occurring in ascending sequences. Compose cycles each element of which is an ascending sequence in this partition and other cycles whose elements are the descending x sequences. The number of ways to do so is clearly $u!v!$. In each case we obtain a permutation π such that $b(\pi)=x_1+\ldots+x_m$.

Now let us arrange the u ascending sequences of the given partition into a linear order, followed by the v descending sequences. The number of ways to do so is $u!v!$. This results in a permutation ϱ of $\{1,\ldots,n\}$ such that

$$a(\varrho) = x_1 + \ldots + x_m.$$

It is easy to see that, conversely, every permutation ϱ is obtained from a unique partition into ascending and descending sequences.

So every partition of $\{1,\ldots,n\}$ into u ascending and v descending sequences corresponds to $u!v!$ permutations π and the same number of permutations ϱ, and $b(\pi)=a(\varrho)$ for each such π and ϱ. This proves the assertion.

It is interesting to remark that this result has important applications in the theory of orthogonal series. [F. Spitzer, *Trans. Amer. Math. Soc.* **82** (1956) 323–339.]

(b) Following the hint, take $x_1 = \ldots = x_m = 1$, $x_{m+1} = \ldots = x_{2m} = -1$ in (a). Let π be a permutation of $\{1,\ldots,2m\}$. This clearly corresponds to the boys and girls forming "rings" to dance. The condition that the number of boys is equal to the number of girls in each "ring" is clearly equivalent to $b(\pi) = 0$. The number of π's with this property is, by (a), the same as the number of permutations ϱ of $\{1, \ldots, 2m\}$ such that $a(\varrho) = 0$. Each such permutation yields a sequence of 1's and (-1)'s such that all sums of the first r of them ($r = 1, \ldots, 2m$) are non-positive. Each such ± 1-sequence corresponds to $(m!)^2$ permutations ϱ with $a(\varrho) = 0$. The number of such ± 1-sequences is, by 1.33b,

$$\frac{1}{m+1}\binom{2m}{m}.$$

So the number of permutations ϱ with $a(\varrho) = 0$ is

$$\frac{1}{m+1}\binom{2m}{m}(m!)^2 = \frac{(2m)!}{m+1}.$$

Dividing this by the total number $(2m)!$ of permutations we obtain the probability as desired.

23. (a) The set off all k-gons can be split into classes, where each class contains k-tuples arising by rotation from each other. We want to determine the number of these classes.

   In a class, there are the n rotated copies of a k-gon K. These are not all different, however; if $c(K)$ denotes the number of rotations which map K onto itself, then the class contains $\frac{n}{c(K)}$ elements. The sum

$$\sum_K c(K)$$

over all convex k-gons counts the class containing K $c(K) \cdot \frac{n}{c(K)} = n$ times. Thus, we want to determine the number

$$\frac{1}{n}\sum_K c(K).$$

Let us consider how many times a given rotation is counted in $\sum c(K)$. If it is the rotation by m, then we get all convex k-gons invariant under this rotation as follows: we split are n-gon into $\frac{n}{(m,n)}$ arcs of length (m,n) and specify $\frac{k(m,n)}{n}$ points on one of these arcs. The other arcs must contain rotated copies of this tuple and therefore we uniquely determined a convex k-gon. Thus, the number of convex k-gons invariant under the rotation by m is

$$\binom{(m,n)}{\frac{k(m,n)}{n}} \quad \text{if } n \mid k(m,n)$$

$$0 \qquad\qquad \text{otherwise}$$

and thus,

$$\frac{1}{n}\sum_{K} c(K) = \frac{1}{n}\sum_{\substack{m \leq n \\ n|k(m,n)}} \binom{(m,n)}{\frac{k(m,n)}{n}},$$

Introducing here $d = \frac{n}{(m,n)}$, we find that $d|(n,k)$ and there are exactly $\varphi(d)$ terms with the same d, and thus

$$\frac{1}{n}\sum_{K} c(K) = \frac{1}{n}\sum_{d|(n,k)} \binom{n/d}{k/d}\varphi(d).$$

(b) Proceeding as before, we consider the number $c(\alpha)$ of rotations keeping a given k-coloration α invariant. Then

$$\frac{1}{n}\sum_{\alpha} c(\alpha)$$

is the number of essentially different k-colorations. If we consider the rotation by m, there will be $k^{(n,m)}$ k-colorations invariant under it and thus,

$$\frac{1}{n}\sum_{\alpha} c(\alpha) = \frac{1}{n}\sum_{m=1}^{n} k^{(n,m)} = \frac{1}{n}\sum_{d|n} \varphi(d)k^{\frac{n}{d}}.$$

24. Let $V_1, \ldots, V_k$ be the orbits of Γ, $|V_i| = n_i$. Let $x \in V_i$ and denote by Γ_x the subgroup fixing x. Then the right cosets of Γ_x in Γ consist of those permutations in Γ which map x onto a given element of V_i. Hence Γ_x has index n_i and

$$|\Gamma_x| = \frac{|\Gamma|}{n_i}.$$

Now let $k_1(\pi)$ be the number of fixed points of the permutation π. Then the average number of fixed points is

$$\frac{1}{|\Gamma|}\sum_{\pi \in \Gamma} k_1(\pi) = \frac{1}{|\Gamma|}\left(\sum_{x \in \{1,\ldots,n\}} |\Gamma_x|\right) = \frac{1}{|\Gamma|}\sum_{i=1}^{k} n_i \frac{|\Gamma|}{n_i} = k.$$

25. Consider

$$\sum_{\pi \in \Gamma}\sum_{\pi(x)=x} w(x) = \sum_{x \in \Omega} w(x)|\Gamma_x|,$$

where Γ_x is the stabilizer of x. Let V_x be the orbit containing x, then

$$|\Gamma_x| = \frac{|\Gamma|}{|V_x|}$$

and thus

$$\sum_{\pi \in \Gamma} \sum_{\pi(x)=x} w(x) = \sum_{x \in \Omega} w(x) \frac{|\Gamma|}{|V_x|} = \sum_{\Theta} \sum_{x \in \Theta} w(x) \frac{|\Gamma|}{|\Theta|} = |\Gamma| \sum_{\Theta} w(\Theta).$$

26. If we define

$$\bar{\pi}(f)(x) = f(\pi(x))$$

for $f : D \to R$ and $\pi \in \Gamma$, then $\bar{\pi}$ can be considered as a permutation of the set Ω of mappings of D into R. By Burnside's lemma 3.24, the number of orbits of Γ on Ω, i.e. the number of essentially different mappings of D into R, is equal to the average number of fixed points of elements of Γ, acting on Ω.

Let $\pi \in \Gamma$. Then $f \in \Omega$ is a fixed point of $\bar{\pi}$ if

$$f(\pi(x)) = f(x) \qquad (x \in D),$$

i.e. if f is constant on the orbits of Γ in D. Let $k(\pi)$ be the number of orbits of π in D, then the number of such mappings f is, obviously, $|R|^{k(\pi)}$ and hence, the average number of fixed points of elements of Γ in Ω is

$$\frac{1}{|\Gamma|} \sum_{\pi \in \Gamma} |R|^{k(\pi)} = F(|R|, \dots, |R|).$$

27. Consider the permutation group $\Gamma_2 = \Gamma \times \Gamma_1$, acting on Ω, defined by

$$((\pi, \varrho)f)(x) = \varrho\left(f(\pi^{-1}(x))\right) \quad (\pi \in \Gamma, \ \varrho \in \Gamma, \ f \in \Omega, \ x \in D).$$

Then we have to determine the number of orbits of Γ_2. By Burnside's lemma 3.24, this is equal to

(1) $$\frac{1}{|\Gamma| \cdot |\Gamma_1|} \cdot \sum_{\pi \in \Gamma} \sum_{\varrho \in \Gamma_1} k_1(\pi, \varrho),$$

where $k_1(\pi, \varrho)$ is the number of fixed points of the permutation $(\pi, \varrho) \in \Gamma_2$, i.e. the number of mappings $f \in \Omega$ such that

(2) $$\varrho(f(\pi^{-1}(x))) = f(x)$$

holds for each $x \in D$. To determine this number, let $k_i(\pi) = k_i$, $k_i(\varrho) = l_i$ ($i = 1, 2, \dots$). Let $C_1, \dots, C_k$ be the cycles of π and $c_i \in C_i$. Now note that the values $f(c_1), \dots, f(c_k)$ uniquely determine f; for by (2),

(3) $$f(\pi^t(c_i)) = \varrho^{-t}(f(c_i))$$

and every element of D can be written as $\pi^t(c_i)$ with some t and i. Also, (3) must hold for $t = |C_i|$:

$$f(c_i) = \varrho^{|C_i|} f(c_i),$$

i.e. the length of the cycle of ϱ containing $f(c_i)$ must divide $|C_i|$.

Conversely, if $f(c_i)$, $i=1,\ldots,k$ are given such that the length of the cycle of ϱ containing $f(c_i)$ divides $|C_i|$, then (3) defines a fixed point of (π, ϱ). To select the image of a point c_i with $|C_i| = m$, we therefore have

$$\sum_{d|m} dl_d$$

possibilities. This can be done independently, hence

$$k_1(\pi, \varrho) = \prod_{m=1}^{|D|} \left(\sum_{d|m} dl_d \right)^{k_m}$$

Thus (1) can be written as

$$\frac{1}{|\Gamma_1|} \frac{1}{|\Gamma|} \sum_{\varrho \in \Gamma_1} \sum_{\pi \in \Gamma} \prod_{m=1}^{|D|} \left(\sum_{d|m} dk_d(\varrho) \right)^{k_m(\pi)} = \frac{1}{|\Gamma_1|} \sum_{\varrho \in \Gamma_1} F(u_1(\varrho), u_2(\varrho), \ldots,),$$

where

$$u_m(\varrho) = \sum_{d|m} dk_d(\varrho).$$

To express this in terms of F and G, observe that

$$F(u_1(\varrho), u_2(\varrho), \ldots) = F\left(\frac{\partial}{\partial z_1}, \frac{\partial}{\partial z_2}, \ldots \right) e^{u_1(\varrho)z_1 + u_2(\varrho)z_2 + \cdots} \Big|_{z_i=0}$$

and here

$$\sum_{m=1}^{\infty} u_m(\varrho)z_m = \sum_{m=1}^{\infty} \sum_{d|m} dk_d(\varrho)z_m = \sum_{d=1}^{\infty} dk_d(\varrho) \sum_{s=1}^{\infty} z_{ds}.$$

Thus

$$\frac{1}{|\Gamma_1|} \sum_{\varrho \in \Gamma_1} F\left(\frac{\partial}{\partial z_1}, \frac{\partial}{\partial z_1}, \ldots \right) e^{u_1(\varrho)z_1 + u_2(\varrho)z_2 + \cdots} =$$

$$= F\left(\frac{\partial}{\partial z_1}, \frac{\partial}{\partial z_2}, \ldots \right) \frac{1}{|\Gamma_1|} \sum_{\varrho \in \Gamma_1} \prod_{d=1}^{\infty} (e^{d(z_d + z_{2d} + \cdots)})^{k_d(\varrho)} =$$

$$= F\left(\frac{\partial}{\partial z_1}, \frac{\partial}{\partial z_2}, \ldots \right) G(e^{z_1 + z_2 + \cdots}, e^{2z_2 + 2z_4 + \cdots}, \ldots) =$$

and so, the number of essentially different mappings is

$$F\left(\frac{\partial}{\partial z_1}, \frac{\partial}{\partial z_2}, \ldots \right) G(e^{z_1 + z_2 + \cdots}, e^{2z_2 + 2z_4 + \cdots}) \Big|_{z_i=0}$$

[N. G. de Bruijn in: *Applied Combinatorial Mathematics*, (E. E. Beckenbach, ed.) Wiley, 1964.]

28. We use the same notation as before. The group Γ_2 keeps invariant the set Ω' of all one-to-one mappings of D into R. We want to determine the number of orbits of Γ_2 in Ω'. As before, this is equal to

$$\frac{1}{|\Gamma| \cdot |\Gamma_1|} \cdot \sum_{\pi \in \Gamma} \sum_{\varrho \in \Gamma_1} k_1'(\pi, \varrho),$$

where $k_1'(\pi, \varrho)$ is the number of those one-to-one mappings of D into R which are kept fixed by (π, ϱ).

Let $C_1, \ldots, C_k$ be the cycles of π. Then, again, it suffices to determine the image of a representative system $\{c_1, \ldots, c_k\}$, $c_i \in C_i$. Moreover, since f is one-to-one, $f(c_i)$ must belong to a cycle of ϱ of length $|C_i|$. Let $C_1, \ldots, C_{k_i}$ be the cycles of π of length i (say), then we have $i \cdot l_i$ choices for the image of $c_1 \in C_1$; $c_2 \in C_2$ cannot be mapped onto the same cycle of and thus, we have $i(l_i - 1)$ choices for the image of c_2. Going on similarly, we will have

$$(il_i)(i(l_i - 1)) \ldots (i(l_i - k_i + 1)) = i^{k_i} l_i (l_i - 1) \ldots (l_i - k_i + 1)$$

possible ways to map $c_1, \ldots, c_{k_i}$. Since the images of cycles of π of distinct length will be automatically disjoint, we can choose them independently and therefore

$$k_1'(\pi, \varrho) = \prod_{i=1}^{|D|} i^{k_i} l_i (l_i - 1) \ldots (l_i - k_i + 1)$$

and hence, the number of essentially different one-to-one mappings of D into R is

$$\frac{1}{|\Gamma|} \frac{1}{|\Gamma_1|} \sum_{\pi \in \Gamma} \sum_{\varrho \in \Gamma_1} \prod_{i=1}^{|D|} i^{k_i(\pi)} k_i(\varrho)(k_i(\varrho) - 1) \ldots (k_i(\varrho) - k_i(\pi) + 1).$$

Now observe that

$$\frac{1}{|\Gamma|} \sum_{\pi \in \Gamma} \prod_{i=1}^{|D|} i^{k_i(\pi)} k_i(\varrho)(k_i(\varrho) - 1) \ldots (k_i(\varrho) - k_i(\pi) + 1) =$$

$$= F\left(\frac{\partial}{\partial z_1}, \frac{\partial}{\partial z_2}, \ldots\right) (1 + z_1)^{k_1(\varrho)} (1 + 2z_2)^{k_2(\varrho)} \ldots \bigg|_{z_i = 0}$$

and thus, the result is

$$\frac{1}{|\Gamma_1|} \sum_{\varrho \in \Gamma_1} F\left(\frac{\partial}{\partial z_1}, \frac{\partial}{\partial z_2}, \ldots\right) (1 + z_1)^{k_1(\varrho)} (1 + 2z_2)^{k_2(\varrho)} \ldots \bigg|_{z_i = 0} =$$

$$= F\left(\frac{\partial}{\partial z_1}, \frac{\partial}{\partial z_2}, \ldots\right) G(1 + z_1, 1 + 2z_2, \ldots) \bigg|_{z_i = 0}.$$

29. Following the hint, consider a $\gamma \in \Gamma$; let $C_1, \ldots, C_k$ be its cycles. We prove that

$$(1) \qquad \sum_{n=0}^{\infty} q_n(\gamma) x^n = \prod_{i=1}^{k} r(x^{|C_i|}).$$

In fact, the coefficient of x^n on the right-hand side is

$$\sum_{\substack{\nu_1, \ldots, \nu_k > 0 \\ \sum \nu_i |C_i| = n}} r_{\nu_1} \cdots r_{\nu_k},$$

and clearly this is the same as the number of mappings satisfying $(*)$ which are constant on each of the sets $C_1, \ldots, C_k$.

Now (1) can be rewritten as

$$\sum_{n=0}^{\infty} q_n(\gamma) x^n = \prod_{j=1}^{|D|} r(x^j)^{k_j(\gamma)}.$$

By Burnside's lemma 3.24,

$$a_n = \frac{1}{|\Gamma|} \sum_{\gamma \in \Gamma} q_n(\gamma)$$

and hence

$$\sum_{n=0}^{\infty} a_n x^n = \frac{1}{|\Gamma|} \sum_{\gamma \in \Gamma} q_n(\gamma) x^n = \frac{1}{|\Gamma|} \sum_{\gamma \in \Gamma} \prod_{j=1}^{|D|} r(x^j)^{k_j(\gamma)} = F(r(x), r(x^2), \ldots).$$

[G. Pólya, *Acta Math.* **68** (1937) 145–254.]

30. Let D be the set of k forints and R, the set of n people to get them. A distribution of the forints is a mapping of D into R. Since all forints are alike, two distributions f, g are essentially the same if there is a permutation π of D such that

$$f(\pi(x)) = g(x) \qquad (x \in D).$$

By 3.26, the number of such distributions is

$$p_k(n, \ldots, n),$$

where $p_k(x_1, \ldots, x_k)$ is the cycle index of S_k. By the solution of 3.12,

$$p_k(n, \ldots, n) = \binom{n+k-1}{k} = \binom{n+k-1}{n-1}.$$

31. Let $|D| = n$, $|R| = N \geq n$. Two mappings $f, g : D \to R$ induce the same partition of R iff

$$g(x) = \varrho(f(x)) \qquad (x \in D)$$

for some permutation ϱ of R. Thus, the number of partitions of D is, by 3.27,

$$P_n = \left(\frac{\partial}{\partial z_1}\right)^n p_N(e^{z_1+z_2+\cdots}, e^{2z_2+2z_4+\cdots}, \ldots)\bigg|_{z_i=0} = \left(\frac{\partial}{\partial z}\right)^n p_N(e^z, 1, \ldots, 1)\bigg|_{z=0},$$

i.e. $\frac{P_n}{n!}$ is the coefficient of z^n in the Taylor series of $p_N(e^z, 1, \ldots, 1)$, for any $N \geq n$. Since by 3.11,

$$\lim_{N \to \infty} p_N(e^z, 1, \ldots, 1) = e^{e^z - 1},$$

$\frac{P_n}{n!}$ is the coefficient of z^n in the power series of e^{e^z-1} as well; i.e.

$$\sum_{n=0}^{\infty} \frac{P_n}{n!} z^n = e^{e^z - 1}.$$

§ 4. Two classical enumeration problems in graph theory

1. We use induction on n. For $n = 1, 2$ the statement is trivial. Since $\sum_{i=1}^{n} d_i = 2n - 2 < 2n$, there is a $d_i = 1$. We may assume that $d_n = 1$. Remove v_n. In any tree under consideration, v_n is adjacent to some v_j, $1 \leq j \leq n-1$ and the removal of v_n results in another tree on $\{v_1, \ldots, v_{n-1}\}$ with degrees $d_1, \ldots, d_{j-1}, d_j - 1, d_{j+1}, \ldots, d_{n-1}$. Conversely, if we are given a tree on $\{v_1, \ldots, v_{n-1}\}$ with degrees $d_1, \ldots, d_{j-1}, d_j - 1, d_{j+1}, \ldots, d_{n-1}$ then joining v_j to v_n we get a tree on $\{v_1, \ldots, v_n\}$ with degrees $d_1, \ldots, d_n$. The number of trees on $\{v_1, \ldots, v_{n-1}\}$ with degrees $d_1, \ldots, d_{j-1}, d_j - 1, d_{j+1}, \ldots, d_{n-1}$ is

$$\frac{(n-3)!}{(d_1-1)! \ldots (d_{j-1}-1)!(d_j-2)!(d_{j+1}-1)! \ldots (d_{n-1}-1)!} =$$
$$= \frac{(d_j-1)(n-3)!}{(d_1-1)! \ldots (d_n-1)!}.$$

This is also valid if $d_j = 1$ since, then it is 0. Thus, the number of trees on $\{v_1, \ldots, v_n\}$ with degrees $d_1, \ldots, d_n$ is

$$\sum_{j=1}^{n-1} \frac{(d_j-1)(n-3)!}{(d_1-1)! \ldots (d_n-1)!} = \left(\sum_{j=1}^{n-1}(d_j-1)\right) \frac{(n-3)!}{(d_1-1)! \ldots (d_n-1)!} =$$
$$= \frac{(n-2)(n-3)!}{(d_1-1)! \ldots (d_n-1)!},$$

which proves the assertion [see B].

2. The number of trees on n points is

$$\sum_{\substack{d_1,\ldots,d_n \geq 1 \\ d_1+\ldots+d_n=2n-2}} \frac{(n-2)!}{(d_1-1)!\ldots(d_n-1)!} = \sum_{\substack{k_1,\ldots,k_n \geq 0 \\ k_1+\ldots+k_n=n-2}} \frac{(n-2)!}{k_1!\ldots k_n!} =$$

$$= (\underbrace{1+\ldots+1}_{n})^{n-2} = n^{n-2} \qquad \text{[A. Cayley; see B].}$$

3. $p_1(x_1,\ldots,x_{n-1},0) = \sum x_1^{d_T(v_1)-1}\ldots x_{n-1}^{d_T(v_{n-1})-1}$, where the summation extends over all trees having an endpoint at v_n. If we set $S = T - v_n$, then S is a tree on $\{v_1,\ldots,v_{n-1}\}$ and conversely, for any tree S on $\{v_1,\ldots,v_{n-1}\}$ we can get $n-1$ trees with an endpoint at v_n by joining v_n to a point of S. For these trees T, we have

$$x_1^{d_T(v_1)-1}\ldots x_{n-1}^{d_T(v_{n-1})-1} = x_j x_1^{d_S(v_1)-1}\ldots x_{n-1}^{d_S(v_{n-1})-1},$$

where j is the index for which v_j is adjacent to v_n. Hence

$$\sum_{\substack{T \\ T-v_n=S}} x_1^{d_T(v_1)-1}\ldots x_n^{d_T(v_n)-1} = (x_1+\ldots+x_{n-1}) \cdot x_1^{d_S(v_1)-1}\ldots x_{n-1}^{d_S(v_{n-1})-1}$$

and summing over all trees S on $\{v_1,\ldots,v_{n-1}\}$, we get

(1) $p_n(x_1,\ldots,x_{n-1},0) = (x_1+\ldots+x_{n-1})p_{n-1}(x_1,\ldots,x_{n-1})$.

Now we prove by induction on n that

(2) $p_n(x_1,\ldots,x_n) = (x_1+\ldots+x_n)^{n-2}$.

By (1), we may assume that

$$p_n(x_1,\ldots,x_{n-1},0) = (x_1+\ldots+x_{n-1})^{n-2}.$$

Set

$$p_n^*(x_1,\ldots,x_n) = (x_1+\ldots+x_n)^{n-2}.$$

Then

$$p_n(x_1,\ldots,x_{n-1},0) = p_n^*(x_1,\ldots,x_{n-1},0)$$

and similarly, if we substitute 0's for any non-empty set of variables, p_n and p_n^* become identical. Hence, using the notation of 2.5,

$$\sigma_k p_n(x_1,\ldots,x_n) = \sigma_k p_n^*(x_1,\ldots,x_n)$$

for $k = 1,\ldots,n$. Now by 2.5,

$$p_n - \sigma_1 p_n + \sigma_2 p_n + \ldots + (-1)^n \sigma_n p_n = 0,$$
$$p_n^* - \sigma_1 p_n^* + \sigma_2 p_n^* + \ldots + (-1)^n \sigma_n p_n^* = 0$$

because p_n, p_n^* are of degree $n-2 < n$. This implies that $p_n = p_n^*$.

The Cayley formula now follows on substituting $x_1 = \ldots = x_n = 1$. [A. Rényi, in: *Combinatorial Str. Appl.* Gordon and Breach, 1970, 355–360.]

4. If we contract each T_i onto a single point v_i $(i = 1, \ldots, r)$, then any tree on V containing $T_1, \ldots, T_r$ is mapped onto a tree on $\{v_1, \ldots, v_r\}$. Let us count how many distinct trees on V are mapped onto a fixed tree T' on $\{v_1, \ldots, v_r\} = V$.

If, for each edge $(v_i, v_j) \in E(T')$, we choose a (T_i, T_j)-edge will get such a tree and, obviously, any tree on V which is mapped onto T' can be obtained in this way. The number of selections of such trees is

$$\prod_{(v_i, v_j) \in E(T')} |V(T_i)| \cdot |V(T_j)| = \prod_{i=1}^{r} |V(T_i)|^{d_T(v_i)}$$

and hence, the number of all trees on V containing $T_1, \ldots, T_r$ is

$$\sum_{T'} \prod_{i=1}^{r} |V(T_i)|^{d_{T'}(v_i)},$$

where the summation extends over all trees on V'. By the preceding formula, this is equal to

$$|V(T_1)| \ldots |V(T_r)| (|V(T_1)| + \ldots + |V(T_r)|)^{r-2} = |V(T_1)| \ldots |V(T_r)| \, |V|^{r-2}$$

[J. W. Moon; see B].

5. (a) Let $b_1, \ldots, b_{n-1}$ be the indices of removed points. Let us see how to determine b_i, if we know the Prüfer code.

b_i is obviously different from $b_1, \ldots, b_{i-1}, a_i$. Also, $b_i \neq a_j$ for $j > i$; for b_i is removed and it cannot be the neighbor of an endpoint at a later step. Conversely, if $k \notin \{b_1, \ldots, b_{i-1}, a_i, \ldots, a_{n-1}\}$, then v_k is an endpoint of $T - \{v_{b_1}, \ldots, v_{b_{i-1}}\}$; otherwise it would be a neighbor of a point removed at a later step. Thus,

(1) $b_i = \min\{k : k \notin \{b_1, \ldots, b_{i-1}, a_i, \ldots, a_{n-1}\}\}.$

Thus, the Prüfer code uniquely determines the numbers b_i. Since (v_{a_i}, v_{b_i}) are the edges of T, the Prüfer code uniquely determines T.

(b) Let $(a_1, \ldots, a_{n-1})$ be any sequence of integers with $1 \leq a_i \leq n$, $a_{n-1} = n$. Define b_i recursively by (1) and join v_{a_i} to v_{b_i} for $i = 1, \ldots, n-1$. We claim that the resulting graph T is a tree with Prüfer code $(a_1, \ldots, a_{n-1})$. Both assertions will follow, if we show that v_{b_i} is an endpoint of the graph $T_i = T - \{v_{b_1}, \ldots, v_{b_{i-1}}\}$ and no point with smaller index is endpoint. We have

$$v_{a_i} \in V(T_i)$$

because $a_i \neq b_1, \ldots, b_{i-1}$ by (1). Thus, v_{b_i} has a neighbor in T_i. v_{b_i} cannot be adjacent to any other point of T_i; for suppose that (v_{a_j}, v_{b_j}) is another edge of T_i adjacent to b_i, then $j > i$ as $v_{b_j} \in V(T_i)$ and either $b_i = b_j$ or $b_i = a_j$, which both contradict (1). Hence v_{b_i} is an endpoint of T_i. This proves that T and all T_i's are trees.

Now suppose that T_i has an endpoint v_k with $k < b_i$. Since k did not come into consideration when defining b_i by (1), it follows that either $k = b_\nu$, $\nu < i$ or $k = a_j$, $j \geq i$. Now the first possibility does not occur because $v_{b_\nu} \in V(T_i)$, so $k = a_j$, $j \geq i$. Since $a_{n-1} = n \geq b_i > k$, we have $j \leq n - 2$. By the argument above, v_{b_j} is an endpoint of T_j and its neighbor is v_{a_j}. But $v_{a_j} = v_k$ is an endpoint of T_i, and therefore it must be an endpoint of T_j too. Hence $V(T_j) = \{v_{a_j}, a_{b_j}\}$ which is impossible as T_j has $n - j + 1 > 3$ points.

The Cayley formula follows immediately: The number of sequences $(a_1, \ldots, a_{n-1})$ with $1 \leq a_i \leq n$, $a_{n-1} = n$ is, obviously, n^{n-2}. [A. Prüfer, *Archiv f. Math. u. Phys.* **27** (1918) 142–144.]

6. If T is a tree on $V = \{v_1, \ldots, v_n\}$ and $e \in E(T)$, then $T - e$ consists of two disjoint trees, which together cover V. This way we get $(n-1)T_n$ such pairs. Let T', T'' be two disjoint trees which together cover V; let $|V(T')| = k$, $|V(T'')| = n - k$, $v_1 \in V(T')$. Then we have $k(n-k)$ possible ways to add an edge joining T' to T'' to obtain a tree on V. Thus the pair (T', T'') arises from $k(n-k)$ trees T. The number of such pairs T', T'' is, obviously,

$$\binom{n-1}{k-1} T_k T_{n-k}$$

and hence,

(1)
$$\sum_{k=1}^{n-1} \binom{n-1}{k-1} T_k T_{n-k} k(n-k) = (n-1) T_n.$$

Since

$$\binom{n-1}{k-1} = \frac{n-1}{n-k} \binom{n-2}{k-1},$$

the result follows. Observe that substituting $T_n = n^{n-2}$ we obtain one of Abel's identities.

To get Cayley's formula note that by changing the index to $n-k$, we get from (1),

(2)
$$\sum_{k=1}^{n-1} \binom{n-1}{k} T_k T_{n-k} k(n-k) = (n-1) T_n$$

and hence, adding (2) to (1),

(3)
$$\sum_{k=1}^{n-1} \binom{n}{k} T_k T_{n-k} k(n-k) = 2(n-1) T_n.$$

Using induction on n, we know

$$T_k = k^{k-2}, \quad T_{n-k} = (n-k)^{n-k-2} \qquad (1 \leq k \leq n-1)$$

and so, the left-hand side of (3) is

$$\sum_{k=1}^{n-1} \binom{n}{k} k^{k-1}(n-k)^{n-k-1} = 2(n-1)n^{n-2}$$

by Abel's identity 1.44. This proves that

$$T_n = n^{n-2}.$$

[O. Dziobek, *Sitzungsber. Berl. Math. G.* **17** (1917) 64–67.]

7. (a) The recurrence relation in the last problem can be written in the form

$$\frac{T_n}{(n-2)!} = \sum_{k=1}^{n-1} \frac{kT_k}{(k-1)!} \frac{T_{n-k}}{(n-k-1)!}$$

and thus

(1) $$\sum_{n=2}^{\infty} \frac{T_n}{(n-2)!} x^{n-2} = \sum_{n=2}^{\infty} \left(\sum_{k=1}^{n-1} \frac{kT_k}{(k-1)!} \frac{T_{n-k}}{(n-k-1)!} \right) x^{n-2}.$$

The right-hand side is the product of the power series

$$\sum_{k=1}^{\infty} \frac{kT_k}{(k-1)!} n^{k-1} \quad \text{and} \quad \sum_{n=1}^{\infty} \frac{T_n}{(n-1)!} x^{n-1}$$

and thus, (1) yields

$$\left(\frac{t(x)}{x} \right)' = t'(x) \frac{t(x)}{x}$$

or, equivalently,

$$\left(\ln \frac{t(x)}{x} \right)' = \frac{\left(\frac{t(x)}{x} \right)'}{\frac{t(x)}{x}} = t'(x),$$

$$\ln \frac{t(x)}{x} = t(x) + C,$$

$$\frac{t(x)}{x} = c \cdot e^{t(x)}.$$

Substituting $x=0$ we get $c=1$, thus

$$x = t(x)e^{-t(x)}.$$

(b) By the Cauchy integral formula we have for $n \geq 1$,

$$T_n = \frac{1}{n} t^{(n)}(0) = \frac{(n-1)!}{2\pi i n} \oint_C \frac{t'(z)}{z^n} dz = \frac{(n-1)!}{2\pi i n} \oint_C \frac{t'(z)}{(t(z)e^{-t(z)})^n} dz,$$

where C is a small circuit around the origin. Since $t'(0) = 1 \neq 0$, the mapping t is a homeomorphism in a small neighborhood of 0. Hence $t(C)$ is a simple closed curve around 0 and hence,

$$
\text{(1)} \qquad T_n = \frac{(n-1)!}{2\pi i n} \oint_{t(C)} \frac{dw}{(we^{-w})^n} = \frac{(n-1)!}{2\pi i n} \oint_{t(C)} \frac{e^{wn}}{w^n} dw.
$$

Here

$$
\frac{1}{2\pi i} \oint_{t(C)} \frac{e^{wn}}{w^n} dw
$$

is the coefficient of x^{n-1} in the power series

$$
e^{xn} = \sum_{k=0}^{\infty} \frac{(xn)^k}{k!}.
$$

Hence

$$
T_n = \frac{(n-1)!}{n} \cdot \frac{n^{n-1}}{(n-1)!} = n^{n-2}.
$$

[G. Pólya, *Acta Math.* **68** (1937) 145–254.]

8. First, let the endpoints $v_1, \ldots, v_{n-l}$ be specified. Then, by 4.1, the result is

$$
\text{(1)} \qquad \sum_{\substack{k_{n-l+1} + \ldots + k_n = n-2 \\ k_i \geq 1}} \frac{(n-2)!}{k_{n-l+1}! \ldots k_n!}.
$$

because $d_1 = \ldots = d_{n-l} = 1$ in this case. Now observe that

$$
\frac{(n-2)!}{k_{n-l+1}! \ldots k_n!}
$$

is the number of ways to partition $n-2$ objects into l labelled classes (i.e. there is a first, second, ... class specified), where the ith class contains k_{n-i+1} elements. Thus, (1) counts the total number of partitions of $n-2$ objects into l labelled non-empty classes. Thus, (1) is equal to $l! \left\{ {n-2 \atop l} \right\}$.

   Since we have not specified the endpoints, they can be chosen in $\binom{n}{l}$ ways. Thus, the number of trees on n points with exactly $n-l$ endpoints is

$$
\binom{n}{l} l! \left\{ {n-2 \atop l} \right\} = \frac{n!}{(n-l)!} \left\{ {n-2 \atop l} \right\}.
$$

[A. Rényi, *Mat. Kut. Int. Közl.* **4** (1959) 73–85.]

9. (a) We may assume without loss of generality that the nth row of A has been removed. Let B be an $(n-1) \times (n-1)$ submatrix of A_0. We claim that

$$
\det B = \begin{cases} \pm 1 & \text{if the subgraph } G' \text{ formed by edges} \\ & \text{corresponding to columns in } B \text{ is a} \\ & \text{spanning tree,} \\ 0 & \text{otherwise.} \end{cases}
$$

We use induction on n. Suppose first that there is a point v_i $(i \neq n)$ which has degree 1 in G'. Then the ith row of B contains exactly one ± 1 (all other entries in this row belonging to B are 0). Expand $\det B$ by this row. The resulting $(n-2) \times (n-2)$ determinant $\det B'$ will correspond to $G' - v_i$ in the same way as $\det B$ does to G'. Since G' is a spanning tree of G iff $G' - v_i$ is a spanning tree of $G - v_i$ and $|\det B| = |\det B'|$ the assertion follows.

Secondly, suppose that no point of G has degree 1, except possibly v_n. Then G' is not a spanning tree. Since $|E(G')| = n-1 < n$, there must be a point having degree 0 in G'. If this point differs from v_n, then B has a 0 row and $\det B = 0$. If this point is v_n, then each column of B contains a 1 and a -1, thus the sum of rows of B is 0, hence $\det B = 0$.

Now use the Binet–Cauchy formula:

$$
\det A_0 A_0^T = \sum (\det B)^2,
$$

where B ranges over all $(n-1) \times (n-1)$ submatrices of A_0. Since by the above,

$$
(\det B)^2 = \begin{cases} 1 & \text{if } B \text{ corresponds to a spanning tree,} \\ 0 & \text{otherwise,} \end{cases}
$$

the assertion follows.

(b) The $(i,j)^{\text{th}}$ entry of $A_0 A_0^T$ $(i \neq j)$ is the inner product of the ith and j^{th} rows of A_0. In any column of A_0 but at most one, one of the ith and j^{th} entry is 0. In the column corresponding to the (i,j)-edge (if any) we have a $+1$ and a -1. Thus the $(i,j)^{\text{th}}$ entry of $A_0 A_0^T$ is -1.

The $(i,j)^{\text{th}}$ entry of $A_0 A_0^T$ is, obviously, the number of non-zero elements in the ith row of A_0, i.e. the degree of v_i. Hence the $(i,j)^{\text{th}}$ entry is

$$
\begin{cases} \text{the degree of } v_i & \text{if } i = j, \\ -1 & \text{if } v_i \text{ and } v_j \text{ are adjacent,} \\ 0 & \text{otherwise.} \end{cases}
$$

(c) The number of trees on n points is the same as the number of spanning trees of K_n, i.e.

$$\begin{vmatrix} n-1 & -1 & \cdots & -1 \\ -1 & n-1 & \cdots & -1 \\ \vdots & \vdots & \ddots & \vdots \\ -1 & -1 & \cdots & n-1 \end{vmatrix} = \begin{vmatrix} 1 & 1 & \cdots & 1 \\ -1 & n-1 & \cdots & -1 \\ \vdots & \vdots & \ddots & \vdots \\ -1 & -1 & \cdots & n-1 \end{vmatrix}$$

$$\underbrace{\qquad\qquad\qquad\qquad}_{n-1}$$

(by adding all rows to the first one)

$$= \begin{vmatrix} 1 & 1 & \cdots & 1 \\ 0 & n & \cdots & 0 \\ \vdots & \vdots & \ddots & \vdots \\ 0 & 0 & \cdots & n \end{vmatrix}$$

(by adding the first row to the others)

$$= n^{n-2}$$

[G. Kirchhoff; see Biggs].

10. Let us orient G in some way to get a digraph $\overrightarrow{G}$ and define $E(G)=\{e_1,\ldots,e_m\}$,

$$a_{ij} = \begin{cases} -\sqrt{x_i x_k} & \text{if } e_j = (v_i, v_k), \\ \sqrt{x_i x_k} & \text{if } e_j = (v_k, v_i), \\ 0 & \text{if } e_j \text{ is not adjacent to } v_i \ (1 \le i \le n, \ 1 \le j \le m), \end{cases}$$

and set

$$A_0 = (a_{ij})_{i=1\ j=1}^{n-1\ m}.$$

Then it is easy to check that

$$D = A_0 A_0^T.$$

Let B be an $(n-1) \times (n-1)$ submatrix of A_0. We claim that

$$\det B = \begin{cases} \pm\sqrt{x_1^{d_T(v_1)}} \cdots \sqrt{x_n^{d_T(v_n)}} & \text{if the columns of } B \text{ correspond to a spanning tree,} \\ 0 & \text{otherwise.} \end{cases}$$

This follows by induction on n in the same way as in the solution of 4.9.

By the Binet–Cauchy formula,

$$\det A_0 A_0^T = \sum (\det B)^2 = \sum_T x_1^{d_T(v_1)} \ldots x_n^{d_T(v_n)},$$

where T ranges over all spanning trees of G. This proves the formula.

11. We have to determine the number of spanning trees of the complete bipartite graph $K_{n,m}$. It follows from 4.9 (or from 4.10 by setting $x_1 = \ldots = x_n = 1$), that

$$T(K_{n,m}) = \begin{vmatrix} \overbrace{m \quad \ldots \quad 0}^{n} & \overbrace{-1 \quad \ldots \quad -1}^{m-1} \\ \vdots \quad \ddots \quad \vdots & \vdots \qquad \vdots \\ 0 \quad \ldots \quad m & -1 \quad \ldots \quad -1 \\ -1 \quad \ldots \quad -1 & n \quad \ldots \quad 0 \\ \vdots \qquad \vdots & \vdots \quad \ddots \quad \vdots \\ -1 \quad \ldots \quad -1 & 0 \quad \ldots \quad n \end{vmatrix} ;$$

add all rows to the first one, then add the first row to each of the last $m-1$ rows, and then we get

$$T(K_{n,m}) = \begin{vmatrix} 1 & 1 & \ldots & 1 & 0 & \ldots & 0 \\ 0 & m & \ldots & 0 & -1 & \ldots & -1 \\ \vdots & & \ddots & \vdots & \vdots & & \vdots \\ 0 & 0 & \ldots & m & -1 & \ldots & -1 \\ 0 & 0 & \ldots & 0 & n & \ldots & 0 \\ \vdots & & & \vdots & \vdots & \ddots & \vdots \\ 0 & 0 & \ldots & 0 & 0 & \ldots & n \end{vmatrix} = m^{n-1} n^{m-1}.$$

12. (a) Let $E(\overline{G}) = \{e_1, \ldots, e_m\}$ and let A_k denote the set of those trees on $V(G)$ containing e_k; we denote by S the set of all trees on $V(G')$. Then, by the inclusion-exclusion principle,

$$(1) \qquad T(G) = \left| S - \bigcup_{k=1}^{m} A_k \right| = \sum_{K \subseteq \{1,\ldots,m\}} (-1)^{|K|} |A_K|,$$

where

$$A_\emptyset = S, \qquad A_K = \bigcap_{k \in K} A_k.$$

We determine $|A_K|$. Set

$$E_K = \{e_k : k \in K\}, \qquad G_K = (V(G), E_K)$$

and let $X_1^{(K)}, \ldots, X_{r_K}^{(K)}$ be the sets of points of components of G_K. If G_K is not a forest, then obviously, $A_K = \emptyset$. Suppose that G_K is a forest. A_K consists of all trees on $V(G)$ containing G_K. By 4.4 the number of such trees is

$$|X_1^{(K)}| \ldots |X_{r_K}^{(K)}| n^{r_K - 2}.$$

Also, $|K| = n - r_K$ and hence, (1) yields

$$T(G) = \sum_{\substack{K \subseteq \{1, \ldots, m\} \\ G_K \text{ is a forest}}} (-1)^{n - r_K} |X_1^{(K)}| \ldots |X_{r_K}^{(K)}| n^{r_K - 2}.$$

For a fixed partition $\{X_1, \ldots, X_r\}$ the number of terms K with $\{X_1^{(K)} \ldots X_{r_K}^{(K)}\} = \{X_1, \ldots, X_r\}$ is, obviously,

$$T(\overline{G}[X_1]) \ldots T(\overline{G}[X_r]);$$

thus, if we sum over all partitions $X_1, \ldots, X_r$ of $V(G)$, we get

$$\sum (-1)^{n-r} T(\overline{G}[X_1]) \ldots T(\overline{G}[X_r]) |X_1| \cdot \ldots \cdot |X_r| n^{r-2}$$

as stated [H. N. V. Temperley; see B].

(b) *First solution.* By 4.9,

$$
T(G) = \begin{vmatrix}
n-2 & -1 & -1 & \cdots & -1 & -1 & -1 & \cdots & -1 \\
-1 & n-2 & 0 & & -1 & -1 & -1 & \cdots & -1 \\
-1 & 0 & n-2 & & -1 & -1 & -1 & \cdots & -1 \\
\vdots & & & \ddots & & & & & \\
-1 & -1 & -1 & & n-2 & 0 & -1 & \cdots & -1 \\
-1 & -1 & -1 & & 0 & n-2 & -1 & \cdots & -1 \\
-1 & -1 & -1 & & -1 & -1 & n-1 & \cdots & -1 \\
\vdots & \vdots & \vdots & & \vdots & \vdots & \vdots & \ddots & \vdots \\
-1 & -1 & -1 & \cdots & -1 & -1 & -1 & \cdots & n-1
\end{vmatrix}.
$$

Add all rows to the first one, then add the first row to all others; we get

$$
T(G) = \left|
\begin{array}{c|ccccc|ccc}
0 & 1 & 1 & 1 & \cdots & 1 & 1 & \cdots & 1 \\
\hline
-1 & n-1 & 1 & 0 & \cdots & 0 & 0 & \cdots & 0 \\
-1 & 1 & n-1 & 0 & \cdots & 0 & 0 & \cdots & 0 \\
\vdots & & & & \ddots & & & & \\
\hline
-1 & 0 & 0 & 0 & \cdots & 0 & n & \cdots & 0 \\
\vdots & \vdots & \vdots & \vdots & & \vdots & \vdots & \ddots & \vdots \\
-1 & 0 & 0 & 0 & \cdots & 0 & 0 & \cdots & n
\end{array}
\right|.
$$

(with braces: $2q-1$ over the middle columns, $n-2q$ over the last columns)

Now add $\frac{1}{n}$ times all columns to the first one, we get

$$
T(G) = \left|
\begin{array}{c|ccc|ccc}
\frac{n-2}{n} & 1 & 1 & \cdots & 1 & \cdots & 1 \\
\hline
0 & n-1 & 1 & & 0 & \cdots & 0 \\
0 & 1 & n-1 & & 0 & \cdots & 0 \\
\vdots & & & \ddots & & & \\
\hline
0 & 0 & 0 & & n & \cdots & 0 \\
\vdots & \vdots & \vdots & & \vdots & \ddots & \vdots \\
0 & 0 & 0 & & 0 & \cdots & n
\end{array}
\right| =
$$

(braces: $2q-1$, $n-2q$)

$$
= \frac{n-2}{n} n^{n-2q} \left| \begin{array}{cc} n-1 & 1 \\ 1 & n-1 \end{array} \right|^{q-1} = (n-2)^q n^{n-q-2}.
$$

Second solution. In the formula of 4.12, it suffices to consider those partitions $(X_1, \ldots, X_r)$ in which every X_i is one-element or consists of two points adjacent in $\overline{G}$. There are $\binom{q}{j}$ ways to select j such classes, thus

$$
T(G) = \sum_{j=0}^{q} \binom{q}{j} (-1)^{n-(n-j)} 2^j n^{n-j-2} = n^{n-q-2}(n-2)^q \quad \text{[L. Weinberg, see B]}.
$$

(c) We shall only give in detail the solution based on (a). Here it suffices to consider partitions $(X_1, \ldots, X_r)$, where X_1 contains v and some j points adjacent to v and all other X_i's are one-element. The number of such partitions $\binom{q}{j}$. Thus we get

$$T(G) = \sum_{j=0}^{q} \binom{q}{j}(-1)^{n-(n-j)}(j+1)n^{n-j-2} = n^{n-2}((1-x)^q x)' \bigg|_{x=\frac{1}{n}} =$$

$$= n^{n-2}\left(1 - \frac{1}{n}\right)^q - n^{n-2}q\left(1 - \frac{1}{n}\right)^{q-1}\frac{1}{n} =$$

$$= (n - q - 1)(n - 1)^{q-1}n^{n-q-2} \quad \text{[P. V. O'Neil; see B]}.$$

13. (a) Draw the tree so that a is the uppermost point, and departing from a we descend while the other endpoints are on the "ground". It is clear that each binary plane tree can be drawn essentially uniquely in this way. The endpoints different from a have a natural ordering $v_1, \ldots, v_k$; going along the "ground" line in this direction, we have a to our left. Note that, from

$$|E(T)| = \frac{1}{2}(k + 1 + 3(2n - k - 1)) = 2n - 1,$$

we have $k = n$.

The tree can be considered as a diagram to calculate the product $x_1 \ldots x_n$. The variables are associated with the endpoints on the ground, and every inner point represents the multiplication of the two products corresponding to the two points immediately below it. Thus the number of binary plane trees with $2n$ points is the same as the number of bracketings of a product of n factors, which is $\frac{1}{n}\binom{2n-2}{n-1}$ by 1.38.

(b) Consider the ± 1-sequence defined in the hint associated with a rooted plane tree. It follows trivially by induction that the sum of the first k entries is equal to our distance from the root after having walked along k edges. This sum is therefore positive, except for $k = 2n - 2$, when it is 0. Conversely, given a sequence of $n-1$ $(+1)$'s and $n-1$ (-1)'s such that the partial sums are positive, there is a unique rooted plane tree with this code. In fact, we can build up the walls and the walk around them step-by-step as follows; we start from the root. Assume that the walls corresponding to the first k entries have been built up, and we are where we have to be after k steps in the walk. If the next entry is $+1$, we attach a new wall and build it for a while, always having it to our left. If the next entry is -1, we do not build but just walk along the wall toward the root (keeping it to our left). It is easy to check that this is a correct "decoding". So the number of plane rooted trees on n points is equal to the number of sequences of $n-1$ $(+1)$'s and $n-1$ (-1)'s such that all partial sums are positive. This number is

$$\frac{1}{n-1}\binom{2n-4}{n-2}$$

by 1.33b. [G. Pólya, *Acta Math.* **68** (1937) 145–254. Note the relation between the results of parts (a) and (b); cf. D. A. Klarner, *J. Comb. Theory* **9** (1970) 401–411.]

14. *First solution.* Draw an arbitrary tree T with point $v_1, \ldots, v_k$. Then any forest F in which $v_1, \ldots, v_k$ are in distinct components and which has k components gives, together with T, a tree containing T_0; and conversely. Thus the result is the same as the number of trees on n points containing a given k-point subtree; this number is, by 4.4,

$$(1) \qquad\qquad E(n, k) = kn^{n-k-1}.$$

Second solution. Let F be one of the forests on $\{v_1, \ldots, v_n\}$ having $k - 1$ components and such that $v_1, \ldots, v_{k-1}$ are in distinct components $T_1, \ldots, T_{k-1}$. Let $v_k \in V(T_i)$ and remove the edge adjacent to v_k which is on the (v_k, v_i)-path. Then we get a forest F' which has k components and $v_1, \ldots, v_k$ are in distinct components.

Let us see how many times a given forest F' occurs. Let F' have components $T_1', \ldots, T_k'$, $v_i \in V(T')$. Then we have to join v_k to a point of $V(T_1' \cup \ldots \cup T_{k-1}')$, which gives $|V(T_1' \cup \ldots \cup T_{k-1}')| = n - |V(T_k')|$ possibilities.

Now, unfortunately, $n - |V(T_k')|$ depends on the special choice of F'. But we can sum this over all permutations of $v_1, \ldots, v_k$. Then a given F' will be counted

$$(k - 1)! \sum_{i=1}^{k} (n - |V(T_i')|) = (k - 1)!(nk - n)$$

times; on the other hand, we get from each forest F with $k - 1$ components ($v_1, \ldots, v_{k-1}$ in distinct components) $k!$ forests F'. Hence

$$k!E(n, k - 1) = (k - 1)!n(k - 1)E(n, k)$$

or, equivalently,

$$(1) \qquad\qquad E(n, k - 1) = \frac{k - 1}{k} nE(n, k).$$

Hence

$$E(n, k) = \frac{k}{k + 1} nE(n, k + 1) = \frac{k}{k + 1} \frac{k + 1}{k + 2} n^2 E(n, k + 2) =$$

$$= \ldots = \frac{k}{k + 1} \frac{k + 1}{k + 2} \cdots \frac{n - 1}{n} n^{n-k} E(n, n).$$

Since $E(n, n) = 1$ by definition, this gives

$$E(n, k) = kn^{n-k-1}.$$

Since

$$T(n) = E(n, 1) = n^{n-2},$$

we have obtained a new proof of the Cayley formula, too [A. Cayley; see B].

15. Select one edge entering each point $x \neq a$. Then we get an arborescence T rooted at a; for we have $n-1$ edges ($n = |V(G)|$) and for each x, we can find an edge $(x_1, x) \in E(T)$; then an edge $(x_2, x_1) \in E(T)$, etc. The same point cannot occur twice in the sequence $x, x_1, x_2, \ldots$, because G is acyclic. Hence there is a unique (a, x)-path in T for every x, and, consequently, T is an arborescence rooted at a.

Conversely, every arborescence rooted at a occurs in this way. Hence the number of such arborescences is

$$\prod_{x \neq a} d^-(x),$$

the $d^-(x)$ is the indegree of x.

16. (a) We use induction on $|E(G)|$. If $|E(G)| < n-1$, then $\Delta(G)$ contains a column which is identically 0. So we may suppose that $|E(G)| \geq n-1$. Also we may assume that no edge enters the root v_n, because such an edge plays no role.

Suppose first that there is a point, say v_1, which is the head of at least two edges. Split the edges entering v_1 into two non-empty classes C_1, C_2. Since any arborescence contains exactly one edge which enters v_1, the number of all (spanning) arborescences of G is equal to the sum of the numbers of arborescences of $G - C_1$ and $G - C_2$. On the other hand, all but one row of $\Delta(G - C_i)$ are the same as those of $\Delta(G)$; the first rows of $\Delta(G - C_1)$ and $\Delta(G - C_2)$ add up to the first row of $\Delta(G)$. Hence

$$\Delta(G - G_1) + \Delta(G - C_2) = \Delta(G).$$

Since by the induction hypothesis, $\Delta(G - C_i)$ is the number of spanning arborescences rooted at v_n in $G - C_i$, the assertion follows.

In the case remaining, exactly one edge enters each point v_i, $1 \leq i \leq n-1$. If G is an arborescence, then we have to show

$$\Delta(G) = 1.$$

To show this we may renumber the points in such a way that, if v_i is joined to v_j, then $i > j$. Then the matrix $\Delta(G)$ has 0's above the diagonal and 1's in the diagonal, whence $\Delta(G) = 1$ follows.

On the other hand, assume that G is not an arborescence. Since each point $\neq v_1$ has indegree 1 and thus, $|E(G)| = n-1$, G must be disconnected; let, say, $v_1, \ldots, v_k$ form a component of G which does not contain v_n. Then the sum of the first k rows of $\Delta(G)$ is 0, hence $\Delta(G) = 0$. This completes the proof.

(b) Replace each edge e of G by x_e parallel edges. Then the number of arborescences in the resulting graph G', rooted at v_n is

$$\sum x_{e_1} \ldots x_{e_{n-1}},$$

where the summation extends over all $(n-1)$-tuples $\{e_1, \ldots, e_{n-1}\}$ of edges which form a spanning arborescence. Let $x_{e_i} = y_\nu$, if $e_i = (v_\nu, v_\mu)$. Then

$$\sum x_{e_1} \ldots x_{e_{n-1}} = \sum_T \prod_{i=1}^{n} y_i^{d_T(v_i)};$$

on the other hand, this is equal to

$$\Delta(G) = \begin{vmatrix} \sum_{j \neq 1} a_{j1} y_j & -a_{12} y_1 & \cdots & -a_{1,n-1} y_1 \\ -a_{21} y_2 & \sum_{j \neq 2} a_{j2} y_j & \cdots & -a_{2,n-1} y_2 \\ \vdots & & & \\ -a_{n-1,1} y_{n-1} & -a_{n-1,2} y_{n-1} & \cdots & \sum_{j \neq n-1} a_{j,n-1} y_j \end{vmatrix}.$$

Since this equality holds for any choice of y_i such that $y_i > 0$ and y_i is an integer, it holds identically [see B].

17. I. It is easy to prove that if π has no fixed points, then no tree is invariant under it. We use here the notation of the center (see 6.21) but the reader will find no difficulty in replacing this by other considerations if he wants to avoid reference to later chapters. If our tree has a center this must be a fixed point of π. If it has a bicenter, π must exchange the two points of this. But then it must exchange the two branches of the tree relative to the edge connecting the bicenters, which is only possible if the tree has an even number of points.

II. Now suppose that x is a fixed point of π. Let $\overrightarrow{T}$ be the arborescence arising by orienting T so that x becomes its root. We need some further information on $\overrightarrow{T}$.

Let $V_1, \ldots, V_m$ be the underlying sets of the cycles of π, $|V_1| \geq |V_2| \geq \ldots \geq |V_m|$, $V_m = \{x\}$.

(i) π is an automorphism of $\overrightarrow{T}$, obviously.

(ii) If $(x, y) \in E(\overrightarrow{T})$, $x \in V_i$, $y \in V_j$, then $|V_i| \mid |V_j|$. In fact, if $|V_i|$ does not divide $|V_j|$, then there is an image $(\pi^k(x), \pi^k(y))$ of (x, y) with $\pi^k(y) = y$ but $\pi^k(x) \neq x$. This means two edges of $\overrightarrow{T}$ enter y, which is impossible, because $\overrightarrow{T}$ is an arborescence.

(iii) If $(x, y) \in E(\overrightarrow{T})$, $x \in V_i$, $y \in V_j$, then $i \neq j$. For there is an edge (z, u) with $u \in V_j$, $z \notin V_j$; a suitable image $(\pi^k(z), \pi^k(u))$ is such that $\pi^k(u) = y$; thus $\pi^k(z) = x \notin V_j$.

(iv) If we contract each V_i into a point v_i, and cancel the arising multiplicities of edges of $\overrightarrow{T}$, then $\overrightarrow{T}$ is mapped onto a certain arborescence $\overrightarrow{T}_1$ on $\{v_1, \ldots, v_m\}$.

In fact, the same argument as in (i) or (ii) shows that all edges of $\overrightarrow{T}$ entering V_i must come from the same V_j. Hence exactly one edge of $\overrightarrow{T}_1$ enters each v_i, $i \neq m$. Since, obviously, each point v_i is accessible from v_m on a path on $\overrightarrow{T}_1$, it follows that $\overrightarrow{T}_1$ is an arborescence rooted at v_m.

(v) $\overrightarrow{T}_1$ is a subgraph of the digraph G_1 in which $(v_i, v_j) \in E(G_1)$ iff $|V_i| \, | \, |V_j|$; by (i).

Now let us determine, how many trees T yield a given arborescence $\overrightarrow{T}_1$. Each such T arises by taking one (V_i, V_j)-edge and its images by π for every $(v_i, v_j) \in E(\overrightarrow{T}_1)$; conversely, if we select one (V_i, V_j)-edge and its images for every $(v_i, v_j) \in E(\overrightarrow{T}_1)$ we get a tree which is invariant under π. The set of all (V_i, V_j)-edges consists of $|V_i|$ orbits (remember $|V_i| \, | \, |V_j|$). Thus the number of trees T belonging to a given arborescence $\overrightarrow{T}_1$ is

$$\sum_{(v_i, v_j) \in E(\overrightarrow{T}_1)} |V_i| = \prod_{i=1}^{m} |V_i|^{d^-_{\overrightarrow{T}_1}(v_i)}$$

and hence, the number of all trees on V invariant under π is

$$\sum_{\overrightarrow{T}_1} |V_1|^{D^-_{\overrightarrow{T}_1}(v_1)} \ldots |V_m|^{D^-_{\overrightarrow{T}_1}(v_m)} .$$

By the preceding problem, this is equal to the determinant

$$\Delta = \begin{vmatrix} \sum_{j \neq 1} a_{j1}|V_j| & -a_{12}|V_1| & \ldots & -a_{1,m-1}|V_1| \\ -a_{21}|V_2| & \sum_{j \neq 2} a_{j2}|V_j| & \ldots & -a_{2,m-1}|V_2| \\ \vdots & & & \\ -a_{m-1,1}|V_{m-1}| & -a_{m-1,2}|V_{m-1}| & \ldots & \sum_{j \neq m-1} a_{j,m-1}|V_j| \end{vmatrix},$$

where

$$a_{ij} = \begin{cases} 1 & \text{if } |V_i| \, | \, |V_j| \quad (1 \le i, j \le m), \\ 0 & \text{otherwise.} \end{cases}$$

This determinant can be evaluated as follows. It has the form

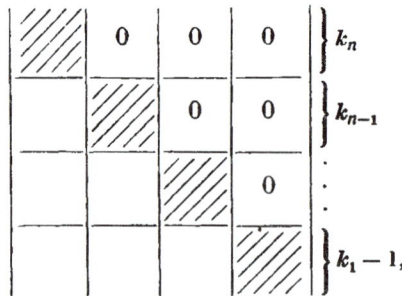

since $|V_i| \big| |V_j|$, $i \geq j$ implies $|V_i| = |V_j|$. Hence Δ is equal to the product of the shaded subdeterminants. Consider the block corresponding to k_i. It has

$$\alpha_i = \sum_{d|i} dk_d - i$$

in the diagonal, so its value is

$$\Delta_i = \begin{vmatrix} \alpha_i & -i & \cdots & -i \\ -i & \alpha_i & \cdots & -i \\ \vdots & & \ddots & \vdots \\ -i & -i & \cdots & \alpha_i \end{vmatrix} = (\alpha_i + i)^{k_i - 1}(\alpha_i - (k_i - 1)i)$$

$$= \left(\sum_{d|i} dk_d \right)^{k_i - 1} \left(\sum_{\substack{d|i \\ d \neq i}} dk_d \right)$$

for $i > 1$ and $k_i \geq 1$, and

$$\Delta_1 = \underbrace{\begin{vmatrix} k_1 - 1 & -1 & \cdots & -1 \\ -1 & k_1 - 1 & & -1 \\ \vdots & & \ddots & \vdots \\ -1 & -1 & \cdots & k_1 - 1 \end{vmatrix}}_{k_1 - 1} = k_1^{k_1 - 2}$$

(one has to check that these formulas give $\Delta_i = 1$ if $k_i = 0$ $(i = 2, \ldots, m)$ and $\Delta_1 = 1$ if $k_1 = 1$). Since

$$\Delta = \Delta_1 \ldots \Delta_n,$$

the assertion is proved.

18. Since the number of rooted trees on n (labelled) points is n^{n-1} by Cayley's formula and at most $(n-1)!$ of these are isomorphic, we have

$$W_n \geq \frac{n^{n-1}}{(n-1)!} = \frac{n^n}{n!} > 2^n \qquad (n \geq 6).$$

Each rooted tree can be drawn in the plane. Join its root to a new point of degree 1 to get a plane tree on $n+1$ points rooted at an endpoint. This correspondence shows that the number of non-isomorphic rooted trees is not larger than the number of essentially different plane trees rooted at an endpoint, which is, by 4.13b,

$$\frac{1}{n}\binom{2n-2}{n-1} < 4^n.$$

19. (a) Following the hint, we write the right-hand side in the form

$$x(1 + x + x^2 + \ldots)^{W_1}(1 + x^2 + x^4 + \ldots)^{W_2} \ldots =$$
$$= x \prod_T (1 + x^{|V(T)|} + x^{2|V(T)|} + \ldots),$$

where T ranges over all rooted trees. The coefficient of x^n in this product is the number of representations of the number n in the form

(1) $$n = 1 + \sum_T \nu_T |V(T)|, \qquad \nu_T \geq 0.$$

Given such a representation, take ν_T copies of T for each rooted tree T and join their roots to a new root. In this way we obtain an n-point rooted tree and conversely, every n-point rooted tree yields a solution of (1). Thus the coefficient of x^n in the product is W_n, which proves the identity. (Convergence questions can be settled easily on the basis of 4.18, and will be ignored.)

(b) The two identities are equivalent by 3.7. (For $x \leq 1$ the series on the right-hand sides are convergent because then $w(x^n) = 0(x^n)$. The first one follows easily from (a):

$$\log \frac{w(x)}{x} = -\sum_{n=0}^{\infty} W_n \log(1 - x^n) = \sum_{n=0}^{\infty} W_n \sum_{k=1}^{\infty} \frac{x^{kn}}{k} = \sum_{k=1}^{\infty} \frac{1}{k} w(x^k),$$

which proves the identity.

To get the second identity directly by the Pólya–Redfield method, set $D = \{1, \ldots, d\}$, $R = \{$ isomorphism types of rooted trees $\}$, $w(T) = |V(T)|$ for $T \in R$, and let Γ be the symmetric group on D. A rooted n-point tree is a d-tuple of "branches", where d is the degree of the root. The branches are themselves rooted trees with total weight $n-1$. Thus the number $W_n^{(d)}$ of rooted n-point trees in which the root has degree d is equal to the number of essentially different mappings of D into R with total weight $n-1$. Hence by 3.29,

$$\sum_{n=1}^{\infty} W_n^{(d)} x^n = x p_d(w(x), w(x^2), \ldots).$$

Summing over all d, we obtain the identity in the problem [A. Cayley].

20. (a) The fact that the radius of the convergence of $w(x)$ satisfies $0 < \tau < 1$ is trivial by 4.18. It follows that

$$\varphi(z) = z \exp\left(\frac{w(z^2)}{2} + \frac{w(z^3)}{3} + \ldots\right)$$

is analytic in the circle of radius $\sqrt{\tau} > \tau$. Write the first identity of 4.19a as

(1) $w(z)e^{-w(z)} = \varphi(z).$

By 4.7 this can also be written as

(2) $w(z) = t(\varphi(z)) = \sum_{n=1}^{\infty} \frac{n^n}{n!} (\varphi(z))^n.$

Consider first the values $0 \le x < \tau$. Since φ has positive coefficients, it follows by (1) that

$$0 < \varphi'(x) = w'(x)(1 - w(x))e^{-w(x)}, \qquad w(x) < 1.$$

Again by (1),

$$\varphi(x) < \frac{1}{e}.$$

By the positivity of the coefficients of φ, we must have $|\varphi(z)| < \frac{1}{e}$ for $|z| \le \tau$ except for $z = \tau$. We must have $\varphi(\tau) = 1/e$, otherwise (2) would define w analytically in a larger circle. Hence w is continuous at τ and $w(\tau) = 1$. Moreover, (2) defines w analytically along the boundary except for $z = \tau$.

Investigating now the behavior in the neighborhood of τ for $0 < x < \tau$, set $y = \tau - x$, $u = 1 - w$. Then (1) can be written in the form

$$\frac{u^2}{2} + \frac{u^3}{3} + \ldots + \frac{u^n}{n(n-2)!} + \ldots = 1 - e\varphi(x) = a_1 y + a_2 y^2 + \ldots,$$

where

$$a_1 = e\varphi'(\tau) > 0.$$

Hence

(3) $u\sqrt{\frac{1}{2} + \frac{u}{3} + \ldots} = \sqrt{y}\sqrt{a_1 + a_2 y + \ldots}$

(the square-roots are to be taken positive if y, u are small positive numbers). The left-hand side is analytic for sufficiently small u and its derivative is non-zero at 0, therefore u can be expressed as an analytic function of the right hand side in a sufficiently small neighborhood of 0. Consequently u is an analytic function of $\sqrt{y}$ for a sufficiently small y. Thus

$$(4) \qquad w(x) = \sum_{k=0}^{\infty} b_k (\tau - x)^{k/2},$$

where $b_0 = 1$ and by (3), $b_1 = -\sqrt{2a_1} < 0$.

(b) By (4) above, we have for $|z| < \tau$

$$(5) \qquad w''(z) = \sqrt{\frac{a_1}{8}}(\tau - z)^{-3/2} + \frac{3b_3}{4}(\tau - z)^{-1/2} + h(z),$$

where $h(z)$ has the following expansion in a neighborhood of τ;

$$(6) \qquad h(z) = \sum_{k=4}^{\infty} \frac{k(k-2)}{4} b_k (\tau - z)^{\frac{k-4}{2}}.$$

(5) implies that $h(z)$ is analytic in the open disc $|z| < \tau$ and even at all points of the boundary except $z = \tau$. (6) imples $h(z)$ is continuous even here. Therefore if we expand $h(z)$ around 0, we have

$$h(z) = \sum_{n=0}^{\infty} c_n z^n$$

where

$$c_n \frac{1}{2\pi i} \oint_{|z|=\tau} \frac{h(z)}{z^{n+1}} dz = O(\tau^{-n}).$$

Hence by (5), the coefficient of x^n in $w''(x)$ is

$$\sqrt{\frac{a_1}{8}}\binom{-\frac{3}{2}}{n}(-1)^n \tau^{-n-\frac{3}{2}} + \frac{3}{4}b_3\binom{-\frac{1}{2}}{n}(-1)^n \tau^{-n-\frac{1}{2}} + O(\tau^{-n}).$$

Here the second term is $0(\tau^{-n})$ as well, while the first term is asymptotic to

$$c\sqrt{n}\tau^{-n}$$

by elementary calculations using the Stirling formula. This proves the assertion of the problem.

21. Let $a_{1,\pi(1)}a_{2,\pi(2)}\cdots a_{n,\pi(n)}$ be an expansion term of per$A$. Observe that the number of 1-factors consisting of a $(u_1, v_{\pi(1)})$-edge, a $(u_2, v_{\pi(2)})$-edge, ..., a $(u_n, v_{\pi(n)})$-edge is exactly

$$a_{1,\pi(1)}a_{2,\pi(2)}\cdots a_{n,\pi(n)}$$

because we can choose a $(u_1, v_{\pi(1)})$-edge, a $(u_2, v_{\pi(2)})$-edge, etc. independently. This has to be summed over all permutations π, i.e. over all expansion terms. This proves the assertion. [see LP]

22. Let a_n denote the number required. We have

(1) $$a_0 = 1, \quad a_1 = 1.$$

Let F be any 1-factor of the ladder with n steps. There are two possibilities. If F starts with a vertical edge at the left end, then there are a_{n-1} possible ways to continue. If F starts with horizontal edges, then it can be one of a_{n-2} 1-factors. Hence

(2) $$a_n = a_{n-1} + a_{n-2}.$$

(1) and (2) show that a_n is the nth Fibonacci number (cf. 1.27).

23. The number of 1-factors in $K_{n,n}$ is, obviously, $n!$ If we have n edges removed from $K_{n,n}$, then we do not want to count those 1-factors of $K_{n,n}$ containing a removed edge. The number of 1-factors containing j given edges is $(n-j)!$. Hence, by inclusion-exclusion, the number under consideration is

$$\sum_{j=0}^{n}(-1)^j\binom{n}{j}(n-j)! = n!\sum_{j=0}^{n}\frac{(-1)^j}{j!}.$$

The sum here is a partial sum of the series for $\frac{1}{e}$ the remainder being less than $\frac{1}{(n+1)!}$. Hence

$$\left| n!\sum_{j=0}^{n}\frac{(-1)^j}{j!} - \frac{n!}{e} \right| < \frac{1}{n+1}.$$

Thus, our number is the integer next to $\frac{n!}{e}$.

24. Let $B = (b_{ij})_{i,j=1}^{n}$, then

(1) $$\det B = \sum \varepsilon(\pi)b_{1,\pi(1)}\cdots b_{n,\pi(n)},$$

where π ranges over all permutations of $\{1,\dots,n\}$ and $\varepsilon(\pi)$ is the sign of π. Suppose that the cycle decomposition of π contains an odd cycle $(i_1, i_2 \dots i_{2k+1})$, $k \geq 1$. Let π' be the permutation obtained by replacing $(i_1 i_2 \dots i_{2k+1})$ by $(i_{2k+1} i_{2k} \dots i_1)$. Then the expansion terms corresponding to π and π' cancel out as B is skew symmetric. Also, if π has a fixed element, then the corresponding expansion term is 0 because B has 0's in the diagonal. Thus it suffices to consider only those permutations π in (1) which decompose into even cycles (and these cycles partition $\{1,\ldots,n\}$). This proves the assertion if n is odd, since then $\det B = (\mathrm{Pf}\,B)^2 = 0$. Call two permutations *equivalent* if their cycle decompositions differ in the direction of cycles only. Then all terms corresponding to equivalent permutations are equal. For a permutation π with even cycles that partition $\{1,\ldots,n\}$, let $r(\pi)$ be the number of cycles and $s(\pi)$ the number of 2-cycles. Then the sum of all terms equivalent to π is

$$(-1)^{r(\pi)}2^{r(\pi)-s(\pi)}b_{1,\pi(1)}\cdots b_{n,\pi(n)}$$

and hence

$$\det B = \sum (-1)^{r(\pi)}2^{r(\pi)-s(\pi)}b_{1,\pi(1)}\cdots b_{n,\pi(n)},$$

where π now ranges over a system of representatives of equivalence classes.

Consider $(\mathrm{Pf}\,B)^2$. If we expand this, each term will have the form

(2)
$$\varepsilon\delta b_{i_1 i_2}\cdots b_{i_{n-1} i_n}b_{j_1 j_2}\cdots_{j_{n-1} j_n},$$

where $\{\{i_1,i_2\},\ldots,\{i_{n-1},i_n\}\}$, $\{\{j_1,j_2\},\ldots,\{j_{n-1},j_n\}\}$ are partitions and ε, δ are the corresponding signs.

Consider the graph on $\{1,\ldots,n\}$ whose edges are (i_{2l-1},i_{2l}) and (j_{2l-1},j_{2l}) $(l=1,\ldots,n)$. This graph consists of s edges which arise in both forms and $r-s$ circuits whose edges are alternatingly (i_{2l-1},i_{2l}) and (j_{2l-1},j_{2l}). We may assume that (i_{2l-1},i_{2l}) and (j_{2l-1},j_{2l}) are either identical or consecutive edges on these circuits. Then set

$$\varrho = \begin{pmatrix} i_1 i_2 \ldots i_n \\ j_1 j_2 \ldots j_n \end{pmatrix}.$$

ϱ is determined up to equivalence by the given term of $(\mathrm{Pf}\,B)^2$.

Also $\varepsilon\delta$ is the sign of ϱ. Hence, (2) is equal to

$$(-1)^{r(\varrho)}b_{1,\varrho(1)}\cdots b_{n,\varrho(n)}.$$

Moreover, we get equivalent permutations for $2^{r(\varrho)-s(\varrho)}$ terms of $(\mathrm{Pf}\,B)^2$. Thus

$$(\mathrm{Pf}\,B)^2 = \sum_{\varrho} (-1)^{r(\varrho)}2^{r(\varrho)-s(\varrho)}b_{1,\varrho(1)}\cdots b_{n,\varrho(n)} = \det B.$$

25. (a) We may assume that n is even, otherwise the assertion is trivial. Observe that each non-zero term in $\mathrm{Pf}\,B$ corresponds to a 1-factor of G and conversely. Moreover, each such term is 1 in absolute value. Thus, the number of 1-factors of G is $\geq |\mathrm{Pf}\,B|$.

(b) Equality holds iff all non-zero terms in $\mathrm{Pf}\,B$ have the same sign. To express this in terms of the orientations of circuits, let

$$\varepsilon b_{i_1 i_2}b_{i_3 i_4}\cdots b_{i_{n-1} i_n},$$

$$\delta b_{j_1 j_2} b_{j_3 j_4} \cdots b_{j_{n-1} j_n} \qquad (\varepsilon, \delta = \pm 1)$$

be two non-zero terms in $\operatorname{Pf} B$. We must have $(v_{i_{2l-1}}, v_{i_{2l}}) \in E(G)$, $(v_{j_{2l-1}}, v_{j_{2l}}) \in E(G)$. Let

$$F = (v_{i_1}, v_{i_2}), \ldots, (v_{i_{n-1}}, v_{i_n}),$$
$$F' = (v_{j_1}, v_{j_2}), \ldots, (v_{j_{n-1}}, v_{j_n})$$

be the two 1-factors of G corresponding to the expansion terms above. $F \cup F'$ decomposes into alternating circuits $C_1, \ldots, C_r$ and some edges belonging to both. Let us orient each C_i in an arbitrary way and let π denote the permutation corresponding to the resulting cycles. Since we have freedom in the ordering of the pairs $(i_1, i_2), \ldots, (i_{n-1}, i_n)$, as well as in the choice of order within the pairs, we may assume that $(v_{i_{2l-1}}, v_{i_{2l}})$, $(v_{j_{2l-1}}, v_{j_{2l}})$ are either identical or consecutive edges on the same (directed) cycle C_i.

Now the ratio, or product, of the two expansion terms of the Pfaffian is

(1) $$(\varepsilon\delta) b_{i_1 i_2} b_{j_1 j_2} b_{i_3 i_4} b_{j_3 j_4} \cdots b_{i_{n-1} i_n} b_{j_{n-1} j_n}.$$

Here $\varepsilon\delta$ is the sign of the permutation

$$\begin{pmatrix} i_1 i_2 \ldots i_n \\ j_1 j_2 \ldots j_n \end{pmatrix} = \pi,$$

i.e.

$$\varepsilon\delta = (-1)^r.$$

On the other hand, the product $b_{i_1 i_2} \ldots b_{j_{n-1} j_n}$ contains a -1 for every edge of $F \cup F'$ whose orientation in G does not agree with the direction on the C_i containing it (for the edges of $F \cap F'$ the corresponding two factors $b_{i_{2l-1} i_{2l}} b_{j_{2l-1} j_2}$ cancel each other out). Let s denote the number of the C_i for which the number of edges oriented in the given direction is even, then (1) is equal to

$$(-1)^r (-1)^{r-1} = (-1)^s.$$

Now it is easy to prove the equivalence of the three conditions.

(i)$\Longrightarrow$(ii): Assume that $|\operatorname{Pf} B|$ is equal to the number of 1-factors. Let C be a circuit which alternates with respect to a 1-factor F, and let

$$F' = F \bigtriangleup E(C).$$

The terms corresponding to F and F' must have the same sign, and so the observations above show that C has an odd number of edges oriented in each direction.

(ii)$\Longrightarrow$(iii): Trivial.

(iii)$\Longrightarrow$(i): If all circuits C of G which alternate with respect to a given 1-factor F_0 have the property that an odd number of their edges is oriented in a given direction, then it follows from the discussion above that for any 1-factor F, the term corresponding to F has the same sign as the term corresponding to F_0, and hence $|\mathrm{Pf}\,B|$ is the number of 1-factors of G. [P.W.Kasteleyn; see B, LP]

26. Write

$$\mathsf{E}(\det B) = \sum_{\pi} \varepsilon(\pi)\mathsf{E}(b_{1,\pi(1)} \cdots b_{n,\pi(n)}).$$

Here the $b_{i,j}$ are random variables with $\mathsf{E}(b_{i,j}) = 0$. If an expansion term contains $b_{i,j}$ where (i,j) is not an edge then it is always 0. If it contains $b_{i,j}$ but not $b_{j,i}$ for some edge (i,j), then this factor is independent of the other factors in the expansion term, and so the expectation of the expansion term is 0. Finally, if an expansion term contains $b_{i,j}$ whenever it contains $b_{j,i}$ (and thus corresponds to a 1-factor) then it is always 1 and so it contributes 1 to $\mathsf{E}(\det B)$. [C. D. Godsil and I. Gutman, in: *Algebraic Methods in Graph Theory*, (eds L. Lovász–V. T. Sós), Coll. Math. Soc. J. Bolyai **25** (1981), 241–249; LP]

27. First we prove the assertion of the hint by induction. If the graph is a tree, there is no bounded face and the assertion is trivial.

Suppose that G is not a tree. Remove an edge of G which is on the boundary of the infinite face and which also belongs to a circuit and orient the rest as required. Put this edge back, then all bounded faces but the one containing this edge have the property that going around their boundary in the positive sense, we find an odd number of edges oriented in the direction of our walk. A suitable orientation of e will guarantee this for the additional bounded face too.

Now we prove that any circuit C which alternates with respect to the given 1-factor F contains an odd number of edges oriented in one way (or the other) around the circuit. G contains an even number $2p$ of points inside the circuit C because F matches these points. Let $|V(C)| = 2k$, and denote by $A_1, \ldots, A_f$ the faces inside C; let q_i be the number of edges passed in the right direction, when going around A_i in the positive sense.

Consider $q_1 + \ldots + q_f$. Note that here the edges inside C are counted exactly once; the edges on C are counted iff they are oriented according to the positive orientation of C. Thus we are interested in the number

$$q = q_1 + \ldots + q_f - m \equiv f - m \pmod 2,$$

where m is the number of edges inside C.

Now consider the graph G' formed by C and those edges and points of G inside C. G' has $f + 1$ faces, $m + 2k$ edges and $2p + 2k$ points. By the Euler polyhedron theorem (cf. also 5.24),

$$f + 1 + 2p + 2k = m + 2k,$$

hence

$$m \equiv f + 1 \pmod 2$$

and

$$q \equiv f - m \equiv 1 \pmod 2$$

as stated [P. W. Kasteleyn; see B, LP].

28. If we consider the orientation of the ladder shown in Fig. 13 we find that the boundary of any bounded face contains an odd number (3) of edges going in the positive direction. Thus, by the preceding solution, any circuit which alternates with respect to a 1-factor contains an odd number of edges oriented one given way around the circuit.

The corresponding matrix B defined as in 4.24 is

$$\begin{pmatrix} A & I \\ -I & -A \end{pmatrix},$$

where

$$A = \begin{pmatrix} 0 & 1 & & & 0 \\ -1 & 0 & & & \\ & & \ddots & & \\ & & & 0 & 1 \\ 0 & & & -1 & 0 \end{pmatrix}, \qquad I = \begin{pmatrix} 1 & & & 0 \\ & \ddots & & \\ 0 & & & 1 \end{pmatrix}.$$

The number of 1-factors of the ladder is given by

$$\det \begin{pmatrix} A & I \\ -I & -A \end{pmatrix}^{1/2} = \det \begin{pmatrix} 1 & I - A^2 \\ -I & 0 \end{pmatrix}^{1/2} = (\det(I - A^2))^{1/2} =$$

$$= (\det(I - A)\det(I + A))^{1/2} = \det(I + A),$$

because

$$I - A = (I + A)^T.$$

It is quite easy to verify by induction that this is the same as the nth Fibonacci number.

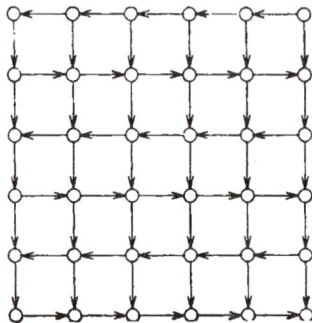

FIGURE 26

29. (a) What we have to enumerate is the number of 1-factors of a $(2n) \times (2n)$ "lattice", i.e. the graph shown in Fig. 26. If we consider the orientation given there, it will satisfy the condition that the boundary of any bounded face contains one edge going one way and three other way. Hence by the solution of 4.27, every circuit which alternates with respect to a given 1-factor has an odd number of edges oriented in a given direction around the circuit. Hence, if we form the corresponding matrix B as in 4.24, the number of 1-factors will be equal to $\sqrt{\det B}$.

Now B can be written in the form

(1)
$$
B = \begin{pmatrix}
A & I & & & & & 0 \\
-I & -A & I & & & & \\
 & -I & \ddots & \ddots & & & \\
 & & \ddots & \ddots & \ddots & & \\
 & & & \ddots & \ddots & \ddots & \\
0 & & & & \ddots & A & I \\
 & & & & & -I & -A
\end{pmatrix}
\underbrace{}_{n}
$$

where

$$
A = \begin{pmatrix}
0 & 1 & & 0 \\
-1 & 0 & \ddots & \\
 & \ddots & \ddots & 1 \\
0 & & -1 & 0
\end{pmatrix}.
$$

If we multiply the first column, then the 3rd and 4th row, then the 4th and 5th column, then the 7th and 8th row, etc. of the partitioned matrix (1) by -1, we get the partitioned matrix

$$
B' = \begin{pmatrix}
-A & I & & & & & 0 \\
I & -A & I & & & & \\
 & I & \ddots & \ddots & & & \\
 & & \ddots & \ddots & \ddots & & \\
 & & & & \ddots & -A & I \\
0 & & & & & I & -A
\end{pmatrix}.
$$

Set

$$
A' = \begin{pmatrix}
0 & 1 & & & 0 \\
1 & 0 & 1 & & \\
 & 1 & \ddots & \ddots & \\
 & & \ddots & \ddots & \ddots \\
 & & & \ddots & \ddots & 1 \\
0 & & & & 1 & 0
\end{pmatrix}, \qquad p_n(\lambda) = \det(A' - \lambda I),
$$

then

$$
\det B = \det B' = \det p_n(A).
$$

By 1.29, the eigenvalues of B are $2\cos\frac{k\pi}{2n+1}$, $k=1,\ldots,2n$. Hence

$$\det p_n(A) = \det \prod_{k=1}^{2n} \left(2\cos\frac{k\pi}{2n+1}I - A\right) = \prod_{k=1}^{2n} \det\left(2\cos\frac{k\pi}{2n+1}I - A\right).$$

Let

$$q_n(\lambda) = \det(A - \lambda I),$$

then, in turn,

$$\det B = \prod_{k=1}^{2n} q_n\left(2\cos\frac{k\pi}{2n+1}\right).$$

The eigenvalues of A can be determined in the same way as those of A' and they turn out to be $2i\cos\frac{k\pi}{2n+1}$, $k=1,\ldots,2n$. Thus

$$\det B = \prod_{k=1}^{2n}\prod_{l=1}^{2n}\left(2\cos\frac{k\pi}{2n+1} - 2i\cos\frac{l\pi}{2n+1}\right) =$$

$$= 2^{4n^2}\prod_{k=1}^{2n}\prod_{l=1}^{2n}\left|\cos\frac{k\pi}{2n+1} - i\cos\frac{l\pi}{2n+1}\right| =$$

$$= 2^{4n^2}\prod_{k=1}^{2n}\prod_{l=1}^{2n}\left(\cos^2\frac{k\pi}{2n+1} + \cos^2\frac{l\pi}{2n+1}\right)^{1/2} =$$

$$= 2^{4n^2}\prod_{k=1}^{n}\prod_{l=1}^{n}\left(\cos^2\frac{k\pi}{2n+1} + \cos^2\frac{l\pi}{2n+1}\right)^{2}.$$

Hence,

$$a_n = \sqrt{\det B} = 2^{2n^2}\prod_{k=1}^{n}\prod_{l=1}^{n}\left(\cos^2\frac{k\pi}{2n+1} + \cos^2\frac{l\pi}{2n+1}\right) =$$

$$= \prod_{k=1}^{n}\prod_{l=1}^{n}\left(4\cos^2\frac{k\pi}{2n+1} + 4\cos^2\frac{l\pi}{2n+1}\right).$$

(b) The polynomial with roots $4\cos^2\frac{k\pi}{2n+1}$, $k=1,\ldots,n$ is

$$f(x) = x^n - \binom{2n-1}{1}x^{n-1} + \binom{2n-2}{2}x^{n-2}\ldots$$

Observe that a_n is the resultant of $f(x)$ and $f(-x)$. By the Sylvester form of the resultant,

$$
a_n = \left|
\begin{array}{cccccc}
1 & -\binom{2n-1}{1} & \binom{2n-2}{2} & \cdots & & \\[2mm]
0 & 1 & -\binom{2n-1}{1} & \cdots & & \\[2mm]
\vdots & & & \ddots & & \\[1mm]
0 & & & & & \cdot \\[2mm]
\hdashline
1 & \binom{2n-1}{1} & \binom{2n-2}{2} & \cdots & & \\[2mm]
0 & 1 & \binom{2n-1}{1} & \cdots & & \\[2mm]
\vdots & & & \ddots & & \\[1mm]
0 & & & & &
\end{array}
\right|
\begin{array}{l} \\ \left.\rule{0pt}{18mm}\right\} n \text{ rows} \\ \\ \left.\rule{0pt}{18mm}\right\} n \text{ rows.} \end{array}
$$

$$\underbrace{}_{2n \text{ columns}}$$

Add the $(n+k)^{\text{th}}$ row to the kth $(k=1,\ldots,n)$, then divide the first n rows by 2 and subtract the kth row from the $(n-k)^{\text{th}}$. The resulting determinant is

$$
a_n = 2^n \left|
\begin{array}{cccccc}
\binom{2n}{0} & 0 & \binom{2n-2}{2} & 0 & \cdots & \\[2mm]
0 & \binom{2n}{0} & 0 & \binom{2n-2}{2} & \cdots & \\[2mm]
\vdots & & & \ddots & & \\[1mm]
\hdashline
0 & \binom{2n-1}{1} & 0 & \binom{2n-3}{3} & \cdots & \\[2mm]
0 & 0 & \binom{2n-1}{1} & 0 & \cdots & \\[2mm]
\vdots & & & & \ddots &
\end{array}
\right|
\begin{array}{l} \left.\rule{0pt}{12mm}\right\} n \text{ rows} \\ \\ \left.\rule{0pt}{12mm}\right\} n \text{ rows.} \end{array}
$$

$$\underbrace{}_{2n \text{ columns}}$$

Suppose that, e.g., n is even. Then the last column looks like this:

$$a_n = 2^n \left|
\begin{array}{cccccc}
\binom{2n}{0} & 0 & \binom{2n-2}{2} & 0 & \cdots & 0 \\[1em]
0 & \binom{2n}{0} & 0 & \binom{2n-2}{2} & \cdots & 0 \\[1em]
\vdots & & \ddots & & & \vdots \\[1em]
0 & \cdots & & \binom{2n}{0} & \cdots & \binom{n}{n} \\[1em]
0 & \binom{2n-1}{1} & 0 & \binom{2n-3}{3} & \cdots & 0 \\[1em]
\vdots & & \ddots & & & \vdots \\[1em]
0 & \cdots & & \binom{2n-1}{1} & \cdots & 0
\end{array}
\right| \begin{array}{l} \left.\begin{array}{c} \\[3em] \end{array}\right\} n \text{ rows} \\[1em] \left.\begin{array}{c} \\[3em] \end{array}\right\} n \text{ rows.} \end{array}$$

$$\underbrace{\hphantom{xxxxxxxxxxxxxxxxxxxxxxxxxxxxxxx}}_{2n \text{ columns}}$$

Thus expanding by the first and last columns we get

$$a_n = 2^n \left|
\begin{array}{ccccc}
\binom{2n}{0} & 0 & \binom{2n-2}{2} & 0 & \cdots \\[1em]
0 & \binom{2n}{0} & 0 & \binom{2n-2}{2} & \cdots \\[1em]
\vdots & & \ddots & & \\[1em]
\hdashline
\binom{2n-1}{1} & 0 & \binom{2n-3}{3} & 0 & \cdots \\[1em]
0 & \binom{2n-1}{1} & 0 & \binom{2n-3}{3} & \cdots \\[1em]
\vdots & & \ddots & &
\end{array}
\right| \begin{array}{l} \left.\begin{array}{c} \\[2.5em] \end{array}\right\} n-2 \text{ rows} \\[1em] \left.\begin{array}{c} \\[2.5em] \end{array}\right\} n \text{ rows.} \end{array}$$

Now this determinant is the direct product of two determinants of the form

$$
\left. \begin{vmatrix}
\dbinom{2n}{0} & \dbinom{2n-2}{2} & \dbinom{2n-4}{4} & \cdots \\[2ex]
0 & \dbinom{2n}{0} & \dbinom{2n-2}{2} & \cdots \\[2ex]
\vdots & & \ddots \\[2ex]
\end{vmatrix} \right\} \; \frac{n-2}{2} \text{ rows}
$$

$$
\left. \begin{vmatrix}
\dbinom{2n-1}{1} & \dbinom{2n-3}{3} & \dbinom{2n-5}{5} & \cdots \\[2ex]
0 & \dbinom{2n-1}{1} & \dbinom{2n-3}{3} & \cdots \\[2ex]
\vdots & & \ddots \\[2ex]
\end{vmatrix} \right\} \; \frac{n}{2} \text{ rows.}
$$

This proves the assertion. The case of odd n follows similarly.

(c) We have

$$
\frac{\log a_n}{n^2} = \frac{1}{n^2} \sum_{k=1}^{n} \sum_{l=1}^{n} \log \left(4 \cos^2 \frac{k\pi}{2n+1} + 4 \cos^2 \frac{l\pi}{2n+1} \right).
$$

The right-hand side tends to

(1) $$
\frac{4}{\pi^2} \int_0^{\pi/2} \int_0^{\pi/2} \log(4 \cos^2 x + 4 \cos^2 y) \, dx \, dy = c \approx 1,17.
$$

Thus,

$$
\frac{\log a_n}{n^2} \to c \approx 1,17.
$$

[P. W. Kasteleyn; see LP, and E. W. Montroll, in: *Applied Combinatorial Mathematics*, (E. E. Beckenbach, ed.) Wiley, 1964; see this last reference for a series expansion of c]

30. (a) Let G denote the $(2n-1) \times (2n-1)$ lattice; its points are (i,j) $(0 \le i \le 2n-2,\ 0 \le j \le 2n-2)$. Call point (i,j) *black* if i and j are both even; *green* if one of them is odd; *red* if both are odd. The black points form an $n \times n$ lattice graph H. Let H be any spanning tree of H. Let $a = (2n-2, 2n-2)$ and $x \ne a$, a black point. Then there is a well-defined first edge on the path in T connecting x to a and this contains a green point x'.

Let y be a red point. In the lattice of red points, there is a unique path not crossing T which connects y to the outside boundary of the lattice G and on this, there is a first edge which contains a green point. Let y' be this green point.

The pairs (x, x'), (y, y') form, as is easily verified, a 1-factor of $G-a$ (Fig. 27).

Conversely, let F be a 1-factor of $G-a$. Consider the set T of those edges of H which contain an edge of F. These form a spanning tree. In fact, the number of edges of F adjacent to black points is n^2-1, so T contains $n^2-1=|V(H)|-1$ edges; it suffices to prove they do not form circuit. Suppose, by way of contradiction, that they form a circuit C. The number of points of G inside C is odd (this easily follows, e.g. by induction on the length of C) and so, F cannot match them, a contradiction.

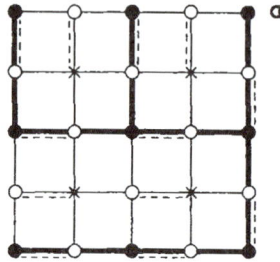

FIGURE 27

Thus we have established a one-to-one correspondence between spanning trees of H and 1-factors of $G-x$.

(b) Consider G embedded in the plane and let G^* be its dual. Then $G \cup G^*$ can be considered as a graph H on $|V(G)|+|V(G^*)|+|E(G)|$ vertices and $4|E(G)|$ edges. Fix vertices $a \in V(G)$ and $b \in V(G^*)$, and consider any spanning tree T of G. This corresponds to a tree T' in G. Moreover, those edges of G^* not crossing T form a spanning tree of G^*, which corresponds in H to a tree T''. Clearly, T' and T'' are vertex-disjoint and together cover all vertices of H.

Observe that every vertex of T' at an even distance from a has degree 2; hence $T'-a$ has a unique 1-factor F'_T. Similarly, $T''-b$ has a unique 1-factor F''_T, and $F_T = F'_T \cup F''_T$ is a 1-factor of $H - \{a,b\}$. It is easy to see (similarly as in part (a)) that every 1-factor of $H - \{a,b\}$ arises from a unique spanning tree of G by this construction. [H. N. V. Temperley, in: *Combinatorics*, London Math. Soc. Lecture Notes Series **13** (1974) 202–204].

31. Orient the edge (u_i, v_j) from u_i to v_j and then identify u_i and v_i $(i=1,\ldots,n)$, our graph is thereby mapped onto the transitive tournament T_n.

If M is a k-element matching in G, then M is mapped onto a set M' of edges of T_n such that at most one edge of M' enters and leaves any given point. Hence M' consists of disjoint directed paths; if we consider the points of T_n not touched by M' as one-element paths, we can observe that M corresponds to a system of $n-k$ disjoint directed paths which cover all points of T_n. Conversely, the edges of a system of $n-k$ disjoint directed paths covering $V(T_n)$ correspond to a k-element matching of G.

Now let $P_1, \ldots, P_{n-k}$ be disjoint directed paths which cover $V(T_n)$. Then $\{V(P_1), \ldots, V(P_{n-k})\}$ is a partition of $V(T_k)$. Each partition of $V(T_k)$ into $n-k$

classes arises uniquely in this way; if $\{V_1, \ldots, V_{n-k}\}$ is a partition of $V(T_n)$ and P_i is the unique Hamiltonian path of the (transitive) subtournament induced by V_i, then $V(P_i) = V_i$ and $P_1, \ldots, P_{n-k}$ are disjoint directed paths that cover $V(T_n)$.

Hence, there is a one-to-one correspondence between k-element matchings of G and partitions of $V(T_n)$ into $n - k$ (non-empty) classes. Thus the number in question is $\left\{ {n \atop n-k} \right\}$.

32. Let F_n denote the number of such permutations. By the assumption, $\pi(n) = n$ or $\pi(n) = n-1$. The number of permutations π with the desired property which fix n is F_{n-1}. The number of those with $\pi(n) = n-1$ is F_{n-2}, because, then it necessarily exchanges n and $n-1$ (nothing else can be mapped onto n). Thus

$$F_n = F_{n-1} + F_{n-2}.$$

Since $F_0 = F_1 = 1$, F_n is the nth Fibonacci number.

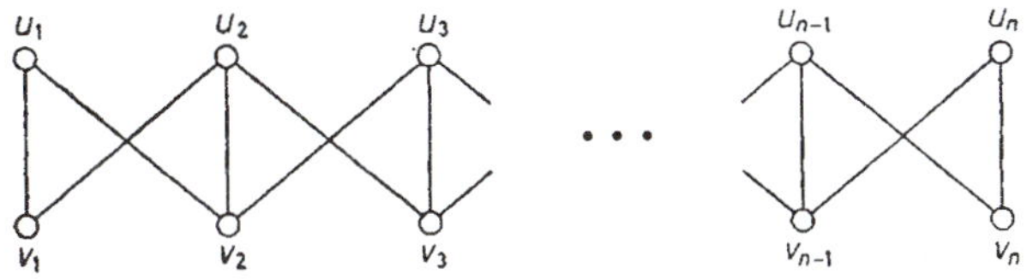

FIGURE 28

One can formulate the problem as follows. Let $\{v_1, \ldots, v_n\}$, $\{u_1, \ldots, u_n\}$ be two disjoint sets of points. Join u_i to v_j iff $|i - j| \leq 1$ (Fig. 28). If π is any admissible permutation, then the pairs $(i, \pi(i))$ form a 1-factor of the resulting graph G and conversely. Thus we want to know the number of 1-factors of the graph G. Since G is isomorphic to the graph in Fig. 1 (problem 4.22) it follows that this number is the nth Fibonacci number.

33. It is easy to see that

$$a_n = \text{per} \begin{pmatrix} 1 & 1 & 1 & & & & & \\ 1 & 1 & 1 & 1 & & & & 0 \\ 1 & 1 & 1 & & \ddots & & & \\ & 1 & & \ddots & & \ddots & & \\ & & \ddots & & \ddots & & 1 & \\ 0 & & & \ddots & & 1 & 1 & \\ & & & & 1 & 1 & 1 \end{pmatrix}}_{n}$$

Expanding this by the first row we get

$$a_n = a_{n-1} + b_{n-1} + c_{n-1},$$

where

$$b_n = \operatorname{per} \begin{pmatrix} 1 & 1 & 1 & & & & & & \\ 1 & 1 & 1 & 1 & & & & \Large 0 & \\ 0 & 1 & 1 & 1 & \ddots & & & & \\ & 1 & 1 & 1 & & \ddots & & \\ & & 1 & & & & & 1 \\ \Large 0 & & & \ddots & & & 1 & 1 \\ & & & & & 1 & 1 & 1 \end{pmatrix},$$

$$\underbrace{}_{n}$$

$$c_n = \operatorname{per} \begin{pmatrix} 1 & 1 & 1 & & & & & & \\ 1 & 1 & 1 & 1 & & & & \Large 0 & \\ 0 & 1 & 1 & 1 & \ddots & & & & \\ & 0 & 1 & 1 & & \ddots & & \\ & & 1 & & & & & 1 \\ \Large 0 & & & \ddots & & & 1 & 1 \\ & & & & & 1 & 1 & 1 \end{pmatrix}.$$

$$\underbrace{}_{n}$$

To get a complete recurrence relation we also expand b_n, c_n by their first columns:

(2) $b_n = a_{n-1} + b_{n-1};$

but with c_n we get a new permanent:

(3) $c_n = b_{n-1} + d_{n-1},$

where

$$d_n = \operatorname{per} \begin{pmatrix} 1 & 1 & 0 & & & & & & \\ 1 & 1 & 1 & 1 & & & & \Large 0 & \\ 0 & 1 & 1 & 1 & 1 & & & & \\ & 1 & 1 & 1 & & \ddots & & \\ & & 1 & & & & & 1 \\ \Large 0 & & & \ddots & & & 1 & 1 \\ & & & & & 1 & 1 & 1 \end{pmatrix}.$$

$$\underbrace{}_{n}$$

Expand d_n too:

(4) $d_n = a_{n-1} + e_{n-1},$

where

$$e_n = \text{per} \begin{pmatrix} 1 & 1 & 1 & & & & & \\ 0 & 1 & 1 & 1 & & & \Large 0 & \\ 0 & 1 & 1 & 1 & 1 & & & \\ & & 1 & 1 & 1 & & \ddots & \\ & & & 1 & & & & 1 \\ & & & & \ddots & & 1 & 1 \\ \Large 0 & & & & & 1 & 1 & 1 \end{pmatrix}.$$

$$\underbrace{\hspace{6cm}}_{n}$$

Trivially,

(5) $$e_n = a_{n-1}$$

(1)–(5) yield recurrence relations for a_n, b_n, c_n, d_n, e_n, starting with the following values for $n=2$: $a_2=b_2=c_2=d_2=2$, $e_2=1$. Set

$$A = \begin{pmatrix} 1 & 1 & 1 & 0 & 0 \\ 1 & 1 & 0 & 0 & 0 \\ 0 & 1 & 0 & 1 & 0 \\ 1 & 0 & 0 & 0 & 0 \\ 1 & 0 & 0 & 0 & 1 \end{pmatrix}, \qquad \mathbf{v}_n = \begin{pmatrix} a_n \\ b_n \\ c_n \\ d_n \\ e_n \end{pmatrix},$$

then

$$\mathbf{v}_n = A\mathbf{v}_{n-1} = \ldots = A^{n-2}\mathbf{v}_2 = A^{n-1}\mathbf{v}_1 = A^n\mathbf{v}_0,$$

where

$$\mathbf{v}_0 = \begin{pmatrix} 1 \\ 0 \\ 0 \\ 0 \\ 0 \end{pmatrix}, \qquad \mathbf{v}_1 = \begin{pmatrix} 1 \\ 1 \\ 0 \\ 1 \\ 1 \end{pmatrix}$$

by convention. Hence,

$$\sum_{n=0}^{\infty} t^n \mathbf{v}_n = \left(\sum_{n=0}^{\infty} t^n A^n \right) \mathbf{v}_0 = (I - tA)^{-1}\mathbf{v}_0.$$

We are interested in the first entry of $(I-tA)^{-1}\mathbf{v}_0$ i.e. the upper left entry of $(I-tA)^{-1}$. By Cramer's rule this is equal to

$$f(t) = \frac{\begin{vmatrix} 1-t & 0 & 0 & 0 \\ -t & 1 & -t & 0 \\ 0 & 0 & 1 & -t \\ 0 & 0 & 0 & 1 \end{vmatrix}}{\begin{vmatrix} 1-t & -t & -t & 0 & 0 \\ -t & 1-t & 0 & 0 & 0 \\ 0 & -t & 1 & -t & 0 \\ -t & 0 & 0 & 1 & -t \\ -t & 0 & 0 & 0 & 1 \end{vmatrix}} = \frac{1-t}{t^5 - 2t^3 - 2t + 1}.$$

[D. H. Lehmer, in: *Comb. Theory Appl.* Coll. Math. Soc. J. Bolyai **4**, Bolyai–North-Holland (1970) 755–770.]

34. Expanding the permanent in the hint by its first row we get

$$u_{n,p} = p \cdot u_{n-1,p}.$$

Hence

$$u_{n,p} = pu_{n-1,p} = \ldots = p^{n-p}u_{p,p} = p^{n-p}p!\,.$$

35. We want to determine

(1)

$$\text{per} \begin{pmatrix} 1 & \cdots & & 1 & 0 & \cdots & 0 \\ & \vdots & & & & \ddots & \vdots \\ & 1 & & & & & 0 \\ 0 & & & & & & 1 \\ \vdots & \ddots & & & & & \vdots \\ 0 & \cdots & 0 & 1 & \cdots & & 1 \end{pmatrix} \begin{matrix} \left.\vphantom{\begin{matrix}1\\1\\1\end{matrix}}\right\} p \\ \\ \left.\vphantom{\begin{matrix}1\\1\\1\end{matrix}}\right\} n-p \end{matrix} \quad ,$$

$$\underbrace{\quad\quad}_{p} \underbrace{\quad\quad}_{n-p}$$

i.e. the number of those expansion terms of

$$\text{per} \begin{pmatrix} 1 & \cdots & 1 \\ \vdots & & \vdots \\ 1 & \cdots & 1 \end{pmatrix},$$

$$\underbrace{\quad\quad\quad}_{n}$$

which do not contain any entry from the two triangles which contain 0's in (1). By inclusion-exclusion,

$$a(n,p) = \sum_{U,V} (-1)^{|U|+|V|} K_{U \cup V},$$

where U is a subset of the entries in the triangle in the lower left corner and V is a set of entries in the triangle in the upper right corner, and $K_{U \cup V}$ is the number of expansion terms containing $U \cup V$. Obviously, it suffices to consider those sets U, V in which no two entries are in a row or column.

We have

$$K_{U \cup V} = (n - |U \cup V|)!\,.$$

Thus our number is

$$a(n,p) = \sum_{U,V} (-1)^{|U|+|V|}(n - |U| - |V|)! = \sum_{k=0}^{p} \sum_{l=0}^{p} (-1)^{k+1}(n - k - l)! \sum_{\substack{|U|=k \\ |V|=l}} 1.$$

By 4.31, the number of sets U under consideration with $|U|=k$ is $\left\{ \begin{matrix} p+1 \\ p+1-k \end{matrix} \right\}$ and,

similarly, we have $\left\{ \begin{matrix} p+1 \\ p+1-l \end{matrix} \right\}$ sets V. Thus the result is

$$a(n,p) = \sum_{k=0}^{p}\sum_{l=0}^{p}(-1)^{k+l}(n-k-l)!\left\{ \begin{matrix} p+1 \\ p+1-l \end{matrix} \right\}\left\{ \begin{matrix} p+1 \\ p+1-k \end{matrix} \right\},$$

and so

$$\frac{a(n,p)}{(n-2p)!} =$$

$$= \sum_{k=0}^{p}\sum_{l=0}^{p}(-1)^{k+l}\left\{ \begin{matrix} p+1 \\ p+1-l \end{matrix} \right\}\left\{ \begin{matrix} p+1 \\ p+1-k \end{matrix} \right\}(n-k-l)(n-k-l-1)\ldots(n-2p+1).$$

36. If an expansion terms of the permanent contains k 1's from the upper left block, then it contains $n-k$ 1's from the upper right block, $n-k$ 1's from the lower left block and k 1's from the lower right block. If we select the $n-k$ 1's from the upper right block and $n-k$ 1's from the lower left block, then there are $(k!)^2$ possible ways to select the rest.

Observe that the $(n-k)$-tuples of 1's, no two in a row or column, out of the upper right $n \times n$ block corresponds to $(n-k)$-element matchings of the graph defined in 4.31; hence their number is $\left\{ \begin{matrix} n \\ k \end{matrix} \right\}$. Similarly for the lower left corner. Thus the number of selections for a fixed k is

$$\left\{ \begin{matrix} n \\ k \end{matrix} \right\}^2 (k!)^2,$$

which proves the assertion. [K. Vesztergombi, *Studia Sci. Math. Hung.* **9** (1974) 181–185.]

§ 5. Parity and duality

1. (a) Summing the degrees of a graph, we get every edge counted twice (once for each of its endpoints). Thus, *the sum of degrees is twice the number of edges.*

As $3+3+3+3+5+6+6+6+6+6$ is odd (there is an odd number of odd summands), this sequence is not degree-sequence of any graph. This gives generally that *the number of odd degrees is even.*

(b) Let $d_1,\ldots,d_k$ be the degrees of the "upper" points, $r_1,\ldots,r_e$ the degrees of "lower" points of a bipartite graph. Then both $d_1+\ldots+d_k$ and $r_1+\ldots+r_e$ give the number of edges, hence

(1) $$d_1 + \ldots + d_k = r_1 + \ldots + r_e.$$

Now suppose that the degrees are as given in the problem. Let $d_1=5$ (say). Then the right-hand side of (1) is divisible by 3, while the left-hand side is not. Thus, this sequence is not the degree-sequence of any bipartite graph (it is, however, the degree sequence of a graph).

(c) Since the point with degree 9 is adjacent to all other points, it must in particular be adjacent to both points of degree 1. Hence, the point with degree 8 cannot be adjacent to either of the two points of degree 1; but then there are only 7 more points left and so, it cannot have degree 8 [cf. problem 7.51].

2. Assume that there is a k-regular simple graph on n points. By the solution of 5.1a, $k \cdot n$ is twice the number of edges, thus

(1) $$k \cdot n \text{ is even};$$

also trivially

(2) $$k \leq n - 1.$$

We show that, if (1) and (2) are satisfied, there is a k-regular simple graph on n points.

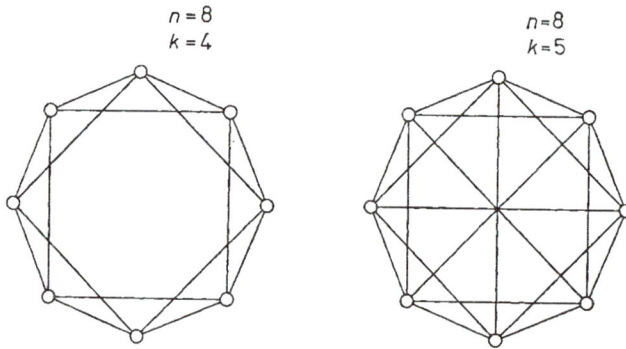

$$n = 8$$
$$k = 4$$

$$n = 8$$
$$k = 5$$

FIGURE 29

Case 1. k is even. Consider the vertices of a regular n-gon and join each of them to its neighbors, 2nd neighbors, etc., $(k/2)^{\text{th}}$ neighbors. Since $k < n$, this gives no multiple edges (Fig. 29).

Case 2. k is odd. Then by (1), n is even. Consider the graph of degree $k-1$ on n points constructed in Case 1. By (2), the endpoints of the longest diagonals are not joined; thus, if we add these edges, we get a k-regular simple graph.

3. If the graph G is bipartite and C is a circuit of G, then C, as a subgraph of G, is bipartite as well and hence it is even. Now suppose that G has only even circuits and let P_1, P_2 be (x,y)-paths. We prove by induction on $V(P_1)$ that P_1, P_2 have the same parity. If $P_1 \cup P_2$ is a circuit, then, since it is even, P_1 and P_2 have the same parity.

So suppose that P_1, P_2 have a point z in common. Let z split them into pieces P_1', P_1'' and P_2', P_2'', respectively. The notation may be chosen so that P_1', P_2' are (x,z)-path and hence by the induction hypothesis, they have the same parity. Similarly, P_1'' and P_2'' have the same parity, which proves that so do $P_1 = P_1' \cup P_1''$ and $P_2 = P_2' \cup P_2''$.

To prove 2-colorability we may assume that G is connected. Let $\pi(x,y)$ denote the common parity of (x,y)-paths ($\pi(x,y)=0$ or 1). Let $x_0 \in V(G)$ and set

$$S_i = \{y : \pi(x_0, y) = i\} \quad (i = 0, 1).$$

Then $\{S_0, S_1\}$ is a good bicoloration; for let (u,v) be an edge and $u \in S_0$ (say), and consider an (x_0, u)-path P. If $v \notin V(P)$, then $P + (u,v)$ is an (x_0, v)-path, whence

$$\pi(x_0, v) = \pi(x_0, u) + 1 = 1.$$

If v is on P, and divides it into an (x_0, v)-piece P_1 and a (v, u)-piece P_2, then $P_2 + (u, v)$ is a circuit and hence, it is even. This means that P_2 has odd length. Now

$$0 = \pi(x_0, u) \equiv |E(P)| = |E(P_1)| + |E(P_2)| \equiv \pi(x_0, v) + 1 \pmod{2}.$$

This shows that $\pi(x_0, v) = 1$, i.e. $v \in S_1$.

The statement is not true for digraphs, since the triangle with transitive orientation is not bipartite and has no odd cycles (no cycles at all, in fact).

However, it remains true for strongly connected digraphs. To show this, first one proves that if G has no odd cycles and P is an (x,y)-path while Q is a (y,x)-path, then P, Q have the same parity. This goes exactly like the first part of the proof above. Now let P_1, P_2 be two (x,y)-paths, then choose a (y,x)-path Q (here we use the fact that G is strongly connected), then P_1 and Q have the same parity. But so do P_2 and Q, and therefore P_1 and P_2 have the same parity. The rest of the proof runs exactly as in the undirected case.

4. If there is a potential $p(x)$, the work on any walk is the change in potential between the endpoints. In particular, it is 0 on any closed walk, so on any circuit.

Now suppose that the sum of work on any circuit is 0. We claim it is 0 on any closed walk W. We use induction on the length of W. If W is a circuit the statement is the assumption. So suppose that W uses a point x twice. x splits W into two smaller closed walks, W_1, W_2. The work on W_1 and on W_2 is 0 by the induction hypothesis, but then so is the work on W.

Now if P_1 and P_2 are any two (x,y)-walks we can form a closed walk W by going on P_1 from x to y and then returning on P_2. Since W has total work 0, P_1 and P_2 need the same work.

Now fix a point x_0 and let $p(y)$ be defined as the work needed to go from x_0 to y. By the above, this is independent of the way we go from x_0 to y. It is easy to see that this potential function satisfies the requirements.

We remark that the values of $v(e)$ and $p(x)$ can be from an arbitrary group. Moreover, if this group is such that each element has order 2, the orientation of G plays no role.

5. First let C be an even cycle. We can 2-color C in such a way that each point of C is joined to a point of opposite color, simply coloring its points alternatingly. If C does not span G, we extend this 2-coloration as follows. There is a point x_1 joined to $\tilde{C}$, because G is strongly connected; similarly, select points $x_2, \ldots, x_m$ so that x_i is joined to $C \cup \{x_1, \ldots, x_{i-1}\}$ $(i = 2, \ldots, m)$. Finally, every point of G

not on C becomes one of the x_i's. Now if $C \cup \{x_1, \ldots, x_{i-1}\}$ is already 2-colored, give x_i the color opposite to the color of a point of $C \cup \{x_1, \ldots, x_{i-1}\}$ which x_i is joined to. Obviously, this coloration has the desired property.

Suppose now that there is a 2-coloration of G with the property in consideration, with color classes S_1, S_2. Let G_0 denote the graph consisting of the edges between S_1 and S_2. Since G_0 has positive outdegrees by the assumption, it contains a cycle C. Since G_0 is bipartite, C has even length.

6. Let L be a maximum closed trail in G. If L does not contain all the edges, then there is an edge $e(x,y)$ which has a point in common with L; for otherwise L would be a connected component of G which is impossible because G is weakly connected. We may assume that e has its trail x on L.

Let us start with this edge and walk along edges of $G - E(L)$, using an edge only once as long as we can. We cannot get stuck at any point $z \neq x$; for there are as many edges of $G - E(L)$ leaving it as running into it and any time we use an edge leaving z we had to use one running into it before.

Thus we get stuck at x; now let us go on to traverse L starting with x. This gives us a trail longer than L.

7. (a) The indegree and outdegree of any point are both equal to n. Moreover, G is strongly connected, since

$$(a_1, \ldots, a_k), \ (a_2, \ldots, a_k, b_1), \ (a_3, \ldots, b_2), \ \ldots, \ (a_k, b_1, \ldots, b_{k-1}), \ (b_1, \ldots, b_k)$$

is a directed walk from $(a_1, \ldots, a_k)$ to $(b_1, \ldots, b_k)$. Thus G is Eulerian.

(b) The case $k = 2$ is trivial. Let $k \geq 3$. If (a,b) is an edge of $G_{k-1,n}$, then $a = (x_1, \ldots, x_{k-1})$, $b = (x_2, \ldots, x_k)$ for some $1 \leq x_i \leq n$ by definition. Let us associate the point $(x_1, \ldots, x_k)$ of $G_{k,n}$ with the edge (a,b). It is immediate that this yields an isomorphism between $L(G_{k-1,n})$ and $G_{k,n}$. An Euler trail of $G_{k-1,n}$ yields a Hamiltonian circuit of $L(G_{k-1,n})$.

8. Let $(a_1, \ldots, a_{2k})$ be a Hamiltonian cycle in $G_{k,2}$. Then, by definition of adjacency, $a_1 = (x_1, \ldots, x_k)$, $a_2 = (x_2, \ldots, x_k, x_{k+1})$, $\ldots$, $a_{2k} = (x_{2k}, x_1, \ldots, x_{k-1})$. Let the numbers $x_1, \ldots, x_{2k}$ be associated with the points of the cycle. Then the arcs of length k yield the 01-sequences $a_1, \ldots, a_{2k}$. Since a Hamiltonian cycle contains each point exactly once, the arcs yield each 01-sequence exactly once.

9. If the pieces $L_1, \ldots, L_d$ described in the hint are given, any permutation of them determines an Euler trail, and two of these Euler trails are the same iff the two permutations are the same as cyclic permutations. Thus there are $(d-1)!$ Euler trails obtained from $L_1, \ldots, L_d$. This proves that the total number of Euler trails is divisible by $(d-1)!$ which is even because $d \geq 3$.

10. We cannot get stuck anywhere except at x_0; this follows just as in 5.6. Assume now that there are edges of G not on the trail L we have traversed. L uses every edge incident to x_0; this is obvious for the edges leaving x_0 and follows for the incoming edges by $d^+(x_0) = d^-(x_0)$. If $(x,y) \notin L$, then the (unique) edge of T starting from x does not belong to L either, by $(*)$. Thus, there are edges of T not on L. Let us consider one whose head z is nearest to x_0 (the distance measured on T). Since L does not use all edges running into z, it returns to z less than $d^+(z) = d^-(z)$ times, hence by (*), it does not use the (unique) edge (z, u) of T leaving z. However, u is nearer to x_0 than is z (on T), which is a contradiction.

11. We consider all Euler trails as starting from x_0 along e_0. Let L be an Euler trail, and T the set of those edges (x, y) $(x \neq x_0)$, which satisfy

(**)   every edge starting from x is used earlier than (x, y).

Then T is a spanning arborescence. For we first observe that there is exactly one edge of T starting from every $x \neq x_0$. Moreover, T is acyclic; for if

$$C = (x_1, e_1, x_2, e_2, \ldots, e_n, x_1)$$

is a cycle in T, then e_2 is the last edge incident with x_2 on L and hence e_1 anticipates e_2. Similarly, e_2 anticipates e_3, ..., e_n anticipates e_1, a contradiction. The definition of T assures that L arises from T by the construction of the preceding problem.

If T is given, then we can characterize any Euler trail obtained from it by specifying an ordering of edges of $G - E(T)$, starting from x, for any $x \neq x_0$, and an ordering of the edges $\neq e_0$ starting from x_0; the orderings indicate in which order the Euler trail has to use the edges. Thus there are exactly $(d_0 - 1)! \ldots (d_{n-1} - 1)!$ such Euler trails. Hence, the number of Euler trails is $(d_0 - 1)! \ldots (d_{n-1} - 1)! \times$ (number of spanning arborescences rooted at x_0). [Aardenne–Ehrenfest, de Bruijn; see B.]

12. We cannot get stuck at $x \neq x_0$; this follows just as in 5.6.

x_0 is a "good" point. For if there is an edge (x_0, y) we have not passed from x_0 to y, then we are not stuck; but if every edge (x_0, y) has been used in this direction, then we left x_0 $d(x_0)$ times, hence we entered $d(x_0)$ times, i.e. all edges have been used to enter x_0.

Assume that we meet a "bad" point on our walk and let x be the first one. We enter x on an edge (y, x) and this being the first time we meet x, this will be the edge marked. x is "bad" means we enter x less than $d(x)$ times and hence, by (**) we do not use (x, y) to leave x. However, this contradicts the fact that y is "good".

Thus, each point we meet is a "good" point. By the definition of "good" points, their neighbors are reached during the walk and are, consequently, "good" points as well. Thus, the good points form a connected component of G. Since G is connected, every point is "good". This completes the proof. [Tarry algorithm; see B.]

13. To show the statement of the hint, let C_1 be a circuit of G (G is not a forest as a forest which has an edge has a point of degree 1). Let G_1 be the graph obtained by removing the isolated points of $G - E(C_1)$. Then G_1 has even degrees (each degree is reduced by 2 or 0), thus

$$G_1 = C_2 \cup \ldots \cup C_k$$

by induction, and

$$G = C_1 \cup \ldots \cup C_k.$$

Now assume that G is a graph with even degrees. Then

$$G = C_1 \cup \ldots \cup C_k$$

(we may forget about isolated points). Orienting each C_i cyclically, we get a desired orientation of G.

14. (a) If G has an Euler trail it is, obviously, connected and has even degrees. Conversely, if it is connected and has even degrees, then it can be oriented in such a way that each point has equal indegree and outdegree, by 5.13. The resulting graph has an Euler trail by 5.6. This Euler trail gives an Euler trail of G.

(b) If G' is the graph obtained by adding a new point and joining it to all points of G of odd degree, then G' has even degrees and is connected, thus it has an Euler trail. Removing the new point, this trail decomposes into k edge-disjoint trails covering G.

15. We may assume that G has all degrees at least 4.

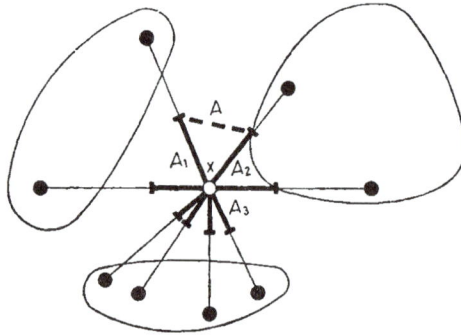

FIGURE 30

Let $x \in V(G)$ and let $A_1, \ldots, A_{2d}$ be short beginning segments of the edges incident with x, in this cyclic order. We may assume that if $G - \{A_1 \cup \ldots \cup A_{2d})$ is disconnected (this is the case, e.g. if there is a loop at x), then A_1, A_2 go to separate components of it. Remove $A_1 \cup A_2$ but connect their endpoints different from x by a new arc A (Fig. 30). The resulting map G' is connected: this follows from the observation that each component of $G - (A_1 \cup \ldots \cup A_{2d})$ is connected to x by at least two A_i's because G is Eulerian. Also, G' has fewer edges than G. Thus by induction, G' has an appropriate Euler trail. Replacing A by $A_1 \cup A_2$ again, we obtain an appropriate Euler trail of G.

16. The statement of the hint is easily verified. As for the assertion of the problem, it is trivial for a tree ($2^{(n-1)-n+1} = 1$, in accordance with the fact that the only "good" subgraph of a tree G is $(V(G), \emptyset)$).

Suppose that G is connected and has M "good" subgraphs. Those "good" subgraphs of $G+e$ which do not contain e are exactly the "good" subgraphs of G and thus their number is M. All we have to do is to show that the number of "good" subgraphs containing e is the same. Let C be a circuit of $G+e$ containing e (which exists because G is connected). Then G_1 is a good subgraph of $G+e$ containing e iff $(V(G), E(G_1) \triangle E(C))$ is a good subgraph not containing e. This correspondence is one-to-one and proves the assertion.

17. (a) We may restrict ourselves to simple graphs as the removal of two parallel edges does not alter the assertion.

If every point has even degree, we take $V_1 = V(G)$, $V_2 = \emptyset$. Suppose that a is a point of odd degree. Let S be the set of its neighbors. Define G_1 by

$$V(G_1) = V(G) - \{a\},$$

$$(x, y) \in E(G_1) \quad \text{if} \quad \begin{cases} (xy) \notin E(G) & \text{if } x, y \in S, \\ (xy) \in E(G) & \text{otherwise.} \end{cases}$$

By induction on n, we may assume that $V(G_1) = W_1 \cup W_2$, where W_1 and W_2 span subgraphs of G_1 with even degrees G_1. Since

$$|S \cap W_1| + |S \cap W_2| = |S| \equiv 1 \pmod 2,$$

we may assume that $|S \cap W_1|$ is even and $|S \cap W_2|$ is odd. Set

$$V_1 = W_1 \cup \{a\}, \quad V_2 = W_2.$$

Then V_1, V_2 span subgraphs of G with even degrees. First, let $x \in V_1$. If $x \notin S$, then its degree is even in $G[V_1]$ obviously. Let $x \in S$. Then

$$d_{G[V_1]}(x) = d_{G_1[W_1]}(x) - d_{G_1[W_1 \cap S]}(x) + d_{G[W_1 \cap S]}(x) + 1 =$$
$$= d_{G_1[W_1]}(x) - d_{G_1[W_1 \cap S]}(x) + (|W_1 \cap S| - 1 - d_{G_1[W_1 \cap S]}(x) + 1 =$$
$$= d_{G_1[W_1]}(x) - 2d_{G_1[W_1 \cap S]}(x) + |W_1 \cap S|,$$

and here each term is even. It follows similarly for $x \in V_2$ that it has even degree in $G[V_2]$.

(b) Let v be a new point, joined to all points of $V(G)$, and let G_1 be the resulting graph. By the preceding exercise, $V(G_1) = U_1 \cup U_2$, where U_1, U_2 span subgraphs with even degrees. Now if $v \in U_2$ (say), then $V_1 = U_1$, $V_2 = U_2 - \{v\}$ is a desired partition of $V(G)$. [T. Gallai, unpublished; W. K. Chen, *SIAM J. Appl. Math.* **20** (1971) 526–529; the proof given here is due to L. Pósa]

(c) We have to find a subset S of $V(G)$ such that every $v \in S$ is connected to an even number of vertices in S and every $v \in V(G) \backslash$ is connected to an odd number of vertices of S. Add a new point u to G and connect it to every vertex with even degree. Apply (a) to the resulting graph G', to get a partition $V_1 \cup V_2 = V(G) \cup \{u\}$ such that $G'[V_i]$ has even degrees. We may assume that $u \in V_2$; then $S = V_1$ satisfies the requirements above.

18. (a) Let S be the set of r-element matchings of the complete graph on $V(G)$. Let $e_1, \ldots, e_m$ be the edges of $\overline{G}$ and A_i the set of those elements of S which contain e_i. We want to determine

$$m_r(G) = \left| S - \bigcup_{i=1}^{m} A_i \right|.$$

By the sieve formula (2.2),

$$m_r(G) = |S| - \sum_{1 \leq i \leq m} |A_i| + \sum_{1 \leq i < j \leq m} |A_i \cap A_j| - \ldots +$$

$$+ (-1)^r \sum_{1 \leq i_1 < \ldots < i_r \leq m} |A_{i_1} \cap \ldots \cap A_{i_r}|;$$

here $|A_{i_1} \cap \ldots \cap A_{i_\nu}|$ is the number of those elements of S which contain $e_{1_i}, \ldots, e_{i_\nu}$. This is non-zero only if $e_{1_i}, \ldots, e_{i_\nu}$ are independent and then it is $\binom{n-2\nu}{2r-2\nu} \cdot (2r-2\nu-1)!!$. Hence,

$$m_r(G) = \sum_{\nu=0}^{r} (-1)^\nu m_\nu(\overline{G}) \binom{n-2\nu}{2r-2\nu} (2r-2\nu-1)!! \qquad ((-1)!! = 1).$$

(b) Let $n = 2k$. By (a),

$$m_k(G) = \sum_{\nu=0}^{k} (-1)^\nu \binom{2k-2\nu}{2k-2\nu} (2k-2\nu-1)!! m_\nu(\overline{G}) \equiv \sum_{\nu=0}^{k} m_\nu(\overline{G}) \quad (\text{mod } 2).$$

Thus

$$m_k(G) \neq 0.$$

(c) Let A be the adjacency matrix of G and consider $\det A$. Each expansion term here which is non-zero and symmetric with respect to the main diagonal corresponds to a 1-factor and vice versa. The other expansion terms correspond to each other in pairs under reflexion in the main diagonal. Thus the number of 1-factors of G has the same parity as $\det A$, and is even iff $\det A$ vanishes over $GF(2)$. This occurs iff the rows of A are linearly dependent over $GF(2)$, i.e. there are elements $g_1, \ldots, g_n$ of $GF(2)$ such that not all of them are 0 and multiplying the rows of A in order by $g_1, \ldots, g_n$, the sum of rows will be 0. Let S be the set of points for which the corresponding g_i is 1, then S has the property that each point is adjacent to an even number of elements of it. Conversely, each such set S yields appropriate coefficients $g_1, \ldots, g_n$. [G.H.C.Little, *Discrete Math.* **2** (1972), 179–181.]

19. Let S be the set of all Hamiltonian paths of the complete digraph on $V(G)$ and let $e_1,\ldots,e_m$ be the edges of $\overline{G}$. Let A_i denote the set of elements of S containing e_i, then we have

$$(*) \qquad h(G) = \sum_{\nu=0}^{n-1} (-1)^\nu \sum_{1\le i_1 < \ldots < i_\nu \le m} |A_{i_1} \cap \ldots \cap A_{i_\nu}|.$$

(the term with $\nu = 0$ is $|S|$). Now, $|A_{i_1} \cap \ldots \cap A_{i_\nu}|$ is the number of Hamiltonian paths of the complete digraph containing $e_{i_1},\ldots,e_{i_\nu}$. This is 0 unless $e_{i_1},\ldots,e_{i_\nu}$ form disjoint paths; in the letter case it is $(n-\nu)!$, since the graph $(V(G);\{e_{i_1},\ldots,e_{i_\nu}\})$ has $n-\nu$ components and any Hamiltonian path through $e_{i_1},\ldots,e_{i_\nu}$ defines an ordering of these components and conversely. Hence,

$$|A_{i_1} \cap \ldots \cap A_{i_\nu}| \text{ is } \begin{cases} 1 & \text{If } \nu = n-1 \text{ and } e_{i_1},\ldots,e_{i_\nu} \text{ form a} \\ & \text{Hamiltonian path} \\ \text{even} & \text{otherwise.} \end{cases}$$

This proves that $h(G) \equiv H(\overline{G}) \pmod 2$.

If G is an undirected graph, then, defining S, e_i, A_i similarly, we have the same formula as $(*)$, but now we have

$$|A_{i_1} \cap \ldots \cap A_{i_\nu}| = \frac{(n-\nu)!}{2} 2^\mu,$$

if $e_{i_1},\ldots,e_{i_\nu}$ form μ vertex-disjoint proper paths. If $n \ge 4$, then this is also even except when $\nu = n-1$ and $e_{i_1},\ldots,e_{i_\nu}$ form a Hamiltonian path. Hence, we can conclude as above. [T. Szele; see B.]

20. If the statement of the hint is true, the assertion of the problem follows easily: by reversing edges we get a transitive tournament, which has one Hamiltonian path. As the parity remained the same, the original tournament had an odd number of Hamiltonian paths.

Thus, it suffices to show that, if T is a tournament, and T' is the tournament obtained from it by reversing an edge e, we have

$$h(T') \equiv h(T) \pmod 2.$$

Let G_1 and G_2 be the digraphs obtained from T by removing e and adding the inverse of e, respectively. Then a simple computation shows that

$$h(G_1) + h(G_2) = h(T) + h(T').$$

Moreover, $\overline{G}_1$ arises from G_2 simply by reversing all edges, whence

$$h(\overline{G}_1) = h(G_2).$$

Finally,

$$h(G_1) \equiv h(\overline{G}_1) \pmod 2$$

by 5.19. Hence

$$h(T) + h(T') = h(G_1) + h(G_2) = h(G_1) + h(\overline{G}_1) \equiv 0 \quad (\text{mod } 2),$$

which proves the assertion. [L. Rédei; see B.]

21. If F_1, F_2 are disjoint 1-factors of G, then $E(G) - F_1 - F_2$ is a 1-factor. Conversely, if we have a partition of $E(G)$ into three 1-factors, then we can choose as F_1 the 1-factor containing e and any one of the two others as F_2. Hence, the number m of pairs (F_1, F_2) of 1-factors with $F_1 \cap F_2 = \emptyset$, $e \in F_1$, $e \notin F_2$ is twice the number of partitions of $E(G)$ into 1-factors. Thus, m is even. Now consider $F_1 \cup F_2$; this consists of even circuits covering $V(G)$. If, conversely, $H \subseteq E(G)$ is the set of edges of a system of k disjoint even circuits covering $V(G)$ and containing e, then H can be decomposed into $H = F_1 \cup F_2$ (F_1, F_2 1-factors, $e \in F_1$, $e \notin F_2$) in exactly 2^{k-1} ways. Thus, if m_k denotes the number of systems of k disjoint even circuits covering $V(G)$ and containing e, then

$$m = m_1 + 2m_2 + \ldots + 2^{k-1}m_k + \ldots$$

Note here that m_1 is the number of Hamiltonian circuits through e and therefore

$$m_1 \equiv m \equiv 0 \quad (\text{mod } 2).$$

[C. A. B. Smith; see B.]

22. We prove the assertion by induction on $|V(G)|$. If $|V(G)| = 4$, it is obvious. Also, we may assume that G is connected.

If there is a double edge in G, then all Hamiltonian circuits must use one of the two and hence the Hamiltonian circuits occur in pairs, differing only in which of the two parallel edges they use. Thus the total number of Hamiltonian circuits is even again.

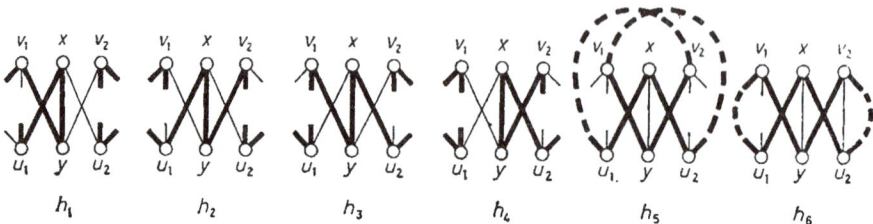

FIGURE 31

So suppose that G is simple. Let (x,y), (x,u_1), (x,u_2), (y,v_1), $(y,v_2) \in E(G)$. (see Fig. 14). Remove x and y and join u_1 to v_1, u_2 to v_2 to get G'; also, join u_1 to v_2 and u_2 to v_1 to get G''.

The Hamiltonian circuits of G may or may not contain (x,y). There are four kinds of Hamiltonian circuits through (x,y) containing (u_1xyv_1), (u_1xyv_2), (u_2xyv_1), (u_2xyv_2), respectively (Fig. 31). Let their numbers be h_1, h_2, h_3, h_4. Also, those Hamiltonian circuits not containing (x,y) may go like $(\ldots u_1xu_2 \ldots v_1yv_2 \ldots)$ or $(\ldots u_1xu_2 \ldots v_2yv_1 \ldots)$. Let the numbers of such Hamiltonian circuits be h_5 and h_6.

The h_1 Hamiltonian circuits defined above are in a one-to-one correspondence with those Hamiltonian circuits of G' containing (u_1, v_1) but not (u_2, v_2). Similarly, h_2, h_3, h_4 are equal to the numbers of Hamiltonian circuits of G'' and G' containing (u_1, v_2), (u_2, v_1), (u_2, v_2) but not the other new edge. The h_5 Hamiltonian circuits of G correspond to those Hamiltonian circuits of G' containing both (u_1, v_1) and (u_2, v_2) and going through them like $(\ldots u_1 v_1 \ldots u_2 v_2 \ldots)$. Similarly, the h_6 Hamiltonian circuits correspond to those Hamiltonian circuits of G'' going through (u_1, v_2) and (u_2, v_1) like $(\ldots u_1 v_2 \ldots u_2 v_1 \ldots)$.

The Hamiltonian circuits of G' not considered so far are those going through (u_1, v_1) and (u_2, v_2) like $(\ldots u_1 v_1 \ldots v_2 u_2 \ldots)$ and those not containing the new edges. Let h_7 and h_8 be the numbers of such Hamiltonian circuits.

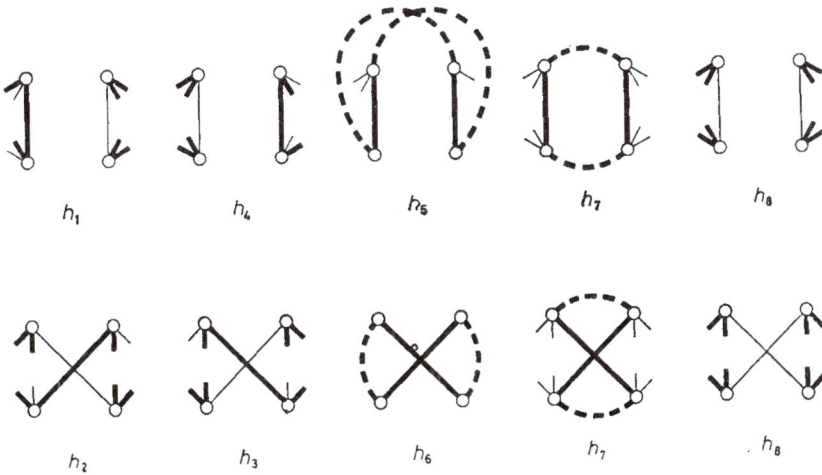

FIGURE 32

In G'', we have not considered so far the Hamiltonian circuits of form $(\ldots u_1 v_2 \ldots v_1 u_2 \ldots)$ and those not containing any of the new edges. The numbers of such Hamiltonian circuits are, as is easily seen, h_7 and h_8, respectively (Fig. 32).

Now the number of Hamiltonian circuits of G is $h_1 + h_2 + h_3 + h_4 + h_5 + h_6 \equiv (h_1 + h_4 + h_5 + h_7 + h_8) + (h_2 + h_3 + h_6 + h_7 + h_8) \equiv 0 \pmod 2$, since here in the brackets we have the numbers of Hamiltonian circuits of G' and G'', respectively. [J. Bosák; see B.]

23. To show the statement of the hint, let F^* be the graph on $V(G^*)$ formed by those edges of G^* which are not crossed by the edges of F. F^* does not contain a cycle, for then the points of F inside this cycle would not be connected to those outside. Moreover, F^* is connected. For if $F^* = F_1 \cup F_2$, $F_1 \cap F_2 = \emptyset$, then let U be the union of faces of G with "capital" in F^*. Then U is not the whole plane (no point of F_2 belongs to it) and therefore, its boundary B is non-empty. Now B consists of certain edges of F, each having a face of U on one side and a face not in U on the other. Therefore, there is no point of degree 1 on the boundary, i.e. it contains a circuit. This is a contradiction.

Hence $F \to F^*$ gives a one-to-one correspondence between the spanning trees of G and G^*. This proves the assertion of the problem.

Remark: Our argument uses some facts of plane topology like the Jordan curve theorem (each simple closed curve divides the plane into two parts) and other similar statements. As our goal is to illuminate the combinatorial content of the problems we assume these facts without proof.

24. Consider a spanning tree F of G and F^* as in 5.23. Then

$$|E(F)| = |V(G)| - 1,$$

$$|E(F^*)| = |V(G^*)| - 1$$

and by the definition of F^*,

$$|E(F)| + |E(F^*)| = |E(G)|.$$

Hence

$$|V(G^*)| = |E(G)| + 2 - |V(G)|.$$

25. Since each face has at least three [four] edges on its boundary and each edge is on the boundary of exactly two faces, we get by Euler's formula

$$2m \geq 3(m - n + 2) \qquad [2m \geq 4(m - n + 2)]$$

or, equivalently,

$$m \leq 3n - 6 \qquad [m \leq 2n - 4].$$

26. We may assume that there are no loops. We use induction on $|E(G)|$.

Let F be a face bounded by the edges $e_1, \ldots, e_m$. We claim that every point x is incident with an even number of them. For let k "corners" in x belong to F, then two of these "corners" cannot have a common edge e because this edge would be a cut-edge (see Fig. 33) and hence a component of $G-e$ would contain exactly one point of odd degree. Thus x is incident with $2k$ edges of the boundary of F.

Now remove the edges $e_1 \ldots, e_m$. The remaining map G' has even degrees, thus its faces can be 2-colored. The faces of G' are the union of F and its neighbors, as well as all the other faces of G. Interchanging the color of F but keeping it everywhere else, we get a 2-coloration of the faces of G.

27. (a) First, 2-color the faces according to 5.26 with red and blue, say. Then, orient each edge so that the red face incident with a given edge e is on its left-hand side. This orientation satisfies the requirement. (For this part, we don't need that G is simple.)

(b) We may assume that G is connected; since it is Eulerian, it is also 2-edge-connected. By a "corner" we mean an ordered pair of edges that are consecutive on the boundary of a face, where the boundary is traversed so that the face is to the left.

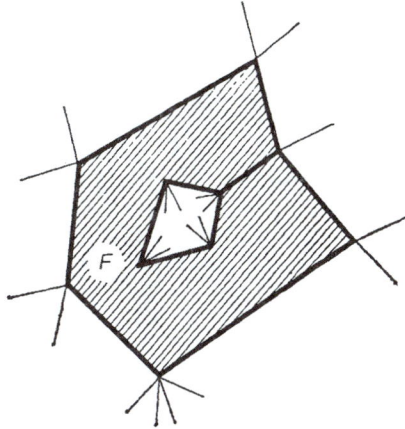

FIGURE 33

Consider any 2-coloration of the edges of G. A face with i edges is incident with at most $\lfloor i/2 \rfloor$ red-blue corners. So if f_i denotes the number of faces with i edges, then the number of red-blue corners is at most

$$f_3 + 2f_4 + 2f_5 + 3f_6 + \ldots \leq f_3 + 2f_4 + 3f_5 + 4f_6 + \ldots$$

Let G have n vertices and m edges; then we know by Euler's Formula that

$$f_3 + f_4 + f_5 + f_6 + \ldots = m - n + 2,$$

and clearly

$$3f_3 + 4f_4 + 5f_5 + 6f_6 + \ldots = 2m.$$

Hence the number of red-blue corners is at most $2m - 2(m-n+1) = 2n-2 < 2n$, and so there must be a vertex incident with at most one red-blue corner. It follows that at this vertex, red edges (and the blue edges as well) are consecutive in the cyclic order of the edges. (In this form, the assertion of the problem extends to all planar graphs.) [Cauchy]

28. (a) Suppose that there were a planar map with even degrees, and all faces triangular except one which is pentagonal. 2-color the faces (5.26) with red and blue, and suppose that the pentagon is red (say). Count the number of edges. Each blue face has 3 edges on its boundary, and this counts each edge exactly once; hence the number of edges is divisible by 3. On the other hand, the red faces have altogether $3k + 5$ edges on their boundaries (k is the number of triangular red faces), which should give the same total number of edges, a contradiction.

(b) The graph G' constructed in the hint has even degrees at every point except possibly z, but then, by 5.1a, z has an even degree as well. Let $a_1 + 1, a_2 + 1, \ldots, a_{2s} + 1$ be the sizes of faces incident with z in a cyclic order. 2-color the faces (Fig. 34). The color containing the face of size $a_1 + 1$ has $(a_1 + 1) + (a_3 + 1) + \ldots +$

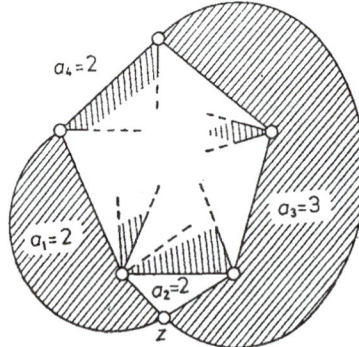

FIGURE 34

$(a_{2s-1}+1)+3k$ edges on its boundary, the other one has $(a_2+1)+\ldots+(a_{2s}+1)+3N$ edges. In both cases we get the total number of edges of G. Hence,

$$a_1 + a_3 + \ldots + a_{2s-1} \equiv a_2 + a_4 + \ldots + a_{2s} \pmod 3.$$

Moreover, we have $\sum_{i=1}^{2s}(a_i - 1) = 5$. It is easy to verify that the only solution of these equations is

$$a_1 = 2, \qquad a_2 = 5$$

(or conversely). Thus, exactly two, neighboring points of the pentagon have odd degrees [T. Gallai].

29. Let the three colors be red, blue, green. Each triangle whose points get different colors has one red-blue edge; any other triangle has 0 or two. Moreover, every red-blue edge is counted twice, in both triangles adjacent to it. So this adds up to an even number, i.e. the number of triangles with 3 different colors is even.

Remark: The statement is a special case of Sperner's lemma in algebraic topology, which is equivalent to a similar statement for n-dimensional triangulations [see e.g. L. S. Pontryagin, *Grundzüge der Kombinatorischen Topologie*, Berlin, 1956, p.73].

30. Consider the coloration defined in the hint. If there is a face with points x, y, z, of color red, blue, green (in this order) and $z \in V_1$, (say), then z, together with the (x,a)-path in V_1, yields a (z,a)-path in V_1, whence z should have been colored red.

Now if there is no (a,c)-path in V_1 and no (b,d)-path in V_2, then c, d are green. Hence, if we add the edge (a,c), we get a triangulation with only the face abc meeting three colors. This contradicts the preceding exercise.

31. (a) True; in fact, A is regular iff $\det A \neq 0$, a fact whose standard proof uses considerations valid in any field.

(b) True; for

$$\det A^T A = \det A^T \det A = (\det A)^2 \neq 0.$$

(c) False; for we may have a vector $\mathbf{u} \neq \mathbf{0}$ with $\mathbf{u}^T \mathbf{u} = 0$ (e.g. if $F = GF(2)$, $\mathbf{u} = \mathbf{e}_1 + \mathbf{e}_2$ or F is the complex field and $\mathbf{u} = (1, i)^T$), and then the transformation A defined by

$$A\mathbf{e}_1 = \mathbf{u}, \quad A\mathbf{e}_2 = \ldots = A\mathbf{e}_n = \mathbf{0} \quad \text{(for some basis } \mathbf{e}_1, \ldots, \mathbf{e}_n)$$

satisfies $A \neq 0$, $\mathbf{y} A^T A \mathbf{x} = (A\mathbf{y})^T A \mathbf{x} = 0$ for any $\mathbf{x}$, $\mathbf{y}$, thus $A^T A = 0$.

(d) False; if $\mathbf{u}^T \mathbf{u} = 0$, $\mathbf{u} \neq \mathbf{0}$, then let $M = \langle \mathbf{u} \rangle$ (the subspace generated by $\mathbf{u}$); then $M \subseteq M^\perp$.

(e) False; for the preceding M, we have $\langle M, M^\perp \rangle = M^\perp \neq V$, as e.g. one of $\mathbf{e}_1, \ldots, \mathbf{e}_n$ is surely non-orthogonal to M.

(f) True; let A be as in the hint. Then

$$\mathbf{x} \in M^\perp \Leftrightarrow \mathbf{x}^T \mathbf{v}_i = 0 \quad (i = 1, \ldots, k) \Leftrightarrow \mathbf{x}^T A \mathbf{e}_i = 0 \quad (i = 1, \ldots, k),$$
$$\Leftrightarrow \mathbf{e}_i^T A^T \mathbf{x} = 0 \quad (i = 1, \ldots, k) \Leftrightarrow A^T \mathbf{x} \in \langle \mathbf{e}_{k+1}, \ldots, \mathbf{e}_n \rangle,$$
$$\Leftrightarrow \mathbf{x} \in (A^T)^{-1} \langle \mathbf{e}_{k+1}, \ldots, \mathbf{e}_n \rangle.$$

Hence

$$M^\perp = (A^T)^{-1} \langle \mathbf{e}_{k+1}, \ldots, \mathbf{e}_n \rangle,$$

which shows that $\dim M^\perp = n - k$.

(g) True; for let $\mathbf{u} \in M$, then $\mathbf{u}$ is orthogonal to every element of $M^\perp$, i.e. $\mathbf{u} \in (M^\perp)^\perp$. Thus

$$M \subseteq (M^\perp)^\perp.$$

On the other hand,

$$\dim(M^\perp)^\perp = n - \dim M^\perp = n - (n - \dim M) = \dim M,$$

which proves the assertion.

32. (a) $\mathbf{x} \in \langle M, M^\perp \rangle^\perp$ iff $\mathbf{x}$ is orthogonal to both M and $M^\perp$, i.e. $\mathbf{x} \in M^\perp \cap (M^\perp)^\perp = M \cap M^\perp$. Hence

$$\langle M, M^\perp \rangle = (M \cap M^\perp)^\perp.$$

Now, let $\mathbf{u} \in M \cap M^\perp$. Then $\mathbf{u} \in M$ and $\mathbf{u} \in M^\perp$, thus $\mathbf{u}^T \mathbf{u} = 0$. This means, however, that the number of 1's in $\mathbf{u}$ is even, which is equivalent to $\mathbf{u}^T \mathbf{j} = 0$. Thus $\mathbf{j} \in (M \cap M^\perp)^\perp = \langle M, M^\perp \rangle$. [T. Gallai, unpublished; W. K. Chen, *SIAM J. Appl. Math.* **20** (1971) 526–529.]

(b) The identity in the hint is obvious, since in $\mathbf{v}^T A \mathbf{v}$ the terms coming from off-diagonal entries of A cancel each other in pairs over $GF(2)$, while in the diagonal entries we have $v_i^2 = v_i$.

Now assume that $\mathbf{a}$ is not in the column space of A. Then there exists a 0-1 vector $\mathbf{v}$ orthogonal to all columns of A but not to $\mathbf{a}$ (by 5.31(g)). So $A\mathbf{v} = 0$ and thus $\mathbf{v}^T A\mathbf{v} = 0$. But by the identity in the hint, this implies that $\mathbf{a}^T\mathbf{v} = 0$, a contradiction.

Considering A as the adjacency matrix of a graph $G = (V, E)$, we obtain the following: given $T \subseteq V$, there exists a set $S \subseteq V$ such that for each $v \in V$,

$$\Gamma(v) \cap S = \begin{cases} \text{even}, & \text{if } v \in S \cup (V \setminus T), \\ \text{odd}, & \text{if } v \in T \setminus S. \end{cases}$$

Choosing T as the set of vertices of odd degree, even degree, and the set of all vertices, respectively, we obtain 5.17 (a), (b) and (c). [N. Alon]

33. (a) We show U_G consists of the cuts. Cuts form a subspace; for the sum of cuts determined by the sets S_1 and S_2, respectively, is the cut determined by $S_1 \triangle S_2$. Each star is a cut. Conversely, the cut determined by S is the sum of stars determined by the points of S.

The star of x is orthogonal to a set A of edges iff the subgraph determined by A has even degree at x. Hence, W_G consists of the sets of edges of subgraphs with even degrees. By the solution of 5.13, W_G is generated by the circuits of G.

If all circuits are even, then $\mathbf{j} = (1, \ldots, 1)$ is orthogonal to every circuit, hence $\mathbf{j} \in (U_G^\perp)^\perp = U_G$. This says by (a) that G is bipartite. Hence 5.3 follows.

To show 5.16, we have to determine $\dim W_G$. It is enough to determine $\dim U_G$ by 5.21f. Let $A_1, \ldots, A_n$ be all the stars. Since

$$A_n = A_1 + \ldots + A_{n-1},$$

we have $U_G = \langle A_1, \ldots, A_{n-1} \rangle$. We show that $A_1, \ldots, A_{n-1}$ are linearly independent. Any linear dependence between them would be of the form

$$A_{i_1} + \ldots + A_{i_k} = 0,$$

as we are working over $GF(2)$. However, $A_{i_1} + \ldots + A_{i_k}$ is a cut determined by a non-empty proper subset of $V(G)$, which is non-empty because G is connected.

Thus $\dim U_G = n - 1$, $\dim W_G = m - n + 1$, q.e.d. 5.17 follows from 5.32 immediately.

(b) The assertion is trivial for disconnected graphs, so suppose that G is connected. The number of decompositions, i.e. the number of solutions of $\mathbf{a} + \mathbf{b} = \mathbf{j}$, $\mathbf{a} \in U_G$, $\mathbf{b} \in W_G$, is clearly equal to $|U_G \cap W_G|$. This is the same as the number of vectors $\mathbf{u} \in U_G$ orthogonal to every element of U_G.

Let A_G be the point-edge incidence matrix of G and A_0 the matrix arising from it by canceling a row. Then the range of A_0^T is U_G. We also know A_0^T is one-to-one. Thus, we are interested in the number of vectors $\mathbf{x}$ such that

$$(A_0\mathbf{x})^T(A_0\mathbf{y}) = \mathbf{x}^T A_0 A_0^T \mathbf{y} = 0$$

for every $\mathbf{y}$. But this holds for every $\mathbf{y}$ iff $A_0 A_0^T \mathbf{x} = 0$. This equation has a unique solution iff $\det A_0 A_0^T \neq 0 \pmod 2$. By 4.9 this means the number of spanning trees is odd.

34. Let $C_1, \ldots, C_{f-1}$ be the finite faces, considered as circuits of G, i.e. elements of W_G. We show that every circuits of G is the sum of some C_i's. Let C be a circuit of G. Each face C_i lies either inside or outside C. Let, say $C_1, \ldots, C_r$ lie inside C. Then
$$C = C_1 + \ldots + C_r;$$
in fact, if e is an edge of C, then exactly one of the two faces incident with e lies inside C, hence $e \in E(C_1 + \ldots + C_r)$. If e is inside (outside) C, then both (none) of these two faces are inside C, hence e does not occur in $C_1 + \ldots + C_r$. On the other hand, if $C_1, \ldots, C_{f-1}$ were dependent, say
$$C_1 + \ldots + C_r = 0,$$
then draw a continuous line from an inner point x of C_1 to the infinity, avoiding the vertices. This line leaves the union of faces $C_1, \ldots, C_r$ at a point which belongs to an edge e of G. Then exactly one of the two faces adjacent to e belongs to $C_1, \ldots, C_r$, hence
$$e \in C_1 + \ldots + C_r,$$
a contradiction.

    Remark: Since the dimension of W_G is $m-n+1$ by the previous problem, we get
$$m = n + 1 = f - 1,$$
i.e. we also obtain Euler's formula at least for 2-connected graphs. Conversely, the use of Euler's formula would have made one half of the proof superfluous.

35. (a) Let $e \in \sum_{i \in I \cup J} C_i$, and say $e \in C_\mu$, $\mu \in I \cup J$. We claim $\mu \in I$. This is clear if $\mu \notin J$. Suppose that $\mu \in J$, then since no other C_ν, $\nu \in I \cup J$ contains e, we have
$$e \in \sum_{i \in J} C_i \subseteq \bigcup_{i \in I} C_i,$$
whence $\mu \in I$. Thus, $e \in \sum_{i \in I} C_i = K$, i.e. $\sum_{i \in I \cup J} C_i \subseteq K$. Since K is a circuit and $\sum_{i \in I \cup J} C_i \neq \emptyset$, we have
$$K = \sum_{i \in I \cup J} C_i.$$
Since K has a unique decomposition into the sum of C_i's, we get
$$I \cup J = I, \quad J \subseteq I.$$

    (b) If $C_1, \ldots, C_f$ are the only circuits, then it is easy to see that they are precisely the blocks of G (not counting cutting edges).

Let $K \neq C_1, \ldots, C_f$ be a circuit such that its representation as the sum of C_i's has a minimum number of terms. Let, say

$$K = C_1 + \ldots + C_r \quad (r \geq 2), \quad I = \{1, \ldots, r\}$$

and, e.g. $C_1 \cap C_2 \neq \emptyset$. Then $C_1 + C_2$ has even degrees and therefore

(1) $C_1 + C_2 = K_1 + \ldots + K_s,$

where $K_1, \ldots, K_s$ are edge-disjoint circuits. At least one of them, say K_1, must be different from $C_1, \ldots, C_f$; otherwise (1) would show that $C_1, \ldots, C_f$ are linearly dependent. Let

$$K_1 = \sum_{i \in J} C_i.$$

By (a), $J \subseteq I$ and hence by the minimality of I, we have $J = I$. Thus $K = K_1$ and so

(2) $K \subseteq C_1 + C_2.$

We claim that $K = C_1 + C_2$. Suppose that there is an edge of C_1 not in $K \cup C_2$ (say). This edge belongs to a C_k, $3 \leq k \leq r$. (2) holds for C_k as well, i.e.

(3) $K \subseteq C_1 + C_k.$

But (2) and (3) imply that an edge of K not in C_1 must belong to both of C_2 and C_k and thus it cannot occur in $\sum_{i=1}^{r} C_i$, a contradiction.

Thus $C_1 + C_2$ is a circuit.

(c) We use induction on the number of edges. Let, say, $C_1 + C_2 = C$ be a circuit. Remove the edges common to C_1 and C_2 and the isolated points that possibly arise. Let G' be the resulting graph.

Consider the system $C, C_3, \ldots, C_f$. It is obvious from the construction that each edge of G' is contained in at most two of them. Moreover, if A is a subgraph of G' with even degrees, then

$$A = \sum_{i \in I} C_i$$

($I \subseteq \{1, \ldots, f\}$), and here either both C_1 and C_2 occur or neither of them, whence A is a linear combination of $C, C_3, \ldots, C_f$. Moreover, $C, C_3, \ldots, C_f$ are, obviously, linearly independent. Hence, by the induction hypothesis, G' can be embedded in the plane so that $C, C_3, \ldots, C_f$ are boundaries of faces. To get G we have to put back $C_1 \cap C_2$, which is a path joining two points of C and thus, it can be done and C_1, C_2 will become the new boundaries.

(d) It is easy to see that a graph is planar iff each block of it is planar. A similar assertion holds for the property that W_G has a basis such that each edge belongs to at most two elements of it. Therefore to prove the equivalence of these two properties, we may restrict ourselves to 2-connected graphs. For these, the necessity of MacLane's conditions of planarity follows by 5.34. We remark that we may require that the elements of the basis should be circuits. For suppose that there exists a basis $A_1, \ldots, A_f$ of W_G such that each edge belongs to at most two of them. A_1 is the union of edge-disjoint circuits. One of these must clearly be linearly independent of $A_2, \ldots, A_n$. Replacing A_1 with this circuit we obtain a basis with the same property. Going on similarly we can replace $A_2, \ldots, A_f$ by circuits.

Now the sufficiency of MacLane's condition follows by (c). [S. MacLane; see W.]

36. Suppose that G is a planar map and let G^* be its dual graph. Then if φ associates with each $e \in E(G)$ the edge of G^* crossing it, will satisfy the requirements of the statement by 5.23.

Conversely, suppose that G^* and φ exist, we show G is planar.

First, we prove that the edges of a star in G^* correspond to the edges of an element of W_G. For let $X \subseteq V(G)$ and suppose that $\varphi(X)$ is the star of a point $x \in V(G^*)$. Let $B_1, \ldots, B_s$ be the branches of G^* relative to x and let A_i be the set of (x, B_i)-edges. Then A_i is a minimal set which meets every spanning tree of G^*. Therefore, $\varphi^{-1}(A_i)$ is a minimal set which is not contained in any spanning tree of G^*, i.e. $\varphi^{-1}(A_i)$ is a circuit. Hence

$$X = \sum_{i=1}^{s} \varphi^{-1}(A_i) \in W_G.$$

Now, let $C_1, \ldots, C_{n^*}$ be the elements of W_G corresponding to the stars of points in G^*. Then trivially each edge of G is contained in exactly two C_i's. Moreover if we set $f = n^* - 1$, $C_1, \ldots, C_f$ will form a basis of W_G. In fact,

$$\dim W_G = e - n + 1 = n^* - 1 = f$$

(from the definition of G^*), and $C_1, \ldots, C_f$ are linearly independent over $GF(2)$, because so are the corresponding stars in G^*.

Thus G is planar by MacLane's criterion. [H. Whitney; see e.g. Wi.]

37. (a) It is trivial that G is 2-connected. Suppose indirectly that $G = G_1 \cup G_2$ with $V(G_1) \cap V(G_2) = \{x, y\}$, $|V(G_i)| \geq 3$. Let P_i be an (x, y)-path in G_i and $H_i = G_i + P_{3-i}$. Then H_i is planar; embed H_i into the plane so that the path P_{3-i} lies on the boundary of the unbounded domain (this can be achieved by inverting the plane with respect to an appropriate circle). Then identify the two points x and y and delete the paths P_i (Fig. 35). This results in an embedding of G in the plane, a contradiction.

(b) Let $(x_0, \ldots, x_m)$ be a longest path in G. x_0 has degree at least 3, and it is adjacent to no point outside this path because of the maximality of it. Hence it has two neighbors x_i, x_j with $1 < i < j$. Then $(x_0, \ldots, x_j)$ is a circuit with a chord (cf. also 6.35).

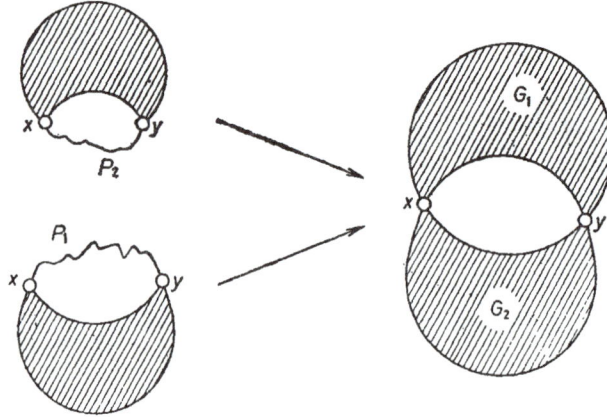

FIGURE 35

(c) Let (x,y) be a chord of a circuit C and choose C so that embedding $G - (x,y)$ in the plane the number of faces inside C is as large as possible. Observe first that there is no point outside C. In fact, let G_0 be a component of $G - V(G)$ and suppose indirectly that G_0 lies outside C. Since G is 3-connected, there must be three points of C adjacent to G_0 and at least two of these, u and v say, are not separated by x and y. Then replacing the (u,v)-arc of e not containing x, y by a (u,v)-path through G_0, we obtain a circuit C' with chord (x,y) and more faces inside.

The same reasoning shows that all chords of C running outside connect inner points of the two (x,y)-arcs of C.

Now let us consider the bridges of C inside C. Call such a bridge flappable if its endpoints do not separate the endpoints of any outside chord of C. It is clear that we can "flap" all these bridges to the outside of C. Among the remaining bridges there must be one containing inner points of both (x,y)-arcs of C, otherwise x and y could be connected inside C and G would be planar. So there is a bridge B inside C and a chord (a,v) outside C such that the endpoints of B separate a from v and x from y on C; also $\{a,v\}$ and $\{x,y\}$ separate each other. This can occur in several ways (Fig. 36):

(a) B contains inner points of the (x,a)-arc and (y,v)-arc (or, symmetrically, of (x,v) and (y,a));

(b) B contains v, an inner point of (x,a) and moreover a point of (y,a) different from a (or any symmetrical situation);

(c) B contains x, y, a, v.

Take a path P connecting two of the mentioned endpoints of B. In case (b) take a path connecting P to the third one. In case (c) take two paths connecting P to the other two endpoints. If these two paths meet each other, then let them have a common initial piece. So we get from case (c) two subcases according to whether the mentioned paths in B form an H or an X (Fig. 37).

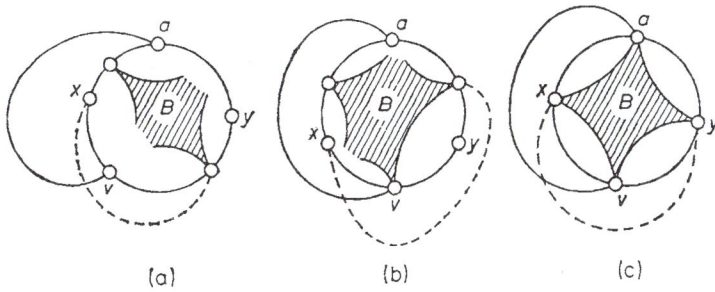

FIGURE 36

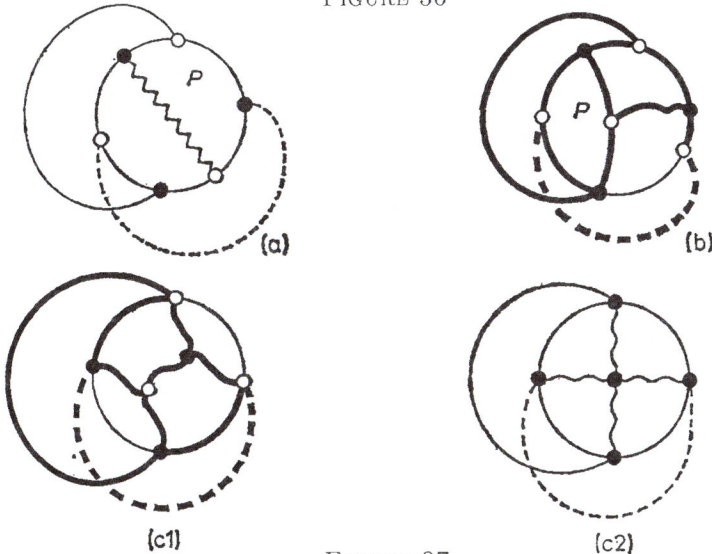

FIGURE 37

In case (a), we see a subdivision of $K_{3,3}$; by the minimality of G, there cannot be other edges or any subdividing points, i.e. $G \cong K_{3,3}$. In cases (b) and (c1) the graph properly contains a subdivision of $K_{3,3}$; this is impossible since each proper subgraph of G is supposed to be planar. In case (c2) we see a subdivision of K_5 and hence $G \cong K_5$.

(d) Suppose that G is planar. Then, obviously, G cannot contain a subdivision of K_5 or $K_{3,3}$. Conversely, suppose that G is not planar. Then G contains a minimal non-planar graph G_0. If we get rid of the points of degree 2 from G_0 (removing them and joining their two neighbors successively), we get another minimal non-planar graph, this time with degrees at least 3. This graph is K_5 or $K_{3,3}$ by (c), thus G_0 is a subdivision of K_5 or $K_{3,3}$ [see S II].

38. We use induction on the number of points. If this is at most 3, the assertion is trivial.

First we show that if G is any planar graph we can introduce new edges to turn all faces into triangles without getting parallel edges. For let us draw new edges as long as we can without getting parallel ones. The graph G has no cutpoints; for if $G = G_1 \cup G_2$ with $V(G_1 \cap G_2) = \{x\}$, then take a point x_i of $G_i - x$ on the boundary of the face which meets both $G_1 - x$ and $G_2 - x$; x_1, x_2 could be connected by a further edge.

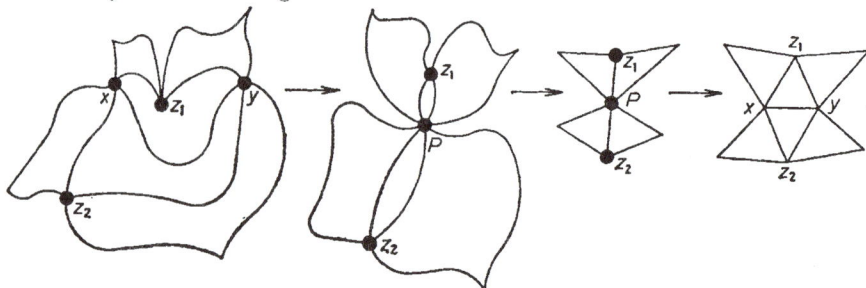

FIGURE 38

So G is 2-connected and, therefore, each face is a circuit. Suppose that C is the boundary of a face with at least 4 vertices and let a, b, c, d be four consecutive points on C. One of the two edges (a,c), (b,d) must be missing; since they both ought to run outside C and therefore, they ought to cross. Suppose that a and c are non-adjacent; then they can be connected by an edge inside C. Hence all faces of G are triangles.

Thus it suffices to prove the assertion for triangulations. Now we find an edge (x,y) which is contained in two triangles only. For let x be a point which is contained inside some triangle T (any point not on the outermost triangle has this property) and choose x, T so that the number of faces inside T is minimal. Let y be any neighbor of x. Now if (x,y) were an edge of three triangles (x,y,z_1), (x,y,z_2), (x,y,z_3), then all these triangles would be properly contained in T and, say, (x,y,z_1) would contain z_3 inside it, contrary to the minimality of T.

So choose an edge (x,y) such that the only two triangles having (x,y) as an edge are the two triangular faces (x,y,z_1), (x,y,z_2) incident with it. Contract (x,y) to a point p and remove one edge from both arising pairs of parallel edges. This way we get a new simple triangulation G_0 and by induction hypothesis, there is a triangulation G_0' with straight edges such that the faces of G_0 and G_0' correspond to each other.

Now consider the edges of G_0' corresponding to (p,z_1) and (p,z_2). They split the angle around p into two angles; one of these contains the edges whose pre-images in G are adjacent to x, the other one those whose pre-images are adjacent to y. Therefore, we can "pull x, y apart" and get an appropriate representation of G with straight edges (Fig. 38). [K. Wagner–I. Fáry; see S.]

§ 6. Connectivity

1. We use induction on $|E(G)|$. If $|E(G)| = 0$, G consists of isolated points and hence, $c(G) = |V(G)|$. Let $e \in E(G)$. Then

$$c(G) \geq c(G - e) - 1,$$

because e connects either two points of the same component of $G-e$, in which case $G-e$ has the same connectivity classes, or two points in different components, in which case $c(G)$ drops by 1. By induction,

$$c(G - e) + |E(G - e)| \geq |V(G - e)| = |V(G)|,$$

hence
$$c(G) + |E(G)| \geq c(G - e) - 1 + |E(G - e)| + 1 =$$
$$= c(G - e) + |E(G - e)| \geq |V(G)|.$$

2. (a) Let H be a component of the graph G^* constructed in the hint. Define

$$\tilde{H} = \bigcup_{(s_i, t_j) \in E(H)} (S_i \cap T_j).$$

Suppose that $(x, y) \in E(G_1 \cup G_2)$, $x \in \tilde{H}$. Let $(x, y) \in E(G_1)$ and $x \in S_i \cap T_j$ (say). Since S_i is a component of G_1, $y \in S_i$. Let $y \in T_{j_0}$. Now $s_i \in V(H)$, $(s_i, t_{j_0}) \in E(H)$, hence $S_i \cap T_{j_0} \subseteq \tilde{H}$, $y \in \tilde{H}$.

Hence, $\tilde{H}$ consists of one or more components of $G_1 \cap G_2$, whence

$$c(G_1 \cup G_2) \geq c(G^*).$$

If $S_i \cap T_j \neq \emptyset$, then it consists of one or more components of $G_1 \cap G_2$, hence

$$c(G_1 \cap G_2) \geq |E(G^*)|.$$

By the previous exercise,

$$c(G_1 \cup G_2) + c(G_1 \cap G_2) \geq |V(G^*)| = c(G_1) + c(G_2).$$

(b) Set $V = V(G_1) \cup V(G_2)$, and add the points of $V - V(G_i)$ to G_i as isolated points. Let G_i' be the resulting graph. Then

$$V(G_1') = V(G_2') = V,$$
$$c(G_i') = c(G_i) + |V| - |V(G_i)|,$$
$$c(G_1' \cap G_2') = c(G_1 \cap G_2) + |V| - |V(G_1 \cap G_2)|,$$
$$c(G_1' \cup G_2') = c(G_1 \cup G_2).$$

By (a),

$$c(G_1' \cup G_2') + c(G_1' \cap G_2') \geq c(G_1') + c(G_2').$$

So

$$c(G_1 \cup G_2) + c(G_1 \cap G_2) = c(G_1' \cup G_2') + c(G_1' \cap G_2')-$$
$$-|V| + |V(G_1 \cap G_2)| \geq c(G_1') + c(G_2') - |V| + |V(G_1 \cap G_2)| =$$
$$= c(G_1) + c(G_2) + |V| - |V(G_1)| - |V(G_2)| + |V(G_1 \cap G_2)| = c(G_1) + c(G_2).$$

3. Assume indirectly that G is disconnected and let G_1 be a component not containing x_n. Let $|G_1| = k$, and $x_{i_1}, \ldots, x_{i_k}$ its points ($1 \leq i_1 < \ldots < i_k < n$). Since the component containing x_n has at least $d_n + 1$ points, we have $k \leq n - d_n - 1$. Moreover,

$$d_k \leq d_{i_k} \leq k - 1,$$

a contradiction [A. Bondy; see B].

4. Suppose first that G_1 contains an odd circuit C and that G_1, G_2 are connected. To follow the hint, let x, $y \in V(G_1)$. If there is an (x, y)-walk in G_1 of length k, then there is one of length $k + 2$ (since we can go on an edge back and forth). Therefore it suffices to show there are both odd and even (x, y)-walks. There is a walk touching C because G_1 is connected; if we add a tour around C to this walk, we get a walk of opposite parity. This proves the statement of the hint. Now let (x, u) and (y, v) be two points of $G_1 \times G_2$. G_2 contains a walk ($u = u_0, u_1, \ldots, u_k = v$). We may assume that this walk is very long. Then by the above, G_1 contains a walk ($x = x_0, x_1, \ldots, x_k = y$) of the same length. Then $((x_0, u_0), (x_1, u_1), \ldots, (x_k, u_k))$ is a walk in $G_1 \times G_2$ connecting (x, u) to (y, v).

Conversely, assume that $G_1 \times G_2$ is connected. Trivially, G_1, G_2 are connected. Suppose that both of G_1 and G_2 are bipartite, and let $V(G_1) = A_1 \cup B_1$, $V(G_2) = A_2 \cup B_2$ be 2-colorations of them. Then no edge of $G_1 \times G_2$ connects $(A_1 \times A_2) \cup (B_1 \times B_2)$ to $(A_1 \times B_2) \cup (B_1 \times A_2)$, a contradiction.

5. Suppose that $G = G_1 \cup G_2$, $V(G_1) \cap V(G_2) = \{x\}$, $|V(G_i)| \geq 2$. Then $1 \leq d_{G_1}(x) \leq k - 1$, on the other hand, $d_{G_1}(y) = k$ for $y \in V(G_1) - \{x\}$. Let $u_1 \ldots, u_r, x$; $v_1, \ldots, v_s$ be the points in the two color classes of G_1. Then

$$|E(G_1)| = d_{G_1}(u_1) + \ldots + d_{G_1}(u_r) + d_{G_1}(x) = d_{G_1}(v_1) + \ldots + d_{G_1}(v_s)$$

whence

$$k \cdot r + d_{G_1}(x) = k \cdot s, \qquad k | d_{G_1}(x),$$

a contradiction.

6. (a) Let $P = (x_0, x_1, \ldots, x_m)$ be a maximum path in G. Suppose that $G - x_0$ is not connected. Let G_1 be a component of $G - x_0$ which does not contain $P - x_0$. Since G is connected, there must be an edge (y, x_0) joining G_1 to x_0. Now $(y, x_0, x_1, \ldots, x_m)$ is a longer path than P.

A cycle shows that the assertion does not hold for strongly connected digraphs.

(b) Let $P = (x_0, x_1, \ldots, x_m)$ be a longest path in G. If $G - x_0 - x_1$ is connected we are finished. Assume not, then there is a point y separated from $x_2, \ldots, x_m$ by x_0 and x_1. Let Q be a (y, P)-path in G. Then Q hits P at x_0 or x_1. However, by the maximality of P, it must hit at x_1. Also, since $Q \cup (P - x_0)$ is a path and has length $\leq |E(P)|$ by the maximality of P, Q consists of a single edge, joining y to x_1. No other edge can leave y, since it cannot go to a point outside P (by the maximality of P) and cannot go to a point of P since x_0, x_1 disconnect P and y. Thus, y is of degree 1.

Now observe that $(y, x_1, \ldots, x_m)$ is a maximum path and therefore the same argument yields a point $z \neq y$ of degree 1 connected to x_1. This contradicts the assumption of the theorem.

(c) Let T be any spanning tree of G; then deleting any two endpoints of T, the graph remains connected. So if G has no non-adjacent non-separating pair of vertices, then the endpoints of T must induce a complete graph.

Now choose T to have a maximum number of endpoints. If T is a path then G is a circuit; if T is a star then G is complete. Suppose that T is neither a path nor a star, and let U be the set of its endpoints. Let v and v' be two endpoints of the tree $T \setminus U$. Since v is not an endpoint of T, it has a neighbor u in U. Add (u, v) and all edges (u, w) $(w \in U \setminus \{u\})$ to the subtree $T \setminus U$, to get a spanning tree T'. The endpoints of T' must form a complete graph again, thus v' is adjacent to all vertices in $U \setminus \{u\}$. But then by the same argument (and using that $|U| > 2$), v must be adjacent to all vertices in U. Adding all edges (v, w) $(w \in U)$ to $T \setminus U$, we get a spanning tree with more endpoints than T, a contradiction.

7. (a) Let $e \in E(T_1) - E(T_2)$. Then $T_1 - e$ has two components and T_2 has an edge f connecting these. $T_1 - e + f$ is a spanning tree having more edges in common with T_2 than T_1. Repeating this we can transform T_1 into T_2.

(b) Let W be a largest common subtree of T_1 and T_2. We prove the assertion by induction on $|V(G) - |V(W)|$. If $|V(W)| \geq V(G)| - 1$ the assertion is trivial. Suppose that $|V(W)| \leq V(G)| - 2$, and let $e_i = (x_i, y_i)$ be an edge of T_i with $x_i \in V(W)$, $y_i \notin V(W)$.

Case 1. $y_1 \neq y_2$. Let T_3 be a spanning tree of G containing $W + e_1 + e_2$. Then by the induction hypothesis T_1 can be transformed into T_3 and T_3 can be transformed into T_2 in the required way, which proves the assertion.

Case 2. $y_1 = y_2$. Let $e_3 = (x_3, y_3)$ be any edge of $G - y_1$ such that $x_3 \in V(W)$, $y_3 \notin V(W)$ (here we use the fact that G is 2-connected). Let T_3 and T_4 be spanning trees of G containing $W + e_1 + e_3$ and $W + e_2 + e_3$, respectively. Then by the induction hypothesis, we can transform T_1 into T_3 into T_4 into T_2, which proves the assertion.

Note that this procedure has the additional property that if T_1 and T_2 have a common subtree W this remains unchanged during the sequence of transformations, and if W consists of a single point this never plays the role of the endpoint x.

8. (a) Let $e = (x_1, x_2)$ be an edge of G and let T_i consist of e and a spanning tree of $G - x_i$. Then by 6.7b, T_1 can be transformed into T_2 through a sequence of intermediate trees in the given way, and by the remark after the solution of 6.7b, we may assume that all these trees contain e. The branch of these trees relative to e containing x_1 has one point in T_1 and $n-1$ points in T_2 and its size changes by at most one at each step. Hence it will have exactly n_1 points in some intermediate spanning tree T. Now the two branches of T relative to e give the partition we wanted.

(b) The assertion is clearly equivalent to the fact that a 2-connected non-bipartite graph G has a spanning tree T such that the (unique) 2-coloration of T has equal color-classes.

Let $C = (x_0, \ldots, x_{2k})$ be an odd circuit in G. Consider a spanning tree W of $G - x_0$ containing $C - x_0$ and the two spanning trees T_1, T_2 of G obtained from W by adding the edges (x_0, x_1) and (x_{2k}, x_0), respectively. By 6.7b and the remark after its solution, we can obtain the tree T_2 from T_1 through a sequence of spanning trees of G, each of which arises from the preceding one by the following operation: we remove the edge adjacent to an endpoint $x \neq x_0$ and connect x to the rest by another edge of G. Let us 2-color each of these trees with red and blue in such a way that x_0 is red. Then if T_1 has k red points, T_2 has $2m - k + 1$ red points, because clearly every point except x_0 has different colors in each of them. Moreover, the number of red points changes by at most one at each step. Hence there is an intermediate tree with exactly m red points. [A. Bondy–L. Lovász; see L. Lovász, *Acta Math. Acad. Sci. Hung.* **30** (1977) 241–251.]

9. If G is strongly connected between a and b, let $\emptyset \neq X \subset V(G)$, $a \in X$, $b \in V(G) - X$. Walking along a (directed) (a,b)-path, at some point we have to leave x. The next edge of the (a,b)-path connects the set X to $V(G) - X$ (in this direction).

Conversely, suppose that there is no (a,b)-path. Let X be the set of those points which are accessible from a along a directed path. Then $a \in X$, $b \notin X$ and there is no edge (x,y) with $x \in X$, $y \notin X$ since any (a,x)-path plus this edge would give an (a,y)-path, which, however, does not exist because $y \notin X$.

10. Let G_0 be the digraph obtained by contracting the red edges and removing the green ones.

Suppose first that there is no directed (x,y)-path in G_0. Then there is a set S_0, $x \in S_0$, $y \notin S_0$ such that no black edge goes from S_0 to $V(G_0) - S_0$. Let S be the co-image of S_0 under the contraction of red edges; then no red edge connects S and $V(G) - S$ and no black edge goes from S to $V(G) - S$, i.e. (ii) is satisfied.

Trivially one obtains that if there is an (x,y)-path in G_0, then (i) is satisfied.

Finally, (i) and (ii) cannot be satisfied simultaneously. Suppose indirectly that there is a path P as in (i) and a set S as in (ii). Then P has a first edge f joining S to $V(G) - S$. f cannot be green since P has no green edges and it cannot be red since no red edge joins S to $V(G) - S$.

Thus f is black. But f cannot be oriented from S to $V(G) - S$ by (ii), neither can it be oriented conversely since it lies on P. [G. J. Minty; see B, ch. 1.]

11. (a) Let F be a minimum set of edges such that $G - F$ is not strongly connected. Then by 6.9, there is a set $\emptyset \subset X \subset V(G)$ with $\delta_{G-F}(X) = 0$. Any edge e in F must join a point of X to a point of $V(G) - X$; otherwise e could be placed back and we would still have $\delta_{G-F+e}(X) = 0$. Therefore, if we put back the edges of F inversely the resulting digraph G' has $\delta_{G'}(X) = 0$, i.e. G' is not strongly connected.

(b) Let F be a minimum set of edges such that G/F is strongly connected, and let G_0 be obtained from G by reversing the edges of F. We claim that G_0 is strongly connected.

First we consider the case $F = \{f\}$. G is not strongly connected by the minimality of F; thus there is a set $X \subset V(G)$, $X \neq \emptyset$ with $\delta_G(X) = 0$. Since G/f is strongly connected, f must join a point $y \in V(G) - X$ to a point $x \in X$. Invert f, and assume indirectly that the resulting digraph G_0 is not strongly connected either. Thus there is a set $Y \subset V(G)$, $Y \neq \emptyset$ such that $\delta_G(Y) = 0$. Again since G/f is strongly connected, $y \in Y$ but $x \in V(G) - Y$.

Now if $X \cap Y \neq \emptyset$, then $\delta_{G/f}(X \cap Y) = 0$ which is impossible. Thus $X \cap Y = \emptyset$. Similarly we obtain $X \cup Y = V(G)$. But then $\delta_G(X) = \delta_G(Y) = 0$ implies that f is an isthmus, a contradiction. This settles the case $|F| = 1$.

The general case follows by an easy induction on $|F|$. Let $f \in F$ and denote by H the digraph obtained from G by reversing f. The digraph $G/(F - \{f\})$ clearly contains no isthmuses, is not strongly connected (by the minimality of F) and contracting f in it we obtain a strongly connected digraph. Hence, if we reverse f the resulting graph $H/(F - \{f\})$ will be strongly connected. Moreover, $F - \{f\}$ is a minimal set with this property. For if H/F_0 were strongly connected for some $F_0 \subset (F - \{f\})$, then so would be $(H/F_0)/f = G/(F_0 \cup \{f\})$ which would contradict the minimality of F as $F_0 \cup \{f\} \subset F$. Thus we may apply the induction hypothesis and conclude that the graph obtained from H by reversing the edges of $F - \{f\}$ is strongly connected. But this digraph is just G_0 [A. Frank].

(c) Suppose that H is a digraph which contains no cycles. Then H has a point x with outdegree 0; otherwise, starting from any point and walking along edges in their direction, we would never get stuck and sooner or later we would complete a cycle. Using induction on $V(H)$, we may assume that the points of $H - x$ can be ordered in such a way that every edge has larger head than tail. Putting x on the top, we obtain such an ordering of $V(H)$.

Now let F be a minimal set of edges such that $H - F$ is acyclic. Then $V(H)$ has an ordering such that every edge of $H - F$ has larger head than tail. If we put back any edge of F the ordering loses this property, because the graph is no longer acyclic. Thus all edges of F go "downward". But then inverting them they will have a larger head than tail, i.e. the graph obtained from G by inverting the edges of F is acyclic. [E. J. Grinberg–J. J. Dambit, *Latv. Mat. E.* **2** (1966) 65–70; T. Gallai, *Theory of Graphs* (P. Erdős–G. O. H. Katona, eds.) Akadémiai Kiadó, 1968, 115–118.]

12. If the tournament T has a Hamiltonian cycle, then it is obviously strongly connected. Now suppose that T is strongly connected. Let $C = (y_1, \ldots, y_k)$ be a maximal cycle in T (this exists, as the acyclic tournament is not strongly connected). Suppose indirectly that C is not a Hamiltonian cycle.

Let x be a point not on C. Suppose that, e.g., $(y_1, x) \in E(T)$. If $(x, y_2) \in E(T)$, then $(y_1, x, y_2, \ldots, y_k)$ is a longer cycle. So $(y_2, x) \in E(T)$. Similarly $(y_i, x) \in E(T)$ for $i = 1, 2, \ldots, k$.

Now, let X be the set of all points x such that $(y_1, x) \in E(T)$. Then $(y_i, x) \in E(T)$ for every $x \in X$ as above. Let $(x, z) \in E(T)$ be an edge with $x \in X$, $z \notin X$ (which exists by $\delta_T(X) > 0$). Then $z \notin C$, and thus, $z \notin X$ implies $(z, y_1) \in E(T)$. Now $(x, z, y_1, \ldots, y_k)$ is a cycle longer than C. [P. Camion; see B.]

13. Let $C = (y_1, \ldots, y_k)$ be a longest cycle which is not a Hamiltonian cycle; this exists as $n \geq 4$. If C has length $|V(T)| - 1$, it misses exactly one point x and so $T - x$ is strongly connected.

Assume indirectly that C has length at most $|V(T)| - 2$. The same argument as in the previous solution yields that, for each $x \in V(T) - V(G)$, either $(y_i, x) \in E(T)$ for each $1 \leq i \leq k$ or $(x, y_i) \in E(T)$ for each $1 \leq i \leq k$. Let X denote the set of points with the first property. Let (x, z) be an edge leaving X. Then

$$(x, z, y_1, \ldots, y_{k-1})$$

is a cycle longer than C but missing a point (y_k, in fact).

To show that there are at least two such points let y_1 have this property. y_1 is contained in some circuit shorter than n; for let X denote the set of points z with $(y_1, z) \in E(T)$, then there is an edge (z, u) leaving X and (y_1, z, u) is a 3-cycle.

Consider now a longest cycle $(y_1, \ldots, y_k)$ with $k < n$; we find as before that $k = n - 1$ and this yields a point $x \neq y_1$ such that $T - x$ is strongly connected.

14. Invert all edges of F. The resulting graph G' has no circuit, therefore the beginning point of a longest path is the tail of every edge incident with it. Inverting the edges of F again, we get that this point satisfies the requirements.

15. We use induction on $|V(T)|$. For $|V(T)| \leq 2$ the assertion is trivial. Let $|V(T)| \geq 3$.

If φ is one-to-one then it is an automorphism. Denoting by T' the subtree formed by the inner points, φ maps T' onto itself. If φ is one-to-one, then $\varphi(T)$ is a proper subtree mapped into itself by φ.

In both cases we are finished by induction.

Remark: The assertion is a degenerate case of the Lefschetz Fixed Point Theorem (see, e.g. E. Spanier, *Algebraic Topology*, McGraw-Hill, 1966).

16. The path connecting two points of the intersection is unique and is contained, therefore, in each of the given subtrees. Thus, it belongs to their intersection.

17. (a) Let x_1, x_2 be the endpoints of Q, $x_i \in P_i$. x_i divides P_i' into two pieces P_i', P_i''. We may assume that P_i' is at least as long as P_i'' and that P_1' is at least as long as P_2'. Then

$$|E(P_1' \cup Q \cup P_2')| > |E(P_1')| + |E(P_2')| \geq 2|E(P_2')| \geq$$
$$|E(P_2')| + |E(P_2'')| = |E(P_2)|,$$

i.e. the path $P_1' \cup Q \cup P_2'$ is longer than P_2, a contradiction.

(b) Let P_1, P_2 be two maximum paths. Their intersection Q is a path by 6.15, hence $P_1 \cup P_2$ has the following form: there are two paths P_1', P_2' starting from one endpoint of Q, and two others: P_1'', P_2'' from the other endpoint, so that

$$P_1 = P_1' \cup Q \cup P_1'', \quad P_2 = P_2' \cup Q \cup P_2''.$$

We have $|E(P_1')| = |E(P_2')|$, because, e.g. $|E(P_1')| > |E(P_2')|$ would imply that the path $P_1' \cup Q \cup P_2''$ is longer than P_2. Similarly, $|E(P_1'')| = |E(P_2'')|$. Since $P_1' \cup P_2'$ is a path, we have $|E(P_1')| \le 1/2|E(P_1)|$ and similarly, $|E(P_1'')| \le 1/2|E(P_1)|$.

This shows that the middle point (points) of P_1 belongs to Q and, hence, to P_2. Since this holds for any P_1, this point (points) is in the intersection of all maximum paths.

Remark: An example of H. J. Walther [WV; Fig. 39] shows that this is not true for every connected graph, as was conjectured by Gallai.

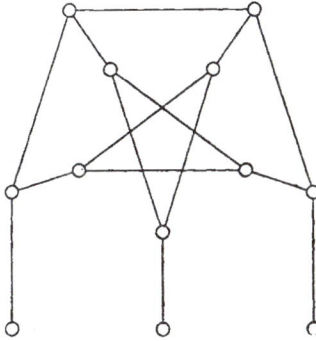

FIGURE 39

18. *First solution.* For $k = 2$ the statement is obvious. Using induction on k, we may assume that $G_1 \cap \ldots \cap G_{k-1} = G_0 \ne \emptyset$. Suppose that $G_0 \cap G_k = \emptyset$. Let P be a (G_0, G_k)-path. Consider an edge (x, y) of P (x nearer G_k). Then obviously $x \notin V(G_0)$, thus $x \notin V(G_i)$ for some $1 \le i \le k-1$. Hence $(x, y) \notin E(G_i)$. Now $G - (x, y)$ is disconnected, and x and y are in distinct components of it. Obviously, G_k is in the component containing x. Since G_i is a connected subgraph of $G - (x, y)$ and meets G_k, it also lies in the same component. But this is impossible because $G_0 \subseteq G_i$ and G_0 has points in the other component.

Second solution. Let x be a point of degree 1 of G, joined to y. Suppose that the assertion holds true for $G - x$.

If none of $G_1, \ldots, G_k$ is the one-point graph with the point x, then $G_1 - x, \ldots, G_k - x$ intersect mutually; for if x is a common point of G_i and G_j, then so is y. Thus by the induction hypothesis $G_1 - x, \ldots, G_k - x$ have a point in common, hence so do $G_1, \ldots, G_k$. If G_1 (say) has only one point x, then x is a common point of $G_1, \ldots, G_k$.

19. To show $d(x, y) + d(y, z) \ge d(x, z)$, consider an (x, y)-path P of length $d(x, y)$ and a (y, z)-path of length $d(y, z)$. $P \cup Q$ contains an (x, z)-path, which obviously has length at most $d(x,y)+d(y,z)$. Thus $d(x,z)$, the minimum length of all such paths, is also $\leq d(x,y)+d(y,z)$.

To show $D(x,y)+D(y,z) \geq D(x,z)$, consider an (x,z)-path P of length $D(x,z)$. Since G is connected, we have a (y,P)-path Q; let t be the endpoint of Q on P. t splits P into two paths P_1, P_2, of lengths l_1 and l_2, say. Let Q have length k. Then $Q \cup P_1$ is an (x,y)-path, hence

$$D(x,y) \geq k + l_1.$$

Similarly,

$$D(y,z) \geq k + l_2,$$

whence

$$D(x,y) + D(y,z) \geq 2k + l_1 + l_2 = 2k + D(x,z) \geq D(x,z).$$

20. In the formulation given in the hint, let us see how many times a given edge e is counted on each of the two sides. $G - e$ consists of two components; let α of the points p_i and β of the points q_i be in one of them. Then e is contained in

$$\alpha(n - \alpha) \quad \text{of the } (p_i, p_j)\text{-paths,}$$
$$\beta(n + 1 - \beta) \quad \text{of the } (q_i, q_j)\text{-paths,}$$
$$\alpha(n + 1 - \beta) + \beta(n - \alpha) \quad \text{of the } (p_i, q_j)\text{-paths.}$$

It suffices to show that

$$\alpha(n - \alpha) + \beta(n + 1 - \beta) \leq \alpha(n + 1 - \beta) + \beta(n - \alpha)$$

or, equivalently,

$$\beta - \alpha \leq (\alpha - \beta)^2,$$

which is true because α, β are integers. [J. B. Kelly, in: *Comb. Structures and their Appl.* Gordon and Breach, 1969, 201–208; cf. 13.16.]

21. (a) First we prove the assertion of the hint. Since all points at distance $\tilde{d}(x)$ from x are, obviously, of degree 1, $\tilde{d}(x)$ decreases at every point. On the other hand, only one point was omitted from the longest path starting from x; hence $\tilde{d}(x)$ decreases by 1.

Now suppose that $|V(G)| \geq 2$. Let Z be the set of points, where $\tilde{d}(x)$ is minimal. If Z contains no points of degree 1, remove all points of degree 1 from G. By the assertion of the hint, this does not alter Z and hence the statement follows by induction. If a point x of degree 1 belongs to Z, look at the neighbor y of x. Obviously, y is strictly nearer than x to any other point. Hence, $\tilde{d}(x)$ can be minimal only if $\tilde{d}(x) = 1$, i.e. G is the tree with two points. In this cases, the statement is obvious.

Note that the center (bicenter) is contained in every maximum path (cf. 6.16b).

(b) Let P be a path of length $\tilde{d}(x)$ starting from x. If P does not contain y or z, then obviously,

$$\tilde{d}(y) = \tilde{d}(x) + 1, \qquad \tilde{d}(z) = \tilde{d}(x) + 1$$

and the assertion follows. If P contains y (say), then

$$\tilde{d}(y) \geq \tilde{d}(x) - 1, \qquad \tilde{d}(z) = \tilde{d}(x) + 1,$$

which proves the statement. [See, e.g. K.]

22. (a) Let k_1, k_2 be the numbers of point in the components of $G - x$ containing y and z, respectively (obviously, these are different components). Moving from x to y, k_1 points approach but $n - k_1$ depart, hence

$$s(x) = s(y) + k_1 - (n - k_1) = s(y) + 2k_1 - n.$$

Similarly,

$$s(x) = s(z) + 2k_2 - n.$$

Summing,

$$2s(x) = s(y) + s(z) + 2(k_1 + k_2 - n) \leq s(y) + s(z) - 2.$$

(b) Assume indirectly that there are two non-adjacent points x, y such that $s(x) = s(y)$ is minimal. Let

$$(x_1 = x, x_2, \ldots, x_p, x_{p+1} = y)$$

be the (x, y)-path. Then

$$s(x_2) \geq s(x).$$

By (a),

$$s(x) + s(x_3) > 2s(x_2) \geq s(x_2) + s(x), \qquad s(x_3) > s(x_2) \geq s(x),$$
$$s(x_2) + s(x_4) > 2s(x_3) > s(x) + s(x_2), \qquad s(x_4) > s(x),$$

and so on, finally we get

$$s(x_{p+1}) = s(y) > s(x_p),$$

a contradiction [K].

(c) Let $x_1, \ldots, x_p$ be the points of a path and $y_1, \ldots, y_q$ other points connected to x_1. Then, if p is even, the center of the tree is $x_{\frac{p}{2}}$; on the other hand, the baricenter is x_1, if $q = \binom{p}{2}$. Thus, the distance of the baricenter and center is $\frac{p}{2} - 1$, which is arbitrarily large indeed.

23. In

$$\sum_{x,y \in V(G)} d(x, y),$$

there are exactly $n - 1$ terms equal to 1; if G is the star, all the other terms are 2; otherwise, they are not less than 2 with at least one of them 3. Hence, the sum is minimal for the star.

To prove the statement given in the hint for the second part of the problem, note that, for an endpoint x of a path P we have

$$(1) \qquad\qquad s(x) = 1 + 2 + \ldots + (n-1).$$

If G is any tree and x an endpoint of it, and $\tilde{d}(x) = d$, we have at least one point of distance 1, 2, ..., d from x, therefore the sum defining $s(x)$ looks like

$$(2) \qquad\qquad s(x) = 1 + 2 + \ldots + d + d_1 + \ldots + d_{n-1-d},$$

where $d_1, \ldots, d_{n-1-d} \le d$. Obviously, $(2) \le (1)$ and equality holds only if $d = n-1$, i.e. G is a path and x its endpoint.

Now we prove by induction on n that

$$\sum_{x,y \in V(G)} d(x,y)$$

is maximal (G ranging over all n-point trees) if G is a path. Let a be a point of degree 1 in G. Then

$$\sum_{x,y \in V(G)} d(x,y) = s(a) + \sum_{x,y \in V(G-a)} d(x,y).$$

Here the first term is maximal if G is a path and a is an endpoint of it. Fortunately, in this case the second term is also maximal by the induction hypothesis.

24. Let $P = (x_1, \ldots, x_{2k-2})$ be a path of length $2k-3$ in G. With any point y not on P, associate a path P_y in the following way. Let Q be the path connecting y to P and having no other point in common with P. If Q is longer than $k-1$, let P_y be the segment of Q of length k, starting from y. Otherwise, let P_y consist of Q and a subpath of P. Such a choice is possible since P has length $2k-3$, hence one of the pieces of P incident with the endpoint of Q is of length $k-1$. Furthermore, let P_{x_i} be a subpath of P incident with x_i for $k \le i \le 2k-2$.

It is easy to see that the P_y's are distinct and that their number is $n-k$.

25. The resulting graph is obviously a spanning tree. Let H be an optimal spanning tree with the maximum number of edges in common with G and let $e_i = (x,y)$ be an edge of G not in H, selected at the ith step. Let P be the path in H connecting x to y; then there is an edge f of P connecting G_i to a point outside it. Since we preferred to choose e_i rather than f, $v(e_i) \le v(f)$. On the other hand, $H - f + e_i$ is a spanning tree and has no more expense but more edges in common with G than H, a contradiction.

Remark: If G_i is the component containing a given point at every step, the graph formed by the edges selected in the first i steps is always connected. [J. B. Kruskal; see S.]

26. Let G, G' be two optimal trees. Let e be an edge of G not in G'. Consider the path in G' connecting the endpoints of e. Some edge of it, f say, connects the two components of $G - e$. Now either $G - e + f$ or $G' - f + e$ has less expense than G.

27. If e, f are contained in the same circuits, then f is contained in no circuit of $G - e$, hence $G - e - f$ is disconnected.

Conversely, if there is a circuit containing f but not e (say), then f is on a circuit of $G - e$, whence $G - e - f$ is connected. This proves the statement of the hint.

Now (a), (b) follow immediately. To show (c), let e be an edge not in the removed equivalence class P, and let $f \in P$. e is not a cut edge of $G - f$, hence there is circuit C containing e but not f. C obviously contains no other edge of P, whence C is a circuit of $G - P$. Obviously C lies in the connected component of $G - P$ containing e. Hence e is not a cut edge of this component.

(d) We show that each component G_0 of $G - P$ is incident with two edges of P. Let (x, y) be an edge of P incident with G_0 ($x \in V(G_0)$, $y \notin V(G_0)$). Let C be a circuit containing (x, y). Let us walk along C, starting from x through (x, y). Upon reaching a point of G_0 again, let us go back to x within G_0. The circuit defined this way contains exactly two edges of P incident with G_0; on the other hand, it contains all edges of P; hence there are exactly two edges of P incident with G_0.

Thus, contracting the components of $G - P$, we get a connected regular graph of degree 2. This is a circuit.

28. We can select a circuit G_1 of G. Assume that $G_1, \ldots, G_j$ are already selected in such a way that G_{i+1} ($1 \le i < j$) is either a path having its endpoints in common with $G_1 \cup \ldots \cup G_i$ or a circuit having one point in common with $G_1 \cup \ldots \cup G_i$. If $G_1 \cup \ldots \cup G_j = G$ we are finished. Otherwise, there is an edge $e = (x, y)$ not in $G_1 \cup \ldots \cup G_j$ but having a point in common with $G_1 \cup \ldots \cup G_j$. Let C be a circuit containing e; start from x walking through e and then along C until we meet $G_1 \cup \ldots \cup G_j$ again; let G_{j+1} be the subgraph consisting of those edges and points we touched. Then, obviously, G_{j+1} is either a path with two endpoints in $G_1 \cup \ldots \cup G_j$, or a circuit C meeting $G_1 \cup \ldots \cup G_j$ in the point x only. Thus we could find a suitable G_{j+1}. G being finite, sooner or later we have to have decomposed the whole G.

29. *First solution.* If we orient G in some way and the resulting $\vec{G}$ is strongly connected, let (x, y) be an edge of $\vec{G}$, directed from x to y. There is a path in $\vec{G}$ connecting y to x; hence (x, y) is contained in a circuit of G (even of $\vec{G}$). Now assume that G is 2-edge-connected. We use induction on the number of edges. Remove an equivalence-class P defined in 6.27. The components of the remaining graph are 2-connected and thus they can be oriented so that they are strongly connected. The edges of P lie on a circuit C; let us orient them in accordance with a cyclic orientation of C. It is easy to see that the resulting digraph is strongly connected.

Second solution. (for the second part): Let $G = G_1 \cup \ldots \cup G_r$ as in 6.28, then we can define an orientation of G by considering an orientation which makes an oriented path or cycle out of each G_i.

Let $a \in V(G_1)$. Then it easily follows by induction on j that any $b \in V(G_j)$ is accessible from a on a directed path and conversely, a is accessible from any other point. By the transitivity of accessibility this proves $\overrightarrow{G}$ is strongly connected. [H. E. Robbins; see B, Wi.]

30. By induction on k, we can find a circuit C_0 such that $e_1, \ldots, e_{k-1}$ are well fitted with C_0. If e_k is well fitted with C_0, we are finished. If both endpoints of e_k are on C_0, e_k and one of the arcs bounded by its endpoints give the desired circuit. Thus we may assume that e has one endpoint x on C_0 and one endpoint y not on C_0.

Contract C_0. The resulting graph G' is obviously 2-edge-connected and by induction, it has a circuit C_1 with which all the edges e_i not on C_0 are well-fitted. Let C_1' be the subgraph of G mapped onto C_1 by the contraction. There are three possibilities.

(1) C_1' is a circuit disjoint from C_0. Then C_1' is the desired circuit.

(2) C_1' is a circuit having one point in common with C_0. Then it must contain e_k (since e_k is well fitted with C_1), hence it does not meet the e_i's on C_0, and has the desired property again.

(3) C_1' is a path connecting two points of C_0. e_k lies again on C_1'. Now C_1' and one of the arcs of C_0 bounded by its endpoints gives a circuit with the desired properties (Fig. 40).

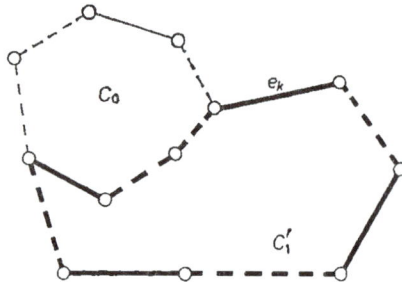

FIGURE 40

31. The only non-trivial statement to prove is that mentioned in the hint. Walk from e_3 in both directions on C_2 till we meet C_1; let x and y be the two points reached (it may happen that x and/or y are endpoints of e_3). x and y are distinct, so they bound two arcs of C_1, one of which contains e_1; this arc and the arc of C_2 bounded by x, y and containing e_3, give a circuit containing e_1 and e_3.

32. (i)$\Rightarrow$(iii): First let e and f be two edges with a common endpoint x. Let y and z be their other endpoints. Since $G - x$ is connected, there is a path in it connecting y to z; this forms with e and f a circuit.

Using 6.31, we get that any two edges are on a circuit; in fact, an equivalence-class of the relation "being on a circuit" contains every edge which is adjacent to its edges; therefore consists of the edges of a component. But G is connected, hence this equivalence-class is the whole graph.

(iii)$\Rightarrow$(ii): Given two points, consider one edge incident with each and a circuit through these two edges.

(ii)$\Rightarrow$(i): Assume that the removal of a point x disconnects G. Let a, b lie in different components of $G-x$. Then no circuit can contain both a and b.

33. The analogue of 6.28 for 2-connected graphs is the following:

(∗) *Every 2-connected graph G has a decomposition (an ear-decomposition) $G = G_1 \cup \ldots \cup G_r$, where G_1 is a circuit and G_{i+1} is a path having exactly its endpoints in common with $G_1 \cup \ldots \cup G_i$.*

It is easy to see that (a) is equivalent to (∗). In fact, if (a) is known, then we can delete the edges and inner points of the path P, find an appropriate decomposition of the remaining graph, and add P as the last "ear". Conversely, if we have an ear-decomposition as in (∗), then G_r can play the role of P in (a). Below we give separate proofs of (a) and (∗).

(a) Let P_1, P_2, P_3 be three paths connecting a and b $(a,b\in V(G))$ and assume that a, b, P_1, P_2, P_3 are chosen so that P_1 is of minimum length. First we show P_1 has inner points of degree 2 only. Assume that (x,y) would be an edge incident with an inner point x of P_1 but not belonging to P_1. Since $G-x$ is connected there is a $(y, P_1\cup P_2\cup P_3-x)$-path Q in $G-x$. The other endpoint of Q may be on P_1 or on $P_2\cup P_3$; in either case we get two points connected by three independent paths, one of them being shorter than P_1, a contradiction with the choice of P_1 (Fig. 41).

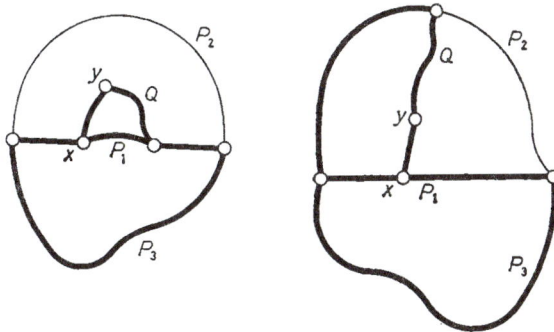

FIGURE 41

Thus the inner points of P_1 are of degree 2. Remove the inner points of P_1. The resulting graph G' is 2-connected. In fact, consider two points a, b of $G'-c$ for some c. There is an (a,b)-path P in $G-c$. If P does not go through P_1, it is an (a,b)-path in $G'-c$. If $P_1\subseteq P$, one of P_2, P_3 does not contain c and we can "avoid" P_1 using it. Thus $G'-c$ is connected.

This proves that P_1 satisfies our requirements.

(b) Let G_1 be a circuit and choose $G_2, \ldots, G_i$ satisfying the requirement above. If $G_1 \cup \ldots \cup G_i \neq G$, we can find an edge (x, y) in $G - E(G_1 \cup \ldots \cup G_i)$ such that $x \in V(G_1 \cup \ldots \cup G_i)$. Since $G - x$ is connected, we can find a $(y, G_1 \cup \ldots \cup G_i - x)$-path Q in $G - x$. Then Q and (x, y) give a G_{i+1}. Therefore, sooner or later we have $G = G_1 \cup \ldots \cup G_r$.

Now $G_1 \cup \ldots \cup G_{r-1}$ is 2-connected, which can be seen very easily. Hence $P = G_r$ satisfies the requirements. [H. Whitney, *Amer. J. Math.* **55** (1933) 236–244; also see G. A. Dirac, *J. Reine Angew. Math.* **228** (1967) 204–216.]

34. Let $G = G_1 \cup \ldots \cup G_r$, where G_1 is a (p, q)-path and G_{i+1} is a path which has exactly its endpoints in common with $G_1 \cup \ldots \cup G_i$. Such a decomposition can be found just as in the first solution of the preceding exercise.

Now order the points of G_1 according to their position on G_1; insert the points of G_2 between them, so that they are ordered in accordance with their positions on G_2 (and thus, lie between the two endpoints of G_2), etc. The ordering obtained finally has the property that each point is adjacent to both earlier and later points.

Now orient any edge, so that its tail anticipates its head in this ordering. Then starting from any edge (x, y), we can find edges (y, y_1), (y_1, y_2), etc. by the properties of the ordering the last of which must end, of course, in q. Similarly, there are edges (x_1, x), (x_2, x_1), etc. the last starting from p. These edges yield a (p, q) path of G containing (x, y). [A. Ádám, A problem of the Schweitzer Competition, *Mat. Lapok*, **22** (1971) 34.]

35. The assertion given in the hint follows from the second solution of problem 6.33; it proves immediately that (i) implies (ii). Conversely, if (ii) holds, let $e = (x, y)$ be an edge such that $G - e$ is 2-connected. Then x and y lie on a circuit of $G - e$ such has e as a chord, a contradiction. [M. D. Plummer, *Trans. Amer. Math. Soc.* **134** (1968) 85–94.]

36. If $G = C$ we have nothing to prove. Otherwise, there is a (C, C)-path in G. For let u be a point of C of degree at least 3, (u, v) an edge not on C, and P_0 a $(v, C - u)$-path in $G - u$. Then $P_0 + (u, v)$ is a (C, C)-path.

Now let P be a (C, C)-path selected as in the hint. R has at least 2 edges by 6.35. Assume that z has degree at least 3. Then it follows as above that z is the endpoint of a (C, C)-path Q. Let w be the other endpoint of Q.

Observe that if P and Q intersect or w is outside R, we can find a circuit such that (x, z) is a chord of it, which contradicts 6.35 (see Fig. 42). However, this means that the arc connecting z and w (in R) is a proper sub-arc of R, which contradicts the choice of P. [M. D. Plummer, ibid.]

37. G' is a forest by 6.36. By the argument in the preceding proof, there is a (G', G')-path Q_1. If G' is connected, then it contains a path Q_2 connecting the endpoints of Q_1. $C = Q_1 \cup Q_2$ is a circuit whose points of degree 2 form an arc of it (these are the inner points of Q_2, in fact).

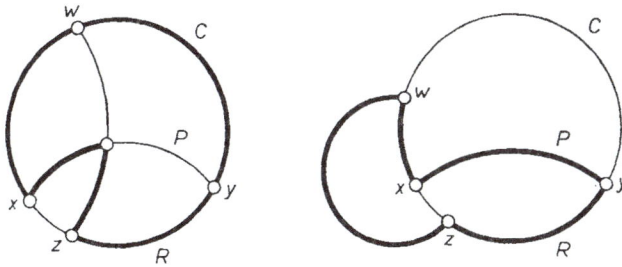

FIGURE 42

Now let P be a (C,C)-path; one of the arcs connecting its endpoints has no points of degree 2. On the other hand, this arc contains a minimal arc of this type and this has a point of degree 2 by the solution of the preceding exercise. [M. D. Plummer, ibid.]

38. Let G_1, G_2 be two isomorphic trees such that every endpoint of them is at distance at least 1000 from their centers, and which have no points of degree 2. Identify the corresponding endpoints of G_1 and G_2. The resulting graph G is 2-connected; for take any two points x, y of it. Let x', y' be the corresponding points in the other tree. We show that even x, x', y, y' (which are not necessarily distinct) lie on a circuit. We may assume that x, $y \in G_1$ and x', $y' \in G_2$. Let P_1 be the path in G_1 connecting x to y. We can lengthen P_1 to a path P which connects two endpoints of G_1. Let P' be the path corresponding to P; then $P \cup P'$ is a circuit which contains x, x', y, y'.

On the other hand, removing an $e \in E(G_1)$ (say) the resulting graph will not even be 2-edge connected, because the edge $e' \in E(G_2)$ corresponding to e will be a cutting edge of it.

39. (a) If G has k edge-disjoint (a,b)-paths, it is obviously k-edge-connected between a and b. To prove the other part, remove edges till the removal of any further edge will destroy k-edge-connectivity between a and b. Then obviously, there will be no edge with head at a or tail at b. Assume first there is an edge e_1 incident neither to a nor to b. Since $G - e_1$ no longer satisfies the conditions, it has a $(k-1)$-element (a,b)-cut C'. Then $C = C' \cup \{e_1\} = \{e_1, \dots, e_k\}$ is a k-element (a,b)-cut and by the choice of e_1, the set S determining C satisfies $|S| \geq 2$, $|V(G) - S| \geq 2$.

Let G_1, G_2 be the graphs obtained by contracting S and $V(G) - S$, respectively; let a' and b' be the images of a in G_1 and b in G_2, respectively.

Obviously, G_1 is k-connected between a' and b, and thus by the induction hypothesis, there are k edge-disjoint (a',b)-paths $P_1, \dots, P_k$. Since the edges going out from a' are only $e_1, \dots, e_k$, we may assume that $e_i \in P_i$. Similarly, there are k edge-disjoint (a,b')-paths $Q_1, \dots, Q_k$ in G_2, $e_i \in Q_i$. Then $P_1 \cup Q_1, \dots, P_k \cup Q_k$ form k-edge-disjoint (a,b)-paths in G.

What is left is the case, when each edge has tail at a or head at b.

If there is an (a,b)-edge, we can remove it and proceed by induction on k, thus we may assume that there is no such edge.

For any $x \neq a$, b, let $k(x)$ be the minimum of the numbers of (a,x)-edges and (x,b)-edges. Then obviously, there are $\sum_{x \neq a,b} k(x)$ edge-disjoint (a,b)-paths. On the other hand, let S be the set of all points x which are connected to b by $k(x)$ edges. Then, the cut determined by $\{a\} \cup S$ has exactly $\sum_{x \neq a,b} k(x)$ edges. Hence

$$\sum_{x \neq a,b} k(x) = k,$$

which proves the assertion.

(b) As suggested in the hint, consider a graph G', which has points a, b and two points x_1, x_2 that each $x \in V(G)$, $x \neq a$, b. Put $a_1 = a_2 = a$ and $b_1 = b_2 = b$. For any edge $e = (x,y) \in E(G)$, G' has the edge $e' = (x_2, y_1)$, moreover, for each $x \in V(G)$, $x \neq a$, b, it has the edge (x_1, x_2). Now

(i) G' is k-edge-connected between a and b iff G is k-connected between a and b;

(ii) G' has k-edge-disjoint (a,b)-paths iff G has k vertex-disjoint (a,b)-paths.

To show (i), consider an (a,b)-cut C in G'. Let A consist of all points x such that $(x_1, x_2) \in C$ and all other edges of C. Then $|A| = |C|$ and A separates a and b in G; for if P is any (a,b)-path in G, then the edges of G' corresponding to edges and inner points of P, form an (a,b)-path P' in G and since P' contains an edge of C, P contains an edge or point of A.

Conversely, if A is a set of edges and points separating a and b in G, then the construction above associates a set C of edges of G' with it, and C will be an (a,b)-cut with $|C| = |A|$. This proves (i).

Now consider k edge-disjoint (a,b)-paths $P_1, \ldots, P_k$ in G'. If $x_i \in P_j$ then, obviously, (x_1, x_2) is an edge of P_j and x_{3-i} is also on P_j. Hence, $P_1, \ldots, P_k$ are vertex-disjoint and even contracting the edges (x_1, x_2) to get G back, we obtain k vertex-disjoint (a,b)-paths of G.

Conversely, if there are k vertex-disjoint (a,b)-paths in G, then the (a,b)-paths of G' associated with them in the natural way are vertex-disjoint, and hence edge-disjoint. This proves (ii).

(i) and (ii) prove the statement of (b), by (a).

(c) The directed graph $\overrightarrow{G}$ obtained as in the hint has the same connectivity and edge-connectivity between any two points as G.

Moreover, the maximum number of edge-disjoint (vertex-disjoint) (a,b)-paths of G and $\overrightarrow{G}$ are the same. For edge[vertex]disjoint (a,b)-paths of G yield such paths in $\overrightarrow{G}$. Conversely, if we have some edge-disjoint (a,b)-paths of $\overrightarrow{G}$, we may assume that they do not use both (x,y) and (y,x); for in this case one can easily find another system of the same number of (a,b)-paths, which contain neither (x,y) nor (y,x) (in the vertex-connectivity case, this difficulty does not even arise). These paths yield edge-disjoint (vertex-disjoint) (a,b)-paths in G. [K. Menger; see any textbook on graph theory.]

40. The graph G' constructed in the hint is k-connected between a and b; since a and b are non-adjacent, it suffices to show that the (a,b)-paths cannot be covered by $k-1$ points; but this is the assumption. Therefore, by Menger's theorem, there are k independent (a,b)-paths in G'; removing a and b from them, we get k (A,B)-paths as desired.

41. To prove the statement formulated in the hint, let us substitute $w(x)$ distinct points for each x, and connect two new points if the original points are connected. Denoting by G' the resulting graph and by A' and B' the sets of points substituted for A and B, respectively, these sets will satisfy the conditions of 6.40. For let X be a set which meets all (A',B')-paths of G'. Choose X to be minimal. If X contains a point x_1 substituted for x, then it contains every other point x_2 substituted for x; since if $X-x_1$ fails to meet some path P, then $P \cap X = \{x_1\}$ and hence if $x_2 \notin X$, replacing x_1 by x_2 in P we get a path disjoint from X. Let X_0 be the set of all points $x \in V(G)$ such that X contains the corresponding points of G', then

$$|X| = \sum_{x \in X_0} w(x),$$

on the other hand X meets all (A,B)-paths of G, hence $|X| \geq k$.

There are, thus, k (A',B')-paths in G' which have no point in common. Contracting the points substituted for x onto x for each $x \in V(G)$, we get k (A,B)-paths in G as required.

Menger's theorem is obtained by putting $w(a) = w(b) = k$ and $w(x) = 1$ elsewhere; 6.40 follows by putting $w(x) = 1$ for every x. A third special case of interest is when $A = \{a\}$ and $w(a) = k$, $w(x) = 1$ for $x \neq a$. Then we get: if every set which does not contain a but meets every (a,B)-path is of cardinality at least k, then there are k (a,B)-paths which have no point in common except a. [G. A. Dirac, *J. London Math. Soc.* **38** (1962) 148–163.]

42. If $R_1, \ldots, R_k$ are chosen in accordance with the hint, we claim that, if the beginning edge of a Q_j $(1 \leq j \leq k+r)$ is not the beginning edge of any R_i, then Q_j is independent of every R_i. For assume that Q_j meets some of them and let x be the first point of Q_j in common with an R_i. Let R_i' be the path consisting of the (a,x)-part of Q_j and the (x,B)-part of R_i. Then

$$|E(R_1 \cup \ldots \cup R_i' \cup \ldots \cup R_k) - E(Q_1 \cup \ldots \cup Q_{k+r})| <$$
$$< |E(R_1 \cup \ldots \cup R_k) - E(Q_1 \cup \ldots \cup Q_{k+r})|,$$

a contradiction.

Now there are at least r of $Q_1, \ldots, Q_{k+r}$ whose beginning edge is not a beginning edge of $R_1, \ldots, R_k$. [H. Perfect, *J. Math. Anal. Appl.* **22** (1968) 96–111.]

43. With $B = V(P_0)$, the conditions of 6.42 are satisfied: we have k independent (a, B)-paths $P_1, \ldots, P_k$ and, on the other hand, suitable beginning sections of $Q_0, \ldots, Q_k$ give $k + 1$ independent (a, B)-paths (no matter where they end, i.e. where they meet P_0). Therefore, we have $k+1$ independent (a, B)-paths $R_1, \ldots, R_k$ and R such that $R_1, \ldots, R_k$ end at b. $R \cup P_0$ contains a $(k+1)^{\text{st}}$ (a, b)-path, independent of $R_1, \ldots, R_k$. [For this and the next problem see G. Hajós, *Theory of Gr. Int. Symp. Rome*, Dunod, Paris — Gordon and Breach, New York, 1967, 147.]

44. Let $c \neq a, b$. Assume indirectly that c is not connected to anyone of a and b. The addition of an (a, c)-edge produces $k + 1$ independent (a, b)-paths; removing this edge again we get k independent (a, b)-paths and a (b, c)-path independent of them. Similarly we get k independent (a, b)-paths and an (a, c)-path independent of them. By 6.43, there are $k + 1$ independent (a, b)-paths in G, which is a contradiction.

We show by a similar argument that, if x is joined to a but not to b and y is joined to b but not to a, then x and y are not adjacent. Assume indirectly that there was an edge e connecting them. Subdivide e by a point c. Then, adding, the edges (a, y) and (b, x) and applying 6.43, we get a contradiction just as above.

This statement means the points connected to both a and b separate them. Hence their number is at least k. On the other hand, through each of these points there goes an (a, b)-path of length 2 and these paths are independent. This proves that the number of points joined to both a and b is k.

Now if G_1 is spanned by a and its neighbors, G_2 by b and its neighbors, then joining two points of G_1 or G_2 cannot produce $k + 1$ independent (a, b)-paths as a and b will still be separated by k points. Thus G_1, G_2 are complete and we are finished.

To deduce Menger's theorem, consider an undirected graph G and $a, b \in V(G)$. We may assume that a, b are independent, since the removal of an (a, b)-edge decreases both the maximum number of independent (a, b)-paths and the connectivity between a and b by 1. Let k denote the maximum number of independent (a, b)-paths; we have to show a and b can be separated by k points. Add edges till k remains the same; the graph we finally get is described above and has a k-element set separating a and b. But the same set separates a and b in G.

45. To disprove (i), consider the graph shown in Fig. 43. Any (a, b)-path starts through e_1, whence any two of them have an edge in common, similarly for (b, a)-path. Taking an (a, b)-path P and a (b, a)-path Q, they connect two opposite corners of the broken rectangle; hence they intersect somewhere. Now each intersection point is a point of the graph and has degree 3, therefore they have an edge in common.

There are two vertex-disjoint paths connecting c to d; hence, if an edge is contained in each (a, b)-path it is either e_1 or f_2. Since, similarly, an edge contained in each (b, a)-path is e_2 or f_1, no edge is contained in all (a, b)-paths and (b, a)-paths.

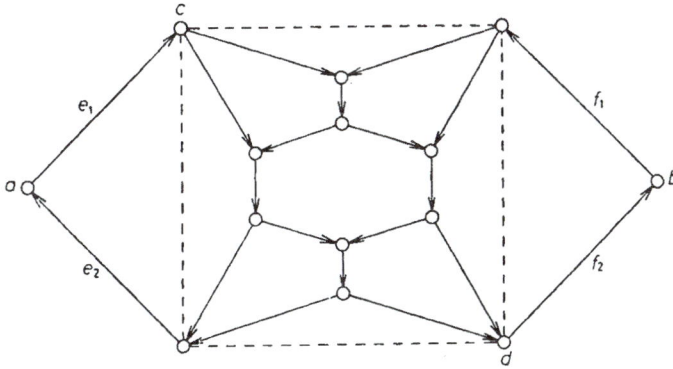

FIGURE 43

This example is not the simplest one. It was given here because contracting e_1, e_2, f_1, f_2, gives a graph G which is 2-connected between the image a' of a and the image b' of b, but any (a,b)-path has an edge in common with any (b,a)-path. This disproves (iii).

Finally, (ii) is obviously false for a graph consisting of k vertex disjoint (a,b)-paths.

46. Let C be an (a,b)-cut, determined by S. Then

$$|C| \geq |C| - |C^*| = \sum_{x \in S}(d_G^+(x) - d_G^-(x)) = k,$$

which proves the statement by Menger's theorem.

47. (i) If there are neither (a,b)-paths, nor (b,a)-paths we have nothing to prove. If P is an (a,b)-path say, then by 6.46, $G - E(P)$ has a (b,a)-path Q which is edge-disjoint from P; a contradiction.

(ii), (iii). Assume that G is k-edge-connected between a and b. Then there are k edge-disjoint (a,b)-paths by 6.39. Remove the edges of these (a,b)-paths, then the remaining graph has k edge-disjoint (b,a)-paths by 6.46. This proves both (ii) and (iii). [A. Kotzig, *Wiss. Z. Martin-Luther Univ. Halle–Wittenberg* **10** (1961–62) 118–125.]

48. (a) Every edge joining $X \cap Y$ to $V(G) - X - Y$ is counted twice on both sides. Those edges joining $X - Y$ to $Y - X$ are counted on the right-hand side only. Any other edge is counted exactly once on each side.

(b) This follows from (a) by substituting $V(G) - Y$ for Y, and taking into account the fact that

$$\delta_G(Y) = \delta_G(V(G) - Y).$$

(c) It is clear that
1° if an edge connects two of $X - Y - Z$, $Y - X - Z$, $Z - X - Y$, $X \cap Y \cap Z$, then it is counted twice on the left-hand side and twice on the right-hand side,

$2°$ if an edge connects one of $X-Y-Z$, $Y-X-Z$, $Z-X-Y$, $X \cap Y \cap Z$ to a point not in these sets, it is counted once on the left-hand side and at least once on the right-hand side,

$3°$ no other edge is counted on the left-hand side.

49. Let $\delta_G(X)=k$, $|X|$ minimal, $x \in X$. If all edges incident with x go to $V(G)-X$, then their number is $\leq \delta_G(X)=k$, so the degree of x is $\leq k$. By the k-edge-connectivity, we have equality here.

Suppose that x is adjacent to $y \in X$. Removing (x,y), the edge-connectivity decreases, hence there is an $\emptyset \neq Z \subset V(G)$ with $\delta_{G-(x,y)}(Z) \leq k-1$. This can only be if $\delta_G(Z)=k$ and Z separates x and y, say $x \in Z$, $y \in V(G)-Z$. We may assume that $X \cup Z \neq V(G)$, otherwise we could consider $V(G)-Z$ instead of Z. By 6.48a,

$$\delta_G(X \cap Z) + \delta_G(X \cup Z) \leq \delta_G(X) + \delta_G(Z).$$

Here $\delta_G(X \cap Z) \geq k$, $\delta_G(X \cup Z) \geq k$, $\delta_G(X) = \delta_G(Z) = k$. Therefore, we must have equality throughout, in particular $\delta_G(X \cap Z) = k$. This contradicts the minimality of X. [W. Mader, *Math. Ann.* **191** (1971) 21–28.]

50. (a) We have to show that if we contract $G-(F_1 \cup \ldots \cup F_m)$ we get a graph G' with not more than $2m$ points. Suppose that we will have p points. These points have degree at least k and thus

$$|E(G')| \geq \frac{k \cdot p}{2}.$$

On the other hand, each edge of G' belongs to one of the cuts $F_1, \ldots, F_m$, otherwise it should be contracted. Thus

$$|E(G')| \leq \sum_{i=1}^{m} |F_i| = m \cdot k,$$

whence

$$p \leq 2m$$

as stated.

51. Let y, z be two neighbors of x. Remove two edges (x,y) and (x,z) and create a new edge (y,z). If the resulting graph G' is not k-edge-connected between two points of $V(G)-x$, then there is a set $\emptyset \neq U \subset V(G)-\{x\}$ such that

$$\delta_{G'}(U) < k.$$

Since G and G' are Eulerian, all cuts in them are even; so k is even and $\delta_{G'}(U)$ as well, whence

(1) $$\delta_{G'}(U) \leq k - 2.$$

Obviously,

$$\delta_{G'}(U) \geq \delta_G(U) - 2 \geq k - 2,$$

whence

$$\delta_G(U) = k, \qquad \delta_{G'}(U) = \delta_G(U) - 2.$$

The latter equality means y, $z \in U$.

Now let y be any neighbor of x and U a maximum set such that $U \subset V(G) - \{x\}$, $y \in U$, $\delta_G(U) = k$. Since

$$\delta_G(U \cup \{x\}) \geq k = \delta_G(U),$$

it follows that U cannot contain all neighbors of x; let $z \notin U$ be a neighbor of x. Now if the pair (y, z) does not satisfy the requirements of our problem, there is a set $V \subset V(G) - \{x\}$ with y, $z \in V$ and $\delta_G(V) = k$. Now by 6.48a,

$$\delta_G(U \cup V) + \delta_G(U \cap V) \leq \delta_G(U) + \delta_G(V) = 2k,$$

and since $U \cap V \neq \emptyset$,

$$\delta_G(U \cap V) \geq k.$$

This implies that

$$\delta_G(U \cup V) \leq k.$$

By the maximality of U, this can only occur if

$$U \cup V = V(G) - \{x\}.$$

Now observe that

$$\delta_G(U - V) + \delta_G(V - U) < \delta_G(U) + \delta_G(V),$$

because in the counting in the solution of 6.48a the edge (x, y) is only counted on the right-hand side. Here

$$\delta_G(U) = \delta_G(V) = k, \quad \delta_G(U - V), \ \delta_G(V - U) \geq k,$$

a contradiction.

52. We use induction on $|V(G)|$; if $|V(G)| = 2$ then we have the initial graph I. Now, suppose that $|V(G)| \geq 3$ and $x \in V(G)$. By the preceding result, we can find two points y_1, z_1 adjacent to x such that if we remove (x, y_1) and (x, z_1) but connect y_1 to z_1, then the resulting graph G_1 is still $(2k)$-edge-connected between any two points of $V(G) - \{x\}$. If x is still not isolated, we can find two points y_2, z_2 adjacent to x in G_1 such that $G_1 - (x, y_2) - (x, z_2) + (y_2, z_2)$ is again $2k$-edge connected between any two points of $V(G) - \{x\}$, etc. Here loops can never arise, because a point with loops could be separated from the other points by less than k edges. Thus we get a $(2k)$-edge-connected, $(2k)$-regular graph G' which arises from G by removing the edges (x, y_1), $(x, z_1), \ldots, (x, y_k)$, (x, z_k) (and then the isolated point x) and creating k new edges $(y_1, z_1), \ldots, (y_k, z_k)$. Now G arises from G' by construction II. [A. Kotzig, *Doctoral dissertation*, Bratislava, 1959; cf. also G. J. Simmons, *Infinite and Finite Sets*, Coll. Math. Soc. J. Bolyai **10**, Bolyai–North-Holland (1974) 1277–1349.]

53. We are going to follow the solution of 6.51 as long as we can. For each pair (y, z) of neighbors of x we find a set U such that $\emptyset \neq U \subset V(G) - \{x\}$, y, $z \in U$, and

$$\delta_G(U) \leq k + 1$$

(because here we do not gain 1 as we did in (1) in the solution of 6.51).

Now let y be a fixed neighbor of x and consider the set H of all sets U like above for all points z. If the assertion is not true, then these sets cover all neighbors of x. Let $U_1, \ldots, U_l$ be a minimum number of sets of H which cover all neighbors of x. Since

$$\delta_G(U_i \cup \{x\}) \geq k \geq \delta_G(U_i) - 1,$$

and x has even degree, it follows that at most $d_G(x)/2$ edges connect x to U_i. Since x is joined to y which belongs to all U_i, it follows that two U_i's cannot contain all neighbors of x. Hence $l \geq 3$.

Consider U_1, U_2, U_3. Since U_1 could not be omitted from H, there is a neighbor of x which is covered by U_1 but not by U_2 or U_3. Similarly we find that $U_2 - U_3 - U_1 \neq \emptyset$ and $U_3 - U_1 - U_2 \neq \emptyset$.

Now use a slight sharpening of 6.48c.

$$\delta_G(U_1 - U_2 - U_3) + \delta_G(U_2 - U_3 - U_1) + \delta_G(U_3 - U_1 - U_2) +$$
$$+ \delta_G(U_1 \cap U_2 \cap U_3) \leq \delta_G(U_1) + \delta_G(U_2) + \delta_G(U_3) - 2.$$

(The -2 comes from the observation that the edge (x, y) is counted 3 times on the right-hand side but only once on the left.) Here all terms on the left-hand side are $\geq k$, but all terms on the right are $\leq k + 1$. Hence

$$4k \leq 3(k + 1) - 2, \quad k \leq 1,$$

a contradiction. (The example of K_4 shows that the assumption on the degree of x cannot be omittted.)

54. (a) Let $\emptyset \neq X \subset V(G)$. Then

$$\delta_G(X) = \delta_{\vec{G}}^+(X) + \delta_{\vec{G}}^+(V(G) - X) \geq k + k = 2k.$$

(b) We prove the assertion by induction on $|V(G)|$. If $|V(G)| = 2$ it is trivial. Let $|V(G)| \geq 3$. We may assume that G is critically $2k$-edge-connected; then by 6.49, there is a point x with $d_G(x) = 2k$. As in the solution of 6.52, we find a $2k$-edge-connected graph G' and k edges $(y_1, z_1), \ldots, (y_k, z_k) \in E(G')$ such that if we subdivide each of these k edges by a point and identify these new points to a point x, we get G (note that now 6.53 must be applied rather than 6.51). Let $\vec{G}'$ be an orientation of G' which is k-edge-connected. This induces an orientation $\vec{G}$ of G: If (y_i, z_i) is directed from y_i to z_i (say), then we direct (y_i, x) from y_i to x and (x, z_i) from x to z_i. It is easy to see that $\vec{G}$ is k-edge-connected. [In a stronger form this was proved by Nash-Williams; see B.]

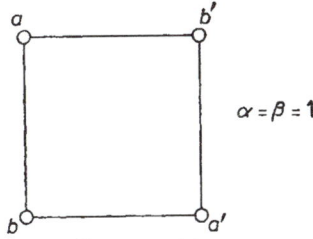

FIGURE 44

55. The necessity of the conditions given is obvious. Consider the graph shown in Fig. 44. Here the conditions the satisfied since a quadrilateral is 2-edge-connected. On the other hand any (a, a')-path has an edge in common with any (b, b')-path.

56. (a) Take two new points u, u'. Connect u to a and b by α and β edges, respectively and, similarly, connect u' to a' and b' by α and β edges, respectively. Condition $(*)$ now implies that the resulting graph G' is $(\alpha+\beta)$-connected between u and u'. Hence, it contains $\alpha+\beta$ edge-disjoint (u, u')-paths and thus G contains $\alpha+\beta$ edge-disjoint $(\{a, b\}, \{a', b'\})$-paths, α of them starting at a and the same number of them ending at a', as stated in the hint. Let $P_1, \ldots, P_{\alpha+\beta}$ be these paths.

Now $G - E(P_1) - \ldots - E(P_{\alpha+\beta})$ has even degrees, so by 5.13 it has an orientation with equal out-, and indegrees. Orienting the paths P_i from their endpoint in $\{a, b\}$ toward their endpoint in $\{a', b'\}$, we get a desired orientation of G.

(b) Let $\overrightarrow{G}$ be an orientation of G as in part (a). First, we show that there are α edge-disjoint (a, a')-paths in $\overrightarrow{G}$. Let $X \subseteq V(\overrightarrow{G}) - \{a'\}$, $a \in X$. If b, $b' \notin X$, then

$$\delta_{\overrightarrow{G}}^{+}(X) \geq \delta_{\overrightarrow{G}}^{+}(X) - \delta_{\overrightarrow{G}}^{-}(X) = \delta_{\overrightarrow{G}}^{+}(a) - \delta_{\overrightarrow{G}}^{-}(a) = \alpha.$$

We argue similarly, if b, $b' \in X$ or $b \in X$, $b' \notin X$. Suppose that $b' \in X$, $b \notin X$. Then we have

$$(1) \qquad \delta_{\overrightarrow{G}}^{+}(X) - \delta_{\overrightarrow{G}}^{-}(X) = (\delta_{\overrightarrow{G}}^{+}(a) - \delta_{\overrightarrow{G}}^{-}(a)) + (\delta_{\overrightarrow{G}}^{+}(b') - \delta_{\overrightarrow{G}}^{-}(b')) = \alpha - \beta.$$

On the other hand

$$(2) \qquad \delta_{\overrightarrow{G}}^{+}(X) + \delta_{\overrightarrow{G}}^{-}(X) = \delta_G(X) \geq \alpha + \beta.$$

Summing (1) and (2) we obtain

$$2\delta_{\overrightarrow{G}}^{+}(X) \geq 2\alpha,$$

which proves that there are α edge-disjoint (a, a')-path $P_1, \ldots, P_\alpha$ in $\overrightarrow{G}$.

Now consider $G' = \overrightarrow{G} - E(P_1) - \ldots - E(P_\alpha)$. In this graph

$$\delta_{G'}^+(b) - \delta_{G'}^-(b) = \beta, \quad \delta_{G'}^+(b') - \delta_{G'}^-(b') = -\beta,$$
$$\delta_{G'}^+(x) = \delta_{G'}^-(x) \quad \text{for} \quad x \neq b,\ b'.$$

Therefore by 6.46, G' contains β edge-disjoint (b,b')-paths $Q_1, \ldots, Q_\beta$. Now $P_1, \ldots, P_\alpha, Q_1, \ldots, Q_\beta$ (more exactly, the undirected paths of G corresponding to these directed paths) form a required set of paths. [T. C. Hu, *Integer Programming and Network Flows*, Addison-Wesley, 1969; this proof is due to C. St. J. A. Nash-Williams.]

57. Let G be a 5-connected planar map with a quadrangular face F; let a, c, b, d be the vertices of F in this cyclic order. Then, obviously, any (a,b)-path intersects any (c,d)-path.

Thus it suffices to find a 5-connected planar graph with a quadrangular face. A little experimentation yields, e.g., Fig. 45.

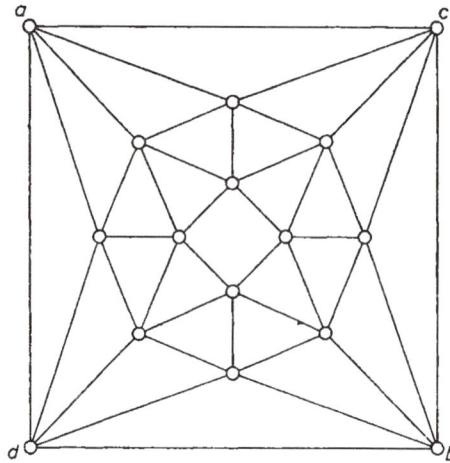

FIGURE 45

Remark: It can be shown [P. Mani–H. A. Jung, see Jung, in: *Comb. Structures Appl.* Gordon and Breach, 1970, 189–191] that all examples are planar. In particular, there are no 6-connected examples. More generally, there exists a function $g(n)$ such that given any $2n$ points $a_1, \ldots, a_n, b_1, \ldots, b_n$ in a $g(n)$-connected graph, there are n vertex-disjoint paths $P_1, \ldots, P_n$, P_i connecting a_i to b_i.

58. *First solution.* C separates a and b; for any (a,b)-path intersects $A \cup B$ (even A) and the first point of it belonging to $A \cup B$ is a point of C.

Similarly, D defined in the hint is a separating set. Hence, $|C| \geq k$, $|D| \geq k$. Now

$$C \cap D \subseteq A \cap B;$$

for if $x \in C \cap D$, then there is an $(a, A \cup B)$-path P_1 connecting x to a and a $(b, A \cup B)$-path connecting x to b. Then $P_1 \cup P_2$ is a path, which meets $A \cup B$ in x only; since each (a, b)-path must meet both A and B, x must belong to both A and B. Thus

$$|A \cap B| \geq |C \cap D| = |C| + |D| - |C \cup D| \geq k + k - |A \cup B| =$$
$$= |A| + |B| - |A \cup B| = |A \cap B|,$$

which shows that we have equality everywhere, in particular $|C| = |D| = k$.

Second solution. Let $P_1, \dots, P_k$ be independent (a, b)-paths (these exist by Menger's theorem), then A (B) contains exactly one point a_i (b_i) from P_i. Let c_i be the one of a_i and b_i nearer a on P_i.

Now $C = \{c_1, \dots, c_k\}$. Obviously, $c_i \in C$. On the other hand, if $c \in C$, then $c \in A \cup B$ and thus $c \in P_i$ for some i. Assume that $c \neq c_i$, then c is on the (c_i, b)-piece of P_i. By definition, there is an (a, c_i)-path not meeting A, which gives with the (c_i, b)-piece of P an (a, b)-path not meeting A, a contradiction.

Thus, $|C| = k$. The fact that it separates a and b follows easily as above [cf. L. Lovász, *Acta Math. Acad. Sci. Hung.* **2** (1970) 365–368.]

59. Let T be as in the hint. It is enough to show that of T separates $G - (x, y)$, then, in fact, it separates x and y; for then, obviously, $T \cap V(H)$ separates x and y in H, i.e. $H - (x, y)$ has a $(k - 1)$-element separating set.

Now if, to the contrary, x, y are in the same component of $G - (x, y) - T$, then, adding (x, y), we see that $G - T$ is still disconnected and this is impossible.

60. (a) 1° In $G - C$, there are no edges between $G_1 \cap G_3$ and the remainder of the graph, which is non-empty. Hence C is a separating set and since $G_1 \cap G_3$ is a proper subgraph of G_1 (as $B \cap V(G_1) \neq \emptyset$), C cannot be a k-element separating set.

2° Let $x \in C$. Then there is an $(a, A \cup B)$-path P connecting a to x. If $x \notin A \cap B$, then $x \notin A$ or $x \notin B$. If $x \notin A$ (say), then P does not intersect A, i.e. x is in the same component G_1 of $G - A$ as a.

3° If $b \in G_2 \cap G_4$, then by 6.58 C has cardinality k.

4°

$$|V(G_1)| > |B \cap V(G_1)| + |V(G_1 \cap G_4)|,$$

while

$$|V(G_4)| = |A \cap V(G_4)| + |V(G_1 \cap G_4)|.$$

Now since

$$|B \cap V(G_1)| + |A \cap B| + |A \cap V(G_3)| \geq |C| > k$$

and

$$|A \cap V(G_4)| + |A \cap B| + |A \cap V(G_3)| = |A| = k,$$

it follows that

$$|B \cap V(G_1)| > |A \cap V(G_4)|,$$

whence

$$|V(G_1)| > |V(G_4)|,$$

which contradicts the minimality of G_1 (Fig. 46).

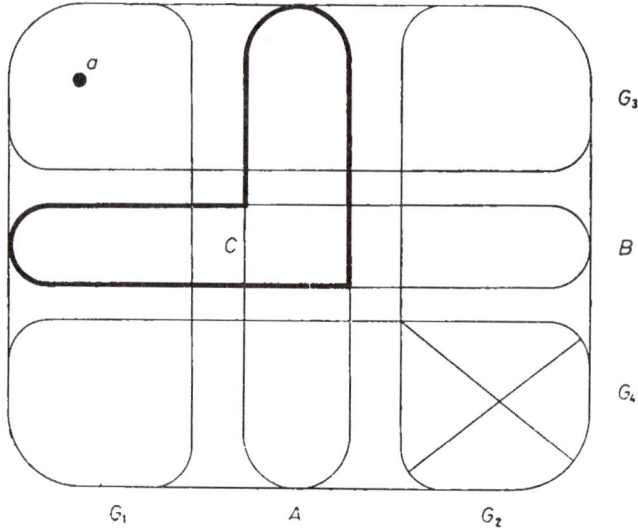

FIGURE 46

To show the second statement, assume first that $V(G_2) \subseteq B$. Since $|V(G_1)| \leq |V(G_2)|$, this immediately implies that $|V(G_1)| \leq k/2$.

Secondly, assume that $V(G_2) \not\subseteq B$. Let, e.g. $G_2 \cap G_3 \neq \emptyset$. Since $(A \cap V(G_3)) \cup (A \cap B) \cup (B \cap V(G_2))$ separates G, if follows that

$$|A \cap V(G_3)| + |A \cap B| + |B \cap V(G_2)| \geq k = |B| =$$
$$= V(G_1)| + |A \cap B| + |B \cap V(G_3)|,$$

whence

$$|V(G_1)| \leq |A \cap V(G_3)|.$$

If $G_2 \cap G_4 \neq \emptyset$ then a similar argument yields that

$$|V(G_1)| \leq |A \cap V(G_4)|,$$

and hence

$$|V(G_1)| \leq \frac{1}{2}|A| = \frac{k}{2}.$$

If $G_2 \cap G_4 = \emptyset$ then a similar argument yields that

$$|V(G_4)| \leq |B \cap V(G_2)|,$$

and so

$$|V(G_1)| \leq |V(G_4)| \leq |B \cap V(G_2)| \leq |B| - |V(G_1)| = k - |V(G_1)|,$$

whence the assertion follows again. [W. Mader, *Arch. Math.* **22** (1971) 333–336; M. E. Watkins, *J. Comb. Th.* **8** (1970) 23–29.]

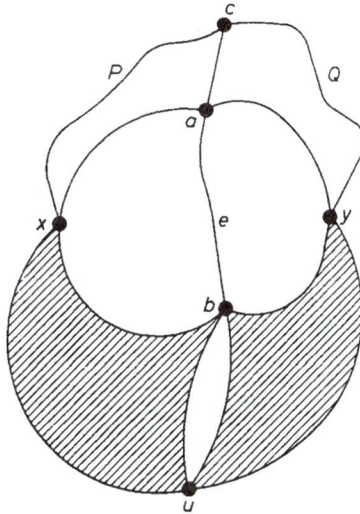

FIGURE 47

    (b) Obviously, G has no multiple edges. We may assume that G is not a $(k+1)$-clique. Then G_1 of (a) exists. Assume that G_1 has at least two points, then G_1 has an edge $e = (x, y)$. $G - e$ has a $(k-1)$-element separating set T. Then $G - T$ is connected but $G - T - e$ is not, whence $G - T - e$ has two components connected by e. Suppose that the component containing x has more than one point, then $T \cup \{x\}$ is a k-element separating set of G, having a point x in common with G_1 but not containing another point, y of it. This is impossible by (a). Thus $G_1 = \{x\}$ and it follows that x has degree k. [R. Halin, *J. Comb. Theory* **7** (1969) 150–154.]

61. (a) Let $\{x, y\}$ be a separating set of $G - e$. Since $G - \{x, y\}$ is connected, e connects different components of $G - e - \{x, y\}$, hence a and b are separated by x, y in $G - e$.

    (b) Let $\{u, v\}$ be a separating set of G/e. We claim one of u, v is the point e is contracted onto. For otherwise $G/e - \{u, v\}$ would arise by contraction of an edge from the connected graph $G - \{u, v\}$ and would be connected. Now if v is this point, then $\{a, b, u\}$ is a separating set. To show it separates x and y, consider a component G' of $G - \{a, b, u\}$. Since G is 3-connected, this component is connected to each of a, b, u and, thus there is an (a, b)-path P_1 with inner points in G'. P_1 contains one of x and y, x say. Considering another component G'' of $G - \{a, b, u\}$, the same argument shows that y must be in G''.

(c) Assume that the component G_1 of $G - e - \{x, y\}$ containing a has a point $c \neq a$. Since G is 3-connected, there is a (c, x)-path P in $G - \{a, y\}$. P does not go through b and u, because b, u and c are separated by a, x, y (even by e, x, and y); hence P is a (c, x)-path in $G - \{a, b, u\}$. Similarly, there is a (c, y)-path Q in $G - \{a, b, u\}$; however, $P \cup Q$ then gives an (x, y)-path in $G - \{a, b, u\}$, which contradicts (b) (Fig. 47).

The degree of a is now 3, a contradiction.

62. First we prove the assertion of the hint. Assume that $G - T$ is disconnected, where $|T| \leq k - 1$. If T does not contain either of the endpoints x, y of e, then $(G - T)/e = G/e - T$ is disconnected, which is a contradiction, because G/e is k-connected. If $x, y \in T$, then G/e can be cut even by the $(k - 2)$-element image of T. Therefore, we may assume that $x \in T$, $y \notin T$ (say).

$G/e - T$ can be obtained from $G - T$ by removing y. $G/e - T$ is connected while $G - T$ is not, which means y was a component of $G - T$. Hence T contains all the neighbors of y, which is impossible, because y has degree at least k.

Now the result follows easily. G/e is 3-connected by the preceding problem. Suppose that it is not minimal, then $G/e - f = (G - f)/e$ is still 3-connected for some $f \in E(G)$. Since the endpoints of e have degree at least 3 in $G - f$, it follows by the above that $G - f$ is 3-connected, contrary to the assumption.

63. Let C be the circuit under consideration. If $|V(C)| = 2$ the statement is void as an edge-critically 3-connected graph cannot have multiple edges.

Assume that $|V(C)| \geq 3$. If C has at most one point of degree 3, it has an edge e whose endpoints have degree at least 4. Then, by 6.26, G/e is a critically 3-connected graph, and the circuit C is mapped by the contraction of e onto a shorter circuit C' which still has at most one point of degree 3. This is a contradiction.

This result — with a more complicated proof — remains valid for critically k-connected graphs. [See W. Mader, *Arch. Math.* **23** (1972) 219–224.]

64. The statement of the hint follows just like (b) in the solution of 6.61. Take a circuit C through b_1 and b_2 in $G - a$ (which is 2-connected so that such circuit exists by 6.32), then one of the (b_1, b_2)-arcs of C contains u_3, the other one contains b_3. Thus C is a circuit through b_1 and b_3 and hence by an argument similar to the one above, it also contains u_2. Thus each circuit of $G - a$ containing two of b_1, b_2, b_3 also contains the third one as well as u_1, u_2, and u_3. Further u_i separates b_{i-1} and b_{i+1}.

It follows that each (u_1, u_2)-path must contain either b_3 or both b_1 and b_2; otherwise we could replace an arc of C by it and get a circuit violating the above assertion. Thus b_3 and b_2 separate u_1 and u_2 in $G - a$, but then they also separate u_1, u_2 in G, a contradiction (Fig. 48). [cf. W. T. Tutte, *Connectivity in Graphs*, Toronto University Press, Toronto, 1966.]

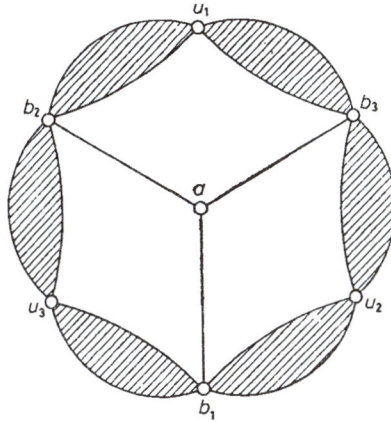

FIGURE 48

65. Construct G as in the hint. I. G is 3-connected. For let us remove 2 points of it. If both are inner points of T, then, since the points of T are of degree at least 3, each component of the remainder of T meets C and this "connects" them. The same argument works if a point of T and a point of C are removed. Finally, if we remove two points of C, the remainder of T is connected and meets all pieces of the remainder of C, thus the resulting graph is connected again.

II. First suppose that $e \in E(C)$, then $G - e$ is at most 2-connected, because the endpoints of e have degree 2 in it. Secondly suppose that $e \in E(T)$. Then $T - e$ has exactly two components T_1, T_2. The endpoints of T_i form an arc A_i of C (the precise proof of this fact is left to the reader). Thus, there are two edges f_1, f_2 connecting A_1 and A_2. $G - e - f_1 - f_2$ is disconnected, showing that $G - e$ is not even 3-edge-connected.

III. Choose the tree T such that some point x has distance 1000 from all endpoints and all inner points have degree 1000.

66. We use induction on k. Let C be a circuit passing through $x_1, \ldots, x_{k-1}$. If $x_k \in V(C)$ we are finished; assume that $x_k \notin V(C)$. Now we distinguish two cases.

(a) Assume that there are k (x_k, C)-paths which have only x_k in common. The points $x_1, \ldots, x_{k-1}$ divide C into $k-1$ arcs, thus one of these arcs (including its endpoints) contains two endpoints of the paths $P_1, \ldots, P_k$. Now adding these two paths to C, but removing the arc between their endpoints (which does not contain any x_i, $1 \leq i \leq k-1$), we get a circuit through $x_1, \ldots, x_k$.

(b) If there are not k (x_k, C)-paths which have no point in common except x_k, then by 6.41 there is a set X, $|X| = k-1$, $x_k \notin X$, which meets each (x_k, C)-path. Since $G - X$ is connected, each point of $C - X$ can be connected to x_k in $G - X$ by a path. This is a contradiction except when $X = V(C) = \{x_1, \ldots, x_{k-1}\}$. In this latter case an argument similar to the one above shows that there are $k-1$

(x_k, C)-paths, having only x_k in common. These paths connect x_k to $x_1, \ldots, x_{k-1}$. Adding two of them, which connect x_k to neighboring points of C, but removing the edge between their endpoints we get a desired circuit again. [G. Dirac; see B.]

67. Suppose that G is a 3-connected graph and e_1, e_2, e_3 are edges of G not on a circuit. We show first that there is a circuit C containing two of them, which does not meet the third one.

Let C_1 be a circuit through e_1 and e_2. If $e_3 = (x,y)$ has no endpoint on C_1, we have the desired circuit. If $x \in V(C_1)$, but $y \notin V(C_1)$, consider an $(y, C_1 - x)$-path P of $G - x$; let this path P end at $z \in V(C_1)$. Then e_3, P and one of the (x,z)-arcs of C_1 give a desired circuit. Finally, if $x, y \in V(C_1)$, then e_3 and one of the (x,y)-arcs of C_1 yield C. Now $C - e_1 - e_2 = P_1 \cup P_2$, where P_1, P_2 are disjoint paths. Consider all $(\{x,y\}, P_1)$ paths of $G - V(P_2)$. No two of them are disjoint, since this would yield a circuit containing e_1, e_2, e_3. Thus, there is a point u meeting each of them. Similarly, there is a point meeting each $(\{x,y\}, P_2)$-path of $G - V(P_1)$. Then $\{u,v\}$ meets all $(\{x,y\}, C)$-paths. Since G is 3-connected, this is only possible if $\{u,v\} = \{x,y\}$, say $u = x$, $v = y$. Thus, any $(\{x,y\}, C)$-path which ends on P_1 (P_2) starts from x (y).

If we show that $G - \{e_1, e_2, e_3\}$ does not contain any (x,y)-path we are through. Assume indirectly that there is such a path T.

Assume first that T does not meet C. Let Q be a (T, C)-path of $G - \{x,y\}$ ending in P_1 (say); then Q and a suitable piece of T would give an $(\{x,y\}, C)$-path starting from y, which is impossible. We reach a similar conclusion when T meets P_1 only. Thus the only case we have to deal with is when T meets both P_1 and P_2. Then it contains a (P_1, P_2)-path R, different from e_1 and e_2. Let $a \in P_1$, $b \in P_2$ be the endpoints of R. Since $G - \{a,b\}$ is connected, there is an $(\{x,y\}, C - \{a,b\})$-path Q_1 in it. Let the endpoints of Q_1 be e.g. x and $z_1 \in P_1$. Let Q_2 be a (y,C)-path in $G - x$, then obviously, the endpoint $z_2 \neq y$ of Q_2 is on P_2. Q_1 and Q_2 are disjoint for otherwise we would get an (x,y)-path of $G - \{e_1, e_2, e_3\}$ disjoint from C.

One of the (z_1, z_2)-arcs of C does not contain a. This arc forms with Q_1, Q_2 and e_3 a circuit C_0, containing e_3 and one of e_1 and e_2, e_2 say. Now there are (e_1, C_0)-paths connecting the same endpoint of e_1 to both component of $C_0 - \{e_2, e_3\}$, a contradiction (Fig. 49).

68. Using induction on k, we may assume that there is an (a,b)-path P which passes through $x_1, \ldots, x_{k-1}$ and assume that $x_k \notin P$. It is impossible to cover all (x_k, P)-paths by k points different from x_k since, if X were a k-element set meeting all of them ($x_k \notin X$), then x_k and $P - X \neq \emptyset$ would belong to different components of $G - X$. Hence by 6.41, there are $k+1$ (x_k, P)-paths $Q_1, \ldots, Q_{k+1}$, having only x_k in common. Since $x_1, \ldots, x_{k-1}$ divide P into k pieces, one of these meets two Q_j's. Then removing the subpath connecting these two endpoints but adding the two Q_j's, we get an (a,b)-path through $x_1, \ldots, x_k$. [M. D. Plummer, in: *Proc. 2nd Louisiana Conf. on Comb.*, Univ. of Manitoba Press, Winnipeg, 1971, 458–472.]

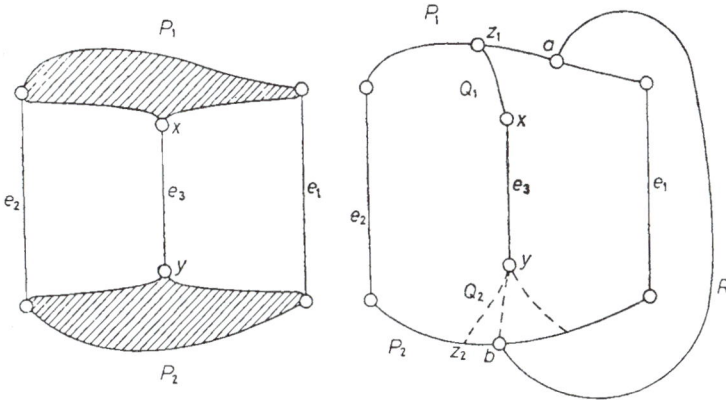

FIGURE 49

69. Let C be the boundary circuit of a face F. Let B be any bridge of C. We show, following the hint, that B contains all points of C. Assume that $x \in V(C)$, $x \notin V(B)$. Let y, z be the two points of $B \cap C$ next to x on C. Then the other (y,z)-arc of C has an inner point; this is trivial, if B is a chord and follows from the fact that B has at least 3 points on C, if B has an inner point. Since $G - \{y, z\}$ is connected, it has a path P connecting the two (y,z)-arcs of C. Let Q be a (y,z)-path in B. Both P and Q run outside F (as F is a face) and so, they intersect by planarity. This means, however, that P belongs to B, a contradiction to the definition of y and z (Fig. 50).

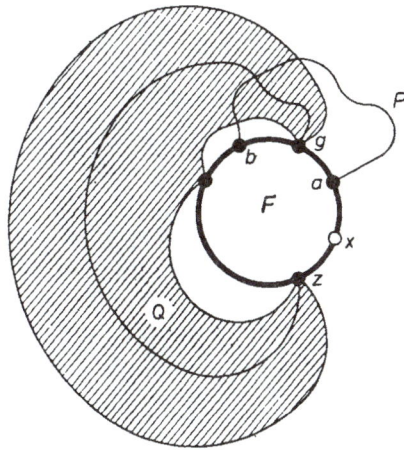

FIGURE 50

The statement just shown implies that C has no chords. Suppose that B_1, B_2 are two bridges. Then $C + B_1$ divides the exterior of F into two regions having at most 2 boundary points on C and B_2 lies inside one of these (as it does not meet B_1). But B_2 has at least 3 points on C, a contradiction.

Since the result of this problem characterizes the boundaries of faces of a planar 3-connected map independently of the embedding, it follows that every isomorphism between two planar 3-connected maps preserves faces. By the Schoenflies Theorem in the plane topology it follows that every isomorphism between two 3-connected maps on the sphere is induced by a homeomorphism of the sphere onto itself. [H. Whitney, *Trans. Amer. Math. Soc.* **34** (1932) 339–362.]

70. Let C be defined as in the hint; we show it has no bridge other than B. As in the previous solution, we show first that each point of C is on B. Suppose that $x \in V(C)$, $x \notin V(B)$. Let y, z be the two points of B next to x on C. As in the previous solution, we find a path P joining the two (y,z)-arcs of C in $G-y-z$. Let a, b be the endpoints of P and let R be an (a,b)-arc of C. Then $P \cup R$ is a circuit such that the bridge of $P \cup R$ containing f properly contains B, a contradiction.

So every point of C is in B. Suppose that B_1 is another bridge of C. Let c, d be two points of B_1 on C. Then a (c,d)-arc of C has an inner point (since B_1 is either a chord or has at least 3 points on C). Denote by R the other (c,d)-arc of C and denote by P a (c,d)-path of B_1, then $R \cup P$ is a circuit through c, d and the bridge of $R \cup P$ containing f properly contains B. A contradiction again. [W. T. Tutte, *Proc. London Math. Soc.* **13** (1963) 743–767.]

71. With the notation of the hint, let P_0 be an (a,b)-path, formed by some edges of $e_1,\dots,e_k$. Since $\{e_1,\dots,e_{k-1}\}$ does not contain any (a,b)-path, there is an (a,b)-cut C_0 such that C_0 contains none of $e_1,\dots,e_{k-1}$. Hence, $C_0 \cap P_0 = \{e_k\}$. Moreover, by definition,

$$\min_{e \in P_0} v(e) = v(e_k), \quad \max_{e \in C_0} v(e) = v(e_k).$$

On the other hand, let C be any (a,b)-cut, then $C \cap P_0 \neq \emptyset$, say $e_i \in C \cap P_0$, $i \le k$. Then

$$\max_{e \in C} v(e) \ge v(e_i) \ge v(e_k)$$

and similarly

$$\min_{e \in P} v(e) \le v(e_k)$$

for any (a,b)-path P, whence

$$\max_P \min_{e \in P} v(e) = v(e_k) = \min_C \max_{e \in C} v(e).$$

[For a more general formulation see J. Edmonds–D. R. Fulkerson, *J. Comb. Theory* **8** (1970), 299.]

72. I. Suppose that φ is a function satisfying $(*)$ and $P = (a, x_1, \dots, x_k, b)$ is any (a,b)-path. Then

$$u(P) = v(a, x_1) + v(x_1, x_2) + \dots + v(x_k, b) \ge$$
$$\ge \varphi(x_1) + (\varphi(x_2) - \varphi(x_1)) + \dots + (\varphi(b) - \varphi(x_k)) = \varphi(b).$$

II. Thus it suffices to construct an (a,b)-path P and a function φ, satisfying
$(*)$, such that $u(P)=\varphi(b)$. We define φ step by step. Put $\varphi(a)=0$. Suppose that
φ is already defined for a set $S\subset V(G)$, $a\in S$ and has the property that for each
point $x\in S$ there is an (a,x)-path in S of value $\varphi(x)$. Consider all edges (x,y)
with $x\in S$, $y\notin S$ and choose one for which $\varphi(x)+v(x,y)$ is minimal. Set $\varphi(y)=$
$\varphi(x)+v(x,y)$. This extends the domain of φ. We show that the property $(*)$ is
preserved. Let (z,u) be any edge spanned by $S\cup\{y\}$, we have to show

$$\varphi(u)-\varphi(z)\le v(z,u).$$

If $u,\ z\in S$ this is obvious. If $z\in S$, $u=y$ then, by the choice of (x,y),

$$\varphi(y)=\varphi(x)+v(x,y)\le\varphi(z)+v(z,u).$$

So let $u\in S$, $z=y$. In this case we claim that $\varphi(u)\le\varphi(y)$, which will prove
the assertion. For this we show, by induction, that a point chosen later has no
smaller φ-value than a point chosen earlier. Suppose indirectly that $\varphi(y)<\varphi(u)$.
Then $\varphi(x)\le\varphi(y)<\varphi(u)$ and hence, by induction, x was chosen earlier than u.
But then instead of u one should have chosen y. So φ satisfies $(*)$. Moreover, by
induction, there is an (a,x)-path with value $\varphi(x)$ and adding (x,y) to this path
we get an (a,y)-path with value $\varphi(y)$.

In particular, we have an (a,b)-path with value $\varphi(b)$ when φ is completely
defined. This proves the assertion. [For most problems of the rest of this paragraph
see FF, Hu, LP.]

73. Since f is an (a,b)-flow, the only non-zero term in the expression given in the
hint is

$$\sum_{e=(a,y)}f(e)-\sum_{e=(y,a)}f(e),$$

which is the value of f. On the other hand, each edge e spanned by S occurs
twice with opposite signs, the edges of C occur with positive sign, the edges of
C^* with negative sign. Thus, the value of f is

$$\sum_{e\in C}f(e)-\sum_{e\in C^*}f(e).$$

74. We follow the hint. Assume that there is an (a,b)-path P in G_1. Let

$$\varepsilon=\min\sum_{e\in P}v_0(e)>0.$$

Now set for $e\in E(G)$,

$$f_1(e)=\begin{cases} f(e)+\varepsilon & \text{if } e\in E(P),\\ f(e)-\varepsilon & \text{if } e'\in E(P),\\ f(e) & \text{otherwise.}\end{cases}$$

Then, obviously, $f_1(e)$ is an (a,b)-flow of value $w(f)+\varepsilon$, a contradiction. Thus, G_1 is not connected between a and b. Hence, by 6.9 there is an $S \subseteq V(E)$ such that $a \in S$, $b \notin S$ and the cut C of G determined by S satisfies

$$f(e) = v(e) \quad \text{if } e \in C,$$
$$f(e) = 0 \quad \text{if } e \in C^*.$$

Hence, by 6.73,

$$w(f) = \sum_{e \in C} f(e) - \sum_{e \in C^*} f(e) = \sum_{e \in C} v(e).$$

Thus

$$\max_f w(f) \geq \min_e \sum_{e \in C} v(e).$$

On the other hand, if f is arbitrary (a,b)-flow with $f \leq v$, and C is any (a,b)-cut, we have

$$w(f) = \sum_{e \in C^*} f(e) - \sum_{e \in C^*} f(e) \leq \sum_{e \in C} f(e) \leq \sum_{e \in C} v(e),$$

which proves the assertion. [C. R. Ford–D. R. Fulkerson; FF.]

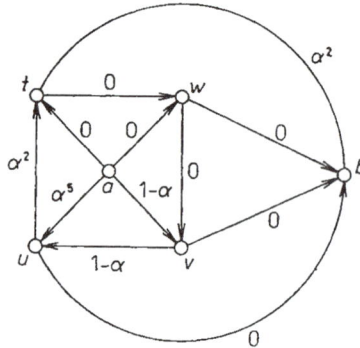

FIGURE 51

75. (a) Consider the network in Fig. 51, where α is the positive root of $\alpha^3+\alpha-1=0$, all capacities are 1 and the values of a flow f_0 are indicated. We start with the flow in Fig. 51, and use the paths $(atwvub)$, $(atuvwb)$, $(avutwb)$, $(avwtub)$, and repeat this cyclically. We consider the change of values of f_k on the quadrilateral $(uvwt)$, since the other edges have a large enough capacity not in interfere.

We claim that after $4m$ steps we will have the flows shown in Fig. 52, and the next 4 increases will be α^{4m+1}, α^{4m+2}, α^{4m+3}, α^{4m+4}. Using induction on

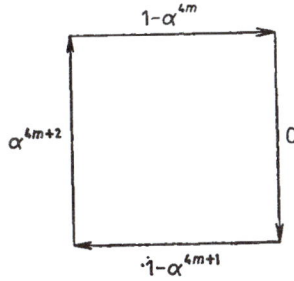

$$1-\alpha^{4m}$$

$$\alpha^{4m+2}$$

$$0$$

$$\cdot 1-\alpha^{4m+1}$$

FIGURE 52

m, it suffices to consider four consecutive steps; so Fig. 53 proves the assertion; the flow values on the edge (a,t) will be partial sums of

$$\alpha - \alpha^2 + \alpha^4 - \alpha^6 + \alpha^8 - \alpha^{10} + \dots,$$

so they will be strictly between 0 and 1; similarly for the other edges at a and b. Now the value of these flows converges to

$$\alpha^2 + (\alpha + \alpha^2 + \dots + \alpha^k + \dots) = \alpha^2 + \frac{\alpha}{1-\alpha} = 1 + \alpha + 2\alpha^2 < 4.$$

On the other hand, the maximum flow clearly has value 4.

(b) Let us form a new digraph G_0 as follows: we orient all edges of P_k and P_{k+1} toward b and take the union of the resulting directed paths P_k and P_{k+1}; but we remove those pairs (x,y) (y,x) of edges where $(x,y) \in P_k$ and $(y,x) \in P_{k+1}$. Those common edges of P_k and P_{k+1}, which are passed in the same sense are taken, however, twice.

Now G_0 has equal outdegrees and indegrees except at a and b. Thus by 6.46 there are two edge-disjoint (directed) (a,b)-paths Q'_1 and Q'_2 in G_0. Let Q_1 and Q_2 be the two corresponding undirected (a,b)-paths in G.

Now observe that Q_1 and Q_2 have the property that, after the kth step, they come into consideration together with P_k, i.e. they have the property that their edges e directed toward a satisfy $f_k(e) > 0$ and those directed toward b satisfy $f_k(e) < v(e)$. This follows immediately from the observation that any edge of Q_i is an edge of P_k or P_{k+1} and Q_i passes it in the same sense.

Thus by the minimality of P_k,

$$|E(Q_i)| \geq E(P_k)|.$$

Since

$$|E(Q_1)| + |E(Q_2)| \leq |E(P_k)| + |E(P_{k+1})|,$$

it follows that

$$|E(P_k)| \leq |E(P_{k+1})|.$$

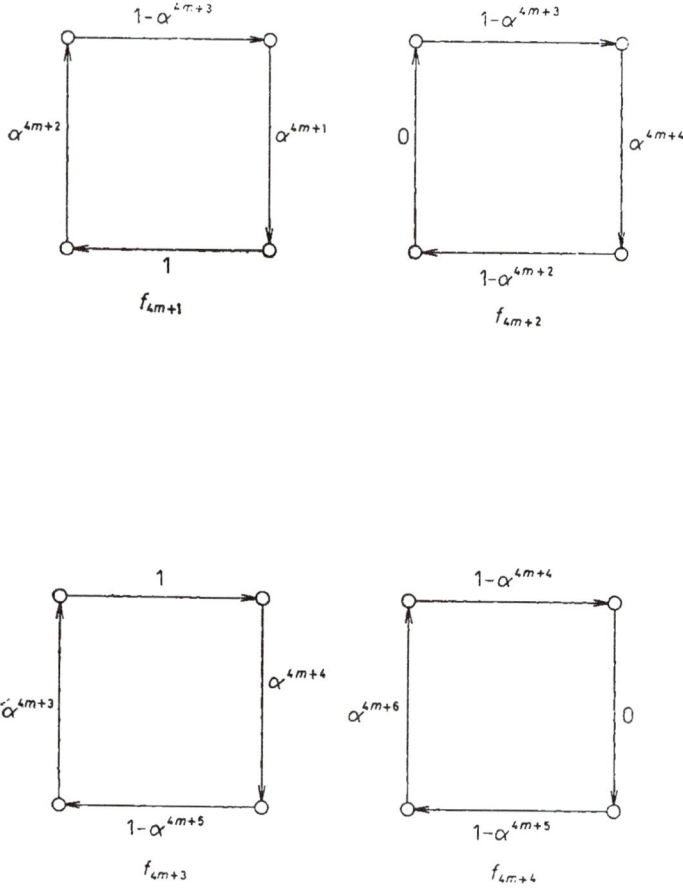

FIGURE 53

(c) Suppose that contrary to the assertion formulated in the hint, $|E(P_{k+l})| =$ $|E(P_k)|$ ($\geq$ follows by (b)). Let l be minimal. Then by (b)

$$|E(P_{k+1})| = \ldots = |E(P_{k+l-1})| = |E(P_k)|$$

and the paths $P_{k+1}, \ldots, P_{k+l-1}$ do not use any edge of P_k or P_{k+l} oppositely to P_k or P_{k+l} by the minimality of l. Hence it follows that those edges e of P_{k+l} not on P_k already satisfy the requirement at the kth step: $f_k(e) > 0$ if e is directed toward a (on P_{k+l}) and $f_k(e) < v(e)$ if e is directed toward b.

Now we get, by an argument similar to the one in (b), two (a,b)-paths Q_1, Q_2 such that $Q_1 \cup Q_2 \subset P_k \cup P_{k+l}$, $Q_1 \cap Q_2 \subseteq P_k \cup P_{k+l}$, and Q_1, Q_2 avoid those edges which are passed by P_k and P_{k+l} oppositely; moreover, Q_1, Q_2 pass each edge in the same direction as P_k or P_{k+l}. By the above, Q_1, Q_2 were candidates for P_k and hence,

$$|E(Q_i)| \geq |E(P_k)|.$$

Moreover, since at least one edge is used by P_k and P_{k+l} in opposite directions,

$$|E(Q_1)| + |E(Q_2)| \leq E(P_k)| + |E(P_{k+l})| - 2.$$

Thus

$$|E(P_{k+l})| \geq |E(P_k)| + 2.$$

This proves the assertion of the hint. To get (c) observe that if $P_k, P_{k+1}, \ldots, P_{k+l}$ do not pass any edge in opposite directions, then the number of those edges e in $P_1 \cup \ldots \cup P_{k+l}$ which satisfy $f_{k+i}(e) = 0$ and are directed toward a (on the path P_{k+j} containing them), or $f_{k+i}(e) = v(e)$ and are directed toward b, increases at each step. Hence

$$l \leq |E(G)| < n^2.$$

Thus, in every n^2 steps the length of P_k increases by at least two. Since it is less than n, the number of steps does not exceed

$$n^2 \cdot \frac{n}{2} = \frac{n^3}{2}.$$

[J. Edmonds and R. M. Karp; see Hu, Ch. 8.]

76. If there is a function as required, we have

$$\sum_x u(x) = \sum_x \left\{ \sum_y f(x,y) - \sum_y f(y,x) \right\}$$

and here each edge occurs twice with opposite signs. This gives a necessary condition

(1) $$\sum_x u(x) = 0.$$

Moreover, define

$$f(a_0, x) = u(x),$$
$$f(x, b_0) = -u(x).$$

Then we obtain an (a_0, b_0)-flow, satisfying $f(e) \leq v(e)$ and of value

$$V = \sum_{u(x)>0} u(x) = \sum_{u(x)<0} (-u(x)).$$

Since such an (a_0, b_0)-flow exists, we must have

$$\sum_{e \in C_1} v(e) \geq V$$

for any (a_0, b_0)-cut C_1 of G_1. Now let C_1 be determined by $S \cup \{a_0\}, \subseteq V(G)$ and let C be the cut of G determined by S, then

$$\sum_{e \in C_1} v(e) = \sum_{e \in C} v(e) + \sum_{\substack{x \in S \\ u(x)<0}} (-u(x)) + \sum_{\substack{x \in V(G) - S \\ u(x)>0}} u(x).$$

Thus,

(2) $$\sum_{e \in C} v(e) \geq \sum_{x \in S} u(x).$$

Conversely, if (1) and (2) are satisfied, then G_1 has a flow $f(e) \leq v(e)$ of value V by 6.77. This flow has, necessarily, $f(e) = v(e)$ for the edges adjacent to a_0 or b_0, and consequently gives the required function, when restricted to G. [D. Gale; see FF.]

77. (a) Let us substitute $v(e)$ parallel edges for each edge e and let G_1 be the resulting graph. Let C be an (a, b)-cut of G and C_1 the (a, b)-cut of G_1, determined by the same set. Then

$$|C_1| = \sum_{e \in C} v(e).$$

Thus putting

$$V = \min_C \sum_{e \in C} v(e),$$

we get $|C_1| \geq V$ and hence, by Menger's theorem, we find V edge-disjoint (a, b)-path in G_1. Let $f(e)$ be the number of edges parallel to e used by these paths, then $f(e)$ is an (a, b)-flow of value V and with integral entries.

We remark that the proof of 6.77 constructs an integer valued flow whenever v is integer valued.

(b) If $v(e)$ is an integer for each e, (a) solves the problem. If $v(e)$ is rational for each e, consider $s \cdot v(e)$, where s is the least common multiple of denominators of the $v(e)$'s, this reduces the problem to the integral case. Finally, if $v(e)$ is arbitrary, take a sequence $v_i \to v$ with rational values $v_i(e)$. Since there are only finitely many (a,b)-cuts, we may assume that the minimum

$$m_i = \min_C \sum_{e \in C} v_i(e)$$

is attained for the same cut C. For each i, there is an (a,b)-flow f_i with

$$0 \le f_i(e) \le v_i(e); \qquad w(f_i) = m_i.$$

As the graph is finite, we may assume that $f_i(e)$ converges for any e. The limit

$$f(e) = \lim_{i \to \infty} f_i(e)$$

is an (a,b)-flow of value

$$\lim_{i \to \infty} m_i = \sum_{e \in C} v(e).$$

This proves the non-trivial part of the Max-Flow-Min-Cut Theorem.

78. Let f' be as in the hint. We show $w(f') = w_1$. Assume that $w(f') > w_1$. Let G' be the subgraph of G determined by those edges e of G for which $f'(e) > 0$. G' is connected between a and b; otherwise we could find an (a,b)-cut C of G which would contain no edge of G', i.e. we would have

$$\sum_{e \in C} f'(e) = 0.$$

On the other hand, the solution of 6.74 yields

$$\sum_{e \in C} f'(e) \ge w(f') > w_1 \ge 0,$$

a contradiction. Thus, we have an (a,b)-path P with $f'(e) > 0$ for each $e \in P$. Subtract 1 from the flow-value of each edge. Then $w(f')$ is decreased but is still $\ge w_1$, and f' is an (a,b)-flow $\le f$. So $w(f') = w_1$.

Now let $f_1 = f'$, $f_2 = f - f'$. Then by the above, f_1, f_2 are (a,b)-flows and

$$0 \le f_1 \le f, \qquad 0 \le f_2 \le f,$$
$$w(f_1) = w_1, \qquad w(f_2) = w(f) - w(f_1) = w_2.$$

§ 7. Factors of graphs

1. First, let F_1 be a system of $\varrho(G)$ edges which cover all points. Since removing any edge from F_1 yields an uncovered point, each edge in F_1 has an endpoint which has degree one in $\langle V(G), F_1 \rangle$. This implies that F_1 consists of disjoint stars.

The number of components of $\langle V(G), F_1 \rangle$ is, obviously, $|V(G)| - |F_1| = |V(G)| - \varrho(G)$. Select an edge from each component of $\langle V(G), F_, \rangle$. Then we have $|V(G)| - \varrho(G)$ independent edges, whence

(1) $\qquad\qquad \nu(G) \geq |V(G)| - \varrho(G), \qquad \nu(G) + \varrho(G) \geq |V(G)|.$

On the other hand, let F_2 be a system of $\nu(G)$ independent edges. Select one edge from each of the $|V(G)| - 2\nu(G)$ points not covered by F_2. Together with F_2, these edges cover all points of G and their number is

$$\nu(G) + (|V(G)| - 2\nu(G)) = |V(G)| - \nu(G),$$

whence

(2) $\qquad\qquad \varrho(G) \leq |V(G)| - \nu(G), \qquad \nu(G) + \varrho(G) \leq |V(G)|.$

(1) and (2) yield the desired equality. [T. Gallai, *Ann. Univ. Sci. R. Eötvös*, Sectio Math. **2** (1959) 133–138. See also LP for most of the problems in this chapter.]

2. *First solution.* Let $\{A, B\}$ be a bicoloration of G, i.e. suppose that $V(G) = A \cup B$, $A \cap B = \emptyset$ and each edge of G joins A to B. Consider two new points a, b. Join a to all points of A and b to all points B. Let G' be the resulting graph. Now observe

(i) a set $S \subseteq V(G)$ separates a and b in G' if and only if S covers all edges of G,

(ii) the maximum number of independent (a, b)-paths is $\nu(G)$; for any $\nu(G)$ independent edges of G yield $\nu(G)$ independent (a, b)-paths and conversely, any k independent (a, b)-paths contain k independent edges of G.

Since, by Menger's theorem, the minimum cardinality of sets separating a and b is equal to the maximum number of independent (a, b)-paths, (i) and (ii) prove $\nu(G) = \tau(G)$. Since by 7.1, $\nu(G) + \varrho(G) = |V(G)|$ and, obviously, $\alpha(G) + \tau(G) = |V(G)|$, the other equation of this problem follows immediately.

Second solution. We show $\nu(G) = \tau(G)$. Since obviously, $\nu(G) \leq \tau(G)$, it suffices to show that $\nu(G) \geq \tau(G)$.

Let G' be a minimal subgraph of G such that $\tau(G') = \tau(G)$. We show G' consists of disjoint edges. This will be sufficient since the number of these edges must be $\tau(G') = \tau(G)$, thus $\nu(G) \geq \tau(G)$.

Suppose indirectly that there is a point x adjacent to two edges $e_1 = (x, y_1)$, $e_2 = (x, y_2)$ of G'. By the minimality of G', there is a set S_i with $|S_i| = \tau(G) - 1$, which covers all edges of $G' - e_i$ ($i = 1, 2$). Obviously, x, $y_i \notin S_i$ but $y_1 \in S_2$, $y_2 \in S_2$. Set $R = S_1 \cap S_2$, $|R| = p$, $\tau(G) = t$, and let G_1 denote the subgraph of G' induced by $(S_1 - R) \cup (S_2 - R) \cup \{x\}$.

Since G_1 is a subgraph of G, it is bipartite. Let $\{A, B\}$ be a bicoloration of G_1 and, say, $|A| \leq |B|$. Then

$$|A| \leq \left\lfloor \frac{|(S_1 - R) \cup (S_2 - R) \cup \{x\}|}{2} \right\rfloor = \left\lfloor \frac{2(t - 1 - p) + 1}{2} \right\rfloor = t - p - 1.$$

We claim that $A \cup R$ covers all edges of G'. In fact, let $e \in E(G')$. If e has an endpoint in R, we are done. We show that, if e is not covered by R, then it belongs to G_1. This is obvious if $e = e_i$, since x, $y_i \in V(G_1)$. If $e \neq e_i$, then both $S_1 - R$ and $S_2 - R$ have to cover e and hence, e has to join $S_1 - R$ to $S_2 - R$. Thus e has both endpoints in $V(G_1)$ again. Since A covers all edges of G_1, it covers e.

Thus, $\tau(G') \leq |A \cup R| = t - p - 1 + p = t - 1$, which contradicts the definition of G'. [D. König; see any textbook on graph theory.]

3. I. Suppose that some point x of F is joined to a point $y \in B_1$. Obviously, $x \in A$. By (**), there is a path P in F connecting x to a point $a \in A_1$. This path starts from a on an edge not in M. Thus by (*), its second edge belongs to M. Its third edge is therefore not in M while by (*) its fourth edge is in M, etc. Hence, it is a path alternating relative to M and its last edge belongs to M. Exchange the edges of $E(P) \cap M$ and $E(P) - M$, then a matching M' arises to which (x, y) can be added. Thus M is not a maximum matching.

II. On the other hand, suppose that no point of F is adjacent to any point in B_1. Let $X = A - V(F)$ and $Y = V(F) \cap B$. We claim $|X \cup Y| = |M|$ and $X \cup Y$ covers all edges.

All points of $X \cup Y$ are covered by M, for suppose not. If $y \in Y = V(F) \cap B$, then by (*) y is covered by M. If $x \in X = A - V(F)$ and $x \in A_1$, then the forest $F' = F \cup \{x\}$ is larger than F, contradicting the assumption that F was maximal relative to M.

Moreover, no edge of M covers two points of $X \cup Y$, for if $(x, y) \in M$ and $y \in V(F) \cap B$, then $x \in V(F)$ by the maximality of F. Hence $|X \cup Y| = |M|$.

Secondly, let (u, v) be any edge not covered by $X \cup Y$ and suppose that $u \in A$, $v \in B$. Then $u \in V(F)$ and $v \notin V(F)$. Moreover, by the assumption, $v \notin B_1$. Thus, there is an edge $(w, v) \in M$, and $w \neq u$ since $|X \cup Y| = |M|$. Here $w \in V(F)$, for otherwise it would follow as in part I, that w would be connected in F to A_1 by a path starting from w through (w, v), whence $v \in V(F)$, a contradiction. But, if $w \in V(F)$, then the edges (u, v) and (w, v) can be added to F, contradicting the maximality of F.

So $\tau(G) \leq |M|$. Since, obviously, $|M| \leq \nu(G) \leq \tau(G)$, it follows that

$$|M| = \nu(G) \quad \text{and} \quad \tau(G) = \nu(G).$$

The first inequality shows that M is a maximum matching. The second proves König's theorem.

The desired algorithm could now be the following: Start with a matching M. Form a maximal forest with properties (*) and (**). Then either we can enlarge M or conclude that it is a maximum matching. In the latter case we also obtain a minimum cover.

We cannot go into details but remark that this algorithm does not need more time than a polynomial of the number of points.

4. (a) *First solution.* What we want to show is that $\nu(G) = |A|$. It is enough to prove, by 7.2, that $\nu(G) = |A|$. Since A covers all edges, $\tau(G) \leq |A|$. Thus, it suffices to show that any set S covering all edges has at least $|A|$ elements. Observe now that if S covers all edges of G, then

$$\Gamma(A - S) \subseteq S \cap B$$

because, if $x \in A - S$ is joined to $y \in B$, then to cover the edge (x, y), S has to contain y. Hence by the assumption of the theorem

$$|S \cap B| \geq |\Gamma(A - S)| \geq |A - S|$$

and thus,

$$|S| = |S \cap A| + |S \cap B| \geq |S \cap A| + |A - S| = |A|.$$

Second solution. We use induction on $|A|$. Let us distinguish two cases:

Case 1. There is a set $A_1 \subset A$, $A_1 \neq \emptyset$, such that $|\Gamma(A_1)| = |A_1|$. Let G_1 be the subgraph of G induced by $A_1 \cup \Gamma(A_1)$ and let $G_2 = G - A_1 - \Gamma(A_1)$. We claim both G_1 and G_2 satisfy the condition of the problem.

Let $X \subseteq A_1$. Then $\Gamma(X) \subseteq \Gamma(A_1)$ and so,

$$\Gamma_{G_1}(X) = \Gamma(X), \quad |\Gamma_{G_1}(X)| = |\Gamma(X)| \geq |X|.$$

On the other hand, suppose that $X \subseteq A - A_1$. Then

$$|\Gamma_{G_2}(X)| = |\Gamma(X \cup A_1)| - |\Gamma(A_1)| \geq |X \cup A_1| - |\Gamma(A_1)| =$$
$$= |X \cup A_1| - |A_1| = |X|.$$

Thus, by the induction hypothesis, there exists a matching F_1 in G_1 which matches A_1 with $\Gamma(A_1)$ and a matching F_2 in G_2 which matches $A - A_1$ with (certain points of) $B - \Gamma(A_1)$. $F_1 \cup F_2$ is a desired matching.

Case 2. $|\Gamma(X)| > |X|$ for every $X \subset A$, $X \neq \emptyset$. Let $x \in A$, $y \in B$ be adjacent points. We claim $G_1 = G - x - y$ satisfies the conditions of the problem. Let $X \subseteq A - \{x\}$; if $X = \emptyset$, then $|\Gamma(X)| = 0 = |X|$, so we may assume that $X \neq \emptyset$. Also $X \neq A$, thus by the assumption, $|\Gamma(X)| > |X|$. So

$$|\Gamma_{G_1}(X)| \geq |\Gamma(X)| - 1 \geq |X|.$$

Hence, by the induction hypothesis, there is a matching F_1 in G_1 which covers all points of $A - \{x\}$. Now $F_1 \cup \{(x, y)\}$ is a matching of the type desired.

(b) If G has a 1-factor, then obviously,

(1) $|\Gamma(X)| \geq |X|$ for each $X \subseteq A$ and $|A| = |B|$.

Conversely, if (1) holds, then (a) implies G has a 1-factor. [P. Hall; B, O, S.]

5. By 7.2, it suffices to show that

$$\tau(G) = |A| - \delta.$$

First let $X \subseteq A$ be a set such that

$$\delta = |X| - |\Gamma(X)|.$$

Set $S = \Gamma(X) \cup (A - X)$. Then S covers all edges of G, and hence

$$\tau(G) \leq |S| = |\Gamma(X)| + |A| - |X| = |A| - \delta.$$

Now let S be a minimum point-cover and $X = A - S$. Then, as before,

$$\Gamma(X) \subseteq S \cap B$$

and hence,

$$|\Gamma(X)| \le |S \cap B| = |S| - |S \cap A| = |S| = |A| + |A - S| =$$
$$= \tau(G) - |A| + |X|,$$

whence

$$\delta \ge |X| - |\Gamma(X)| \ge |A| - \tau(G), \text{ and thus } \tau(G) \ge |A| - \delta.$$

[O. Ore; O, B]

6. (a) By the observation in the hint,

$$|\Gamma(X_1 \cup X_2)| + |\Gamma(X_1 \cap X_2)| \le |\Gamma(X_1) \cup \Gamma(X_2)| + |\Gamma(X_1) \cap \Gamma(X_2)| =$$
$$= |\Gamma(X_1)| + |\Gamma(X_2)| = |X_1| + |X_2| + 2k.$$

On the other hand,

$$|\Gamma(X_1, \cup X_2)| \ge |X_1 \cup X_2| + k \text{ and } |\Gamma(X_1 \cap X_2)| \ge |X_1 \cap X_2| + k,$$

because of (∗) and since $X_1 \cap X_2 \ne \emptyset$. Thus,

$$|\Gamma(X_1 \cup X_2)| + |\Gamma(X_1 \cap X_2)| \ge |X_1 \cup X_2| + |X_1 \cap X_2| + 2k =$$
$$= |X_1| + |X_2| + 2k.$$

Hence equality must hold throughout, in particular

$$|\Gamma(X_1 \cap X_2)| = |X_1 \cap X_2| + k.$$

(b) Let G_1 be defined as in the hint. Let $x \in A$, and let S be the intersection of all sets X with

$$|\Gamma_{G_1}(X)| = |X| + k \text{ and } x \in X.$$

Then, by the repeated application of (a), we see that S itself satisfies

(3) $$|\Gamma_{G_1}(S)| = |S| + k \text{ and } x \in S.$$

(If the assertion is true, $S = \{x\}$, but we do not know this as yet.)

Let y be any point adjacent to x. Since $G_1 - (x, y)$ does not have property (2), we find a set $X \subseteq A$, $X \ne \emptyset$, such that

$$|\Gamma_{G_1 - (x,y)}(X)| < |X| + k.$$

Since, on the other hand, $|\Gamma_{G_1 - (x,y)}(X)|$ can differ from $|\Gamma_{G_1}(X)|$ only by 1, we must have

$$|\Gamma_{G_1}(X)| = |X| + k, \quad x \in X, \quad y \notin |\Gamma_{G_1 - (x,y)}(X)|.$$

Thus, by the definition of S, $S \subseteq X$ and also

(4) $$y \notin \Gamma_{G_1}(S - \{x\}),$$

because

$$\Gamma_{G_1}(S - \{x\}) = \Gamma_{G_1 - (x,y)}(S - \{x\}) \subseteq \Gamma_{G_1 - (x,y)}(X).$$

Now since (4) holds for each neighbor y of x, we get

$$\Gamma_{G_1}(x) \cap \Gamma_{G_1}(S - \{x\}) = \emptyset.$$

Hence

$$|\Gamma_{G_1}(S)| = |\Gamma_{G_1}(x) \cup \Gamma_{G_1}(S - \{x\})| = |\Gamma_{G_1}(x)| + |\Gamma_{G_1}(S - \{x\})|.$$

Here

(5)
$$|\Gamma_{G_1}(x)| \geq k + 1$$

and

$$|\Gamma_{G_1}(S - \{x\})| \begin{cases} = 0 = |S - \{x\}| & \text{if } S - \{x\} = \emptyset. \\ \geq |S - \{x\}| + k \geq |S - \{x\}| & \text{if } S - \{x\} \neq \emptyset. \end{cases}$$

Consequently,

$$|\Gamma_{G_1}(S)| \geq k + 1 + |S - \{x\}| = |S| + k.$$

Comparing this with (1), we find that equality must hold everywhere, in particular in (5). [L. Lovász, *Acta Math. Acad. Sci. Hung.* **21** (1970) 443–446.]

7. (i)$\Rightarrow$(iii). G has a 1-factor, hence $|A| = |B|$. Let $\emptyset \neq X \subset A$ and assume indirectly that $|\Gamma(X)| \leq |X|$. We must then have equality as G has a 1-factor. Now $X \cup \Gamma(X)$ is joined by an edge e to the rest of the graph, because G is connected. Obviously, e joins $\Gamma(X)$ to $A - X$. Now this edge e is not contained in any 1-factor; for if F is a 1-factor, then it contains $|X|$ edges leaving X. But these edges cover all of $\Gamma(X)$ so that e cannot belong to F.

(iii)$\Rightarrow$(ii). We have to show $G - x - y$ has a 1-factor if $x \in A$, $y \in B$. Since $|A - \{x\}| = |B - \{y\}|$, it suffices to show that there is a matching in $G - x - y$ covering $A - \{x\}$. But this is equivalent to saying that, for each $X \subseteq A - \{x\}$,

$$|\Gamma_{G-x-y}(X)| \geq |X|.$$

This is obvious if $X = \emptyset$. Otherwise,

$$|\Gamma_{G-x-y}(X)| \geq |\Gamma(X)| - 1 \geq |X|.$$

(ii)$\Rightarrow$(i). Suppose that G is disconnected. Then we can find a component G_1 such that $|V(G_1) \cap A| \leq |V(G_1) \cap B|$. Let $x \in V(G_1) \cap A$, $y \in B - V(G_1)$. Then $G - x - y$ cannot have a 1-factor, because $V(G_1) \cap B$ is adjacent to $|V(G_1) \cap A| - 1 < |V(G_1) \cap B|$ points only. Thus G is connected.

Now let $e = (x, y)$ be any edge of G. Then $G - x - y$ has a 1-factor, which extends to 1 a-factor of G containing e. Thus each edge of G is contained in a 1-factor. [G. Hetyei, *Pécsi Tan. Főisk. Közl.* (1964) 151–168.]

8. I. First we show that the graphs of form

$$G = G_0 \cup P_1 \cup \ldots \cup P_k,$$

where $P_1, \ldots, P_k$ are chosen as described in the problem, are elementary. They are obviously bipartite and connected.

We use induction on k. Suppose that

$$G' = G_0 \cup P_1 \cup \ldots \cup P_{k-1}$$

is elementary and let x, y be the two endpoints of P_k. Observe first that every 1-factor of G' extends to a 1-factor of G, simply by adding every second edge of P_k, starting with the second edge. Thus, all edges of G' as well as every second edge of P_k belongs to some 1-factor of G.

On the other hand, $G' - x - y$ has a 1-factor F_0. Add to F_0 every second edge of P_k, starting with the first edge. Then we get a 1-factor of G which includes the edges of P_k not used before.

Thus we have shown that G is connected and each edge of it belongs to a 1-factor. This proves by the preceding result that it is elementary.

II. Now suppose that G is any elementary graph. Let F be a 1-factor of G and let G_0 be the graph formed by an edge of F. We choose $P_1, P_2, \ldots, P_k$ one by one so that they have the properties required in the problem and in addition that

(∗) they are alternating paths with respect to F, starting with an edge not in F (we allow P_i to be a single edge not in F).

Suppose that $P_1, \ldots, P_i$ are chosen, and $G' = G_0 \cup \ldots \cup P_i \neq G$. Let e be an edge not in G' but having at least one endpoint x in G'. Since G is connected, such an edge exists. Let F_1 be a 1-factor of G containing e.

Observe that by the additional requirement (∗), G' has the property that $F \cap E(G')$ is a 1-factor of G', and $F_0 = F - E(G')$ does not meet G' at all. Thus, if we consider $F_1 \cup F_0$, one component of it will be a path ending with e. This path P_{i+1} has both endpoints in G' (one of them is x) and has no other point in common with G'. Its end-edges belong to F_1, and hence by reasons of parity, P_{i+1} has odd length and connects two points in different classes of G'. By its definition, it alternates with respect to F in the required manner. Thus, if $G_0 \cup P_1 \cup \ldots \cup P_i \neq G$, we were able to extend this union by P_{i+1}. Sooner or later G will be exhausted, which proves the assertion. [G. Hetyei, ibid.]

9. Represent G in the form

$$G_0 \cup P_1 \cup \ldots \cup P_k$$

as in the preceding problem. Here P_k cannot be a single edge, because

$$G_0 \cup P_1 \cup \ldots \cup P_{k-1}$$

is elementary and therefore cannot be obtained from G by removing a single edge. Thus P_k has an inner point (at least two, in fact) and these have degree 2 [G. Hetyei, ibid.].

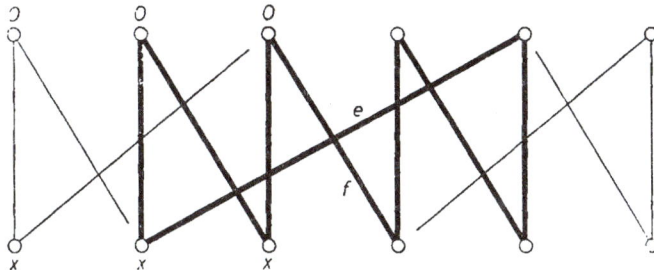

FIGURE 54

The stronger statement that every edge would have an endpoint of degree 2 is false. The graph shown in Figure 54 arises from the construction of 7.8, so it is elementary. It is also minimal. To see this we remark that an elementary graph has degrees at least 2 by 7.7. Removing any edge except e and f, the remaining graph will have a point of degree 1, and so will not be elementary. If e or f is removed, the three points marked by x or O will be joined to the three points only, showing by 7.7 that the remaining graph is not elementary again. However, both endpoints of e have degree 3 [J. Csima].

10. Suppose first that G is r-regular. We show G has a 1-factor. Obviously,

$$|E(G)| = r|A| = r|B|,$$

whence $|A| = |B|$.

Thus, it suffices to show that G contains a matching which covers A or by 7.4, that for each $X \subseteq A$,

$$|\Gamma(X)| \geq |X|.$$

Suppose that $X \subseteq A$ and let us count the edges leaving X. There are, by regularity, $r|X|$ such edges. On the other hand, each of them goes to a point of $\Gamma(X)$ and any point of $\Gamma(X)$ meets at most r of them. Hence,

$$r|X| \leq r|\Gamma(X)|, \quad |X| \leq |\Gamma(X)|.$$

Thus G has a 1-factor.

Let F be a 1-factor of G. $G - F$ is an $(r-1)$-regular bipartite graph and, by induction, it decomposes into the union of $r - 1$ 1-factors. Adding F to them we obtain a desired decomposition.

Now let G be any bipartite graph with maximum degree r. Add new (isolated) points to make the two color-classes of G equal. Then add new edges as long as you can without increasing the maximum degree. The graph G_1 obtained this way must be r-regular; for if there was a point x with degree $d_{G_1}(x) < r$, then the opposite color class would contain such a point y as well and we could add an (x, y)-edge.

Thus, G_1 is an r-regular graph containing G. Hence we may split G_1 into r 1-factors, and thereby produce a decomposition of G into r matchings.

11. Let G_1 denote the bipartite graph obtained as described in the hint. Then

$$d_{G_1}(x) \leq k \qquad (x \in V(G_1))$$

and hence, by the preceding problem, $E(G_1)$ is the union of k matchings $F_1, \ldots, F_k$.

Now identify those points of G_1 which corresponds to the same point of G, then $F_1, \ldots, F_k$ are mapped onto certain subgraphs $F'_1, \ldots, F'_k$. Since F_i is a matching, there is at most one F_i-edge incident with any of the $\left\lceil \frac{d(x)}{k} \right\rceil$ points of G_1 corresponding to $x \in V(G)$. Hence,

$$d_{F_i'}(x) \leq \left\lceil \frac{d(x)}{k} \right\rceil .$$

On the other hand, there are $\left\lfloor \frac{d(x)}{k} \right\rfloor$ points of G_1 corresponding to x which have degree k. There must be an F_i-edge starting from each of these, whence

$$d_{F_i'}(x) \geq \left\lfloor \frac{d(x)}{k} \right\rfloor$$

[D. de Werra; B].

12. By 7.11, $E(G)$ has a partition $G_1 \cup \ldots \cup G_r$ where

$$\left\lfloor \frac{d(x)}{r} \right\rfloor \leq d_{G_i}(x) \leq \left\lceil \frac{d(x)}{r} \right\rceil .$$

for any point x. Since $d(x) \geq r$ by the definition of r, this implies that

$$d_{G_i}(x) \geq 1,$$

i.e. G_i is an edge-cover [R. P. Gupta, *Theory of Gr. Int. Symp. Rome*, Dunod, Paris — Gordon and Breach, New York, (1967) 135–138].

13. Suppose that G_1 as defined in the hint has no 1-factor. Let $\{A, B\}$ be a bicoloration of G, obviously $|A| = |B|$. Thus, if G_1 has no 1-factor, 7.4b implies that there exists an $X \subseteq A$ such that

(1) $|\Gamma_{G_1}(X)| < |X|.$

We count the edges of G leaving X by two different methods. Firstly, the number of these edges is $r \cdot |X|$. Secondly, there are at most $|X| \cdot (n - |\Gamma_{G_1}(X)|)$ edges joining X to $B - \Gamma_{G_1}(X)$ (since these edges are simple edges) and at most $r \cdot |\Gamma_{G_1}(X)|$ edges joining X to $\Gamma_{G_1}(X)$, since any point of $\Gamma_{G_1}(X)$ has degree r. Thus,

$$r \cdot |X| \leq |X| \cdot (n - |\Gamma_{G_1}(X)|) + r|\Gamma_{G_1}(X)|$$

or, equivalently,

(2) $(r - n)|X| \leq (r - |X|)|\Gamma_{G_1}(X)|.$

Now obviously, we are interested in the case $r > n$, otherwise not even the existence of multiple edges follows. Then $|X| < r$ and (1), (2) imply that

$$(r - n)|X| \leq (r - |X|)(|X| - 1)$$

or, equivalently,

$$r \leq (n - |X| + 1) \cdot |X| \leq \left\lfloor \frac{n+1}{2} \right\rfloor \cdot \left\lfloor \frac{n+2}{2} \right\rfloor .$$

Thus we have shown that $r > \left\lfloor \frac{n+1}{2} \right\rfloor \cdot \left\lfloor \frac{n+2}{2} \right\rfloor$ implies the existence of a 1-factor in G_1.

Conversely, let $|A| = |B| = n$, $A' \subset A$, $|A'| = \left\lfloor \frac{n+1}{2} \right\rfloor$, $B' \subset B$ and $|B'| = \left\lfloor \frac{n-1}{2} \right\rfloor$. Join each point of A' to each point of B' by $\left\lfloor \frac{n+2}{2} \right\rfloor$ edges, each point of $A - A'$ to each point of $B - B'$ by $\left\lfloor \frac{n+1}{2} \right\rfloor$ edges; and each point of A' to each point of $B - B'$ by one edge. The resulting graph G is $\left\lfloor \frac{n+1}{2} \right\rfloor \cdot \left\lfloor \frac{n+2}{2} \right\rfloor$-regular. For any $r \leq \left\lfloor \frac{n+1}{2} \right\rfloor \cdot \left\lfloor \frac{n+2}{2} \right\rfloor$, G contains an r-regular spanning subgraph G_r (see 7.10). Observe that G and, consequently, G_r has no 1-factor whose edges have a parallel in G_r. This proves that $r(n) = \left\lfloor \frac{n+1}{2} \right\rfloor \cdot \left\lfloor \frac{n+2}{2} \right\rfloor + 1$.

14. I. We show we cannot get stuck in any situation other than $x = a_1$ and a_1 is joined to some b_i by r parallel edges.

Suppose that we get stuck when $x = b_i$. The reason why we cannot continue is that we have come from a_μ $(1 \leq \mu \leq n)$ and b_i has no neighbor other than a_μ. Since b_i has degree $r-1$, when $x = b_i$, we must have $r - 1$ (a_μ, b_i)-edges. So at the previous step when $x = a_\mu$, there were r (a_μ, b_i)-edges. Suppose that x came to a_μ from $b_j \neq b_i$. Then the edge (a_μ, b_j) was doubled, i.e. the degree of a_μ was at least $r + 2$, which is not allowed.

Similarly it follows that if we get stuck when $x = a_i$, then $i = 1$ and we have the situation described.

II. So it remains to show that the procedure cannot run infinitely. Suppose indirectly that it does. Note that all edges in all graphs arising are parallel to edges of G. Consider those edges of G which x passes infinitely many times. These edges must be passed infinitely often in both directions, otherwise their multiplicities would tend to $+\infty$ or $-\infty$, both impossible.

Consider a point b_i, and suppose that 3 edges (a_μ, b_i), (a_ν, b_i), (a_ϱ, b_i) are used infinitely often $(\mu < \nu < \varrho \leq n)$. Then (a_ϱ, b_i) could never be passed from b_i to a_ϱ, since (a_μ, b_i) or (a_ν, b_i) is always a choice with μ, $\nu < \varrho$. But this is impossible. Similarly it follows that each a_μ is adjacent to at most two infinitely used edges.

Now consider a stage when all the finitely used edges have already been used up. Since each point is adjacent to at most two infinitely used edges, x can only run around a circuit. But then the edges of this circuit are always passed in the same direction, a contradiction.

The procedure thus results in an r-regular graph which is obtained from a subgraph of G by multiplying edges and in which the pair $\{a_1, b_i\}$ spans a component. Repeating it, we finally obtain a graph which arises from a 1-factor of G by taking each edge r times. Thus we obtain a 1-factor of G. Note that it suffices to store the current list of edges plus one vertex; the procedure can be carried out on a blackboard with an eraser and one piece of chalk. [J. Csima; cf. J. Csima–L. Lovász, *Discrete Appl. Math.* **35** (1992) 197–203.]

15. (a) We use simultaneous induction on $|V(G)|$ and k.

Suppose first that G is elementary. Let $x \in A$, $y \in B$ be adjacent. Then the 1-factors of G containing (x, y) correspond to the 1-factors of $G - x - y$. Since (x, y) is contained in some 1-factor, $G - x - y$ has a 1-factor. Moreover, every point of $A - x$ has degree $\geq k - 1$ in $G - x - y$, so $G - x - y$ has at least $(k-1)!$ 1-factors. Thus, each edge of G is contained in at least $(k-1)!$ 1-factors.

Fix an $x \in A$. There are at least k edges incident with x and each of them is contained in at least $(k-1)!$ 1-factors. This gives $k(k-1)! = k!$ 1-factors which are obviously different.

Now suppose that G is not elementary. Then by 7.7, there is an $\emptyset \neq X \subset A$ such that

$$|\Gamma(X)| = |X|.$$

Let F be any 1-factor of G and let F_1 be the set of its edges starting from $A - X$. Then, obviously, the other edges of F cover $\Gamma(X)$, so F_1 matches up $A - X$ with $B - \Gamma(X)$.

Observe that the subgraph G_1 induced by $X \cup \Gamma(X)$ satisfies the same requirements as G: every point of X has degree at least k in G_1, and it has a 1-factor (e.g. $F - F_1$). Therefore, G_1 has at least $k!$ 1-factors by the induction hypothesis. Adding F_1 to these 1-factors, we obtain $k!$ 1-factors of G. [M. Hall, *Bull. Amer. Math. Soc.* **54** (1948) 922–926.]

(b) We use inductions on $m - n + 2$. If this $= 2$, then G is a single circuit and the assertion is trivial. Suppose that $m > n$. By 7.8,

$$G = G_1 \cup P,$$

where G_1 is an elementary bipartite graph and P is an odd path, which has only its endpoints x, y in common with G_1 and x, y belong to different color classes of G_1. Clearly

$$|E(G_1)| - |V(G_1)| + 2 = m - n + 1 < m - n + 2$$

and thus, by the induction hypothesis, G_1 contains at least $m - n + 1$ 1-factors. Each of these can be extended to a 1-factor of G. Moreover by 7.7, $G_1 - x - y$ contains a 1-factor which can also be extended to a 1-factor of G using every second edge of P. Thus we indeed have found $(m-n+1)+1 = m-n+2$ 1-factors in G [cf. L. Lovász–M. D. Plummer, in: *Infinite and Finite Sets*, Coll. Math. Soc. J. Bolyai **10**, Bolyai–North-Holland (1975) 1051–1079].

16. I. Suppose that G has an f-factor F. Then, obviously,

$$\sum_{x \in A} f(x) = |F| = \sum_{x \in B} f(x),$$

which proves (i). Let $X \subset A$ and $Y \subseteq B$. Then there are at most $m(X, Y)$ edges of F joining X to Y and at most $\sum\limits_{y \in B - y} f(y)$ edges of F joining X to $B - Y$. Since there are exactly $\sum\limits_{x \in X} f(x)$ edges of F joining X to B, (ii) follows.

     II. Now suppose that (i) and (ii) are satisfied. Direct all edges from A to B. Take two points a, b and join a to each point of A and each point of B to b by directed edges. Define the capacity $c(e)$ of an edge e of the resulting digraph G_0 by

$$c(e) = \begin{cases} f(x) & \text{if } e = (a, x) \text{ or } e = (x, b), \\ 1 & \text{if } e = (x, y),\ x \in A,\ y \in B. \end{cases}$$

Observe that G_0 has an internal (a, b)-flow with value $\sum\limits_{x \in A} f(x) = \sum\limits_{y \in B} f(y)$, if and only if G has an f-factor. So what we have to verify is, by 6.74 and 6.77, that each (a, b)-cut of G_0 has capacity at least $\sum\limits_{x \in A} f(x)$.

     Let S determine an (a, b)-cut $(a \in S \subset V(G_0))$. Set $X = S \cap A$ and $Y = B - S$. Then the capacity of the cut determined by S is

$$\sum_{x \in A - X} f(x) + \sum_{y \in B - Y} f(y) + m(X, Y) \geq \sum_{x \in A - X} f(x) + \sum_{x \in X} f(x) = \sum_{x \in A} f(x)$$

by (ii). This completes the proof [O. Ore–D. Gale–H. J. Ryser; see O].

17. Suppose that G is embedded into a d-regular bipartite graph G' with color classes A, B. Let $A_1 = A \cap V(G)$, $B_1 = B \cap V(G)$, $A_2 = A - A_1$, $B_2 = B - B_1$. Since there are exactly $d - d(x)$ edges of G' joining x to B_2 for any $x \in A_1$, we have

$$\sum_{x \in A_1} (d - d(x)) \leq d \cdot |B_2|$$

or, equivalently,

$$d \cdot |A_1| - m \leq d \cdot |B_2|.$$

Similarly,

$$d \cdot |B_1| - m \leq d \cdot |A_2|.$$

Summing, we get

$$dn - 2m \leq d(|A_2| + |B_2|) = d(|V(G')| - n),$$

thus

$$|V(G')| \geq 2n - \frac{2m}{d}$$

and, since $|V(G')|$ is an even integer,

$$|V(G')| \geq 2n - 2 \left\lfloor \frac{m}{d} \right\rfloor.$$

Next we show that G can be embedded as an induced subgraph into a d-regular bipartite graph G' with exactly $2n - 2\lfloor \frac{m}{d} \rfloor$ points. Let $\{A, B\}$ be a bicoloration of G. Since $m \leq d \cdot |A|$ and $m \leq d \cdot |B|$, we have

$$|B| = n - |A| \leq n - \left\lfloor \frac{m}{d} \right\rfloor \text{ and } |A| \leq n - \left\lfloor \frac{m}{d} \right\rfloor.$$

So we can construct two disjoint sets $A' \supset A$ and $B' \supset B$ with

$$|A'| = |B'| = n - \left\lfloor \frac{m}{d} \right\rfloor.$$

We add now $(A, B' - B)$-edges and $(B, A' - A)$-edges so that all the degrees in $A \cup B$ become equal to d. This is possible because

$$\sum_{x \in A}(d - d(x)) = |A| \cdot d - m = nd - |B| \cdot d - m \leq$$

$$\leq nd - |B| \cdot d - \left\lfloor \frac{m}{d} \right\rfloor \cdot d = |B'| \cdot d - |B| \cdot d = |B' - B| \cdot d$$

and similarly,

$$\sum_{x \in A}(d - d(x)) \leq |A' - A| \cdot d$$

(and, because we do not mind multiple edges). Finally, we add $(A' - A, B' - B)$-edges to make all degrees d.

18. Considering the "bipartite complement" $\tilde{G}$ of G as defined in the hint, by 7.16 it suffices to show that whenever $X \subseteq A$, $Y \subseteq B$, then

$$(n - 2d)(n - |Y|) + m_{\tilde{G}}(X, Y) \geq (n - 2d) \cdot |X|$$

or, equivalently,

$$(n - 2d)(|X| + |Y| - n) \leq m_{\tilde{G}}(X, Y).$$

Since this inequality is symmetric in X, Y we may assume that $|X| \geq |Y|$. Moreover,

$$m_{\tilde{G}}(X, Y) = |X| \cdot |Y| - m_G(X, Y),$$

and so, it suffices to show that

(1) $$(n - 2d)(|X| + |Y| - n) \leq |X| \cdot |Y| - m_G(X, Y).$$

Here $$m_G(X, Y) \leq d \cdot |Y|$$

and so, (1) would follow from

$$(|X| + d - n)(|Y| + 2d - n) \geq d(2d - n).$$

If $|Y| \le n - 2d$, then $|Y| + 2d - n \le 0$ and so,

$$(|X| + d - n)(|Y| + 2d - n) \ge (n + d - n)(|Y| + 2d - n) \ge d(2d - n).$$

If $n - 2d \le |Y| \le |X| \le n - d$ and $|X| \ge d$, then $|X| + d - n \le 0$, thus

$$(|X| + d - n)(|Y| + 2d - n) \ge (|X| + d - n)d \ge (2d - n)d.$$

If $n - 2d \le |Y|$ and $n - d \le |X|$, then recalling also that $d \le n/2$, we have

$$(|X| + d - n)(|Y| + 2d - n) \ge 0 \ge d(2d - n).$$

Finally, if $|X| \le d$, we use

$$m_G(X, Y) \le |X| \cdot |Y|$$

and thus it suffices to show that

$$(n - 2d)(|X| + |Y| - n) \le 0,$$

which is clear, since $|Y| \le |X| \le d \le \frac{n}{2}$.

19. The transformation of the problem given in the hint is easily verified (cf. 4.21).

   Suppose that G has no 1-factor, then by 7.4b, there is a set $X \subseteq A = \{u_1, \ldots, u_n\}$ with $|\Gamma(X)| < |X|$. Let, say, $X = \{u_1, \ldots, u_k\}$ and $B - \Gamma(X) = \{v_1, \ldots, v_l\}$. Thus $k + l > n$, and $a_{ij} = 0$ for $1 \le i \le k$, $1 \le j \le l$. Now

$$\sum_{i=1}^{n} a_{ij} = 1,$$

because A is doubly stochastic, hence

$$\sum_{j=1}^{l} \sum_{i=1}^{n} a_{ij} = l,$$

On the other hand,

$$\sum_{j=1}^{l} \sum_{i=1}^{n} a_{ij} = \sum_{i=k+1}^{n} \sum_{j=1}^{l} a_{ij} \le \sum_{i=k+1}^{n} \sum_{j=1}^{n} a_{ij} = n - k,$$

which contradicts $k + l > n$.

20. (a) It is clear from the remark in the hint.

   (b) Suppose that G has no 1-factor. Then $\det(a_{ij}) = 0$. Let, say the first $k + 1$ columns form a minimal linearly dependent set of columns. Since the rank of the matrix A' formed by these $k + 1$ columns is k, it contains a $k \times k$ non-sigular submatrix. We may assume that the first k rows and columns form this submatrix.

Since the first $k+1$ columns are linearly dependent, we have numbers $\lambda_1, \ldots, \lambda_{k+1}$ such that

$$(1) \qquad \sum_{j=1}^{k+1} \lambda_j a_{ij} = 0 \qquad (i = 1, \ldots, n).$$

Since $\lambda_\mu \neq 0$ by the minimality of k, we may divide by any λ_μ and determine the numbers $\lambda_\nu / \lambda_\mu$ by Cramer's rule from the first k rows. Thus $\lambda_\nu / \lambda_\mu$ belongs to the field generated by $\{a_{ij} : 1 \leq i \leq k, \ 1 \leq j \leq k+1\}$.

We claim that $a_{ij} = 0$ if $k+1 \leq i \leq n$ and $1 \leq j \leq k+1$. We show, e.g. that $a_{k+1,1} = 0$. We have

$$a_{k+1,1} = -\frac{1}{\lambda_1} \sum_{j=2}^{k+1} \lambda_j a_{k+1,j}$$

from (1). Here the right-hand side is in the field generated by $\{a_{ij}; (i,j) \neq (k+1,1)\}$. By the algebraic independence of the a_{ij}'s this is only possible if $a_{k+1,1} = 0$.

Setting $X = \{a_{k+1}, \ldots, a_n\}$, we have

$$\Gamma(X) \subseteq \{b_{k+2}, \ldots, b_n\}, \qquad |\Gamma(X)| < |X|,$$

which proves the non-trivial part of 7.4b. [J. Edmonds, *J. Res. Natl. B. Standards* **71B** (1967), 241–245.]

21. If the conditions imposed upon y are linearly independent the answer is clearly $m - n$, where $n = |V(G)|$; if $n - 1$ of them are linearly independent but the nth one is not, the dimension is $m - n + 1$. We claim the first case occurs, if G is not bipartite and the second one, if it is.

To prove this, consider a non-trivial linear relation among the left-hand sides of the constraints (1). We want to show that this can exist only, if G is bipartite and that in this case it is essentially unique. So let

$$\sum_{v \in V(G)} \lambda_v \sum_{e_i \ni v} y_i$$

be identically 0, i.e.

$$\lambda_v + \lambda_u = 0$$

for every edge (u, v). Hence using the fact that G is connected we conclude that $|\lambda_v| = \text{const} = \lambda$ and points with $\lambda_v = \lambda$ are only connected to points with $\lambda_v = -\lambda$ and vice verse. Hence G is bipartite. Since the 2-coloration of a connected bipartite graph is unique, we also see that the linear relation among the left-hand sides of the constraints (1) is unique up to a scalar factor. Thus in the bipartite case there is the unique relation

$$\sum_{v \in A} \sum_{e_i \ni v} y_i = \sum_{v \in B} \sum_{e_i \ni v} y_i,$$

where $\{A, B\}$ is the bicoloration of G. [H. Saks, *Wiss. Z. Tech. Hochschule Ilmenau* **12** (1966), 7–12.]

22. Following the hint, let $(x, y) \in F$. If there are at least 3 edges joining $\{x, y\}$ to $\{u, v\}$, then there are two of them which are independent, e.g. (x, u) and (y, v). Now $F = \{(x, y)\} \cup \{(x, u), (y, v)\}$ is a bigger matching than F, a contradiction. So each edge of F is joined to $\{u, v\}$ by at most two edges. Since u and v are not joined to each other or to any point not on the edges of F (otherwise F would not be maximal), this implies that

$$d(u) + d(v) \le 2|F| \le 2(n - 1).$$

This, however, contradicts the hypothesis that $d(y)$ and $d(v)$ are at least n (cf. also problem 10.19).

23. Let X be the set of points covered by F_0. Following the hint, let F be a maximum matching of G containing as many edges of F_0 as possible. We claim F covers X. Suppose indirectly that there exists an $x \in X$ not covered by F. Let (x, y) be the edge of F_0 incident with x. Since $F \cup \{(x, y)\}$ is not a matching by the maximality of F, there is an $e \in F$ incident with y. Set $F' = F - \{e\} \cup \{(x, y)\}$, then F' is a matching of the same size as F, i.e. it is a maximum matching. Moreover, it has more edges in common with F_0 than F, a contradiction.

24. (a) For $k = 1$, a single edge satisfies the requirements. Suppose that G_{k-1} is a graph with a unique 1-factor and with degrees at least $k-1$. Let G'_{k-1} be another disjoint copy of it, and also consider two new points x, y. Join x to all points of G_{k-1}, y to all points of G'_{k-1}, and x to y. Let G_k be the resulting graph (Fig. 55).

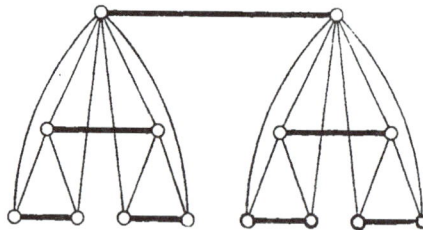

FIGURE 55

Every point of $V(G_{k-1}) \cup V(G'_{k-1})$ has greater degree than in G_{k-1} or G'_{k-1}, hence it has degree at least k. Obviously, this holds for x and y. Moreover, G_k has a unique 1-factor. For let F be any 1-factor of G_k. By reasons of parity it must contain the cut-edge (x, y). Then, however, the rest of it forms a 1-factor of

G_{k-1} and a 1-factor of G'_{k-1}. Since these are unique, F is uniquely determined. So G_k has at most one 1-factor. But the one described (composed of the 1-factors of G_{k-1}, G'_{k-1} and of (x,y)) is indeed a 1-factor of G_k. Thus G_k has a unique 1-factor.

(b) *First solution.* Suppose that G is a minimum counterexample. Let F be its unique 1-factor and let $e_1 \in E(G) - F$. Consider the equivalence-class defined in 6.27 containing e_1. If it consists only of e_1, then $G - e_1$ is 2-edge-connected and we get a contradiction of the minimality of G. Suppose that it has other edges $e_2, \ldots, e_k$, $k \geq 2$.

Let G_1 be any component of $G - \{e_1, \ldots, e_k\}$. By 6.27, it is incident with exactly two edges of $\{e_1, \ldots, e_k\}$. If these two edges do not belong to F, we get by induction that G_1 has two distinct 1-factors which together with $F - E(G_1)$, yield two distinct 1-factors of G. Thus at least one of the two edges of $\{e_1, \ldots, e_k\}$ incident with G_1 belongs to F. Since this is true for any component of $G - \{e_1, \ldots, e_k\}$, at least half of the edges $\{e_1, \ldots, e_k\}$ belong to F. Since, in turn, this holds for any equivalence class, at least half of the edges of G belong to F. Hence $|E(G)| \leq |V(G)|$. Since G is 2-edge-connected, this is possible only, if G is a circuit. However, an (even) circuit has two 1-factors.

Second solution. Suppose that G is 2-edge-connected. Let F be a 1-factor of G. Let C be a circuit such that any edge of F either lies on C or is completely disjoint from C. Such a circuit exists by 6.30. Now C must be alternating relative to F (since F is a 1-factor). Thus, "switching" on C (i.e. replacing the edges of $E(C) \cap F$ by the edges of $E(C) - F$, we get another 1-factor. [A. Kotzig, *Mat. Fyz. Časopis* **9** (1959) 73–91; L. Beineke–M. D. Plummer, *J. Comb. Th.* **2** (1967) 285–289.]

(c) The proof is by induction on $|E(G)|$.

The sharpening of (b) formulated in the hint follows from (b) trivially. It suffices to prove this for connected graphs, since if G has a unique 1-factor, so do its components. Double all edges of G not in F. This cannot produce any new 1-factors, so the resulting graph G' has a cut-edge by (b). This edge must belong to F, since it is not doubled.

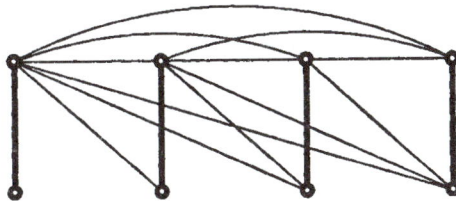

FIGURE 56

Let $e = (x,y)$ be a cut-edge of G, belonging to the unique 1-factor F of G, then $G - x - y$ has a unique 1-factor. By the induction hypothesis, $E(G - x - y) \leq$

$(n-1)^2$. Moreover, since (x, y) is a cut-edge, no point is joined to both x and y. Hence, at most $1 + (2n - 2) = 2n - 1$ edges are adjacent to one of x and y. Thus,

$$|E(G)| \leq (n-1)^2 + 2n - 1 = n^2.$$

The result is sharp, as shown by graphs of structure similar to those in Fig. 56. [G. Hetyei, *Pécsi Tan. Főisk. Közl.* (1964) 151–168.]

25. Let $x_1, \ldots, x_\delta$ be the points not covered by a maximum matching F, $\delta = \delta(G) = |V(G)| - 2\nu(G)$. Let $F = \{e_1, \ldots, e_\nu\}$, and let f_i be an edge adjacent to x_i $(i = 1, \ldots, \delta)$. Then each f_i joins x_i to a point covered by F. Moreover, no two f_i's go to different endpoints of the same e_j, since this would make it possible to enlarge F. Therefore, we can select an endpoint u_j of e_j $(j = 1, \ldots, \nu)$ so that $\{u_1, \ldots, u_\nu\}$ covers all $f_1, \ldots, f_\delta$. Since $d(u_j) \leq d$, we have

$$\delta \leq (d-1)\nu \quad \text{or} \quad n - 2\nu \leq (d-1)\nu,$$

and hence

$$\nu \geq \frac{n}{d+1}.$$

26. Following the hint, observe that if x and y are adjacent, then $\nu(G - x - y) < \nu(G)$, because any matching of $G - x - y$ can be extended using (x, y).

Let x, y be two points at distance $k > 1$, and let z be any inner point of a minimum (x, y)-path. Suppose indirectly that $\nu(G - x - y) = \nu(G)$. Let F be any maximum matching in $G - x - y$. Also, consider a maximum matching F_0 in $G - z$; then $|F| = |F_0| = \nu(G)$ (ν for brevity). Observe that

(1) F must cover z, otherwise, it would be a ν-element matching of $G - x - z$, which contradicts the induction hypothesis $\nu(G - x - z) < \nu$, and

(2) similarly, F_0 must cover x and y.

Now $F \cup F_0$ is the union of some disjoint alternating paths, circuits and common edges of F and F_0. There is an alternating path P_1 starting from x and such a path P_2 starting from y. If $P_1 = P_2$, then replace the edges of $F \cap E(P_1)$ by the edges of $F_0 \cap E(P_1)$ in F; this way we get a $(\nu + 1)$-element matching (since both end-edges of P_1 belong to F_0), which is a contradiction. Thus, $P_1 \neq P_2$. At least one of P_1 and P_2 does not contain z, say $z \notin V(P_1)$. Now replace the edges of $F_0 \cap E(P_1)$ by the edges of $F \cap E(P_1)$ in F_0. The resulting matching F' has $|F'| \geq |F|$ (because P_1 starts from x with an F_0-edge and so, if the last edge of P_1 is in F_0, then $|F'| > |F|$, otherwise $|F'| = |F|$). Moreover, z is still not covered by F' (since $z \notin V(P_1)$) and thus F' is a matching of cardinality $\geq \nu$ in $G - y - z$, again a contradiction.

Thus, $\nu(G - x - y) < \nu(G)$ for every x, y. Let F be any maximum matching of G. Then there is at most one point not covered by F, since if there were two such points x, y, then we would have

$$\nu(G - x - y) \geq |F| = \nu(G),$$

a contradiction. Moreover, F cannot be a 1-factor for if it were, then

$$\nu(G - x) \leq \frac{1}{2}|V(G - x)| < \frac{1}{2}|V(G)| = |F| = \nu(G),$$

which contradicts one of the hypotheses. Hence,

$$\nu(G) = |F| = \frac{1}{2}(|V(G)| - 1),$$

which proves the assertion. [T. Gallai, *MTA Mat. Kut. Int. Közl.* **8** (1963) 135–139.]

27. (a) Let $X \subseteq V(G)$ and F, a maximum matching of G. Let $G_1, \ldots, G_k$ be the odd components of $G - X$. Let, say, $G_1, \ldots, G_i$ be those containing a point not covered by F. From every G_j, $i < j \leq k$, there must be an edge of F going to X by parity. Since these edges have to go to different points of X, we have

$$k - i \leq |X|.$$

On the other hand, each of $G_1, \ldots, G_i$ contains an uncovered point, so

(1) $$\delta(G) \geq i \geq k - |X| = c_1(G - X) - |X|.$$

It remains to show — and this is the crux of the argument — that there exists a set X for which equality holds in (1). We show this by induction on $|V(G)|$.

Suppose first that there exists an $x \in V(G)$ such that $\nu(G - x) < \nu(G)$. Obviously, $\nu(G - x) = \nu(G) - 1$ and hence

$$\delta(G - x) = |V(G - x)| - 2\nu(G - x) = |V(G)| - 2\nu(G) + 1 = \delta(G) + 1.$$

Thus by the induction hypothesis, there exists an $X' \subseteq V(G - x)$ such that

$$\delta(G) + 1 = c_1(G - x - X') - |X'|.$$

Now set $X = X' \cup \{x\}$. Then

$$\delta(G) = c_1(G - X) - |X|.$$

Secondly, suppose that $\nu(G - x) = \nu(G)$ for every $x \in V(G)$. Let $G_1, \ldots, G_k$ be the components of G, and $x \in V(G_i)$. Then

$$\nu(G) = \nu(G_1) + \ldots + \nu(G_k) = \nu(G - x) =$$
$$= \nu(G_1) + \ldots + \nu(G_i - x) + \ldots + \nu(G_k),$$

hence $\nu(G_i - x) = \nu(G_i)$. By the preceding problem, this implies that $|V(G_i)|$ is odd and $\delta(G_i) = 1$. Thus,

$$\delta(G) = \sum_{i=1}^{k} \delta(G_i) = k = c_1(G),$$

and $X = \emptyset$ satisfies the requirement [C. Berge; see B].

(b) G has a 1-factor if and only if $\delta(G)=0$, i.e. by (a), if and only if

(2) $$\max_X\{c_1(G-X)-|X|\}=0.$$

Since $c_1(G)-|\emptyset|\geq 0$, the maximum is always ≥ 0 and so, (2) is equivalent to

$$c_1(G-X)-|X|\leq 0$$

for every $X\subseteq V(G)$ as stated [W. T. Tutte; see B].

28. (a) Let V_1 be the set of those points of G joined to every other point and $V_1=V(G)-V_1$. We want to show that if a, b, $c\in V_2$, and b is adjacent to both a and c, then a and c are adjacent.

Suppose that this is not the case. Since $b\in V_2$, there is a fourth point d which is not adjacent to b. By the maximality assumption on G, $G+(a,c)$ has a 1-factor F_1 and $G+(b,d)$ has a 1-factor F_2. Obviously, $(a,c)\in F_1$, $(b,d)\in F_2$, $(b,d)\notin F_1$ and $(a,c)\notin F_2$.

Consider $F_1\cup F_2$. This consists of disjoint alternating circuits and common edges of F_1 and F_2. (a,c) is on an alternating circuit C_1 and similarly, (b,d) is on an alternating circuit C_2.

Case 1. $C_1\neq C_2$. Then replace the edges of $E(C_1)\cap F_1$ by the edges of $E(C_1)\cap F_2$ in F_1 and get a 1-factor of G, a contradiction.

Case 2. $C_1=C_2$. Starting from b through (b,d) and walking along C_1, sooner or later we must hit (a,c). Without loss of generality we may suppose that we hit a before c.

So we have a (b,a)-path starting with the F_2-edge (b,d) and ending with another F_2-edge, since $(a,c)\in F_1$. Thus, $K=P+(a,b)$ is a circuit alternating relative to F_2. Replacing the edges of $E(K)\cap F_2$ by the edges of $E(K)-F_2$ we get a 1-factor of G from F_2. This is again a contradiction.

Thus we have shown that adjacency is an equivalence relation on V_2, i.e. V_2 decomposes into disjoint complete graphs.

(b) We only prove the non-trivial part of 7.27b, which says that if G has no 1-factor, then there is an $X\subseteq V(G)$ with

$$c_1(G-X)>|X|.$$

If $|V(G)|$ is odd, $X=\emptyset$ has this property; so we may assume that $|V(G)|$ is even. Let us add edges to G as long as we do not produce any 1-factor. When we have to stop we will have a graph G' on the same point-set as G with the property that G' has no 1-factor, but joining any two non-adjacent points of G' produces a graph with a 1-factor.

By (a), G' has the following form: if V_1 is the set of those points connected to every other point, then $G'-V_1$ consists of disjoint complete subgraphs $G_1,\ldots,G_k$.

We claim there are at least $|V_1|+1$ odd complete graphs among $G_1,\ldots,G_k$. Suppose not. Select a 1-factor of each even G_i, select a maximum matching of each odd G_i, this misses exactly one point. Select independent edges matching these points with some points in V_1. Since $|V(G)|$ is even, there are an even number of points of V_1 still uncovered. But we may match them arbitrarily with each other, because V_1, obviously, spans a complete subgraph of G'. So we get a 1-factor, and hence a contradiction.

Thus, there are at least $|V_1|+1$ odd components of $G'-V_1$. When the edges of $E(G')-E(G)$ are removed, these odd components may fall apart, but each of them yields at least one odd component of $G-V_1$. Hence

$$c(G-V_1) > |V_1|,$$

i.e. $X=V_1$ is an appropriate choice for X.

Remark: The reader may easily verify the fact (which was not needed here) that $G_1,\ldots,G_k$ are all odd and $k=|V_1|+2$. [G. Hetyei, *Pécsi Tanárképző Főiskola Közl.* 1974; L. Lovász, *J. Comb. Th.* **19** (1975) 269–271.]

29. (a) By Tutte's theorem, we have to show that

$$c_1(G-X) \le |X|$$

for every $X \subseteq V(G)$.

Let G_1 be any odd component of $G-X$. Since G is 2-connected, there are at least 2 edges joining G_1 to X. There cannot be exactly two such edges, however, for this would imply the sum of the degrees of G_1 is $3|V(G_1)|-2$, an odd number, which is impossible. So there are at least 3 edges joining G_1 to X. Thus, there are at least $3c_1(G-X)$ edges coming from the odd components of $G-X$ to X. But each point of X is adjacent to at most 3 of them, so their number is at most $3|X|$. Thus

$$3c_1(G-X) \le 3|X|, \quad c_1(G-X) \le |X|$$

as required.

(Note that only 2-edge-connectivity of G was used. It is trivial to see though that, if a 3-regular graph is 2-edge-connected then it is also 2-connected.) [J. Petersen; for a proof not using Tutte's theorem, see K, S.]

(b) If e is a cut-edge of a 3-regular graph, then e is contained in every 1-factor (if any). In fact, the two components of $G-e$ are odd (they contain exactly one point of even degree). Thus, $G-e$ has no 1-factor, so each 1-factor, if any, contains e.

It therefore suffices to construct a simple cubic graph with three separating edges meeting at some point: any 1-factos of such a graph must contain all three of them, which is impossible. Figure 57 shows such a construction (the reader may wish to verify that this is the smallest such example).

30. By Tutte's theorem (7.27b), we only have to show that

$$c_1(G'-X) \le |X|$$

for any $X \subseteq V(G)$. Suppose that this does not hold for some X. Since $|V(G)|$ is even and so

$$c_1(G'-X) \equiv |V(G')-X| \equiv |X| \pmod 2$$

we have

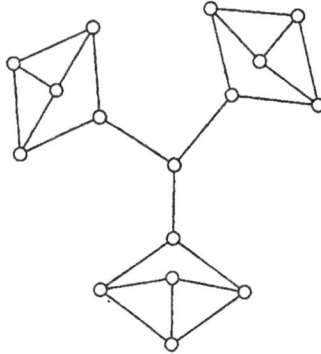

FIGURE 57

(1)
$$c_1(G' - X) \geq |X| + 2.$$

Let $G_1,\ldots,G_m$ be the odd components, $G_{m+1},\ldots,G_M$ the even components of $G' - X$. For $1 \leq i \leq M$, let α_i (β_i) be the number of edges of $E(G) - E(G')$ joining G_i to X (to the other G_j's), and let γ_i denote the number of edges of G' joining G_i to X. Then $\alpha_i + \beta_i + \gamma_i$ is the total number of edges of G leaving G_i, which is at least $k-1$.

Moreover, if G_i is an odd component, then $\alpha_i + \beta_i + \gamma_i > k-1$; for $\alpha_i + \beta_i + \gamma_i = k-1$ would mean that the sum of degrees of the subgraph of G induced by $V(G_i)$ is $|V(G_i)| - k + 1 = k(|V(G_i)| - 1) + 1 \equiv 1 \pmod 2$. Hence

(2)
$$\sum_{i=1}^{m} \alpha_i + \sum_{i=1}^{m} \beta_i + \sum_{i=1}^{m} \gamma_i \geq k \cdot m.$$

The number of removed edges between two of $G_1,\ldots,G_M$ and X is $\sum_{1}^{M} \alpha_i + \frac{1}{2}\sum_{1}^{M} \beta_i$, whence

$$2\sum_{i=1}^{M} \alpha_i + \sum_{i=1}^{M} \beta_i \leq 2(k-1).$$

The total number of edges entering X is $\sum_{1}^{M} \alpha_i + \sum_{1}^{M} \gamma_i$, whence

$$\sum_{i=1}^{M} \alpha_i + \sum_{i=1}^{M} \gamma_i \leq k \cdot |X|.$$

Adding the last two inequalities,

$$3\sum_{i=1}^{M} \alpha_i + \sum_{i=1}^{M} \beta_i + \sum_{i=1}^{M} \gamma_i \leq k \cdot (|X| + 2) - 2.$$

Comparing this with inequality (2), we get

$$km \leq \sum_{i=1}^{M} \alpha_1 + \sum_{i=1}^{M} \beta_i + \sum_{i=1}^{M} \gamma_i \leq 3 \sum_{i=1}^{M} \alpha_i + \sum_{i=1}^{M} \beta_i + \sum_{i=1}^{M} \gamma_i \leq$$

$$\leq k \cdot (|X| + 2) - 2 < k(|X| + 2)$$

or, equivalently,

$$m < |X| + 2$$

which contradicts (1). [J. Plesnik, *Mat. Čas. Slov. Akad. Vied.* **22** (1972) 310–318.]

31. Let F be a maximum matching which misses a given point x. Let y be a point adjacent to x and let F_0 be a 1-factor of $G - y$. $F \cup F_0$ consists of disjoint alternating circuits, common edges of F and F_0 and an alternating (x,y)-path P. Now $P_0 = P + (x,y)$ is an odd circuit which contains $\frac{1}{2}(|V(P_0)| - 1)$ edges of F.

Now suppose that $P_0, \ldots, P_i$ have been chosen so that for all $j < i$, P_{j+1} is an odd $(P_0 \cup \ldots \cup P_j, P_0 \cup \ldots \cup P_j)$-path or an odd circuit having exactly one point in common with $P_0 \cup \ldots \cup P_j$ and P_{j+1} alternates relative to F (starting and ending with an edge not in F). If P_{j+1} happens to be an odd circuit, then we adopt the convention that it starts and ends at its common point with $P_0 \cup \ldots \cup P_j$. Also suppose that $P_0 \cup \ldots \cup P_i \neq G$. Let (a,b) be any edge not in $P_0 \cup \ldots \cup P_i$ such that $a \in V(P_0 \cup \ldots \cup P_i)$ (such an edge exists since G is connected).

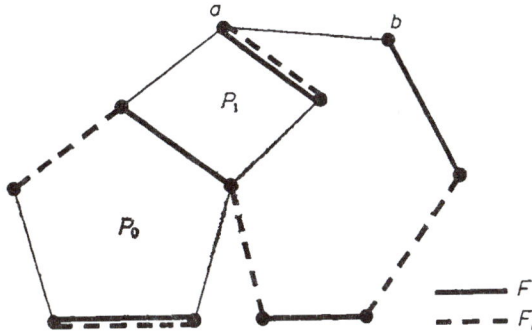

FIGURE 58

If $b \in V(P_0 \cup \ldots \cup P_i)$, we may take $P_{i+1} = (a,b)$. So suppose that $b \notin V(P_0 \cup \ldots \cup P_i)$. Let F_i be a 1-factor of $G - b$. $F \cup F_i$ consists of disjoint alternating circuits, common edges of F and F_i, and an alternating (b,x)-path Q. Traverse Q from b to the first point c of $P_0 \cup \ldots \cup P_i$. The last edge traversed then belongs to F_i, because such F-edge meeting $P_0 \cup \ldots \cup P_i$ belongs to this union by the construction. Let Q' be the piece of Q between b and c, and $P_{i+1} = Q' + (a,b)$. Then P_{i+1} satisfies the requirements (Fig. 58).

Thus if $P_0 \cup \ldots \cup P_i \neq G$, we can choose a P_{i+1}. Sooner or later we exhaust G, and the required decomposition is obtained. [L. Lovász, *Studia Sci. Math. Hung.* **7** (1972) 279–280; also cf. the Edmonds Matching Algorithm 7.34.]

32. (a) Set $A = A_G$, $C = C_G$, $\ldots$; $A' = A_{G-x}$, $C' = C_{G-x}$, $\ldots$ It suffices to prove that $D' = D$, since then C is the set of points not adjacent to $D = D'$ and so is the same as C'.

Observe that since $G - x$ contains no maximum matching of G, $\nu(G - x) = \nu(G) - 1$. Hence if M is any maximum matching of G, $M - x$ is a maximum matching of $G - x$. Since for each $y \in D$ there is a maximum matching of G avoiding y and this yields such a maximum matching of $G - x$, it follows that $y \in D'$, i.e. $D \subseteq D'$.

Conversely, let $y \notin D$ and suppose indirectly that $y \in D'$. Then there is a maximum matching M' of $G - x$ avoiding y. Let M be a maximum matching of G avoiding a neighbor $z \in D$ of x. Consider $M \cup M'$. Since y is covered by M, but not by M', the component of $M \cup M'$ containing y is a path P starting from y with an edge of M. P cannot end with an edge of M', since then replacing the edges of $M \cap E(P)$ by the edges of $M' \cap E(P)$ we would get a maximum matching of G avoiding y. Thus P ends with an edge of M. Now replace the edges of M' on P with the edges of $M \cap E(P)$. This way we get a matching larger than M'. Therefore, this new matching cannot lie in $G - x$. This implies that P ends at x. But now $(M - E(P)) \cup (E(P) \cap M') \cup \{(x, z)\}$ is a maximum matching avoiding y, a contradiction again.

(b) Let $G_1 \ldots, G_t$ be those components of $G - A$ contained in D. Since no edge connects D to C, these components partition D. Let H be the subgraph induced by C.

If we remove the points of A one by one, (a) implies that the sets C, D do not change. Hence every maximum matching of $G - A$ will cover all points of H, but each point of D will be missed by some maximum matching of $G - A$. Every maximum matching of $G - A$ consists of a maximum matching M_0 of H and maximum matchings M_i of G_i, $i = 1, \ldots, t$. Since every point of H must be covered, M_0 is a 1-factor of H. Moreover, since each point of G_i is missed by some maximum matching of G_i, G_i is factor-critical by 7.26.

(c) If M is any minimum matching of G then by the argument at the beginning of the proof, $M - A$ is a maximum matching of $G - A$. In part (b), we have already seen what $M - A$ must look like.

(d) By (b),

$$\nu(G - A) = \frac{1}{2}(|V(G)| - |A| - t).$$

By the remark above concerning removal of points of A one at the time,

$$\nu(G - A) = \nu(G) - |A|.$$

Hence the equality follows.

(e) Suppose that $c_1(G-X) \geq |X|$ holds for every X. In particular, $c_1(G-A) \geq |A|$. On the other hand $c_1(G-A) = t$. Hence by (d),

$$\nu(G) \geq \frac{1}{2}|V(G)|,$$

i.e. G has a 1-factor. (The other direction of Tutte's theorem is trivial.) [T. Gallai, *MTA Mat. Kut. Int. Közl.* **8** (1963) 373–395; J. Edmonds, *Canad. J. Math.* **17** (1965) 449–467.]

(f) See Fig. 59.

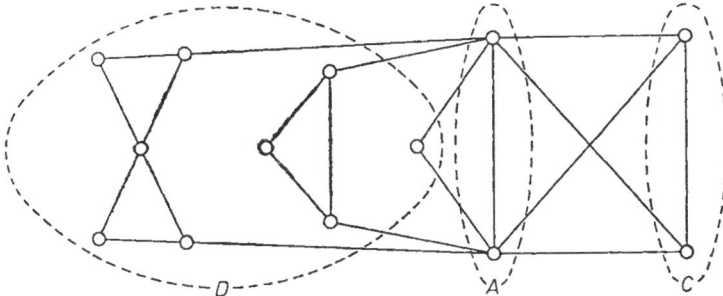

FIGURE 59

33. I. Suppose that M' is not a maximum matching in G'. Then G' has a matching M_0 with $|M_0| > |M'|$. M_0 is a matching in G which contains at most one point of C. Hence we can add k edges of C to M_0 to get a matching M_1 with

$$|M_1| = |M_0| + k > |M'| + k = |M|,$$

a contradiction.

II. Now suppose that M' is a maximum matching of G'. Consider the sets D' A', C' defined for G' as in 7.32. Let c be the image of C in G'. Since M' avoids c, $c \in D'$. Hence if we "blow up" c, the component of $G'-A'$ containing c becomes a component of $G-A'$ containing C, also odd. Hence

$$c_1(G-A') = c_1(G'-A')$$

and

$$\nu(G) \leq \frac{1}{2}\{|V(G)| - c_1(G-A') + |A'|\} =$$

$$= \frac{1}{2}\{|V(G)| - c_1(G'-A') + |A'|\} =$$

$$= \frac{1}{2}\{|V(G')| - c_1(G'-A') + |A'|\} + k = |M'| + k = |M|,$$

showing that M is a maximum matching in G.

34. (a) I. A forest with properties $(*)$–$(**)$ certainly exists; e.g. let the forest consist of the points not covered by M and no edges. Let F be a maximal such forest. Then no outer point can be adjacent to $V(G) - V(F)$, for by the properties of F, $(G - V(F)) \cap M$ is a 1-factor of $G - V(F)$ (since all the uncovered points must be roots in F) and so, if $a \in V(G) - V(F)$ is adjacent to the outer point b of F and $(a, c) \in M$, then (a, b) and (a, c) could be added to F.

II. If two outer points, x and y say, in distinct components of F are adjacent, then consider the paths P and Q connecting a and y to the roots of their components in F. Then $R = P \cup Q \cup \{(a, y)\}$ is a path alternating relative to M and if we replace $M \cap E(R)$ by $E(R) - M$, we get a matching larger than M (Fig. 60).

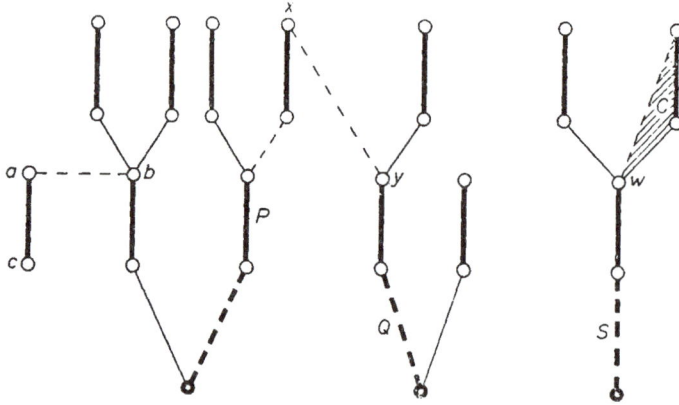

FIGURE 60

III. Now suppose that two outer points u, v of the same component of F are adjacent. Then they form with certain edges of F an odd circuit C. Let w be the point of C nearest the root r of this component of F ($w = r$ may occur). "Switching" on the alternating path connecting w to r, we get a new matching M_0 such that M_0 contains $\frac{1}{2}(|V(C)| - 1)$ edges of C and no other edge of M_0 meets C. By the preceding result, if we contract C, then $M_0' = M_0 - E(C)$ will be a maximum matching in the resulting graph G' iff M_0 (or, equivalently, M) is a maximum matching of G. Moreover, if we find in G' a matching larger than M_0' we can easily extend it to a matching of G larger than M.

IV. Finally, suppose that the outer points of G are independent. Let A be the set of inner points. Then the outer points will be isolated in $G - A$; their number is

$$c_1(G - A) = |M \cap E(F)| + (|V(G)| - 2|M|),$$

while

$$|A| = |M \cap E(F)|.$$

Hence

$$|M| = \frac{1}{2}(|V(G)| + |A| - c_1(G - A)).$$

But by 7.27 the right-hand side is an upper bound for $\nu(G)$. Thus M is a maximum matching.

The desired algorithm is now clear by summarizing the above results. At each step, we have a matching M of G and a forest $F \subset G$ satisfying (*) and (**). We consider the edges adjacent to outer points of F

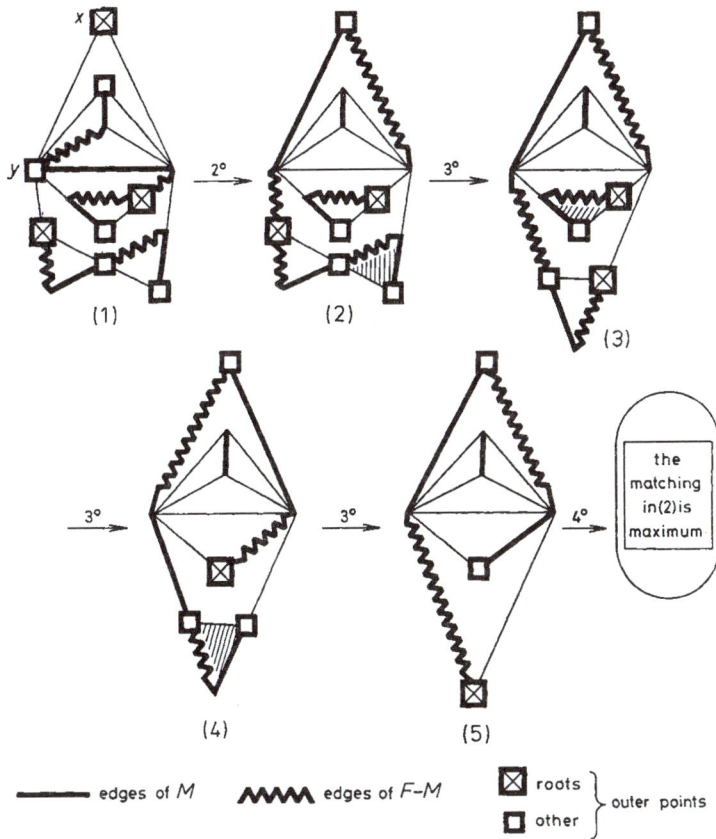

FIGURE 61

1° If there is an edge connecting an outer point of F to a point of $V(G)-V(F)$, we enlarge F as in part I.

2° If there is an edge connecting two outer points in *different* components of F, we enlarge M as in part II.

3° If there is an edge connecting two outer points in the *same* component of F, we contract an odd circuit as in part III and try to enlarge the resulting matching of the resulting graph. We can either conclude that M is maximal or enlarge it.

$4°$ If all edges adjacent to outer points connect them to inner points, we conclude that M is a maximum matching.

Remarks. 1. The above algorithm is efficient in the sense that it performs in $O(n^4)$ steps. In fact, let $f(n)$ denote the least upper bound on the number of steps the algorithm needs to check whether a given matching M of an n-point graph is maximum and replace it by a larger one if not. It takes $O(n^2)$ steps to find a maximal forest with properties $(*)$ and $(**)$. (We carry a list of certain outer points. At each step we check whether or not the first point on the list is adjacent to any point outside F. If so, we enlarge F, and put the new outer point on the end of the list. If not, we cancel this outer point from the list. One such step needs the inspection of $O(n)$ adjacencies. At each step either $V(F)$ or the number of canceled outer points increases, so the number of steps is $O(n)$.) Then we have to carry out one of the steps $2°$, $3°$, $4°$. Since this takes, in the worst case when step $3°$ must be carries out, at most $f(n-2)+O(n^2)$ steps, we get that

$$f(n) = f(n-2) + O(n^2),$$

whence $f(n) = O(n^3)$. Since the size of the matching is increased at most $O(n)$ times, the whole algorithm finishes in no more than $O(n^4)$ steps.

2. The following observation is very significant: If we carry out step $3°$, the forest F is mapped onto a forest of the same kind for the resulting graph. Using this one can improve the upper bound on the length of the algorithm to $O(n^3)$. The same observation and a more careful analysis of the algorithm would enable us to avoid the use of 7.33 (and through this, the use of the Gallai–Edmonds Structure Theorem). We also remark that the sets A_G, C_G, D_G can be determined and several previous results (e.g. 7.26, 7.27, 7.31, 7.32) can be deduced from the algorithm. The reader may find it interesting and instructive to work out the details.

(b) See Fig. 61 (cf. also 7.32f). [J. Edmonds, *Canad. J. Math.* **17** (1965) 449–467.]

35. Suppose that G has no 1-factors. Consider the sets A, C, D as in 7.32. Let $G_1,\ldots,G_t$ be components of $G[D]$, then by 7.32 they are factor-critical. 7.31 implies (it is also easy to see directly) that if $|V(G_i)| > 1$, then G_i contains an odd circuit. Hence, at most one of $G_1,\ldots,G_t$ has more than one point. Let, say,

$$|V(G_1)| \geq 1, \quad |V(G_2)| = |V(G_3)| = \ldots = |V(G_t)| = 1.$$

Now there are exactly k edges joining G_i to A for $i=2,\ldots,t$. Since each point in A is adjacent to at most k of these, we get

$$k(t-1) \leq k \cdot |A|$$

or, equivalently,

$$t \le |A| + 1.$$

Now since

$$t = c_1(G - A) \equiv |V(G - A)| \equiv |A| \pmod 2,$$

this implies that

$$t \le |A|, \quad \nu(G) = \frac{1}{2}\{|V(G)| - t + |A|\} \ge \frac{1}{2}|V(G)|,$$

whence G has a 1-factor, a contradiction. [D. R. Fulkerson–A. J. Hoffman–M. H. McAndrew, *Canad. J. Math.* **17** (1965) 166–177.]

36. (a) Let x be as in the hint, and (x,y) any edge incident with x. y must be covered by some edge e of F, otherwise $F + (x,y)$ would be a larger matching. Now $F - e + (x,y)$ is a maximum matching containing (x,y).

(b) We use induction on $|E(G)|$. By 7.24b we know that G has two distinct 1-factors F_1, F_2. Let C be a circuit in $F_1 \cup F_2$. If some point of C is of degree 2 we are done, since then the two edges adjacent to it belong to the 1-factors F_1 and F_2, respectively.

So suppose that each point of C has degree at least 3. We claim that $G-e$ is 2-connected for some $e \in E(C)$. This will settle the problem; for $G-e$ has a 1-factor (e belongs to exactly one of F_1 and F_2) and so by the induction hypothesis, there is a point x such that each edge of $G-e$ adjacent to x, belongs to some 1-factor of $G-e$. Since e belongs to a 1-factor, x will have the same property in G.

Suppose, indirectly, that $G-e$ has a cut-point x_e for each $e \in E(C)$. Clearly $x_e \in V(C)$. Let e be chosen in such a way that x_e is as close to e on C as possible. Obviously, x_e is not an endpoint of e.

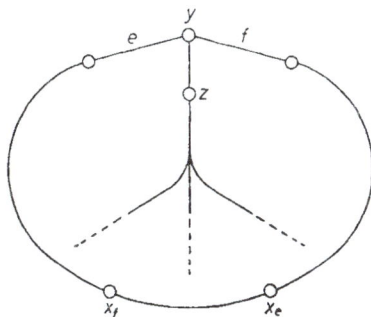

FIGURE 62

Let f be the edge of C adjacent to e on the shorter (e, x_e)-arc of C (Fig. 62). Then by the choice of e, x_f must lie on the other (e, x_e)-arc. Now the common point y of e and f is adjacent to a third edge (y, z). Since $G-y$ is connected, z is joined to some point of $C-y$ by a path P. Now it is seen that no matter where P hits C, we get a contradiction to the fact that both $\{e, x_e\}$ and $\{f, x_f\}$ separate G. [J. Zaks, *Combin. Structures Appl.*, Gordon and Breach, 1970, 481–488.]

37. I. Let w be any 2-matching and choose an independent set $X \subseteq V(G)$. Set

$$\overline{w}(x) = \sum_{\substack{e \in E(G) \\ e \ni x}} w(e), \quad \delta_X = \sum_{x \in X} (2 - \overline{w}(x)).$$

The total w-weight of edges going from X to $\Gamma(X)$ is

$$2|X| - \delta_X,$$

on the other hand, it is $\leq 2|\Gamma(X)|$, whence

$$\delta_X \geq 2|X| - 2|\Gamma(X)|.$$

Thus,

$$2|w| = \sum_{x \in V(G)} \overline{w}(x) = 2|V(G)| - \sum_{x \in V(G)} (2 - \overline{w}(x)) \leq$$

$$\leq 2|V(G)| - \sum_{x \in X} (2 - \overline{w}(x)) = 2|V(G)| - \delta_X \leq 2(|V(G)| - |X| + |\Gamma(X)|).$$

This proves that the maximum size of a 2-matching is

$$\leq |V(G)| - \max_{X \text{ indep}} (|X| - |\Gamma(X)|).$$

II. First we remark that

$$\max_{X \text{ indep}} \{|X| - |\Gamma(X)|\} = \max_{\text{all } X} \{|X| - |\Gamma(X)|\}.$$

In fact, for an arbitrary $X \subseteq V(G)$ define $X' = X - \Gamma(X)$. Then X' is independent, and $\Gamma(X') \subseteq \Gamma(X) - X$. Hence

$$|X| - |\Gamma(X)| = |X'| - |\Gamma(X) - X| \leq |X'| - |\Gamma(X')|,$$

showing that the maximum of $|X| - |\Gamma(X)|$ is attained by some independent set X.

Let G_0 be the bipartite graph defined in the hint. Observe that for $X \subseteq V(G)$, $|\Gamma_{G_0}(X)| = |\Gamma(X)|$. Therefore, we have

$$\delta = \max_{X \subseteq V(G)} \{|X| - |\Gamma_{G_0}(X)|\} = \max_{X \subseteq V(G)} \{|X| - |\Gamma(X)|\}.$$

By 7.5, G_0 has a set F of $|V(G)| - \delta$ independent edges. For $e = (u, v) \in E(G)$ let $w_0(e)$ denote the number of edges of F among (v', u') and (u', v'); so $w_0(e) = 0$ 1 or 2. Moreover, for any fixed v, there is at most one edge of F leaving v and at most another one leaving v'. Hence

$$\overline{w}_0(v) = \sum_{\substack{e \in E(G) \\ e \ni v}} w_0(e) \leq 2,$$

i.e. w_0 is a 2-matching. Moreover,

$$|w_0| = \sum_{e \in E(G)} w_0(e) = |F| = |V(G)| - \delta,$$

whence

$$\max_{w \text{ a 2-matching}} |w| \geq |w_0| = |V(G)| - \delta = |V(G)| - \max_{X \subseteq V(G)} \{|X| - |\Gamma(X)|\}.$$

[cf. W. T. Tutte, *Proc. Amer. Math. Soc.* **4** (1953) 922–931.]

38. I. Observe that a 2-matching of size $|V(G)|$ is necessarily maximum. Therefore, if G has a 1-factor F, then taking the edges of F with weight 2, we get a maximum 2-matching containing the maximum 1-matching F.

II. Now let G be factor-critical and suppose that $|V(G)| > 1$, Let F be any maximum matching of G. The solution of 7.31 shows that there exists an odd circuit P_0 such that F has $\frac{1}{2}(|V(P_0)|-1)$ edges on P_0; i.e. $F - E(P_0)$ is a 1-factor of $G - V(P_0)$. Set

$$w(e) = \begin{cases} 1 & \text{if } e \in E(P_0), \\ 2 & \text{if } e \in F - E(P_0), \\ 1 & \text{otherwise,} \end{cases}$$

then w is a 2-matching of size $|V(G)|$, containing F.

III. Now suppose that G is not factor-critical and has no 1-factor. Consider the Gallai–Edmonds decomposition of G (7.32). Let $G_1, \ldots, G_t$ be the components of $G[D_G]$; let, say,

$$|V(G_1)| = \ldots = |V(G_k)| = 1 < |V(G_{k+1})|, \ldots, V(G_t)|,$$
$$V(G_i) = \{y_i\}, \quad (i = 1, \ldots, k), \quad Y = \{y_1, \ldots, y_k\}.$$

Also set, as before

$$\delta = \max_{X \subseteq A(G)} \{|X| - |\Gamma(X)|\}.$$

Since, therefore,

$$|\Gamma(X)| \geq |X| - \delta$$

holds for every $X \subseteq Y$, we get from 7.5, applied with the bipartite graph formed by the edges incident with Y, that there are $|Y| - \delta = k - \delta$ independent edges joining Y to $\Gamma(Y) \subseteq A$. We may assume that $y_1, \ldots, y_{k-\delta}$ can be matched with $k - \delta$ points of A.

By 7.23, there exists a *maximum* matching F of G covering $y_1, \ldots, y_{k-\delta}$. We define a maximum 2-matching w containing F as follows: let $G_{k+1}, \ldots, G_m$ be joined to A by some edge of F ($m \leq t$). By 7.32, $F \cap E(G_i)$ is a maximum matching in G_i ($i = m+1, \ldots, t$), and hence, as in part II of the proof, we can define a 2-

matching w_i of G_i of size $|V(G_i)|$ containing $F \cap E(G_i)$, for $i = m+1, \ldots, t$. Now, for each $e \in E(G)$, let

$$w(e) = \begin{cases} w_i(e), & \text{if } e \in E(G_{m+1}) \cup \ldots \cup E(G_t), \\ 2, & \text{if } e \in F - E(G_{m+1}) - \ldots - E(G_t), \\ 0, & \text{otherwise.} \end{cases}$$

Then, obviously, w is a 2-matching which contains F. We have to show w is maximal. Observe that exactly two edges of w contain each point except $y_{k-\delta+1}, \ldots, y_k$, which are contained in no edge of w. Thus

$$w = |V(G)| - \delta,$$

which shows by 7.37 that w is a maximum 2-matching.

39. Let L be an Euler trail of G. Consider every second edge of L. Since the total number of edges is even, this is possible. Then the considered edges form a d-factor.

40. Let $\overrightarrow{G}$ be a pseudosymmetric orientation of G (see 5.13), set $V(G) = \{v_1, \ldots, v_n\}$ and define a bipartite graph G_0 by setting

$$V(G_0) = \{v_1, \ldots, v_n, \ v_1', \ldots, v_n'\},$$
$$E(G_0) = \{(v_i, v_j') : (v_i, v_j) \in E(\overrightarrow{G})\}.$$

Observe that G_0 is a d-regular bipartite graph and G arises from it by identifying v_i and v_i' for each $1 \le i \le n$.

Let us color the edges of G_0 by d colors so that each color forms a 1-factor (this is possible by 7.10). Identifying v_i and v_i', this yields a d-coloration of the edges of G such that each color forms a 2-factor. (Note the difference between this and the construction in the proof of 7.37; there two edges of the bipartite graph were mapped onto the same edge of G, therefore we could only conclude that a matching of the bipartite graph yields a 2-*matching* of the graph; here no two edges are mapped onto the same edge, therefore a 1-factor of the bipartite graph yields a 2-*factor* of the graph.)

41. First suppose that all degrees are even. If the number of edges is even, the same proof as in 7.39 gives a desired subgraph; i.e. consider an Euler trail and take every second edge of it. On the other hand, if such a subgraph F exists, then

$$2|E(F)| = \sum_{x \in V(G)} d_F(x) = \sum_{x \in V(G)} \frac{d_G(x)}{2} = |E(G)|,$$

whence $|E(G)|$ is even.

Now suppose that there are points with odd degree. Let v be a new point and join v to all points of G of odd degree. Adding a loop at v if necessary, we thereby construct a graph G' with even degrees and with an even number of edges. Thus, again by 7.39, G' has a spanning subgraph F' with degrees

$$d_{F'}(x) = \frac{1}{2}d_{G'}(x) \qquad (x \in V(G')).$$

Now set $F = F' - v$. Then for any $x \in V(G)$,

$$d_F(x) = d_{F'}(x) = \frac{1}{2}d_{G'}(x) = \frac{1}{2}d(x) \qquad \text{if } d(x) \text{ is even,}$$

$$d_F(x) = d_{F'}(x) = \frac{1}{2}d_{G'}(x) = \frac{d(x)+1}{2} \qquad \text{if } d(x) \text{ is odd and } (x,v) \notin E(F),$$

$$d_F(x) = d_{F'}(x) - 1 = \frac{d(x)-1}{2} \qquad \text{if } d(x) \text{ is odd and } (x,v) \in E(F).$$

So all connected graphs have this property except Eulerian graphs with an odd number of edges.

42. (a) Since a graph with odd degrees has an even number of points, G must have an even number of points.

Conversely, suppose that $|V(G)|$ is even, say $V(G) = \{x_1, \ldots, x_{2m}\}$. Let P_i be a path connecting x_i to x_{i+m} $(i = 1, \ldots, m)$, and set

$$F = \sum_{i=1}^{m} E(P_i) \quad (\text{mod } 2).$$

Then for each $x \in V(G)$,

$$d_F(x) \equiv \sum_{i=1}^{m} d_{P_i}(x) \equiv 1 \quad (\text{mod } 2),$$

and F is a spanning subgraph of the type desired.

(b) Let $x \in V(G)$, $d_G(x) \geq 4$. Let us consider two edges (x,y), (x,z) of G such that $G - (x,y) - (x,z)$ is connected. (There exist two such edges, for let y, z lie in distinct components of $G - x$ if $G - x$ is not connected. Then $G - (x,y) - (x,z)$ is connected, if $G - x$ is connected, because $d_G(x) \geq 3$ and also if $G - x$ is disconnected since each component of $G - x$ is joined to x by at least two edges.) Take a new point x' and connect it to x, y and z. Contracting the edge (x,x') we recover G.

Repeating the above procedure we get, in a finite number of steps, a 3-regular graph G', for the sum

$$\sum_{x \in V(G)} (d_G(x) - 3)$$

decreases at each step. G is a contraction of G' and G' is 2-edge-connected (and thus, as remarked in 7.29a, 2-connected).

Hence G' contains a 1-factor F by 7.29a. Then $F' = E(G') - F$ is a 2-factor. Contracting the new edges to get G from G', F' is mapped onto a subgraph of G with positive even degrees.

43. (a) Replace each point x by a set M_x of $f(x)$ independent points. If $(x, y) \in E(G)$, then each point of M_x is joined to each point of M_y, otherwise no edge connects M_x to M_y. It is straightforward to check that the resulting graph has a 1-factor iff G has an f-matching.

(b) Subdivide each edge by two points and define $f(x) = 1$ at the new points. Then the obtained graph G_0 has an f-factor iff G has an f-factor. Each edge of G_0 has at least one endpoint x with $f(x) = 1$; therefore, G_0 has an f-factor iff it has an f-matching. Now if we replace each point x of G_0 by $f(x)$ independent points as in (a), we get a graph G' which has a 1-factor iff G has an f-factor. [W. T. Tutte, *Canad. J. Math.* **6** (1954) 347–352.]

44. Let $\overrightarrow{G}'$ be the digraph defined in the hint; i.e. set $V' = \{x' : x \in V(G)\}$, $V'' = \{x'' : x \in V(G)\}$, $V(\overrightarrow{G}') = V' \cup V''$ and $E(\overrightarrow{G}') = \{(x', y'') : (x, y) \in E(G)\}$. Let G' arise by forgetting about the orientations. Define

$$h(y) = \begin{cases} g(x) & \text{if } y = x', \\ f(x) & \text{if } y = x'' \end{cases} \quad (x \in V(G)).$$

Obviously, G has a desired factor if and only if G' has an h-factor. By 7.16, this is so if and only if

$$\sum_{x \in V'} h(x) = \sum_{y \in V''} h(y)$$

and for every $X \subseteq V'$, $Y \subseteq V''$,

$$\sum_{x \in X} h(x) \leq \sum_{y \in Y} h(y) + m_{G'}(X, V'' - Y).$$

In other words, if and only if

$$\sum_{x \in V(G)} f(x) = \sum_{x \in V(G)} g(x)$$

and for all $X, Y \subseteq V(G)$,

$$\sum_{x \in X} g(x) \leq \sum_{y \in Y} f(y) + m_G(X, V(G) - Y),$$

where $m_G(X, Z)$ denotes the number of edges with tail in X and head in Z.

45. Following the hint, let G_1 be the graph obtained by subdividing the edges of G and extend f to $V(G_1)$ as described. Observe that G has a desired orientation if and only if G_1 has an f-factor.

G_1 is, obviously, bipartite with color classes $V(G)$ and $V(G_1) - V(G)$. Thus by 7.16, G_1 has an f-factor if and only if

$$(1) \qquad\qquad \sum_{x \in V(G)} f(x) = |E(G)|$$

and
(2) for each $X \subseteq V(G)$, $Y \subseteq E(G)$ the number if incidences between X and Y is not less than

$$\sum_{x \in X} f(x) - |E(G) - Y|.$$

Obviously, it suffices to require this for the special case when there is no incidence between X and Y, but Y contains all edges not incident with X. Then (2′) says that

$(2')$ at least $\displaystyle\sum_{x \in X} f(x)$ edges are incident with X, for every $X \subseteq V(G)$.

(1) and (2′) are necessary and sufficient conditions for G being orientable in the desired way.

46. (a) We use induction on the number of points. Suppose first that G is a circuit, $V(G) = \{x_1, \ldots, x_n\}$ and $E(G) = \{(x_1, x_2), \ldots, (x_n, x_1)\}$. If $g(x) > 0$ for every $x_i \in V(G)$, we can take $F = (V(G), \emptyset)$. So suppose that, e.g., $g(x_1) = 0$. Then put (x_1, x_2) into F. Going along the cycle, put (x_i, x_{i+1}) into F whenever $(x_{i-1}, x_i) \in F$ and $g(x_i) = 1$ or $(x_{i-1}, x_i) \notin F$ and $g(x_i) = 0$. Deciding this way about every edge we end up with 1 or 2 F-edges at x_1, thus $d_F(x_1) \neq g(x_1) = 0$. The other points satisfy the requirement, because F has been constructed that way.

Now suppose that G is not a single circuit. Then we find a point $x \in V(G)$ such that $G_1 = G - x$ is connected and has a circuit. Define g_1 on $V(G_1)$ as follows: if $g(x) > 0$, we set $g_1 = g$; if $g(x) = 0$, we select a neighbor y of x and let

$$g_1(z) = \begin{cases} g(z) & \text{if } z \neq y, \\ g(y) - 1 & \text{if } z = y. \end{cases}$$

By the induction hypothesis, G_1 has a spanning subgraph F_1 with degree different from G_1. In case $g(x) > 0$, $F = F_1$ is a desired subgraph of G. If $g(x) = 0$, $F = F_1 \cup \{(x, y)\}$ has the required property.

(b) I. Suppose that both (i) and (ii) hold; let F be a spanning subgraph with degrees $\neq g(x)$ and $\overrightarrow{G}$ an orientation with indegrees $g(x)$. Reverse the orientation of the edges of F. Since the resulting digraph is acyclic, it has a point x with indegree 0. However, reversing the edge of F again, x will have indegree $g(x)$, whence it is adjacent to $g(x)$ edges of F, a contradiction.

II. Let G be any tree, we show either (i) or (ii) is satisfied. Let x be a point of degree 1. If $g(x) < 0$ or $g(x) > 1$, then (ii) is satisfied. For add a loop to x, apply (a), and then remove the loop. We get a spanning subgraph having degree $\neq g(y)$ at each $y \neq x$ and of course, its degree is different from $g(x)$ at x.

Thus we may assume that $g(x)=0$ or $g(x)=1$. Suppose that, e.g., $g(x)=1$ (the other case follows similarly). If $G-x$ has a spanning subgraph F with degrees $\neq g(y)$ for each $y\in V(G)-\{x\}$, the same subgraph (with x as an isolated point) satisfies the requirements for G; thus (b) holds again. Suppose, therefore, that no such F exists. Then, by the induction hypothesis, $G-x$ has an orientation with indegrees $g(y)$ for each $y\in V(G)-x$. Orient the edge adjacent to x toward x. Then the resulting orientation shows that (i) holds. [L. Lovász, *Periodica Math. Hung.* **4** (1973) 121–123.]

47. The necessity is obvious as a tree on n points has $n-1$ edges. Now suppose that $d_1+\ldots+d_n=2n-2$. We use induction on n. The case $n\leq 2$ is trivial, so suppose that $n\geq 3$. Then $d_1=1$, since $d_1\geq 2$ would imply $d_1+\ldots+d_n\geq 2n>2n-2$. Also $d_n>1$ as $d_n=1$ would imply $d_1+\ldots+d_n=n<2n-2$. Since

$$d_2+\ldots+d_{n-1}+(d_n-1)=2n-4=2(n-1)-2,$$

there is a tree T with degrees $d_2,\ldots,d_{n-1},d_n-1$. Add a new point and join it to the point of T with degree d_n-1; then the resulting tree has the desired degrees [see B].

48. Again the necessity is obvious. We prove the sufficiency using induction on $\sum_{i=1}^{n} d_i$. We distinguish two cases.

I. $d_{n-2}<d_n$. Then d_n-1 is a largest element in $d_1,d_2,\ldots,d_{n-2},d_{n-1}-1,d_n-1$ and so what we have to verify is

$(1')$ $\qquad d_1+\ldots+d_{n-2}+(d_{n-1}-1)+(d_n-1)\equiv 0\pmod 2,$

$(2')$ $\qquad d_1+\ldots+d_{n-2}+d_{n-1}-1\geq d_n-1,$

which are obvious from (1) and (2).

II. $d_{n-2}=d_n$. Then $d_{n-1}=d_n$. Obviously, $(1')$ holds and

$(2'')$ $\qquad d_1+\ldots+d_{n-3}+(d_{n-1}-1)+(d_n-1)\geq d_{n-2},$

provided $d_{n-2}\geq 2$. If $d_{n-2}=1$, then the left-hand side of $(2'')$ is odd by (1), and is therefore at least 1 trivially.

Thus $d_1,\ldots,d_{n-2},d_{n-1}-1,d_n-1$ satisfy the assumptions of the problem and by induction, there exists a graph without loops on n points realizing this sequence. Joining the points with degree $d_{n-1}-1$ and d_n-1 by a new edge, we get a graph with degrees $d_1,\ldots,d_n$ [see B].

49. Let K be the simple digraph without loops defined by $V(K)=\{v_1,\ldots,v_n\}$, $E(K)=\{(v_i,v_j):i\neq j\}$. The existence of a digraph with the desired properties is equivalent to the existence of a subgraph of K with given outdegrees and indegrees. By 7.44, this is equivalent to (1) and (2), since the number of edges of K connecting $\{v_i;i\in I\}$ to $\{v_j;j\notin J\}$ is $|I|(n-|J|)-|I-J|$. (Remember here that K has no loops!)

Now if $f_1 \leq \ldots \leq f_n$ and $g_1 \leq \ldots \leq g_n$, then for every I, J with $|I| = k$, $|J| = l$, we have

$$\sum_I f_i \leq \sum_{i=n-k+1}^n f_i,$$

as well as

$$\sum_{j \in J} g_j + k(n-l) - |I - J| \geq \sum_{j=1}^l g_j + k(n-l) - \min(k, n-l).$$

Thus (2) can be replaced by

(2')
$$\sum_{i=n-k+1}^n f_i \leq \sum_{j=1}^l g_j + (n-l)k - \min(k, n-l).$$

For (2') is, of course, a special case of (2) (with $I = \{n-k+1, \ldots, n\}$ and $J = \{1, \ldots, l\}$), so it is necessary; and moreover as we have seen (2') implies (2) under the assumption $f_1 \leq \ldots \leq f_n$, $g_1 \leq \ldots \leq g_n$ [see B].

50. Suppose first that there exists a simple graph G with degrees $d_1, \ldots, d_n$ and let v_i be the point with degree d_i. Also set $p = n - d_n$. We claim G can be chosen so that v_n is adjacent to each of $v_1, \ldots, v_{n-1}$ (but to no other point). For let us choose G among all graphs with $d_G(v_i) = d_i$ such that v_n is adjacent to as many points in $\{v_p, \ldots, v_{n-1}\}$ as possible. Suppose that v_n is not adjacent to v_k, for some $p \leq k \leq n-1$. Then v_n must be adjacent to some v_ν, $1 \leq \nu \leq p-1$, since $d_n = n - p$. Since $d_\nu \leq d_k$, there must be a point v_m, $m \neq k$, ν, adjacent to v_k but not to v_ν. Now remove the edges (v_k, v_m) and (v_ν, v_n), but add (v_k, v_n) and (v_ν, v_m). This way a graph with the same degrees is obtained in which v_n is adjacent to more points in $\{v_p, \ldots, v_{n-1}\}$ than in G, a contradiction.

Thus we have a graph G with degrees $d_1, \ldots, d_n$ in which all of $v_p, \ldots, v_{n-1}$ are adjacent to v_n. Removing v_n we obtain a graph with degrees $d_1, \ldots, d_{p-1}, d_p - 1, d_{p+1} - 1, \ldots, d_{n-1} - 1$.

Conversely, suppose that the numbers d'_k are degrees of a simple graph G'. We can then add v_n trivially. [V. Havel, *Čas. Pest. Mat.* **80** (1955) 477–480.]

51. (a) Following the hint, choose H with the given out- and indegrees such that the number of 2-cycles in H is maximal. Let $\tilde{H}$ denote the subgraph formed by those edges $(x, y) \in E(H)$ for which $(y, x) \notin E(H)$.

Now $\tilde{H}$ contains no even cycle. For suppose that $(x_1, \ldots, x_{2k})$ is an even cycle of $\tilde{H}$. Then remove $(x_2, x_3), \ldots, (x_{2k}, x_1)$; but add $(x_2, x_1), (x_4, x_3), \ldots, (x_{2k}, x_{2k-1})$. We thereby get a graph with the same out- and indegrees but with more 2-cycles.

The same argument shows that $\tilde{H}$ cannot have any closed directed trail with an even number of edges. Since $\tilde{H}$ has equal indegrees and outdegrees, it is the union of edge-disjoint cycles (cf. 5.6). By the above, these are odd and no two can have a point in common, because they would then form a trail with an even number of edges.

Suppose that there are two cycles C_1, C_2 in the above cycle-decomposition of $\tilde{H}$, $C_1 = (x_0, \ldots, x_{2k})$ and $C_2 = (y_0, \ldots, y_{2l})$. If $(x_0, y_0) \in E(H)$, then $(x_0, y_0) \notin E(\tilde{H})$ by the above, and so $(y_0, x_0) \in E(H)$ and $C_1 \cup C_2 \cup \{(x_0, y_0), (y_0, x_0)\}$ is a trail with an even number of edges, whence we get a contradiction as above. So $(x_0, y_0), (y_0, x_0) \notin E(H)$. Then remove

$$(x_0, x_1), (x_2, x_3), \ldots, (x_{2k}, x_0), (y_0, y_1), (y_2, y_3), \ldots, (y_{2l}, y_0),$$

but add

$$(x_2, x_1), \ldots, (x_{2k}, x_{2k-1}), (y_2, y_1), \ldots, (y_{2l}, y_{2l-1}), (x_0, y_0), (y_0, x_0).$$

We again get a simple digraph with the same degrees and more 2-cycles.

Thus $\tilde{H}$ contains at most one cycle. If it contains exactly one, then this is odd and $\sum_{i=1}^{n} d_i = |E(H)|$ is odd, a contradiction of the assumption. Thus $\tilde{H}$ has no edge; i.e. H consists of pairs of opposite edges. Replacing each such edge by a single undirected edge we get a simple graph with degrees $d_1, \ldots, d_n$. [cf. D. R. Fulkerson–A. J. Hoffman–M. H. McAndrew, *Canad. J. Math.* **17** (1965) 166–177.]

(b) By (a), the existence of a simple graph with degrees $d_1, \ldots, d_n$ is equivalent to the existence of a simple digraph H on $\{v_1, \ldots, v_n\}$ such that $d_H^+(v_i) = d_H^-(v_i) = d_i$ $(i = 1, \ldots, n)$ (under the necessary assumption that $d_1 + \ldots + d_n$ is even). By 7.49, such a digraph exists iff for every $k, l \in \{1, \ldots, n\}$,

$$(2') \qquad \sum_{i=n-k+1}^{n} d_i \leq \sum_{j=1}^{l} d_j + k(n-l) - \min(k, n-l).$$

It suffices to require $(2')$ for the case $k + l \leq n$. In fact, if $k + l > n$, then $(2')$ is equivalent to

$$(2'') \qquad \sum_{i=l+1}^{n} d_i = \sum_{j=1}^{n-k} d_j + k(n-l) - \min(k, n-l),$$

which can be also derived from $(2')$ with $n-l$ and $n-k$ playing the roles of k and l, respectively.

Now if $k + l \leq n$, then

$$\sum_{j=1}^{l} d_j + k(n-l) \geq \sum_{j=1}^{n-k} \min(d_j, k) + k^2.$$

So if

(2)
$$\sum_{i=n-k+1}^{n} d_i \le \sum_{j=1}^{n-k} \min(d_j, k) + k^2 - k,$$

then (2′) is satisfied. On the other hand, (2) is a special case of (2′) when we take $m = \max\{j : d_j \le k\}$ and $l = \min\{n - k, m\}$. So (2) is a necessary condition [P. Erdős, T. Gallai; see B].

52. Suppose first that $d_1, \ldots, d_n$ are the degrees of a connected graph. Then 7.51 implies (1), while (2) and (3) are trivial.

Suppose, conversely, that $d_1, \ldots, d_n$ satisfy (1), (2) and (3). By (1), there is a simple graph G with degrees $d_1, \ldots, d_n$. Choose G so that it has the least possible number of components.

We claim that G is connected. Suppose not, then by (3), one of its components contains a circuit. Let G_1 be the component and (x, y) an edge of G_1 on a circuit. Let G_2 be any other component and (u, v) any edge of G_2 (G_2 has an edge since it is not an isolated point by (2)). Then $G - \{(x, y), (u, v)\} + \{(x, u), (y, v)\}$ is a graph on the same set, with the same degrees and with fewer components, a contradiction [P. Erdős, T. Gallai; B].

53. Let G be a graph with a 1-factor F and with points $v_1, \ldots, v_n$ such that $d(v_i) = d_i$. Then, obviously $G - F$ has degree sequence $d_1 - 1, \ldots, d_n - 1$. To prove the converse, let G be any simple graph on $V = \{v_1, \ldots, v_n\}$ with $d_G(v_i) = d_i$ and G' another simple graph on V with $d_{G'}(v_i) = d_i - 1$. Moreover, among all such pairs of graphs, choose a pair with $|E(G) - E(G')|$ minimal. We claim $G' \subseteq G$. Suppose not, then there are points incident with edges of $E(G') - E(G)$. Let x be one adjacent to the maximum number k of edges of $E(G') - E(G)$. Then x is contained in $k + 1$ edges if $E(G) - E(G')$. Let $(x, y_1), \ldots, (x, y_{k+1}) \in E(G) - E(G')$ and $(x, z) \in E(G') - E(G)$.

Since $d_G(z) > d_{G'}(z)$, there is a point v such that $(z, v) \in E(G) - E(G')$. Choose an $1 \le i \le k + 1$ such that $y_i \ne v$. Then $(v, y_i) \in E(G)$. For if $(v, y_i) \notin E(G)$, then removing (x, y_i) and (z, v) from G, but adding (x, z) and (v, y_i), we get a graph with the same degrees as G, but with more edges in common with G'. Similarly, $(v, y_i) \notin E(G')$, for if $(v, y_i) \in E(G')$ we could remove (v, y_i) and (x, z) from G', but add (x, y_i) and (z, v) to G' and get a graph with the same degrees as G', but having more edges in common with G. Thus, $(v, y_i) \in E(G) - E(G')$.

Now, if $v \ne y_1, \ldots, y_{k+1}$, then v is adjacent to at least $k + 2$ edges of $E(G) - E(G')$ $((v, z), (v, y_1), \ldots, (v, y_{k+1}))$ so by hypothesis it is adjacent to at least $k + 1$ edges of $E(G') - E(G)$, a contradiction of the choice of x. If $v = y_1$ (say) then, again, it is adjacent to at least $k + 2$ edges of $E(G) - E(G')$ (in this case $(v, z), (v, x), (v, y_2), \ldots, (v, y_{k+1})$ are these edges), and we get a contradiction as before. [S. Kundu, *Discrete Math.* **6** (1973) 367–376; L. Lovász, *Periodica Math. Hung.* **5** (1974) 149–151.]

§ 8. Independent sets of points

1. Let S be a maximal independent set. Then any point $x \in V(G) - S$ is joined to a point in S. Since a point of S can only be joined to at most d points of $V(G) - S$, we have

$$|V(G)| - |S| = |V(G) - S| \leq d|S|,$$

i.e.

$$|S| \geq \frac{1}{d+1}|V(G)|.$$

2. Define $S_1, S_2, \ldots$ as in the hint. Then

$$|S_i| \leq \alpha(G[S_{i+1} \cup S_i]) \leq \alpha(G - S_1 - \ldots - S_{i-1}) = |S_i|,$$

whence

$$\tau(G[S_{i+1} \cup S_i]) = |S_{i+1}|.$$

Hence by König's theorem 7.2, we have $|S_{i+1}|$ independent edges in $G[S_i \cup S_{i+1}]$, which is a matching F_i of S_{i+1} into S_i.

Now $F_1 \cup F_2 \cup \ldots$ consists of $|S_2|$ disjoint paths, which cover all points except $|S_1| - |S_2|$ of the points of S_1. Taking these as one-point paths, we get $|S_1|$ paths covering $V(G)$.

3. If G consists of independent points, the assertion is obvious. So suppose that this is not the case. We use induction on $\alpha(G)$.

Let P be a longest path in G and let x be an endpoint of P. If x has degree 1, remove x and its neighbor y. Observe that $\alpha(G - x - y) < \alpha(G)$, since any independent set of $G - x - y$ can be completed with x. Hence, $G - x - y$ can be covered by $\alpha(G) - 1$ disjoint circuits, edges and points, and adding the edge xy, we get a desired cover of G.

So suppose that x has degree at least 2. Each point adjacent to x is on P, because P is maximal. Let y be the farthest point on P adjacent to x and let C be the circuit formed by the (x, y)-piece of P and the edge (x, y). Observe again that $\alpha(G - V(C)) < \alpha(G)$, because C contains all neighbors of x. Hence $G - V(C)$ can be covered by $\alpha(G) - 1$ disjoint circuits, edges and points, and adding C, we get a desired cover of G again. [L. Pósa, *MTA Mat. Kut. Int. Közl.* **8** (1963) 355–361.]

4. We use induction on $|V(G)|$. For $|V(G)| = 1$ the assertion is trivial.

Consider a set S such that there are disjoint paths $P_1, \ldots, P_{|S|}$ starting from points of S and covering all points; suppose, moreover, that S is minimal. What we have to show is $|S| \leq \alpha(G)$. Suppose that this is not the case, then, obviously,

S is not independent, i.e. there are x_1, $x_2 \in S$ with $(x_1, x_2) \in E(G)$. Let, say, P_i be the path starting from x_i. Then P_1 is not a single point, since then the paths

$$P_2' = P_2 + (x_1, x_2), \; P_3, \; \ldots, \; P_{|S|}$$

would cover every point, which is impossible, because they are disjoint and start from the points of $S - \{x_1\}$. Thus, P_1 has a second point z. Consider now $G' = G - x_1$, $S' = S - x_1 \cup \{z\}$ and the paths

$$P_1' = P_1 - x_1, \; P_2, \; P_3, \; \ldots, \; P_{|S|}.$$

They start from the points of S', cover $V(G')$, and they are disjoint. Hence, by the induction hypothesis, there is an $S_0 \subseteq S'$ with $|S_0| \leq \alpha(G') \leq \alpha(G) < |S|$ such that certain disjoint paths, starting from the points of S_0, cover G. Now if $z \in S_0$, then adding x_1 to the path starting from z, we get $|S_0| < |S|$ disjoint paths starting from a proper subset of S and covering G, which is a contradiction. A similar argument works if $z \notin S_0$ but $x_2 \in S_0$. Finally, suppose that x_2, $z \notin S_0$. Then $|S_0| \leq |S| - 2$, thus we may add $\{x_1\}$ as a path to the system to get $|S_0| + 1 < |S|$ disjoint paths, starting from a subset of S and covering all points of G. This is a contradiction again [T. Gallai; see B].

5. (a) If S is a maximal independent set, then each $x \notin S$ must be adjacent to a $y \in S$. By the symmetry assumption, this means that both (x, y) and (y, x) are edges of the digraph.

(b) Let x be a point with indegree 0 (the existence of such a point follows from the fact that the digraph G is acyclic); let T be the set of neighbors of x. Obviously, $G_0 = G - x - T$ is acyclic, so by induction on the number of points it has a (unique) kernel S_0. We claim that $S = \{x\} \cup S_0$ is a kernel of G. It is independent since G_0 contains no neighbor of x; moreover, any $y \in V(G) - S$ is either $\in V(G_0) - S_0$ in which case it can be reached from S_0, or $\in T$, in which case it can be reached from x on an edge of G.

Now if S' is any other kernel of G, then $x \in S'$ (since x cannot be reached from any point) and hence, $T \cap S' = \emptyset$. Let $S_0' = S' \cap V(G_0)$, then S_0' is a kernel of G_0 and hence by the uniqueness of S_0, $S_0' = S_0$. Thus,

$$S' = S_0' \cup \{x\} = S_0 \cup \{x\} = S.$$

(c) Let G_0 be a strongly connected component of G such that no edge of G enters G_0. By 5.3, G_0 is bipartite; let $S_0 \neq \emptyset$ be one of its color classes, and let T be the set of all neighbors of S_0 in $V(G) - V(G_0)$. Using induction on $|V(G)|$ we may assume that $G' = G - V(G_0) - T$ has a kernel S_1. Set

$$S = S_1 \cup S_0.$$

Then S is, obviously, independent. Further let $x \in V(G) - S$. If $x \in V(G') - S_1$, then it can be reached from S_1 through an edge, because S_1 is a kernel of G'. If $x \in T$, it is adjacent to a $y \in S_0$ by the definition of T and the edge joining them goes from y to x by the definition of G_0. Finally, let $x \in V(G_0) - S_0$. Since G_0 is strongly connected and $S_0 \neq \emptyset$, we have an edge (x,z) of G_0 leaving x. Since S_0 has been one of the color classes of G_0, $z \in S_0$. This proves that S is a kernel of G [see B].

6. Let x have the highest outdegree among the points of T. We claim any other point y can be reached from x on a path of length at most 2.

Suppose that y is a point that cannot be reached. Then

(1) the edge joining x and y has tail in y,

(2) whenever (x,z) is an edge, (y,z) is an edge (otherwise, (xzy) would be an (x,y)-path of length at most 2).

Hence, for each edge leaving x, there is an edge leaving y and (1) yields another edge leaving y. This means that the outdegree of y is larger than the outdegree of x, a contradiction.

7. Let $x \in V(G)$ and let T be the set of all points, which can be reached from x on an edge. Set $G' = G - T - x$ and suppose that, by the induction hypothesis, that G' contains an independent set S' such that every point of $G' - S'$ can be reached on a path of length at most 2 from S'. Now, we distinguish two cases:

Case 1. $S' \cup \{x\}$ is independent. Then set $S = S' \cup \{x\}$ and observe that a $y \in V(G') - S'$ can be reached from S' on a path of length at most 2, while a $y \in T$ can be reached from x even on an edge.

Case 2. $S' \cup \{x\}$ is not independent, i.e. there is a $z \in S'$ adjacent to x. Since $z \notin T$, $(z,x) \in E(G)$. Now set $S = S'$. If $y \in V(G') - S'$, it can be reached from S' in at most 2 steps by the definition of S'. If $y \in T$, it can be reached on zxy. [V. Chvátal–L. Lovász, in: *Hypergraph Seminar*, Lecture Notes in Math. 411, Springer (1974) p. 175.]

8. (a) To be precise, we define a *winning move* as a point such that, if a player occupies it, he can win (no matter what his opponent does). E.g. a point with outdegree 0 is a winning move, since there is no response at all. Now it follows from the definition that

(1) *winning moves are independent points*; for if x is a winning move and $(x,y) \in E(G)$ then, by definition, if the first player occupies x, he can win; this happens, in particular, if his opponent occupies y at the next step, so y is not a winning move (here we use the assumption that the graph is acyclic and hence, once y is occupied, it does not matter whether x has been occupied);

(2) *any point x, which is not a winning move is joined to a winning move.* Since by definition, if the first player occupies x, the second player can still win, i.e. he has a response y which is a winning move (again, we use that G is acyclic).

Thus, the set of winning moves is the "dual kernel", i.e. the kernel of the graph obtained by inverting the arrows. So the first player can win by choosing a point of dual kernel at each step.

(b) Assume that the first player cannot win if he starts with x_0; this means that the second player has a response x_1 such that the first player cannot win after this move. In this case, let the first player imagine he is the second player and his opponent has opened the game with x_0; so respond with x_1 and follow the second player's previous strategy. Since x_0 has indegree 0, it will never happen that his opponent makes a move which was illegal in the previous game (the only such move would be occupying x_0). Thus he wins.

(c) Assume that first that for each strong component G_i of G $(i = 1, \ldots, k)$, the second player has a winning strategy. Then he can follow this rule: whenever A moves in a component G_i, he responds in the same component G_i with the move determined by his winning strategy for this particular component. So sooner or later A has no move in this component G_i, and he either loses or has to move to another component G_j. Clearly, they have not played in G_j earlier and therefore B can consider the move of A as an opening move of the game in G_j, and respond according to his strategy in G_j, etc.

Conversely, assume that there are strong components of G, where the first player has a winning strategy. Let G_i be such a component with the property that no such component can be reached from G_i on a path (if for each such component there is another one accessible on a path we can make a closed walk through more than one component, contradicting the definition of component). Let the first player start in G_i according to the winning strategy in G_i. Then he can force his opponent to make a move outside G_i first. However, this move is in a component, where the second player has a winning strategy and so the first player can force his opponent to leave this component first (or to lose), etc.

9. Suppose first that G has a 1-factor $\{e_1, \ldots, e_r\}$. Then the second player can win by following this rule: whenever the first player occupies an endpoint of an e_i, his response is to occupy the other endpoint of the same edge. So the first player has to choose from a new edge of the 1-factor at each step and the response of the second player is always legitimate.

Now suppose that G has no 1-factor. Let $\{e_1, \ldots, e_v\}$ be a maximum matching of G and x, a point not covered by $e_1, \ldots, e_v$. We claim that the first player wins using the following strategy: he opens with x and whenever his opponent occupies an endpoint of an e_i he responds by occupying the other one.

What we have to show is that the second player can never make a move to occupy a point that is not an endpoint of an edge e_i. Suppose that at the jth move, the second player first has a move y_j that is not an endpoint of $e_1, \ldots, e_v$. Let $x, x_1, \ldots, x_{j-1}$ be the previous moves of the first player and $y_1, \ldots, y_{j-1}$, the previous moves of the second player. Then (x_i, y_i) is an edge of the maximum matching $\{e_1, \ldots, e_v\}$; say $(x_i, y_i) = e_i$ $(i = 1, \ldots, j-1)$. Now observe that $\{(x, y_1), \ldots, (x_{j-1}, y_j), e_j, \ldots, e_v\}$ is a larger matching, a contradiction. [cf. W. N. Anderson, *J. Comb. Theory* B **17** (1972) 234–239.]

10. For $k = 1$ the statement is evident. Suppose that

$$|T_1 \cup \ldots \cup T_{k-1}| + |T_1 \cap \ldots \cap T_{k-1}| \geq 2\alpha(G).$$

In fact, we show that

$$|T_1 \cup \ldots \cup T_k| + |T_1 \cap \ldots \cap T_k| \geq |T_1 \cup \ldots \cup T_{k-1}| + |T_1 \cap \ldots \cap T_{k-1}|$$

or, equivalently,

$$|T_1 \cup \ldots \cup T_k - T_1 \cup \ldots \cup T_{k-1}| \geq |T_1 \cap \ldots \cap T_{k-1} - T_1 \cap \ldots \cap T_k|.$$

Now set

$$A = T_1 \cup \ldots \cup T_k - T_1 \cup \ldots \cup T_{k-1},$$
$$B = T_1 \cap \ldots \cap T_{k-1} - T_1 \cap \ldots \cap T_k,$$
$$C = T_k - A = T_k \cap (T_1 \cup \ldots \cup T_{k-1}).$$

Then $|A| + |C| = |T_k| = \alpha(G)$. On the other hand, $B \cup C$ is independent. In fact, suppose that $x, y \in B \cup C$. Obviously, B, C are independent, so we may assume that $x \in B$, $y \in C$. Then $y \in T_1 \cup \ldots \cup T_{k-1}$, say $y \in T_1$. But also $x \in T_1$ and hence, x and y are non-adjacent. Thus $B \cup C$ is independent and hence,

$$|B \cup C| \leq \alpha(G) = |A \cup C|,$$

i.e.

$$|B| \leq |A|.$$

[A. Hajnal; see B.]

11. It suffices to consider the case $k = 1$, since the general case is trivial by induction. Observe that

$$\Gamma(S) \subseteq \Gamma(X) - T_1.$$

Hence

$$|\Gamma(X)| - |\Gamma(S)| \geq |\Gamma(X) \cap T_1| = |T_1| - |S| - |T_1 - X - \Gamma(X)| =$$
$$= \alpha(G) - |S| + |X| - |T_1 \cup X - \Gamma(X)|.$$

Here $T_1 \cup X - \Gamma(X)$ is, obviously, independent, and has, therefore, at most $\alpha(G)$ points. Thus

$$|\Gamma(X)| - |\Gamma(S)| \geq |X| - |S|,$$

which proves the assertion. [LP]

12. Let T be an $(\alpha(G) + 1)$-element independent set in $G - (x, y)$. Since T is not independent in G, we have $x, y \in T$. Now $T - y$ is an $\alpha(G)$-element independent set in G containing x, and $T - x$ is another one missing it.

If x, y do not form a component of G, we can find a z joined to one of them, say to x. Then $G - (x, z)$ contains an $(\alpha(G) + 1)$-element independent set T. Obviously, $x, z \in T$. If $y \in T$, then $T - x$ contains y but not x; if $y \notin T$, then $T - z$ contains x but not y. Finally, if x, y are adjacent, then $y \notin T$ and so $T - x$ misses both x and y.

13. Let G' be the graph obtained from G by replacing x by x_1 and x_2 as in the hint. Since $G' - x_2 \cong G$, we have $\alpha(G') \geq \alpha(G)$. On the other hand, no independent set of G contains both x_1 and x_2 and so, it yields an independent set of G' of the same size. Hence, $\alpha(G') = \alpha(G)$.

Now let $e \in E(G')$. Suppose first that $e = (x_1, x_2)$. Let T be a maximum independent set of G containing x. Then $T - \{x\} \cup \{x_1, x_2\}$ is an $(\alpha(G)+1)$-element independent set in $G - e$.

Suppose that $e = (x_1, y)$ $(y \neq x_2)$. Let T be an $(\alpha(G)+1)$-element independent set in $G - (x, y)$. Then $T - \{x\} \cup \{x_1\}$ is an $(\alpha(G)+1)$-element independent set in $G - (x_1, y)$. Similarly, if $e = (x_2, y)$ or (z, y) $(z, y \neq x_1, x_2)$, we get $\alpha(G-e) > \alpha(G)$. [W. Wessel, *Coll. Math. Soc. J. Bolyai* **4** (1970) 1123–1139.]

14. Suppose that we have a G such that G_0 is an induced subgraph of G and $\alpha(G-e) > \alpha(G)$ for $e \in E(G_0)$. Remove edges of G as long as you do not increase $\alpha(G)$; then you never remove any edge from G_0 and so, the α-critical graph you get contains G_0 as an induced subgraph.

So it suffices to construct such a G. Let $E(G_0) = \{e_1, \ldots, e_m\}$ and $\alpha = \alpha(G_0)$. Let $S_1, \ldots, S_m$ be $(\alpha-1)$-element sets disjoint from each other and $V(G_0)$. Set $V(G) = V(G_0) \cup S_1 \cup \ldots \cup S_m$. Let, moreover, each $x \in S_j$ be joined to all other points except the points of S_j and the endpoints of e_j. Thus, we get a graph G.

Now $\alpha(G) \geq \alpha$, because S_1, together with an endpoint of e_1, form an α-element independent set. On the other hand, if T is an independent set and it contains a point of S_j, then it contains at most S_j and one endpoint of e_j; if $T \subseteq V(G_0)$, then $|T| \leq \alpha(G_0)$. Hence $\alpha(G) = \alpha$.

Moreover, $G - e_j$ contains the $(\alpha+1)$-element independent set $S_j \cup e_j$ $(j = 1, \ldots, m)$.

15. (a) Consider a $(6t+3)$-cycle. This is α-critical. Substitute a K_{r-1} for every third point of it. Then the resulting graph is r-regular, connected and α-critical by 8.13.

(b) It is trivial that the tetrahedron is α-critical and that the octahedron and cube are not.

Using the fact that the automorphism group of the icosahedron is vertex- and edge-transitive, it is easy to verify that its independence number is 3 but this increases to 4 if any edge is removed. Thus the icosahedron is α-critical.

Note that two opposite vertices of the icosahedron, though non-adjacent, never occur in the same maximum independent set. This shows that the assertion of problem 8.12 could not be extended in this direction.

Finally, to show that the dodecahedron is not α-critical note that it contains 8 independent vertices (Fig. 63). Remove an edge and consider an independent set in the remaining graph. This contains at most three vertices of the two faces adjacent to the removed edge and at most 2 vertices of any other face. Each element of the independent set belongs to three faces. Therefore the cardinality of this independent set is no larger than

$$\frac{2 \cdot 3 + 10 \cdot 2}{3} = 8.66,$$

showing that the independence number is still 8.

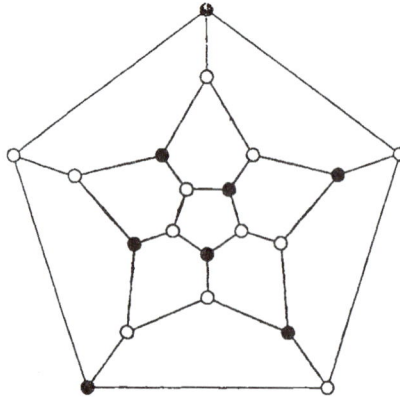

FIGURE 63

16. Let S_1, S_2 be as in the hint. Obviously $x \in S_1 \cap S_2$, $y_i \in S_i$, $y_i \notin S_{3-i}$. Since $S_1 \triangle S_2 \cup \{x\}$ induces a bipartite graph, it contains an independent set T with

$$|T| \geq \frac{1}{2}|S_1 \triangle S_2 \cup \{x\}| = |S_1 - S_2| + \frac{1}{2},$$

i.e. $|T| \geq |S_1 - S_2| + 1$. Now note that $T \cup (S_1 \cup S_2) - \{x\})$ is independent; for let u, $v \in T \cup ((S_1 \cap S_2) - \{x\})$; we show they are non-adjacent. If u, $v \in T$ or u, $v \in S_1 \cap S_2$, we are done. Let, say, $u \in T$, $v \in S_1 \cap S_2$. Since $T \subseteq S_1 \cup S_2$, we have, e.g. $u \in S_1$. So u, v are in S_1 and $v \neq x$, y_1, hence they are non-adjacent. But

$$|T \cup ((S_1 \cap S_2) - \{x\})| = |T| + |S_1 \cap S_2| - 1 \geq$$
$$\geq |S_1 - S_2| + 1 + |S_1 \cap S_2| - 1 = |S_1| > \alpha(G),$$

a contradiction.

17. Let x be a point of the α-critical graph G, adjacent to y_1 and y_2. Let S_i be an $(\alpha(G)+1)$-element independent set in $G - (x,y_i)$. As before, x, $y_i \in S_i$, $y_i \notin S_{3-i}$. Let $G' = G[S_1 \triangle S_2 \cup \{x\}]$. If G' is bipartite, we get a contradiction exactly as above. So G' contains an odd circuit. A minimal odd circuit C is chordless. Since, obviously, $G' - (x,y_1)$ is bipartite with bipartition $\{S_1 - S_2, S_2 - S_1 \cup \{x\}\}$, C must go through (x,y_1). similarly, C goes through (x,y_2). [Andrásfai; L. Beineke–F. Harary–M. D. Plummer; C. Berge; see B, LP.]

18. Following the hint, let (x,y_1), (x,y_2) be two edges and suppose that y_1, y_2 are in different components of $G - S$. Then G contains a chordless (odd) circuit through (x,y_1) and (x,y_2). This has another point z of S because S separates y_1 and y_2 and z is joined to x because S spans a complete graph. Thus C is not chordless, a contradiction.

19. (a) Let T be an independent set in G. If T contains at most one of x_1, x_2, then $T \cap (V(G_1) - \{x\})$ and $T \cap V(G_2)$ are independent sets in G_1 and G_2, respectively, whence $|T| \leq \alpha(G_1) + \alpha(G_2)$. If x_1, $x_2 \in T$, then $(T \cap V(G_1) - \{x\}) \cup \{x\}$ and

$T \cap V(G_2) - \{x_1\}$ are independent sets in G_1 and G_2, respectively, and we reach the same conclusion.

On the other hand, let T_1 be a maximum independent set of G_1 containing x and T_2 an $(\alpha(G_2)+1)$-element independent set of $G_2 - (y_1, y_2)$ (thus containing y_1 and y_2). Then $T_1 - \{x\} \cup T_2$ is an $(\alpha(G_1) + \alpha(G_2))$-element independent set of G. Hence $\alpha(G) = \alpha(G_1) + \alpha(G_2)$.

Let $e \in E(G_1)$. Let T_1 be an $(\alpha(G_1)+1)$-element independent set in $G_1 - e$. If $x \in T_1$, let T_2 be an $(\alpha(G_2)+1)$-element independent set of $G_2 - (y_1, y_2)$ and $T = T_1 - \{x\} \cup T_2$. If $x \notin T_1$, let T_2 be a maximum independent set of G_2 missing y_1 and y_2 and $T = T_1 \cup T_2$. In both cases, T is an $(\alpha(G)+1)$-element independent set in $G - e$.

Let $e \in E(G_2)$, $e \neq (y_1, y_2)$. Let T_2 be an $(\alpha(G_2)+1)$-element independent set in $G_2 - e$. If y_1, $y_2 \notin T_2$, let T_1 be a maximum independent set of G_1 missing x and $T = T_1 \cup T_2$. If $y_1 \in T_2$ (say), let z be a neighbor of x_2 in $V(G_1)$, let T_1 be an $(\alpha(G_1)+1)$-element independent set in $G_1 - (x, z)$ and $T = T_1 - \{x\} \cup T_2$. In both cases T is an $(\alpha(G)+1)$-element independent set in $G - e$. This proves G is α-critical.

(b) Let G be a connected α-critical graph, $\alpha(G) = \alpha$. By the preceding exercise, G is 2-connected. Suppose that G is not 3-connected, then $G = G_1 \cup G_2$ with $V(G_1) \cap V(G_2) = \{x, y\}$, $|V(G_i)| \geq 3$. By the preceding problem again, x and y are non-adjacent.

For each $X \subseteq \{x, y\}$ and $i = 1, 2$, denote the maximum size of an independent set $T \subseteq V(G_i)$ with $T \cap \{x, y\} = X$ by a_X^i. We set $a^i = a_\emptyset^i$. Then clearly

$$(1) \qquad a_X^1 + a_X^2 < \alpha + |X|, \qquad a_X^i \leq a^i + |X|.$$

Let $(x, z) \in E(G)$, $z \in V(G_1)$. Then $G - (x, z)$ contains an $(\alpha+1)$-element independent set T. Clearly, x, $z \in T$. If $y \notin T$, we get from this

$$(2) \qquad \begin{cases} a_x^1 + a_x^2 \geq \alpha + 1, \\ a^1 + a_x^2 \geq \alpha + 1. \end{cases}$$

Similarly if $y \in T$, we obtain

$$(3) \qquad \begin{cases} a_{xy}^1 + a_{xy}^2 \geq \alpha + 2, \\ a_y^1 + a_{xy}^2 \geq \alpha + 2. \end{cases}$$

Thus either (2) or (3) holds. Similarly we obtain that one of the following systems of inequalities holds:

$$(4) \quad \begin{cases} a_x^1 + a_x^2 \geq \alpha + 1, \\ a_x^1 + a^2 \geq \alpha + 1, \end{cases} \qquad (5) \quad \begin{cases} a_{xy}^1 + a_{xy}^2 \geq \alpha + 2, \\ a_{xy}^1 + a_y^2 \geq \alpha + 2. \end{cases}$$

Now (2) and (4) cannot hold simultaneously since (1) implies

$$(a^1 + a_x^2) + (a_x^1 + a^2) = (a^1 + a^2) + (a_x^1 + a_x^2) \leq 2\alpha + 1.$$

Similarly, (3) and (5) cannot hold simultaneously. Thus, choosing the indices appropriately we get that (2) and (5) hold. Interchanging the role of x and y we get that either

(6) $\begin{cases} a_y^1 + a_y^2 \geq \alpha + 1, \\ a^1 + a_y^2 \geq \alpha + 1, \end{cases}$ and (7) $\begin{cases} a_{xy}^1 + a_{xy} \geq \alpha + 2, \\ a_{xy}^1 + a_{xy}^2 \geq \alpha + 2, \end{cases}$

or

(8) $\begin{cases} a_y^1 + a_y^2 \geq \alpha + 1, \\ a_y^1 + a^2 \geq \alpha + 1, \end{cases}$ and (9) $\begin{cases} a_{xy}^1 + a_{xy}^2 \geq \alpha + 2 \\ a_x^1 + a_{xy}^2 \geq \alpha + 2 \end{cases}$

hold. But (2), (5) and (8), (9) cannot hold simultaneously; in fact, they would imply

$$(a^1 + a_x^2) + (a_y^1 + a^2) + (a_{xy}^1 + a_y^2) + (a_x^1 + a_{xy}^2) \geq 4\alpha + 6;$$

but the left-hand side is equal to

$$(a^1 + a^2) + (a_x^1 + a_x^2) + (a_y^1 + a_y^2) + (a_{xy}^1 + a_{xy}^2) \leq 4\alpha + 4$$

by (1), a contradiction. Thus (2), (5), (6), (7) hold but (3), (4), (8), (9) do not. Since the first inequalities in (3), (4), (8), (9) also occur in (2), (5), (6), (7) their second inequalities must fail to hold, i.e.

(10) $\begin{cases} a_x^1 + a^2 \leq \alpha & a_x^1 + a_{xy}^2 \leq \alpha + 1, \\ a_y^1 + a^2 \leq \alpha & a_y^1 + a_{xy}^2 \leq \alpha + 1. \end{cases}$

Since by (1) and (6)

$$\alpha \geq a^1 + a^2 \geq a^1 + a_y^2 - 1 \geq \alpha.$$

we have

$$a^1 + a^2 = \alpha, \qquad a_y^2 = a^2 + 1.$$

But, then by (6) and (10),

$$\alpha + 1 \leq a_y^1 + a_y^2 = a_y^1 + a^2 + 1 \leq \alpha + 1,$$

whence

$$a_y^1 = \alpha - a^2 = a^1.$$

Similarly we deduce that

$$a_x^1 = a^1, \quad a_x = a^2 + 1, \quad a_{xy}^1 = a^1 + 1, \quad a_{xy} = a^2 + 1.$$

Now denote by $\tilde{G}_1$, $\tilde{G}_2$ the graphs obtained from G_1 by connecting x to y and from G_2 by identifying x and y, respectively. We claim these two graphs are α-

critical. Since G arises from them by the operation in (a), this will prove the assertion.

We have

$$\alpha(G_1) = \max(a^1, a^1_x, a^1_y) = a^1,$$

$$\alpha(G_2) = \max(a^2, a^2_{xy} - 1) = a^2.$$

Let $e \in E(G_1)$. If $e = (x,y)$, then $\alpha(G_1 - e) \geq a^1_{xy} > a^1$; if $e \neq (x,y)$, then $e \in E(G)$. Let T be an $(\alpha+1)$-element independent set in $G - e$, $T \cap \{x,y\} = X$. Then

$$|T \cap V(G_2)| \leq a^2_X$$

and thus by (1)

$$|T \cap V(G_1)| \geq \alpha + 1 - a^2_X + |X| = a^1_X + 1.$$

Now, if $X \neq \{x,y\}$, then $T \cap V(G_1)$ is an independent set in $G_1 - e$ of size $a^1_X + 1 > a^1$; if $X = \{x,y\}$, then $T - \{x\}$ is such a set of size $a^1_{xy} > a^1$. This proves G_1 is α-critical. For G_2 this follows similarly. [T. Gallai–M. D. Plummer–W. Wessel; see W. Wessel, *Manuscripta Math.* **2** (1970) 309–334.]

20. Let $T_1, \ldots, T_k$ be all maximum independent sets of G. Then, by 8.12,

$$T_1 \cap \ldots \cap T_k = \emptyset, \quad T_1 \cup \ldots \cup T_k = V(G).$$

Thus, by 8.10,

$$|V(G)| = |T_1 \cup \ldots \cup T_k| = |T_1 \cup \ldots \cup T_k| + |T_1 \cap \ldots \cap T_k| \geq 2\alpha(G)$$

[P. Erdős, T. Gallai; B].

21. Let $T_1, \ldots, T_k$ be the maximum independent sets of G. Then, as we have seen,

$$T_1 \cap \ldots \cap T_k = \emptyset.$$

Thus setting $S = T_1 \cap \ldots \cap T_k \cap X = \emptyset$, we get by 8.11

$$|\Gamma(X)| - |X| \geq |\Gamma(S)| - |S| = 0.$$

[A. Hajnal; see B, LP]

22. Let $S = X \cap T_1 \cap \ldots \cap T_k$, where $T_1, \ldots, T_k$ are as in the hint. Then $x \in S$. Let $S' = S - \{x\}$. We claim that $\Gamma(S') \cap \Gamma(x) = \emptyset$. Suppose that there is a v adjacent to x and to a point u in S'. Let T be an $(\alpha(G)+1)$-element independent set in $G - (x,v)$, then x, $v \in T$ and hence $u \notin T$. Moreover, $T - \{v\}$ is a maximum independent set of G containing x and so, $T - \{v\} = T_i$ for some i. But then $u \notin T_i$, hence $u \notin S$, a contradiction.

Thus $|\Gamma(S)| = |\Gamma(S')| + |\Gamma(x)|$; by 8.11 we have

$$|\Gamma(S)| - |S| \leq |\Gamma(X) - |X|$$

whence

$$d_G(x) = |\Gamma(x)| = |\Gamma(S)| - |\Gamma(S')| = |\Gamma(S)| - |S| - (|\Gamma(S')| - |S'| + 1 \leq$$
$$\leq \Gamma(X) - |X| + 1$$

[L. Lovász, L. Surányi; see LP]

23. (a) Let $x \in V(G)$ and let T be an independent set of size $\alpha(G)$ containing x. By the preceding problem,

$$d_G(x) \leq |\Gamma(T)| - |T| + 1 = |V(G) - T| - |T| + 1 = n - 2\alpha(G) + 1$$

[A. Hajnal; see B, LP]

(b) By part (a), the number m of edges satisfies

$$m \leq \frac{1}{2}n(n - 2\alpha(G) + 1) \leq \frac{1}{2}(n - \alpha(G))(n - \alpha(G) + 1) = \binom{n - \alpha(G) + 1}{2}.$$

[Erdős, Hajnal and Moon; LP]

24. Let S, T be maximum independent sets with $S \cap T \neq \emptyset$. Obviously

$$\Gamma(S \cap T) \subseteq V(G) - S - T.$$

On the other hand if $x \in S \cap T$, we have

$$|V(G)| - 2\alpha(G) + 1 = d_G(x) \leq |\Gamma(S \cap T)| - |S \cap T| + 1$$

by 8.22, so

$$|\Gamma(S \cap T)| \geq |S \cap T| + |V(G)| - 2\alpha(G) =$$
$$= |V(G)| + |S| + |T| - |S \cap T| = |V(G)| - |S \cup T|,$$

whence we have $\Gamma(S \cap T) = V(G) - S - T$.

Now let a be an arbitrary point of $V(G)$, and $b_1, \ldots, b_k$ ($k = |V(G)| - 2\alpha(G) + 1$) be the neighbors of a. Let S_i be an $(\alpha(G) + 1)$-element independent set of $G - (a, b_i)$ and set $T_i = S_i - a$. Then, obviously, T_i is a maximum independent set of G, and $b_i \in T_i$, $b_i \notin T_j$ for $j \neq i$. Thus, $T_i \cap T_j$ contains no neighbors of a. This means by the statement proved above that $T_i \cap T_j = \emptyset$. Hence

$$|V(G)| \geq 1 + \sum_{i=1}^{k} |T_i| = 1 + (|V(G)| - 2\alpha(G) + 1)\alpha(G)$$

or, equivalently,

$$(\alpha(G) - 1)(|V(G)| - 2\alpha(G) - 1) \leq 0.$$

Now if $\alpha(G) = 1$, we have a complete graph. If $\alpha(G) > 1$ and $|V(G)| \leq 2\alpha(G) + 1$, then the degree is at most 2 and so we have K_2 or a cycle. The latter is α-critical only if it is odd.

25. If $|V(G)| = 2\alpha(G)$, we have $d_G(x) = 1$ by 8.22 and so, $G \cong K_2$. If $|V(G)| = 2\alpha(G) + 1$, we have $d_G(x) \leq 2$ for every point x, thus G is a circuit or a path. It is easily seen that only odd circuits are α-critical among these.

Now consider those graphs which are α-critical and satisfy $|V(G)| = 2\alpha(G) + 2$. They have $d_G(x) \leq 3$ for every point x. Observe that if we subdivide an edge of them by two points we get a graph from the same class; because this operation is a special case of the operation in 8.19 when G_1 is a triangle and G_2 is the given graph. Conversely, if we have such a graph and a point x of degree 2 in it (no point can have degree 1, since then its neighbor would be a one-element cutset, contradicting e.g. 8.18), then the two neighbors of x form a 2-element cutset and by part b of 8.19 we get that, contracting the two edges adjacent to x, an α-critical graph arises. This graph G' has degrees at most 3 as well since simple computation shows that $|V(G')| - 2\alpha(G') = 2$. Therefore, one of the neighbors of x in G must have been of degree 2 and so, G arises from G' by subdividing an edge by two points. Thus, it suffices to consider those graphs with degree 3.

By the previous problem, a 3-regular graph with $|V(G)| = 2\alpha(G) + 2$ is K_4. Hence all graphs G which are connected, α-critical and satisfy $|V(G)| = 2\alpha(G) + 2$ arise from K_4 by subdividing some edges by an even number of points. [B. Andrásfai, in: *Theory of Gr. Int. Symp. Rome*, Dunod, Paris–Gordon and Breach, New York, 1967, 9–19.]

Remark: It is known [L. Lovász, in: *Combinatorics*, Coll. Math. Soc. Bolyai **15**, Bolyai–North-Holland (1977); see also LP] that for each δ there exists a finite number of α-critical graphs G with $2\alpha(G) + \delta$ points such that all the other such graphs arise from them by subdividing some edges by an even number of points.

26. Remove edges till finally an α-critical graph G' with $\alpha(G') = \alpha(G)$ arises. G' does not have isolated points since for such a point x we would have $\alpha(G - x) \leq \alpha(G' - x) = \alpha(G) - 1$. Similarly, no component of G' consists of two points; since if x, y constitute a component of G', then

$$\alpha(G - \{x, y\}) \geq \alpha(G' - \{x, y\}) = \alpha(G) - 1.$$

Hence, for each component G_0 of G', we have $|V(G_0)| \geq 2\alpha(G_0) + 1$ by 8.25. Since $|V(G)| = 2\alpha(G) + 1$, we have only one component and so by 8.25, this is an odd circuit.

So G' is an odd circuit. We show $G' = G$. Suppose that there is an edge of G not in G'. This edge, together with an arc of G, forms an odd circuit C and the rest of G, an arc of odd length, has a 1-factor F. Now $G'' = C \cup F$ has $\alpha(G'') = \alpha(G)$ and we get a contradiction with the above. [V. G. Vizing–L. S. Melnikov, *Diskret. Analiz* **19** (1971) 11–14.]

27. As before, let G' be an α-critical spanning subgraph of G with $\alpha(G') = \alpha(G)$. We see as before that G has no component, which is an isolated point or a single edge. Moreover, each point of G' has degree greater than 2. For suppose that x is adjacent in G' to y and z only. Then $\alpha(G' - \{x, y, z\}) < \alpha(G)$, because we could add x to any independent set of $G - \{x, y, z\}$ to get an independent set of G. Thus,

$$\alpha(G - \{x, y, z\}) \leq \alpha(G' - \{x, y, z\}) < \alpha(G),$$

a contradiction.

By 8.23, we have $|V(G_0)| \geq 2\alpha(G_0) + 2$ for each component G_0 of G. This implies that

$$|V(G)| \geq 2\alpha(G) + 4,$$

if G is not connected; so we may assume that G is connected. By 8.25, $G \cong K_4$ or

$$|V(G)| \geq 2\alpha(G) + 3.$$

[E. Szemerédi, *Comb. Theory Appl.* Coll. Math. Soc. J. Bolyai **4** Bolyai–North-Holland (1970), 1051–1053.]

§ 9. Chromatic number

1. Let $x \in V(G)$. Since every point of $G - x$ has degree at most k, by using induction we can $(k+1)$-color it. x is adjacent to at most k points, therefore we can find a color, which does not occur among the neighbors of x. If we give x this color, we get a good coloration of G.

2. The first statement is trivial: Consider any (good) coloration of G with $\chi(G)$ colors, let V_1 be the union of $k < \chi(G)$ color classes and V_2 the rest. The $\chi(G)$-coloration we started with induces a k-coloration of $G[V_1]$ and a $(\chi(G) - k)$-coloration of $G[V_2]$: thus

$$\chi(G[V_1]) \leq k, \qquad \chi(G[V_2]) \leq \chi(G) - k.$$

On the other hand, $G[V_1]$ cannot be colored by less than k colors, since such a coloration, together with the coloration of $G[V_2]$ we already have, would give a coloration of G by less than $\chi(G)$ colors. Thus

$$\chi(G[V_1]) = k$$

and similarly

$$\chi(G[V_2]) = \chi(G) - k.$$

Now suppose that G is not a clique, then it contains a maximal clique H_1. Let $V_1 = V(H_1)$, $V_2 = V(G) - V_1$ and $H_2 = G[V_2]$. We claim that

$$\chi(H_1) + \chi(H_2) > \chi(G).$$

Since H_1 is a clique, $\chi(H_1) = |V_1| = k$ (say). We have to show that $\chi(H_2) > \chi(G) - k$.

Suppose indirectly that H_2 admits a $(\chi(G)-k)$-coloration, using colors $1, \ldots,$ $\chi(G)-k$. Use colors $\chi(G)-k+1, \ldots, \chi(G)$ to color H_1. We are going to show that we can get rid of one of these k colors, color $\chi(G)$, say. To this end, look at the set T of those points of color 1, which are neighbors of the point x with color $\chi(G)$. Since H_1 is a maximal complete subgraph, each point $t \in T$ is non-adjacent to a certain point $t' \in V_1$. Now re-color the graph giving to each $t \in T$ the color of t'. This is another legitimate coloration, because the color of t' does not occur anywhere else, and the new points of this color are independent. Now observe that in this new coloration x is the only point of color $\chi(G)$ and has no neighbor of color 1. So if we re-color x by 1, we get a coloration by $\chi(G) - 1$ colors. This is a contradiction.

3. Let α_i be a $\chi(G_i)$-coloration of G_i ($i = 1, 2$). We may assume that $V(G_1) = V(G_2)$, since this can always be achieved by adding isolated points to the two graphs. Then

$$\alpha(x) = (\alpha_1(x), \alpha_2(x))$$

defines a coloration of $G_1 \cup G_2$ with $\chi(G_1) \cdot \chi(G_2)$ colors, which is easily seen to be legitimate.

4. Let α, β be as in the hint. We show that β uses each color at least twice. Suppose that, e.g., 1 occurs only once. Let x be the point of β-color 1, and let S be the color-class of x with respect to α. Let β' be obtained from β by re-coloring each point of S by 1. Then the "old" common color classes of α and β remain untouched, but we get a new color class S in common, a contradiction. [T. Gallai, *Mat. Kut. Int. Közl.* **8** (1964) 373–385.]

5. (a) By induction on k, we may suppose that

$$\chi(G - V_k) \leq n - |V_k| - (k - 1) + 1.$$

Let α be a coloration of $G - V_k$ with colors $1, \ldots, n - |V_k| - k + 2$, and extend α to a coloration of G by coloring the points of V_k with colors $n - |V_k| - k + 3, \ldots, n - k + 2$. We want to re-color the graph, so as to get rid of one of these colors.

By the assumption, each V_i ($1 \leq i \leq k - 1$) contains a point y_i, which is non-adjacent to a certain point x_i of V_k. There must be a color which occurs in the set $\{y_1, \ldots, y_{k-1}\}$ only, since the number of points outside this set is only $n - k + 1$. Let, e.g. 1 be this color. Now re-color the points y_i such that $\alpha(y_i) = 1$ with the color of x_i. It is clear that the resulting coloration is good and uses $n - k + 1$ colors only. [R. P. Gupta, in: *Theory of Graphs*, Proc. Int. Coll. Rome, Gordon and Breach, 1969.]

(b) Consider a $\chi(\overline{G})$-coloration of $\overline{G}$ with color classes $V_1, \ldots, V_k$ ($k = \chi(\overline{G})$). Then any two V_i's are connected by an edge in $\overline{G}$, since otherwise they could be merged. Thus G satisfies the conditions of (a) and so,

$$\chi(G) \leq n - k + 1 = n - \chi(\overline{G}) + 1.$$

Hence the first inequality. The second is trivial by 9.3:

$$\chi(G) \cdot \chi(\overline{G}) \geq \chi(G \cup \overline{G}) = \chi(K_n) = n.$$

[E. A. Nordhaus–J. W. Gaddum, *Amer. Math. Monthly* **63** (1956) 175–177.]

6. Let $V(K_m) = \{x_1, \ldots, x_m\}$. Suppose that T is an independent set of $G \oplus K_m$. Then T contains at most one point of $\{(y, x_1), \ldots, (y, x_m)\}$ ($y \in V(G)$) since this is a complete graph. Therefore, $|T| \leq |V(G)|$. If equality holds, then exactly one of $\{(y, x_1), \ldots, (y, x_m)\}$ belongs to T. Define then an m-coloration of G by giving y the color i, if $(y, x_i) \in T$. This is a legitimate coloration since, if y, z both have color i, then (y, x_i), $(z, x_i) \in T$ and so are non-adjacent, but then so are y, z.

Conversely, suppose that G is m-colorable, and consider a coloration α of it with colors $1,\dots,m$. Then the set $\{(y, x_{\alpha(y)}); y \in V(G)\}$ is an independent subset of $G \oplus K_m$ of cardinality $|V(G)|$ [V. G. Vizing; B].

7. (a) Let α be a $\chi(G_1)$-coloration of G_1. Define a coloration β of $G_1 \times G_2$ by $\beta(x,y) = \alpha(x)$ ($x \in V(G_1)$, $y \in V(G_2)$). This is a legitimate coloration, for if (x_1,y_1) is joined to (x_2,y_2), then x_1 is joined to x_2 (and y_1 is joined to y_2), so $\alpha(x_1) \neq \alpha(x_2)$. Thus $\chi(G_1 \times G_2) \leq \chi(G_1)$. Similarly, $\chi(G_1 \times G_2) \leq \chi(G_2)$.

(b) Since the points of $\{(x,x) : x \in V(G)\}$ induce a subgraph of $G \times G$ isomorphic to G, we have $\chi(G \times G) \geq \chi(G)$ and by (a), we have equality.

(c) Suppose that $k = \chi(G \times K_n) < n$, we show that $k \geq \chi(G)$. Let α be a k-coloration of $G \times K_n$. Since $k < n$, we have a repetition in the set $\{\alpha(v,x); x \in V(K_n)\}$ for every $v \in V(G)$; let $\beta(v)$ be this more than once occurring color. We claim β is a good k-coloration of G. Suppose that $(v_1, v_2) \in E(G)$. There are x_i, x_i' with $\alpha(v_i, x_i) = \alpha(v_i, x_i') = \beta(v_i)$ ($i = 1, 2$). We assume that $x_1 \neq x_2$. Then (v_1, x_1) and (v_2, x_2) are adjacent in $G \times K_n$, so

$$\beta(v_1) = \alpha(v_1, x_1) \neq \alpha(v_2, x_2) = \beta(v_2).$$

(d) Let α be an n-coloration of $G \times K_n$. If, for every v, there is a color which occurs twice among $\{\alpha(v,x) : x \in V(K_n)\}$, then we get a good n-coloration of G as in the previous solution and this contradicts $\chi(G) > n$. So we have a $v \in V(G)$ such that $\alpha(v,x)$, $x \in V(K_n)$ are different.

We claim that the same holds for each neighbor u of v, and, in fact, $\alpha(u,x) = \alpha(v,x)$ for each $x \in V(K_n)$. This follows since we have $\alpha(u,x) \neq \alpha(v,y)$ for $x \neq y$ and all colors but $\alpha(v,x)$ occur among $\alpha(v,y)$, $y \neq x$. So $\alpha(u,x) = \alpha(v,x)$ and as G is connected, this will hold for any u and v. Hence, the n-coloration α is the one induced by the (unique) n-coloration of K_n.

(e) If $n = 2$, a bipartite graph G has $2^{c(G)-1}$ 2-colorations, since each component has a unique 2-coloration (up to exchanging colors) and these can be combined in $2^{c(G)-1}$ ways.

So suppose that $n \geq 3$. We claim that K_n^m has exactly m n-colorations (those induced by the n-colorations of the factors). We use induction on m. Let α be an n-coloration of $K_n^m = K_n^{m-1} \times K_n$.

Suppose first that there is a point $v \in V(K_n^{m-1})$ such that all points (v,x), $x \in V(K_n)$, have the same color. Then it follows as in the previous solution that $\alpha(v,x)$ is independent of v (because K_n^{m-1} is connected for $n \geq 3$ by 6.4), i.e. it is induced by an n-coloration of the last factor. So suppose that there are two points x_v, $x_v' \in V(K_n)$ for each $v \in V(K_n^{m-1})$ such that $\alpha(v, x_v) = \alpha(v, x_v')$. If $\alpha(v,x)$ is independent of x, then it is induced by a coloration of K_n^{m-1} and we are finished by induction. So suppose that there is a $v_0 \in V(K_n^{m-1})$ and x_0, $x_0' \in$

$V(K_n)$ such that $\alpha(v_0, x_0) \neq \alpha(v_0, x_0')$. Now define

$$\beta(v) = \begin{cases} \alpha(v_0, x_0) & \text{for } v = v_0, \\ \alpha(v, x_v) & \text{otherwise,} \end{cases}$$

$$\beta'(v) = \begin{cases} \alpha(v_0, x_0') & \text{for } v = v_0, \\ \beta(v) & \text{otherwise.} \end{cases}$$

Then it is easy to check that β, β' are legitimate n-colorations of K_n^{m-1} and they differ only at v_0. But we know all n-colorations of K_n^{m-1} from the induction hypothesis and there are no two such β, β' among them. This contradiction proves the assertion. [I. Tomescu] [D. L. Greenwell–L. Lovász, *Acta Math. Acad. Sci. Hung.* **25** (1974) 335–340.]

8. (a) The complete graph on S.

(b) Take $k - |S| + 1$ new points and join them to each other and to S.

(c) Take $k - 1$ new points and join them to each other and to S.

(d) Take an odd circuit C of length $\geq |S|$; join each point of C to a point of S, so that each point of S occurs. Take, furthermore, $k-3$ other points $u_1, \ldots, u_{k-3}$ and join them to each other, to the points of C and to a given point $s_0 \in S$.

No k-coloration of the resulting graph G induces $\{S\}$. For if all points of S are colored the same, then $u_1, \ldots, u_{k-3}$ are colored by $k - 3$ other colors (being adjacent to s_0), and we can use only two colors to color C, which is impossible. Let us consider a k-coloration of S, where at least two colors occur. Let us color $u_1, \ldots, u_{k-3}$ by colors different from the color of s_0. Now at each point of C there are at most $k - 2$ colors excluded (the colors of $u_1, \ldots, u_{k-3}$ and the color of its neighbor in S) and so, there are still two legal colors. Moreover, these two colors are not the same at every point. We show that this implies that we can extend the coloration to C.

Let $C = (x_1, \ldots, x_m)$ (we do not need in this argument the fact that C is odd). Also we may assume that the two colors permitted at x_1 and x_m are not the same, i.e. there is a color α_1 permitted at x_1 but not at x_m. Let α_2 be a color permitted at x_2 different from α_1 (since two colors are permitted at x_2, there is such a color) and similarly, if $\alpha_1, \ldots, \alpha_i$ are defined, then let α_{i+1} be a color permitted at x_{i+1} different from α_i. Then

$$\alpha(x_i) = \alpha_i \qquad (i = 1, \ldots, m)$$

defines a good coloration of C. In fact, x_i, x_{i+1} have different colors by definition and so do x_1 and x_m since the color α_m is one of the colors permitted at x_m while the color of x_1 is not.

(e) *First solution.* First, let a partition P be given. We construct a graph G_P such that $S \subseteq V(G_P)$ and the k-colorations of G_P induce all partitions of S except P. Let $P = \{X_1, \ldots, X_r\}$. Let us associate a new point with each $(r-1)$-subset $A = \{a_1, \ldots, a_{r-1}\}$ such that $a_i \in A_i$ $(i = 1, \ldots, r-1)$, and connect it to $a_1, \ldots, a_{r-1}$. Let T be the set of these new points. Now note the following two properties of the resulting graph:

(1) If we k-color it so that the coloration induces the partition P on S, then $T \cup X_r$ is necessarily monochromatic.

(2) If we k-color S so that the corresponding partition is different from P, then the coloration can be extended over T so that $T \cup X_r$ will not be monochromatic.

Now attach a graph to $T \cup X_r$ such that a k-coloration of $T \cup X_r$ can be extended over this graph iff $T \cup X_r$ is not monochromatic (see part (d)). Then the resulting graph G_P has the desired property.

If any set $\{P_1, \ldots, P_N\}$ of partitions of S is given, construct the graph G_P for each partition $P \neq P_1, \ldots, P_N$. We may assume that the graphs G_P have only the points of S in common. Then $G = \cup G_P$ is a graph with the desired properties.

Note the additional property that S is an independent set of points.

Second solution. We only describe the construction without details. Consider K_k^N. By 9.7, this graph has exactly N k-colorations $\alpha_1, \ldots, \alpha_N$. Let β_i be a k-coloration of S inducing the partition P_i $(i = 1, \ldots, N)$, and $\gamma_i = \alpha_i \cup \beta_i$ (this is a k-coloration of $S \cup V(K_k^N)$). Add all edges between S and K_k^N to the graph which leave the colorations γ_i legitimate. It is easy to verify that the k-colorations of the resulting graph are just $\gamma_1, \ldots, \gamma_N$, and thus it satisfies our requirements.

9. Suppose first that G has no cycles. Let $\alpha(x)$ denote the maximum length of a path starting from x. Then α is a coloration with colors 0, 1, ..., $m-1$. For suppose that there were adjacent points x, y with the same color i; let say $(x,y) \in E(G)$. Let P be a path of length i starting from y; then P does not go through x, because G has no cycles. Therefore, $(x,y) + P$ is a path of length $i+1$ starting from x, hence $\alpha(x) \geq i+1$, a contradiction. Observe that in this coloration, no path joins points of the same color.

Now let $\{e_1, \ldots, e_k\}$ be a minimum set of edges whose removal destroys all cycles. Then $G' = G - \{e_1, \ldots, e_k\}$ is acyclic and has the m-coloration constructed above. Since $G' + e_i$ is not longer acyclic, it has a cycle through e_i. Hence, G' has a path connecting the endpoints of e_i. By our remark above, this implies that the endpoints of e_i have different colors. Thus, the m-coloration of G' is a good m-coloration of G as well [T. Gallai, B. Roy; see B].

10. Assume that we have a coloration with the mentioned property. Then every circuit must go "up" and "down" equally often, i.e. has the same number of edges oriented in each of the two directions.

Conversely, suppose that the circuits have this property. Let W be any closed walk, we claim that W has the same property. We use induction on $|E(W)|$. If W is a circuit this is trivial; otherwise let x be a point, which occurs twice on W. Then x divides W into two shorter closed walks, and we are finished by induction.

Define the "expense" of a walk as the number of edges used in the right direction less the number of edges used backwards. Let x_0 be a point. For any point a, the expenses of all (x_0, a)-walks are the same, because otherwise, walking from x_0 to a on one walk and coming back on the other the expense of this walk would be non-zero. Color a by the expense of any (x_0, a)-walk, then the coloration has the desired property.

11. If G is k-colorable with colors $1,\ldots,k$, direct (a,b) from a to b, if the color b has a higher number than the color of a. It is easy to check that this orientation has the property stated. Conversely, assume that $\overrightarrow{G}$ is an appropriate orientation of G. Let us verify first the statements of the hint.

(a) Observe that by the assumption on the orientation, if I walk around any circuit of G, I cannot gain. Thus, if I remove a "loop" from any (a,b)-walk, I obtain an (a,b)-walk with no more expense. The number of (a,b)-paths being finite, we have one with minimal expense.

(b) Suppose that a and b are adjacent and say $(a,b) \in E(G)$. Let W_a, W_b be optimal (x_0,a)- and (x_0,b)-walks with expenses $\varepsilon(a)$ and $\varepsilon(b)$, respectively. Then $W_a+(a,b)$ is an (x_0,b)-walk with expense $\varepsilon(a)-1$, thus $\varepsilon(b) \le \varepsilon(b)-1$. Conversely, $W_b+(a,b)$ is an (x_0,a)-walk with expense $\varepsilon(b)+k-1$, hence $\varepsilon(a) \le \varepsilon(b)+k-1$. Thus,

$$1 \le \varepsilon(a) - \varepsilon(b) \le k - 1$$

as stated.

Now fix a point x_0 and define the color of a as the minimum expense of (x_0,a)-walks (mod k). Thus we use colors 0, 1, $\ldots$, $k-1$. This coloration is legal, because if a, b are adjacent points then by (b) above, the minimum expenses of (x_0,a) and (x_0,b)-walks belong to different residue classes mod k. [G. J. Minty, *Am. Math. Monthly* **69** (1962) 623–624.]

12. (a) Let x_0 be a point with $|C(x_0)| > d_G(x_0)$; suppose that $x_0,\ldots,x_i$ ($0 \le i \le n-1$) are defined, then let x_{i+1} be a point adjacent to one of them (which exists, because G is connected). So we get the sequence $(x_0,\ldots,x_{n-1})$.

Now color x_{n-1} with one of the colors of $C(x_{n-1})$; suppose that $x_{n-1},\ldots,x_{n-i}$ are colored, then choose a color from $C(x_{n-i-1})$, which is different from the colors or previously colored neighbors of x_{n-i-1}. This is possible, if $i \ne n-1$ since then x_{n-i-1} has fewer than $d_G(x_{n-i-1})$ neighbors previously colored and $|C(x_{n-i-1})| \ge d_G(x_{n-i-1})$. But it is also possible for $i = n-1$, because x_0 has $d_G(x_0)$ neighbors and $|C(x_0)| > d_G(x_0)$. Thus we get a desired coloration.

(b) We can find two neighboring points a, b with $C(a) \ne C(b)$. We may assume that $C(a) \not\supseteq C(b)$. Since $G-b$ is connected, we can arrange its points in a sequence $(x_0=a,\ldots,x_{n-2})$ so that x_j is adjacent to some x_i with $i < j$ for any $1 \le j \le n-2$. Let $b = x_{n-1}$, it also has this property.

Let us give x_{n-1} a color not in $C(a)$ (which exists, because $C(a) \not\supseteq C(b)$). Going backwards as in the previous solution, we get $x_{n-2},\ldots,x_1$ colored. However, we can find a suitable color for $x_0 = a$ as well; since it has $d_G(a) = |C(a)|$ neighbors but one of these, b, has a color α, which does not come into consideration since $\alpha \notin C(a)$. Therefore, we can find a color in $C(a)$, which is different from the color of the neighbors of a.

13. (a) x_0, x_{n-2}, x_{n-1} with the property given in the hint exist, as otherwise the relation of being adjacent would be an equivalence relation and G would be complete.

Since G is 3-connected, $G - x_{n-2} - x_{n-1}$ is connected and its points can be arranged in a sequence $x_0, \ldots, x_{n-3}$ so that x_j is joined to an x_i with $i < j$ ($j = 1, \ldots, n-3$).

Now color x_{n-1}, x_{n-2} with color 1. Working backwards again, we can give each x_i ($i \geq 1$) one of the colors $1, \ldots, k$, since x_i has less than k neighbors colored previously. Reaching x_0, we can find a color for it too, since although x_0 has k previously colored neighbors, two of these, x_{n-1} and x_{n-2}, have the same color.

(b) We may assume that G is 2-connected; for otherwise, if we can k-color the blocks of it, we can put together these colorations in the obvious way (permuting the colors, if necessary, to get the same color for the cutpoints). We may also assume that G is k-regular otherwise 9.12a applies.

Now suppose that G is not 3-connected, i.e. $G = G_1 \cup G_2$ with $V(G_1) \cap V(G_2) = \{x_1, x_2\}$, $|V(G_i)| \geq 3$. Since $k \geq 3$, G_1 and G_2 have more than three points. Also, we may assume that $d_{G_1}(x_1) > 1$ otherwise we could replace x_1 by its neighbor in G_1. Similarly, we may assume that $d_{G_2}(x_2) > 1$. Then $G_1 + (x_1, x_2)$ and $G_2 + (x_1, x_2)$ have points of degree at most k and so, they can be k-colored (by induction or by 9.12a, whichever you prefer). In these k-colorations x_1, x_2 get different colors, since they are adjacent. Therefore, by permuting the colors we can achieve that x_1 gets the same color in both graphs and so does x_2. Now the two k-colorations can be put together to get a k-coloration of G. [R. L. Brooks; see any textbook on graph theory.]

14. (a) As noted in the hint, we have to show that "to be non-adjacent" is an equivalence-relation on $V(G)$. This will imply that there is a partition $\{V_1, V_2, \ldots\}$ of $V(G)$ such that two points are adjacent iff they belong to distinct classes. The number of classes will be not more than k since G cannot contain a complete $(k+1)$-graph; the fact that there cannot be less than k classes follows from the maximality of G.

The symmetry and reflexivity of the relation "to be non-adjacent" is trivial. Thus, what we have to show is that $(a, b) \notin E(G)$, $(b, c) \notin E(G)$ imply $(a, c) \notin E(G)$. By the maximality of G, there is a finite subgraph H_1 of G such that $H_1 + (a, b)$ is not k-colorable. Clearly $a, b \in V(H_1)$. Similarly, there exists a finite subgraph H_2 of G with the property that $H_2 + (b, c)$ is not k-colorable. Consider $H_1 \cup H_2 + (a, c)$; we claim that this graph is not k-colorable. Suppose indirectly that it has a k-coloration α. Then $\alpha(a) = \alpha(b)$, othewise α would be a k-coloration of $H_1 + (a, b)$. Similarly $\alpha(b) = \alpha(c)$. Now $\alpha(a) = \alpha(c)$ and this is impossible as a and c are adjacent in $H_1 \cup H_2 + (a, c)$. Thus $H_1 \cup H_2 + (a, c)$ is not k-colorable and so, it cannot be a subgraph of G. Thus a, c are non-adjacent in G.

(b) We may suppose that G is simple. Consider all graphs G' such that $V(G') = V(G)$, and all finite subgraphs of $G' \cup G$ are k-colorable. Since this property depends on the finite subgraphs G' only (i.e. G' has this property iff all finite subgraphs have), there is a maximal simple graph G_0 with this property. Clearly $G_0 \supseteq G$. Thus by (a), $V(G)$ has a partition $V_1, \ldots, V_k$ such that two points are adjacent in G iff they belong to distinct classes. Clearly this partition is a good coloration of G. [P. Erdős–N. G. De Bruijn, *Indag. Math.* **13** (1961) 371–373; this proof is due to L. Pósa.]

(c) Let $e = (x, y) \in E(G)$. Then the set

$$F_e = \{\alpha : \alpha(x) \neq \alpha(y)\}$$

of k-colorations of $V(G)$ is closed, since its complement is the union of the open sets

$$\{\alpha : \alpha(x) - \alpha(y) = j\} \qquad (j = 1, \ldots, k).$$

The set of legal colorations is just

$$\bigcap_{e \in E(G)} F_e,$$

which is, therefore, closed as well.

To prove the Erdős–de Bruijn theorem, note that by Tihonov's Lemma the product space $\underset{v \in V(G)}{\times} T_v$ is compact. The assumption that each finite subgraph of G is k-colorable translates into the assumption that the intersection of any finite number of sets F_e is non-empty. The proposition that G is k-colorable is equivalent to the assertion that the intersection of all of them is non-empty. But this follows by the Riesz Intersection Theorem.

15. I. Clearly $\mathcal{K}$ contains some complete graphs. Let K_{k+1} be the least complete graph in $\mathcal{K}$. We claim that $\mathcal{K}$ is precisely the class of non-k-colorable graphs. First let $G \in \mathcal{K}$ and suppose indirectly that G is k-colorable. Then the k-coloration of G defines a homomorphism of it into K_k, thus by (i) $K_k \in \mathcal{K}$, a contradiction.

II. We now show that every non-k-colorable graph belongs to $\mathcal{K}$. Suppose that G would be a counterexample. We may suppose that G is such that connecting any two non-adjacent points of G by a new edge, we get a graph in $\mathcal{K}$. We claim that "to be non-adjacent" is an equivalence-relation on $V(G)$. Suppose indirectly that $(a, b) \notin E(G)$, $(b, c) \notin E(G)$ but $(a, c) \in E(G)$ (the other conditions of an equivalence-relation are trivially fulfilled). By the maximality of G, $G + (a, b) \in \mathcal{K}$, $G + (b, c) \in \mathcal{K}$. Hence by (ii) $G \in \mathcal{K}$, a contradiction.

Thus "to be non-adjacent" is an equivalence-relation and hence, $V(G)$ has a partition $\{V_1, \ldots, V_m\}$ such that two points are adjacent iff they belong to distinct classes. Since G is not k-colorable, $m \geq k + 1$. But then G contains a K_{k+1}, a contradiction as $K_{k+1} \in \mathcal{K}$ but $G \notin \mathcal{K}$.

16. It is straightforward to show that operations (α), (β), (γ) yield non-k-colorable graphs if the initial graphs are not k-colorable.

Let $\mathcal{K}$ denote the class of all graphs arising by repeated application of (α), (β), (γ) from the complete $(k+1)$-graph. We claim that $\mathcal{K}$ satisfies the conditions of 9.15. (i) is trivial. Let H be a graph, a, b, $c \in V(H)$ such that $H + (a, b) \in \mathcal{K}$, $H + (b, c) \in \mathcal{K}$. Let H' be a graph isomorphic with H and denote by x' the point of H' corresponding to $x \in V(H)$. Then $H' + (b', c') \in \mathcal{K}$. Let us identify b and b' and connect a to c'; then identify x with x' ($x \in V(H)$) and cancel the multiplicities of edges to get a simple graph. From $H+(a,b)$ and $H'+(b',c')$, by operations (γ), (β) we obtain H this way. This proves that $H \in \mathcal{K}$.

Thus $\mathcal{K}$ consists of all non-r-colorable graphs for some r. Obviously, $r = k$ [G. Hajós; see B, S].

17. (a) By 5.3, a graph is 2-colorable iff it contains no odd circuit. Hence the critically 3-chromatic graphs are precisely the odd circuits.

(b) Assume that for the sake of simplicity that n is odd; the even case could be treated similarly. Consider a complete bipartite graph $K_{n,n}$ with color classes A, B. Take two disjoint odd circuits C_A, C_B of length n and connect each point of C_A to a point of A, each point of C_B to a point of B by independent edges. Let G denote the resulting graph (Fig. 64).

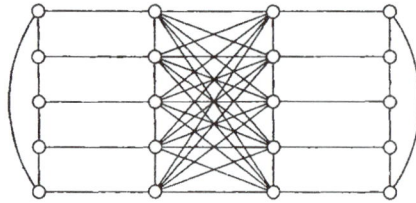

FIGURE 64

The graph G is not 3-colorable. In fact, every 3-coloration of $K_{n,n}$ colors one of the sets A, B monochromatically. But then this coloration does not extend to the corresponding odd circuit.

On the other hand, G is χ-critical. For let $e = (x,y)$ be any edge. If $x \in A$, $y \in B$, then color x and y red, all points of $A - \{x\}$ blue, all points of $B - \{y\}$ green. In this coloration neither A nor B is monochromatic, hence it extends to a 3-coloration of the whole graph $G - e$. We obtain by similar reasoning that $G - e$ is 3-colorable for every edge e.

Since G has more than n^2 edges, the desired graph is found. [B. Toft, *Studia Sci. Math. Hung.* **5** (1970) 461–470.]

(c) Again suppose that for the sake of simplicity that n is odd. Take two disjoint n-circuits C_1, C_2 and connect each point of C_1 to each point of C_2. The resulting graph is not 5-colorable, for we need 3 colors to color C_1 and 3 different colors to color C_2. It is straightforward to find 5-colorations of the graphs $G - e$, $e \in E(G)$, thus G is χ-critical. It trivially fulfills the degree requirement. [G. A. Dirac, *J. London Math. Soc.* **27** (1952) 85–92.]

18. Set $\chi(G) = k$. We show first that $\chi(G') > k$. Suppose indirectly that there is a k-coloration α of G'. Let $\alpha(y) = 1$; define a coloration β of $V(G)$ by

$$\beta(x) = \begin{cases} \alpha(x) & \text{if } \alpha(x) \neq 1, \\ \alpha(x') & \text{if } \alpha(x) = 1. \end{cases}$$

Then β is a legitimate coloration of G. In fact, if $(x,z) \in E(G)$, then one of them, say x, has $\alpha(x) \neq 1$, so $\beta(x) = \alpha(x)$. Now no matter whether $\beta(z) = \alpha(z)$ or $\alpha(z')$, $\beta(z)$ will be different from $\beta(x)$ since both z, z' are neighbors of x. β uses only $k-1$ colors, which is a contradiction.

Now let $e \in E(G')$, we show that $\chi(G' - e) \geq k$. If $e \in E(G)$, then take a $(k-1)$-coloration of $G - e$, color the points x' $(x \in V(G))$ with color k and y with color 1. If $e = (u, v')$, take a $(k-1)$-coloration β of $G - (u, v)$. Define

$$\alpha(z) = \begin{cases} \beta(z) & \text{if } z \in V(G) - \{v\}, \\ k & \text{if } z = v \text{ or } z = y, \\ \beta(x') & \text{if } z = x'. \end{cases}$$

Finally, if $e = (y, v')$, let β be a $(k-1)$-coloration of $G - v$ and define

$$\alpha(z) = \begin{cases} \beta(z) & \text{if } z \in V(G) - \{v\}, \\ k & \text{if } z = v, \, v' \text{ or } y, \\ \beta(x) & \text{if } z = x'. \end{cases}$$

In all cases we have found a k-coloration of $G' - e$. Thus G' is χ-critical and $\chi(G') = k + 1$ as stated [J. Mycielsky; see S].

19. (a) Assume that G_0 is a subgraph of a 4-chromatic χ-critical graph G. Then $G - \{x_3, x_4\}$ is 3-colorable. Consider any 3-coloration of it. Then x_1, x_5 have two different colors; therefore both x_2, x_4 have the third color and so, x_3, x_4 have different colors. But this implies our coloration is a good coloration of G as well, which is a contradiction.

(b) Suppose that $G_0 \subseteq G$, where G is χ-critical and $\chi(G) = k + 1$. Let $e \in E(G_0)$. Then $G - e$ has a $(k-1)$-coloration α. Since this is not a good coloration of G, the endpoints of e have the same color. However, this means that α yields a k-coloration of G_0/e.

Conversely, suppose that $\chi(G_0/e) \leq k$ for every $e \in E(G)$. We show that G_0 can be embedded into a critically $(k+1)$-chromatic graph even as an induced subgraph. The assumption means that for each $e \in E(G)$, there exists a k-coloration of $V(G)$, which associates the same color with the endpoints of e, but different colors with the endpoints of any other edge of G_0. Let P_e be the partition of S induced by α_e. By 9.8, we can find a graph G such that $S \subseteq V(G)$ and the k-colorations of G induce the partitions P_e $(e \in E(G_0))$ of S and no other partition. Set $G' = G \cup G_0$.

Now $\chi(G') > k$. For if there existed a k-coloration of G', this would induce a P_e on S, which is impossible as e joins two points in the same class of P_e. Therefore, G' contains a $(k+1)$-chromatic χ-critical graph G''. All we have to show is that to obtain G'' from G' we do not have to remove any edge of G_0. Suppose indirectly that $e \in E(G_0)$, $e \notin E(G'')$. Let α_e be a k-coloration of G inducing P_e on S. By the definition of P_e, α_e is a good coloration of $G'' \subset G \cup (G_0 - e)$. This contradicts $\chi(G'') = k + 1$. [D. Greenwell–L. Lovász, *Acta Math. Acad. Sci. Hung.* **25** (1974) 335–340.]

20. The first assertion is trivial: for any edge e of G, $G-e$ is k-colorable and this yields a k-coloration of $G'-e$. Hence either G' itself is k-colorable or it is critically $(k+1)$-chromatic.

If $k=2$, then the critically 3-chromatic graphs are the odd circuits. Splitting any point of these, we obtain a 2-chromatic graph (a path).

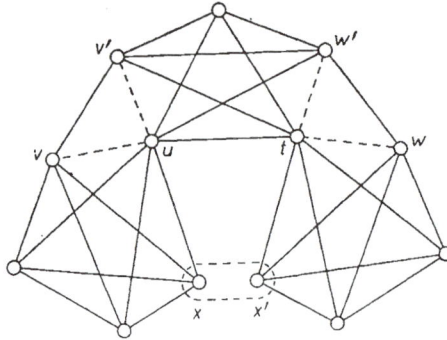

FIGURE 65

We show by example that for $k \geq 3$ the second possibility can, in fact, occur. Put 3 copies of K_{k+1} together as in Hajós' construction shown in Fig. 65. Then by 9.16 (or trivially) the resulting graph G' is not k-colorable. On the other hand, even if we identify the points x and x', the resulting graph will be critically $(k+1)$-chromatic. For let us remove any edge e, for example from the K_{k+1} containing x. We can then color the points of this K_{k+1} is such a way that only the endpoints of e have the same color; in particular u and v are differently colored. Color the points of the other K_{k+1} (containing x') such that only the colors of t and w are the same. Color v' with the color of u. The remaining $k-2$ points can be colored with the $k-2$ colors different from the colors of v' w' and u. Similar arguments yield a k-coloration of $G-e$, when e is any other edge.

21. Assume indirectly that same $m \leq k-1$ edges $e_1,\ldots,e_m$ separate G into two pieces G_1, G_2. We may assume that this is a minimal cutset, so that each e_i connects G_1 to G_2. Consider a k-coloration of G_1 with color-classes $T_1,\ldots,T_k$ and a k-coloration of G_2 with color-classes $S_1,\ldots,S_k$ (these exist by the criticality of G).

We want to match each T_i with some S_i such that no edge connects T_i and S_i. Since there are only $m \leq k-1$ edges to consider, there certainly exists an S_i not connected to T_1 by an edge. We can clearly select this S_i such that either T_1 or S_i should be incident with one of the edges $e_1,\ldots,e_m$. Let this index be $1'$. Then we have to match the remaining $k-1$ classes T with the remaining $k-1$ classes S, and the number of edges e_i between these classes is now $\leq m-1 \leq k-2$. So we can proceed in the same way as above. (Actually what we have shown is that removing $k-1$ edges of $K_{k,k}$, the remaining bipartite graph has a 1-factor. This would of course follow from the König–Hall theorem immediately.)

Now the partition $\{T_1 \cup S_1, \ldots, T_k \cup S_k\}$ is a k-coloration of G, a contradiction. [G. Dirac; B.]

22. Suppose first that G is separable, i.e. $G = G_1 \cup G_2$ with $V(G_1 \cap G_2) = \{x\}$, $|V(G_i)| \geq 2$. Since G is χ-critical, $\chi(G_i) < \chi(G) = k+1$. Consider a k-coloration α_i of G_i. By permuting the colors we may assume that $\alpha_1(x) = \alpha_2(x)$. Then putting α_1 and α_2 together we obtain a k-coloration of G, a contradiction.

Now suppose that $G = G_1 \cup G_2$ with $|V(G_i)| \geq 3$, $V(G_1 \cap G_2) = \{x, y\}$. Let α_i be a k-coloration of G_i $(i = 1, 2)$. It is impossible that $\alpha_1(x) \neq \alpha_1(y)$ and $\alpha_2(x) \neq \alpha_2(y)$; since then, by permuting colors, we could assume that $\alpha_1(x) = \alpha_2(x)$, $\alpha_1(y) = \alpha_2(y)$ and so, we could put α_1, α_2 together to get a k-coloration of G. We reach a contradiction similarly, if $\alpha_1(x) = \alpha_1(y)$ and $\alpha_2(x) = \alpha_2(y)$. So we have that, choosing the indices approximately, any k-coloration of G_1 paints x, y with the same color, but no k-coloration of G_2 does so. Hence, $G_1' = G_1 + (x, y)$ is not k-colorable, and neither is the graph G_2' obtained from G_2 by identifying x and y.

We claim that G_1', G_2' are χ-critical. Obviously, $\chi(G_1') = \chi(G_2') = \chi(G)$. Let $e \in E(G_1')$. Now $G - e$ is k-colorable; from above, this coloration associates different colors with x and y (G_2 forces this) and therefore, it yields a k-coloration of $G_1' - e$. Thus G_1' is χ-critical. Similarly, G_2' is χ-critical.

Note that the k-colorability of G_2 does not follow from the fact that G_2' is critically $(k+1)$-chromatic (cf. 9.20). Therefore, we can summarize the result as follows:

Every critically $(k+1)$-chromatic graph G, which is not 3-connected arises by the following construction. We take two critically $(k+1)$-chromatic graphs G_1', G_2'. We remove an edge (x_1, y_1) of G_1'. We split a point of G_2' into two points x_2, y_2 in such a way that the resulting graph is k-colorable. Finally, we identify x_1 with x_2 and y_1 with y_2. It is easy to verify that the converse is true as well: every graph arising by this construction is critically $(k+1)$-chromatic. [G. A. Dirac, *J. Reine Angew. Math.* **214/215** (1964) 43–52.]

23. Let α_i be a k-coloration of $G - V(G_i)$. Then α_i induces a partition P_i of S into at most k classes. We claim that $P_i \neq P_j$ for $i \neq j$. For if $P_i = P_j$ then, by permuting colors, we could arrange for α_i and α_j to color S the same way. Now use α_i on $G - V(G_i)$ and α_j on G_i (where it is defined, as $i \neq j$) to get a k-coloration of G. This contradicts $\chi(G) = k+1$. Thus the P_i's are different and hence $N \leq B_{m,k}$.

To construct a χ-critical graph G having a set $S \subseteq V(G)$ with $|S| = m$ that separates it into exactly $p = B_{m,k}$ components, let $P_1, \ldots, P_p$ be the partitions of S into at most k classes. For each i, 9.8 yields a graph G_i containing S such that the k-colorations of G_i induce all partitions of S into at most k classes except P_i. We may assume that $V(G_i) \cap V(G_j) = S$. Let $G = G_1 \cup \ldots \cup G_p$.

G is in general not χ-critical. But certainly, it is not k-colorable, for a k-coloration of G would induce a partition P_i of S and this contradicts the definition of G_i. So we can consider a χ-critical subgraph G' of G with $\chi(G') = k+1$. We show that G' has the desired property: it suffices to verify

(1) $S \subseteq V(G')$;

(2) G' meets every $G_i - S$.

To show (1), suppose that $a \in S$ but $a \notin V(G')$. Let, say, $P_1 = \{\{a\}, S - \{a\}\}$, $P_2 = \{S\}$. Let α_1 be a k-coloration of G_1, which induces the partition P_2 of S and let α_i be a k-coloration of G_i inducing P_1 for $i = 2, \ldots, p$. Since P_1 and P_2 coincide on $S - a$, we may assume that the α_i's can be put together to get a k-coloration of $G - a$; therefore, $G' \subseteq G - a$ is k-colorable, a contradiction.

To show (2), suppose that G' does not meet $G_i - S$. Let α_j be a k-coloration of G_j inducing P_i of S ($j \neq i$). Then the α_j's can be put together to get a k-coloration of $G - (V(G_i) - S)$. Since G' is a subgraph of this, G' is k-colorable, a contradiction.

24. Let $(x, y) \in E(G)$. Since G is χ-critical, $G - (x, y)$ has a k-coloration α. Since G itself is not k-colorable, $\alpha(x) = \alpha(y) = 1$ (say). Consider another color i and the graph H_i induced by those points of colors 1 and i. Let H_i' be the component of H_i containing x. Then we claim that $y \in V(H_i')$; for otherwise, we could switch colors 1 and i in $V(H_i')$; the arising coloration α' is legitimate and has $\alpha'(x) \neq \alpha'(y)$, i.e. it is a good coloration of G, which is impossible. So there are (x, y)-paths in H_i; let P_i be a shortest one. Since P_i contains alternating points with colors 1 and i, it is of even length; by minimality, it spans no chord. Since each edge of P_i connects a point of color 1 to a point of color i, the paths $P_2, \ldots, P_k$ must be edge-disjoint. This proves the assertion. We remark that 9.21 follows easily from this.

25. (a) We have to find the minimum number of colors needed to color the edges of K_n so that edges with an endpoint in common have different colors. Since $L(K_n)$ contains a complete $(n-1)$-graph, its chromatic number is at least $n-1$. Realizing K_n by the edges and diagonals of a regular n-gon, we can color parallel segments with the same color, giving an n-coloration of $L(K_n)$. So

$$n - 1 \leq \chi(L(K_n)) \leq n.$$

We show that the lower bound gives equality for even n and the upper bound gives equality for odd n.

First, assume that n is odd. Then the red color must miss at least one vertex, and the $n-1$ edges incident with this vertex need $n-1$ further colors; so at least n colors are needed.

Second, assume that n is even. Delete any vertex v of K_n, and consider the $(n-1)$-coloration of $K_{n-1} = K_n - v$ as described above. In this coloration, every vertex u of K_{n-1} misses exactly one color, and each color is missed by exactly one vertex of K_{n-1}. Thus we can color each edge uv by the color missed by u.

(b) Now we have to color the edges of K_n so that edges with the same color must share an endpoint. We can order the vertices and "color" the edges by their smaller endpoint. This coloration uses $n-1$ colors; we can gain one for $n \geq 3$ by observing the the last three edges may get the same color. So

$$\chi(\overline{K}_n) \leq n - 2.$$

We show that equality holds. In any legitimate coloration, edges with the same color form a star or a triangle. Let, say, vertices $1, \ldots, n-k$ be centers of the stars used. Then the edges spanned by vertices $n-k+1, \ldots, n$ must be colored so that the colors form triangles. Hence the number of colors used is at least

$$n - k + \frac{1}{3}\binom{k}{2} \geq n - 2,$$

by elementary arguments.

(c) Translating the problem, we have to color the edges of $\overrightarrow{K}_n$ by the minimum number of colors so that pairs of edges forming a 2-path, or a 2-cycle, get different colors. First, assume that we have such a coloration, with t colors. Associate with each vertex v the set A_v of colors of edges leaving v. Then for any two vertices u and v, $A_u \not\subseteq A_v$ (the color of uv occurs in A_u but not in A_v). Hence by Sperner's Theorem 13.21,

(1) $$n \leq \binom{t}{\lfloor t/2 \rfloor}.$$

We show that, conversely, if (1) holds then $L(\overrightarrow{K}_n)$ is t-colorable. We label the vertices of $\overrightarrow{K}_n$ by the $\lfloor t/2 \rfloor$-element subsets of $\{1, \ldots, t\}$; let A_u be the label of vertex u. With every edge uv, we associate any element of $A_u \setminus A_v$ as its color. Trivially, this defines a good coloring of $L(\overrightarrow{K}_n)$.

Thus the chromatic number of $L(\overrightarrow{K}_n)$ is the least t for which (1) holds.

26. (a) Let $\chi(L(G)) = k$ and consider a k-coloration α of $L(G)$. Then α is a k-coloration of the edges of G such that the edges (x, y) and (y, z) cannot have the same color. For each point u, let A_u be the set of colors occurring among the edges leaving u. Then $A_u \neq A_v$ for every edge (u, v), since the color of (u, v) is contained in A_u but not in A_v. So we have defined a coloration of G with 2^k colors.

To show that equality holds for an appropriate reorientation of G, let $k = \lceil \log_2 \chi(G) \rceil$, and color the vertices of G with the subsets of $\{1, \ldots, k\}$. Let A_u be the color of u. For each edge (u, v), choose an element j from $A_u \triangle A_v$, use it to color the edge, and reverse the edge if $j \in A_v$. If G' denotes the graph obtained from G after these reversals, then we have a good coloration of $L(G')$.

(b) First, assume that $\chi(G) \leq \binom{k}{\lfloor k/2 \rfloor}$. Then we can color the vertices of G by the $\lfloor k/2 \rfloor$-element subsets of $\{1, \ldots, k\}$, and hence we can obtain a k-coloration of $L(G)$ just as in the solution of 9.25(c).

Second, assume that $\chi(L(G)) \leq k$. This means that the edges of G are colored with k colors so that any two edges forming a 2-path or a 2-cycle have different colors. As in the solution of 9.25(c), we consider the set A_u of colors of edges leaving vertex u of G. Just as there, it follows that if (u, v) is an edge of G, then A_u is not a subset of A_v (and vice versa); but now A_u may contain A_v if u and v

are non-adjacent. However, we may use 13.20 instead of Sperner's Theorem: the subsets of $\{1,\ldots,k\}$ can be split into $\binom{k}{\lfloor k/2 \rfloor}$ chains. We use the chain containing A_u as the color of u. Then if u and v are adjacent, they do not belong to the same chain, and so they get different colors. [A. Hajnal]

27. (a) Let G_3 be a 5-circuit and, if G_r is defined, let G_{r+1} be the graph obtained by the construction of problem 9.18. Then it follows by the assertion of this problem that G_r is an r-chromatic graph (χ-critical too) and it is easy to check that it has no triangles.

(b) Let G_3 be a 7-circuit. Suppose that G_n ($n \geq 3$) is defined such that $\chi(G_n) \geq n$. Take a set S of $n(|V(G_n)| - 1) + 1$ points. With each $|V(G_n)|$-element subset F of S, we associate a copy of G_F of G_n (these copies should be disjoint), and join it to the points of F by a matching. Let G_{n+1} be the resulting graph, then $\chi(G_{n+1}) > n$. Suppose that G_{n+1} could be n-colored, then some $|V(G_n)|$ points of S would get the same color 1; let F be the set of these points. As remarked in the hint, the points of G_F cannot get the color 1, so they are, in fact, $(n-1)$-colored; this, however, contradicts the definition of G_n.

It is immediate that G_n contains no 3-, 4-, or 5-circuits. [W. T. Tutte (Blanche Descartes); see S.]

(c) *First solution.* Let $\overrightarrow{G}_{r,n}$ be the r-times iterated line-graph of T_n, the transitive tournament. By the preceding problem,

$$\chi(\overrightarrow{G}_{r,n}) \geq \underbrace{\log\log\ldots\log}_{r} n.$$

Now let $G_{r,n}$ be $\overrightarrow{G}_{r,n}$ with the orientations ignored. We show that $G_{r,n}$ has no odd circuits of length at most $2r+1$. We use induction on r. Let C be a closed path in $G_{r,n}$ of length $2s+1$. Since $\overrightarrow{G}_{r,n}$ is acyclic, C decomposes into arcs P_1, Q_1, P_2, Q_2, ..., P_t, Q_t such that $P_1,\ldots,P_t$ are directed paths in $\overrightarrow{G}_{r,n}$ going in the same direction around C and $Q_1,\ldots,Q_t$ are directed paths going in the opposite direction; moreover, they follow each other on the walk as listed.

Since $\overrightarrow{G}_{r,n}$ is acyclic, it is easy to see that the points of P_i (or Q_i) correspond to the edges of a path P_i' (or Q_i') in $\overrightarrow{G}_{r-1,n}$. Then the last edge of P_1' coincides with the last edge of Q_2' (with respect to orientation in G_r), etc. Thus, if we remove these last edges the rest of $P_1', Q_1', \ldots, P_t', Q_t'$ forms a walk of length $2s + 1 - 2t$. Clearly, this walk contains an odd circuit of length $l \leq 2s+1-2t \leq 2s-1$. By the induction hypothesis $l \geq 2r$, whence $2s+1 > 2r+1$ as stated.

Second solution. Consider the surface of the k-dimensional unit ball. Connect two points of it, if their distance, measured on the surface, is at least $(1-1/l)\pi$. The resulting graph is not k-colorable. In fact, a theorem of Borsuk asserts that in any k-coloration of the surface of the k-dimensional ball, one of the color-classes has diameter 2; this color-class then contains adjacent points.

On the other hand, let $(x_0, \ldots, x_{2p})$ be any odd circuit in the graph. If d denotes the spherical distance, then $d(x_i, x_{i+1}) \geq (1-(1/l))\pi$, whence $d(x_i, x_{i+2}) \leq 2\pi/l$, and hence by induction $d(x_0, x_{2p}) \leq 2p\pi/l$. On the other hand, this distance is at least $(1-(1/l))\pi$. Hence $2p+1 \geq l$. So our graph contains no short odd circuits.

This graph is infinite but by the Erdős–de Bruijn theorem 9.14, it has a finite subgraph with similar properties. [P. Erdős–A. Hajnal, *Ann. N. Y. Acad. Sci.* **157** (1970) 115–124.]

28. (a) Let, say, $I_1 = [b, a]$ be the interval whose endpoint a is the first from left. The system $I_2, \ldots, I_n$ determines the graph $G - x_1$. By induction, $\chi(G - x_1) = \omega(G - x_1) \leq \omega(G)$. Observe, moreover, that if, x_i is adjacent to x_1, i.e. I_i meets I_1, then $a \in I_i$ (since otherwise the right endpoint of I_i would lie to the left of a). Since at most $\omega(G)$ intervals contain a (including I_1), the degree of x_1 is at most $\omega(G) - 1$. Therefore, we can extend the $\omega(G)$-coloration of $G - x_1$ to x_1. Hence, $\chi(G) \leq \omega(G)$. Since, obviously, $\chi(G) \geq \omega(G)$, we are done.

(b) Let $I_1, I_2, \ldots, I_r$ denote the intervals containing a. Then $x_1, x_2, \ldots, x_r$ form an independent set in $\overline{G}$. Consider $\overline{G} - \{x_1, \ldots, x_r\}$. If $x_{i_1}, \ldots, x_{i_q}$ form a clique in $\overline{G} - \{x_1, \ldots, x_r\}$, then $I_{i_1}, \ldots, I_{i_q}$ are disjoint intervals not containing a. By the definition of a, $I_{i_1}, \ldots, I_{i_q}$ are to the right of a and therefore I_1 is disjoint from them. Hence $\{x_1, x_{i_1}, \ldots, x_{i_q}\}$ is a clique in $\overline{G}$. This proves that $\chi(\overline{G} - \{x_1, \ldots, x_r\}) \leq \omega(\overline{G}) - 1$. Thus, $\overline{G} - \{x_1, \ldots, x_r\}$ can be colored by $\omega(\overline{G}) - 1$ colors and we can use the same new color for $x_1, \ldots, x_r$. Hence $\chi(\overline{G}) \leq \omega(\overline{G})$. The equality is obvious as before.

(c) Let x_2, x_3 be the neighbors of x_1 on the circuit. Then, as we have already seen, I_2, I_3 contain a and therefore, they meet each other. Thus, (x_2, x_3) is a chord of the circuit (unless $C = (x_1, x_2, x_3)$). [G. Hajós, *Intern. Math. Nachr.* **11** (1957) Sondernummer 65.]

29. (a) If G_1 is a component of $G - S$, then it is adjacent to all points of S, since otherwise a set smaller than S (the set of neighbors of G_1) would separate G.

Let $u, v \in S$ and G_1, G_2 two components of $G - S$. Let P_i be a minimal path, which connects u, v through G_i. Then P_i has no chords by minimality, except possibly (u, v). Set $C = P_1 \cup P_2$, then C is longer than 3 and therefore, C has a chord. This cannot join an inner point of P_1 to an inner point of P_2 as these points are in distinct components of $G - S$. Hence, it must join two points of the same P_i. This is only possible if it is (u, v), Thus, u, v are adjacent. [G. Dirac]

(b) First, let G be defined through the subtrees as formulated, and consider a circuit C of G. We may assume that C is a Hamiltonian circuit.

Let us remove points of degree 1 from T as long as the remainders of $F_1, \ldots, F_n$ have at least one point. We cannot change the graph G, i.e. if $F_i \cap F_j \neq \emptyset$ and x is a point of degree 1, then $(F_i - x) \cap (F_i - x) \neq 0$; since, if x is a common point of F_i and F_j, then certainly so is its neighbor. When we get stuck, we have an F_i

with one point; let F_1 be this, say. Let F_2, F_3 be the two neighbors of F_1 on C, then F_2, F_3 both contain the (unique) point of F_1, and hence, they intersect. So (x_2, x_3) is a chord of C.

Conversely, suppose that G is a graph without chordless circuits of length at least 4; we show by induction on $|V(G)|$ that it is representable in the mentioned way. Let S be a minimum cutset of G, then by the previous result, S induces a complete graph H. Let $G = G_1 \cup G_2$, $G_1 \cap G_2 = H$, G_1, $G_2 \neq H$. By the induction hypothesis, G_1, G_2 can be represented in the given way, through certain subtrees of trees T_1 and T_2, respectively. Let $F_1, \ldots, F_{|S|}$; $F_1', \ldots, F_{|S|}'$ be the subtrees corresponding to the points of S in the representation of G_1 and G_2, respectively. By 6.18, $F_1, \ldots, F_{|S|}$ have a point v in common and similarly, $F_1', \ldots, F_{|S|}'$ have a point w in common. Set

$$T = T_1 \cup T_2 + (v, w), \quad F_i'' = F_i \cup F_i' + (v, w), \qquad (i = 1, \ldots, |S|),$$

then the subtrees F_i'' and those corresponding to points of $G_1 - S$ and $G_2 - S$ yield a representation of G.

(c) We give here proofs using the representation (b). As we have seen in the first part of that proof, we may remove points of degree 1 from T, and still get a system of trees representing the same G until finally, we have a tree F_1 with one point. Let x_1 be the corresponding point of G. Then $G - x_1$ has $\chi(G - x_1) = \omega(G - x_1) \leq \omega(G)$. Moreover, the neighborhood of x_1 forms a complete graph and so, x_1 has degree at most $\omega(G) - 1$. Thus coloring $G - x_1$ by $\omega(G)$ colors we can find a color for x_1.

To show $\chi(\overline{G}) = \omega(\overline{G})$, remove x_1 and all its G-neighbors $x_2, \ldots, x_r$. Any set which forms a clique in $\overline{G} - \{x_1, \ldots, x_r\}$, i.e. which is independent in $G - \{x_1, \ldots, x_r\}$ can be completed with x_1. So $\omega(\overline{G} - \{x_1, \ldots, x_r\}) \leq \omega(\overline{G}) - 1$. Color $\overline{G} - \{x_1, \ldots, x_r\}$ by $\omega(\overline{G}) - 1$ colors by induction; then using a further color for $x_1, \ldots, x_r$ we get the desired coloration of $\overline{G}$. [A. Hajnal–J. Surányi; C. Berge; see B.]

30. Let G have m isolated points $x_1, \ldots, x_m$. By 7.2, $V(G - \{x_1, \ldots, x_r\})$ can be covered by

$$\alpha(G - \{x_1, \ldots, x_m\}) = \alpha(G) - m$$

edges. So $V(G)$ can be covered by $\alpha(G)$ edges and points of G. We may assume that these edges are disjoint, since if two have an endpoint in common, we can replace one of them by its other endpoint. Then these $\alpha(G)$ edges and points of G yield $\alpha(G) = \omega(\overline{G})$ independent subsets of $\overline{G}$ covering the points, i.e. an $\omega(\overline{G})$-coloration of G. Hence, $\chi(\overline{G}) \leq \omega(\overline{G})$. Since the converse is true as well, we have equality.

31. Let H_i be a maximum complete subgraph of G_i $(i = 1, 2)$, then $V(H_1) \times V(H_2)$ spans a complete subgraph of $G_1 \cdot G_2$ of size $|V(H_1)| \cdot |V(H_2)| = \omega(G_1) \cdot \omega(G_2)$. Hence $\omega(G_1 \cdot G_2) \geq \omega(G_1) \cdot \omega(G_2)$.

On the other hand, let α_i be a $\chi(G_i)$-coloration of G_1 $(i = 1, 2)$. Define a coloration α by

$$\alpha(x_1, x_2) = (\alpha_1(x_1), \alpha_2(x_2)) \qquad (x_i \in V(G_i)),$$

then it is easy to verify that α is a good coloration of $G_1 \cdot G_2$ and uses $\chi(G_1) \cdot \chi(G_2)$ colors. Therefore $\chi(G_1 \cdot G_2) \leq \chi(G_1) \cdot \chi(G_2)$. Thus,

$$\chi(G_1 \cdot G_2) \leq \chi(G_1) \cdot \chi(G_2) = \omega(G_1) \cdot \omega(G_2) \leq \omega(G_1 \cdot G_2)$$

and, the converse inequality being trivial, we have equality as stated.

32. (a) The exact formulation of the hint is this. Let $\alpha(x)$ denote the maximum length of a chain $x = x_1 < x_2 < \ldots < x_\alpha$. Then $1 \leq \alpha(x) \leq \omega(G)$, because $\omega(G)$ is the maximum size of a totally ordered subset, i.e. a chain. Moreover, α is a legitimate coloration. For suppose that x and y are adjacent, say $x > y$. Then a chain $x = x_1 < x_2 < \ldots < x_{\alpha(x)}$, of maximum length $\alpha(x)$ can be extended over y, showing $\alpha(y) \geq \alpha(x) + 1$.

(b) Define a directed graph $\overrightarrow{G}$ by orienting the edge (x, y) from x to y if $x < y$. Then, obviously, $\omega(\overrightarrow{G}) = \alpha(\overrightarrow{G})$. By 8.4, we can cover the points of $\overrightarrow{G}$ by $\alpha(\overrightarrow{G})$ disjoint paths. A path of $\overrightarrow{G}$ corresponds to a chain in P. So, it spans a complete subgraph of G; i.e. an independent set in $\overline{G}$. Thus, $\overline{G}$ can be partitioned into $\alpha(\overrightarrow{G}) = \omega(\overline{G})$ independent sets, i.e. $\chi(\overline{G}) \leq \omega(\overline{G})$. The converse is, as always, trivial. [This is an equivalent form of Dilworth' theorem; see B.]

33. In 9.28, 29, 30, 32, the graphs considered are such that all induced subgraphs of them belong to the same class, thus they satisfy $\chi(G) = \omega(G)$ as well. However, we do not get perfect graphs from 9.31 in general. To show this, let G_1, G_2 be two 3-paths, then $G_1 \cdot G_2$ contains a chordless 7-cycle as shown in Fig. 66.

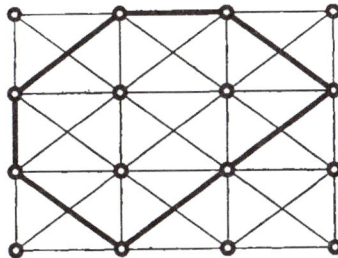

FIGURE 66

34. Suppose first that G is perfect. Let G' be any induced subgraph of G. Let S be a color-class of a $\chi(G')$-coloration of G'. Then $\chi(G' - S) = \chi(G') - 1$ and so, S meets all maximum cliques of G'.

Suppose that, conversely, every induced subgraph of G has an independent set, which meets all maximum cliques. We prove that $\chi(G') = \omega(G')$ for the induced subgraphs by induction on $|V(G')|$. Let S be an independent set, which meets all maximum cliques of G'. Since S is independent,

$$\chi(G') \leq \chi(G' - S) + 1$$

and by its definition,

$$\omega(G') \geq \omega(G' - S) + 1.$$

The two right-hand sides are equal by the induction hypothesis, thus $\chi(G') \leq \omega(G')$. The converse inequality is trivial.

35. We may substitute the graphs G_x for the points of G one by one, and it suffices to show that perfectness is preserved at each step. So we may assume that a non-trivial G_{x_0} is substituted for one point x_0, while the other points are not touched (or we may say one-point graphs are substituted for them).

Also, it suffices to show that the maximum cliques of G' can be covered by an independent set; for the induced subgraphs of G' arise in the same way as G' and so, the previous exercise yields that G' is perfect. Now let us $\chi(G)$-color G and take the color-class S containing x_0. Also, take an independent subset S_1 of G_{x_0}, which meets all maximum cliques of G_{x_0}.

Then clearly $(S-x_0) \cup S_1$ is independent in G'. We show it covers all maximum cliques of G'. Let T be a maximum clique of G'. If T does not meet G_{x_0}, then it is a maximum clique of G and so $S-x_0$ covers it. If T meets G_{x_0}, then, obviously, it contains a maximum clique of G_{x_0} and S_1 covers it. [L. Lovász, *Discrete Math.* **2** (1972) 253–267.]

36. If P is a partition of $V(G)$ into independent sets, then, obviously, it can arise in

$$\lambda(\lambda - 1) \ldots (\lambda - |P| + 1)$$

ways as a partition induced by λ-colorations of G; this clearly also holds, if $|P| > \lambda$. Hence

(1) $$P_G(\lambda) = \sum_P \lambda(\lambda - 1) \ldots (\lambda - |P| + 1),$$

where the summation is over all partitions of $V(G)$ into independent sets. Thus $P_G(\lambda)$ is a polynomial indeed.

Obviously, each term in (1) has degree at most $n = |V(G)|$ and the only term with this degree belongs to the partition into one-element sets. Thus, if one-element sets are independent, i.e. G has no loops, then $P_G(\lambda)$ is of degree n and λ^n has coefficient 1.

If G has a loop, then it has no good coloration at all, so $P_G(\lambda) = 0$. [Concerning chromatic polynomials see O, B, Wi; R. C. Read, *J. Comb. Theory* **4** (1968) 52–71; W. T. Tutte, *Comb. Structures and their Appl.* Gordon and Breach, 1970, 439–453.]

37. To get all λ-colorations of G, consider all mappings of $V(G)$ into $\{1,\dots,\lambda\}$. We have to exclude those, which map the two endpoints of an edge onto the same point. The inclusion-exclusion formula (2.2) yields

$$P_G(\lambda) = \sum_{T \subseteq E(G)} (-1)^{|T|} p(T,\lambda),$$

where $p(T,\lambda)$ denotes the number of those mappings which identify the endpoints of each edge of T. Note that a mapping does this if and only if it maps each component of the graph $\langle V(G), T \rangle$ onto a single number; hence, $p(T,\lambda) = \lambda^{c(T)}$ and the formula is proved.

38. If a λ-coloration of $G - e$ associates different colors with the endpoints of e, it is a good λ-coloration for G, and conversely. Thus $P_{G-e}(\lambda) - P_G(\lambda)$ is the number of those λ-colorations of $G-e$, which associate the same number with the endpoints of e. Now note that such a λ-coloration of $G-e$ yields a λ-coloration of G/e and conversely. Hence,

$$P_{G-e}(\lambda) - P_G(\lambda) = P_{G/e}(\lambda)$$

for every natural number λ. Since we have polynomials on both sides, this must hold for every λ.

39. Since K_n has only one partition into independent sets, the formula in the solution of 9.36 implies that

$$P_{K_n}(\lambda) = \lambda(\lambda - 1)\dots(\lambda - n + 1).$$

Let F be a tree and x a point of degree 1 of F. Any λ-coloration of $F-x$ can be extended to x in $\lambda - 1$ ways; hence

$$P_F(\lambda) = (\lambda - 1)P_{F-x}(\lambda).$$

Since the chromatic polynomial of the one-point tree is λ, we have

$$P_F(\lambda) = \lambda(\lambda - 1)^{n-1}.$$

Let C_n be a circuit of length n, and let e be an edge of it. Then

$$P_{C_n}(\lambda) = P_{C_n-e}(\lambda) - P_{C_n/e}(\lambda).$$

Now note that $C_n - e$ is a path on n points, while C_n/e is an $(n-1)$-circuit C_{n-1}. Thus,

$$P_{C_n}(\lambda) = \lambda(\lambda - 1)^{n-1} - P_{C_{n-1}}(\lambda)$$

and repeating this reduction,

$$P_{C_n}(\lambda) = \lambda(\lambda - 1)^{n-1} - \lambda(\lambda - 1)^{n-2} + \dots + (-1)^{n-2}\lambda(\lambda - 1),$$

since $P_{C_1}(\lambda) = 0$. Summing this series,

$$P_{C_n}(\lambda) = (\lambda - 1)^n + (-1)^n(\lambda - 1).$$

Finally, let W_n be the wheel with n points. Color the center arbitrarily, then the rim has to be colored by the remaining $\lambda - 1$ colors. Thus

$$P_{W_n}(\lambda) = \lambda P_{C_{n-1}}(\lambda - 1) = \lambda(\lambda - 2)^{n-1} + \lambda(-1)^{n-1}(\lambda - 2).$$

40. Let $G_1, \dots, G_m$ be the components of G. Then we can combine λ-colorations of them arbitrarily to get a λ-coloration of G, thus

$$P_G(\lambda) = P_{G_1}(\lambda) \dots P_{G_m}(\lambda).$$

Now suppose that G is connected and let $B_1, \dots, B_m$ be its blocks. We may assume that B_m is a block, which has only one point x_0 in common with the rest and set $G' = B_1 \cup \dots \cup B_{m-1}$.

Take a λ-coloration α_0 of G'. Now those and only those λ-colorations of B_m yield a λ-coloration of G, which color x_0 with $\alpha(x_0)$. Since this only means permutation of colors, the number of such λ-colorations of B_m is

$$\frac{1}{\lambda}P_{B_m}(\lambda).$$

Hence

$$P_G(\lambda) = P_{G'}(\lambda)\frac{1}{\lambda}P_{B_m}(\lambda)$$

and by induction we obtain

$$P_G(\lambda) = \frac{1}{\lambda^{m-1}}P_{B_1}(\lambda) \dots P_{B_m}(\lambda).$$

41. Let G be the graph obtained from the system $\{[a_1, b_1], [a_2, b_2], \dots, [a_n, b_n]\}$ of intervals. We may assume that $b_1 < b_2 < \dots < b_n$. Let x_i be the point of G corresponding to $[a_i, b_i]$. Then, as observed and used before, the neighbors of x_1 form a complete graph. Hence, if we take any λ-coloration of $G - x_1$, this excludes exactly $d(x_1)$ colors for x_1 (we assume that λ is large). Thus,

$$P_G(\lambda) = (\lambda - d(x_1))P_{G-x_1}(\lambda).$$

Let μ_i denote the number of intervals containing b_i, then $d(x_1) = \mu_1 - 1$, so we may write

$$P_G(\lambda) = (\lambda - \mu_1 + 1)P_{G-x_1}(\lambda).$$

Continuing in a similar manner, we get

$$P_G(\lambda) = (\lambda - \mu_1 + 1)(\lambda - \mu_2 + 1)\dots(\lambda - \mu_n + 1)$$

(we have to remark that the first j intervals do not contain b_{j+1} by definition, so if we remove them, b_{j+1} will be contained in the same number μ_{j+1} of intervals).

Note that 9.28 is an immediate consequence; if $\lambda \geq \omega(G) = \max \mu_i$, then $P_G(\lambda) \neq 0$, so there exists a λ-coloration.

42. We use induction on $|E(G)|$. For the edgeless graph on n points we have $P_G(\lambda) = \lambda^n$. We may assume that G is simple since reducing multiple edges to simple ones does not influence $P_G(\lambda)$. So G/e has no loops. By 9.38,

$$P_G(\lambda) = P_{G-e}(\lambda) - P_{G/e}(\lambda)$$

for every edge e of G. By the induction hypothesis

$$P_{G-e}(\lambda) = \lambda^n - a'_{n-1}\lambda^{n-1} + a'_{n-2}\lambda^{n-2} - \cdots$$

$$P_{G/e}(\lambda) = \lambda^{n-1} - a''_{n-2}\lambda^{n-2} + a''_{n-3}\lambda^{n-3} - \cdots,$$

where $a'_i, a''_i \geq 0$. From the recurrence relation we get

$$P_G(\lambda) = \lambda^n - (a'_{n-1} + 1)\lambda^{n-1} + (a'_{n-2} + a''_{n-2})\lambda^{n-2} - \cdots,$$

which proves the assertion. It is seen that $a_{n-1} = a'_{n-1} + 1$, thus it follows by induction that a_{n-1} is the number of edges of G (provided G is simple).

To find an interpretation for a_1, consider the formula given in 9.37. The linear terms correspond to those subgraphs, which are connected spanning subgraphs; thus, $(-1)^{n-1}a_1$ is the difference between the numbers of connected spanning subgraphs with an even and an odd number of edges, respectively. It is remarkable that $a_1 \geq 0$ implies an inequality between these two numbers. The reader may find it interesting to give a direct proof.

43. We use induction of $|E(G)|$. If the graph is a tree the statement is true by 9.40 (and the binomial theorem). Suppose that G is a simple connected graph, which is not a tree, then it has an edge e such that $G-e$ is connected. Obviously so is G/e and they both have no loops. By the previous solution (and with the notation as there),

$$a_i \geq a'_i > 0.$$

Moreover, $a'_{n-1} < a'_{n-2} < \cdots < a'_{\lfloor n/2 \rfloor}, a''_{n-2} < \cdots < a''_{\lfloor (n-1)/2 \rfloor}$ by the induction hypothesis, thus

$$a_{n-1} < \cdots < a_{\lfloor n/2 \rfloor}.$$

44. By 9.36,

$$P_G(\lambda) = \sum_P \lambda(\lambda - 1)\ldots(\lambda - |P| + 1),$$

where P ranges over all partitions of $V(G)$ into independent sets. If $\lambda > n-1$ all terms are positive.

45. We may suppose that G is connected. We show $(-1)^{n-1}P_G(\lambda) > 0$, if $0 < \lambda < 1$. The statement is true for trees. Let e be an edge of G such that $G-e$ is connected. Then

$$(-1)^{n-1}P_{G-e}(\lambda) > 0, \quad (-1)^{n-2}P_{G/e}(\lambda) > 0$$

by the induction hypothesis, so

$$(-1)^{n-1}P_G(\lambda) = (-1)^{n-1}\{P_{G-e}(\lambda) - P_{G/e}(\lambda)\} =$$
$$= (-1)^{n-1}P_{G-e}(\lambda) + (-1)^{n-2}P_{G/e}(\lambda) > 0.$$

46. (a) If G is connected, 9.43 implies 0 is a simple root of P_G. Thus by 9.40, the multiplicity of the root 0 is equal to the number of components. Let G be 2-connected, we show by induction on $|E(G)|$ that 1 is a simple root; we show that, in fact,

$$(-1)^n\frac{P_G(\lambda)}{\lambda - 1} > 0.$$

To this end let us first remark if G has an edge, then 1 is a root by definition; so if G is connected but not 2-connected and $G \neq K_2$, then 1 is a multiple root, which follows immediately from 9.40. Thus, if e is any edge of G,

$$(-1)^n\frac{P_{G-e}(\lambda)}{\lambda - 1}\bigg|_{\lambda=1} \geq 0,$$
$$(-1)^n\frac{P_{G/e}(\lambda)}{\lambda - 1}\bigg|_{\lambda=1} \geq 0,$$

since, if $G-e$ and/or G/e are 2-connected, it follows from the induction hypothesis, otherwise the left-hand side is 0.

Since for $|V(G)| \leq 3$ the assertion is trivially true, we may assume that $|V(G)| \geq 4$. We are going to show that one of $G-e$, G/e is 2-connected. Suppose that G/e is separable; clearly only the image of e can be a cutpoint of it, whence the two endpoints x, y of e form a cutset. Let P_1, P_2 be (x,y)-paths through two components of $G - x - y$. Then $P_1 \cup P_2$ is a circuit and e is a chord of it. But as in 6.35, this implies $G-e$ is 2-connected.

(b) Let $\sigma(G)$ denote the number of acyclic orientations of G. Consider an acyclic orientation of $G-e$. This can be extended to an acyclic orientation of G in one or two ways according as it is an acyclic orientation of G/e. Thus,

$$\sigma(G) = \sigma(G - e) + \sigma(G/e).$$

Hence the result follows by induction and 9.38. [R. Stanley, *Discrete Math.* **5** (1973), 171–178.]

47. (a) By 9.38,

$$P_{G_1}(\lambda) = P_{G_1-e}(\lambda) - P_{G_1/e_1}(\lambda) = P_G(\lambda) - P_{G_1/e_1}(\lambda),$$
$$P_G(\lambda) = P_{G_1}(\lambda) + P_{G_1/e_1}(\lambda),$$

and similarly

$$P_G(\lambda) = P_{G_2}(\lambda) + P_{G_2/e_2}(\lambda);$$

thus,

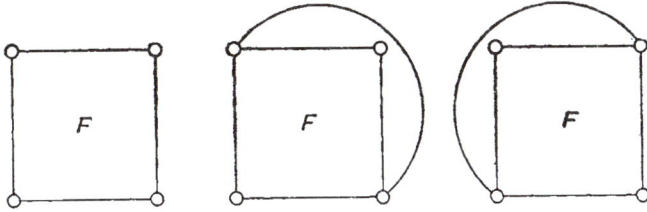

FIGURE 67

$$P_{G_1}(\lambda) + P_{G_1/e_1}(\lambda) = P_{G_2}(\lambda) + P_{G_2/e_2}(\lambda).$$

(b) Consider the three graphs G is Fig. 67, if there is a linear relation, then it is satisfied by them. So

$$a\lambda(\lambda - 1)(\lambda^2 - 3\lambda + 3) + b\lambda(\lambda - 1)(\lambda - 2)^2 + c\lambda(\lambda - 1)(\lambda - 2)^2 = 0,$$
$$a\lambda(\lambda - 1)(\lambda - 2)^2 + b\lambda(\lambda - 1)(\lambda - 2)^2 + c\lambda(\lambda - 1)(\lambda - 2)(\lambda - 3) = 0,$$
$$a\lambda(\lambda - 1)(\lambda - 2)^2 + b\lambda(\lambda - 1)(\lambda - 2)(\lambda - 3) + c\lambda(\lambda - 1)(\lambda - 2)^2 = 0.$$

If this system of equations has a non-trivial solution, its determinant is 0. But its determinant is

$$\lambda^3(\lambda - 1)^3(\lambda - 2)^2 \begin{vmatrix} \lambda^2 - 3\lambda + 3 & \lambda - 2 & \lambda - 2 \\ (\lambda - 2)^2 & \lambda - 2 & \lambda - 3 \\ (\lambda - 2)^2 & \lambda - 3 & \lambda - 2 \end{vmatrix} =$$
$$= \lambda^3(\lambda - 1)^3(\lambda - 2)^2(\lambda^2 - 3\lambda + 1).$$

So if $\lambda \neq 0$, 1, 2, $\frac{3 \pm \sqrt{5}}{2}$, there is no linear relation (in particular, there is none, which would hold for every λ). For $\lambda = 0$, 1, 2, we have trivial relations like $P_{G_1}(\lambda) = 0$.

(c) We use induction on $|E(G)|$. If $V(G) = V(F)$, we have only three graphs to check, namely those in Fig. 67 (since multiple edges can be ignored and if G has a loop, both sides are 0). Fortunately, there are only four numbers to calculate, namely setting $\xi = \tau + 1 = \frac{3 + \sqrt{5}}{2}$, we have

$$\xi(\xi - 1)(\xi^2 - 3\xi + 3) = (\xi^2 - \xi) \cdot 2 = 4\xi - 2,$$
$$\xi(\xi - 1)(\xi - 2)^2 = (\xi^2 - \xi)(\xi^2 - 4\xi + 4) = (2\xi - 1)(3 - \xi) = \xi - 1,$$
$$\xi(\xi - 1)(\xi - 2)(\xi - 3) = \xi(\xi - 3) \cdot (\xi - 1)(\xi - 2) =$$
$$= (\xi^2 - 3\xi)(\xi^2 - 3\xi + 2) = -1$$

(we have made use of the fact throughout that $\xi^2 = 3\xi - 1$). With these values, and the chromatic polynomials determined in the last solution, it is easy to check that the relation holds as stated.

Now suppose that $V(G) \neq V(F)$. If there are only isolated points in G besides the points of F, they only mean that the equations already verified are multiplied by a power of ξ. So suppose that there is an edge e with at most one endpoint on F. Set $G' = G - e$, $G'_1 = G' + e_1$, $G'_2 = G' + e_2$ and similarly, $G'' = G/e$, $G''_1 = G'' + e_1$, $G''_2 = G'' + e_2$. Then by the induction hypothesis,

$$P_{G'}(\xi) = \xi(P_{G'_1}(\xi) + P_{G'_2}(\xi)),$$
$$P_{G''}(\xi) = \xi(P_{G''_1}(\xi) + P_{G''_2}(\xi)),$$

Subtracting we get the desired equality by 9.38.

48. Let $q(x) = x^2 - 3x + 1$; the roots of $q(x)$ are $\xi = \frac{3+\sqrt{5}}{2}$ and $\xi' = \frac{3-\sqrt{5}}{2}$. Then $q \nmid P_G$, because by 9.45, ξ' is not the root of any chromatic polynomial. Since q is irreducible over the rationals, this implies it has no root in common with P_G, i.e. $P_G(\xi) \neq 0$.

49. (a) We use the notations of 9.47. By 9.38 and 9.47c,

$$P_{G_1}(\tau+1) + P_{G_2}(\tau+1) + P_{G_1/e_1}(\tau+1) + P_{G_2/e_2}(\tau+1) =$$
$$= 2(\tau+1)(P_{G_1}(\tau+1) + P_{G_2}(\tau+1)),$$

whence

(1) $P_{G_1/e_1}(\tau+1) + P_{G_2/e_2}(\tau+1) = (2\tau+1)(P_{G_1}(\tau+1) + P_{G_2}(\tau+1)) =$
$$= \tau^3(P_{G_1}(\tau+1) + P_{G_2}(\tau+1)).$$

Also by 9.47a,

(2) $P_{G_1/e_1}(\tau+1) - P_{G_2/e_2}(\tau+1) = P_{G_1}(\tau+1) - P_{G_2}(\tau+1).$

and multiplying (1) and (2) together we obtain

(3) $P^2_{G_1/e_1}(\tau+1) - P^2_{G_2/e_2}(\tau+1) = \tau^3(P^2_{G_1}(\tau+1) - P^2_{G_2}(\tau+1)).$

Now suppose that the maps G_2, G_1/e_1, G_2/e_2 satisfy the identity given in the problem. Then (3) yields upon multiplying by $(\tau+2)\tau^{3n-13}$,

$$P_{G_1/e_1}(\tau+2) - P_{G_2/e_2}(\tau+2) = (\tau+2)\tau^{3n-10}P^2_{G_1}(\tau+1) - P_{G_2}(\tau+2).$$

By 9.47a, this implies G_1 also satisfies the identity.

Now let G_1 be a minimal triangulation, which does not satisfy the identity and suppose that G_1 has a point x of degree as large as possible. Let (x, y, z) be any triangle adjacent to x, and let (x', y, z) be the second triangle with side (y, z). Then $x' = x$. In fact, if $x' \neq x$, then letting $G = G_1 - (y, z)$ and $G_2 = G + (x, x')$, $G_1/(y, z_j)$, $G_2/(x, x')$ and G_2 will satisfy the identity by the minimality assumption on G_1 and the maximality assumption on $d_{G_1}(x)$. Hence by the above, G_1 also satisfies the identity, a contradiction.

The above argument shows that, if a face is adjacent to x, then so are all the neighboring faces. Hence all faces are adjacent to x. This implies that x is adjacent to all other points and $G - x$ has no circuits. Since a triangulation is always 2-connected (if x were a cut-point, then consider a face which meets two components of $G - x$; the fact that this face is a triangle implies that there is an edge connecting two components of $G - x$, a contradiction), $G - x$ is a tree.

Now if we want to λ-color G, we can give x any of the λ colors and $(\lambda - 1)$-color the remaining points. Hence

$$P_G(\lambda) = \lambda P_{G-x}(\lambda - 1) = \lambda(\lambda - 1)(\lambda - 2)^{n-2}$$

and

$$P_G(\tau + 2) = (\tau + 2)(\tau + 1)\tau^{n-2} = (\tau + 2)\tau^n$$

as $\tau + 1 = \tau^2$. Also,

$$(\tau + 2)\tau^{3n-10}P_G^2(\tau + 1) = (\tau + 2)\tau^{3n-10}(\tau + 1)^2\tau^2(\tau - 1)^{2(n-2)} =$$
$$= (\tau + 2)\tau^{3n-10}\tau^4\tau^2\tau^{-2(n-2)} = (\tau + 2)\tau^n$$

as $\tau - 1 = \tau^{-1}$. Thus G satisfies the identity, a contradiction.

(b) We use induction on the number of points and "backwards" induction on the number of edges; we may suppose that there are no 2-gonal faces. If we have a triangulation, (a) implies the assertion. Otherwise, we have a face with more than 3 points. Put in a diagonal e into this face. Then we have, by the induction hypothesis,

$$P_{(G+e)/e}(\tau + 2) > 0,$$
$$P_{G+e}(\tau + 2) > 0.$$

Now by 9.38,

$$P_G(\tau + 2) = P_{(G+e)/e}(\tau + 2) + P_{G+e}(\tau + 2) > 0.$$

50. We use induction on $|V(G)|$. We may assume that G is simple. By 5.25a, the number of edges of G is at most $3|V(G)| - 6$, therefore G has a point x of degree at most 5. If the degree of x is at most 4, we can remove it, 5-color the rest and extend this coloration to x. So suppose that $d(x) = 5$. x has two neighbors, which are non-adjacent, for otherwise we would get a K_6 in the graph. Let y, z be two such neighbors of x. Contract the edges (x, y) and (x, z) and consider the resulting graph G. This has no loops, thus by the induction hypothesis, it is 5-colorable. Let us consider the corresponding 5-coloration of G. This is not legitimate, because x, y and z have the same color; but only the edges (x, y) and (x, z) are monochromatic. Now note that x has only 3 other neighbors, so we can find a color, which is different from the colors of all neighbors of x, and re-color x with this color.

51. Suppose first that the faces are 4-colorable; then it is easy to check that the 3-coloration of the edges defined in the hint is a good one.

Secondly, let us consider a 3-coloration of the edges. Let G_1 be the map formed by red and blue edges and G_2 the map formed by red and green edges. Clearly, G_1 and G_2 consists of disjoint circuits and hence, the faces of G_1 can be 2-colored with red and green; also the faces of G_2 can be 2-colored with "light" and "dark". Now use "light red", "light green", "dark red" and "dark green" to color the faces of G; each face F of G is contained in a face F_1 of G_1 and in a face F_2 of G_2 and color it "light red", if F_1 is red and F_2 is light etc. This defines a good coloration of the faces of G.

Remark: One could also prove the second half by using 5.4 [P. G. Tait; see S, OF, Wi].

52. If the faces are 4-colorable the edges are 3-colorable by 9.51; let 1, 2, 3 be their colors. Consider the assignment described in the hint. Go around the boundary circuit of a face F counterclockwise, and consider the "clock" in Fig. 68. Whenever we pass through a point with value $+1$, the color of the next edge will be the next mark clockwise; if we pass through a point with -1, the color of the next edge is the next mark counterclockwise. Since going around the whole circuit we get back to the same color we started with, the sum of $+1$'s and -1's is divisible by 3.

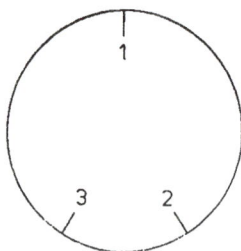

FIGURE 68

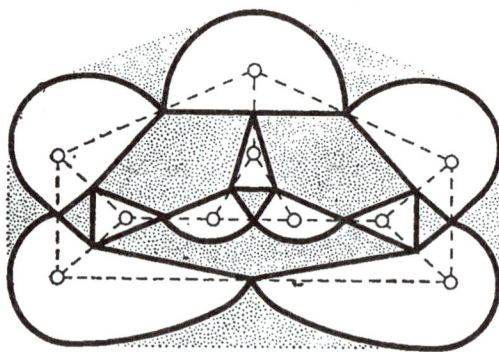

FIGURE 69

Conversely, suppose that we have an assignment of $+1$'s and -1's as required. Consider $L(G)$, this is also planar and has two kinds of faces: For each $x \in V(G)$ it has a triangular face F_x surrounding x and for each face F of G, it has a face F', which lies in F and has the same number of edges (Fig. 69).

Orient $L(G)$ by orienting the triangles F_x clockwise, and define the work $v(e)$ on an edge $e \in E(L(G))$ as the value assigned to the corresponding point of G.

So the three edges of a triangular face of $L(G)$ corresponding to a point of G have the same work. Thus, the sum of work of edges around any face is $\equiv 0$ (mod 3). But, then this holds for any circuit C. For go around each face inside this circuit clockwise and write up the sum of the work. This is 0. On the other hand, the sum of these sums is exactly the work around C (clockwise), because the inside edges are counted twice, and in fact, in different direction. So the work needed to go around C is 0. By 5.4, we can find a potential $p(x)$ of $V(L(G))$ such that for each edge (x, y), $v(x, y) \equiv p(y) - p(x)$ (mod 3). Obviously, we may assume that p takes only the values 0, 1, 2. Then $p(x)$ is a good 3-coloration of $L(G)$. [P. J. Heawood, see S, OF.]

53. The faces inside the Hamiltonian circuit C can be 2-colored. To show this, note that the corresponding piece of the dual graph G^* contains no circuit, for such a circuit would isolate a point of G from C, which is clearly impossible. Thus, the piece of G^* inside C is a forest, and this is 2-colorable.

Similarly, 2 other colors suffice to color the faces outside C.

54. (a) We use induction on $|V(G)|$. If this number is at most 4 the assertion is obvious.

Let $d_G(x) \le 4$. By the induction hypothesis, $G - x$ has a 4-coloration α. If $d_G(x) \le 3$, this 4-coloration can obviously be extended to x. So suppose that $d_G(x) = 4$. Let y, z, u, v be the neighbors of x. We think of G as embedded in the plane and assume that the edges (x, y), (x, z), (x, u), (x, v) leave x in this cyclic order.

If one of the colors does not occur among $\alpha(y)$, $\alpha(z)$, $\alpha(u)$, $\alpha(v)$, we can color x with this color. So suppose that $\alpha(y) = 1$, $\alpha(z) = 2$, $\alpha(u) = 3$, $\alpha(v) = 4$.

Consider the subgraph spanned by colors 1 and 3 and the component G_1 of it containing y. If $u \notin V(G_1)$, we can exchange the colors of points of G_1 and get a 4-coloration of $G - x$, where 1 does not occur among the colors of y, z, u, v. Then, as before, x can get color 1. So we may suppose that $u \in V(G_1)$ i.e. there is a (y, u)-path P_1 whose points have only colors 1 and 3.

Similarly, we may assume that there is a (z, v)-path P_2 having only colors 2 and 4. But since G is planar, P_1 and P_2 cross, and their common point cannot have any of the colors 1, 2, 3, 4, a contradiction (Fig. 70) [G. Dirac; see S, OF. This method is called *Kempe chaining*].

(b) Suppose first that all faces are triangles. Then it is easy to see that we must have the graph of the icosahedron which, in fact, admits a 4-coloration (Fig. 71).

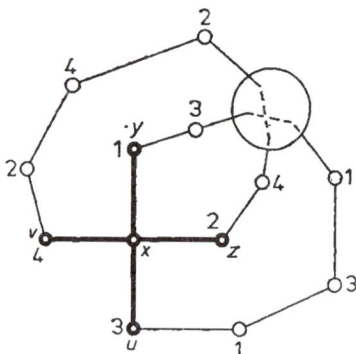

FIGURE 70

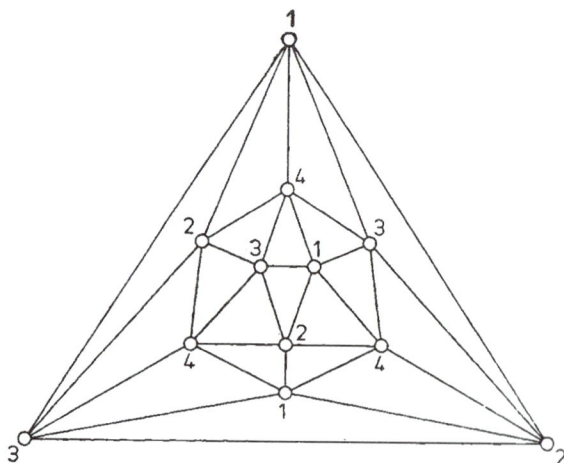

FIGURE 71

We may assume that the graph is 3-connected, this follows by an argument similar to the one in the solution of 9.13b.[†] If F is a non-triangular face, then F has two points x, y which are at distance 2 on the boundary C of F; but, then x, y are not adjacent in G since otherwise, they would separate the two arcs of C connecting them.

Identify x and y, and cancel the arising multiplicities of edges. The resulting graph G' has at least one point of degree at most 4: this is the common neighbor of x and y. It has degrees at most 5 except that the image x' of x (and y) has higher degree.

If G' is 4-colorable clearly so is G. So suppose that G' is not 4-colorable and consider a critically 5-chromatic subgraph G'' of it. The argument in part (a)

[†] Note the analogy between 9.13a–b and this problem.

shows that G'' can have no point of degree at most 4. Hence G'' is a proper subgraph of G' and, moreover, all edges of G' incident with G'' must be incident with x'; otherwise, their endpoints in G'' would have degree at most 4. But, then x' is a cutting point which is impossible, because as G is 3-connected, G' is 2-connected [J. M. Aarts and J. de Groot; see S].

55. Clearly, we may assume that G is 2-connected; otherwise, add another edge joining two blocks of G, not parallel to the old edges.

Consider a face F of G, which is not a triangle, and let $C = (x_1, \ldots, x_k)$ be its boundary circuit. Either (x_1, x_3) or (x_2, x_4) is not an edge of G, for otherwise they ought to cross each other outside C. Say $(x_1, x_3) \notin E(G)$. Then add (x_1, x_3) to G; we have broken F into with smaller faces this way. Carrying on, we can extend G into a simple triangulation. If this triangulation can be colored in the desired way, i.e. so that no face of the triangulation is monochromatic, then no face of G is monochromatic.

Thus, we may assume that G is a triangulation. Then G^* is 3-regular and since G is simple, G^* is 2-edge-connected.

By Petersen's theorem (7.29), G^* has a 1-factor F. Let Φ be the set of corresponding edges of G. Then Φ contains exactly one edge from each face of G. Therefore, the faces of $G - \Phi$ are quadrilaterals, i.e. by the dual of 5.26, $G - \Phi$ is bipartite. Consider the 2-coloration of $G - \Phi$; this 2-coloration meets the requirements for G. [M. Schäuble, *Beiträge zur Graphentheorie*, Teubner, Leipzig (1968) 137–142.]

56. Suppose that every point has even degree. Let us 2-color the faces (5.26) with red and blue and orient the edges so, that they have red on their right side. Let C be any circuit. Count the number of edges on and inside C in two different ways. Let α edges of C have red face incident from the inside and β edges from the outside. Let, moreover, k red and l blue faces lie inside C. Then the number of edges on and inside C is

$$3k + \beta,$$

and also

$$3l + \alpha.$$

Therefore, the work needed to go around C counterclockwise is

$$\alpha - \beta = 3k - 3l \equiv 0 \pmod{3}.$$

This implies, by 5.4, that there is a potential $p(x)$ defined on $V(G)$ such that if $(x, y) \in E(G)$, then $p(y) - p(x) \equiv 1 \pmod{3}$. This $p(x)$ yields a 3-coloration of G [H. Whitney; OF, Wi].

57. Let $(x_1, y_1), \ldots, (x_n, y_n)$ be the intersection points. We may suppose that the direction of axes is chosen so that $x_1, \ldots, x_n$ are different, say $x_1 < x_2 < \ldots < x_n$. Now 3-color the points in the above order. When we get to (x_i, y_i) at most two neighbors of it have previously been colored and so, we can find a color for it. Thus we can 3-color the graph [H. Sachs].

§ 10. Extremal problems for graphs

1. We may assume that G is 2-connected. For if $G = G_1 \cup G_2$ with $|V(G_1) \cap V(G_2)| \le 1$, $|V(G_i)| \ge 2$, then one of G_1, G_2 satisfies

$$|E(G_i)| > \frac{3}{2}(|V((G_i)| - 1),$$

and we can use induction on n.

So suppose that G is 2-connected. Clearly, it is not a single circuit, so by 6.33(b), it has an ear-decomposition $G = P_0 \cup P_1 \cup \ldots \cup P_k$ $(k \ge 1)$, where P_0 is a circuit and P_{i+1} is a $(P_0 \cup \ldots \cup P_i, P_0 \cup \ldots \cup P_i)$-path for $i = 0, \ldots, k-1$. Then $P_0 \cup P_1$ is a Θ-graph.

The sharpness of the result is shown by any graph whose blocks are triangles.

2. (a) Let $P = (x_0, x_1, \ldots, x_p)$ be a longest path in G. x_0 is adjacent to two points other than x_1; since P is maximal these two points must lie on P; let, say, x_0 be adjacent to x_i and x_j, $2 \le i < j \le p$. Then $(x_0, \ldots, x_j)$ is a circuit in G which has the chord (x_0, x_i) [J. Czipszer].

(b) We use induction on n. The assertion is clear for $n = 4$. If G has degrees at least 3, then it contains a circuit with a chord by (a).

Thus we may assume that G has a points with degree at most 2. Then $G - x$ contains at least $2n - 3 - 2 = 2(n-1)$ edges and thus, by the induction hypothesis, $G - x$ contains a circuit with a chord. But, then so does G.

$K_{2,n-2}$ shows that the result of (b) is sharp [L. Pósa].

3. (a) We prove by induction on $|V(G)|$ the assertion of the hint.

I. Assume first that G has a point x of degree at most 2. If x has degree 1, then $G - x$ has at most one point of degree at most 2 and we are done by the induction. If x has degree 2, i.e. it is adjacent to two points y, z and $(y,z) \notin E(G)$, then remove x and join y to z. The resulting graph has degrees at least 3 and thus, it contains a subdivision of K_4. If we subdivide (y,z) by x, we get G back; hence G too contains a subdivision of K_4. If y, z are adjacent, then we may assume that $d_G(y) = d_G(z) = 3$ (otherwise, $G - x$ satisfies the assumption). If the third neighbors of y and z are the same point w, then $G - x - y - z - w$ satisfies the assumption; if y is adjacent to u, z is adjacent to v (x, y, z, u, v being distinct) then contracting x, y, z to a single points x', we get a graph G' in which only x' has degree at most 2. Hence G' contains a subdivision of K_4. Since subdividing the edge (x', u) of G; we get a subgraph of G, G also contains a subdivision of K_4 (see Fig. 72).

II. Now assume that every point of G has degree at least 3. We may assume that G has a point x of degree 3, otherwise we could remove edges as long as all but one of the points of G have degree at least 3. Let u, v and w be the points adjacent to x. If u, v and w are mutually adjacent, then x, u, v and w spans a K_4. Thus, we may suppose that $(u,v) \notin E(G)$. Remove x and join u to v. The resulting graph G' has degrees at least 3 except possibly that w has degree 2.

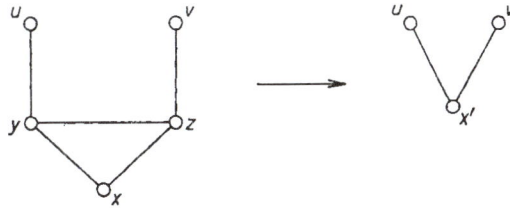

FIGURE 72

Thus G' contains a subdivision of K_4 by induction. Since subdividing the edge (u, v) by x we get a subgraph of G, G also contains a subdivision of K_4.

(b) If every point of G has degree at least 3, we are done by (a). If there is a point of degree 2, we can remove it and proceed by induction.

The graph shown in Fig. 73 has exactly two points of degree 2 and $2n - 3$ edges, it does not contain a subdivision of K_4 (prove!). Thus (a) and (b) are sharp. [G. Dirac, *Math. Nachr.* **22** (1960) 61–85.]

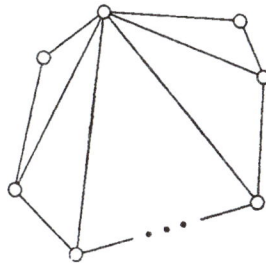

FIGURE 73

4. First we prove, as suggested in the hint, that G does not contain a subdivision H of K_4 and an edge (u, v) disjoint from it. Suppose that it does. Consider the component G_1 of $G - V(H)$ containing (u, v). Trivially G_1 is a tree. Let x, y be two endpoints of G_1. Then x (respectively y) must have two neighbors p_1, p_2 (respectively q_1, q_2) in H. No two of p_1, p_2, q_1, q_2 can be contained in the "subdivided star" of one of the principal points of H as center or inner point of the "rays", for such a position would yield two disjoint circuits (Fig. 74a). But then, as is easily verified, p_1, p_2, q_1, q_2 must be the principal points of H and we again get two disjoint circuits as shown in Fig. 74b.

Now we have to distinguish 5 cases.

Case 1. Assume there is in G a subdivision H of K_4 and a point v of degree at least 4 not in H. A similar argument to the one above shows that v has degree exactly 4 and is adjacent to the four principal points x_1, x_2, x_3, x_4 of H. Exchanging the role of x_i and v we can use the assertion of the hint and see that x_i must be adjacent to x_j ($j \neq i$) and v, and no other point. Hence $G = K_5$.

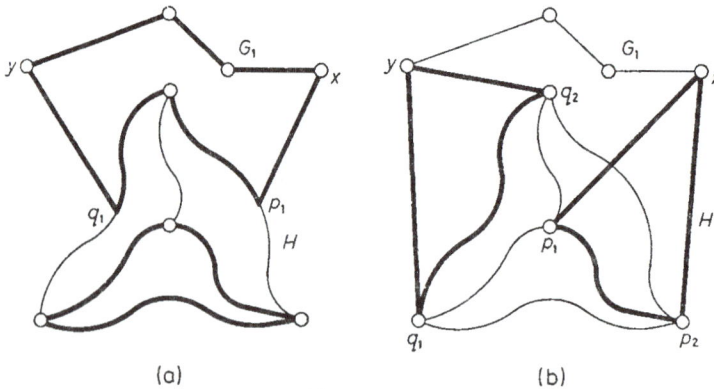

FIGURE 74

Case 2. Assume that G contains a subdivision H of points K_4 and a point v of degree 3 not in H but connected to three principal points x_1, x_2, x_3 of H. Interchanging the role of v and the fourth principal point x_4 we get that $\Gamma(x_4) = \{x_1, x_2, x_3\}$. We show that every edge (r, s) of G contains one of x_1, x_2, x_3. Suppose not, then r, $s \neq x_1$, x_2, x_3, x_4, v. But x_1, x_2, x_3, x_4 and v, together with any one of the subdivided edges (x_1, x_2), (x_2, x_3), (x_1, x_3), form a subdivision of K_4. Hence (r, s) must meet each of these subdivided edges at inner points, which is impossible. Thus x_1, x_2, $_3$ is fact cover all edges. But this clearly means we have graph (iii).

Case 3. There is a subdivision H of K_4 and a point not in H of degree 3, connected to at most two principal points of H. By the assertion of the hint, the neighbors of v must be in H and it is easy to see that at least two of them, x_1 and x_2 say, must be principal points. The third one, being not principal must be an inner point p of the subdivided (x_3, x_4)-edge. Interchanging the roles of x_3 and v we get $\Gamma(x_3) = \{x_1, x_2, p\}$ and similarly, $\Gamma(x_4) \triangleq \{x_1, x_2, p\}$. We claim that x_1, x_2, p cover all edges; then we shall be able to conclude as before. Since x_1, x_2, x_3, x_4, p, v include a subdivision of K_4, each edge must meet one of these points. But those meeting x_3, x_4 or v also meet x_1, x_2 or p. This settles this case.

Case 4. So we may assume that each subdivision of K_4 is a spanning subgraph. By 10.3a, G contains a subdivision H of K_4. Let x_1, x_2, x_3, x_4 be its principal points. Assume that this subdivided K_4 can be chosen so that x_1 has degree at least 4. Let $x_1, u)$ be an edge adjacent to x_1 but not in H. Since H is a spanning subgraph, u is a point of H; trivially u must lie on the circuit C of $H - x_1$. Replacing the subdivided edge (x_1, x_i) $(i = 2, 3, 4)$ by (x_1, u), we must get spanning subdivided K_4's. Hence x_1 is adjacent to x_2, x_3 and x_4. Any point of C can, trivially, be adjacent to x_1 only (except for its neighbors on C). Thus G is a wheel.

Case 5. Only one more case is left, when each subdivided K_4 in G is a spanning subgraph and its principal points have degree 3 in G. If $G = K_4$, then it is a wheel. So suppose that H has a chord e, which must then connect inner points of two subdivided edges of H. Clearly these subdivided edges must be disjoint, e.g. (x_1, x_2) and (x_3, x_4). Now G contains no further point since removing such a point the rest of $H + e$ would still contain a subdivision of K_4, contrary to the assumption. Hence G has six point, a trivial case.

Remark: By similar arguments we could determine all (not necessarily simple) graphs, which contain no two disjoint circuits. Given any such graph, we can remove points of degree one and "smooth out" points of degree 2, so it suffices to describe these graphs with degrees at least 3. The result is the following: the three types of graphs in the problem with the modification that the spokes of the wheel and the edges in the 3-element class of $K_{3,n-3}$ may be multiple; and there is one new class: take a forest F, and a point x with possibly some loops at x, and connect x to F arbitrarily. [G. A. Dirac, *Canad. Math. Bull.* **8** (1965) 459–463; L. Lovász, *Mat. Lapok* **16** (1965) 289–299.]

5. (a) By 10.3b, G contains a subdivision of K_4. If this subdivision is proper, i.e. at least one of the subdivided edges has an inner point we easily find a subdivision of $K_{2,3}$. So we may suppose that G contains four pairwise adjacent points x, y, z, u. Since $n \geq 5$, there is at least one more point v. Since G is 2-connected there are two independent paths connecting v to two distinct points of x, y, z, u. Now x, y, z, u, v are principal points of a subdivision of $K_{2,3}$ (Fig. 75).

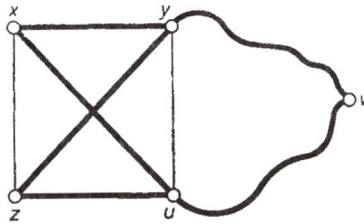

FIGURE 75

(b) Since $|E(G)| \geq 3n - 5$, 5.25(a) implies that G is non-planar and hence, by Kuratowski's theorem 5.39d, G contains a subdivision of $K_{3,3}$ or K_5. In the first case we are done, so suppose that G contains a subdivision of K_5 with principal points $x_1, \ldots, x_5$. As in part (a), we distinguish whether or not this subdivision is proper.

1° Suppose that the subdivided (x_1, x_2)-edge Q (say) contains inner points. $G - \{x_1, x_2\}$ being connected, we have a path P joining an inner point y of Q to a point z outside Q. z may have three essentially different positions but in either case, we find a subdivision of $K_{3,3}$ easily (see Fig. 76, p. 426).

2° Suppose that $x_1, \ldots, x_5$ are mutually adjacent. Since $n \geq 6$, we must have a further point v. Since G is 3-connected, there are 3 independent paths connecting v to x_1, x_2, x_3 (say). Then again a subdivision of $K_{3,3}$ is found trivially (Fig. 77).

(c) We claim that if a graph has at least $2n - 2$ edges and contains no subdivision of $K_{3,2}$, then it is connected and its blocks are K_4's. We use induction on n.

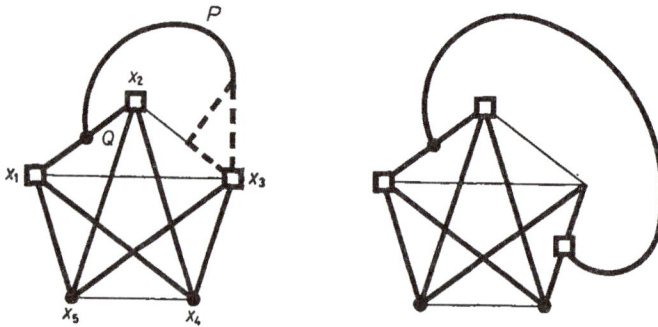

FIGURE 76

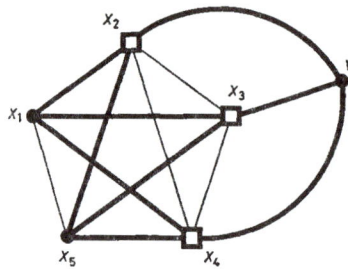

FIGURE 77

Suppose first that a graph G with $2n-2$ edges is not connected, i.e. $G = G_1 \cup G_2$, $G \cap G_2 = \emptyset$. Let $|V(G_1)| = n_i > 0$ $(i = 1, 2)$. Then

$$|E(G_1)| + |E(G_2)| \geq 2n - 2 = (2n_1 - 2) + (2n_2 - 2) + 2,$$

whence, e.g.

$$|E(G_1)| \geq 2n_1 - 1.$$

Thus, by induction, G_1 is connected and all blocks of it are K_4's. But, then $|E(G_1)| = 2n_1 - 2$, a contradiction. So G is connected.

If G is 2-connected the assertion follows from (a). So suppose that it has a cut-point x. Let $G = G_1 \cup G_2$. $V(G_1 \cap G_2) = \{x\}$, $|V(G_i)| = n_i < n$. Then

$$|E(G_1)| + |E(G_2)| \geq 2n - 2 = (2n_1 - 2) + (2n_2 - 1).$$

As above, we get a contradiction, if

$$|E(G_i)| > 2n_i - 2$$

for any i. Thus it follows that

$$|E(G_1)| = 2n_1 - 2, \quad |E(G_2)| = 2n_2 - 2$$

and hence by the induction hypothesis, G_1, G_2 are connected graphs whose blocks are K_4's. Thus so is G.

A similar argument shows that, if G is a graph with n points and $3n-5$ edges containing no subdivision of $K_{3,3}$, then G is 2-connected and it is composed of complete 5-graphs by the following recursive rule. We take a graph G (already constructed) and a complete 5-graph K_5, and identify an edge of G with an edge of K_5 (see Fig. 79a).

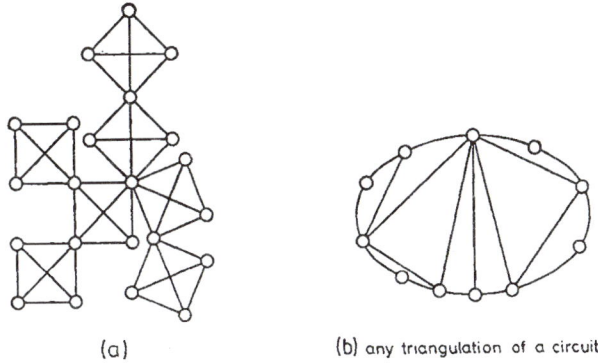

(a) (b) any triangulation of a circuit

FIGURE 78

Thus the answer to (c) is: A graph with n points and

$$2n - 1 \text{ edges if } n \equiv 1 \pmod 3,$$
$$2n - 2 \text{ edges if } n \equiv 0,\ 2 \pmod 3$$

contains a subdivision of $K_{2,3}$; a graph with n points and

$$3n - 4 \text{ edges if } n \equiv 2 \pmod 3,$$
$$3n - 5 \text{ edges if } n \equiv 0,\ 1 \pmod 3$$

contains a subdivision of $K_{3,3}$. These bounds are best possible as shown by the graphs in Figs 78 and 79 [L. Pósa].

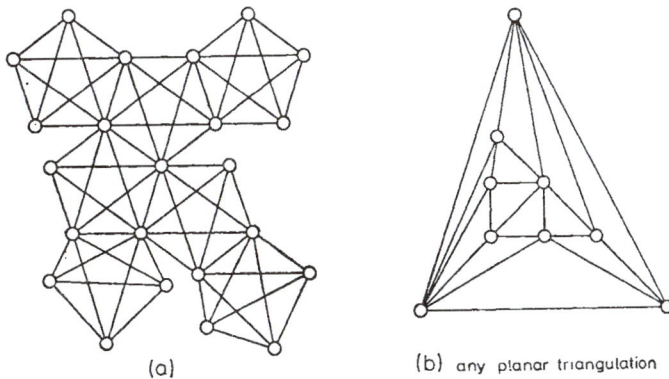

(a) (b) any planar triangulation

FIGURE 79

6. We show: (a') If G_0 has degrees at most 3, then any graph contractible onto G_0 contains a subdivision of G_0; (b') If G_0 has a point with degree at least 4 this is no longer true.

(a') It suffices to show on the one hand that, if G/e $(e = (u,v) \in E(G))$ contains a subdivision H of G_0, then so does G. If the point x of G/e, which is the image of e is not a principal point of H, then this is clear; suppose that it is a principal point. By the assumption, there are at most 3 subdivided edges of G starting from x. If we pull u and v apart again, some of these will start from u and some will start from v. We may assume that at most one starts from v; then adding e to this subdivided edge and considering u as a principal point we find a subdivision of G_0 in G.

(b') We show that there is a graph G_1, contractible onto G_0, with degrees at most 3. Let x be a point of G_0 with degree at least 4. Let us split x into d points $x_1, \ldots, x_d$ of degree 1 and connect these by a circuit. Do so for each point x with degree at least 4, and denote the resulting graph by G_1. Then G_1 has degrees at most 3 and is contractible onto G_0, If G_0 has a point with degree at least 4, then, clearly, G_1 contains no subdivision of G_0.

A 4-connected counterexample in the case of K_5 is $K_{4,4}$.

7. Let $a \in V(G)$. Contract edges and cancel and arising multiplicities repeatedly so that

(∗) the contracted edge is adjacent to the image of a,

(∗∗) $\dfrac{|E(G')|}{|V(G')|} \geq \dfrac{m}{n}$ holds for the resulting graphs.

Suppose that we get stuck with the graph G_0. Let a_0 be the image of a in G_0. It follows from (∗) above that the subgraph G_1 of G mapped onto a_0 is connected and $G - V(G_1) = G_0 - a_0$.

Let x be any point adjacent to a_0 in G_0 (i.e. adjacent to $V(G_1)$ in G). Let ϑ denote the number of points of G_0 adjacent to a_0 and x (this is the degree of x in the subgraph of G induced by the neighbors of G_1). Try to contract (a_0, x) and cancel one edge from each of the arising ϑ parallel pairs. Then (∗∗) must fail to hold for the resulting graph G_0'. Since

$$|V(G_0')| = |V(G_0)| - 1$$
$$|E(G_0')| = |E(G_0)| - 1 - \vartheta,$$

we have

$$\frac{|E(G_0)| - \vartheta - 1}{|V(G_0)| - 1} < \frac{m}{n}$$

or, equivalently,

$$\vartheta > |E(G_0)| - 1 - \frac{m}{n}(|V(G_0)| - 1).$$

Since

$$\frac{|E(G_0)|}{|V(G_0)|} \geq \frac{m}{n},$$

we have

$$\vartheta > \frac{m}{n}|V(G_0)| - 1 - \frac{m}{n}(|V(G_0)| - 1) = \frac{m}{n} - 1,$$

which proves the assertion [W. Mader; see W].

8. We use induction on m. For $m = 3$ the assertion is clear. Let $m \geq 4$. We may assume that G is connected.

By the preceding problem, G contains a connected subgraph G_1 such that the subgraph G_2 induced by the neighbors of G_1 has degrees $> \frac{|E(G)|}{n} - 1 \geq 2^{m-3} - 1$, so $\geq 2^{m-3}$. Hence

$$|E(G_2)| \geq 2^{m-4}V(G_2),$$

and thus, by the induction hypothesis, G_2 can be contracted onto K_{m-1}. Carrying out the same contraction on G and also contracting G_1, we obtain a graph which contains K_m [ibid].

9. Induction on k: for $k = 1$ the assertion is trivial.

Again, let G_1 be a connected subgraph as in 10,7, i.e. let the subgraph G_2 induced by the neighbors of G_1 have degrees $\geq 2^k$. Then

$$|E(G_2)| \geq 2^{k-1}|V(G_2)|$$

and thus, G_2 contains a subdivision H of $F - e$ for any $e \in E(F)$. Let x_1, x_2 be the principal points of H corresponding to the endpoints of e. Since $x_i \in V(G_2)$, it is adjacent to some point $y_i \in V(G_1)$ $(i = 1, 2)$. Since G_1 is connected it contains a (y_1, y_2)-path P. Then

$$H' = H + (x_1, y_1) + (x_2, y_2) + P$$

is a subdivision of F contained in G. [ibid.]

10. I. Let x_1 be a point of the 3-regular graph G with girth 4, and let x_2, x_3, x_4 be its neighbors. Then x_2, x_3, x_4 must be independent, because G contains no triangle. Let x_5, c_6 be two further neighbors of x_2. Thus $|V(G)| \geq 6$; equality holds only of x_3, x_4 are also adjacent to x_5 and x_6, i.e. $G = K_{3,3}$.

II. Let $(x_1, \ldots, x_p)$ be the least circuit in G, $p \geq 5$. Let y_i be the third neighbor of x_i. By the minimality of the circuit considered, y_i does not lie on it and $y_i \neq y_j$ for $i \neq j$. Hence $|V(G)| \geq 2p \geq 10$. If equality holds, then $p = 5$. The only points at distance at least 4 from y_1 are y_3 and y_4, so these must be the two further neighbors of y_1 besides x_1. Similarly, y_2 is adjacent to y_4 and y_5, and y_3 to y_5, i.e. we get the Petersen graph (Fig. 80).

11. I. Assume first that g is odd. Let $x_0 \in V(G)$, and denote by S_i, the set of points at distance i from x_0 $\left(i = 0, \ldots, \frac{g-1}{2}\right)$. From each point x of S_i there is exactly one edge to S_{i-1}, since two such edges would yield two paths from x to

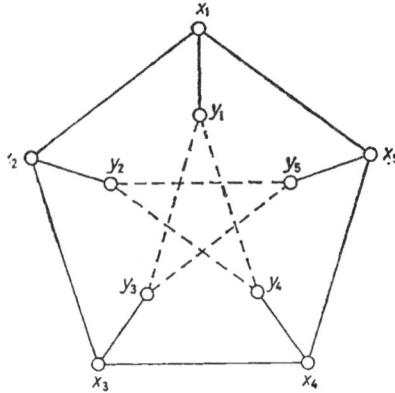

FIGURE 80

x_0 of length i, and these would form a circuit of length less than g. Thus $|S_{i+1}| = (r-1) \cdot |S_i| \left(i=1, \ldots, \frac{g-2}{3} \right)$, and hence

$$|V(G)| \geq |S_0| + |S_1| + \ldots + \left| S_{\frac{g-1}{2}} \right| =$$

$$= 1 + r + r(r-1) + \ldots + r(r-1)^{\frac{g-3}{2}}.$$

II. Let g be even and consider two adjacent points x, y. Denote by S_i the set of points at distance i from the set $\{x, y\}$ $(i=1, \ldots, g/2-1)$. The result follows by the same counting as above. [W. T. Tutte; P. Erdős–H. Sachs; see S.]

12. Assume that we have constructed $G(r,g)$ and $G(r',g-1)$, where $r'=|V(G(r,g))|$. Split each point of $G(r',g-1)$ into r' points of degree one and identify these r' points with the points of a copy of $G(r,g)$. The resulting graph G' is, obviously, $(r+1)$-regular. We claim that it has girth g. If we consider a minimum circuit in a copy of $G(r,g)$, this has length g; hence G' has girth at most g. Any other circuit in copies of $G(r,g)$ has length at least g. Let C be a circuit of length s not in the copies of $G(r,g)$. Contract each copy of $G(r,g)$, then G' is mapped onto $G(r'g-1)$. C is mapped onto a non-empty subgraph with even degrees, hence the image of C contains a circuit C'. Now C contains at least one edge[†] of a copy of $G(r,g)$ and hence, C' has less than s edges. Thus $s-1 \geq g-1$, $s \geq g$. Thus $G'=G(r+1,g)$.

Now since $G(r,2)$ as well as $G(2,g)$ exists trivially (r parallel edges and a g-gon, respectively) we get that $G(3,3)$, $G(4,3)$, …, $G(3,4)$; $G(4,4)$, …, ; …, ; $G(3,g)$, $G(4,g)$, … also exist. [P. Erdős–H. Sachs; see S, WV.]

13. (a) Suppose indirectly that a, $b \in V(G)$ are at distance $> g$. Remove a and b and add r new edges, matching the r neighbors of a with the r neighbors of b. The resulting graph G' is obviously, r-regular.

---

† At least $s/2$ such edges, in fact; but we do not use this observation.

We show that G' has girth at least g. Let C be a circuit in G'. If C does not contain any of the r new edges, then it is a circuit in G and thus, the length of C is at least g. So suppose that C contains some new edges.

Now consider a path P in $G - \{a,b\}$ joining two endpoints of the new edges. If P joins a neighbor of a to a neighbor of b, then, since a, b are at distance at least $g+1$, P has length at least $g-1$. If P joins two neighbors of a (or b), then with two further edges it forms a circuit of G through a (or b) and hence, it has length at least $g-2$.

Now if C contains only one new edge, then it contains a path joining the two endpoints of this edge. By the above argument, this path has at least $g-1$ edges and thus, C has at least g edges.

On the other hand, if C contains at least two new edges, then it also contains at least two paths connecting endpoints of these. Hence its length is at least

$$2 + 2(g - 1) \ge g.$$

Thus G' is r-regular and has girth at least g. Since $|V(G')| < |V(G)|$, this is a contradiction.

(b) I. Assume that $r = 2l$. Let $x \in V(G)$. Remove x and join the $2l$ neighbors of it by new independent edges. The resulting graph G' is clearly r-regular. Since $|V(G')| < |V(G)|$, G' must have girth less than g, i.e. it must contain a circuit C with length less than g. Obviously, C contains some (at least one) new edges; let $e_1, \ldots, e_s$ be these edges and let them split C into arcs $P_1, \ldots, P_s$. We observe that P_i forms a circuit in G when we add the two new edges joining its endpoints to x, thus $|E(P_i)| \ge g - 2$. Hence

$$g > |E(C)| \ge s(g - 2) + s = s(g - 1).$$

Thus $s = 1$, i.e. C consists of a new edge e_1 and a path P_1 connecting its endpoints, such that $P_1 \subseteq G - \{x\}$. It also follows that P_1 has $g - 2$ edges, i.e. joining its endpoints to x we get a circuit of length g in G.

II. Assume that $r = 2l + 1$. If $f = 2$ or 3, then G consists of r parallel edges and $G = K_{r+1}$ respectively. Thus we may assume that $g \ge 4$.

Let $(x,y) \in E(G)$. Remove x and y from G and pair up the $2l$ remaining neighbors of x as well as the $2l$ remaining neighbors of y. The resulting graph G' is r-regular again. Since $|V(G')| < |V(G)|$, G' contains a circuit C of length less than g. Again, let $e_1, \ldots, e_s$ be the new edges on C and let them divide C into arcs $P_1, \ldots, P_s$. Note that

$$|E(P_i)| \ge \begin{cases} g - 2 & \text{if } P_i \text{ joins two neighbors of } x \text{ or two neighbors of } y. \\ g - 3 & \text{if } P_i \text{ connects a neighbor of } x \text{ to a neighbor of } y. \end{cases}$$

Thus

$$g > |E(C)| \ge s(g - 3) + s = s(g - 2),$$

whence $s = 1$ as $g \ge 4$. Now we conclude as before that P_1 is contained in a circuit of G of length g, which passes through x or y.

(c) Let $x \in V(G)$. by (a), any point of G can be reached from x by a path of length at most g. But there are

$$\leq r(r-1)^{j-1}$$

paths of length j starting from x (some of them may terminate at the same point; some may come back to x) and thus,

$$|V(G)| \leq 1 + r + r(r-1) + \ldots + r(r-1)^{g-1} =$$
$$= 1 + \frac{r}{r-2}((r-1)^g - 1) < \frac{r}{r-2}(r-1)^g$$

[ibid.].

14. Take a maximal graph G', which arises from G by adding independent edges and which has girth at least g. Assume indirectly that G' has a point u of degree r. Then for reasons of parity, it has another such point v.

Let S be the set of points of G' at distance at most $g-1$ from $\{u,v\}$. Each point w of G' outside S must be of degree $r+1$, otherwise the edge (u,w) could be added to G'. Since by a computation similar to that in the solution of 10.11 we have

$$|S| \leq 2(1 + r + \ldots + r^{g-1}) < 2r^g \leq \frac{1}{2}|V(G)|,$$

if follows that there are two points x, y outside S connected by an edge of $E(G')-E(G)$. Now $G' - (x,y) + (x,u) + (x,v)$ is a graph with girth at least g, arising from G by addition of independent edges, and larger than G', a contradiction.

15. (a) Let U be the set of point and W the set of lines of the projective plane over $GF(p)$. Connect $u \in U$ to $w \in W$ iff $u \in w$. The resulting bipartite graph is clearly $(p+1)$-regular and has $2(p^2 + p + 1)$ points. To show it has girth 6, note that it contains no quadrilaterals; in fact, if (u_1, w_1, u_2, w_2) were a circuit, then the lines w_1, w_2 would have two points in common u_1, u_2. It contains no circuits of length 3 or 5, because it is bipartite. Any triangle in the plane yields a hexagon in the graph.

(b) Consider the hypersurface $\mathcal{F} = \{x : x^T x = 0\}$ in the 4-dimensional projective space over $GF(p)$. First prove some geometric properties of $\mathcal{F}$.

(i) $\mathcal{F}$ contains no plane. Suppose indirectly that π is a plane contained in $\mathcal{F}$ and let u, $v \in \pi$. Then $u^T u = 0$, $v^T v = 0$ and $(u+v)^T(u+v) = 0$ (since u, v and $u+v$ are vectors representing points of π). Hence it follows that $u^T v = 0$. Hence if V denotes the linear subspace of the 5-dimensional vector space formed by homogeneous coordinate quintuples of points of π, we have $V \subseteq V^\perp$. But this implies that $\dim V^\perp \geq \dim V = 3$, contradicting 5.31.

(ii) There are exactly $p+1$ lines of $\mathcal{F}$ through each point b of $\mathcal{F}$. Let $\Sigma = \{x : b^T x = 0\}$ be the "tangent hyperplane" at b. Each line through b is contained in Σ; for if v is a point of such a line then as in (i), it follows that $b^T v = 0$. Conversely, if $v \in \mathcal{F} \cap \Sigma$, then all points of the line space spanned by b and v belong to $\mathcal{F} \cap \Sigma$.

Let π be any plane in Σ avoiding b. Then $\pi \cap \mathscr{F}$ is a conic section, which is non-degenerate since if it contained a line, then this, together with b, would give a plane contained in $\mathscr{F}$, contradicting (i). So $\pi \cap \mathscr{F}$ has $p+1$ points as is well known. Now the lines in $\mathscr{F}$ through b are exactly those lines connecting b to the points of $\pi \cap \mathscr{F}$, so their number is $p+1$.

Let us note that these considerations together with (i) also imply that no three lines in $\mathscr{F}$ form a triangle.

(iii) $|\mathscr{F} = p^3 + p^2 + p + 1$. Observe that given any line Λ and point b on $\mathscr{F} - \Lambda$, there is a unique line on $\mathscr{F}$ containing b and meeting Λ. In fact, the tangent hyperplane Σ at b cannot contain Λ, so Σ intersects Λ in a single point a, and then the line ab is this unique line. Thus, if we consider any line Λ, any point of $\mathscr{F}$ is incident with exactly one of these lines. Hence $|\mathscr{F} - \Lambda| = p^2(p+1)$, which proves the assertion.

Since each line contains $p+1$ points and each point is incident with $p+1$ lines of $\mathscr{F}$, the number of lines of $\mathscr{F}$ is the same. Let $\mathscr{L}$ denote the set of lines of $\mathscr{F}$.

Now form a bipartite graph on $\mathscr{F} \cup \mathscr{L}$ by connecting $b \in \mathscr{F}$ to $\Lambda \in \mathscr{L}$ iff $b \in \Lambda$. The resulting bipartite graph is $(p+1)$-regular by (ii) and has $2(p^3 + p^2 + p + 1)$ points by (iii). It contains no 4- or 6-circuit, since a 4-circuit would correspond to two lines on $\mathscr{F}$ meeting in two points and a 6-circuit to 3 lines on $\mathscr{F}$ forming a triangle, both impossible.

16. (a) If Z represents all circuits, then $V(G) - Z$ spans a forest and hence, it spans at most $n - |Z| - 1$ edges. Since each point has degree at least 3, there are $3(n - |Z|)$ edges leaving the points of $V(G) - Z$. The edges spanned by this set are counted here twice nut still we have at least

$$3(n - |Z|) - 2(n - |Z| - 1) = n - |Z| + 2$$

edges connecting $V(G) - Z$ to Z. On the other hand, a point of Z is incident with at most d edges, hence the number of $(V(G) - Z, Z)$-edges is at most $d \cdot |Z|$. Thus,

$$d \cdot |Z| \geq n - |Z| + 2$$

or, equivalently,

$$|Z| \geq \frac{n+2}{d+1}.$$

(b) We may suppose that $g \geq 3$, as $g = 1, 2$ are trivial.

Let d denote the maximum degree in G. The same counting as in 10.11 yields

$$n \geq 1 + d + 2d + \ldots + 2^{\lfloor \frac{g-3}{2} \rfloor} \cdot d = 1 + d\left(2^{\lfloor \frac{g-1}{2} \rfloor} - 1\right).$$

Now by (a), any set Z representing all circuits satisfies

$$|Z| \geq \frac{n+2}{d+1} \geq \frac{3 + d\left(2^{\lfloor \frac{g-1}{2} \rfloor} - 1\right)}{d+1} \geq \frac{3 + 3\left(2^{\lfloor \frac{g-1}{2} \rfloor} - 1\right)}{3+1} \geq \frac{3}{8} \cdot 2^{g/2}.$$

[H.–J. Voß, M. Simonovits; see WV].

17. Let $G_1, G_2, \ldots, G_\nu$ be defined as in the hint; $G_{\nu+1} \neq \emptyset$. One may assume that the girth of each G_i is at most g, otherwise we could consider G_i instead of G. Hence $|V(C_1) \cup \ldots \cup V(G_\nu)| \leq \nu g$.

On the other hand, it is easy to verify that $V(C_1) \cup \ldots \cup V(C_\nu)$ represents all circuits of G. Thus part (b) of the preceding problem yields

$$|V(C_1) \cup \ldots \cup V(C_\nu)| \geq \frac{3}{8} 2^{g/2}.$$

Hence

$$\nu \geq \frac{3}{8g} 2^{g/2},$$

as stated [ibid.].

18. (a) $\nu = 1$ means that ant two circuits of G have a point in common. If G is not simple, i.e. it contains a loop or two parallel edges, then all circuits are covered by 1 or 2 points, respectively. So we may assume that G is simple. Then by 10.4, G is one of the graphs displayed in Fig. 8. The circuits of K_5 are covered by any 3 points of it; those of the wheel are covered by the center and any one point of the rim; those of the third example ($K_{3,n-3}$ with some edges added in the 3-element class) are covered by any two points of the 3-element class [B. Bollobás, L. Pósa].

(b) We use induction on $V(G)$. We may assume that G has no points of degree 1 or 2, because these could be removed and "smoothed out", respectively. Remove the points of a shortest circuit. By 10.16b, the length of this satisfies

$$g \leq 4 \log_2 \tau.$$

Let τ' denote the minimum number of points of covering all circuits of the remaining graph. Clearly

$$\tau \leq \tau' + g.$$

So using the induction hypothesis,

$$\nu \geq \frac{\tau'}{4 \log \tau'} + 1 \geq \frac{\tau'}{4 \log \tau} + 1 \geq \frac{\tau - g}{4 \log \tau} + 1 \geq \frac{\tau}{4 \log \tau}.$$

(c) Let G be a 3-regular graph with girth g and minimum number n of points. By 10.13c,

$$n \leq 3 \cdot 2^g, \qquad \text{or} \qquad g \geq \log_2 \frac{n}{3}.$$

By 10.16a, we need at least $n/4$ points to represent all circuits, i.e. $n \leq 4\tau$. Furthermore,

$$\nu \leq \frac{n}{g} \leq \frac{n}{\log_2(n/3)} \leq \frac{4\tau}{\log_2 \tau}.$$

[P. Erdős–L. Pósa; this proof is due to H.–J. Voß, in: *Theory of Graphs* (Akadémiai Kiadó, Budapest 1968) and M. Simonovits, *Acta Math. Acad. Sci. Hung.* **18** (1967) 191–206; also see WV.]

19. Consider $G_1 = G - x$. Since G is 2-connected, G_1 is connected.

1° If G_1 is 2-connected, then let x_1 be any point adjacent to x. Since $d_{G_1} \geq k - 1$ for every $z \neq x_1$, y (for every $z \neq y$, in fact) there is an (x_1, y)-path of length at least $k - 1$ in G_1. Adding (x, x_1) we get an (x, y)-path of length at least k.

2° Assume that G_1 is not 2-connected and let $G_1 = A \cup B$, $|V(A) \cap V(B)| = 1$. Choose the notation so that $y \in V(A)$ and suppose that B is minimal. Then B is, obviously, 2-connected. Let $\{y_1\} = V(B) \cap V(A)$. Since G is 2-connected, x must be adjacent to a point $x_1 \in V(B) - \{y_1\}$.

Now each point $z \neq y_1$ of B has degree at least $k - 1$, hence B contains an (x_1, y_1)-path P of length at least $k - 1$. Also, let P_2 be a (y_1, y)-path in A; then

$$P = P_1 + P_2 + (x, x_1)$$

is an (x, y)-path in G of length at least k.

20. Let X be defined as in the hint. Obviously, all path arising by repeated deformation from P have the same points, thus $X \subseteq V(P)$. Let $X = \{x_0, x_{i_1}, \ldots, x_{i_t}, x_m\}$ $(0 < i_1 < \ldots < i_t < m)$. We claim that

$$\Gamma(X) \subseteq \{x_1, x_{m-1}, x_{i_\nu \pm 1} : \nu = 1, \ldots, t\} \cup X$$

(whence $|\Gamma(X) - X| \leq 2t + 2 = 2|X| - 2$).

Let $u \in \Gamma(X)$, $u \notin X$. Then u is adjacent to a point $v \in X$. Since, by definition, v is an endpoint of a maximum path whose points are the points of P, u must belong to P. Let $u = x_i$.

By the definition of v, there are paths $P_0 = p, P_1, \ldots, P_s$ such that P_{j+1} arises by deformation from P_j $(j = 0, \ldots, s-1)$ and v is an endpoint of P_s. If both edges (x_i, x_{i-1}) and (x_i, x_{i+1}) belong to P_s, then let, say, (x_i, x_{i+1}) be the first edge of the (U, v)-arc of P_s. Then x_{i+1} is an endpoint of a path arising from P_s by deformation, i.e. $x_{i+1} \in X$ and we are finished.

So suppose, e.g., that $(x_{i+1}, x_i) \notin E(P_s)$. Then there is an index j, $0 \leq j \leq s-1$ such that $(x_{i+1}, x_i) \in E(P_j)$ but $(x_{i+1}, x_i) \notin E(P_{j+1})$. Since P_{j+1} arises from P_j by deformation, this can only happen if one of x_{i+1}, x_i is the endpoint of P_{j+1}. Since $x_i \notin X$, we must have $x_{i+1} \in X$ and we are done again.

Thus we know that

$$|\Gamma(X) - X| \leq 2|X| - 2,$$

whence, by the assumption, $|X| \geq k + 1$. Let $X_1 \subset X$, $|X_1| = k$. Then by the assumption

$$|\Gamma(X_1) - X_1| \geq 2|X_1| - 1 = 2k - 1$$

and since

$$X_1 \cup \Gamma(X_1) \subseteq V(P),$$

we have

$$|V(P)| \geq |X_1| + |\Gamma(X_1) - X_1| \geq k + 2k - 1 = 3k - 1.$$

The assertion is sharp for any k as is shown by any graph consisting of disjoint complete $(3k-1)$-graphs. [L. Pósa, *Discrete Math.* **14** (1976) 359–364.]

21. (a) Suppose indirectly that G has no Hamiltonian circuit. Let us add edges to G as long as we do not form a Hamiltonian circuit. Since condition (a) remains valid (by are way, so do (b), (c), (d)) we may suppose that G is already saturated, i.e. adding any new edge to it, a graph containing a Hamiltonian circuit arises. Let $(x,y) \notin E(G)$, and let H be a Hamiltonian circuit of $G+(x,y)$. Then G contains a Hamiltonian path $P = (z_1 = x, z_2, \ldots, z_n = y)$ connecting x to y.

Let $z_{i_1}, \ldots, z_{i_k}$ be the neighbors of x on P, $2 = i_1 < i_2 < \ldots < i_k \leq n$. Then y cannot be adjacent to $z_{i_\nu - 1}$ $(1 \leq \nu \leq k)$; otherwise, $(z_1, \ldots, z_{i_\nu - 1}, z_n, z_{n-1}, \ldots, z_{i_\nu})$ would be a Hamiltonian circuit. Hence

$$d_G(y) \leq n - 1 - k = n - 1 - d_G(x) \leq n - 1 - \frac{n}{2} < \frac{n}{2},$$

a contradiction.

We next assume that (d) holds. We again may assume that G contains no Hamiltonian circuit but joining any two non-adjacent points by an edge a Hamiltonian circuit arises. Let x_k, x_l be a non-adjacent pair with $k+l$ maximal $(k < l)$. Then x_k is adjacent to $x_{l+1}, \ldots, x_n$, hence

(1) $$d_k \geq n - l;$$

also, x_l is adjacent to $x_{k+1}, \ldots, x_{l-1}, x_{l+1}, \ldots, x_n$, i.e.

(2) $$d_l \geq n - k - 1.$$

The same argument as used in (a) yields

(3) $$d_k + d_l \leq n - 1.$$

Now from (2) and (3),

(4) $$d_k \leq n - 1 - d_l \leq (n-1) - (n-k-1) = k.$$

Set $m = d_k$. Then by (4), $m \leq k$ and thus, $d_m \leq d_k = m$. Also, (3) implies that

$$m = d_k < \frac{n}{2}.$$

Thus by the assumption,

$$d_{n-m} \geq n - m = n - d_k \geq d_l + 1.$$

It follows that

$$n - d_k = n - m > l,$$

or, equivalently,

$$d_k < n - l,$$

which contradicts (1).

If (b) holds, then, obviously, (d) holds. Suppose that (c) holds, and let $d_k \leq k < \frac{n}{2}$, $l = n - k$. Then by (c), either $d_l \geq l$ or $d_k + d_l \geq l + k = n$, whence $d_l > l$. Thus (d) holds again and the existence of a Hamiltonian circuit follows [see B].

22. Let G be a simple graph, which does not contain a Hamiltonian circuit through the edges in F. We may assume that, adding any edge to G, we already have such a Hamiltonian circuit (because in the complete graph on $V(G)$, there is a

Hamiltonian circuit F, as F consists of disjoint paths). Let $(x,y) \notin E(G)$. Then there is a Hamiltonian circuit in $G + (x,y)$ through F, i.e. there is a Hamiltonian path P in G through F connecting x to y. Let $P = (x = z_1, z_2, \ldots, z_n = y)$. Let x be adjacent to $z_{i_1}, \ldots, z_{i_k}$ $(2 = i_1 < \ldots < i_k < n)$. As before, it follows that $z_{i_\nu - 1}$ $(1 \le \nu \le k)$ can be adjacent to y only if $(z_{i_\nu - 1}, z_{i_\nu}) \in F$. Thus, at most $n - 1 - k + q$ points are adjacent to y. Hence

(1) $$d_G(y) + d_G(x) \le n + q - 1.$$

This is clearly impossible.

Remark: One could give analogous generalizations as in the preceding problem [see B].

23. Add a point y and connect it to all points. The resulting graph on $2n+2$ points has degree at least $n+1$ and has a Hamiltonian circuit by Dirac's theorem 10.21a. Removing y we still have a Hamiltonian path $(x_0, \ldots, x_{2n})$ in G. Suppose that G has no Hamiltonian circuit, then we have the following rule: *if x_0 is adjacent to x_i, then x_{2n} is adjacent to x_{i-1}.* Since the degrees of x_0 and x_{2n} are n, it must hold that if x_0 *is not adjacent to* x_i *then* x_{2n} *is adjacent to* x_{i-1}.

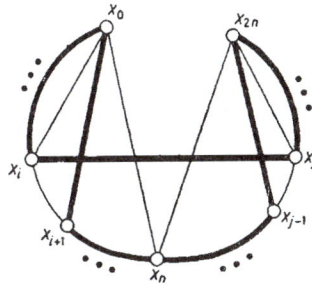

FIGURE 81

Suppose first that x_0 is adjacent to $x_1, \ldots, x_n$ and x_{2n} is adjacent to $x_n, \ldots, x_{2n-1}$. There is an i, $1 \le i \le n$ such that x_i is not adjacent to x_n; then x_i is adjacent to x_j for some $n < j \le 2n - 1$, because $d_G(x_i) = n$. Then the circuit $(x_i, x_{i-1}, \ldots, x_0, x_{i+1}, \ldots, x_{j-1}, x_{2n}, \ldots, x_j)$ is a Hamiltonian circuit (Fig. 81).

Now let $1 \le i \le 2n - 1$ be such that x_{i+1} is adjacent to x_0 but x_i is not. By the above argument, x_{i-1} is adjacent to x_{2n}. Thus G contains a $(2n)$-circuit $(x_{i-1}, \ldots, x_0, x_{i+1}, \ldots, x_{2n})$.

Let $C = (y_1, \ldots, y_{2n})$ be a $(2n)$-circuit in G, and let y_0 be the last point. Since C is a maximum circuit in G, y_0 cannot be adjacent to two neighboring points of C; so it must be adjacent to every second point in C, to $y_1, y_3, \ldots, y_{2n-1}$, say. Replacing y_{2i} by y_0 we get another maximum circuit and so, y_{2i} must also be adjacent to $y_1, y_3, \ldots, y_{2n-1}$. Now we observe that y_1 is adjacent to $y_0, y_2, \ldots, y_{2n}$, i.e.

$$d_G(y_1) \geq n + 1,$$

a contradiction. [C. St. J. A. Nash-Williams, *Proc. Amer. Math. Soc.* **17** (1966) 466–467.]

24. Let $x, y \in V(G)$, we prove that they can be connected by a Hamiltonian path. We may assume that $(x, y) \in E(G)$, since connecting x, y by a new edge influences neither the assumption nor the conclusion.

Subdivide the edge (x, y) by a new point z. Then it is easy to see that the resulting graph G' has a Hamiltonian circuit, if and only if G contains a Hamiltonian path connected x to y.

Now the degrees of G' are

$$2 \leq d_2 \leq \ldots \leq d_{n+1},$$

where $d_2 \leq \ldots \leq d_{n+1}$ are the degrees of G; hence $d_2 \geq \frac{n+1}{2}$ by the assumption. Since $|V(G')| = n+1$, Pósa's condition (b) in 10.21 is satisfied and hence, G' has a Hamiltonian circuit. Thus, G contains a Hamiltonian path connecting x to y [see B].

25. Let $C = (x_1, \ldots, x_m)$ be a longest cycle in G. Then $|V(C)| > \frac{n}{2}$. In fact, let $Q = (z_0, \ldots, z_p)$ be a longest (directed) path and $z_{i_1}, \ldots, z_{i_k}$ $(i_1 < \ldots < i_k)$ be those points of G for which $(z_{i_\nu}, z_0) \in E(G)$ (they obviously lie on Q). Then $i_k \geq k \geq n/2$ and thus, the cycle $(z_0, \ldots, z_{i_k})$ has length $> n/2$.

Suppose that C is not Hamiltonian and let $P = (y_0, \ldots, y_l)$ be a longest path in $G - V(C)$. We have $(u, y_0) \in E(G)$ for at least $\frac{n}{2} - l$ points u not in P. All these points belong to C; let us denote them by $x_{i_1}, \ldots, x_{i_t}$, $t \geq (n/2) - l$. Similarly, there are points $x_{j_1}, \ldots, x_{j_s}$, $s \geq (n/2) - l$ such that $(y_l, x_{j_\nu}) \in E(G)$, $1 \leq \nu \leq s$.

Now observe that if $x_{i_\nu} \neq x_{i_\mu}$, then the (x_{i_ν}, x_{i_μ})-arc of C must be at least as long as $l + 2$; otherwise, this arc of C would be replaced by $(x_{i_\nu}, y_0) + P + (y_l, x_{j_\mu})$. Thus if we consider an arc A_ν of C, starting at $x_{i_\nu + 1}$ $(x_{m+1} = x_1)$ and having length l, $(1 \leq \nu \leq t)$, then

$$x_{j_\mu} \notin \bigcup_{\nu=1}^{t} V(A_\nu) \qquad (1 \leq \mu \leq s).$$

But it is easy to see that $\left| \bigcup_{\nu=1}^{t} V(A_\nu) \right| \geq t + l$ and thus

$$s \leq m - (t + l)$$

or, equivalently,

$$m \geq s + t + l \geq \left(\frac{n}{2} - l \right) + \left(\frac{n}{2} - l \right) + l = n - l.$$

But P has $l + 1$ points out of $n - m$, thus

$$l + 1 \leq n - m, \quad m \leq n - l - 1,$$

a contradiction [C. St. J. A. Nash-Williams; see B].

26. Let C be a minimum circuit in G. Suppose indirectly that C is not a Hamiltonian circuit. Then $G - V(C)$ is non-empty; let G_1 be a component of it. Let $x_1, \ldots, x_s$ be those points of C adjacent to G_1. Observe that no two x_i are neighboring points on C by the maximality of C. This implies that $\{x_1, \ldots, x_s\}$ separates G and hence, $s \geq k$. Going around C in a given direction, let $y_1, \ldots, y_s$ be the points following $x_1, \ldots, x_s$.

Now we claim that $y_1, \ldots, y_s$ are independent. Suppose indirectly that y_i and y_j are adjacent, then remove (x_i, y_i) and (x_j, y_j) from C and add (y_i, y_j) and an (x_i, x_j)-path through G_1; the resulting circuit is longer than C, a contradiction.

Also, no y_i is adjacent to G_1 and thus, selecting a $y_0 \in V(G_1)$, the set $S = \{y_0, y_1, \ldots, y_s\}$ will be independent. But

$$|S| = s + 1 \geq k + 1,$$

a contradiction [P. Erdős–V. Chvátal; see B].

27. (a) Let $P = (x_0, x_1, \ldots, x_m)$ be a longest path. Then all neighbors of x_0 are on P; since $d_G(x_0) \geq k$, one of these neighbors is x_i with $k \leq i \leq m$. Then $C = (x_0, \ldots, x_i)$ is a circuit of length $i + 1 \geq k + 1$.

(b) Let P be a longest path as above. Suppose first that there are points x_i, x_j such that $i < j$, x_i is adjacent to x_m and x_j is adjacent to x_0. We may assume that $j - i$ is minimal among such pairs of indices. Let $C = (c_0, \ldots, x_i, x_m, x_{m-1}, \ldots, x_j)$. If $j = i + 1$, then C has length $m + 1$ and must be a Hamiltonian circuit, since otherwise there would be a point outside C connected to C, which would yield a path longer than P. So we may assume that $j \geq i + 2$. Then $x_{i+1}, \ldots, x_{j-1}$ are not adjacent to x_0 or x_m, hence C contains x_m, all neighbors of x_m and all points x_ν for which $x_{\nu+1}$ is adjacent to x_0, except x_{j-1}. These points are distinct and hence C has at least $2k$ points.

So suppose that the last point x_i adjacent to x_0 comes before the first point x_j is adjacent to x_m (possibly $x_i = x_j$). Since G is 2-connected, there are two disjoint paths P_1, P_2 connecting the circuits $C_1 = (x_0, \ldots, x_i)$ and $C_2 = (x_j, \ldots, x_m)$. We may assume that one of them starts at x_i; for otherwise we can walk on the path $(x_i, \ldots, x_j)$ till we hit either C_2 or P_ν and replace an appropriate piece of P_ν by this path. Similarly, we may assume that one of P_1, P_2 ends at x_j. It may be that the same P_ν ends at x_i or not; but in both cases we get a circuit which contains x_0, x_m and all neighbors of them as shown in Fig. 82. This circuit is longer than $2k$. [G. A. Dirac, *Proc. London Math. Soc.* **2** (1952) 69–81.]

28. For $n \leq k$ the assertion is void, so suppose that $n > k$. If G has a point x with degree at most $k/2$, $G - x$ has more than $\frac{(n-2)k}{2}$ edges and so, by the induction hypothesis, $G - x$ contains a circuit of length greater than k. Thus we may assume that every point of G has degree at least $\frac{k+1}{2}$.

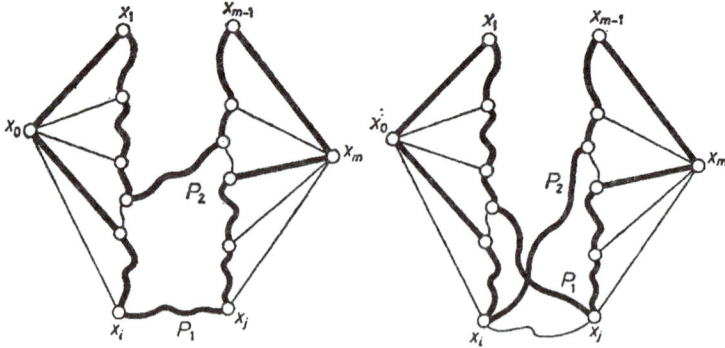

FIGURE 82

Also, if G is not 2-connected, say $G = G_1 \cup G_2$ with $|V(G_1) \cap V(G_2)| \leq 1$ then, by

$$|E(G_1)| + |E(G_2)| > \frac{k(n-1)}{2} \geq \frac{k}{2}(|V(G_1)| + |V(G_2))| - 2),$$

it follows that for $i = 1$ or 2,

$$|E(G_i)| > \frac{k}{2}(|V(G_i)| - 1)$$

and induction works again. Thus we may suppose that G is 2-connected. Then 10.27 implies that G contains a circuit of length at least $k+1$.

The result is sharp for graphs whose blocks are K_k's. [P. Erdős–T. Gallai, *Acta Math. Acad. Sci. Hung.* **10** (1959) 337–356.]

29. Let $P = (0, 1, \ldots, N)$ be any path in G. Let R_1 be a (P, P)-path with endpoints $i_1 = 0$ and j_1 such that j_1 is maximal. Suppose that the (P, P)-paths $R_1, \ldots, R_k$ have been selected, R_ν has endpoints $i_\nu < j_\nu$ and $j_k < N$. Let R_{k+1} be a (P, P)-path connecting $\{0, \ldots, j_k - 1\}$ to $\{j_k + 1, \ldots, N\}$ (such a path exists as G is 2-connected) and let j_{k+1} be as large as possible. If $j_k = N$, we stop; let s be this value of k.

It follows from the definition that

$$0 < j_1 < j_2 < \ldots < j_s = N.$$

Also,

$$i_{k+1} \geq j_{k-1},$$

since otherwise R_{k+1} could have been chosen for R_k. Thus we have

$$0 = i_1 < i_2 < j_1 \leq i_3 < j_2 \leq \ldots \leq i_s < j_{s-1} < j_2.$$

Similar reasoning shows that $R_1, \ldots, R_s$ are independent paths. Thus $P \cup R_1 \cup \ldots \cup R_s$ has the structure indicated in the hint.

Suppose that $s = 2p + 1$ is odd (the even case can be treated similarly and is left to the reader). Then the arc between i_{p+1} and j_{p+1} is not longer than $l - 1$

since it forms a circuit with R_{p+1}. Similarly, the (i_p, i_{p+1})-arc and (j_{p+1}, j_{p+2})-arc of P form a circuit with R_p, R_{p+1}, R_{p+2} and the (j_p, i_{p+2})-arc of P (this latter may degenerate; Fig. 83) and hence their total length is at most $l-3$. Similarly we get the sum of lengths of the (i_q, i_{q+1})-arc and the (j_{2p+1-q}, j_{2p+2-q})-arc is at most $l-(p+1-q)$. Thus

$$|E(P)| \leq (l-1) + (l-3) + \ldots = \frac{l^2}{4},$$

which proves the assertion [H.-J. Voß, G. Dirac, see WV].

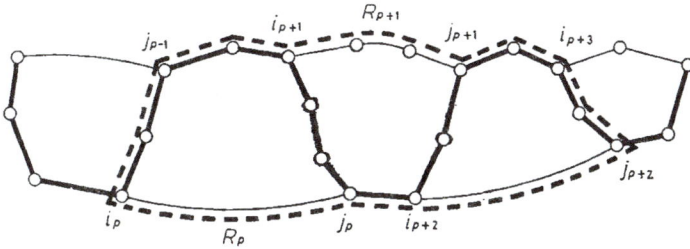

FIGURE 83

30. Let $x, y \in V(G)$, $(x,y) \in (EG)$. Any $z \in V(G) - \{x,y\}$ is adjacent to at most one of x, y. Hence

$$(d(x) - 1) + (d(y) - 1) \leq n - 2, \quad d(x) + d(y) \leq n.$$

Sum this over all edges (x,y), then we get each $d(x)$ exactly $d(x)$ times on the left-hand side, i.e.

$$\sum_x d^2(x) \leq n \cdot |E(G)|.$$

Here

$$\sum_x d^2(x) \geq \frac{1}{n} \left(\sum_x d(x) \right)^2 = \frac{4}{n}|E(G)|^2,$$

thus

$$\frac{4}{n}|E(G)|^2 \leq n|E(G)|, \qquad |E(G)| \leq \frac{n^2}{4}.$$

[W. Mantel, *Wiskundige Opgaven* **10** (1906) 60–61. Other simple proofs can be obtained by specializing the solutions of Turán's theorem 10.34 and 10.35.]

31. Let x be any point of G. Then no two neighbors of x can be adjacent, hence

$$d(x) \leq \alpha(G).$$

Now let S be a minimum point cover of G. Then each edge of G is represented by a point in S, thus

$$|E(G)| \leq \sum_{x \in S} d(x) \leq \alpha(G) \cdot |S| = \alpha(G) \cdot \tau(G).$$

Since

$$\alpha(G) \cdot \tau(G) \leq \left(\frac{\alpha(G) + \tau(G)}{2}\right)^2 = \left(\frac{n}{2}\right)^2 = \frac{n^2}{4},$$

the result in the preceding problem also follows from this.

32. Let us call a triple $\{x, y, z\}$ *bad* if it does not span a triangle of G or $\overline{G}$. Then the number of bad triples such that exactly one edge spanned by them contains the point x is $d(x)(n-1-d(x))$, so

$$\sum_x d(x)(n - 1 - d(x))$$

counts the bad triples twice. The number of triangles in G and $\overline{G}$ together is, therefore,

$$\binom{n}{3} - \frac{1}{2} \sum_{x \in V(G)} d(x)(n - 1 - d(x)).$$

(a) If G is k-regular this formula reduces to

$$\binom{n}{3} - \frac{1}{2} nk(n - 1 - k)$$

as stated.

(b) We have

$$d(x)(n - 1 - d(x)) \leq \left(\frac{n-1}{2}\right)^2$$

and hence, the number of triangles in G and $\overline{G}$ is at least

$$\binom{n}{3} - \frac{n}{2}\left(\frac{n-1}{2}\right)^2 = \frac{n(n-1)(n-5)}{24}.$$

Note: This is positive for $n > 5$, which is a very special case of Ramsey's theorem (§ 14). [A. W. Goodman, *Amer. Math. Monthly* **66** (1959) 778–783.]

33. Let $(x,y) \in E(G)$. Then at least $d(x)+d(y)-n$ points are adjacent to both x and y, i.e. at least this many triangles contain (x,y). Hence

$$\frac{1}{3} \sum_{(x,y)\in E(G)} (d(x)+d(y)-n)$$

estimates the total number of triangles from below. Since $d(x)$ occurs in this sum exactly $d(x)$ times, the number of triangles on G is at least

$$\frac{1}{3}\left(\sum_{x\in V(G)} d(x)^2 - nm\right).$$

By the Cauchy–Schwartz inequality, this is

$$\geq \frac{1}{3}\left(\frac{(\sum_x d(x))^2}{n} - nm\right) = \frac{4m}{3n}\left(m - \frac{n^2}{4}\right).$$

[ibid.]

34. We use induction on m. For $m=1$ the assertion is void.

Let H be a complete k-graph in G (such a subgraph exists by induction on k,m or let us "saturate" G by adding edges as long as no complete $(k+1)$-graph is produced, and consider this saturated graph rather than G). If G contains no complete $(k+1)$-graph, each $x \in V(G)-V(H)$ is adjacent to no more than $k-1$ points in H. Hence the graph $G_1 = G - V(H)$ has more than

$$\binom{k}{2}m^2 - (k-1)(m-k) - \binom{k}{2} = \binom{k}{2}(m-1)^2$$

edges. By the induction hypothesis, this implies that G_1 contains a complete $(k+1)$-graph and hence, so does G.

The equality is attained when $V(G) = V_1 \cup \ldots \cup V_k$, $|V_i| = m$ and two points are adjacent iff they belong to different sets V_j.

If $n=mk+r$ then we claim that the most number of edges in a simple graph without triangles is attained when $G = H_{n,k}$, $V(H_{n,k}) = V_1 \cup \ldots \cup V_k$, $|V_1| = \ldots = |V_r| = m+1$, $|V_{r+1}| = \ldots = |V_k| = m$ and again, two points are adjacent iff they belong to different classes. This number of edges is

$$\binom{r}{2}(m+1)^2 + r(k-r)m(m+1) + \binom{k-r}{2}m^2.$$

This claim is trivially true if $m=0$ and can be proved by induction on m in the same way as above [P. Turán; see B, S].

35. Let x be a point with maximum degree and denote by G_0 the subgraph induced by the neighbors of x. Then there is a graph H_0 on $V(G_0)$, which is $(k-1)$-chromatic and satisfies

$$d_{H_0}(z) \geq d_{G_0}(z) \qquad (z \in V(G_0));$$

this follows by induction on k since G_0 contains no complete k-graph. Define H on $V(G)$ by connecting all points of $V(G) - V(G_0)$ to all points of $V(G_0)$ and also all pairs of points adjacent in H_0.

Obviously, H is k-chromatic. Also

$$d_H(y) = |V(G_0)| = d_G(x) \geq d_G(y) \qquad (y \notin V(G_0)),$$
$$d_H(y) = |V(G)| - |V(G_0)| + d_{H_0}(y) \geq$$
$$\geq |V(G)| - |V(G_0)| + d_{G_0}(y) \geq d_G(y) \qquad (y \in V(G_0)).$$

Note that Turán's theorem is an immediate consequence. For let $A_1, \ldots, A_k$ be the color-classes of H, $|A_i| = a_i$. Then $a_1 + \ldots + a_k = n$ and

$$|E(G)| = \frac{1}{2} \sum_{x \in V(G)} d(x) \leq \frac{1}{2} \sum_{x \in V(H)} d_H(x) = |E(H)| \leq \sum_{1 \leq i < j \leq k} a_i a_j.$$

The right-hand side is maximal if the a_i;s are as nearly equal as possible, i.e. $a_1 = \ldots = a_r = m+1$, $a_{r+1} = \ldots = a_k = m$. This gives Turán's bound. [P. Erdős, *Mat. Lapok* **21** (1970) 249–251.]

36. (a) Suppose that G is a simple graph on n points containing no 4-circuit. Let us count those pairs $\{x,y\}$ of points both adjacent to a third point z.

For a fixed z, we count $\binom{d(z)}{2}$ such pairs. On the other hand, each pair $\{x,y\}$ is counted at most once since if it were counted both with z_1 and with z_2, then (x, z_1, y, z_2) would form a quadrilateral. Hence

$$\sum_{z \in V(G)} \binom{d(z)}{2} \leq \binom{n}{2}.$$

Here by Jensen's inequality,

$$\sum_{z \in V(G)} \binom{d(z)}{2} \geq n \binom{\frac{2|E(G)|}{n}}{2} = |E(G)| \frac{2|E(G)| - n}{n}$$

and thus

$$|E(G)|^2 - \frac{n}{2}|E(G)| \leq \frac{n^3 - n^2}{4},$$

whence

$$|E(G)| \leq \sqrt{\frac{n^3}{4} - \frac{3n^2}{16}} + \frac{n}{4} = \frac{n}{4}(1 + \sqrt{4n - 3}) \sim \frac{1}{2} n^{3/2}.$$

(b) Let the points of $V(G)$ be the points of the projective space over $GF(p)$, and connect points $[x, y, z]$ and $[u, v, w]$ iff $xu + yv + zw = 0$ (this means, if they are conjugate with respect to the conic section $x^2 + y^2 + z^2 = 0$; we use homogeneous coordinates).

The points adjacent to any given point u form a line of the geometry (the polar of u), which goes through u iff u is on the conic $x^2 + y^2 + z^2 = 0$. Hence G has $p+1$ points of degree p and the remaining p^2 points of degree $p+1$. Thus

$$|E(G)| = \frac{1}{2}(p(p+1) + (p+1)p^2) = \frac{1}{2}p(p+1)^2 = \frac{n-1}{4}(\sqrt{4n-3} - 1).$$

On the other hand, G contains no 4-circuit. In fact, if u, v are two points, then their polars meet in a single point (since they are trivially distinct lines), so any two points have at most one common neighbor. [I. Reiman, *Acta Math. Acad. Sci. Hung.* **9** (1959) 269–279.]

37. Let $m = |E(G)|$, $V(G) = \{x_1, \ldots, x_n\}$, $d_G(x_i) = d_i$. Since G contains no $K_{r,r}$, any given r-tuple in $V(G)$ is contained in the neighborhood of at most $r-1$ points. The neighborhood of x_i contains $\binom{d_i}{r}$ r-tuples. Thus

$$\sum_{i=1}^{n} \binom{d_i}{r} \le (r-1)\binom{n}{r}.$$

We are going to estimate the left-hand side from below using Jensen's inequality. Unfortunately, $\binom{x}{r}$ is convex only for $x \ge r - 1$. but set

$$f(x) = \begin{cases} \binom{x}{r} & \text{if } x \ge r-1, \\ 0 & \text{if } x \le r-1, \end{cases}$$

then $f(x)$ is convex and, since d_i is an integer, we have

$$\sum_{i=1}^{n} \binom{d_i}{r} = \sum_{i=1}^{n} f(d_i) \ge n \cdot f\left(\frac{d_1 + \ldots + d_n}{n}\right) = nf\left(\frac{2m}{n}\right).$$

Now, if $\frac{2m}{n} < r-1$, we have nothing to prove. Suppose that $\frac{2m}{n} \ge r-1$, then we have

$$(r-1)\binom{n}{r} \ge \sum_{i=1}^{n} \binom{d_i}{r} \ge nf\left(\frac{2m}{n}\right) = n\binom{2m/n}{r}.$$

Here

$$(r-1)\binom{n}{r} < (r-1)\frac{n^r}{r!}, \quad \binom{2m/n}{r} > \frac{\left(\frac{2m}{n} - r + 1\right)^r}{r!},$$

$$((2m/n) - r + 1)^r < (r-1)n^{r-1},$$

$$2m < \sqrt[r]{r-1} \cdot n^{2-\frac{1}{r}} + (r-1)n < C \cdot n^{2-\frac{1}{r}}.$$

[K. Zarankiewicz's problem; T. Kővári–V. T. Sós–P. Turán, *Colloqu. Math.* **3** (1954) 50–57.]

38. (a) We use induction on k. For $k = 1$ the assertion is true. Let $k \geq 2$ and $s = \left\lceil \frac{1}{\varepsilon} t \right\rceil$. If n is large enough then, by the induction hypothesis, we find k disjoint s-sets $A_1, \ldots, A_k$ such that any two points in distinct sets A_i are adjacent.

Now let W denote the set of those points in $U = V(G) - A_1 - \ldots - A_k$ adjacent to at least t points in each A_i. Then there are at least

$$(|U| - |W|)(s - t) \geq (|U| - |W|)(1 - \varepsilon)s = (n - ks - |W|)(1 - \varepsilon)s$$

edges missing between U and $A_1 \cup \ldots \cup A_k$. On the other hand, there are at most $\left(\frac{1}{k} - \varepsilon \right)$ edges missing from a given point and hence, there are at most

$$ks\left(\frac{1}{k} - \varepsilon \right) n = (1 - k\varepsilon)sn$$

edges missing between U and $A_1 \cup \ldots \cup A_k$. Thus

$$(n - ks - |W|)(1 - \varepsilon)s \leq (1 - k\varepsilon)sn,$$

whence

$$|W| \geq \frac{k}{1 - \varepsilon} n - ks.$$

Thus W gets large if n is large.

Select t points adjacent to $w \in W$ from each A_i, for all $w \in W$. If

$$|W| > \binom{s}{t}^k (t - 1),$$

then we necessarily select the same t-tuples for t distinct points in W. But these t points together with the k t-tuples belonging to them form a desired subgraph.

(b) Remove a point with degree less than $\left(1 - \frac{1}{k} + \frac{\varepsilon}{2} \right) \cdot |V(G)|$ (if any); do so again for the resulting graph and so on. Suppose that we get stuck, i.e. we get a graph H with all degrees at least $\left(1 - \frac{1}{k} + \frac{\varepsilon}{2} \right) |V(H)|$. Then if $N = |V(H)|$ is large enough, the preceding problem implies the assertion. Thus it suffices to show that N cannot be too small (and that we get stuck at all, i.e. $N \neq 0$). In the construction of H we removed altogether at most

$$\sum_{j=N+1}^{n} j \left(1 - \frac{1}{k} + \frac{\varepsilon}{2} \right) = \left(\binom{n+1}{2} - \binom{N+1}{2} \right) \left(1 - \frac{1}{k} + \frac{\varepsilon}{2} \right) \leq$$

$$\leq \left(\binom{n}{2} - \binom{N}{2} \right) \left(1 - \frac{1}{k} + \frac{\varepsilon}{2} \right) + (n - N)$$

edges. The remainder has at most $\binom{N}{2}$ edges, thus,

$$|E(G)| = \left(1 - \frac{1}{k} + \varepsilon\right)\binom{n}{2} \leq \left(1 - \frac{1}{k} + \frac{\varepsilon}{2}\right)\left(\binom{n}{2} - \binom{N}{2}\right) + (n - N) + \binom{N}{2}$$

or, equivalently,

$$\frac{\varepsilon}{2}\binom{n}{2} \leq \left(\frac{1}{k} - \frac{\varepsilon}{2}\right)\binom{N}{2} = (n - N).$$

Hence N gets large of n is large and the assertion follows.

(c) Set $\chi(G_0) = k+1$. Consider Turán's graph $H_{n,k}$ as in the solution of 10.34. This is k-chromatic and hence it does not contain a G_0. Hence

$$M(n, G_0) \geq |E(H_{n,k})| \sim \frac{n^2}{2}\left(1 - \frac{1}{k}\right),$$

i.e.

$$\liminf \frac{M(n, G_0)}{n^2} \geq \frac{1}{2}\left(1 - \frac{1}{k}\right).$$

Now suppose indirectly that there were arbitrarily large graphs G containing no copy of G_0 with

$$\frac{|E(G)|}{n^2} > \frac{1}{2}\left(1 - \frac{1}{k}\right) + \varepsilon.$$

Then by (b), an appropriately large one would contain $k+1$ disjoint sets of cardinality $|V(G_0)|$ such that any two points in two different such sets are adjacent. But then G_0 is a subgraph of this subgraph already, a contradiction. [P. Erdős–A. H. Stone, *Bull. Amer. Math. Soc.* **52** (1946) 1089–1091. This proof is a specialization of P. Erdős–M. Simonovits, *Studia Sci. Math. Hung.* **1** (1966) 51–57.]

39. Set $n_0 = |V(G_0)|$. First we prove the assertion formulated in the hint, more exactly:

(∗) If $G \neq H_{n,k}$ is a simple graph with $n \geq N(k, n_0)$ points, $|E(H_{n,k})|$ edges and G contains no G_0, then it contains an induced subgraph G_1 on $n_1 = n - 2kn_0$ points such that

$$|E(G_1)| - |E(H_{n,k})| > |E(G)| - |E(H_{n,k})|.$$

In fact $N = N(k, n_0)$ is chosen so that a graph on $n \geq N$ points and with $|E(H_{n,k})|$ edges should contain k disjoint classes $A_1, \ldots, A_k$, $|A_i| = 2n_0$ such that any two points in different sets A_i are adjacent. Such an N exists by 10.38b since

$$|E(H_{n,k})| \sim \binom{k}{2}\left(\frac{n}{k}\right)^2 > \left(1 - \frac{1}{k-1} + \frac{1}{k^2}\right)\frac{n^2}{2}.$$

Now for each $x \notin A_1 \cup \ldots \cup A_k$, there must be at least $2n_0$ points of $A_1 \cup \ldots \cup A_k$ non-adjacent to x. Otherwise, x would be adjacent to at least one point in each class and to at least n_0 points in all but one of the classes. Hence we could select a set $B_i \subseteq A_i$, $|B_i| = n_0$ such that x is adjacent to all points in, say $B_1 \cup \ldots \cup B_{k-1}$

and to at least one point in B_k. Since $G_0 - e$ is k-chromatic, the subgraph induced by $B_1 \cup \ldots \cup B_k \cup \{x\}$ contains G_0, a contradiction.

Also observe that equality can hold only if x is adjacent to all points of all but one A_i and to no point of this A_i. Suppose that the equality holds for every x. Then the x's can be divided into k classes $C_1, \ldots, C_k$, $x \in C_i$ being adjacent to all points of A_j, $j \neq i$ but to no point of Z_i. No two points in the same C_i can be adjacent since then we could find a G_0 in G in the same way as above. Hence, G is k-chromatic. Among k-chromatic graphs, obviously, $H_{n,k}$ has the most edges, which contradicts $G \neq H_{n,k}$ and $|E(G)| \geq |E(H_{n,k})|$.

Thus some x is adjacent to less than $(2k-2)n_0$ points of $A_1 \cup \ldots \cup A_k$. Hence the removal of $A_1 \cup \ldots \cup A_k$ destroys less than $M = 4\binom{k}{2}n_0^2 + (n - 2kn_0)(2k-2)n_0$ edges. The removal of $2n_0$ points from each class of $H_{n,k}$ destroys exactly M edges and results in $H_{n_1,k}$. Thus

$$|E(G - A_1 - \ldots - A_k)| - |E(H_{n_1,k})| > |E(G)| - |E(H_{n,k})|.$$

This proves $(*)$.

Now suppose indirectly that there is a simple graph $G \neq H_{n,k}$ with $n \leq 2kn_0\binom{N}{2} + N$ points, at least $|E(H_{n,k}|$ edges and containing no G_0. Set $n_i = n - 2kin_0$. Select induced subgraphs $G \supseteq G_1 \supseteq G_2 \ldots$ such that $|V(G_i)| = n_i$ and $|E(G)| - |E(H_{n,k})| < |E(G_1)| - |E(H_{n_1,k})| < \ldots$. Then

$$(1) \qquad\qquad |E(G_i)| \geq |E(H_{n_i,k})| + i$$

Let us see when the sequence $G_1, G_2, \ldots$ stops. Suppose that this happens at the tth step. By $(*)$ we can find a G_{t+1} unless $G_t \cong H_{n_t,k}$ or $|E(G_t)| < |E(H_{n_t,k})|$ or $|V(G_t)| < N$. (1) rules out the first two possibilities immediately. But also, if

$$|V(G_t)| = n_t = n - 2ktn_0 < N$$

then

$$t > \frac{n - N}{2kn_0} \geq \binom{N}{2}$$

and so,

$$|E(G_t)| \geq t > \binom{N}{2} > \binom{n_t}{2},$$

a contradiction. [M. Simonovits, in: *Theory of Graphs* (P. Erdős–G. Katona, ed.) Akadémiai Kiadó, Budapest, 1968, 270–319.]

40. (a) Let $A_1, \ldots, A_{N_k}$ be the complete k-graphs and $B_1, \ldots, B_{N_{k-1}}$ be the complete $(k-1)$-graphs in G. Let A_i be contained in a_i complete $(k+1)$-graphs; also, let B_i be contained in b_i complete k-graphs.

Let x be a point not in A_i which does not form a complete $(k+1)$-graph with A_i. Then x is non-adjacent to some point $y \in A_i$. Let $U_1, \ldots, U_{k-1}$ be those k-

subsets of $A_i \cup \{x\}$ containing $\{x, y\}$. Then any pair (A_i, U_j) consists of a complete k-graph and a non-complete k-graph, and has $|A_i \cap U_j| = k - 1$. So

$$\sum_{i=1}^{N_k} (k-1)(n-k-a_i)$$

is a lower bound for the number of pairs (A, U) such that A is a complete k-graph, U is non-complete k-graph and $|A \cap U| = k - 1$.

On the other hand, the number of such pairs (A, U) is, obviously,

$$\sum_{i=1}^{N_{k-1}} b_i (n - k + 1 - b_i).$$

Thus we have

(1) $$\sum_{i=1}^{N_k} (k-1)(n-k-a_i) \leq \sum_{i=1}^{N_{k-1}} b_i(n-k+1-b_i).$$

We know that

(2) $$\sum_{i=1}^{N_k} a_i = (k+1)N_{k+1}, \qquad \sum_{i=1}^{N_{k-1}} b_i = kN_k.$$

The left-hand side of (1) is equal to

$$(k-1)(n-k)N_k - (k^2-1)N_{k+1},$$

while the right-hand side can be estimated by Jensen's inequality

$$\sum_{i=1}^{N_{k-1}} b_i(n-k+1-b_i) \leq N_{k-1} \frac{\sum\limits_{i=1}^{N_{k-1}} b_i}{N_{k-1}} \left(n-k+1 - \frac{\sum\limits_{i=1}^{N_{k-1}} b_i}{N_{k-1}} \right) =$$

$$k(n-k+1)N_k - \frac{k^2 N_k^2}{N_{k-1}},$$

whence (a) follows. [J. W. Moon–L. Moser, *Mat. Kut. Int. Közl.* **7** (1962) 283–286.]

(b) Estimate (∗) in the hint follows from (a) by a straightforward induction. Now from (∗),

$$N_k \geq \prod_{i=1}^{k} \frac{\vartheta-i+1}{\vartheta} \frac{n}{i} = \binom{\vartheta}{k}\left(\frac{n}{\vartheta}\right)^k.$$

41. (a) Any triple of points of T_n span either a 3-cycle or a transitive triangle. If two edges have a common trail their points form a transitive triangle and each transitive triangle contains such pair of edges. Hence the number of transitive triangles is

$$\sum_{i=1}^{n} \binom{d_i}{2} \geq \binom{\frac{n-1}{2}}{2} = \frac{n(n-1)(n-3)}{8}$$

$\left(\text{by Jensen's inequality, since } \sum d_i = \binom{n}{2}\right)$. Thus the number of 3-cycles is

$$\leq \binom{n}{3} - \frac{n(n-1)(n-3)}{8} = \frac{(n+1)n(n-1)}{24} = \frac{1}{4}\binom{n+1}{3}.$$

(b) Let $x \in V(T)$, and denote by X and Y the sets of points y such that $(y,x) \in E(G)$ or $(x,y) \in E(G)$, respectively. There must be an edge joining Y to X, otherwise T would be strongly connected. This edge forms, with x, a 3-cycle.

We prove that the number of 3-cycles is at least $n-2$ by induction on n. For $n=3$ this is obvious.

By 6.13 we can find a point x such that $T-x$ is strongly connected; thus, by the induction hypothesis, $T-x$ contains at least $n-3$ 3-cycles. As observed above, there is at least one 3-cycle incident with x; hence T contains at least $n-2$ 3-cycles.

Remark: Both assertions are essentially sharp: if n is odd, then K_n has an orientation such that each point has outdegree $\frac{n-1}{2}$ by 5.13 and this tournament has exactly $\frac{1}{4}\binom{n+1}{3}$ 3-cycles. On the other hand, consider a transitive tournament and invert the edge e connecting its "first" and "last" point. Then we obtain a strongly connected tournament in which each 3-cycle contains e, thus their number is $n-2$ [see S, M].

42. First we show by induction on n that for each $3 \leq k \leq n$, there is a k-cycle through each point x. For $n=k$ this is true since T contains a Hamiltonian circuit by 6.11. Suppose that $k<n$.

By 6.13, there is a point $y \neq x$ such that $T-y$ is strongly connected. Thus $T-y$ contains a k-cycle through x and hence, so does T.

Now it follows by an easy induction like the one in part (b) of the preceding solution that T contains at least $n-k+1$ k-cycles for $3 \leq k \leq n$. Hence the total number of cycles is

$$\geq \sum_{k=3}^{n}(n-k+1) = \sum_{\nu=1}^{n-2} \nu = \binom{n-1}{2}.$$

The assertion is sharp for the tournament obtained from a transitive one by inverting the edges of the Hamiltonian path [see M].

43. We already know there is a Hamiltonian path (see 5.20). Let $e_1, \ldots, e_{\lfloor n/2 \rfloor}$ be independent edges. A Hamiltonian path whose $(2i-1)^{\text{st}}$ edge is e_i $(i=1,\ldots,\lfloor n/2 \rfloor)$ is uniquely determined. Hence, the number of Hamiltonian paths is not greater than the number of ways to select $\lfloor n/2 \rfloor$ independent edges (and order them).

We have $\binom{n}{2}$ choices for e_1; $\binom{n-1}{2}$ choices for e_2, etc. Thus the number of ways to select $e_1, \ldots, e_{\lfloor n/2 \rfloor}$ is

$$\binom{n}{2}\binom{n-2}{2}\cdots\left(\binom{n-2\lfloor n/2 \rfloor + 2}{2}\right) = \frac{n!}{2^{\lfloor n/2 \rfloor}}. \quad [\text{M}].$$

44. (a) We may assume that $n = 2^{k-1}$ and want to prove T contains a transitive subtournament with k points, using induction on k. For $k = 1$ the assertion is trivial.

Suppose that $k > 1$ and choose an $x \in V(T)$. Let X and Y denote the sets of points z such that $(z,x) \in E(T)$ and $(x,z) \in E(T)$, respectively. Since $|X| + |Y| = 2^{k-1} - 1$, we may assume, e.g. that $|X| \geq 2^{k-2}$. Then, by the induction hypothesis, X spans a transitive subtournament T_1 with $k-1$ points. Together with x, T_1 yields a transitive subtournament with k points.

(b) We prove the assertion by induction on k. For $k = 1$ it is obvious again. Suppose that $k > 1$. Define

$$f(x) = \begin{cases} \prod\limits_{j=0}^{k-2}\left(\frac{x+1}{2^j} = 1\right) & \text{if } x \geq 2^{k-2} - 1, \\ 0 & \text{if } x \leq 2^{k-2} - 1. \end{cases}$$

Then $f(x)$ is a convex function.

Let $V(T) = \{x_1, \ldots, x_n\}$, x_i having outdegree d_i. By the induction hypothesis, there are at least $f(d_i)$ transitive $(k-1)$-tournaments spanned by the d_i points x_i is joined to; this gives $f(d_i)$ transitive k-tournaments with sources in x_i. Thus the number of transitive k-tournaments is

$$\geq \sum_{i=1}^{n} f(d_i) \geq n \cdot f\left(\frac{n-1}{2}\right)$$

$$\left(\text{and since } \frac{n-1}{2} \geq 2^{k-1} - 1\right)$$

$$\geq n \prod_{j=0}^{k-2}\left(\frac{\frac{n-1}{2}+1}{2^j} - 1\right) = n \prod_{j=1}^{k-1}\left(\frac{n+1}{2^j} - 1\right),$$

which proves the assertion.

(c) This time we use induction on n to prove the assertion of the hint. For $n \leq k$ the assertion is true. Let $n > k$. By 6.13, there is a point x such that $T - x$ is strongly connected. Thus, there are at least $\binom{n-3}{k-2}$ k-tuples in $T - x$ containing a 3-cycle. Also, by the solution of 10.40, there is a 3-cycle through x and there are $\binom{n-3}{k-3}$ further k-tuples containing this triangle. This gives

$$\binom{n-3}{k-3} + \binom{n-3}{k-2} = \binom{n-2}{k-2}$$

k-tuples containing a 3-cycle as stated.

Since a transitive k-tournament contains no 3-cycle, their number is

$$\leq \binom{n}{k} - \binom{n-2}{k-2}.$$

Note that equality holds for the tournament in the solution of 10.41 [see M].

§ 11. Spectra of graphs

1. An eigenvector of A_G assigns numbers $x_1, \ldots, x_n$ to the point $1, \ldots, n$ in such a way that the sum of values assigned to neighbors of i is λx_i. Now the graphs in consideration have such a simple structure that it is easy to guess such assignments.

For the complete graph we have $(1, \ldots, 1)$, yielding the eigenvalue $(n-1)$; we also have vectors of form $(1, 0, \ldots, 0, -1, 0, \ldots, 0)$, which are linearly independent and yield the eigenvalue (-1) with multiplicity $n-1$. Since there are n eigenvalues altogether, this is all.

For the star, we look for eigenvectors, which associate the value 1 with the center. If λ is the corresponding eigenvalue we must have $1/\lambda$ associated with each of the other points, so

$$(n-1)\frac{1}{\lambda} = \lambda, \qquad \lambda = \pm\sqrt{n-1}.$$

This gives two eigenvectors and two eigenvalues. To get the rest, we have to consider eigenvectors associating 0 with the center. Then the eigenvalue belonging to it must be 0. It follows that the eigenvectors have arbitrary numbers at other points, the only restriction being that the sum of entries must be 0 [as it has to give $0 \times$(number in the center)]. There are $n-2$ linearly independent such vectors, thus we have all eigenvalues again.

More generally, if $G = K_{n,m}$, then we get two eigenvectors by placing $\sqrt{m}$ on the color class of size n and $\pm\sqrt{n}$ on the other. This yields the eigenvalues $\pm\sqrt{mn}$. The remaining eigenvalues are 0, since the adjacency matrix has only two different rows.

Finally, if $G = C_n$ is an n-circuit with the vertices labelled in a natural order, then let

$$x_\mu = \varepsilon^\mu, \qquad \mathbf{x} = \begin{pmatrix} x_1 \\ \vdots \\ x_n \end{pmatrix},$$

where ε is any n^{th} root of unity. Then we have

$$Ax = (x_\nu')_{\nu=1}^n,$$

where

$$x_\nu' = x_{\nu-1} + x_{\nu+1} = \varepsilon^{\nu-1} + \varepsilon^{\nu+1} = \left(\varepsilon + \frac{1}{\varepsilon}\right)\varepsilon^\nu = (\varepsilon + \bar\varepsilon)x_\nu.$$

Thus $\varepsilon + \bar\varepsilon$ is an eigenvalue and considering the n roots of unity, we get all of them. Hence the eigenvalues of C_n are

$$2,\ 2\cos\frac{2\pi}{n},\ 2\cos\frac{4\pi}{n},\ \ldots,\ 2\cos\frac{2(n-1)\pi}{n}.$$

2. (a) $\mathbf{j} = (1,\ldots,1)$ is an eigenvector of G with eigenvalue d, the degree of G. This follows immediately, because G is regular. Similarly, $\mathbf{j}$ is an eigenvector of $\overline{G}$ with eigenvalue $n - 1 - d$.

Now let $\mathbf{v}_1 = \mathbf{j}, \mathbf{v}_2, \ldots, \mathbf{v}_n$ be a basis of the n-space consisting of orthogonal eigenvectors of G. We claim that $\mathbf{v}_1, \ldots, \mathbf{v}_n$ are eigenvectors of $\overline{G}$. For

$$A_{\overline{G}}\mathbf{j} = (n - 1 - d)\mathbf{j},$$

which, if written out, means only that $\overline{G}$ is $(n - d - 1)$-regular. For the other eigenvectors we have

$$A_{\overline{G}}\mathbf{v}_i = (J - I - A_G)\mathbf{v}_i = J\mathbf{v}_i - \mathbf{v}_i - A_G\mathbf{v}_i = -\mathbf{v}_i - A_G\mathbf{v}_i = (-1 - \lambda_i)\mathbf{v}_i$$

since, by $\mathbf{j}\mathbf{v}_i = 0$, $J\mathbf{v}_i = 0$. Thus $\mathbf{v}_i$ is an eigenvector indeed and thus, if $\lambda_1 \geq \ldots \geq \lambda_n$ are the eigenvalues of G, then $n - 1 - \lambda_1$, $-1 - \lambda_2$, $\ldots$, $-1 - \lambda_n$ are the eigenvalues of $\overline{G}$.

(b) The two equalities in the hint are easily verified. We use now the well-known fact that if C is an $(m \times n)$ matrix and D is an $(n \times m)$ matrix, then

$$\det(xI - CD) = x^{m-n}\det(xI - DC).$$

Let G have m edges and n points. Then

$$\det(xI - A_{L(G)}) = \det((x + 2)I - B_G^T B_G) =$$

$$= (x + 2)^{m-n}\det((x + 2)I - B_G B_G^T) = (x + 2)^{m-n}\det((x + 2 - d)I - A_G).$$

Thus if $m \geq n$, the eigenvalues of $L(G)$ are

$d - 2 + \lambda$ for each eigenvalue λ of G, with the same multiplicity, and
$\quad -2$ $\qquad\qquad\qquad\qquad\qquad$ with multiplicity $m - n$.

(If $-d$ is an eigenvalue of G, then further -2's occur among the eigenvalues of the first type.) If $m < n$, then we have to remove $n - m$ numbers -2 from the numbers listed first to get the spectrum of $L(G)$ [H. Sachs; see Biggs].

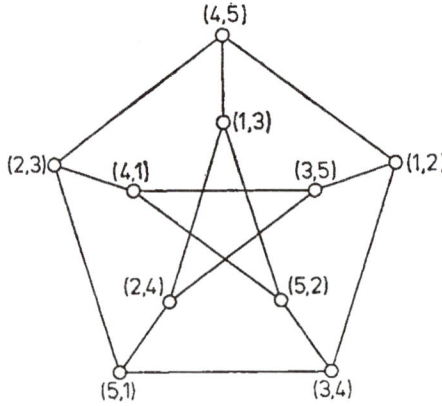

FIGURE 84

(c) Figure 84 shows how to assign the edges of the complete 5-graph to the points of the Petersen graph in such a way that adjacent edges correspond to non-adjacent points, and vice versa. Hence the Petersen graph is $\overline{L(K_5)}$, and by 11.1 and (a), (b), its eigenvalues are $-2, -2, -2, -2, 1, 1, 1, 1, 1, 3$.

3. The expansion of $\det(\lambda I - A_T)$ by its first k rows contains only two non-zero terms: One, where the $1^{\text{st}}, 2^{\text{nd}}, \ldots, k^{\text{th}}$ columns go with the first k rows, the other, where the $1^{\text{st}}, \ldots, (k-1)^{\text{st}}, (k+1)^{\text{st}}$ columns do so (any other combination yields a column of zeros in one subdeterminant or the other). Now the first term yields $p_{T_1}(\lambda) p_{T_2}(\lambda) = p_{T-e}(\lambda)$; the second term is the product to two determinants p_1, p_2, which do not correspond to characteristic polynomials directly. But if we expand D_1 (the subdeterminant composed of the $1^{\text{st}}, \ldots, k^{\text{th}}$ rows and $1^{\text{st}}, \ldots, (k-1)^{\text{st}}, (k+1)^{\text{st}}$ columns) by its last column we get $p_{T_1 - x_k}(\lambda)$; similarly, $D_2 = p_{T_2 - x_{k+1}}(\lambda)$ and so, taking the sign into consideration, the second term is

$$-p_{T_1 - x_k}(\lambda) p_{T_2 - x_{k_1}}(\lambda) = p_{T - x_k - x_{k+1}}(\lambda).$$

[A. Mowshowitz, *J. Comb. Th.* B **12** (1972) 177–193; L. Lovász–J. Pelikán, *Periodica Math. Hung.* **3** (1973) 175–182.]

4. We use induction on $|E(T)|$.

The statement is true for forests with no edges. Let $e = (x, y) \in E(T)$. By the previous exercise,

$$p_T(\lambda) = p_{T-e}(\lambda) - p_{T-x-y}(\lambda)$$

and by the induction hypothesis,

$$p_{T-e}(\lambda) = \lambda^n - a_1' \lambda^{n-2} + a_2' \lambda^{n-4} - \ldots,$$

$$p_{T-x-y}(\lambda) = \lambda^{n-2} - a_1'' \lambda^{n-4} + a_2'' \lambda^{n-6} - \ldots,$$

where a'_k (a''_k) is the number of k-element matchings in $T-e$ ($T-x-y$, resp.). Thus

$$p_T(\lambda) = \lambda^n - (a'_1 + 1)\lambda^{n-2} + (a'_2 + a''_1)\lambda^{n-4} - (a'_3 + a''_2)\lambda^{n-6} + \dots$$

To prove the formula it suffices to show that $a'_k + a''_{k-1} = a_k$ (and $a'_1 + 1 = a_1$). Since a_1 is the number of edges of T, a'_1 is the number of edges of $T-e$, $a'_1 + 1 = a_1$ obviously. Moreover, a'_k is the number of k-element matchings in $T-e$, i.e. the number of those k-element matchings of T not containing e. On the other hand, the $(k-1)$-element matchings of $T-x-y$ are in an obvious one-to-one correspondence with those k-element matchings of T containing e. Thus, $a'_k + a''_{k-1}$ gives the total number of k-element matchings in T as stated [ibid.].

5. (a) By 11.4,

$$p_{P_n}(\lambda) = \lambda^n - a_1\lambda^{n-2} + a_2\lambda^{n-4} - \dots,$$

where a_k is the number of k-element matchings in P_n; this means the number of ways to select the k non-neighboring edges from $n-1$, which is $\binom{n-k}{k}$ by 1.31.

(b) In 1.29 the recurrence relation

$$p_{P_n}(\lambda) = p_{P_{n-1}}(\lambda) - p_{P_{n-2}}(\lambda)$$

was deduced and formula (b) was proved by it (note that this recurrence relation is a consequence of 11.3 as well). This yields a simple inductive proof of (a) whose details are left to the reader. Also in 1.29 the eigenvalues of the path were found to be

$$2\cos\frac{k\pi}{n+1}, \qquad k = 1, \dots, n.$$

[ibid.]

For (d), the recurrence given in the hint expresses that $\mathbf{y}$ is an eigenvector; if we set $y_{-1} = y_{t+1} = 0$ then it will hold for $0 \le k \le t$. Hence, as in 1.29, we can express y_k in the form $\alpha x_1^k + \beta x_2^k$, where $x_1, x_2 = (\lambda \pm \sqrt{\lambda^2 - 4D})/2D$ are the roots of the equation

$$Dx^2 - \lambda x + 1 = 0.$$

From $y_{-1} = 0$ we get that $\beta = -(x_2/x_1)\alpha$, and hence we my assume (by scaling) that $y_k = x_1^{k+1} - x_2^{k+1}$. So $y_{t+1} = 0$ implies that $x_1 = \varepsilon x_2$, where ε is a $(t+2)^{\text{nd}}$ root of unity; $\mathbf{y} \ne 0$ implies that $\varepsilon \ne 1$. Solving this equation for λ, we get $\lambda = (\delta + 1/\delta)\sqrt{D}$, where $\delta = \varepsilon^{1/2}$ is a $2(t+2)^{\text{nd}}$ root of unity. This gives

$$\lambda = 2\sqrt{D}\cos\left(\frac{jn}{t+2}\right) \qquad (j = 1, \dots, 2t+4).$$

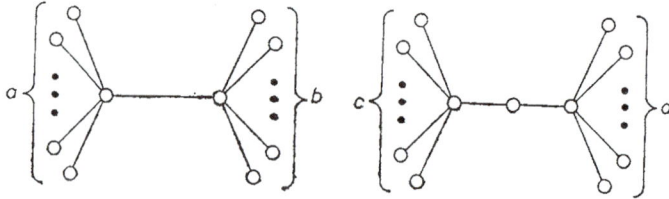

FIGURE 85

6. Trees with no 3 independent edges have one of the forms shown in Fig. 85. Let T_1 be one from the first class, T_2 be one from the second class. Let a_k, a_k' denote the number of k-element matchings in T_1, T_2 respectively. Then

$$a_1 = a + b + 1, \qquad a_1' = c + d + 2,$$
$$a_2 = ab, \qquad a_2' = cd + c + d.$$

If we find values a, $b > 0$, c, $d > 0$ such that $a_1 = a_1'$, $a_2 = a_2'$, then we are done as clearly this also implies that T_1, T_2 have the same number of points ($a_1 + 1 = a_1' + 1$). A little experimentation yields, e.g. the solution $a = b = u^2 + u + 1$, $c = u^2$, $d = (u+1)^2$. (In fact, almost every tree has a "pair" with the same spectrum; see A. Schwenk, in *New Directions in the Theory of Graphs*, Acad. Press, 1973, 275–307.)

7. (a) The adjacency matrix of G arises from the adjacency matrix of G_1 by replacing the 1's by I, the $|V(G_2)|$-dimensional identity matrix, and putting A_{G_2} into the diagonal. Hence $\lambda I - A_G$ arises by substituting $-I$ for the -1's in $\lambda I - A_{G_1}$ and $\lambda I - A_{G_2}$ for the λ's:

$$\underbrace{\begin{pmatrix} \lambda & 0 & -1 & -1 \\ 0 & \lambda & 0 & 0 \\ -1 & 0 & \lambda & -1 \\ -1 & 0 & -1 & \lambda \end{pmatrix}}_{\lambda I - A_{G_1}} \underbrace{\begin{pmatrix} \lambda I - A_{G_2} & 0 & -I & -I \\ 0 & \lambda I - A_{G_2} & 0 & 0 \\ -I & 0 & \lambda I - A_{G_2} & -I \\ -I & 0 & -I & \lambda I - A_{G_2} \end{pmatrix}}_{\lambda I - A_G}.$$

Since $\lambda I - A_{G_2}$ and I commute, this yields

$$p_G(\lambda) = \det(\lambda I - A_G) = \det p_{G_1}(\lambda I - A_{G_2}).$$

Let $\lambda_1, \ldots, \lambda_n$ be the eigenvalues of G_1, $\mu_1 \ldots, \mu_m$ be those of G_2. Then

$$p_G(\lambda) = \det \prod_{i=1}^{n} (\lambda I - A_{G_2} - \lambda_i I) = \prod_{i=1}^{n} \det((\lambda - \lambda_i)I - A_{G_2}) =$$
$$= \prod_{i=1}^{n} p_{G_2}(\lambda - \lambda_i) = \prod_{i=1}^{n} \prod_{j=1}^{m} (\lambda - \lambda_i - \mu_j).$$

Hence the eigenvalues of G are $\lambda_i + \mu_j$ ($1 \leq i \leq n$, $1 \leq j \leq m$).

(b) Let $A'_G = A_G + I$. With this notation, if x_i, $y_i \in V(G_i)$, then

$$\{((x_1, x_2), (y_1, y_2)) \in E(G_1 \cdot G_2) \text{ or } (x_1, x_2) = (y_1, y_2) \Leftrightarrow$$
$$((x_1, y_1) \in E(G_1) \text{ or } x_1 = y_1) \text{ and } ((x_2, y_2) \in E(G_2) \text{ or } x_2 = y_2)\},$$

which means that $A'_{G_1 \cdot G_2}$ is the Kronecker product of A'_{G_1} and A'_{G_2}.

The eigenvalues of A'_{G_1} and A'_{G_2} are $\lambda_1 + 1, \ldots, \lambda_{n_1} + 1, \ldots, \mu_1 + 1, \ldots, \mu_{n_2} + 1$, where $\lambda_1, \ldots, \lambda_{n_1}$; $\mu_1, \ldots, \mu_{n_2}$ are the eigenvalues of G_1 and G_2, respectively. Thus the eigenvalues of $A'_{G_1 \cdot G_2}$ are

$$(\lambda_i + 1)(\mu_j + 1) \qquad (1 \le i \le n_1, \ 1 \le j \le n_2)$$

and the eigenvalues of $G_1 \cdot G_2$ are

$$\lambda_i \mu_j + \lambda_i + \mu_j \qquad (1 \le i \le n_1, \ 1 \le j \le n_2).$$

A similar argument shows that the eigenvalues of the weak direct product are $\lambda_i \mu_j$. [D. Cvetković, *Univ. Beograd, Publ. Elektrotehn. Fak. Ser. Mat. Fiz.* **354–356** (1971) 1–50.]

8. The ith coordinate of $A\mathbf{v}$ is

$$\sum_{j=1}^{n} a_{ij} v_j = \sum_{(i,j) \in E(G)} \chi(\gamma_{1,j}) = \sum_{(i,j) \in E(G)} \chi(\gamma_{i,j}) \chi(\gamma_{1,i}).$$

Take here a term and set $\gamma_{i,j}(1) = k$ (i.e. $\gamma_{i,j} = \gamma_{1,k}$). We claim that $(i,j) \in E(G)$ if and only if $(1,k) \in E(G)$. For consider $\gamma_{1,i}$. Then $\gamma_{1,i}(1) = i$ by definition. On the other hand,

$$\gamma_{1,i}(k) = \gamma_{1,i}(\gamma_{i,j}(1)) = \gamma_{i,j}(\gamma_{1,j}(1)) = j$$

(by the commutativity of Γ) and so, $(1,k)$ is mapped to (i,j) by $\gamma_{i,j}$. Thus, they are both edges or non-edges.

Thus, if j ranges over all neighbors of i, the point $k = \gamma_{i,j}(1)$ ranges over all neighbors of 1 (it is different for different points j, because Γ is regular). Hence, the formula for the ith coordinate of $A\mathbf{v}$ can be rewritten as

$$\left\{ \sum_{(1,k) \in E(G)} \chi(\gamma_{1,k}) \right\} \chi(\gamma_{1,i})$$

showing that $\mathbf{v}$ is an eigenvector with eigenvalue

$$\sum_{(1,k) \in E(G)} \chi(\gamma_{1,k}).$$

We remark that, since a commutative group Γ has exactly $|\Gamma|$ distinct characters (and they all have $\chi(1) = 1$, so that they do not give parallel vectors), the eigenvalues found above are all the eigenvalues of G. (For the case of non-commutative groups see L. Lovász, *Periodica Math. Hung.* **6** (1975) 191–195.)

9. *First solution.* Let us represent the points of Q_n by 01-vectors of length n, two being adjacent iff they differ at exactly one place. For each 01-vector $(a_1,\ldots,a_n)$, the mapping

$$\alpha_{a_1,\ldots,a_n} : (x_1,\ldots,x_n) \mapsto (x_1 + a_1,\ldots,x_n + a_n)$$
$$((x_1,\ldots,x_n) \in V(Q_n); \text{ addition mod } 2)$$

is an automorphism of Q_n. Geometrically, there are those automorphisms of Q_n preserving the directions of edges; they are reflexions in certain affine subspaces. They form a commutative, regular group Γ, which is (abstractly) isomorphic with $(Z_2)^n$.

Now the characters can be obtained this way: let $1 \le i_1 < i_2 < \ldots < i_k \le n$, and

$$\chi_{i_1,\ldots,i_k}(\alpha_{a_1,\ldots,a_n}) = (-1)^{a_{i_1}+\ldots+a_{i_k}}.$$

It is easy to verify that there are characters and since their number is $2^n = |\Gamma|$, they are all the characters. Thus, we get an eigenvalue as follows:

$$\lambda_{i_1,\ldots,i_k} = \sum_{((a_1,\ldots,a_n),(0,\ldots,0)) \in E(Q_n)} \chi_{i_1,\ldots,i_k}(\alpha_{a_1,\ldots,a_n}) =$$

$$= \sum_{((a_1,\ldots,a_n),(0,\ldots,0)) \in E(Q_n)} (-1)^{a_{i_1}+\ldots+a_{i_k}}.$$

By the definition of Q_n, the sequences $(a_1,\ldots,a_n)$ considered here are those having exactly one non-zero entry. Hence

$$\lambda_{i_1,\ldots,i_k} = n - 2k.$$

Thus, the eigenvalues of Q_n are

$$n, \ n-2, \ \ldots, \ -n; \text{ the multiplicity of } n-2k \text{ is } \binom{n}{k}.$$

Second solution. Q_n is the Cartesian product of n copies of K_2. Hence by 11.7a, its eigenvalues arise in the form $\lambda_1 + \ldots + \lambda_n$, where $\lambda_i = \pm 1$ are eigenvalues of K_2. We conclude as before.

10. The mappings

$$\varphi_\nu : \varepsilon^k \to \varepsilon^{k+\nu}$$

are automorphisms of G, forming a cyclic group Γ of order p, which acts transitively on $V(G)$. The characters of Γ are given by

$$\chi_t(\varphi_\nu) = \varepsilon^{t\nu} \qquad (t = 0,\ldots,p-1; \quad \nu = 0,\ldots,p-1),$$

where $\varepsilon = e^{\frac{2\pi i}{p}}$. Thus by 11.8, the eigenvalues of G are given by

$$\lambda_t = \sum_{j=1}^{k} \varepsilon^{ta_j} \qquad (t = 0,\ldots,p-1),$$

where 1 is adjacent to $\varepsilon^{a_1},\ldots,\varepsilon^{a_k}$. Similarly, the eigenvalues of H are

$$\mu_t = \sum_{j=1}^{k} \varepsilon^{tb_j} \qquad (t = 0, \ldots, p-1),$$

where $\varepsilon^{b_1}, \ldots, \varepsilon^{b_k}$ are the neighbors of 1. Since G and H must have the same spectra, we have

$$\lambda_1 = \mu_t \qquad (0 \le t \le p-1)$$

or, equivalently,

$$\sum_{j=1}^{k} (\varepsilon^{a_j} - \varepsilon^{tb_j}) = 0.$$

Let $\bar{a}$ denote the residue of a mod p, $0 \le \bar{a} \le p-1$. Then ε is a root of the polynomial

$$f(x) = \sum_{j=1}^{k} (x^{\bar{a}_j} - x^{\overline{tb_j}}).$$

Since the minimal polynomial of ε is $1 + x + \ldots + x^{p-1}$, it follows that

$$1 + x + \ldots + x^{p-1} | f(x).$$

Since $\deg f \le p-1$, we get;

$$f(x) = c(1 + x + \ldots + x^{p-1}).$$

From $f(0) = 0$, we deduce that $c = 0$, i.e. $f(x) = 0$. But this means that in $f(x)$ all terms cancel out, i.e.

$$\{\bar{a}_j : 1 \le j \le k\} = \{\overline{tb_j} : 1 \le j \le k\}$$

proving the assertion. [D. Z. Djoković, *Acta Math. Acad. Sci. Hung.* **21** (1970) 267–277; B. Elspas–J. Turner, *J. Comb. Th.* **9** (1970) 297–307.]

11. If $\det A \ne 0$, we have a non-zero expansion term of $\det A$; say $a_{1,i_1}, \ldots, a_{n,i_n} \ne 0$, where $(i_1, \ldots, i_n)$ is a permutation of $(1, \ldots, n)$. Thus $\nu, i_\nu) \in E(G)$ $(i = 1, \ldots, n)$. If say, $1, \ldots, k$ $(k \ge n/2)$ form one of the two color-classes of G, then the edges

$$(1, i_1), \ldots, (k, i_k)$$

are independent and their number is $k \ge n/2$, thus they form a 1-factor of G.

12. Let $G_1, \ldots, G_m$ be vertex-disjoint subgraphs as shown in Fig. 10, which cover all points (so $m = n/6$). Let S be the set of endpoints of $G_1 \cup \ldots \cup G_m$ and $T = V(G) - S$. Then each point is adjacent to two points of S and to one point of T. Hence we set

$$x_i = \begin{cases} 2 & \text{if } i \in T, \\ -1 & \text{if } i \in S \end{cases} \qquad \mathbf{x} = \begin{pmatrix} x_1 \\ \vdots \\ x_n \end{pmatrix},$$

we have $A\mathbf{x} = 0$, showing that $\lambda = 0$ is an eigenvalue. [H. Sachs. *Wiss. Z. Tech. Hochschule Ilmenau*, **19** (1973) 83–99.]

13. We may assume that G' is a spanning subgraph of G; otherwise, we can add the remaining points of G to G' as isolated points and this only adds 0's to the spectrum of G', thus does not change the maximum eigenvalue of it.

Let $\mathbf{v}$ be an eigenvector of $A_{G'}$, with eigenvalue λ_1' and with non-negative entries. We may assume that $|\mathbf{v}| = 1$, then

$$\lambda_1' = \mathbf{v}^T A_{G'} \mathbf{v} \le \mathbf{v}^T A_G \mathbf{v}$$

as $\mathbf{v}$ has non-negative entries and $A_G \ge A_{G'}$. On the other hand,

$$\lambda_1 = \max_{|\mathbf{x}|=1} \mathbf{x}^T A_G \mathbf{x} \ge \mathbf{v}^T A_G \mathbf{v} \ge \lambda_1'.$$

14. (a) I. Let $\mathbf{j}$ be the vector of length n all of whose entries are 1. Then $A_G \mathbf{j}$ is a vector whose entries are the degrees and so,

$$A_G \mathbf{j} \ge d\mathbf{j}, \quad \mathbf{j}^T A_G \mathbf{j} \ge d\mathbf{j}^T \mathbf{j} = dn.$$

Thus

$$\lambda_1 = \max_{\mathbf{x} \ne 0} \frac{\mathbf{x}^T A_G \mathbf{x}}{\mathbf{x}^T \mathbf{x}} \ge \frac{\mathbf{j}^T A_G \mathbf{j}}{\mathbf{j}^T \mathbf{j}} \ge d.$$

II. Let G' denote the star of a point of maximum degree. Then, by 11.1, the maximum eigenvalue of G' is $\sqrt{D}$. By 11.13,

$$\lambda_1 \ge \sqrt{D}.$$

III. We have $A_G \mathbf{j} \le D\mathbf{j}$ by the definition of D.

Let $\mathbf{v}$ be an eigenvector belonging to D; we may assume that the maximum coordinate of it, say the first one, is equal to 1. Then

$$\lambda_1 \mathbf{v} = A_G \mathbf{v} \le A_G \mathbf{j} \le D\mathbf{j},$$

so considering the first coordinates,

$$\lambda_1 \le D.$$

IV. Suppose that we have equality here. Then $A_G \mathbf{v}$ and $A_G \mathbf{j}$ have the same first coordinate D, i.e.

$$\sum_{(1,i) \in E(G)} v_i = D.$$

Since $v_i = 1$, this can only happen of each i adjacent to 1 has the same value $v_i = 1$. Then the above argument applies to these points and yields that their neighbors j also have $v_j = 1$ and so on; since G is connected, we conclude that $v_i = 1$ for every point. Thus $\mathbf{v} = \mathbf{j}$, and then

$$A_G \mathbf{j} = \lambda_1 \mathbf{j} = D\mathbf{j}$$

yields that G is D-regular.

(b) The two identities in the hint are easily seen: If $\lambda_1 \geq \ldots \geq \lambda_n$ are the eigenvalues of G, then $\sum_{i=1}^{n} \lambda_i = \operatorname{Tr} A_G = 0$, $\sum_{i=1}^{n} \lambda_i^2 = \operatorname{Tr} A_G^2 = 2m$. Hence

$$\lambda_1^2 = (\lambda_2 + \ldots + \lambda_n)^2 \leq (n-1)(\lambda_2^2 + \ldots + \lambda_n^2) = (n-1)(2m - \lambda_1^2),$$

whence the upper bound follows. The lower bound follows by the argument of I.

(c) G is a subgraph of a rooted complete $(D-1)$-ary tree T. The largest eigenvalue of T is $2\sqrt{D-1}\cos(n/(t+2)) < 2\sqrt{D-1}$, by 11.5(d). Thus $\lambda_1 < 2\sqrt{D-1}$.

15. The formula

$$A_{L(G)} = B^T B - 2I$$

follows easily.

Since $B^T B$ is positive semidefinite, all of its eigenvalues are non-negative. Hence, the eigenvalues of $A_{L(G)}$ are ≥ -2. Moreover, if $|V(G)| < |E(G)|$, then

$$r(B^T B) = r(B) \leq |V(G)| < |E(G)|$$

($r(X)$ is the rank of the matrix X). So, $B^T B$ has at least one 0 eigenvalue, i.e. $A_{L(G)}$ has at least one -2 eigenvalue. [A. J. Hoffman, in *Beiträge zur Graphentheorie*, Teubner (1968) 75–80.]

16. Let α be an automorphism of G; define a matrix $P = (p_{ij})$ by

$$p_{ij} = \begin{cases} 1 & \text{if } \alpha(i) = j, \\ 0 & \text{otherwise.} \end{cases}$$

Then P is a permutation matrix, i.e. each row and column of it contains exactly one 1. Hence $P^{-1} = P^T$.

The fact that α is an automorphism is equivalent to the relation

$$P^{-1} A_G P = A_G.$$

For if $A_G = (a_{ij})$, then the $(i,j)^{\text{th}}$ entry of $P^{-1} A_G P$ is

$$\sum_{\nu} \sum_{\mu} p_{\nu i} a_{\nu \mu} p_{\mu j} = a_{i',j'},$$

where i', j' are determined by $\alpha(i') = i$, $\alpha(j') = j$. Thus α maps the pair (i',j') onto the pair (i,j), and it as an automorphism iff

$$a_{i'j'} = a_{ij} \qquad (i,j = 1, \ldots, n)$$

as stated.

Now A_G is symmetric, hence $A_G = T^{-1} L T$, where

$$L \begin{pmatrix} \lambda_1 & & & 0 \\ & \lambda_2 & & \\ & & \ddots & \\ 0 & & & \lambda_n \end{pmatrix}$$

and T is an orthogonal matrix. Thus

$$P^{-1}T^{-1}LTP = T^{-1}LT,$$

or

$$L(TPT^{-1}) = (TPT^{-1})L.$$

Multiplying a matrix by L from the left (from the right) means to multiply its ith row (i^{th} column) by λ_i for each $1 \leq i \leq n$. Since $\lambda_i \neq \lambda_j$ for $i \neq j$ by the assumption, this yields that the matrix TPT^{-1} is a diagonal matrix, say

$$TPT^{-1} = \begin{pmatrix} t_1 & & 0 \\ & \ddots & \\ 0 & & t_n \end{pmatrix}.$$

Here, however, P and T are orthogonal matrices and thus, so is TPT^{-1}. Hence $t_i = \pm 1$. But then

$$TP^2T^{-1} = (TPT^{-1})^2 = I, \qquad P^2 = I.$$

Now it is easy to see that P^2 is the permutation matrix associated with α^2, whence $\alpha^2 = 1$. [A. Mowshowitz, in *Proof Techn. in Graph Th.* Academic Press, 1969, 109–110; M. Petersdorf–H. Sachs, in *Combin. Th. Appl.* Coll. Math. Soc. J. Bolyai **4**, Bolyai–North-Holland, (1970) 891–907. For directed graphs cf. C. Y. Chao, *J. Comb. Th.* **3** (1971) 301–302.]

17. Since by 11.16, all elements of Γ are of order 2, Φ is commutative $((xy)^2 = 1 = (x^2y^2)$ and hence, $xy = yx)$. Suppose that Γ is transitive on the points, then it follows that it is regular[†]. Thus 11.8 applies and we get that any eigenvalue of G is of the form

$$\lambda = \sum_{(1,i) \in E(G)} \chi(\gamma_{1,i}),$$

where χ is the character of Γ. Moreover, since $\chi(\gamma)^2 = \chi(\gamma^2) = \chi(1) = 1$ for each γ, the summands here are ± 1's and hence,

$$-d \leq \lambda \leq d, \qquad \lambda \equiv d \pmod{2}.$$

This leaves exactly $d+1$ values for λ. Since there are $|V(G)|$ eigenvalues, all different, we have $|V(G)| \leq d+1$. Since G is simple, it must be a complete graph, but this is impossible again except that $G \cong K_2$, since K_n has $n-1$ equal eigenvalues (see 11.1) [ibid.].

[†] Suppose that $\varphi \in \Gamma$ fixes a point x. Then by $\varphi(\alpha(x)) = \alpha(\varphi(x)) = \alpha(x)$, it follows that it fixes every point $\alpha(x)$, $\alpha \in \Gamma$, Since Γ is transitive, $\varphi = 1$.

18. Let G_1, G_2 be the subgraphs induced by a maximum clique and a maximum independent set, respectively. Then A_{G_1}, A_{G_2} are symmetric submatrices of A_G. Since G_1 has eigenvalues

$$\omega(G) - 1, \underbrace{-1, \ldots, -1}_{\omega(G)-1},$$

we have, by the theorem mentioned in the hint,

$$\lambda_{\omega(G)} \geq -1, \qquad \lambda_{n-\omega(G)+2} \leq -1.$$

Similarly, G_2 has eigenvalues $0,\ldots,0$ and therefore

$$\lambda_{\alpha(G)} \geq 0, \qquad \lambda_{\tau(G)+1} \leq 0.$$

19. (a) Suppose first that $-\lambda_1$ is an eigenvalue.

Let $\mathbf{w} = (w_1,\ldots,w_n)^T$ be the eigenvector belonging to $-\lambda_1$ and set $\mathbf{w}' = (|w_1|,\ldots,|w_n|^T)^T$, $A_G = (a_{ij})$. Then

$$(1) \qquad |\mathbf{w}^T A_G \mathbf{w}| = |-\lambda_1 \mathbf{w}^T \mathbf{w}| = \lambda_1 \mathbf{w}^t \mathbf{w};$$

on the other hand,

$$\mathbf{w}^T A_G \mathbf{w} = \left| \sum_i \sum_j a_{ij} w_i w_j \right| \leq \sum_i \sum_j a_{ij} |w_i||w_j| =$$

$$(2) \qquad = \mathbf{w}'^T A_G \mathbf{w}' \leq \lambda_1 \mathbf{w}'^T \mathbf{w}' = \lambda_1 \mathbf{w}^T \mathbf{w},$$

(since λ_1 is the maximum of $\mathbf{u}^T A_G \mathbf{u}/\mathbf{u}^T \mathbf{u}$).

Comparing (1) and (2), we find that we must have equality everywhere in (2). Thus, $\mathbf{w}'$ is an eigenvector belonging to λ_1 and since G is connected, no entry of $\mathbf{w}'$ is 0. Moreover, all non-zero terms in

$$\sum_i \sum_j a_{ij} w_i w_j = -\lambda_1 \mathbf{w}^T \mathbf{w}$$

must have the same sign and since the sum is negative, all summands are negative. Thus, if $a_{ij} \neq 0$, w_j have different signs. So, if

$$S_1 = \{i : w_i > 0\} \quad S_2 = \{i : w_i < 0\},$$

then $\{S_1, S_2\}$ is a 2-coloration of G.

Conversely, let G be bipartite and let $\mathbf{w} = (w_1,\ldots,w_n)$ be the eigenvector belonging to the eigenvalue λ_1. Let S_1, S_2 be a 2-coloration of G and set

$$w_i' = \begin{cases} w_i & \text{if } i \in S_1, \\ -w_i & \text{if } i \in S_2, \end{cases}$$

$$\mathbf{w}' = (w_i',\ldots,w_n')^T.$$

Then the ith coordinate of $A_G \mathbf{w}'$ is

$$\sum_j a_{ij} w'_j = -\sum_i a_{ij} w_j = -\lambda_1 w_i = -\lambda_1 w'_i,$$

if $i \in S_1$ (since all j's here for which $a_{ij} \neq 0$ have $j \in S_2$), and similarly,

$$\sum_j a_{ij} w'_i = \sum_j a_{ij} w_j = \lambda_1 w_i = -\lambda_1 w'_j,$$

if $i \in S_2$. Thus $\mathbf{w}'$ is an eigenvector and the corresponding eigenvalue is $-\lambda_1$. The same idea proves the "only if" part of (b) too.

(b) Suppose that the spectrum is symmetric with respect to the origin. If G is connected (a) proves the theorem. We are going to reduce the general case to this, using induction on the number of components.

Note that the spectrum of G is the union of the spectra of its components (in the same sense that, if λ occurs on the spectra of components with multiplicity $m_1, \ldots, m_r$, then it has multiplicity $m_1 + \ldots + m_r$ in the spectrum of G). Let λ_1 be the largest eigenvalue and let G_1 be a component such that $-\lambda_1$ is an eigenvalue of G_1. Then the largest eigenvalue of G_1 is at least $|-\lambda_1| = \lambda_1$ and, since it cannot be greater than this, it is λ_1. Hence, applying (a) to G_1 we get that G_1 is bipartite. Hence, its spectrum is symmetric with respect to the origin. We obtain the spectrum of $G - G_1$ by removing the spectrum of G_1 from the spectrum of G. Both these spectra being symmetric, so is the spectrum of $G - G_1$. By induction, $G - G_1$ is bipartite. So is G_1 as we have seen and therefore, so is G.

20. We use induction on the number of points.

Set $k = \lfloor \lambda_1 \rfloor + 1$. Let x be a point of minimum degree d. By 11.14, $d \leq \lambda_1$, so $d \leq k-1$. Let λ'_1 be the maximum eigenvalue of $G - x$. By 11.13, $\lambda'_1 \leq \lambda_1$, so by the induction hypothesis

$$\chi(G - x) \leq \lambda'_1 + 1 \leq \lambda_1 + 1, \quad \text{so} \quad \chi(G - x) \leq k.$$

Let us k-color $G - x$. Since $d = k - 1$, this coloration can be extended to x, and so, $\chi(G) \leq k \leq \lambda_1 + 1$ [H. S. Wilf; see Biggs].

21. (a) It follows from $\chi(G) = k$ that for a suitable choice of indices, A_G can be written in the form

$$\begin{pmatrix} 0 & A_{12} & \cdots & A_{1k} \\ A_{21} & 0 & & A_{2k} \\ \vdots & \vdots & \ddots & \\ A_{k1} & A_{k2} & & 0 \end{pmatrix},$$

where A_{ij} is an $m_i \times m_j$ matrix and, of course, $A_{ji} = A_{ij}^T$, here m_i is the number of points with color i.

Let $\mathbf{v}$ be an eigenvector belonging to λ_1. Let us break $\mathbf{v}$ into pieces $\mathbf{v}_1, \ldots, \mathbf{v}_k$ of length $m_1, \ldots, m_k$, respectively. Set

$$\mathbf{w}_i = \left.\begin{pmatrix} |\mathbf{v}_i| \\ 0 \\ \vdots \\ 0 \end{pmatrix}\right\} m_i \text{ entries}, \quad \mathbf{w} = \begin{pmatrix} \mathbf{w}_1 \\ \vdots \\ \mathbf{w}_k \end{pmatrix}.$$

Let B_i be any orthogonal matrix such that

$$\mathbf{w}_i B_i = \mathbf{v}_i \qquad (i = 1, \ldots, k),$$

and

$$B = \begin{pmatrix} B_1 & & & 0 \\ & B_2 & & \\ & & \ddots & \\ 0 & & & B_k \end{pmatrix}.$$

Then $B\mathbf{w} = \mathbf{v}$ and

$$B^{-1}AB\mathbf{w} = B^{-1}A\mathbf{v} = \lambda_1 B^{-1}\mathbf{v} = \lambda_1 \mathbf{w},$$

so $\mathbf{w}$ is an eigenvector of $B^{-1}AB$.

Moreover, $B^{-1}AB$ has the form

$$\begin{pmatrix} 0 & B_1^{-1}A_{12}B_2 & \ldots & B_1^{-1}A_{1k}B_k \\ B_2^{-1}A_{21}B_1 & 0 & & B_2^{-1}A_{2k}B_k \\ \vdots & & \ddots & \vdots \\ B_k^{-1}A_{k1}B_1 & B_k^{-1}A_{k2}B_2 & \ldots & 0 \end{pmatrix}.$$

Pick the entry in the upper left corner of each of the k^2 submatrices $B_i^{-1}A_{ij}B_j$ ($A_{ii} = 0$), these form a $k \times k$ submatrix D. Now observe that

$$\mathbf{u} = \begin{pmatrix} |\mathbf{v}_1| \\ \vdots \\ |\mathbf{v}_k| \end{pmatrix}$$

is an eigenvector of D; for $\mathbf{w}$ is an eigenvector of $B^{-1}AB$ and has 0 entries on places corresponding to those rows and columns of $B^{-1}AB$, which are to be deleted to get D. Moreover, the eigenvalue belonging to $\mathbf{u}$ is λ_1.

Now let $\mu_1 \geq \ldots \geq \mu_k$ be the eigenvalues of D. Since D has 0's in its main diagonal,

$$\mu_1 + \ldots + \mu_k = 0.$$

On the other hand, λ_1 is an eigenvalue of D and so

$$\lambda_1 \leq \mu_1,$$

while by the Interlacing Eigenvalue Theorem (quoted in the proof of 11.18),

$$\lambda_n \le \mu_k, \dots, \lambda_{n-k+2} \le \mu_2.$$

Thus

$$\lambda_n + \dots + \lambda_{n-k+2} \le \mu_k + \dots + \mu_2 = -\mu_1 \le -\lambda_1.$$

[A. J. Hoffman, in: *Graph Th. Appl.* Academic Press (1970) 79–91.]

(b) Let the k-coloration of G have a_i points of the ith color $(i = 1, \dots, k)$. Then the a_i rows of the matrix A_G corresponding to the ith color are equal, hence the rank of A_G is at most k and $n - k$ of the eigenvalues are 0; moreover if we denote by $\mathbf{w}_i$ the vector with

$$\mathbf{w}_i = (w_{ij}),$$
$$w_{ij} = \begin{cases} 1 & \text{if the } j^{\text{th}} \text{ point has color } i \\ 0 & \text{otherwise,} \end{cases}$$

then the eigenvectors belonging to the $n - k$ 0's are orthogonal to every $\mathbf{w}_i$. We look now for eigenvectors of the form

$$(1) \qquad\qquad\qquad \mathbf{w} = \sum_{i=1}^{k} x_i \mathbf{w}_i;$$

these will be different from the above mentioned ones and if we find k of them (i.e. k eigenvalues belonging to them) we have all. (1) is an eigenvector with eigenvalue λ if and only if

$$(2) \qquad\qquad\qquad \sum_{\nu \ne i} a_\nu x_\nu = \lambda x_i,$$

i.e. if λ is an eigenvalue and $\begin{pmatrix} x_1 \\ \vdots \\ x_k \end{pmatrix}$ is a corresponding eigenvector of

$$A' = \begin{pmatrix} 0 & a_2 & a_3 & \dots & a_k \\ a_1 & 0 & a_3 & \dots & a_k \\ \vdots & & & & \\ a_1 & a_2 & a_3 & \dots & 0 \end{pmatrix}.$$

We claim that every eigenvalue of A' except the largest one is non-positive. In order to show this write (2) as

$$(3) \qquad\qquad\qquad \sum_{\nu=1}^{k} a_\nu x_\nu = (\lambda + a_i) x_i.$$

If $\lambda = -a_i$ for some i, then we have nothing to show. Otherwise,

$$\sum_{\nu=1}^{k} a_\nu x_\nu \neq 0$$

and hence, we may assume

(4) $$\sum_{\nu=1}^{k} a_\nu x_\nu = 1.$$

Then (3) yield

$$x_i = \frac{1}{\lambda + a_i} \qquad (i = 1, \ldots, k),$$

whence by (4)

(5) $$\sum_{\nu=1}^{k} \frac{a_\nu}{\lambda + a_\nu} = 1.$$

Now it is easy to see that (5) has exactly one positive root. Thus all but one of the eigenvalues of A' are non-positive.

Thus, if $\lambda_1, \ldots, \lambda_{k-1}$ are the k least eigenvalues of A_G and Λ is the largest one, then $\lambda_1, \ldots, \lambda_{k-1}, \Lambda$ are exactly the eigenvalues of A'. Since A' has 0's in its diagonal,

$$\lambda_1 + \ldots + \lambda_{k-1} + \Lambda = 0,$$

which proves the assertion.

22. The identity in the hint is easily verified by comparing coefficients on both sides. Now the inequality follows from the observation that $\sum_i x_i = 0$ defines the subspace H orthogonal to the eigenvector $\mathbf{1}$ of A_G and hence invariant under A_G. So the eigenvalues of the quadratic form $\mathbf{x}^T(dI - A_G)\mathbf{x}$, restricted to this subspace, are $d - \lambda_n \geq \ldots \geq d - \lambda_2$.

23. Let A be a maximum independent set of vertices. Define a vector $\mathbf{y}$ by

$$y_i = \begin{cases} n - \alpha, & \text{if } i \in A, \\ -\alpha, & \text{if } i \in V(G) \setminus A. \end{cases}$$

Then $\sum_i y_i = 0$, $\sum_i y_i^2 = \alpha(n - \alpha)n$, and

$$\sum_{(i,j) \in E} (y_i - y_j)^2 = \alpha dn^2,$$

since every edge between A and $V(G) \setminus A$ contributes n^2 to the sum, while the other edges contribute 0. Hence by 11.22,

$$\alpha(n - \alpha)n(d - \lambda_n) \geq \alpha dn^2,$$

whence the assertion follows. [cf. L. Lovász, *IEEE Trans. Inf. Theory* **25** (1979), 1–8.]

24. Let $G = G_1 \cup \ldots \cup G_m$, where G_i is a complete bipartite graph. We add all points of $V(G) - V(G_i)$ as isolated points to G_i. Then

$$A_G = A_{G_1} + \ldots + A_{G_m}, \quad \mathbf{x}^T A_G \mathbf{x} = \sum_{i=1}^{m} \mathbf{x}^T A_{G_i} \mathbf{x}.$$

Denote by $\{A_i, B_i\}$ the 2-coloration of G_i. Then

$$\mathbf{x}^T A_{G_i} \mathbf{x} = \sum_{\nu \in A_i} \sum_{\mu \in B_i} x_\nu x_\mu = \frac{1}{4} \left[(\sum_{\nu \in A_i} x_\nu + \sum_{\mu \in B_i} x_\mu)^2 - (\sum_{\nu \in A_i} x_\nu - \sum_{\mu \in B_i} x_\mu)^2 \right]$$

and thus

$$\mathbf{x}^T A_G \mathbf{x} = \sum_{i=1}^{m} \frac{1}{4} (\sum_{\nu \in A_i} x_\nu + \sum_{\mu \in B_i} x_\mu)^2 - \sum_{i=1}^{m} \frac{1}{4} (\sum_{\nu \in A_i} x_\nu - \sum_{\mu \in B_i} x_\mu)^2.$$

Thus $\mathbf{x}^T A_G \mathbf{x}$ is the sum of m positive and m negative squares. Hence, as is well known, $m \geq k$ and $m \geq l$ both follow. To give an argument, let U denote the subspace of vectors $\mathbf{x}$ such that

$$\sum_{\nu \in A_i} x_\nu = \sum_{\mu \in B_i} x_\mu \quad (i = 1, \ldots, m)$$

and V the subspace spanned by the eigenvectors with negative eigenvalue. Then clearly $U \cap V = 0$ and so

$$l = \dim V \leq n - \dim U \leq n - (n - m) = m.$$

[R. L. Graham–H. O. Pollak, in: *Graph Th. Appl.* (Lecture Notes in Math. 303, Springer 1972) 99–110.]

25. Let k be the number of distinct eigenvalues of G and assume indirectly that the diameter of G is at least k. Let $f(x) = (x - \lambda_1) \ldots (x - \lambda_k) = x^n + a_{n-1} x^{n-1} + \ldots + a_0$, where $\lambda_1, \ldots, \lambda_k$ are all the distinct eigenvalues of G. As is well known

(1) $f(A_G) = A_G^k + a_{n-1} A_G^{k-1} + \ldots + a_0 I = 0.$

Now let i and j be two points at distance k (these exist as the diameter of G is at least k). Observe that the $(i,j)^{\text{th}}$ entry of A_G^m is the number of (i,j)-walks of length m; because it is

$$\sum_{i_1=1}^{n} \sum_{i_2=1}^{n} \cdots \sum_{i_{m-1}=1}^{n} a_{i i_1} a_{i_1 i_2} \cdots a_{i_{m-1} j}$$

and this term is 1 if $(i, i_1, \ldots, i_{n-1}, j)$ is a walk and 0 otherwise.

Thus, the $(i,j)^{\text{th}}$ entry in A_G^k is positive, but it is 0 in all the other term in (1), which is a contradiction [see Biggs].

26. We consider the following three statements:

(i) G is connected and d-regular.

(ii) There exists s polynomial f such that $f(A_G) = J$;

(iii) Every eigenvector of A_G is an eigenvector of J, i.e. either parallel or orthogonal to $\mathbf{j}$.

(ii)⇔(iii). Let $\mathbf{u}$ be any eigenvector of A_G; it is an eigenvector of any polynomial of A_G, in particular of J.

(iii)⇔(i). Since the eigenvectors of A_G span the whole space, one of them must be $\mathbf{j}$. Let d denote the eigenvalue belonging to $\mathbf{j}$, then

$$A_G\mathbf{j} = d\mathbf{j}$$

expresses exactly the fact that G is d-regular.

Suppose that G is disconnected, say $G = G_1 \cup G_2$, $G_1 \cap G_2 = \emptyset$. Define $\mathbf{u} = (u_i)$ by

$$u_i = \begin{cases} 1 & \text{if } i \in V(G_1), \\ 0 & \text{otherwise.} \end{cases}$$

Then $\mathbf{u}$ is an eigenvector of A_G but not of J, a contradiction.

(i)⇔(ii). Since G is D-regular, $\mathbf{j}$ is an eigenvector with eigenvalue d. Let $\mathbf{u}_1 = \mathbf{j}, \mathbf{u}_2, \ldots, \mathbf{u}_n$ be a basis of eigenvectors of A_G, with corresponding eigenvalues $d = \lambda_1, \ldots, \lambda_n$. As is well-known, $\mathbf{j}^T\mathbf{u}_i = 0$ $(2 \le i \le n)$.

Consider a polynomial $f(x)$ such that

$$f(d) = n, \quad f(\lambda_i) = 0 \quad (2 \le i \le n).$$

Since G is connected, d is a single eigenvalue and this definition makes sense. Then

$$f(A_G)\mathbf{j} = n\mathbf{j}, \quad f(A_G)\mathbf{u}_i = 0 \quad (2 \le i \le n),$$

and therefore, the action of $f(A_G)$ coincides with the action of J on a basis; hence, $f(A_G) = J$. [A. J. Hoffman, *Amer. Math. Monthly* **70** (1963) 30–36.]

27. From the basic properties of projective planes, we get

$$A_L^2 = \begin{pmatrix} J + qI & 0 \\ 0 & J + qI \end{pmatrix},$$

where q is the order of the plane. Hence the eigenvalues of A_L^2 are: $(q+1)^2$ (twice) and $q(2(q^2 + q)$ times). Now the spectrum of L is symmetric with respect to the origin (see 11.19), hence L has eigenvalues

$q + 1$	with multiplicity	1,
$\sqrt{q}$	"	$q^2 + q,$
$-\sqrt{q}$	"	$q^2 + q,$
$-q - 1$	"	1 .

[A. J. Hoffman, *Proc. Amer. Math. Soc.* **16** (1965) 297–302.]

28. (a) (i) expresses the fact that each diagonal entry of A_G^2 is d; (ii) expresses the fact that the off-diagonal entries are b. Hence

$$A_G^2 = bJ + (d-b)I.$$

Hence the eigenvalues of A_G^2 are $nb+d-b,\ d-b,\dots,d-b$. Note that G is connected by (ii) and is not bipartite, since again by (ii), it contains triangles. Hence 11.19 implies that $-\sqrt{nb+d-b}$ is not an eigenvalue of G. Thus $\sqrt{nb+d-b},\ \pm\sqrt{d-b}$ are the different eigenvalues of G. Since by (i), d is the largest eigenvalue of G, we must have

(1) $$\sqrt{nb+d-b} = d,\ \ n = \frac{d^2-d}{b}+1.$$

Suppose that G has k eigenvalues equal to $\sqrt{d-b}$ and $n-1-k$ eigenvalues equal to $-\sqrt{d-b}$. Since the sum of eigenvalues of G is 0, we have

$$-(n-1-2k)\sqrt{d-b}+d = 0.$$

Hence $\sqrt{d-b}$ is rational, consequently integral; say $\sqrt{d-b}=r$. Then

(2) $$d = r^2 + b$$
and
$$r(n-1-2k) = r^2 + b.$$

Hence
(3) $$r|b.$$
Also, by (1),
$$b|d^2 - d = r^4 + 2r^2 b + b^2 - r^2 - b$$
thus
(4) $$b|r^2(r^2-1).$$

(b) Now if $b=1$ (3) implies that $r=1$ and hence, by (2), $d=2$. Thus from (1) we get $n=3$, i.e. $G=K_3$.

Note that (3), (2), and (1) imply that $n\le b^2(b+2)$ and hence there is a finite number of such graphs for fixed b; (4) shows that one need not even consider all divisors of b in (3) [S. S. Shrikhande, see Biggs].

(c) Let the points of G be the non-vertical lines of the affine plane over $GF(2^k)$, and let us connect two of them if they intersect and the intersection point is on one of 2^{k-1} prescribed vertical lines $L_1,\dots,L_{2^{k-1}}$. Set $L=L_1\cup\dots\cup L_{2^{k-1}}$. If two non-vertical lines E, F do not intersect in L, then there are $2^{k-1}(2^{k-1}-1)$ other non-vertical lines meeting both E, F in L. If E, F intersect in a point $p\in L$, then there are $2^k - 2$ other non-vertical lines through p and $(2^{k-1} - 1)(2^{k-1} - 2)$ non-vertical lines avoiding p, which meet both E, F in L. This gives again

$$2^k - 2 + (2^{k-1} - 1)(2^{k-1} - 2) = 2^{k-1}(2^{k-1} - 1)$$

common neighbors of E, F. Thus G has the desired property.

29. (a) Let $\mathbf{x}$ be an eigenvector of A belonging to the eigenvalue λ_2 with $\sum_i x_i^2 = 1$. Since $\mathbf{1}$ is an eigenvector belonging to $\lambda_1 \neq \lambda_2$, it is orthogonal to the eigenvector $\mathbf{1}$ belonging to λ_1, i.e., we have $\sum_i x_i = 0$. We may assume that x_1 is the smallest entry of $\mathbf{x}$, x_k is the largest entry of $\mathbf{x}$, and $x_2, \ldots x_{k-1}$ correspond to the vertices on a shortest path from 1 to k $(k \leq D + 1)$. From $\sum_i x_i = 0$ it follows that $x_1 < 0$ and $x_k > 0$. From $\sum_i x_i^2 = 1$ it follows that $\max\{|x_1|, |x_k|\} \geq 1/\sqrt{n}$. We may choose the sign of $\mathbf{x}$ so that $x_k \geq 1/\sqrt{n}$. Then (using the solution of 11.22),

$$d - \lambda_2 = \sum_{(i,j) \in E} (x_i - x_j)^2 \geq \sum_{i=1}^{k-1} (x_{i+1} - x_i)^2$$

$$\geq \frac{1}{k-1} \left(\sum_{i=1}^{k-1} (x_{i+1} - x_i) \right)^2 = \frac{1}{k-1} (x_k - x_0)^2 > \frac{1}{Dn}.$$

(b) The matrix $A_G^2 - dI$ may be viewed as the adjacency matrix of a regular graph H of degree $d^2 - d$. We show that H is connected and its diameter is at most D. Consider two vertices x and y. Clearly there exists a walk $W = (x_0 = x, x_1, \ldots, x_{2m} = y)$ of even length in G connecting them (since G is non-bipartite). Consider the shortest such walk. It suffices to show that $m \leq D$, since then $(x_0, x_2, \ldots x_{2m})$ is a walk in H of length at most D connecting x and y. Assume (by way of contradiction) that $m > D$. Let P be a shortest (x, x_m)-path and Q, a shortest (x_m, y)-path in G, and let p and q be the length of P and Q, respectively. If $p \equiv m \pmod q$ then replacing the first half of W by P we obtain a shorter walk of even length from x to y. So we must have $p \equiv m - 1 \pmod 2$. Similarly, $q \equiv m - 1 \pmod 2$. But then $P \cup Q$ is a walk from x to y of even length shorter than W, a contradiction.

The largest eigenvalue of H is $d^2 - d$, which must have multiplicity 1 as H is connected (this also follows by 11.19(a)). The other eigenvalues of H are $\lambda_i^2 - d$ $(i = 1, \ldots, n)$, and so the second largest is either $\lambda_2^2 - d$ or $\lambda_n^2 - d$. In either case (a) implies that

$$d^2 - \lambda_n^2 \geq \frac{1}{Dn},$$

which implies the assertion as $0 < d - \lambda_n \leq 2d$.

30. We may assume that $\bar{y} = 0$. Following the hint, we have

$$\sum_{(i,j)\in E} |y_i - y_j| = \sum_{i=1}^{n-1} \delta_G(S_i)(y_{i+1} - y_i),$$

where $S_i = \{1, \dots, i\}$. By the definition of Φ, this implies

$$\sum_{(i,j)\in E} |y_i - y_j| \geq \Phi \sum_{i=1}^{n-1} \min\{i, n-i\}(y_{i+1} - y_i) = \Phi \left(\sum_{i>n/2} y_i - \sum_{i\leq n/2} y_i \right)$$

$$= \Phi \sum_i |y_i|.$$

31. (a) Let $n = |V|$ and $S \subseteq V$ with $0 < p = |S| \leq n/2$ and such that $\Phi = \delta_G(S)/p$. Let **x** denote the vector which is $n-p$ on S and $-p$ on $V \setminus S$. Then

$$\sum_i x_i = 0, \qquad \sum_i x_i^2 = p(n-p)n,$$

and so by 11.22,

$$d - \lambda_2 \leq \sum_{(i,j)\in E} (x_i - x_j)^2 \bigg/ \sum_i x_i^2 = \delta_G(S) \frac{n}{p(n-p)} \leq \frac{2}{p}\delta_G(S) = 2\Phi.$$

(b) Let **x** be a unit length eigenvector belonging to λ_2, such that $x_1 \geq x_2 \geq \dots \geq x_n$. Let $\bar{x}$ be any median of x. Setting $z_i = (\max\{0, x_i - \bar{x}\})$ and choosing the sign of **x** appropriately, we may assume that

$$\sum_i z_i^2 \geq \frac{1}{2} \sum_i (x_i - \bar{x})^2 = \frac{1}{2} \sum_i x_i^2 - \bar{x}\sum_i x_i + \frac{n}{2}\bar{x}^2 = \frac{1}{2} + \frac{n}{2}\bar{x}^2 \geq \frac{1}{2}.$$

By 11.30,

$$\sum_{(i,j)\in E} |z_i^2 - z_j^2| \geq \Phi \sum_i z_i^2.$$

On the other hand, using the Cauchy–Schwartz inequality,

$$\sum_{(i,j)\in E} |z_i^2 - z_j^2| \leq \left(\sum_{(i,j)\in E} (z_i - z_j)^2 \right)^{1/2} \left(\sum_{(i,j)\in E} (z_i + z_j)^2 \right)^{1/2}.$$

Here the second factor can be estimated as follows:

$$\sum_{(i,j)\in E} (z_i + z_j)^2 \leq 2 \sum_{(i,j)\in E} (z_i^2 + z_j^2) = 2d \sum_i z_i^2.$$

Combining these inequalities, we obtain

$$\sum_{(i,j)\in E}(z_i - z_j)^2 \geq \left(\sum_{(i,j)\in E}|z_i^2 - z_j^2|\right)^2 \Big/ \sum_{(i,j)\in E}(z_i + z_j)^2$$

$$\geq \Phi^2 \left(\sum_i z_i^2\right)^2 \Big/ 2d\sum_i z_i^2 = \frac{\Phi^2}{2d}\sum_i z_i^2 \geq \frac{\Phi^2}{4d}.$$

Since

$$\sum_{(i,j)\in E}(x_i - x_j)^2 \geq \sum_{(i,j)\in E}(z_i - z_j)^2,$$

from here we can conclude by the equation $d - \lambda_2 = \sum_{(i,j)\in E}(x_i - x_j)^2$. [A. Sinclair—M. Jerrum, *Proc. 20th ACM STOC* (1988), 235–244; cf. also N. Alon, *Combinatorica* **6** (1986) 83–96.]

32. (a) By 11.8, the eigenvalues of G are

$$\lambda_k = \sum_{j=1}^{d} \varepsilon^{a_j k},$$

where $\varepsilon = \exp(2\pi i/n)$, $a_1, \ldots, a_d$ are the neighbors of vertex 0, and $0 \leq k \leq n-1$. The largest eigenvalue is $\lambda_0 = d$. We show that there is a different one close to this. By Dirichlet's Theorem on simultaneous diophantine approximation, there exists an integer $1 \leq q < n$, and integers $p_1, \ldots, p_d$ such that

$$\left|q\frac{a_i}{n} - p_i\right| \leq n^{-1/d} \qquad (i = 1, \ldots, d).$$

This implies that the argument of $\varepsilon^{a_i q}$ is between $-2\pi n^{1/d}$ and $2\pi n^{-1/d}$, and so its real part is at least $\cos(2\pi n^{-1/d})$. Thus

$$\lambda_q > d\cos(2\pi n^{-1/d}) > d(1 - 2\pi^2 n^{-2/d})$$

and so

$$d - \lambda_q \leq 2\pi^2 dn^{-2/d}.$$

By 11.31(b), this implies that $\Phi(G) \leq 2\sqrt{2}\pi dn^{-1/d} < 10dn^{-1/d}$.

(b) For each pair i, j of points, select a shortest path P_{ij} connecting them. Let $\mathcal{P}$ denote the family of these paths and all their images under automorphisms of G. The total number of paths in $\mathcal{P}$ is $\binom{n}{2}g$, where g is the number of automorphisms of G. Moreover, $\mathcal{P}$ contains exactly g paths connecting any given pair of points.

We claim that every edge occurs in at most $Dg(n-1)$ paths of $\mathcal{P}$. In fact, if an edge e occurs in p paths then so does every image of e under the automorphisms, and there are at least $n/2$ distinct images by vertex-transitivity. This gives $pn/2$ edges, but the total number of edges of paths in $\mathcal{P}$ is at most $Dg\binom{n}{2}$, which proves the claim.

Now let $S \subseteq V(G)$, $|S| = s \leq |V(G)|/2$. The number of paths in $\mathcal{P}$ connecting S to $V(G) \setminus S$ is exactly $gs(n-s)$. On the other hand, this number is at most $\delta_G(S) \cdot Dg(n-1)$, and hence

$$\delta_G(S) \geq \frac{gs(n-s)}{Dg(n-1)} = \frac{s}{D} \cdot \frac{n-s}{n-1} \geq \frac{s}{2D}.$$

[cf. L. Babai, in: *Proc. 23. Ann. ACM Symp. on Theory of Computing*, 164–174.]

33. (a) Let $E(G) = M_1 \cup M_2 \cup M_3$, where M_1, M_2 and M_3 are three random 1-factors on $\{1, \ldots, 2n\}$. Let $p(n)$ be the probability that G is a simple 3-regular graph, i.e., that these 1-factors are mutually disjoint. We express $p(n)$ by the sieve formula (2.2(a)). For each edge e, we consider three kinds of bad choices of M_1, M_2 and M_3: those where $e \in M_1 \cap M_2$, $e \in M_2 \cap M_3$, and $e \in M_1 \cap M_3$. For $X_1, X_2, X_3 \subseteq E(K_{2n})$, let $p(X_1, X_2, X_3)$ denote the probability that $X_1 \subseteq M_2 \cap M_3$, $X_2 \subseteq M_1 \cap M_3$, and $X_3 \subseteq M_1 \cap M_2$. Then by the sieve formula,

$$p(n) = \sum_{X_1, X_2, X_3} (-1)^{|X_1| + |X_2| + |X_3|} p(X_1, X_2, X_3).$$

Now $p(X_1, X_2, X_3) = 0$ unless $X_1 \cup X_2 \cup X_3$ is a matching; in that case

$$p(X_1, X_2, X_3) =$$
$$= \frac{(2n - 2|X_1 \cup X_2| - 1)!!(2n - 2|X_2 \cup X_3| - 1)!!(2n - 2|X_1 \cup X_3| - 1)!!}{(2n - 1)!!^3}.$$

(We have used the simple fact that the number of 1-factors on $2n$ vertices is $(2n-1)!! = (2n)!/2^n n!$.) If we fix the cardinalities $t_1 = |X_1 \setminus X_2 \setminus X_3|$, $t_{12} = |(X_1 \cap X_2) \setminus X_3|$, $t_{123} = |X_1 \cap X_2 \cap X_3|$, and analogously t_2, t_3, t_{13} and t_{23}, and set $t = t_1 + \cdots + t_{123} = |X_1 \cup X_2 \cup X_3|$, then the number of choices for X_1, X_2 and X_3 is

$$\frac{\binom{2n}{2}\binom{2n-2}{2} \cdots \binom{2n-2t+2}{2}}{t_1! t_2! t_3! t_{12}! t_{13}! t_{23}! t_{123}!} = \frac{(2n)(2n-1) \cdots (2n-2t+1)}{2^t t_1! t_2! t_3! t_{12}! t_{13}! t_{23}! t_{123}!},$$

and hence

$$p(n) = \sum_{t_1, \ldots, t_{123} \geq 0} (-1)^{t_1 + t_2 + t_3 + t_{123}} \frac{(2n)(2n-1) \cdots (2n-2t+1)}{2^t t_1! t_2! t_3! t_{12}! t_{13}! t_{23}! t_{123}!}$$
$$\times \frac{(2n - 2t + 2t_1 - 1)!!(2n - 2t + 2t_2 - 1)!!(2n - 2t + 2t_3 - 1)!!}{(2n-1)!!^3}.$$

This sum is majorized, uniformly in n, by the convergent series

$$\sum_{t_1,\dots,t_{123}\geq 0} \frac{1}{t_1!t_2!t_3!t_{12}!t_{13}!t_{23}!t_{123}!} = e^7,$$

and hence we can let $n\to\infty$ term-by-term. But

$$\lim_{n\to\infty} (2n)(2n-1)\cdots(2n-2t+1)\cdot$$
$$\cdot\frac{(2n-2t+2t_1-1)!!(2n-2t+2t_2-1)!!(2n-2t+2t_3-1)!!}{(2n-1)!!^3}$$
$$= \begin{cases} 1, & \text{if } t_{12}=t_{23}=t_{13}=t_{123}=0, \\ 0, & \text{otherwise.} \end{cases}$$

Thus

$$\lim_{n\to\infty} p(n) = \sum_{t_1,t_2,t_3\geq 0} (-1)^{t_1+t_2+t_3}\frac{1}{2^t t_1!t_2!t_3!} = e^{-3/2}.$$

This implies that $0.1 < p(n) < 0.9$ if n is large enough.

(b) Let $(S,\bar{S})$ be a partition of $V(G)$; set $k=|S|$ and assume that $k \leq n$. For convenience, assume that k is even. Let $p < k/1000$, $p \equiv k \pmod 2$. The probability that M_1 contains p edges connecting S to $\bar{S}$ is

$$\binom{k}{p}\binom{2n-k}{p}\frac{p!(k-p-1)!!(2n-k-p-1)!!}{(2n-1)!!} = 2^p\binom{n}{p}\binom{n-p}{(k-p)/2}\Big/\binom{2n}{k}.$$

Using that $\binom{n}{k/2} > \binom{n-p}{(k-p)/2}\binom{p}{p/2}$ (cf. 1.42(c)), this is bounded by

$$\frac{2^p\binom{n}{p}\binom{n}{k/2}}{\binom{2n}{k}\binom{p}{p/2}} < \frac{p\binom{n}{p}\binom{n}{k/2}}{\binom{2n}{k}}.$$

Since the right hand side is monotone increasing with p for $p < n/2$, the probability that M_1 contains at most $q = k/1000$ such edges is at most

$$q^2\binom{n}{q}\binom{n}{k/2}\Big/\binom{2n}{k}.$$

The probability that neither M_1 nor M_2 nor M_3 contains more than q such edges is the cube of this, and so the probability that there exists a partition $(S,\bar{S})$ with $|S|=k$ for which this holds is at most

$$\alpha = \binom{2n}{k}\left[q^2\binom{n}{q}\binom{n}{k/2}\Big/\binom{2n}{k}\right]^3.$$

Using that $\binom{2n}{k} \geq \binom{n}{k/2}^2$ (cf. again 1.42(c)), we get

$$\alpha \leq q^6 \binom{n}{q}^3 \Big/ \binom{n}{k/2} = q^2 \binom{n}{q}^3 \Big/ \binom{n}{500q} \; .$$

Thus the probability that the conductance is smaller than $1/1000$ is at most of order

$$\gamma(n) = \sum_{k \leq n} k^6 \binom{n}{\lfloor k/1000 \rfloor}^3 \Big/ \binom{n}{\lfloor k/2 \rfloor} \; .$$

This can be shown to tend to 0 as $n \to \infty$ using Stirling's formula; an argument with less computation is the following. Each term is monotone decreasing in n, and tends to 0 as $n \to \infty$. Hence the sum is majorized by

$$\sum_{k \leq n} k^2 \frac{\binom{k}{\lfloor k/1000 \rfloor}^3}{\binom{k}{\lfloor k/2 \rfloor}} < \sum_{k=1}^{\infty} k^2 \frac{\binom{k}{\lfloor k/1000 \rfloor}^3}{\binom{k}{\lfloor k/2 \rfloor}} \; ,$$

which is a convergent series. Therefore we can take the limit term-by-term and see that $\gamma(n) \to 0$ as $n \to \infty$.

Thus, if n is large enough, G is 3-regular and has conductance at least $1/1000$, with probability at least 0.1. [M. Pinsker; an explicit construction was given by G. A. Margulis; for a particularly powerful construction of expanders see A. Lubotzky–R. Phillips–P. Sarnak, *Combinatorica* **8** (1988), 261–277.]

34. The equation in the hint is trivial. We denote by $\mathbf{p}^{(k)}$ the vector $(p_i^{(k)})_{i=1}^n$ and by M, the matrix defined by

$$(M)_{ij} = \begin{cases} 1/d(i), & \text{if } (i,j) \in E(G), \\ 0, & \text{otherwise.} \end{cases}$$

Then the equation in the hint can be written as $\mathbf{p}^{(k+1)} = M^T \mathbf{p}^{(k)}$, and hence

$$\mathbf{p}^{(k)} = (M^T)^k \mathbf{p}^{(0)}.$$

Let D denote the diagonal matrix with $(D)_{ii} = 1/d(i)$, then $M = DA_G$. If G is d-regular, then $M = (1/d)A_G$.

35. (a) Since

$$\sum_{i \in \Gamma(j)} \frac{1}{d(i)} \cdot \frac{d(i)}{2m} = \frac{d(j)}{2m},$$

we see that the distribution $p_i^{(0)} = d(i)/(2m)$ is stationary.

(b) By 11.34, a distribution on $V(G)$ is stationary iff it is a left eigenvector of M with eigenvalue 1. Since $(d(i)/(2m))$ defines a positive eigenvector of M, it follows by the Frobenius–Perron theory that 1 is the largest eigenvector of M. Since M is an irreducible matrix, the Frobenius–Perron theory implies that the eigenvector belonging to the largest eigenvalue is unique up to a scalar; but since we are considering probability distributions, the scaling is also uniquely determined.

(c) By the Frobenius–Perron theory, every eigenvalue of M is in the interval $[-1, 1]$; if G is non-bipartite then it follows by an argument similar to the solution of 11.19 that -1 is not an eigenvalue. Hence it follows (considering e.g. the Jordan normal form of M) that $(M^T)^k \to \mathbf{p}\mathbf{q}^T$ $(k \to \infty)$, where $\mathbf{p}$ and $\mathbf{q}$ are the left and right eigenvectors of M belonging to the eigenvalue 1, normalized so that $\mathbf{p}^T\mathbf{q} = 1$. We may assume that $\mathbf{q} = \mathbf{1}$, then $\mathbf{p}$ is just the stationary distribution. Hence for any starting distribution $\mathbf{p}^{(0)}$, we have

$$(M^T)^k \mathbf{p}^{(0)} \to \mathbf{p}\mathbf{1}^T \mathbf{p}^{(0)} = \mathbf{p}.$$

If G is bipartite, and v_0 is chosen from one color class, then v_k is always in the same color class for k even and in the opposite, when k is odd. So the distribution of v_k does not tend to the stationary distribution.

36. Assume that $j > i$ and write

$$\mathsf{P}(v_i = x, \; v_j = y) = \mathsf{P}(v_i = x)\mathsf{P}(v_j = y | v_i = x).$$

The part of the random walk after v_i can be viewed as a random walk starting at v_i. Hence we have

$$\mathsf{P}(v_j = y | v_i = x) = \mathsf{P}(v_{j-i} = y | \; v_0 = x) = p_{xy}^{(j-i)},$$

where $p_{xy}^{(j-i)}$ is the probability that a random walk starting at x will be at y after $j - i$ steps. By 11.35(c), $p_{xy}^{(t)} \to p_y$ (as $t \to \infty$), where $\mathbf{p}$ is the stationary distribution. Thus we have a t_0 such that if $j - i > t_0$ then $|p_{xy}^{(j-i)} - p_y| < \varepsilon/2$ (we may choose t_0 valid for all x and y). Then we also have, for $j \geq t_0$,

$$|\mathsf{P}(v_j = y) - p_y| = \left| \sum_z [\mathsf{P}(v_j = y | v_0 = z) - p_y]\mathsf{P}(v_0 = z) \right|$$

$$\leq \sum_z |p_{zy}^j - p_y|\mathsf{P}(v_0 = z) \leq \frac{\varepsilon}{2} \sum_z \mathsf{P}(v_0 = z) = \frac{\varepsilon}{2}.$$

Thus,

$$|\mathsf{P}(v_i = x, \; v_j = y) - \mathsf{P}(v_i = x)\mathsf{P}(v_j = y)| =$$
$$\mathsf{P}(v_i = x)|\mathsf{P}(v_j = y | v_i = x) - \mathsf{P}(v_j = y)|$$
$$\leq \mathsf{P}(v_i = x)\Big(|\mathsf{P}(v_j = y | v_i = x) - p_y| + |p_y - \mathsf{P}(v_j = y)|\Big) < \mathsf{P}(v_i = x)\varepsilon \leq \varepsilon.$$

37. (a) To verify the equality in the hint, let

$$\xi_i = \begin{cases} 1, & \text{if } v_i = x, \\ 0, & \text{otherwise.} \end{cases}$$

Then $\nu_t(x) = \sum_{i=0}^{t-1} \xi_i$, and so

$$E(\nu_t(x)) = \sum_{i=0}^{t-1} E(\xi_i) = \sum_{i=0}^{t-1} P(v_i = x).$$

Thus, we have by 11.35(c),

$$E(\nu_t(x)/t) = \frac{1}{t}\sum_{i=0}^{t-1} P(v_i = x) = \frac{1}{t}\sum_{i=0}^{t-1} p_x^i \to \frac{d(x)}{2m} \qquad (t \to \infty).$$

(b) The equality in the hint is verified similarly as in (a). Using it, we have

$$D^2(\nu_t(x)) = \sum_{i=0}^{t-1}\sum_{j=0}^{t-1}[P(v_i = x, v_j = x) - P(v_i = x)P(v_j = x)]$$

$$\leq \sum_{i=0}^{t-1}\sum_{j=0}^{t-1}|P(v_i = x, v_j = x) - P(v_i = x)P(v_j = x)|$$

$$= \sum_{i,j: |i-j|<t_0} |P(v_i = x, v_j = x) - P(v_i = x)P(v_j = x)|$$

$$+ \sum_{i,j: |i-j|\geq t_0} |P(v_i = x, v_j = x) - P(v_i = x)P(v_j = x)|.$$

Consider any $\varepsilon > 0$ and let t_0 be chosen as in 11.36. Then it follows that

$$D^2(\nu_t(x)) < 2t_0 t + t^2\varepsilon,$$

and hence

$$D^2(\nu_t(x)/t) \leq \frac{2t_0}{t} + \varepsilon < 2\varepsilon$$

as soon as $t > 2t_0/\varepsilon$.

38. (a) Let f denote the expected number of steps before returning to u. Assume that the random walk returns to u the kth time after τ_k steps. Then $E(\tau_k - \tau_{k-1}) =$

f and hence $E(\tau_k) = kf$. Moreover, for any integer $t \geq 0$, $\tau_k \leq t$ if and only if $\nu_t = \nu_t(u) \geq k$, and so for any real number $x > 0$,

$$\frac{\tau_k}{k} \leq x \quad \text{if and only if} \quad \nu_{\lfloor kx \rfloor}/kx \geq \frac{1}{x}.$$

We know by 11.37(a) and (b) that the probability of this event tends to 0 if $x < 2m/d(u)$ and tends to 1 if $x > 2m/d(u)$ (if $k \to \infty$). Hence $E(\tau_k/k) \to 2m/d(u)$, and so $f = 2m/d(u)$.

(b) A random walk with N steps passes through v about $Nd(v)/(2m)$ times, and each time it has a chance of one in $d(v)$ to step to u, so it will pass through vu about $N/(2m)$ times. To make this argument precise, let ρ_t denote the number of times the random walk passes through the edge vu from v to u during the first t steps, and $\nu_t = \nu_t(v)$. Also set $d = d(v)$. Let

$$\xi_t = \begin{cases} 1, & \text{if } v_t = v \text{ and } v_{t+1} = u, \\ 0, & \text{otherwise,} \end{cases}$$

and

$$\eta_t = \begin{cases} 1, & \text{if } v_t = v, \\ 0, & \text{otherwise.} \end{cases}$$

Then trivially

$$E(\xi_t - \eta_t/d) = 0.$$

Moreover, for any walk $u_0 = u, u_1, \ldots, u_t$, we have

$$E\left(\xi_t - \eta_t/d \mid v_0 = u_0, \ldots, v_t = u_t\right) = 0$$

(in other words, the sequence $\xi_t - \eta_t/d$ is a martingale). Thus,

$$E(\rho_t) - E(\nu_t/d) = \sum_{i=0}^{t-1} E(\xi_i - \eta_i/d) = 0,$$

and so

$$E(\rho_t/t) = \frac{1}{d}E(\nu_t/t) \longrightarrow \frac{1}{d}\frac{d}{2m} = \frac{1}{2m}.$$

Moreover, we have

$$(\rho_t - E(\rho_t))^2 = ([\rho_t - \nu_t/d] + [\nu_t/d - E(\nu_t/d)])^2 \leq 2[\rho_t - \nu_t/d]^2 + 2[\nu_t/d - E(\nu_t/d)]^2,$$

and so

$$D^2(\rho_t) \leq 2E([\rho_t - \nu_t/d]^2) + 2E([\nu_t/d - E(\nu_t/d)]^2).$$

The second term is just $D^2(\nu_t/d)$, and by 11.37(b), it tends to 0 when divided by t. To estimate the first term, write

$$E([\rho_t - \nu_t/d]^2) = E\left(\left[\sum_{i=0}^{t-1}(\xi_i - \eta_i/d)\right]^2\right)$$

$$= \sum_{i=1}^{t-1} E((\xi_i - \eta_i/d)^2) + 2 \sum_{0 \le i < j \le t-1} E((\xi_i - \eta_i/d)(\xi_j - \eta_j/d)).$$

By the martingale property,

$$E((\xi_i - \eta_i/d)(\xi_j - \eta_j/d)) = \sum_s E((\xi_j - \eta_j/d) \mid \xi_i - \eta_i/d = s) P(\xi_i - \eta_i/d = s) = 0,$$

and so

$$E([\rho_t - \nu_t/d]^2) = \sum_{i=1}^{t-1} E((\xi_i - \eta_i/d)^2) = E\left(\sum_{i=1}^{t-1} \xi_i^2 - \frac{2}{d} \sum_{i=1}^{t-1} \xi_i \eta_i + \frac{1}{d^2} \sum_{i=1}^{t-1} \eta_i^2 \right).$$

Since $\xi_i^2 = \xi_i \eta_i = \xi_i$ and $\eta_i^2 = \eta_i$, this implies that

$$E([\rho_t - \nu_t/d]^2) = \frac{d-2}{d} E(\rho_t) - \frac{1}{d^2} E(\nu_t) = \frac{d-1}{d^2} E(\nu_t) \sim \frac{d-1}{d^2} \frac{dt}{2m} = \frac{(d-1)t}{2md}.$$

This implies that $D^2(\rho_t/t) \to 0$ as $t \to \infty$. From here the assertion follows as in part (a).

39. Let a denote the mean access time from vertex 1 to vertex 2 in K_n. Starting at 1, with probability $1/(n-1)$, we step to 2 right away; with probability $(n-2)/(n-1)$, we are back to an identical situation. Hence

$$a = \frac{1}{n-1} \cdot 1 + \frac{n-2}{n-1} \cdot (1+a),$$

whence

$$a = n - 1.$$

Concerning the path, observe that the mean access time from the $(n-1)^{\text{st}}$ vertex to the nth of a path with n vertices is one less than the expected mean return time of a random walk starting at the last vertex. By 11.37(a), this number is $2n - 3$.

Now let b_n denote the mean access time from the first point of a path with n vertices to the last. If we start at the first vertex, then in order to reach the nth, we have to reach the $(n-1)^{\text{st}}$; this takes, on the average, b_{n-1} steps. From here, we have to get to the last, which takes, on the average, $2n-3$ steps. This yields the recurrence

$$b_n = b_{n-1} + 2n - 3,$$

whence $b_n = (n-1)^2$.

40. (a) If u is an endpoint and v is the midpoint of a path of length 2, then $a(u,v) = 1$ while $a(v,u) > 1$, since with probability $1/2$, we need more than 1 step to get from v to u.

Consider a d-regular graph that has a cutpoint u; let $G = G_1 \cup G_2$ where $V(G_1) \cap V(G_2) = \{u\}$, $|V(G_i)| > 1$. Let v be a vertex of G_1 different from u. Then the mean access time from v to u is the same as the mean access time from v to u in G_1, which is independent of G_2. On the other hand, walking from u to v we may, with probability $d_{G_2}(u)/d$, step to a vertex of G_2, and then we have to walk until we return to u; the expected time before this happens is, by 11.34, at least $2|E(G_2)|/d_{G_2}(u)$. So $\alpha(u,v) > 2|E(G_2)|/d$, which can be chosen larger than $\alpha(v,u)$.

(b) If $W = (v_0 = u, v_1, \ldots, v_N = v)$ is a walk from u to v then $W' = (v_N, \ldots, v_0)$ is a walk from v to u. Moreover, the probability that a random walk starting at u traverses W is

$$\mathsf{P}(W) = \frac{1}{d(v_0)} \frac{1}{d(v_1)} \cdots \frac{1}{d(v_{N-1})},$$

which is the same as the probability that a random walk starting at v traverses W', since $d(v_0) = d(v_N)$.

Now the probability that a random walk starting at u reaches v without returning to u is the sum of $\mathsf{P}(W)$ over those walks W from u to v that do not pass through u or v. Since W has this property if and only if W' does, the assertion follows.

The same argument shows that in general, the ratio of the given probabilities is $d(v)/d(u)$.

41. Let τ be the first time when a random walk starting at u returns to u and σ the first time when it returns to u after visiting v. By 11.34, $\mathsf{E}(\tau) = 2m/d(u)$ and by definition, $\mathsf{E}(\sigma) = \kappa(u,v)$. Clearly $\tau \leq \sigma$ and the probability of $\tau = \sigma$ is exactly the probability q that the random walk reaches v before returning to u. Moreover, if $\tau < \sigma$ then after the first τ steps, we have to walk from u until we reach v and then return to u. Hence $\mathsf{E}(\sigma - \tau) = (1 - q)\kappa(u,v)$, and thus

$$\kappa(u, v) = \mathsf{E}(\sigma) = \mathsf{E}(\tau) + \mathsf{E}(\sigma - \tau) = \frac{2m}{d(u)} + (1 - q)\kappa(u, v),$$

which implies that

$$q = \frac{2m}{d(u)\kappa(u, v)}.$$

42. (a) Starting a random walk at u, walk until v is visited; then walk until w is visited; then walk until u is reached. Call this random sequence a $uvwu$-tour. The expected number of steps in a $uvwu$-tour is $a(u,v)+a(v,w)+a(w,u)$. On the other hand, we can express this number as follows. Let $W = (u_0, u_1, \ldots, u_N = u_0)$ be a closed walk. The probability that we have walked exactly this way is

$$\mathsf{P}(W) = \prod_{i=0}^{N-1} \frac{1}{d(u_i)},$$

which is independent of the starting point and remains the same if we reverse the order.

Let $a(W)$ denote the number of ways this closed walk arises as a $uvwu$-tour, i.e., the number of occurrences of u in W where we can start W to get a $uvwu$-tour (note that the same value would be obtained by considering v or w instead of u). We shall show that the number of ways the reverse closed walk $W' = (u_N, u_{N-1}, \ldots, u_0 = u_N)$ arises as a $uwvu$-tour is also $a(W)$. Since the expected length of a $uvwu$-tour is $\sum_W p(W) a(W) |W|$, this will prove the identity in the problem. (It will also follow that $a(W)$ is 1 or 2.)

Call an occurrence of u in the closed walk W "forward good" if starting from u and following the walk until v occurs, then following it until w occurs, then following it until u occurs, we traverse the whole walk exactly once. Call this occurrence "backward good" if this holds with the orientation of W as well as the role of v and w reversed. Clearly $a(W)$ is the number of "forward good" occurrences of u, so it suffices to verify that for every closed walk W, the number of "forward good" occurrences of u is the same as the number of "backward good" occurrences. (Note that a "forward good" occurrence need not be "backward good".)

Assume that W arises as a $uvwu$-tour at least once; say $u_0 = u$, $u_i = v$, and $u_j = w$ $(0 < i < j < N)$, where $W_1 = \{u_1, \ldots, u_{i-1}\}$ does not contain v, $W_2 = \{u_{i+1}, \ldots, u_{j-1}\}$ does not contain w and $W_3 = \{u_{j+1}, \ldots, u_{N-1}\}$ does not contain u. Assume first that W_2 does not contain u either. Then u_0 is the only "forward good" occurrence of u, and the last occurrence of u in W_1 is the only "backward good" occurrence.

Second, assume that W_2 contains u. Similarly, we may assume that W_3 contains v and W_1 contains w. Let u_t be the last occurrence of u in W_2. It is easy to check that u_t is "backward good". So we see that if W arises as a $uvwu$-tour then it also arises as a $uwvu$-tour.

Assume now that $a(W) > 1$. Then there must be a second "forward good" element, and it is easy to check that this can only be the first occurrence u_s of u on W_2; it also follows that all occurrences of v on W_2 must come before u_s, and similarly, all occurrences of w on W_3 must come before the first occurrence of v on W_3, and all occurrences of u on W_1 must come before the first occurrence of w on W_1. But in this case there are exactly two "forward good" and exactly two "backward good" occurrences of u. So $a(W) = a(W') = 2$.

(b) Assume that u precedes v in the ordering suggested in the hint. Then $a(u, t) - a(t, u) \le a(v, t) - a(t, v)$ and hence $a(u, t) + a(t, v) \le a(v, t) + a(t, u)$. By (a), this is equivalent to saying that $a(u, v) \le a(v, u)$. [Tetali and Winkler]

43. First we show that if $(u, v) \in E(G)$, then the mean commute time between u and v, i.e., the expected number of steps before the random walk starting at u visits v and returns to u, is at most $2m$. This follows if we show that the expected number of steps before the edge (u, v) is passed from v to u is $2m$. This, in turn, follows by 11.38(b).

Now assume that $r > 1$. Let $(u_0 = u, u_1, \ldots, u_r = v)$ be a shortest path between u and v. The mean commute time between u and v is clearly at most

$$a(u_0, u_1) + a(u_1, u_2) + \cdots + a(u_{r-1}, u_r) + a(u_r, u_{r-1}) + \cdots + a(u_1, u_0)$$
$$= [a(u_0, u_1) + a(u_1, u_0)] + [a(u_1, u_2) + a(u_2, u_1)] + \cdots + [a(u_{r-1}, u_r) + a(u_r, u_{r-1})].$$

Here $a(u_i, u_{i+1}) + a(u_{i+1}, u_i)$ is just the mean commute time between u_i and u_{i+1}, and so it is at most $2m$. Hence the assertion follows.

44. (a) The identity in the hint is obvious. Expressing it in matrix notation, we get that $F = J + MB - B$ is a diagonal matrix. Let $\mathbf{p}$ denote the stationary distribution, then

$$F^T \mathbf{p} = J\mathbf{p} + B^T (M - I)^T \mathbf{p} = J\mathbf{p} = \mathbf{1},$$

whence by 11.35(a),

$$(F)_{ii} = \frac{1}{p_i} = \frac{2m}{d(i)}.$$

Thus $F = 2mD$ as claimed.

(b) Obviously, $F_{ii} = 1 + \frac{1}{d(i)} \sum_{k \in \Gamma(i)} a(k, j)$ is just the expected time needed to return to node i.

(c) *First solution.* The identity in the hint is easily verified by considering the first step when the walk visits j. Hence we get for the generating functions $f_{ij}(x) = \sum_{t=0}^{\infty} p_{ij}^{(t)} x^t$ and $g_{ij}(x) = \sum_{t=0}^{\infty} q_{ij}^{(t)} x^t$ that

$$f_{ij}(x) = g_{ij}(x) f_{jj}(x).$$

Now here

$$f_{ij}(x) = \sum_{t=0}^{\infty} \mathbf{e}_i^T \left(\frac{1}{d} A\right)^t \mathbf{e}_j x^t = \sum_{t=0}^{\infty} \sum_{k=1}^{n} w_{ki} w_{kj} \left(\frac{\lambda_k x}{d}\right)^t = \sum_{k=1}^{n} \frac{w_{ki} w_{kj}}{1 - \lambda_k x/d},$$

and so

$$g_{ij}(x) = \sum_{k=1}^{n} \frac{w_{ki} w_{kj}}{d - \lambda_k x} \bigg/ \sum_{k=1}^{n} \frac{w_{kj}^2}{d - \lambda_k x} = \frac{1}{nd} + \sum_{k=2}^{n} \frac{w_{ki} w_{kj}}{d - \lambda_k x} \bigg/ \sum_{k=1}^{n} \frac{w_{kj}^2}{d - \lambda_k x}.$$

Now $a(i, j) = g_{ij}'(1)$; from this the formula in the problem follows by direct computation.

Second solution. Let $B' = -nd \sum_{k=2}^n \frac{1}{d-\lambda_k} \mathbf{w}_k \mathbf{w}_k^T$. Then

$$(I - M)B' = -nd \sum_{k=2}^n \frac{1}{d - \lambda_k}(I - M)\mathbf{w}_k \mathbf{w}_k^T = -n \sum_{k=2}^n \mathbf{w}_k \mathbf{w}_k^T.$$

Since $\mathbf{w}_1, \ldots, \mathbf{w}_n$ form an orthonormal basis, we have $\sum_{k=1}^n \mathbf{w}_k \mathbf{w}_k^T = I$ and so

$$(I - M)B' = -n(I - \mathbf{w}_1 \mathbf{w}_1^T) = J - nI = (I - M)B.$$

Thus $(I - M)(B - B') = 0$, and so every column of $B - B'$ is in the null space of $M - I$. But this null space is spanned by the vector $\mathbf{1}$, and so $B - B'$ has the same entry through each column. Since the diagonal entries of B are 0, we know that B arises from B' by subtracting each diagonal entry from all other entries in its column.

45. (a) We have

$$a(i,j) + a(j,l) + a(l,i) = nd \sum_{k=2}^n \frac{w_{kj}^2 - w_{ki}w_{kj}}{d - \lambda_k} + \frac{w_{kl}^2 - w_{kj}w_{kl}}{d - \lambda_k} + \frac{w_{ki}^2 - w_{kl}w_{ki}}{d - \lambda_k}$$

$$= nd \sum_{k=2}^n \frac{w_{kj}^2 + w_{kl}^2 + w_{ki}^2 - w_{ki}w_{kj} - w_{kj}w_{kl} - w_{kl}w_{ki}}{d - \lambda_k}.$$

This value is clearly independent of the order of i, j and l.

(b) Let $\mathbf{0} = (0, \ldots, 0)$ and $\mathbf{1} = (1, \ldots, 1)$ represent two antipodal vertices of the k-cube. By 11.9, we get an eigenvector $\mathbf{w}_b$ for every 0–1 vector $\mathbf{b} \in \{0,1\}^k$, defined by $\mathbf{w}_b(\mathbf{x}) = (-1)^{\mathbf{b} \cdot \mathbf{x}}$. The correspoding eigenvalue is $k - 2\mathbf{b} \cdot \mathbf{1}$. Normalizing $\mathbf{w}_b$ and substituting in 11.44(c), we get that

$$a(\mathbf{0}, \mathbf{1}) = k \sum_{j=1}^k \binom{k}{j} \frac{1}{2j}(1 - (-1)^j).$$

To find the asymptotic value of this expression, we substitute $\binom{k}{j} = \sum_{p=0}^{k-1} \binom{p}{j-1}$, and get

$$a(\mathbf{0}, \mathbf{1}) = k \sum_{j=1}^k \sum_{p=0}^{k-1} \frac{1}{2j} \binom{p}{j-1}(1 - (-1)^j)$$

$$= k \sum_{p=0}^{k-1} \frac{1}{2(p+1)} \sum_{j=1}^n \binom{p+1}{j}(1 - (-1)^j)$$

$$= k \sum_{p=0}^{k-1} \frac{2^p}{p+1} = 2^{k-1} \sum_{j=0}^{k-1} \frac{1}{2^j} \frac{k}{k-j} \sim 2^k.$$

(It is easy to see that the exact value is always between 2^k and 2^{k+1}.)

46. (a) By definition and by 11.44(c), the mean commute time is

$$a(i,j) + a(j,i) = nd \sum_{k=2}^{n} \frac{(w_{ki} - w_{kj})^2}{d - \lambda_k}.$$

(b) Using the result of (a), we have

$$a(i,j) + a(j,i) \leq \frac{nd}{d - \lambda_2} \sum_{k=2}^{n} (w_{ki} - w_{kj})^2 = \frac{nd}{d - \lambda_2}(\mathbf{w}_i - \mathbf{w}_j)^2 = \frac{2nd}{d - \lambda_2},$$

and similarly,

$$a(i,j) + a(j,i) \geq \frac{2nd}{d - \lambda_n} \geq n,$$

since $\lambda_n \geq -d$.

The example in 11.40 shows that the mean access time between two vertices of a d-regular graph can remain bounded for arbitrary large n.

47. (a) Let τ_i denote the first time when i vertices have been visited. So $\tau_1 = 0 < \tau_2 = 1 < \tau_3 < \ldots < \tau_n$. Now $\tau_{i+1} - \tau_i$ is the number of steps while we wait for a new vertex to occur — an event with probability $(n-i)/(n-1)$, independently of the previous steps. Hence

$$\mathsf{E}(\tau_{i-1} - \tau_i) = \frac{n-1}{n-i},$$

and so the mean cover time is

$$\mathsf{E}(\tau_n) = \sum_{i=1}^{n-1} \mathsf{E}(\tau_{i+1} - \tau_i) = \sum_{i=1}^{n-1} \frac{n-1}{n-i} \approx n \log n.$$

(b) Let $(u_0 = u, u_1, \ldots, u_p)$ be a walk visiting all vertices of G; clearly we have such a walk with $p \leq 2n - 3$. The mean cover time is then at most

$$a(u_0, u_1) + a(u_1, u_2) + \cdots + a(u_{p-1}, u_p) \leq 2pm < 4nm.$$

48. (a) For two vertices $x \neq v$, let $A(x,v)$ denote the event that our random walk hits x before v (the starting vertex u is fixed). Clearly,

$$\mathsf{P}(A(x,v)) + \mathsf{P}(A(v,x)) = 1.$$

We have for $v \in V(G) \setminus \{u\}$,

$$\mu(u,v) = 1 + \sum_{x \in V \setminus \{u,v\}} \mathsf{P}(A(x,v)),$$

and thus

$$\sum_{v \in V \setminus \{u\}} \mu(u,v) = (n-1) + \sum_{x,v \in V \setminus \{u\}} \mathsf{P}(A(x,v)) = (n-1) + \binom{n-1}{2} = \binom{n}{2}.$$

This proves the assertion of the hint, and thereby the assertion of the problem.

(b) Assume, for simplicity, that $n = 2k+1$ is odd. Then, following the hint, the time β when we reach more than half of the vertices is the $(k+1)^{\text{st}}$ largest of the α_v. Hence

$$\sum_v \alpha_v \geq (k+1)\beta,$$

and so

$$b = \mathsf{E}(\beta) \leq \frac{1}{k+1} \sum_v \mathsf{E}(\alpha_v) \leq \frac{n}{k+1} a < 2a.$$

49. Let β_1 be the number of steps before we see more than half of the vertices; let β_2 be the number of steps after that before we see more than half of the rest, etc. By 11.48(b), $\mathsf{E}(\beta_1) \leq 2a$; the same argument can be applied to show that $\mathsf{E}(\beta_i) \leq 2a$. Let $p = \lfloor \log_2 n \rfloor$. After $\beta_1 + \cdots + \beta_p$ steps no vertex remains unvisited, so the mean cover time is at most

$$\mathsf{E}(\beta_1 + \cdots + \beta_p) = \mathsf{E}(\beta_1) + \cdots + \mathsf{E}(\beta_p) \leq 2ap.$$

[P.Mathews, tech. report TR 234, Dept. of Stat., Stanford, 1985]

50. Just as in the solution of 11.44(c), we have

$$p_{ij}^t = \sum_{k=1}^n w_{ki} w_{kj} \left(\frac{\lambda_k}{d}\right)^t.$$

Here the term corresponding to $k=1$ is just $1/n$, so we have

$$\left| p_{ij}^t - \frac{1}{n} \right| = \left| \sum_{k=2}^n w_{ki} w_{kj} \left(\frac{\lambda_k}{d}\right)^t \right| \leq \left(\frac{\lambda}{d}\right)^t \sum_{k=2}^n |w_{ki} w_{kj}|.$$

Here the last sum is less than 1, since the vectors $(w_{ki})_{k=1}^n$ and $(w_{kj})_{k=1}^n$ are unit vectors.

51. *First solution.* Since Q_k is bipartite, we consider its "square", i.e., the graph H with adjacency matrix $A_{Q_k}^2$. H has two isomorphic connected components, and we consider the component H' containing v_0. The sequence $v_0, v_2, v_4, \ldots$ is a random walk on H'. The eigenvalues of H' are $k^2, (k-2)^2, \ldots$ (all non-negative), and so by 11.44A, we have

$$|\mathsf{P}(v_t = u) - 2^{1-k}| < \left(\frac{(k-2)^2}{k^2}\right)^{t/2} < e^{-2t/k} < \frac{1}{100} 2^{1-k}.$$

(More careful computation shows that $t = o(k \log k)$ steps would suffice; see P. Diaconis and D. Strook, The Annals of Appl. Prob. 1 (1991), 36–61.)

Second solution. Consider another random walk $(w_0, w_1, \ldots, w_t)$ on Q_k, where w_0 is uniformly distributed over the color class of v_0. We can "couple" these random walks as follows. As before, we consider the vertices of Q_k as 0-1 vectors of length k; then a step in the random walk consists of flipping a randomly chosen coordinate. At a given step i, let $1 \leq j_1 < \ldots < j_s \leq k$ be those coordinates where v_i and w_i differ. Since v_i and w_i are in the same color class, s is even. If v_{i+1} arises by flipping a coordinate in which v_i and w_i agree, then we flip this same coordinate also in w_i. If v_{i+1} arises from v_i by flipping coordinate j_r, then we flip in w_i coordinate j_{r+1} (subscript modulo s).

The important fact is that viewing w_i in itself, it is a legitimate random walk. Since it starts from the uniform distribution on a regular graph, each w_i, and in particular w_t, is uniformly distributed.

On the other hand, the "coupling" rule above implies that once a coordinate is flipped in $(v_0, v_1, \ldots)$ it becomes equal to the corresponding coordinate of w_i, and this remains so forever. In particular, if all coordinates are flipped at least once in $(v_0, \ldots, v_t)$, then $v_t = w_t$. The probability of the event A that not all coordinates are flipped is at most $(k-1)^t/k^{t-1} < ke^{-t/k}$. Hence for any vertex u in the appropriate color class,

$$|\mathsf{P}(v_t = u) - 2^{1-k}| = |\mathsf{P}(v_t = u) - \mathsf{P}(w_t = u)|$$
$$\leq \mathsf{P}(w_t \neq v_t) \leq \mathsf{P}(A) \leq ke^{-t/k} < \frac{1}{100} 2^{1-k}.$$

52. (a) Let $v \neq s, t$, and consider a random walk starting at v. Let B denote the event that we hit s before t. For every $u \in \Gamma(v)$, let A_u denote the event that the first step of the random walk is to u. Then

$$\phi(v) = \mathsf{P}(B) = \sum_{u \in \Gamma(v)} \mathsf{P}(B \mid A_u)\mathsf{P}(A_u) = \sum_{u \in \Gamma(v)} \phi(u)\frac{1}{d_G(v)}.$$

(b) For every edge $(u, v) \in E(G)$, the current from u to v is $\phi(v) - \phi(u)$ by Ohm's Law. By Kirchhoff's Law the total current entering $v \neq s, t$ is 0, *i.e.*,

$$\sum_{u \in \Gamma(v)} (\phi(v) - \phi(u)) = 0.$$

Rearranging and dividing by $d_G(v)$, we get the equation in the definition of harmonic functions.

(c) Let $\phi(u)$ denote the position of vertex u in equilibrium. Then any edge (u,v) pulls its endpoint u with force $|\phi(v)-\phi(u)|$. This force points in the positive direction if and only if $\phi(v) > \phi(u)$, so we may write it as the signed number $\phi(v)-\phi(u)$. Thus the fact that u is in equilibrium corresponds to the equation

$$\sum_{u\in\Gamma(v)} (\phi(v) - \phi(u)) = 0,$$

which proves the assertion.

(d) Assume that ϕ and ψ are harmonic functions with poles s and t, such that $\phi(s)=\psi(s)=1$ and $\phi(t)=\psi(t)=0$. Then $\eta=\phi-\psi$ is a harmonic function with poles s and t such that $\eta(s)=\eta(t)=0$. Assume that $\eta\neq 0$, and let v be a vertex with $a=|\eta(v)|$ maximum. We may assume that $\eta(v)>0$. Since $\eta(v)$ is the average of the numbers $\eta(u)$, $u\in\Gamma(v)$, it follows that $\eta(u)=a$ for all $u\in\Gamma(v)$. Since G is connected, repetition of this argument yields that $\eta(s)=a$, a contradiction.

53. (a) By 11.52(b), $\phi_{st}(v)$ is the voltage of v if we put a current through G from s to t, where the voltage of s is 0 and the voltage of t is 1. The total current through the network is $\sum_{u\in\Gamma(s)} \phi_{st}(u)$, and so the resistance

$$R_{st} = \left(\sum_{u\in\Gamma(t)} \phi_{st}(u)\right)^{-1}.$$

On the other hand, 11.52(a) says that $\phi_{st}(u)$ is the probability that a random walk starting at u visits s before t, and hence $\frac{1}{d(t)}\sum_{u\in\Gamma(t)} \phi_{st}(u)$ is the probability that a random walk starting at t hits s before returning to t. By 11.40, this probability is $2m/d(t)\kappa(s,t)$. Hence

$$\frac{1}{R_{st}} = \sum_{u\in\Gamma(t)} \phi_{st}(u) = \frac{2m}{\kappa(s,t)},$$

and so $R_{st} = \kappa(s,t)/(2m)$. [A. K. Chandra–P. Raghavan–W. L. Ruzzo–R. Smolensky–P. Tiwari, Proc. 21st ACM STOC (1989), 574–586. See also C.St.J.A.Nash-Williams, Proc. Cambridge Phil. Soc. **55** (1959) 181–194.]

(b) Clearly, the force with which the graph pulls the nails at 0 and 1 is

$$\sum_{u\in\Gamma(t)} \phi_{st}(u) = \sum_{v\in\Gamma(s)} (1 - \phi(v)) = \frac{1}{R_{st}}.$$

The energy of a spring stretched to length h is $h^2/2$, hence the energy in the equilibrium position is

$$\sum_{(u,v)\in E(G)} \frac{1}{2}(\phi(u) - \phi(v))^2 = \sum_u \frac{d(u)}{2}\phi(u)^2 - \sum_{(u,v)\in E(G)} \phi(u)\phi(v)$$

$$= \frac{1}{2} \sum_u \phi(u) \sum_{v \in \Gamma(u)} (\phi(u) - \phi(v)) = \frac{1}{2} \sum_{v \in \Gamma(s)} (1 - \phi(v)) = \frac{1}{2R_{st}}.$$

One may note that the energy is a positive definite quadratic form of the positions, and so there is a unique minimizing position, which is the equilibrium.

(c) We may assume that $V(G) = \{1, 2, \ldots, n\}$, where $s = 1$ and $t = n$. Let $L = (\ell_{ij})$ denote the matrix defined by

$$\ell_{ij} = \begin{cases} d(i), & \text{if } i = j, \\ -1, & \text{if } (i, j) \in E(G), \\ 0, & \text{otherwise.} \end{cases}$$

By 11.52(a) and by Ohm's Law, the current through edge (i, j) is $\phi_{1n}(j) - \phi_{1n}(i)$, and hence Kirchhoff's Law can be expressed as

$$\sum_{i \in \Gamma(j)} (\phi_{1n}(j) - \phi_{1n}(i)) = (L\phi_{1n})_j = 0 \qquad (j \neq 0, n).$$

Moreover, the total current through the network is

$$\sum_{i \in \Gamma(1)} (\phi_{1n}(1) - \phi_{1n}(i)) = (L\phi_{1n})_0,$$

and so $R_{1n} = 1/(L\phi_{1n})_n$. Let us drop the last entry from ϕ_{1n} (which is 0) and also from each row and column of L, to get a vector $\hat{\phi}$ and a matrix $\hat{L}$. Then we have

$$\hat{L}\hat{\phi} = \frac{1}{R_{1n}} e_1,$$

where $e_1 = (1, 0, \ldots, 0)^T$. Solving this system of equations by Cramer's rule, we get that

$$\hat{\phi}(1) = \frac{1}{R_{1n}} \frac{\det(\tilde{L})}{\det(\hat{L})},$$

where $\tilde{L}$ is obtained from $\hat{L}$ by deleting the first row and column. Since we know that $\hat{\phi}(1) = \phi_{1n}(1) = 1$, we get that

$$R_{1n} = \frac{\det(\tilde{L})}{\det(\hat{L})}.$$

But $\det(\hat{L}) = T(G)$ and $\det(\tilde{L}) = T(G')$ by 4.9(a) and (b). (It is easy to extend the formulas in this problem to the case when the resistances of the edges or, equivalently, the constants in Hooke's Law are not all 1.)

54. It suffices to prove that deleting an edge from a graph G cannot increase the energy of the equilibrium configuration in 11.53(b). Clearly, deleting an edge while keeping the positions of the vertices fixed cannot increase the energy. If we let the new graph find its equilibrium then the energy can only further decrease.

55. (a) is just a restatement of 11.54.

(b) Let G be a path on n vertices, with endpoints a and b. Let $s = a$ and let t be the unique neighbor of s. Then the access time from s to t is 1. On the other hand, if we add the edge (a, b) then with probability $1/2$, we have to make more than one step, so the access time from s to t will be larger than one (in fact, it is $n-1$; cf. 11.56 (a)).

(c) Consider a random walk W' on G' starting at s, and follow it until t is reached. Associate with this a random walk W on G as follows: as long as the edge (a, b) is not used by W', we follow W'. The first time W' uses (a, b), we "uncouple" the random walks and let W follow its random choices independently from W'. We also follow W until it reaches t.

Then at any case W is at least as long as W'. This is clear if t is reached before they are "uncoupled". If this is not the case, then W' reaches t in exactly one step after they are "uncoupled" (since $a = t$), while W has to walk at least one step.

Thus the expected length of W' is not larger than the expected length of W.

56. (a) is just a restatement of 11.38(a).

(b) By 11.44(c), we have

$$\sum_s a(t, s) = nd \sum_{k=2}^{n} \sum_s w_{ks}^2 \frac{1}{d - \lambda_k} - nd \sum_{k=2}^{n} \sum_s w_{ks} \frac{w_{kt}}{d - \lambda_k}.$$

Here the second term is 0, since the eigenvector $\mathbf{w}_k$ is orthogonal to the eigenvector **1**. Also, we have $\sum_s w_{ks}^2 = 1$, and so

$$\sum_s a(t, s) = nd \sum_{k=2}^{n} \frac{1}{d - \lambda_k}.$$

We have $d + \sum_{k=2}^{n} \lambda_k = \operatorname{Tr} A_G = 0$, and hence

$$\sum_{k=2}^{n} (d - \lambda_k) = nd.$$

By the inequality between the arithmetic and harmonic means,

$$\frac{1}{n-1} \sum_{k=2}^{n} \frac{1}{d - \lambda_k} \geq \frac{n-1}{\sum_{k=2}^{n}(d - \lambda_k)} = \frac{n-1}{nd},$$

and hence

$$\sum_s a(t, s) \geq (n-1)^2.$$

Since $a(t, t) = 0$, this proves the assertion.

(c) Again by 11.44(c), we have

$$\sum_s a(s,t) = n^2 d \sum_{k=2}^n \frac{1}{d-\lambda_k} w_{kt}^2.$$

By the inequality between arithmetic and harmonic means (considering the w_{kt}^2 as weights), we have

$$\frac{\sum_{k=2}^n \frac{1}{d-\lambda_k} w_{kt}^2}{\sum_{k=2}^n w_{kt}^2} \geq \frac{\sum_{k=2}^n w_{kt}^2}{\sum_{k=2}^n (d-\lambda_k) w_{kt}^2}.$$

Now here

$$\sum_{k=2}^n w_{kt}^2 = \sum_{k=1}^n w_{kt}^2 - \frac{1}{n} = 1 - \frac{1}{n},$$

and

$$\sum_{k=2}^n (d-\lambda_k) w_{kt}^2 = \sum_{k=1}^n (d-\lambda_k) w_{kt}^2 = d - \sum_{k=1}^n \lambda_k w_{kt}^2 = d - (A_G)_{t,t} = d.$$

Thus

$$\sum_s a(s,t) = n^2 d \sum_{k=2}^n \frac{1}{d-\lambda_k} w_{kt}^2 \geq n^2 d \left(\frac{n-1}{n}\right)^2 \frac{1}{d} = (n-1)^2,$$

which proves the assertion.

57. (a) Let $V(G) = \{1,\ldots,n\}$ in this cyclic order. Let p denote the probability that starting at 1, we hit 3 before hitting 2 (note that this is equivalent to saying that we start at 1 and see every vertex before seeing 2). Assume now that we start at 1, and want to find the probability that we see every vertex before i. Follow the walk until a neighbour of i is encountered; say, $i-1$. Then i is the last vertex seen if and only if we see $i+1$ before i. By symmetry, the probability of this is p.

So every vertex different from 1 has the same chance p of being the last hit. Thus $(n-1)p = 1$, which proves the assertion.

It is trivial that every complete graph also has this property.

(b) For each neighbor x of u, let L_x be the event that in a random walk starting at x, the last vertex visited is v; further, let L_x' be the event that the next-to-last vertex visited is v and the last is u. If we begin our random walk at

u and the first step is to x, then the last vertex visited is v if and only if either L_x or L'_x occurs. Since these events are disjoint, this gives

$$\pi(u,v) = \frac{1}{d(u)} \sum_{x \in \Gamma(u)} (\mathsf{P}(L_x) + \mathsf{P}(L'_x)) = \frac{1}{d(u)} \sum_{x \in \Gamma(u)} \pi(x,v) + \frac{1}{d(u)} \sum_{x \in \Gamma(u)} \mathsf{P}(L'_x).$$

By the assumption that $\{u,v\}$ is not a cutset, we see that L'_x has positive probability: it is possible to start at x, visit all vertices of $G \setminus \{u,v\}$, then walk to v in $G \setminus u$, and then walk to u. Hence $\mathsf{P}(L'_x) > 0$ and so

$$\pi(u,v) > \frac{1}{d(u)} \sum_{x \in \Gamma(u)} \pi(x,v).$$

Thus $\pi(u,v) > \pi(w,v)$ for at least one $w \in \Gamma(v)$.

(c) Assume that G is a connected graph, different from the circuits and complete graphs. By 6.6(c), there exists a pair $\{u,v\}$ of non-adjacent vertices such that $G \setminus \{u,v\}$ is connected. Thus (b) implies that $\pi(x,y)$ cannot be the same for all x and y. [L.Lovász and P. Winkler]

58. To prove the identity in the hint, let $T = T(w_0, w_1, \dots)$ denote the tree generated by the random walk $(w_0 = u, w_1, \dots)$, and $T_1 = T(w_1, w_2, \dots)$. Fix any spanning tree S. Let $u_1 \dots, u_r$ be the neighbors of u in S, and let B_i denote the connected component of $S - u$ containing u_i.

We claim that $T = S$ if and only if $w_1 = u_i$ for some i and $T_1 = S - (u, u_i) \cup (u,v)$ for some v with $v^S = u_i$.

Assume first that $S = T$. Then clearly $w_1 = u_i$ for some i. Moreover, T_1 has at least $(n-2)$ edges in common with S: the only edge that may not be present is (u, u_i), and this is replaced by the first edge (v,u) through which u is entered. We must have $v^S = u_i$; in fact, consider the first edge (x,y) of the random walk leaving B_i ($x \in B_i$, $y \notin B_i$). We cannot have $y \in B_j$ ($j \neq i$), since this would imply that $(x,y) \in T$ by the definition of T. So $y = u$, which implies that $x = v$ and $u_i = v^T$. This proves the "only if" part of our claim. The converse is verified along the same lines.

Thus we have

$$P(S,u) = \mathsf{P}(T = S) = \sum_{i=1}^{r} \mathsf{P}(w_1 = u_i) \sum_{v: v^T = u_i} \mathsf{P}(T_1 = S - (u,u_i) \cup (u,v)),$$

which gives the identity in the hint.

Now choose S and u so that $p = P(S,u)$ is maximum. Then it follows from the identity that $P(S - (u, v^S) \cup (u,v), v^S) = p$ is also maximum for every $v \in \Gamma(u)$. It suffices to show that this operation can transform (S,u) into any other pair (S', u'), where S' is a spanning tree and u' is a vertex.

Note that if $(u,v) \in E(S)$ then (S,u) is transformed to (S,v). So the question whether or not (S',u') can be reached by such transformations is independent of u'.

Let $\mathcal{H}$ denote the set of all spanning trees S' into which S can be transformed. Let S_0 be any spanning tree, let $S' \in \mathcal{H}$, let F be a common subtree of S_0 and S', and assume that S' and F are chosen so that $|V(F)|$ is maximum. If $F = S_0$ then we are done, so suppose that $F \neq S_0$. Then there is an edge $(x,y) \in E(S_0)$ such that $x \in V(F)$ and $y \notin V(F)$. Applying the transformation with "root" y, we see that $S'' = S' - (x',y) \cup (x,y) \in \mathcal{H}$, where x' is the first point after x on the (x,y)-path in S'. But S'' have S_0 have the subtree $F \cup (x,y)$ in common, contradicting the maximality of F. (See D. J. Aldous, *SIAM J. Disc. Math.* 3 (1990), 450–465.)

59. Set $t = 8d^2 n^3 \log n$, and let $H = (h_1, \ldots, h_t)$ be randomly chosen from $\{1, \ldots, d\}^t$. For a fixed G, starting point, and labelling, the walk defined by H is just a random walk; so the probability p that H is not a traverse sequence is the same as the probability that a random walk of length t does not visit all vertices.

By 11.47(b), the expected time needed to visit all vertices is at most $2dn^2$. Hence (by Markov's Inequality) the probability that after $4dn^2$ steps we have not seen all vertices is less than $1/2$. Since we may consider the next $4dn^2$ steps as another random walk, etc., the probability that we have not seen all vertices after t steps is less than $2^{-t/(4dn^2)} = n^{-2nd}$.

Now the total number of d-regular graphs G on n vertices, with the ends of the edges labelled, is less than n^{dn} (less than n^d choices at each vertex), and so the probability that H is not a traverse sequence for one of these graphs, with some starting point, is less than

$$n n^{nd} n^{-2nd} < 1.$$

So at least one sequence of length t is a universal traverse sequence. [R. Aleliunas–R. M. Karp–R. J. Lipton–L. Lovász–C. W. Rackoff, in: *Proc. 20th Ann. Symp. on Foundations of Computer Science*, 218–223.]

§ 12. Automorphisms of graphs

1. Figure 84 in the solution of 11.2 shows how to associate lines of K_5 with the points of the Petersen graph P so that adjacent points are associated with non-adjacent lines. So $P \cong \overline{L(K_5)}$ and the automorphism group of P is the same as that of $L(K_5)$. The automorphism group of $L(K_5)$ is isomorphic with the automorphism group of K_5. For obviously, each automorphism of K_5 induces an automorphism of its line-graph. Conversely, let α be an automorphism of $L(K_5)$. Observe that any 4-element clique in $L(K_5)$ corresponds to a star in K_5 and vice versa, so $L(K_5)$ has exactly five 4-element cliques, and any point in $L(K_5)$ is uniquely characterized as the intersection of some two of them. Thus,

α induces a permutation of the five 4-element cliques of $L(K_5)$ and this induces an automorphism of K_5.

Thus, the automorphism group of P is isomorphic with the automorphism group of K_5, which is S_5. [R. Frucht, *Comment. Math. Helv.* **9** (1936–37) 217–223.]

2. (a) Each automorphism of this graph is a congruence of the dodecahedron. This follows from the observation that for any mapping of the star of a point onto the star of any other point there is a unique automorphism extending it; also, there is a unique congruence of the dodecahedron and so, the automorphism must be the same as the one induced by this congruence.

This observation also shows that the dodecahedron has 120 congruences.

Consider the cubes shown in Fig. 16 in the hint. Observe that each diagonal of any face is an edge of a unique cube and each cube contains exactly one diagonal of each face. Thus, the number of such cubes is exactly 5. Let $Q_1,\ldots,Q_5$ be these cubes. Each congruence α of the dodecahedron induces a permutation $\overline{\alpha}$ of $Q_1,\ldots,Q_5$. Suppose that $\overline{\alpha} = 1$ for some α. Consider any point x of the dodecahedron. This is contained in exactly two cubes Q_i, Q_j. These two cubes have only one more point in common, namely the point x' diametrically opposite to x. Thus $\alpha(x) = x$ or x'. Moreover, if $\alpha(x) = x$ for some x, then $\alpha(y)$ must be a neighbor of x for every neighbor y of x, so $\alpha(y) = y$. Thus it follows that α is either the identity or the reflection ϱ about the centre of the dodecahedron.

The number of congruences of the dodecahedron is 120. By the above, $\alpha \to \overline{\alpha}$ is a homomorphism of the congruence group Γ into S_5, and the image has 60 elements. The only 60-element subgroup of S_5 is A_5, so this is the image of Γ. Also the above considerations show that the mapping is monomorphic on the subgroup Γ_0 of direct (orientation preserving) congruences, hence this subgroup (which clearly has index 2) is isomorphic with A_5. Also it follows that $\{1, \varrho\}$ is a normal subgroup, whence ϱ commutes with every element of Γ. Hence

$$\Gamma = \Gamma_0 \times \{1, \varrho\} \cong A_5 \times C_2.$$

(b) It follows as for the dodecahedron that each automorphism of the cube is, in fact, induced by a congruence and that the whole group is the direct product of Z_2 and the group of direct congruences of the cube. Each direct congruence of the cube induces a permutation of the 4 main diagonals. It is also easy to see that, if a direct congruence maps every main diagonal onto itself, then it is the identity. Therefore, the group of direct congruences of the cube is a subgroup of S_4. A straightforward counting shows that the cube has 24 direct congruences (each edge is mapped onto any other edge by exactly two direct congruences) and hence, the group of direct congruences is S_4.

The tetrahedron clearly has automorphism group S_4. The octahedron is the dual of the cube; since its automorphisms must be induced by congruences in the same way as in the previous cases, its automorphism group is the same as that of the cube, i.e. $Z_2 \times S_4$. Similarly, the automorphism group of the icosahedron is isomorphic with $Z_2 \times A_5$.

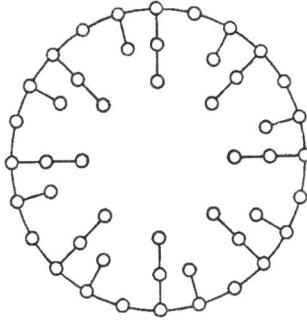

FIGURE 86

3. A graph like the one in Fig. 86 has only the rotations as automorphisms. [R. Frucht, *Compositio Math.* **6** (1938) 239–250.]

4. The right multiplication by any fixed group element g, i.e. the mapping $g_i \mapsto g_i g$ is an automorphism: if g_i is joined to g_j by an edge of color k (i.e. $g_i g_j^{-1} = g_k$), then $(g_i g)(g_j g)^{-1} = g_k$ and hence, $g_i g$ is joined to $g_j g$ by an edge of color k.

Conversely, the automorphism of G arises in this way. For let α be any automorphism of G and set $g = \alpha(1)$. We claim that $g_i g = \alpha(g_i)$. Since $(g_i g)g^{-1} = g_i$, $g_i g$ is joined to g by an edge of color i and g_i is the only point with this property. On the other hand, g_i is joined to 1 by an edge of color i by definition, and since α is an automorphism this implies that $\alpha(g_i)$ is joined to $\alpha(1) = g$ by an edge of color i. Hence, $\alpha(g_i) = g_i g$.

Since it is easy to see that the multiplication of elements of Γ is the same as multiplication between the corresponding automorphisms of G, we conclude that the automorphism group of G is isomorphic with Γ [R. Frucht, ibid.].

5. We may assume that $|\Gamma| \geq 2$. The previous problem yields a colored digraph $\overrightarrow{G}_0$ with automorphism group $\cong \Gamma$.

If $g_i, g_j \in V(\overrightarrow{G}_0) = \Gamma$ are joined by an edge of color k, then connect them by a path of length $k+2$, with paths of length 1 attached to each inner point of it except for the inner point next to g_j, where we attached a path of length 2 (Fig. 87). We do this for each pair g_i, g_j and then remove the directed edges. We denote the resulting graph by G.

Observe first that each automorphism of $\overrightarrow{G}_0$ induces a unique automorphism of G. It suffices to show that, conversely, if α is an automorphism of G, then α is induced by some automorphism of $\overrightarrow{G}_0$. The points of $V(\overrightarrow{G}_0)$ are exactly those points of G, which are neither cutpoints nor endpoints. Hence, $V(\overrightarrow{G}_0)$ is invariant under $A(G)$.

The new paths with the attached shorter paths are the components of $G - V(\overrightarrow{G}_0)$, hence α maps them onto each other. A path of length k must be mapped

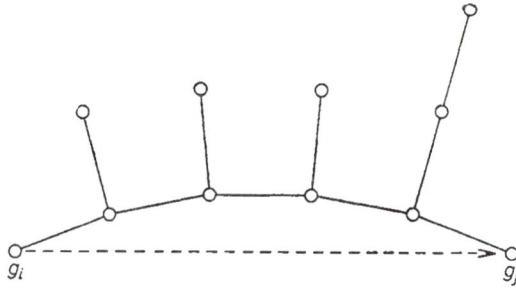

FIGURE 87

onto a path of length k and the end of the path with the longer attached path must be mapped onto the corresponding end. Hence, if g_i is connected to g_j by an edge of color k, then so is $\alpha(g_i)$ to $\alpha(g_j)$. Thus, α yields an automorphism of $\overrightarrow{G}_0$ as stated [R. Frucht, ibid.].

6. Let $V_i = \{(\gamma, i) : \gamma \in \Gamma\}$ ($i = 1, 2$). We may consider Γ to act on $V_1 \cup V_2$ according to

$$\delta(\gamma, i) = (\gamma\delta, i) \qquad (\gamma, \delta \in \Gamma, i = 1, 2).$$

Let $h_1, \ldots, h_m$ be a minimal set of generators of Γ. Assume first that $m \geq 2$. Connect $(1, 1)$ to $(1, 2)$, $(h_1, 2), \ldots, (h_m, 2)$, $(h_1, 1)$. Also connect $(h_1, 2)$ to $(h_2, 2)$ to $\ldots$ to $(h_m, 2)$. Take all images of these edges under Γ and denote the resulting graph by G.

Clearly G admits all elements of Γ as automorphisms. Let us see, if it has any other.

Let α be an automorphism of G and assume first that it keeps V_1, V_2 invariant. By multiplying by a suitable element of Γ, we may achieve $\alpha(1, 1) = (1, 1)$. The point $(1, 1)$ is adjacent to $(1, 2)$, $(h_2, 2), \ldots, (h_m, 2)$ in V_2. α must keep this set invariant.

We claim that these points induce the following subgraph: $(1, 2)$ is isolated, $(h_1, 2), \ldots, (h_m, 2)$ form a path P. Clearly the edges of P are in G but we have to show no other edge runs between these points. More exactly we show: Γ *does not map any pair* $((h_i, 2), (h_{i+1}, 2))$ *onto a pair* $((h_r, 2), (h_s, 2))$ *unless* $\{r, s\} = \{i, i+1\}$ *and it does not map it onto* $((1, 2), (h_r, 2))$ *at all.* For assume that

$$\gamma((h_i, 2)(h_{i+1}, 2)) = ((h_r, 2), (h_s, 2)), \gamma \in \Gamma.$$

Then $\gamma = h_i^{-1} h_r = h_{i+1}^{-1} h_s$. If one of r, s is different from both h_i and h_{i+1}, then it can be expressed from this relation in contradiction with the minimality of generating set $\{h_1, \ldots, h_m\}$. So $\{r, s\} = \{i, i+1\}$. The other assertion above follows similarly.

Thus it follows that α must keep $(1, 2)$ fixed, and that there are only two possibilities for α to act on $(h_1, 2), \ldots, (h_m, 2)$: either α is the identity on this set or $\alpha(h_i, 2) = (h_{m-i+1}, 2)$. We can rule out the second possibility as follows.

$(h_1, 2)$ is adjacent to $(h_1, 1)$, which is a neighbor of $(1, 2)$. We show $(h_m, 2)$ is not adjacent to any neighbor of $(1, 1)$ in V_1. In fact, $(1, 1)$ has two (not necessarily different) neighbors $(h_1, 1)$ and $(h_1^{-1}, 1)$ in V_1. If $(h_1^{\pm 1}, 1)$ were adjacent to $(h_m, 2)$, then we would have $(h_1^{\pm 1}, 1) = \gamma(1, 1)$, and $(h_m, 2) = \gamma(h_r, 2)$ or $(h_m, 2) = \gamma(1, 2)$ for some $\gamma \in \Gamma$ and $1 \le r \le m$. But this would imply $\gamma = h_1^{\pm 1}$ and either $\gamma = h_r^{-1} h_m$ or $\gamma = hm$, which are both impossible by the minimality of the generating set $\{h_1, \ldots, h_m\}$.

So we have shown that if α fixes $(1, 1)$, it also fixes its neighbors in V_2. We show now that α acts on V_1 and V_2 in the same way, i.e. if $\alpha(g, 1) = (g', 1)$, then $\alpha(g, 2) = (g', 2)$. In fact, $g \alpha g'^{-1}$ keeps $(1, 1)$ fixed which, as above, implies that it fixes $(1, 2)$. But this means that $\alpha(g, 2) = (g', 2)$. Thus α fixes $(h_1, 1), \ldots, (h_m, 1)$. It follows then that if α fixes $(g, 1)$, is also fixes $(gh_i, 1)$ for each $i = 1, \ldots, m$. Since $\{h_1, \ldots, h_m\}$ generates Γ it follows that α fixes every element of V_1. As we have seen, this implies that it fixes every element of V_2, i.e. $\alpha = 1$. This proves that those automorphisms of G keeping V_1 and V_2 invariant are only the elements of Γ.

To rule out those automorphisms of G, which would not keep V_1 and V_2 invariant we want to arrange for the degrees in these sets to be different. If G itself fulfils this we are done. Otherwise, take the complement in V_1. Since the degrees of the subgraph induced by V_1 in G are at most 2, in the complement they will become at least $n - 1 - 2 > 2$. Thus they increase, so after complementation they will not be equal to the degrees in V_2. Since the complementation in V_1 does not influence those automorphisms keeping V_1 invariant, we get the desired graph.

In the case when $m = 1$, i.e. Γ is cyclic, take the graph shown in Fig. 88. The points on the rim have degree 5, the points inside have degree 3, so each automorphism maps the boundary circuit onto itself. The inner points prevent reflections. [L. Babai, *Can. Math. Bull.* **17** (1974) 467–470.]

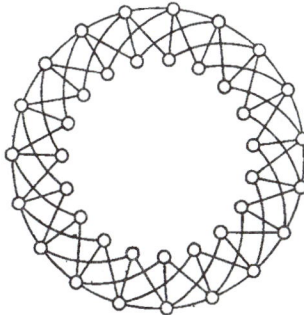

FIGURE 88

7. (a) Suppose indirectly that a tournament T has an even number of automorphisms. By Cauchy's theorem T has an automorphism α of order 2. Let (xy) be a 2-cycle of α. Since γ is a tournament, either (x, y) or (y, x) is an edge of T, but not both. Say $(x, y) \in E(T)$, then α maps (x, y) onto $(y, x) \notin E(T)$, a contradiction.

(b) Let us first note the following fact. Let V be a set such that the group Γ of odd order acts regularly on V. Let us consider the orbits of Γ on the set E of ordered pairs (x,y), $x \neq y$, x, $y \in V$. No element of Γ can map (x,y) onto (y,x); for such a permutation would contain an even cycle and hence, it would be of even order.

Thus, the orbits of Γ on E decompose into pairs $\{S,S'\}$ such that $S' = \{(x,y): (y,x) \in S\}$. Selecting one orbit from each such pair $\{S,S'\}$, the union of these orbits forms a tournament T, which admits all elements of Γ as automorphisms. Now let us take two sets $V_i = \{(\alpha,i): \alpha \in \Gamma\}$ $(i=1,2)$, and define the action of $\alpha \in \Gamma$ on $V_1 \cup V_2$ by

$$\alpha(\gamma,i) = (\gamma\alpha,i) \ (\gamma \in \Gamma, i = 1,2).$$

Let $h_1,\ldots,h_m$ be a minimal generating set of Γ.

Let us connect $(\gamma,1)$ to $(\delta,2)$ if $\delta = \gamma$ or $\delta = \gamma \cdot h_i (1 \leq i \leq m)$, and $(\delta,2)$ to $(\gamma,1)$ otherwise. Take any tournament on V_1 invariant under Γ. To define the tournament on V_2 let us observe that no pair $((h_i,2),(h_j,2))$ or $((h_i,2),(1,2))$ or $((1,2),(h_i,2))$ is mapped onto such a pair by any element $\gamma \in \Gamma$, $\gamma \neq 1$; for, e.g.

$$\gamma((h_i,2),(h_j,2)) = ((h_\mu,2),(h_\nu,2))$$

would imply

$$h_i = h_\mu h_\nu^{-1} h_j,$$

which implies, by the minimality of $\{h_1,\ldots,h_m\}$, that either $i = \nu$, $j = \mu$ and thus $(h_i h_j)^2 = 1$ (which has been excluded, because Γ has odd order) or $i = \mu$, $j = \nu$ and thus $\gamma = 1$. So we may take the edges $((1,2),(h_i,2))$ $(i=1,\ldots,m)$ and $((h_i,2),(h_j,2))$ $(1 \leq i < j \leq m)$ and all images of them under Γ and, as noted at the beginning, one of each remaining pair of orbits of Γ on E, and obtain a tournament on V_2 invariant under Γ. Thus we have defined a tournament T on $V_1 \cup V_2$ invariant under Γ.

We claim that $A(T) = \Gamma$. Let $\alpha \in A(T)$, we want to show $\alpha \in \Gamma$. First observe that the points in V_1 have outdegree

$$\frac{n-1}{2} + m + 1,$$

while the points in V_2 have outdegree

$$\frac{n-1}{2} + n - m - 1,$$

which are different, because n is odd. Therefore, α must keep V_1 and V_2 invariant.

We may assume that $\alpha(1,1) = (1,1)$. Then the set of points in V_2 accessible from $(1,1)$ on an edge is also invariant under α. This set is $(1,2)$, $(h_1,2),\ldots,(h_m,2)$, which is a transitive tournament. Therefore α must fix all these points. Hence it follows as in the preceding solution that α must fix every point i.e. $\alpha = 1$. [J. W. Moon, *Canad. J. Math.* **16** (1964) 485–489; this construction is due to L. Babai.]

8. Instead of the paths shown in Fig. 87, use the paths in Fig. 89 to connect g_i to g_j if they are joined by a directed edge of color k; where the 6-point configuration is repeated k times. In this way, we get a graph G_1 which has automorphism group Γ (this follows exactly as before). However, G_1 is not 3-regular as the points of $V(G_0)$ have degree $2|\Gamma|$.

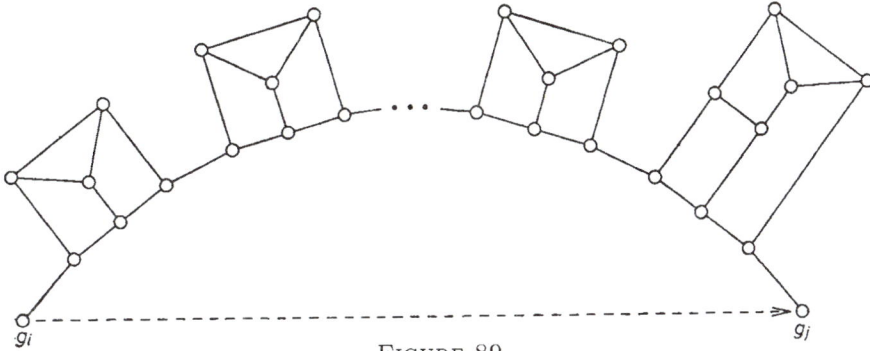

FIGURE 89

Take any $g_i \in V(G_0)$. Observe that there are exactly two of the strings of Fig. 86 of length k attached to it, one with each kind of ending $(k = 1, \ldots, |\Gamma|)$. Split g_i into $2|\Gamma|$ points of degree 1; they can be denoted by $a_{1,6}, \ldots, a_{|\Gamma|,6}$, $a_{1,8}, \ldots, a_{|\Gamma|,s}$, where $a_{k,j}$ is the starting point of a string in Fig. 86 of length k and with a j-point configuration next to $a_{k,j}$. Now join the $a_{k,j}$'s by a circuit C_i in the above order. Do the same for each $i = 1, \ldots, |\Gamma|$. Clearly, each automorphism of G_1 induces an automorphism of the resulting graph G. Conversely, let α be an automorphism of G. The only triangles in G are those in the strings of Fig. 86, so α maps them onto each other; also the triangles in the 6-point configurations have different neighborhoods than the triangles in the 8-point configurations and 8-point-configurations onto 6-point configurations and 8-point-configurations onto 8-point configurations. Thus, α maps the rest, i.e. the points on circuits C_i onto each other. Since the C_i's are the component of the subgraph spanned by $V(C_1) \cup \ldots \cup V(C_{|\Gamma|})$, α maps each C_i onto some C_j. Hence, α induces a (unique) automorphism of the graph obtained by contracting each C_i. But this graph is G_1.

Thus the automorphisms of G and G_1 are in a $1-1$ correspondence, which is obviously an isomorphism. Hence, the automorphism group of G is isomorphic to Γ. Since G is 3-regular, we are finished. [R. Frucht, *Canad. J. Math.* **1** (1949) 365–378. This proof is due to L. Babai. For regular graphs of any degree see G. Sabidussi, *Canad. J. Math.* **9** (1957) 515–525.]

9. Let $x_1, \ldots, x_n$ be as in the hint, and denote by Γ_i the group of those automorphisms fixing $x_1, \ldots, x_i$. Then $\Gamma_1 = A(G)$, because x_1 is the only point of degree 2 and so each automorphism must fix it; $\Gamma_n = \{1\}$. We show that the index $|\Gamma_{i-1} : \Gamma_i|$ is 1 or 2; this will imply the assertion.

Look at the images of x_i under $\Gamma_{i-1}(i \geq 2)$. Let $j < i$ be such that x_j is adjacent to x_i. Then x_j has at most one other neighbor in $\{x_i, \ldots, x_n\}$. Since each $\gamma \in \Gamma_{i-1}$ must map x_i onto such a neighbor of x_j, it follows that either x_i is fixed by all automorphisms in Γ_{i-1} or it has one other image. The index $|\Gamma_{i-1} : \Gamma_i|$ is, accordingly, 1 or 2.

Remark: Every group of order 2^k arises as the automorphism group of such a graph [cf. L. Babai–L. Lovász, *Studia Sci. Math. Hung.* **8** (1973) 141–150].

10. Select an $e \in E(G)$ and orient it arbitrarily. Let f be any other element in the orbit of e under $A(G)$ and let α be the unique automorphism mapping e onto f. Now orient f in such a way that α should preserve the orientation of e, if $\alpha \in \Gamma$ and conversely, otherwise. The remaining edges of G we orient in such a way that every $\gamma \in A(G)$ preserves their orientations (this can be done as $A(G)$ acts semiregularly on $E(G)$).

The resulting digraph $\overrightarrow{G}$ admits the elements of Γ as automorphisms by the construction. On the other hand, let α be any automorphism of $\overrightarrow{G}$. Then, clearly, α is an automorphism of G, i.e. $\alpha \in A(G)$. Since $\alpha(e)$ must be oriented like e, it follows that $\alpha \in \Gamma$ [cf. J. Nešetřil, *Monatsh. f. Math.* **76** (1972) 323–327].

11. It is easy to check that in 12.5, if we remove an edge e, which connects a point of degree 1 to a point of degree 2, we obtain a graph with no automorphisms. Taking the complement we have a connected graph G_2 and an edge $e = (u, v) \in E(G_2)$ such that $A(G_2) = \{1\}$, $A(G_2 - e) \cong \Gamma_2$. Also the construction in 12.5 has the property that its automorphism group acts semiregularly on the points, i.e. no automorphism $\alpha \neq 1$ fixes any point. Thus we have a graph G_1 such that $A(G_1) \cong \Gamma_1$ and only the identity fixes a certain $x \in V(G_1)$.

Now construct G as follows. Let its points be the pairs (x_1, x_2), $x_i \in V(G_i)$. Moreover, connect (x_1, x_2) to (y_1, y_2), if either $(x_1, y_1) \in E(G_1)$ or $x_1 = y_1$ and $(x_2, y_2) \in E(G_2)$. (G is called the lexicographic product of G_1 and G_2.) Also let us color those edges $((x_1, x_2), (y_1, y_2))$ of G with $x_1 = y_1$ red and the rest black. Now $A(G) \cong \Gamma_1$. For obviously, each automorphism of G_1 induces an automorphism of G. On the other hand, let $\alpha \in A(G)$. Since G_2 is connected, the red coloration of edges forces that α must map each class $V_{x_1} = \{(x_1, x_2) : x_2 \in V(G_2)\}$ onto such a class $V_{x_1'}$. From the fact that G_2 has no automorphism except the identity, it follows that $\alpha(x_1, x_2) = (x_1', x_2)$, i.e. α is induced by an automorphism of G_1.

On the other hand, let us remove the edge $f = ((x, u), (x, v))$. Then for each $\gamma \in A(G_2 - e)$ the mapping $\overline{\gamma}$ defined by

$$\overline{\gamma}(x_1, x_2) = \begin{cases} (x, \gamma(x_2)) & \text{if } x_1 = x, \\ (x_1, x_2) & \text{if } x_1 \neq x. \end{cases}$$

is an automorphism of $G - f$. Conversely, if α is any automorphism of $G - f$, then it follows as above that α maps V_{x_1} onto $V_{x_1'}$. But then α must map V_x onto itself (since V_x induces fewer edges than any other V_{x_1}). Therefore, α induces an automorphism $\tilde{\alpha}$ of G_1, which fixes x, whence $\tilde{\alpha} = 1$, i.e. $\alpha(V_{x_1}) = V_{x_1}$ for every

$x_1 \in V(G_1)$. Since V_{x_1} induces a graph with no proper automorphism for $x_1 \neq x$, we conclude that $\alpha = \bar{\gamma}$ for some $\gamma \in A(G_2 - e)$. Thus $A(G - f) \cong \Gamma_2$.

We still have to get rid of the colorings of edges. This can easily be done by replacing each black edge by a path of length of length N, N large [L. Babai].

12. (a) Let us remark first that there is a tree T such that each orbit of Γ contains exactly one point of T. For choose a maximal tree T_0 such that each orbit contains at most one point of T_0. If $V_0 = \bigcup_{\gamma \in \Gamma} \gamma(V(T_0)) = V(G)$ we are done. Suppose that $V_0 \neq V(G)$, then, since G is connected, there is a point $x \notin V_0$ adjacent to some $y \in V_0$. Let $y \in \gamma(T_0)$, then $\gamma(T_0) + (x, y)$ is a larger tree meeting each orbit in at most one point.

So let the tree T meet each orbit in exactly one point. Let us contract each $\gamma(T)$, $\gamma \in \Gamma$. The resulting graph G' is clearly connected, and Γ acts regularly on it. [L. Babai, *Acta Math. Acad. Sci. Hung.* **24** (1973) 215–221; cf. G. Sabidussi, *Proc. AMS* **9** (1958) 800–804.]

(b) We use induction on $|V(G)|$. We may suppose that $\Gamma \neq \{1\}$. Let $e \in E(G)$ such that not both endpoints of e are fixed points of Γ.

Let G' be the subgraph of G formed by those edges of G of form $\gamma(e)$, $\gamma \in \Gamma$. Then G' is a subgraph of G on which Γ acts edge-transitively. Moreover, different elements of Γ have different effects on $V(G')$; for the set N of those permutations $\gamma \in \Gamma$ which have $\gamma|_{V(G')} = 1$ is a normal subgroup of Γ; by the simplicity of Γ, $N = \Gamma$ or $N = \{1\}$. However, $N = \Gamma$ would imply that all points of $V(G')$, in particular both endpoints of e, are fixed points of Γ, a contradiction. Thus if G' is connected we are done.

Suppose that G' is disconnected; let $G_1, \ldots, G_k$ be its components. Then

$$\{V(G_1), \ldots, V(G_k)\} \cup \{\{x\} : x \in V(G) \setminus V(G')\}$$

is a partition of $V(G)$ invariant under Γ. Let us contract each G_i onto one point. The resulting graph G'' is clearly connected. Each automorphism $\gamma \in \Gamma$ induces an automorphism $\tilde{\gamma} \in A(G'')$. As above, the simplicity of Γ implies that either $\tilde{\gamma} = 1$ for each $\gamma \in \Gamma$ or $\gamma \mapsto \tilde{\gamma}$ is one-to-one. But the former possibility cannot occur since some $\gamma \in \Gamma$ maps G_1 onto G_2 and for this $\tilde{\gamma} \neq 1$. Hence $A(G'')$ contains a subgroup isomorphic with Γ. We are finished by induction. [L. Babai, *Discrete Math.* **8** (1974) 13–20.]

13. If a permutation group is commutative and transitive, then it is regular (see the footnote to the solution of 11.17).

If we fix any x_0 then, by the regularity of $A(G)$, each $x \in V(G)$ can be written uniquely as $\alpha(x_0)$, $\alpha \in A(G)$. So the mapping φ with $\varphi(\alpha(x_0)) = \alpha^{-1}(x_0)$ is well defined and is a permutation of $V(G)$. Moreover, φ is an automorphism of G, since if $(x, y) \in E(G)$, $x = \alpha(x_0)$, $y = \beta(x_0)$, $(\alpha, \beta \in A(G))$ then, since α is an automorphism, $(x_0, \alpha^{-1}(\beta(x_0))) = (x_0, \beta(\alpha^{-1}(x_0))) \in E(G)$, whence $(\beta^{-1}(x_0), \alpha^{-1}(x_0)) \in E(G)$. Since φ fixes x_0, we get, by the regularity of $A(G)$, that $\varphi = 1$. Thus, $\alpha^{-1} = \alpha$ for each $\alpha \in A(G)$, i.e. every element of $A(G)$ is of order 2. It is well known from group theory that this implies that $A(G)$ is the direct product of cyclic groups of order 2. [C. Y. Chao, *Proc. AMS* 15 (1964) 291–292; G. Sabidussi, *Monatsh. f. Math.* **68** (1964) 426–438.]

(b) Let $_n$ be the n-cube. We may assume that the vertices of Q_n are 01-vectors of length n and two such vectors are adjacent iff they differ at exactly one place. Denote by e_I the characteristic vector of $I \subseteq \{1,\dots,n\}$. If we switch the coordinates belonging to a subset $I \subseteq \{1,\dots,n\}$, we get an automorphism α_I of Q_n. The group Γ_0 of all automorphisms α_I is isomorphic with $\underbrace{Z_2 \times \dots \times Z_2}_{n}$. However, Q_n has other automorphisms, e.g. exchanging the first two coordinates. To exclude these, let us modify Q_n by joining two vectors by k edges if they differ at the kth place only. It is obvious that α_I as defined above is still an automorphism of the resulting graph Q_n'. We show that Q_n' has no other automorphisms. If it had we could assume that it had one, which fixes $(0,\dots,0)$. However, if α is an automorphism, which fixes $(0,\dots,0)$, then it also fixes $(1,0,\dots,0)$ since this is the only point joined to $(0,\dots,0)$ by exactly one edge. Similarly, it fixes $(0,\dots,0,1,0,\dots,0)$ for any position of the 1. Thus if α fixes a point it fixes all neighbors of it. Q_n' being connected, α must fix every point, i.e. $\alpha = 1$, a contradiction.

(c) We want to introduce new edges to Q_n such that the arising simple graph G still admits every α_I as an automorphism, but has no other automorphism. Let H be a graph on the set of vectors $\{\mathbf{e}_1,\dots,\mathbf{e}_n\}$, $\mathbf{e}_i = (0,\dots,0,1,0,\dots,0)$.

If $(\mathbf{e}_i,\mathbf{e}_j) \in E(H)$, then $\alpha_I(\mathbf{e}_i) = (\mathbf{e}_{I\triangle\{i\}})$, $\alpha_I(\mathbf{e}_j) = \mathbf{e}_{i\triangle\{j\}}$. Thus if we add all edges $(\mathbf{e}_I,\mathbf{e}_J)$ to Q_n such that $I\triangle J = \{i,j\}$ with $(\mathbf{e}_i,\mathbf{e}_j) \in E(H)$, we obtain a graph G invariant under Γ_0.

Let us see what properties H must have to ensure that G has no other automorphism.

Suppose that $A(G) \neq \Gamma_0$. Since Γ_0 is transitive, we have an $\alpha \in A(G)$, $\alpha \neq 1$ with $\alpha(0,\dots,0) = (0,\dots,0)$. Let H_1 be the graph induced by the neighbors of $(0,\dots,0)$ in G (this is, of course, not H since $(0,\dots,0)$ has neighbors other than $\mathbf{e}_i$). The points of H_1 are $\mathbf{e}_1,\dots,\mathbf{e}_n$ and all points $\mathbf{e}_{\{i,j\}}$ with $(\mathbf{e}_i,\mathbf{e}_j) \in E(H)$. Two points of form $\mathbf{e}_i$ are adjacent iff they are adjacent in H. $\mathbf{e}_{\{i,j\}}$ is adjacent to $\mathbf{e}_{\{\mu,\nu\}}$ iff $(\mathbf{e}_i,\mathbf{e}_j)$, $(\mathbf{e}_\mu,\mathbf{e}_\nu)$ are adjacent edges, e.g. $j = \nu$ and $(\mathbf{e}_i,\mathbf{e}_\mu) \in E(G)$.

Thus H_1 arises from H by taking a new point p_e for each edge $\mathbf{e}$ of H, connecting it to the endpoints of $\mathbf{e}$ and connecting p_e to p_e' iff $\mathbf{e}$, $\mathbf{e}'$ are edges of a triangle in H.

Note that if H contains no triangles, then each point of $V(H_1) - V(H)$ has degree 2. If H has no point of degree 1, then these are the only points of degree 2. Thus if in addition H is asymmetric, so is H_1.

If this can be achieved we are done, since it means that α fixes the neighbors of $(0,\dots,0)$ in G, and G being connected, it follows that it fixes every point, a contradiction. So all we need is an asymmetric triangle-free graph with no point of degree 1.

Figure 90 shows such a graph H for $n \geq 8$. [W. Imrich, *Monatsh. f. Math.* **73** (1969) 341–347. For graphs with non-abelian regular automorphism group see L. A. Nowitz–M. E. Watkins, *Canad. J. Math.* **24** (1972) 993–1018.]

14. Let k be the edge-connectivity of G, and $X \subseteq V(G)$ a minimal set with $\delta_G(X) = k$.

By the minimality of X, $|X| \leq |V(G)|/2$. Let x_1, $x_2 \in X$ and $\alpha \in A(G)$ be such that α moves x_1 to x_2. Let $\alpha(X) = X'$. Then $\delta_G(X') = k$ (as α is an automorphism) and $X \cap X' \neq \emptyset$ and $X \cup X' \neq V(G)$ (as $|X| = |X'| \leq |V(G)|/2$). Hence by 6.48a, $\delta_G(X \cap X') = k$. Since X has been minimal, this is only possible, if $X \cap X' = X$, i.e. $X = X'$.

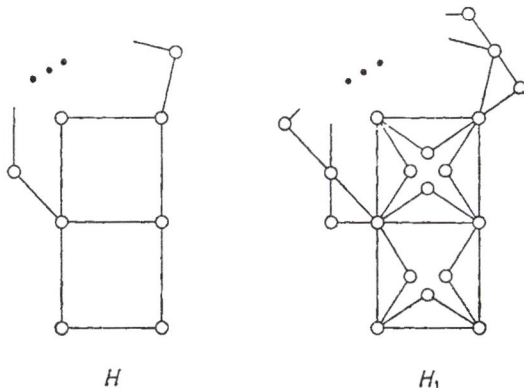

H H_1

FIGURE 90

Thus, those automorphisms which keep X invariant act transitively on X. This implies that each point of X has the same number a of neighbors in X. Obviously,

$$a \leq |X| - 1,$$

and since each point of X is adjacent to $r - a$ points outside X,

$$(r - a)|X| = k.$$

Thus $k = (r - a)|X| \geq (r - a)(a + 1) \geq r.$

Since obviously $r \geq k$, we have $r = k$ as stated.

15. (a) Let $T \subseteq V(G)$, $|T| = 3$, G_1 a component of $G - T$, $X = V(G_1)$ and suppose that T, G_1 are chosen so that $|X|$ is minimal.

Consider all images of X under automorphisms of G. By the transitivity of $A(G)$, they will cover all points. Moreover, they are disjoint (if different). For if $\alpha(X) \cap \beta(X) \neq \emptyset$, $\alpha(X) \neq \beta(X)$ then $\alpha(T)$ meets $\beta(X)$ but does not contain it, which contradicts 6.60(a). This also shows that T is the union of certain images of X. Since $|T| = 3$, this is only possible if $|X| = 1$ (in which case we are finished) or $|X| = 3$.

To exclude this latter case, observe that X is joined by edges to T only. So if $T = \alpha(X)$ for some α, then T is joined to $\alpha(T) = \alpha^2(X)$ only; however, T is in fact joined to X and to some other points too (being a cutset).

The statement is false for 4 instead of 3 as is shown by Fig. 91.

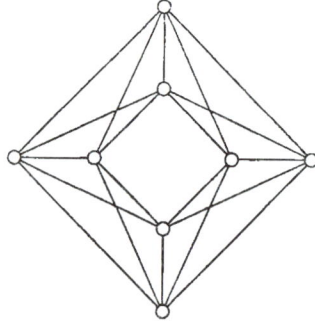

FIGURE 91

(b) Again let T be a minimum cutset of G and G_1 a component of $G - T$, chosen so that $|X|$ is minimal, where $X = V(G_1)$. It follows as before that T is the union of certain images of X which are disjoint, and T cannot be an image of X (under automorphisms of G). Hence,

$$|T| \geq 2|X|.$$

Since any $x \in X$ can be joined to the points of $T \cup (X - \{x\})$ only, we have

$$r \leq |T| + |X| - 1 \leq |T| + \frac{1}{2}|T| - 1 = \frac{3}{2}|T| - 1$$

or, equivalently,

$$|T| \geq \frac{2}{3}(r + 1).$$

This estimate is sharp if $r + 1$ is divisible by 3, as is shown by the strong direct product of a circuit and a complete $(r + 1)/3$-graph (Fig. 88 for $r = 5$). [M. E. Watkins, *J. Comb. Th.* **8** (1970) 223–226; W. Mader, *Arch. d. Math.* **21** (1970) 331–336.]

(c) Again let T be a minimum cutset of G and G_1 a component of $G - T$. Assume that T and G_1 are chosen so that $|V(G_1)|$ is minimal. We want to show that G_1 consists of a single point. Suppose indirectly that G_1 has an edge e. Let f be an edge connecting G_1 to T, and α an automorphism mapping f onto e. Then $\alpha(T)$ meets G_1 and does not contain it, a contradiction with 6.60a.

16. Suppose indirectly that the edge $e = (x, y)$ of G does not occur in any 1-factor (G itself may or may not have 1-factors). Then $G - x - y$ has no 1-factor and therefore, by Tutte's theorem, there exists an $X_1 \subseteq V(G) - \{x, y\}$ such that $G - x - y - X_1$ has more than $|X_1|$ odd components. For reasons of parity, the number of odd components of $G - x - y - X_1$ is $\geq |X_1| + 2$ (since $|V(G)|$ is even). So if we set $X = X_1 \cup \{x, y\}$, then $G - X$ has $\geq |X|$ odd components.

Let $T_1, \ldots, T_k$ be the components of $G - X (k \geq |X|)$ and let G be r-regular. By 12.14, G is r-edge-connected and hence, there must be at least r edges joining T_i to X ($i = 1, \ldots, k$). This means that at least kr edges enter X. But X has only $|X| \leq k$ points to receive them and a point of X can receive at most r. Moreover x and y can receive at most $r-1$ as they are incident with e too. Thus, the points of X cannot receive all edges entering X, a contradiction.

17. Γ is regular (see the footnote to the solution of 11.17). Let $x \in V(G)$ and $\gamma \in \Gamma$ be such that $(x, \gamma(x)) \in E(G)$. Then $(x, \gamma(x), \ldots, \gamma^{r-1}(x))$ is a circuit in G (where r is the order of γ).

Now let Γ_1 be a maximal subgroup of Γ such that the elements of $\Gamma_1(x)$ form a circuit $(x, x_1, \ldots, x_m)$ ($|\Gamma_1| = m + 1$). We claim that $\Gamma_1(x) = V(G)$, which will prove the assertion. Suppose that $\Gamma - \Gamma_1 \neq \emptyset$, then since G is connected, there is an edge (a, b) with $a \in \Gamma_1(x)$, $b \notin \Gamma_1(x)$. Then $a = \gamma(x)$ ($\gamma \in \Gamma_1$) and $(x, \gamma^{-1}(b))$ is an edge as well. Clearly $\gamma^{-1}(b) \notin \Gamma_1(x)$; let $\gamma^{-1}(b) = \delta(x)$, $\delta \in \Gamma - \Gamma_1$. Let p be the least integer with $\delta^p \in \Gamma_1$.

By the commutativity of Γ, $(y, \delta(y)) \in E(G)$ for any $y \in V(G)$; in fact, $y = \gamma(x)$ for some $\gamma \in \Gamma$ and then $(y, \delta(y)) = (\gamma(x), \delta(\gamma(x))) = (\gamma(x), \gamma(\delta(x))) = \gamma(x, \delta(x)) \in E(G)$. Also the commutativity implies that $\Gamma_2 = \Gamma_1 \cup \Gamma_1 \delta \cup \ldots \cup \Gamma_1 \delta^{p-1}$ is a subgroup. If p is even, then

$$(x, x_1, \ldots, x_m, \delta(x_m), \ldots, \delta(x_1), \delta^2(x_1), \ldots, \delta^2(x_m), \ldots,$$
$$\delta^{p-1}(x_m), \ldots, \delta^{p-1}(x_1), \delta^{p-1}(x), \ldots, \delta(x))$$

is a circuit formed by the points of $\Gamma_2(x)$, a contradiction (Fig. 92a). If p is odd we have the circuit

$$(x, x_1, \ldots, x_m, \delta(x_m), \ldots, \delta(x_1), \delta^2(x_1), \ldots, \delta^2(x_m), \ldots,$$
$$\delta^{p-1}(x_m), \delta^{p-1}(x), \delta^{p-2}(x), \ldots, \delta(x)),$$

a contradiction again (Fig. 92b).

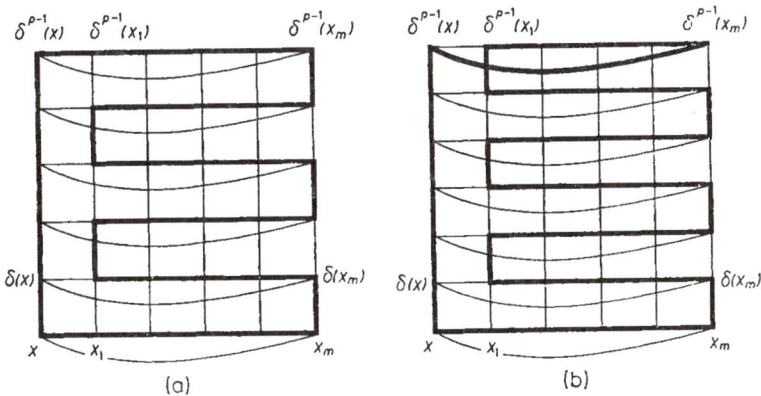

FIGURE 92

Remark: The condition that Γ is abelian cannot be dropped, as is shown by the Petersen graph [J. Pelikán].

18. It suffices to describe *connected* graphs with the desired properties; the disconnected ones are disjoint unions of arbitrary numbers of the same connected examples.

We distinguish two cases. If $\Gamma = A(G)$ acts point-transitively, then by 12.15b, G is at least 2-connected unless $G = K_2$. Moreover, if G is not 3-connected, then it is 2-regular, i.e., a single circuit.

Assume that G is r-regular ($r \geq 3$) and 3-connected. By 6.69 it follows that G has an essentially unique embedding into the sphere and every automorphism maps faces onto faces. By edge-transitivity, either all faces have the same size or there are only two types of faces, k-gons and l-gons, say. In the first case G can only be one of the five platonic bodies, the proof of which is found in any geometry book.

In the second case, the edge-transitivity implies that each edge must be adjacent to just one k-gon and one l-gon. Hence each point is adjacent to some a k-gons and a l-gons. Clearly $a \geq 2$ and by 5.26, $|E(G)| = a \cdot n \leq 3n - 6$, whence $a = 2$.

The total number of k-gons is, clearly, $\frac{2n}{k}$, the number of l-gons is $\frac{2n}{l}$, thus by Euler's formula

$$n + \frac{2n}{k} + \frac{2n}{l} = 2n + 2.$$
$$\frac{1}{k} + \frac{1}{l} = \frac{1}{2} + \frac{1}{n}.$$

We may suppose that $3 \leq k < l$, whence either $k = 3$, $l = 4$, $n = 12$ or $k = 3$, $l = 5$, $n = 30$. It is easy to see that the corresponding graphs are the linegraphs of the cube and dodecahedron, respectively.

Now suppose that Γ is not transitive on the points. Then clearly $A(G)$ has two transitivity classes on $V(G)$, U and W, say. Each edge connects two points in different classes.

If the minimum degree of G is 1, it is a star. If the minimum degree of G is 2, it arises by subdividing each edge of a (not necessarily simple) planar graph G_1 by one point. This graph G_1 has an automorphism group transitive on both the points and the edges. Hence G_1 arises from one of the graphs determined above by multiplying each edge by the same number. Thus G is one of the graphs described in the hint.

So suppose that the minimum degree of G is at least 3. Then G is 3-connected by 12,15c, and thus by 6.69 it has an essentially unique embedding into the sphere. From the edge-transitivity it again follows that $A(G)$ is either transitive on the faces or there are two types of faces. In the first case consider the dual graph G^*. This has an automorphism group transitive on both the points and the edges, and is therefore one of the graphs determined above.

So we may suppose that $A(G)$ is not transitive on the faces either. Let each edge be adjacent to a k-gonal and an l-gonal face. Let the degrees in U and W be *a* and *b*, respectively. Since the faces containing a given point are alternatingly in the two transitivity classes, *a*, *b* are even and therefore at least 4. Similarly *k*, $l \geq 4$. But then the number of points is at most $|E(G)|/2$, the number of faces is at most $|E(G)|/2$, contradicting Euler's theorem. So the last case cannot occur. Thus the list given in the hint is verified.

19. Let Γ be a non-cyclic simple group and assume that $A(G) \cong \Gamma$, where G is planar. We show that this implies $\Gamma \cong A_5$, even if coloring the points is allowed. Let G be a minimum graph with $A(G) \cong \Gamma$. Then G is connected. For, suppose that G has components $G_1, \ldots, G_k$. Then each automorphism of G induces a permutation of the components. Those automorphisms inducing the identity form a normal subgroup of $A(G)$. Since $A(G)$ is simple, either no automorphism induces the identity or all of them do. In the first case the components of G are asymmetric and hence $A(G)$ is a direct product of symmetric groups (permuting the isomorphic components), which is never a non-cyclic simple group. So each automorphism leaves the components invariant. Let G_1 be a component, which is not asymmetric, then those automorphisms keeping the points of G_1 fixed form a normal, proper subgroup of $A(G)$. Since $A(G)$ is simple it follows that only the identity does so, i.e. $A(G) \cong A(G_1)$, a contradiction with the minimality of G.

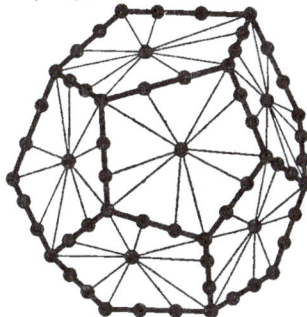

FIGURE 93

Furthermore $A(G)$ has no fixed points. For removing a fixed point and coloring its neighbors with a new color we obtain a smaller colored graph with the same automorphism group.

Now let us contract an appropriate connected subgraph of G onto a graph G_0 as in 12.12b such that $A(G_0)$ contains a subgroup $\Gamma' \cong A(G)$, which acts edge-transitively. Observe that the construction in 12.12b is such that if $A(G)$ has no fixed points, neither does Γ'. G_0 is clearly planar and so it is one of the graphs described in the preceding hint. The stars are ruled out since Γ' is fixed point free. The remaining examples are the platonic bodies or graphs, which are easily seen to have the same automorphism groups as the platonic bodies. Using 12.2 it is easy to see that the automorphism groups as the platonic bodies have no non-cyclic simple subgroups other than A_5.

Remark: Figure 93 shows a planar graph with automorphism group A_5. [L. Babai, in: *Infinite and Finite Sets*, Coll. Math. Soc. J. Bolyai **10** Bolyai–North-Holland (1975) 29–84.]

20. (a) Using 12.12a, we may assume that Γ acts regularly. Then it is easy to see that G' contains the Cartesian product of two $(p-1)$-paths and an edge. (In fact, it contains the Cartesian product of three p-circuits.) Let $G \supset P \oplus P' \oplus Q$, $P = (x_1, \ldots, x_p)$, $P' = (y_1, \ldots, y_p)$, $Q = (z_1, z_2)$. It suffices to show that $P \oplus P' \oplus Q$ can be contracted onto K_p. To this end contract the subgraphs spanned by $P \oplus \{y_i\} \oplus \{z_1\}$ and $\{x_i\} \oplus P' \oplus \{z_2\}$ (which are clearly connected). This results in a graph isomorphic to $K_{p,p}$ which, in turn, is trivially contractible onto K_p (Fig. 94).

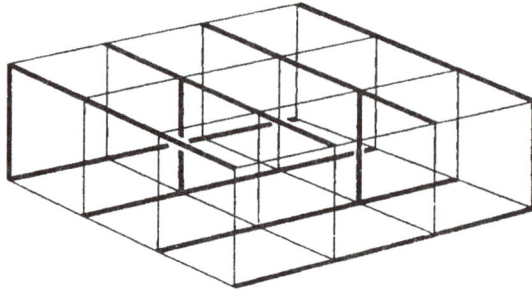

FIGURE 94

(b) For reasons similar to those in the solution of the previous problem we may assume that the action of Γ' is edge-transitive. By (a) we may suppose that the action of γ is not semiregular. Let $\gamma_0 \in \Gamma$ have fixed points. Since G is connected it has two adjacent points x_0, y_0 such that $\gamma_0(x_0) = x_0$, but $\gamma_0(y_0) \neq y_0$. Then y_0, $\gamma_0(y_0), \ldots, \gamma^{p-1}(y_0)$ are different neighbors of x_0. Hence $d_G(x_0) \geq p$.

If $A(G)$ is point-transitive, then each point has degree at least p and we are done by Mader's theorem 10.8. So suppose that $A(G)$ is not point-transitive. By the edge-transitivity, it follows then that Γ' has two transitivity classes, namely $U = \Gamma'(x_0)$ and $W = \Gamma'(y_0)$. Each edge connects U to W. It is trivial that the degrees in U are all at least p and those in W are all at least 2. If the degrees in W are at least $\sqrt[3]{p}$, then we can still conclude as before by 10.8. So suppose that the degrees in W are less than $\sqrt[3]{p}$.

Form a graph G_1 on $V(G_1) = U$ by connecting two points if they are at distance 2 in G. Each $\gamma \in \Gamma'$ induces a $\bar\gamma \in A(G_1)$. Clearly $\gamma \mapsto \gamma'$ is a homomorphism and since Γ' is simple, it is either a monomorphism or maps onto 1. However, the latter is impossible since Γ' acts transitively on U. Moreover, γ_0 has fixed points, but is not the identity, thus as before, it follows that the degrees of G_1 are at least p.

For each edge $e = (u, v) \in E(G_1)$, let $f(e)$ be one of the common neighbors of u, v in G. Clearly $f(e) \in V$. A point $y \in V$ occurs at most

$$\binom{d_G(y)}{2} \leq \binom{\sqrt[3]{p}}{2} < \frac{1}{2} p^{2/3}$$

times as $f(e)$ ($e \in E(G_1)$) and so, the number of points of form $f(e)$ is at least

$$\frac{2|E(G)|}{p^{2/3}} \geq \frac{p \cdot |U|}{p^{2/3}} = |U| \sqrt[3]{p}.$$

For each point x of the form $f(e)$ select an edge $(u_x, v_x) = e_x \in E(G_1)$ such that $f(e_x) = x$ and contract the edge (u_x, x) of G.

Identify those points of W not of form $f(e)$ with an arbitrary neighbor of them. In this way G is mapped onto a subgraph G' of G_1, which contains all edges of the form e_x (and possibly more). So

$$|E(G')| \geq |U| \sqrt[3]{p}.$$

Thus by 10.8 G' can be contracted onto K_m.

(c) Let N be a large enough number; we show that the alternating group A_N is not isomorphic to the automorphism group of any graph from $\mathcal{K}$, even, if coloring of the points is considered. More exactly we show that if a graph G (with possibly colored points) has automorphism group A_N, then a suitable subgraph of it can be contracted onto K_m provided $N \geq 3p$, where p is a prime greater than 8^m. Since for large enough m, $\mathcal{K}$ certainly does not contain any graph contractible onto K_m, this will prove the assertion.

For reasons similar to those in the solution of 12.19, we may suppose that G is connected and $A(G)$ has no fixed points. Then disregarding the coloration of the points, we obtain a graph, which satisfies the conditions of (b). This implies that G can be contracted onto K_m. [L. Babai, *Discrete Math.* **8** (1974) 13–20.]

21. Let G be a graph with $A(G) \cong \Gamma$, $V(G) \cap \Omega = \emptyset$. On the basis of the construction in 12.5 we may assume that $A(G)$ acts on G semiregularly, i.e. no element $\alpha \neq 1$ has a fixed point.

Also by the construction in 12.5 we may assume that $|V(G)| \geq |\Gamma| \cdot |\Omega|$; then $A(G)$ has at least $|\Omega|$ orbits on $V(G)$. Finally, by taking the complement of the construction in 12.5 we can arrange for each degree in G to be greater than $|\Gamma|$. Let $\gamma \to \overline{\gamma}$ be an isomorphism of Γ onto $A(G)$. Now let $\Omega_1, \ldots, \Omega_k$ be the orbits of Γ on Ω and select a $y_i \in \Omega_i (i = 1, \ldots, k)$. Let $x_1, \ldots, x_k \in V(G)$ be points in different orbits. For every $\gamma \in \Gamma$ and $1 \leq i \leq k$, connect $\gamma(y_i)$ to $\overline{\gamma}(x_i)$. The resulting graph G' admits all permutations γ' as automorphisms, where

$$\gamma'(x) = \begin{cases} \gamma(x) & \text{if } x \in \Omega, \\ \overline{\gamma}(x) & \text{if } x \in V(G). \end{cases}$$

For, clearly γ' preserves adjacency in $V(G)$ and if $(\gamma_0(y_i), \overline{\gamma}(x_i))$ is any new edge, then $\gamma'(\gamma_0(y_i), \overline{\gamma}(x_i)) = ((\gamma_0\gamma)(y_i), (\overline{\gamma_0\gamma})(x_i))$ is also a new edge.

On the other hand, G' has no other automorphisms. For let $\alpha \in A(G')$. Then α preserves $V(G)$, as the set of points with degree greater than $|\Gamma|$. Therefore $\alpha|_{v(G)} = \overline{\gamma}|_{V(G)}$ for some $\gamma \in \Gamma$. Let $y \in \Omega$, then $y = \delta(y_i)$ for some $\delta \in \Gamma$ and $1 \leq i \leq k$. Since α is an isomorphism, $\alpha(y)$ is adjacent to $\alpha(\overline{\delta}(x_i)) = (\overline{\delta\gamma})(x_i)$. But the only point of Ω adjacent to $(\overline{\delta\gamma})(x_i)$ is $(\delta\gamma)(y_i)$. So $\alpha(y) = (\delta\gamma)(y_i) = \gamma(y)$. Thus we conclude that $\alpha = \gamma'$.

This proves that G' has the desired properties [I. Z. Bouwer, *J. Comb. Th.* **6** (1969) 378–386; Z. Hedrlin–A. Pultr, *Illinois J. Math.* **10** (1966) 392–405.]

22. The assertion of the hint easily follows from these two observations:

(a) If η is any endomorphism of the Petersen graph and C is a pentagon in it, then η is one-to-one on C. For the image of C contains an odd circuit (it cannot be 2-colorable as this would yield a 2-coloration of C), and the shortest odd circuit in the Petersen graph is pentagon. Thus $\eta(C)$ contains a pentagon and so η is one-to-one on C.

(b) Any two points of the Petersen graph are contained in a pentagon.

Therefore, no endomorphism can identify two distinct points. Thus, the endomorphism semigroup is $\cong S_5$ by 12.1.

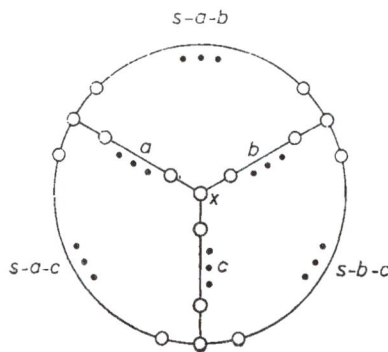

FIGURE 95

23. Consider the graph shown in Fig. 95, where $s > a+b+c$ is odd. This graph G has three circuit of length s and one circuit of length $3s-2(a+b+c) > s$. Therefore, any endomorphism η of G must be one-to-one on the s-circuits. In particular, the three neighbors of x lie pairwise on circuits of length S, thus $\eta(x)$ must have three distinct neighbors, which also lie pairwise on such circuits. Hence $\eta(x) = x$. It follows trivially that if a, b, c are distinct, then η must keep the three "spokes" fixed. Also, the arc of length $s-a-c$ on the "rim" cannot be mapped onto any other arc with the same endpoints, for there are only two other paths with the same parity, namely the paths of length $c+b+(s-a-b)$ and $a+b+(s-b-c)$, but these are longer. Thus $\eta = 1$. [Z. Hedrlin–A. Pultr, *Monatsh. f. Math.* **69** (1965) 318–322.]

24. (a) Let $\eta_1, \ldots, \eta_n$ be the elements of $\sum$. Choose these as points of a colored digraph G. Connect η_i to η_j by a directed edge of color k if $\eta_i = \eta_k \eta_j$. Since we only have a semigroup it may happen that certain pairs are connected by several edges of different colors or by no edge at all.

For any η_m, defining $\bar{\eta}_m(\eta_j) = \eta_j \eta_m$, we obtain an endomorphism of G; for if an edge of color k connects η_i to η_j, i.e. $\eta_i = \eta_k \eta_j$, then $\eta_i \eta_m = \eta_k \eta_j \eta_m$, i.e.

an edge of color k connects $\bar{\eta}_m(\eta_i)$ to $\bar{\eta}_m(\eta_j)$. Moreover, the $\bar{\eta}_m$'s are distinct endomorphisms of G; for if $\bar{\eta}_m(\eta_i) = \bar{\eta}_l(\eta_i)$ for each i, then in particular, $\bar{\eta}_m(1) = \bar{\eta}_l(1)$, i.e. $m = l$.

We claim G has no other endomorphisms. Let $\varepsilon \in \mathrm{End}\,(G)$ and set $\eta_m = \varepsilon(1)$. We claim that $\varepsilon = \bar{\eta}_m$. Let $\eta_i \in V(G)$. Then η_i is connected to 1 by an edge of color i, as $\eta_i = \eta_i \cdot 1$. Therefore, $\varepsilon(\eta_i)$ is connected to $\varepsilon(1) = \eta_m$ by an edge of color i, since ε is an endomorphism. Hence

$$\varepsilon(\eta_i) = \eta_i \eta_m, \text{ i.e. } \varepsilon = \bar{\eta}_m.$$

(b) First we observe that there is an arbitrarily large number of simple graphs, which are rigid and have no homomorphisms into each other. Just take a large enough s and different choices of a, b, c with $a + b + c = s - 1$ in the construction in the solution of 12.20. Let $G_1, \ldots, G_N$ be these rigid graphs, where $N = 2|\sum|$. Each point of each G_i is contained in an odd circuit of length s, which is shortest in each of them.

Now we replace each edge (x, y) of color k in G (constructed in part(a)) by a chain as shown in Fig. 96, where the two edges incident with a G_i are supposed to be attached to it at the same two points in each case. Loops are replaced similarly, except that x and y coincide.

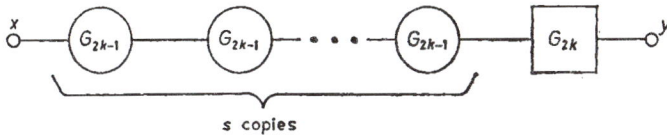

s copies

FIGURE 96

The resulting graph G' clearly has endomorphisms $\tilde{\eta}_i$ corresponding to each endomorphism $\bar{\eta}_i$ of G. We claim it has no other endomorphisms. Since G' has no other s-circuits or shorter circuits than those in the G_i's it follows that any $\varepsilon \in \mathrm{End}(G)$ maps the points of G_i's onto points of G_i's. Since these are rigid and cannot be mapped into each other, it follows that ε maps each G_i into another (or the same) copy of the same graph. The points of G are adjacent to points in different G_i's. Therefore, $\varepsilon(V(G)) \subseteq V(G)$. It is also easy to see that if x and y are connected in G by an edge of color k, i.e. they are connected in G' by a chain of G_{2k-1}'s and G_{2k}'s (see Fig. 92), then so are $\varepsilon(x)$ and $\varepsilon(y)$, and there is a unique way to map the (x, y)-chain onto the $(\varepsilon(x), \varepsilon(y))$-chain. Therefore, $\varepsilon = \tilde{\eta}_i$ for some i. [Z. Hedrlin–A. Pultr, *Monatsh. f. Math.* **68** (1964) 213–217; see also Z. Hedrlin–J. Lambek, *J. of Alg.* **11** (1969) 195–212.]

25. (a) Suppose that End $(G) = \{1, \alpha, \omega\}$, where ω is a 0-element: $\alpha\omega = \omega\alpha = \omega$, 1 is the identity and $\alpha^2 = 1$. Let us remove the edge $e = (x, y)$. If ω does not remain an endomorphism, there exists an edge $(x_1, y_1) \in E(G - e)$ with $\omega(x_1) =$

$x, \omega(y_1) = y$. Similarly, if α does not remain an endomorphism, then there is an edge $(x_2, y_2) \in E(G - e)$ with $\alpha(x_2) = x$, $\alpha(y_2) = y$. But then

$$x_2 = \alpha(x) = \alpha(\omega(x_1)) = \omega(x_1) = x, \quad y_2 = y,$$

which means that (x_2, y_2) is not an edge of $G - (x, y)$, a contradiction.

(b) Consider the construction in the solution of 12.21b, but choose s such that we have one more rigid graph G_{N+1} of girth s having no endomorphism into $G_1, \ldots, G_N$ and conversely. If we add G_{N+1} as a new component to G' we obtain a graph G'' with End $(G'') \cong$ End $(G') \cong \sum$. However, if we connect the point 1 of $V(G)$ to G_{N+1} by an edge, the resulting graph is rigid. [L. Babai–J. Nešetřil.]

§ 13. Hypergraphs

1. Consider G_H as defined in the hint. Obviously, G_H is connected. Moreover, observe that a circuit of H corresponds to a circuit of G_H and conversely. Thus H has no circuits, iff G_H is a tree, which is, in turn, equivalent to

(**) $$|E(G_H)| = |V(G_H)| - 1.$$

Obviously, $|E(G_H)|$ is the sum of degrees in W, i.e.

$$|E(G_H)| = \sum_{E \in E(H)} |E|.$$

On the other hand,

$$|V(G_H)| = |V(H)| + |E(H)|.$$

Thus (**) is the same as

$$\sum_{E \in E(H)} |E| = |V(H)| + |E(H)| - 1,$$

which is the same as (**) [see B].

2. (a) We first verify the assertion of the hint. Suppose that $H' = (H - \{E\}) \setminus E$ is connected. Let F be an edge such that $|E \cap F|$ is maximal. We claim that each point of $E - F$ has degree 1 in H. Let $x \in E \cap F$, $y \in E - F$ (if such x or y does not exist we have nothing to prove), and assume indirectly that $y \in E'$ for some $E' \in E(H)$, $E' \neq E$. By the assumption $F - E$, $E' - E$ are non-empty, thus by the connectivity of H' there exist a sequence $a_0, \ldots, a_k \in V(H) - E$ of points and $E_0 = E'$, $E_1, \ldots, E_{k+1} = F$ of edges such that $a_i \in E_i \cap E_{i+1}$ $(i = 0, \ldots, k)$. If k is as small as possible, we clearly have that $a_i \notin E_j$ unless $j = i$ or $i + 1$. Then the circuit $(y, E_0, a_0, \ldots, E_k, a_k, E_{k+1}, x, E)$ is not balanced, a contradiction.

Now to prove the problem we use induction on $|E(H)|$. We may assume that H is connected, otherwise we may apply the induction hypothesis with one of its components.

Let $x \in V(H)$ and suppose that x is not contained in every edge (if no such point exists the assertion is trivial). Choose an $E \in E(H)$ such that $x \notin E$ and the connected component H_1 of $(h - \{E\}) \setminus E$ containing x is maximal. If $(H - \{E\}) \setminus E$ is connected, we already know the assertion is true; so suppose that there are edges of H, which do not meet $V(H_1)$ other than E and let H_2 be the hypergraph formed by all edges F not meeting $V(H_1)$. Clearly $E \in E(H_2)$ and $|E(H_2)| \geq 2$. Let T be the set of those points in E which belong to edges not in H_2. We claim that every $F \in E(H_2)$ contains T. Suppose that there are $F \in E(H_2)$ and $t \in T - F$. Then consider $(H - \{F\}) \setminus F$. In this hypergraph the component containing x clearly contains $V(H_1)$ and also t, a contradiction.

Thus $\bigcap_{F \in E(H_2)} F \supseteq T$. Clearly H_2 is totally balanced (each partial hypergraph of a totally balanced one is totally balanced) and so, there are $E_1, F_1 \in E(H_2)$ such that all points of $E_1 - F_1$ have degree 1 in H_2. Since the edges of H_1 meet in T only, these points have degree 1 in H.

(b) We use induction on $|E(H)|$. Let E_0, F_0 be two edges such that each point of $E_0 - F_0$ has degree 1. Consider the hypergraph H' arising from H by the removal of E_0 and the points of $E_0 - F_0$.

Obviously, this is totally balanced and satisfies $|E_1 \cap E_2| \leq p$ for any two E_1, $E_2 \in E(H')$. So we have, by induction,

$$\sum_{E \in E(H')} (|E| - p) \leq |V(H')| - p.$$

Now observe that $|E_0 \cap F_0| \leq p$ by the assumption. So $|E_0 - F_0| \geq |E_0| - p$ and so,

$$\sum_{E \in E(H)} (|E| - p) = |V(H')| - p + |E_0| - p \leq$$

$$\leq |V(H')| - p + |E_0 - F_0| = |V(H)| - p.$$

3. Consider the points of P to be linearly ordered. Let $C = (x_1, P_1, \ldots, x_k, P_k)$ be a circuit in H, $k \geq 3$. We may assume that x_1 is neither the least nor the largest point in C, i.e. $x_i < x_1$, $x_j > x_1$ for some i, j. We may suppose that $i < j$; then there is an $i \leq \mu < j$ such that

$$x_\mu < x_1 < x_{\mu+1}.$$

Now P_μ contains x_μ and $x_{\mu+1}$ and since it is a path, it contains x_1. Thus C is balanced.

4. If H has no circuit of length at least 3, then it is totally balanced and thus it satisfies $(*)$ in 13.2b with $p = 2$. But $(*)$ simplifies to

$$|E(H)| \leq |V(H)| - 2,$$

a contradiction. [L. Lovász, *Beiträge zur Graphentheorie*, Teubner (1968) 99–106.]

5. (a) The given condition is obviously necessary. Suppose that it is satisfied, we show H has a system of distinct representatives.

Define a bipartite graph G on the set of points $V(H) \cup E(H)$, by joining $v \in V(H)$ to $E \in E(H)$ iff $v \in E$. Then, for any $X \subseteq E(H)$,

$$|\Gamma(X)| = |\bigcup_{E \in X} E| \geq |X|$$

by the assumption, so by 7.4, G has a matching, which covers all points in $E(H)$. The endpoints of edges of this matching in $V(H)$ form a system of distinct representatives of H [P. Hall; see B, Mi].

(b) Again it is trivial that the given condition is necessary. We show it is sufficient. Form the bipartite graph G as above. Then the condition means that

(1) $$|\Gamma(X)| \geq |X| + 1$$

holds for $\emptyset \neq X \subseteq E(H)$. By 7.6, there is a subgraph G' of G on the same set of points such that each point of $E(H)$ has degree 2 and G' also satisfies (1). For each $E \in E(H)$, let $f(E)$ be the set of points in $V(H)$ adjacent to E in G'. Then $|f(E)| = 2$, and (1) implies

$$|\bigcup_{E \in X} f(E)| \geq |X| + 1$$

for every $\emptyset \neq X \subseteq E(H)$. Hence the pairs $f(E)$ form the edges of a forest on $V(H)$. [L. Lovász, *Acta Math. Acad. Sci. Hung.* **21** (1970) 443–446.]

6. *First solution.* Suppose that there exists a function $f : E(H) \to V(H)$ such that $f(E) \in (E)$, f is one-to-one and $f(e(H)) \supseteq T$. Then, setting $h = f^{-1}|_T$, we can observe that

(a) H has a system of distinct representatives,

(b) h defines a system of distinct representatives of the hypergraph H_T^*, defined by $V(H_T^*) = E(H)$, $E(H_T^*) = \{U_x : x \in t\}$, where $U_x = \{E \in E(H) : x \in E\}$.

Conversely, suppose that g and h are mappings defining systems of distinct representatives of H and H_T^*, respectively. We claim H has a system of distinct representatives containing T.

For each $x \in T$, $x \notin R(g)$ consider the sequence x, $h(x)$ $g(h(x))$, $h(g(h(x)))$, ... until we get stuck; this clearly happens when $g(h \ldots g(h(x))) \in T$. Let H_1 be the set of edges in these sequences. Define

$$f(E) = \begin{cases} h^{-1}(E) & \text{for } E \in H_1, \\ g(E) & \text{for } E \in E(H) - H_1. \end{cases}$$

Then f defines a system of distinct representatives. Clearly, $f(E) \in E$. If $f(E) = F(E')$, then we must have $E \in H_1$, $E' \notin H_1$ (or conversely) and $g(E') = h^{-1}(E)$; but this means that $g(E')$ and E' are the elements preceding E in the chain containing it, i.e. $E' \in H_1$, a contradiction. Finally, every $x \in T$ belongs to $R(f)$.

For if x belongs to one of the chains, then $x = f(h(x))$; if x does not belong to the chains, then $x \in R(g)$ and $g^{-1}(x) \in H_1$, so $x = f(g^{-1}(x))$.

Thus (a) and (b) together are a necessary and sufficient condition for the existence of a system of distinct representatives containing T. By the preceding problem, they can be rewritten in the form

(a') $|V(H')| \geq |E(H')|$ for every partial hypergraph H';

(b') every $T' \subseteq T$ meets at least $|T'|$ edges [This is a version of Berstein's theorem on equicardinality of sets; see B, Mi].

Second solution. Set $F = V(H) - T$, and add F as a new edge $|V(H)| - |E(H)|$ times to H. Then it is easy to verify that the resulting hypergraph H' has a system of distinct representatives iff H has one containing T. Applying the Hall condition (preceding problem) to H' we obtain a condition, which is equivalent to (a'), (b') above. Details are left to the reader.

7. Take two new points a, b. Define a digraph G on the set $\{a\} \cup \{b\} \cup E(H_2) \cup E(H_1) \cup V(H_1)$ as follows: join a to all elements of $E(H_1)$; join each $E \in E(H_1)$ to each $x \in E$; join each $x \in V(H_1) = V(H_2)$ to all $F \in E(H_2)$ containing x; finally, join each element of $E(H_2)$ to b.

Observe that H_1 and H_2 have a common system of distinct representatives iff G has m independent (a, b)-paths. By Menger's theorem, this is equivalent to the property that every set of points of G separating a and b has at least m points.

Now let $X \cup Y \cup Z$ separate a and b, $X \subseteq E(H_1)$, $Y \subseteq V(H_1)$, $Z \subseteq E(H_2)$. Let H_1', H_2' be the hypergraphs formed by the edges in $E(H_1) - X$ and $E(H_2) - Z$, respectively. Obviously, Y must contain $V(H_1') \cap V(H_2')$ and does not have to contain any other point. So the fact that every set of points separating a and b has cardinality at least k is equivalent to saying that

$$|E(H_1) - E(H_1')| + |E(H_2) - E(H_2')| + |V(H_1') \cap V(H_2')| \geq m$$

for every two partial hypergraphs H_1', H_2' of H_1 and H_2, respectively. This is the same as the formula given in the problem [R. Rado, B, Mi].

8. (a) For a fixed $E_0 \in E(H)$, the sets

$$E_0 \triangle F (F \in E(H))$$

are distinct, as $E_0 \triangle F_1 = E_0 \triangle F_2$ implies that
$$F_1 = (E_0 \triangle F_1) \triangle E_0 = (E_0 \triangle F_2) \triangle E_0 = F_2.$$

(b) Let H_1 be the hypergraph obtained from $H \setminus x$ by removing one of each double edge. Obviously, $H \setminus x$ has $|E(H')|$ double edges, where H' is defined as in the hint. So setting

$$m_1 = |E(H_1)|, m_2 = |E(H')|,$$

we have

$$m_1 + m_2 = m.$$

By induction on $|V(H)|$, we find m_1 pairs $\tilde{E}_1$, $\tilde{F}_1;\ldots;\tilde{E}_{m_1}$, $\tilde{F}_{m_1}$ such that $\tilde{E}_i$, $\tilde{F}_i \in H_1$ and $\tilde{E}_1 - \tilde{F}_1,\ldots,\tilde{E}_{m_1} - \tilde{F}_{m_1}$ are distinct. Similarly, we find m_2 pairs $E_1^*,F_1^*;\ldots;E_{m_2}^*4$, $F_{m_2}^*$ such that E_i^*, $F_i^* \in H'$ and $E_1^* - F_1^*,\ldots,E_{m_2}^* - F_{m_2}^*$ are distinct.

There are sets E_i, $F_i \in H(1 \le i \le m_1)$ such that $\tilde{E}_i = E_i$ or $\tilde{E}_i = E_i - \{x\}$, $\tilde{F}_i = F_i$ or $\tilde{F}_i = F_i - \{x\}$. Then, obviously, $E_i - F_i$ $(i=1,\ldots,m_1)$ are distinct.

For each $E_j^*(1 \le j \le m_2)$, we will denote by E_j' one of E_j^* and $E_j^* \cup \{x\}$ (both are edges of H), by the following rule. If $E_j^* - F_j^*$ does not occur among $E_i - F_i$ $(i=1,\ldots,m_1)$, then let $E_j' = E_j^*$. If $E_j^* - F_j^* = E_{i_0} - F_{i_0}$, then let $E_j' = E_j^* \cup \{x\}$. Then $E_j^* - F_j^*$ will be different from the sets $E_i - F_i$; for if $E_j' - F_j^* = E_i - F_i$, then $E_i - F_i = E_{i_0} - F_{i_0}$ which is impossible. Thus, if we consider the differences $E_1 - F_1,\ldots,E_{m_1} - F_{m_1}$, $E_1' - F_1^*,\ldots,E_{m_2}' - F_{m_2}^*$, they are distinct and their number is $m_1 + m_2 = m$. [J. Marica–J. Schönheim, *Canad. Math. Bull.* **12** (1969) 635–637.]

(c) If $m = 2$ the assertion is obviously true. Suppose that $m \ge 3$ and use induction on m.

Let $x \in V(H)$; we may assume that x belongs to at least one edge of E_0, but not to all. Suppose first that x has degree 1. Then $H_1 = H - x$ has $m-1$ edges, so there are at least $m-1$ edges representable in one of the forms $E \cap F$, $E \cup F$ (E, $F \in E(H_1)$, $E \ne F$). None of these sets contains x. Thus, the set $E_0 \cup F$ yields an mth set of the desired form for an arbitrary F.

Secondly, let the degree of x be between 2 and $m-2$ (inclusively). Then consider

$$H_1 = H - x, \quad H_2 = H - E(H_1).$$

They both have at least two edges, so, by the induction hypothesis, there are at least $|E(H_i)|$ sets representable in the form $E \cup F$ or $E \cap F$, $E \ne F \in E(H_i)$, ($i = 1,2$). Moreover, $x \notin E \cap F$, $x \notin E \cup F$, if E, $F \in E(H_1)$ but $x \in E \cap F$, $E \cup F$, if E, $F \in E(H_2)$. Thus we have at least $|E(H_1)| + |E(H_2)| = m$ sets of the required form altogether.

Finally, if x belongs to all edges but E_1, then set $H_2 = H - \{E_1\}$. H_2 has at least two edges, hence there are at least $|E(H_2)| = m - 1$ sets of the form $E \cap F$ or $E \cup F$, $E \ne F \in E(H_2)$. All these sets contain x, thus $E_1 \cap F$ yields an mth for any $F \in E(H_2)$. [D. E. Daykin–L. Lovász, *J. London Math. Soc.* **12** (1976) 225–230.]

9. If H_1, H_2 are defined as in the hint, then (by induction on $|V(H)|$) we may assume that there are permutations σ_i of $E(H_i)$ $(i=1,2)$ such that $E \cap \sigma_i(E) = \emptyset$ for each $E \in E(H_i)$.

Now define σ as follows: Let $E \in E(H)$. If $x \notin E$, then let

$$\sigma(E) = \begin{cases} \sigma_1(E) & \text{if } \sigma_1(E) \cup \{x\} \notin E(H), \\ \sigma_1(E) \cup \{x\} & \text{if } \sigma_1(E) \cup \{x\} \in E(H). \end{cases}$$

If $x \in E$, then let

$$\sigma(E) = \sigma_2(E - \{x\}).$$

It is easy to check that $E \cap \sigma(E) = \emptyset$ for each E. We show that σ is a permutation. Let $E \in H$. If $x \in E$, then $E = \sigma(\sigma_1^{-1}(E - \{x\}))$. If $x \notin E$ but $E \cup \{x\} \in E(H)$, then $E = \sigma(\sigma_1^{-1}(E))$. Finally, if $E \cup \{x\} \notin E(H)$, then $E = \sigma(\sigma_2^{-1}(E))$. So each edge of H is the image of some edge under σ, i.e. σ is a permutation. [P. Erdős–J. Herczog–J. Schönheim, *Israel J. Math.* **8** (1970), 408–412.]

10. (a) *First solution.* We may assume that $n \geq 2$. Then there is a point v contained in some edge but not in all. $X = \{v\}$ satisfies the requirement of the assertion in the hint for $k = 1$.

Suppose that we have a set X' such that $|X'| = k - 1$ and $H_{X'}$ has at least k distinct edges. If $H_{X'}$ has more than k distinct edges, then so does $H_{X' \cup \{y\}}$ for any $y \in V(H') - X'$, and we can set $X = X' \cup \{y\}$. Thus we may assume that $H_{X'}$ has exactly k distinct edges. Since $k < n$, $H_{X'}$ has two distinct edges E, F with $E \cap X' = F \cap X'$. Let $y \in E \triangle F$, and $X = X' \cup \{y\}$. Obviously, H_X has more distinct edges that $H_{X'}$, as $E \cap X \neq F \cap X$. Thus, we have (for $k = n - 1$) a set $X = V(H) - \{x\}$ such that $H_X = H \setminus x$ has n distinct edges.

Second solution. Let x be any point. If $H \setminus x$ has fewer distinct edges than H, then there must be an $E_x \in E(H)$ such that $x \in E_x$ and $F_x = E_x - \{x\} \in E(H)$. Let us fix one such pair (E_x, F_x) of edges for each point x (if, for some x, no such pair exists we are done).

Define a graph G on $V(G) = E(H)$ by joining E to F iff $E = E_x$, $F = F_x$ for some x. This simple graph G has n points and n edges, so it contains a circuit. Let $(E_1, \ldots, E_k)$ form a circuit in $G (E_i \in E(H), k \geq 3)$ and let $E_1 = E_x$, $E_k = F_x$ (say). Then $x \in E_1$, $x \notin E_k$ and thus, there is an index j such that $x \in E_j$, $x \notin E_{j+1}$. But E_j and E_{j+1} are adjacent, so $\{E_j, E_{j+1}\} = \{E_y, F_y\}$ for some y. Since E_y and F_y differ at the element y only, we must have $y = x$ and $E_j = E_y - E_x$, $E_{j+1} = F_y = F_x$. This is, however, a contradiction as $(E_j, E_{j+1}) \neq (E_1, E_k)$. [J. A. Bondy, *J. Comb. Theory* B **12** (1972) 201–202.]

(b) Suppose indirectly that $H \setminus x$ contains at most $m - 2$ distinct edges. Construct a graph G on $E(H)$ as follows. For every $x \in V(H)$, select two pairs $E_x \supseteq F_x$ and $E'_x \supseteq F'_x$ of edges such that $E_x - F_x = E'_x - F'_x = \{x\}$. Such pairs exist otherwise $H \setminus x$ has at least $m - 1$ distinct edges. Now connect the pairs (E_x, F_x) and also the pairs $E'_x, F'_x)$ in G. Thus

$$|V(G)| = m, \quad |E(G)| = 2n.$$

1° G is bipartite. In fact

$$V_1 = \{ E \in E(H) : |E| \text{ is odd}\},$$
$$V_2 = \{ E \in E(H) : |E| \text{ is even}\}$$

defines a 2-coloration of it.

2° No two points of G are connected by three independent paths. Suppose indirectly that E, E' are two points of G connected by three independent paths P_1, P_2, P_3. Let $x \in E - E'$ (say). Then, on each path P_i we find a last point E_i containing x. So if F_i is the point following E_i on P_i, then $x \in E_i - F_i$. Since E_i,

F_i are adjacent they differ at one point only, i.e. $(E_i, F_i) = (E_x, F_x)$ or (E'_x, F'_x). This cannot hold for all three values of i. This proves $2°$.

Now we estimate the number of edges in G. It follows from $2°$ that each block in G is a single edge or a single circuit. Let G have c components, a cutting edges and b other blocks, of sizes $k_1, \ldots, k_b$. By $1°$, $k_i \geq 4$. Also

$$m = |V(G)| = c + a + \sum_{i=1}^{b}(k_i - 1),$$

$$2n = |E(G)| = a + \sum_{i=1}^{b} k_i.$$

Thus

$$m \geq 1 + a + \sum_{i=1}^{b}(k_i - 1) \geq 1 + \frac{3}{4}a + \sum_{i=1}^{b}\left(\frac{3}{4}k_i\right) = 1 + \frac{3}{4}(2n) = 1 + \frac{3}{2}n,$$

a contradiction [B. Bollobás].

(c) We use induction on n. Let H_1, H_2 be as in the hint. If

$$|E(H_1)| > \sum_{i=0}^{k-1}\binom{n-1}{i}$$

then, by the induction hypothesis, it has a set X of n points such that each subset of X can be written as $X \cap E$, $E \in E(H_1)$. Then, obviously, the same set X will be good for H as well. Thus we may assume that

$$|E(H_1)| \leq \sum_{i=1}^{k-1}\binom{n-1}{i}.$$

Then

$$|E(H_2)| = |E(H) - E(H_1)| > \sum_{i=0}^{k-1}\binom{n}{i} - \sum_{i=0}^{k-1}\binom{n-1}{i} = \sum_{i=0}^{k-2}\binom{n-1}{i}$$

and so, by the induction hypothesis, H_2 contains a set X' of $k-1$ elements such that each subset of X' can be written in the form $X' \cap E$, $E \in E(H_2)$. Now set $X = X' \cup \{x\}$. If $Y \subseteq X'$, then $Y = X' \cap E = X \cap E$ for some $E \in E(H_2) \subseteq E(H)$. If $Y \subseteq X$, $x \in Y$, then we know $Y - \{x\} = X' \cap E$ for some $E \in E(H_2)$. Now by definition of H_2, $E = F - \{x\}$ for some $F \in E(H)$, $x \in F$. Then $X \cap F = Y$. Thus we have shown that every subset Y of X can be represented in the form $X \cap E$, $E \in E(H)$. [N. Sauer, *J. Comb. Theory* **133** (1972) 145-147; the largest k for which the hypergraph H contains a set X of points such that every subset of X occurs as $X \cap A$, $A \in E(H)$, is called the *Vapnik–Cervonenkis dimension* of the hypergraph and plays an important role in statistics, the theory of learning, and computational geometry.]

11. We use induction on n. Let $x \in V$ and

$$\mathscr{A}' = \{A \in E(H) : x \in A\}, \quad \mathscr{A}'' = \{A \in E(H) : x \notin A\},$$

$$\mathscr{B}' = \{B \in E(K) : x \in B\}, \quad \mathscr{B}'' = \{B \in E(K) : x \notin B\}.$$

Then clearly

$$E(H) = \mathscr{A}' \cup \mathscr{A}'', \quad E(K) = \mathscr{B}' \cup \mathscr{B}'', \quad E(H) \cap E(K) = (\mathscr{A}' \cap \mathscr{B}') \cup (\mathscr{A}'' \cap \mathscr{B}'').$$

By the induction hypothesis,

$$|\mathscr{A}' \cap \mathscr{B}'| \le \frac{1}{2^{n-1}} |\mathscr{A}'| \cdot |\mathscr{B}'|$$

and omitting x from all members of $\mathscr{A}''$ and $\mathscr{B}''$, we similarly get

$$|\mathscr{A}'' \cap \mathscr{B}''| \le \frac{1}{2^{n-1}} |\mathscr{A}''| \cdot |\mathscr{B}''|.$$

So we know that

$$|E(H) \cap E(K)| \le \frac{1}{2^{n-1}} (|\mathscr{A}'| \cdot |\mathscr{B}'| + |\mathscr{A}''| \cdot |\mathscr{B}''|)$$

and it suffices to prove that

$$|\mathscr{A}'| \cdot |\mathscr{B}'| + |\mathscr{A}''| \cdot |\mathscr{B}''| \le \frac{1}{2}(|\mathscr{A}'| + |\mathscr{A}''|)(|\mathscr{B}'| + |\mathscr{B}''|).$$

This can be re-written as

$$(*) \qquad (|\mathscr{A}'| - |\mathscr{A}''|)(|\mathscr{B}'| - |\mathscr{B}''|) \le 0.$$

Observe that, if $A \in \mathscr{A}''$, then $A - \{x\} \in \mathscr{A}'$, thus $|\mathscr{A}'| \ge |\mathscr{A}''|$. Analogous reasoning gives $|\mathscr{B}'| \le |\mathscr{B}''|$. This proves $(*)$. [D. Kleitman, *J. Comb. Theory* **1** (1966) 153–155.]

12. (a) Let $f(X)$ denote the number of edges contained in X or disjoint from X, for $X \subseteq V(H)$. Set $|V(H)| = n$. Then

$$\sum_{X \subseteq V(H)} f(X) = \frac{|E(H)|}{2^{r-1}},$$

because each given edge E is counted 2^{n-r+1} times (there are 2^{n-r} sets X disjoint from E and 2^{n-r} sets X containing E). Thus,

$$\frac{1}{2^n} \sum_{X \subseteq V(H)} f(X) = |E(H)| 2^{n-r+1}.$$

The left-hand side being the average of $f(X)$, there must be a set X below average, i.e. such that

$$f(X) \le \frac{|E(H)|}{2^{r-1}}$$

as stated.

(b) Let $\varphi: V(H) \to \{1, \ldots, r\}$ be an r-coloration of $V(H)$ and denote by $\|\varphi\|$ the number of those edges E all of whose points are differently colored by φ. Then

$$\sum_{\varphi} \|\varphi\| = r! r^{n-r} |E(H)|,$$

because there are $r!$ ways to color a given edge E and r^{n-r} ways to color the rest (n is again $|V(H)|$). Thus the average of $\|\varphi\|$ is

$$\frac{1}{r^n} \sum_{\varphi} \|\varphi\| = \frac{r!}{R^r} |E(H)|.$$

Since there is a φ such that $\|\varphi\|$ is at least equal to the average, this proves the assertion.

13. Let $d(x)$ denote the degree of x in H. Then, for any $E \in E(H)$,

$$\sum_{x \in E} d(x) = \sum_{F \in E(H)} |E \cap F| = r + \sum_{F \in E(H)-\{E\}} |E \cap F| \le r + (m-1)k.$$

Sum this over all edges E, then

$$\sum_{E \in E(H)} \sum_{x \in E} d(x) = \sum_{x \in V(H)} d(x)^2 = |V(H)| \frac{\sum_{x \in V(H)} d(x)^2}{|V(H)|} \ge$$

$$\ge |V(H)| \left(\frac{\sum_{x \in V(H)} d(x)}{|V(H)|} \right)^2 = \frac{\left(\sum_{E \in E(H)} |E| \right)^2}{|V(H)|} = \frac{m^2 r^2}{|V(H)|}.$$

Hence

$$\frac{m^2 r^2}{|V(H)|} \le m(r + (m-1)k)$$

or, equivalently,

$$|V(H)| \ge \frac{m r^2}{r + (m-1)k}.$$

[K. Corrádi, *Problem at the Schweitzer Competition, Mat. Lapok* **20** (1969) 159–162.]

14. Supposing $m > n$, we have $m - d(x) > n - d(x) \ge n - |E|$ for any pair x, E such that $x \notin E$, and hence

$$\frac{d(x)}{m - d(x)} < \frac{|E|}{n - |E|}$$

(because $|E| \ne 0, n$). Summing for every pair $x \notin E$, we obtain

$$\sum_{x} (m - d(x)) \frac{d(x)}{m - d(x)} < \sum_{E} (n - |E|) \frac{|E|}{n - |E|}$$

or, equivalently,

$$\sum_x d(x) < \sum_E |E|,$$

which is not the case, because here we have equality. The contradiction proves that $m \leq n$ [cf. P. Crawley–R. P. Dilworth, *Algebraic Theory of Lattices*, Prentice-Hall, 1973, proof of 14.2].

15. (a) If $\emptyset \in E(H)$, then clearly $|E(H)| = 1$ and we are done. So suppose that $\emptyset \notin E(H)$. Clearly $V(H) \notin E(H)$. Let $x \in v(H)$, $E \in E(H)$, $x \notin E$. By the preceding problem, it suffices to prove that

$$d(x) \leq |E|.$$

Let $E_1, \ldots, E_d$ be those edges containing x. Then $E_1, \ldots, E_d$ meet E by the assumption and their intersections with E are disjoint since they cannot have any common element other than x. Thus, $|E| \geq d = d(x)$ as stated.

(b) Let $V(H) = \{x_1, \ldots, x_n\}$, $E(H) = \{E_1, \ldots, E_m\}$,

$$a_{ij} = \begin{cases} 1 & \text{if } x_i \in E_j, \\ 0 & \text{otherwise}, \end{cases}$$

$$a_j = \begin{pmatrix} a_{1j} \\ a_{2j} \\ \vdots \\ a_{nj} \end{pmatrix}.$$

We claim that $a_1, \ldots, a_m$ are linearly independent. This clearly also proves that $m \leq n$. Suppose that

(1)
$$\sum_{j=1}^m \eta_j \mathbf{a}_j = \mathbf{0};$$

we show every η_j is 0. Multiplying (1) by $\mathbf{a}_k$, we get

$$\sum_{j=1}^m \eta_j \mathbf{a}_j \mathbf{a}_k = 0.$$

Since by assumption $\mathbf{a}_i \mathbf{a}_k = \lambda$ for $i \neq k$, this yields

$$\lambda \sum_{j=1}^m \eta_j + \eta_k (\mathbf{a}_k^2 - \lambda) = 0.$$

If $\mathbf{a}_k^2 = |E_k| = \lambda$, then E_k is contained in every other edge, which is excluded except when $E_k = \emptyset$, in which case $m = 1 \leq n$. So suppose that $\mathbf{a}_k^2 > \lambda$ for every k. Then

(2)
$$\eta_k = \frac{-\lambda}{\mathbf{a}_k^2 - \lambda} \sum_{j=1}^m \eta_j$$

and so

$$\sum_{k=1}^{m} \eta_k = \left(\sum_{j=1}^{m} \eta_j \right) (-\lambda) \sum_{k=1}^{m} \frac{1}{a_k^2 - \lambda}$$

or, equivalently,

$$\left(\sum_{j=1}^{m} \eta_j \right) \left(1 + \lambda \sum_{k=1}^{m} \frac{1}{a_k^2 - \lambda} \right) = 0.$$

Since the second factor is strictly positive we derive

$$\sum_{j=1}^{m} \eta_j = 0$$

and hence by (2), $\eta_k = 0$ as stated [This is a generalized version of Fisher's inequality; see Hall].

16. Let $x \in V(H)$, and let a and b denote the degrees of x in H and K, respectively. Then x is counted in

(1) $$a(m' - b) + b(m - a)$$

terms of the left-hand side and

(2) $$a(m - a) + b(m' - b)$$

terms of the right-hand side. Their difference is

$$a(m' - b) + b(m - a) - a(m - a) - b(m' - b) =$$
$$= (a - b)^2 - (m' - m)(a - b) \geq 0,$$

since $a - b$ is an integer and $m' - m = 0$ or 1. [J. B. Kelly, *Combin. Structures and Appl.* Proc. Calgary Conference, Gordon and Breach, 1970, 201–207.]

17. (a) As in the hint, assume that $x \in F_1, \ldots, F_d$, $x \notin F_{d+1}, \ldots, F_m$. Define two hypergraphs H, K on $V = \bigcup_{i=1}^{m} F_i$, by letting $E(H)$ and $E(K)$ consist of $m - d$ copies of $F_1 - \{x\}, \ldots, F_d - \{x\}$ and d copies of $F_{d+1}, \ldots, F_m$, respectively. Then $|E(H)| = |E(K)| = d(m - d)$. Also we have

$$\sum_{A \in E(H)} \sum_{B \in E(K)} |A \triangle B| = (2k - 1)d^2(m - d)^2,$$

$$\sum_{\{A, A'\} \subseteq E(H)} |A \triangle A'| = 2k(m - d)^2 \binom{d}{2},$$

$$\sum_{\{B, B'\} \subseteq E(K)} |B \triangle B'| = 2kd^2 \binom{m - d}{2}.$$

Thus, by the inequality in 13.16,

$$(2k-1)d^2(m-d)^2 \geq 2k\left[(m-d)^2\binom{d}{2} + d^2\binom{m-d}{2}\right],$$

whence the inequality of the problem follows after cancellations.

(b) Replacing F_i by $F_i \triangle F_m$ the conditions remain unchanged. Thus we may suppose that $|F_1| = \ldots = |F_{m-1}| = 2k$, $F_m = \emptyset$. Then, if d is the degree of any point, then by (a)

$$d(m-d) \leq km.$$

Assume first that $k+2 \leq d \leq m-k-2$. Then

$$km \geq d(m-d) \geq (k+2)(m-k-2),$$

i.e.

$$m \leq \left\lfloor \frac{(k+2)^2}{2} \right\rfloor \leq k^2+k+2,$$

and we are finished. Hence we may suppose that $d \leq k+1$ or $d \geq m-k-1$ holds for each degree d.

Let $x_1, \ldots, x_l$ be those points with degree at least $m-k-1$. Suppose first that $l \geq k+1$. Clearly there is at most one F_j containing all elements of $\{x_1, \ldots, x_{k+1}\}$. Thus, $m-2$ F_j's miss at least one point of $\{x_1, \ldots, x_{k+1}\}$; one misses $k+1$ of them (the empty set F_m) and one may contain all of them. Thus, there are at most $k(m-2)+k+1$ incidences between these points and the sets F_j. On the other hand, the number of such incidences is at least $(m-k-1)(k+1)$. Thus

$$k(m-2)+k+1 \geq (m-k-1)(k+1).$$

Whence

$$m \leq k^2+k+2$$

and we are finished again. So we may assume that there are $l \leq k$ points $x_1, \ldots, x_l$ only with degree at least $m-k-1$. By our previous assumption, the remaining points have degree at most $k+1$.

There is a set F_j, $1 \leq j \leq m-1$ such that F_j contains less than k elements of $\{x_1, \ldots, x_l\}$. Otherwise, we would have $l=k$ and $x_1, \ldots, x_k$ would be common elements of $F_1, \ldots, F_{m-1}$. But, then the sets $F_1 - \{x_1, \ldots, x_k\}, \ldots, F_{m-1} - \{x_1, \ldots, x_k\}$ are disjoint (by $F_i \triangle F_j = 2k$) and so we would have a situation which we have excluded.

Thus we may assume that e.g. F_1 contains $k+1$ (or more) points whose degree are at most $k+1$. Hence the number of incidences between points in F_1 and the sets $F_1, F_2, \ldots, F_m$ is at most $(k-1)(m-1)+(k+1)^2$. On the other hand, this number is exactly $2k+(m-1)k$, since each of $F_2, \ldots, F_{m-1}$ intersects F_1 in exactly k elements. Thus

$$2k+(m-2)k \leq (k-1)(m-1)+(k+1)^2, \quad m \leq k^2+k+2$$

as stated.

Equality holds if k is such that there exists a finite projective plane of order k. Let us add $k-1$ further points and let $F_1, \ldots, F_{k^2+k+1}$ be the lines of the plane together with these $k-1$ points; let $F_{k^2+k+2} = \emptyset$. Then these sets satisfy our conditions. [M. Deza, *J. Comb. Theory* **16** (1974) 166-167.]

18. (a) Arrange all subsets of $V(H)$ into pairs $(X, V(H)-X)$. Then we will have 2^{n-1} pairs and $E(H)$ can contain at most one member of each. Hence the assertion follows.

(b) *First solution.* Let H' be the hypergraph with $V(H') = V(H)$ and $E(H') = \{X : X \subseteq E \in E(H)\}$. Consider a permutation σ of the edges of H' for which $E \cap \sigma(E) = \emptyset (E \in E(H'))$; such a σ exists by 13.9. The edges $\sigma(E)$, $E \in E(H)$ cannot be edges of H since $E \cap \sigma(E) = \emptyset$, while any two edges of H intersect. Thus

$$(1) \qquad\qquad |E(H')| \geq 2|E(H)|.$$

Observe that no two edges of H' cover $V(H)$; for if $E, F \in E(H')$ have $E \cup F = V(H)$, then choose $E_1, F_1 \in E(H)$ so that $E \subseteq E_1$, $F \subseteq F_1$ and then $E_1 \cup F_1 = V(H)$, a contradiction. Hence $E(H')$ contains at most one of X, $V(H) - X$ for any $X \subseteq V(H)$ and, therefore,

$$(2) \qquad\qquad |E(H')| \leq 2^{n-1}.$$

(1) and (2) prove the assertion. [J. Schönheim, *Combinatorics*, London Math. Soc. Lecture Notes Series **13** (1974) 139–140.]

Second solution. Consider the hypergraph H_1 with $V(H_1) = V(H)$, $E(H_1) = \{X : X = E - F, E, F \in E(H)\}$, and set $H_2 = H \cup H_1$. No edge of H_1 is an edge of H, since $E - F = E_0 (E, F, E_0 \in E(H))$ would imply $F \cap E_0 = \emptyset$, which is not allowed. Moreover, $|E(H_1)| \geq |E(H)|$ by 13.8b. Thus

$$(3) \qquad\qquad |E(H_2)| \geq 2|E(H)|.$$

Now no two edges of H_2 cover $V(H)$, because each edge of H_2 is contained in an edge of H. Hence, as before,

$$(4) \qquad\qquad |E(H_2)| \leq 2^{n-1},$$

and (3), (4) imply the assertion of the problem. [D. E. Daykin–L. Lovász, *J. London Math. Soc.* **12** (1976) 225–230.]

Third solution. Let $\mathscr{A} = \{X \subseteq V(H) : X \subseteq E \in E(H)\}$, $\mathscr{B} = \{X \subseteq V(H) : X \supseteq E \in E(H)\}$. Then by (a), $|\mathscr{A}| \leq 2^{n-1}$ and $|\mathscr{B}| \leq 2^{n-1}$. By 13.11,

$$|E(H)| \leq |\mathscr{A} \cap \mathscr{B}| \leq 1/2^n |\mathscr{A}| \, |\mathscr{B}| \leq 2^{n-2}$$

[D. Kleitman].

19. We may assume that each one-element set belongs to $E(H)$, as well as $\emptyset$ and $V(H)$. If $E[\neq V(H)$ or $\emptyset]$ is any edge of H, then there is a unique minimal edge E_1 properly containing E; for if E_1, E_2 were two such edges, then we would have $E_1 \subset E_2$, $E_2 \subset E_1$ or $E_1 \cap E_2 = \emptyset$ by the assumption; however, any of these is

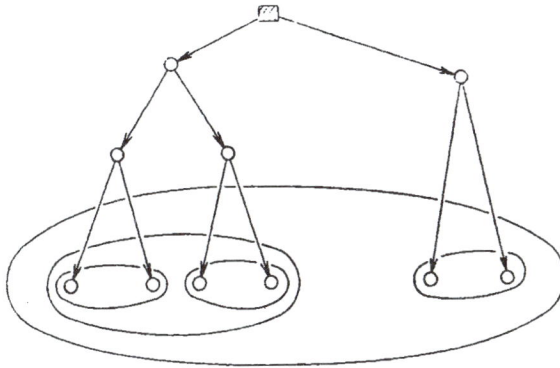

FIGURE 97

impossible. If $E = \emptyset$, then it is contained in n minimal members of H; $E = V(H)$ is contained in none. This way we count $m + n - 2$ edges.

Consider any $E \in E(H)$, $|E| \geq 2$ and consider the maximal edges properly contained in E. They cover E and are disjoint, so their number is at least 2. So E is counted at least twice. If $|E| = 1$, then E is counted once; $E = \emptyset$ is not counted. Hence

$$m + n - 2 \geq 2(m - n - 1) + n,$$

or, equivalently,

$$m \leq 2n.$$

This number can be attained; in fact, if G is any arborescence with outdegrees 2 at all points except n endpoints and, if for each point x of G, we consider the set E_x of endpoints accessible from x, then the hypergraph formed by edges E_x and the empty set has n points and $2n$ edges (Fig. 97).

20. For $n = 1$ the problem is obvious. We use induction on n. Suppose that we have the chains $K_1, \ldots, K_t$ of 2^S, where $|S| = n$ and add an $(n+1)^{\text{st}}$ point x. For each chain $K_i = \{E_r, \ldots, E_{n-r}\}$ we define two new chains

$$K_i' = \{E_r, E_r \cup \{x\}, E_{r+1} \cup \{x\}, \ldots, E_{n-r} \cup \{x\}\}$$

and

$$K_i'' = \{E_{r+1}, \ldots, E_{n-r}\} \text{ (provided } |K_i| \geq 2).$$

Obviously, both K_i' and K_i'' are symmetric chains of subsets of $S \cup \{x\}$. Moreover, any subset X of $S \cup \{x\}$ belongs to exactly one of them. For if $X \subseteq S \cup \{x\}$ and $x \in X$, then let $X - \{x\} \in K_i$; we will have $X \in K_i'$. If $X \subseteq S$, then let $X \in K_i$; if X is the first member of K_i, then $X \in K_i'$; if X is a later member, then $X \in K_i''$. Similarly one sees that no X belongs to more than one K_i' or K_i''.

Thus the chains K_i', K_i'' yield a decomposition of all subsets of $S \cup \{x\}$ into symmetric chains and the induction is complete. [N. G. de Bruijn–C. van E. Tengbergen–D. Kruijswijk, *Nieuw Arch. Wiskunde* **23** (1949–51) 191–193.]

21. (a) *First solution*: Let $K_1, \ldots, K_t$ be the chains of subsets of $V(H)$ defined in 13.20. Since each subset X with $|X| = \lfloor n/2 \rfloor$ belongs to exactly one K_i and conversely, we have $t = \binom{n}{\lfloor n/2 \rfloor}$. Any K_i contains at most one edge of H, as any two edges of K_i are comparable (one contains the other). Hence $|E(H)| \le t$. [E. Sperner, *Math. Z.* **27** (1928) 544–548. This proof is due to de Bruijn, C. A. van E. Tengbergen and D. Kruijswijk, ibid.]

Second solution: Let us consider those permutations $(x_1, \ldots, x_n)$ of $V(H)$ for which $\{x_1, \ldots, x_k\} \in E(H)$ for some k. For any edge E with $|E| = k$, we get $k!(n-k)!$ such permutations by putting the elements of E in the first k places and the other elements in the last $(n-k)$ places. *No permutation is counted more than once*; for if $(x_1, \ldots, x_n)$ is a permutation, then it cannot happen that $E_1 = \{x_1, \ldots, x_{k_1}\}$ a $E_2 = \{x_1, \ldots, x_{k_2}\}$ are both edges since then one of E_1, E_2 would contain the other.

Thus we count $k!(n-k)!$ permutations with any edge E $(k = |E|)$, and, altogether, we count at most $n!$ permutations. Since

$$k!(n-k)! = \frac{n!}{\binom{n}{k}} \ge \frac{n!}{\binom{n}{\lfloor n/2 \rfloor}} = \left\lfloor \frac{n}{2} \right\rfloor! \left\lfloor \frac{n+1}{2} \right\rfloor!,$$

this implies that

$$|E(H)| \cdot \left\lfloor \frac{n}{2} \right\rfloor! \left\lfloor \frac{n+1}{2} \right\rfloor! \le n!,$$

$$|E(H)| \le \frac{n!}{\left\lfloor \frac{n}{2} \right\rfloor! \left\lfloor \frac{n+1}{2} \right\rfloor!} = \binom{n}{\lfloor \frac{n}{2} \rfloor}.$$

[D. Lubbell, *J. Comb. Th.* **11** (1966) 299.]

(b) Let H be a clutter on n points with exactly $\binom{n}{\lfloor n/2 \rfloor}$ edges. Then we must have equivality in all estimations used in the second proof of part (a). In particular, if $E \in E(H)$, $|E| = k$, then

$$k!(n-k)! = \left\lfloor \frac{n}{2} \right\rfloor! \left\lfloor \frac{n+1}{2} \right\rfloor!, \text{i.e.} \frac{n}{\binom{n}{k}} = \frac{n!}{\binom{n}{\lfloor n/2 \rfloor}},$$

whence $k = \lfloor \frac{n}{2} \rfloor$ or $\lfloor \frac{n+1}{2} \rfloor$.

Now there are two cases:

$1°$ $n = 2p$. Then $|E| = p$ for any $E \in E(H)$ and so, $E(H)$ consists of all p-tuples of $V(H)$.

$2°$ $n = 2p+1$. Then $|E| = p$ or $p+1$ for any $E \in E(H)$. We claim that every edge has the same cardinality, i.e. $E(H)$ consists of all p-tuples or of all $(p+1)$-tuples.

If $(x_1, \ldots, x_n)$ is any permutation of the points, then this permutation must be counted in the second solution; i.e. either $\{x_1, \ldots, x_p\}$ or $\{x_1, \ldots, x_{p+1}\}$ is an edge. Thus, if $X \subset Y \subset V(H)$, $|X| = p$, $|Y| = p+1$, then one of X, Y is an edge.

Now suppose that there is an edge $E \in E(H)$ with $|E| = p+1$ and a set $X \subset V(H)$, $|X| = p+1$, which is not an edge (if all $(p+1)$-tuples are edges then, obviously, no p-tuple is an edge, and we have nothing to prove). We can find such a pair E, X with $|E \cap X| = p$; for let $E = \{x_1, \ldots, x_{p+1}\}$, $E \cap X = \{x_i, \ldots, x_{p+1}\}$, $X = \{x_i, \ldots, x_{i+p}\}$. Then consider the sets $\{x_v, \ldots, x_{v+p}\}$ $(\nu = 1, \ldots, i)$. There is a last one among these, which is an edge, say $\{x_v, \ldots, x_{v+p}\}$. Then we can consider $\{x_v, \ldots, x_{v+p}\}$ instead of E and $\{x_{v+1}, \ldots, x_{v+1+p}\}$ instead of X.

Now since $E \cap X \subset E$, $E \cap X \notin E(H)$); on the other hand, $E \cap X \subset X$, $|E \cap X| = p$, $|X| = p+1$ and so by our remark above, $E \cap X \in E(H)$, a contradiction.

Thus $E(H)$ consists of all p-tuples or of all $(p+1)$-tuples of $V(H)$.

22. Suppose that A, B are maximum antichains with m elements and define

$$A \vee B = \{x \in A \cup B : x \not< y \quad \text{for every } y \in A \cup B\},$$
$$A \wedge B = \{x \in A \cup B : x \not> y \quad \text{for every } y \in A \cup B\}.$$

Then $A \vee B$, $A \wedge B$ are antichains by their definition. Moreover

$$(1) \qquad\qquad (A \vee B) \cap (A \wedge B) \supseteq A \cap B.$$

In fact, if $x \in A \cap B$, then there is no $y \in A \cup B$ such that $x < y$ for if $y \in A$ (say), then $x \in A$ also and A is an antichain. Thus, $x \in A \vee B$ and similarly, $x \in A \wedge B$.

$$(2) \qquad\qquad (A \vee B) \cup (A \wedge B) = A \cup B.$$

The inclusion $\subset$ is obvious. Let $x \in A \cup B$, say $x \in A$. If $x \notin A \vee B$, then there is a $y_1 \in A \cup B$ such that $x < y_1$. Obviously, $y_1 \notin A$; so $y_1 \in B$. Similarly, if $x \notin A \wedge B$, then there exists a $y_2 \in B$ with $y_2 < x$. Now $y_2 < y_1$, a contradiction. Thus $x \in (A \vee B) \cup (A \wedge B)$.

From (1) and (2) we deduce that

$$(3) \qquad |A \vee B| + |A \wedge B| = |(A \vee B) \cup (A \wedge B)| +$$
$$+ |(A \vee B) \cap (A \wedge B)| \geq |A \cup B| + |A \cap B| = |A| + |B| = 2m.$$

Since A, B were maximum antichains we have

$$(4) \qquad\qquad |A \vee B| \leq m, \quad |A \wedge B| \leq m.$$

Now from (3) and (4) we see that equality must hold throughout, in particular in (4). This proves that $A \vee B$, $A \wedge B$ are maximum antichains.

It is easy to verify that the set $\mathscr{L}$ of maximum antichains forms a lattice with respect to the operations $\wedge$ and $\vee$. Let E be the unity of $\mathscr{L}$, then E is a maximum antichain which is, obviously, invariant under the automorphisms of $(S, \leq)$.

Now consider $S = 2^{V(H)}$ and the partially ordered set $(S, \subseteq)$. $E(H)$ is an antichain in $(S, \subseteq)$, so we want to determine the maximum size of an antichain. By the above, there is a maximum antichain $\mathscr{E}$, which is invariant under the automorphisms of $(S, \subseteq)$. Let $E \in \mathscr{E}$, $|E| = k$. Any permutation of $V(H)$ induces an automorphism S, and these permutations map E onto all k-element subsets. Thus, all k-element subsets belong to $\mathcal{E}$. Obviously, $\mathcal{E}$ cannot have any other member, thus

$$|\mathcal{E}| = \binom{n}{k} \leq \binom{n}{\lfloor \frac{n}{2} \rfloor}.$$

[R. P. Dilworth, *Combin. Analysis*, Proc. Symp. Appl. Math. AMS (1960) 85.]

23. (a) The hypergraph H with $V(H) = \{1,2,3,4\}$, $E(H) = \{\{1,2,3\}, \{3,4\}, \{1,4\}, \{2\}\}$ is cross-cutting but contains no other cross-cutting hypergraph, in particular, no cross-cutting clutter (Fig. 98).

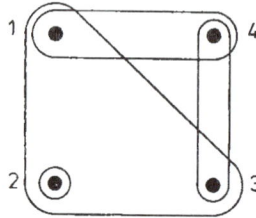

FIGURE 98

(b) Let H be a minimal cross-cutting hypergraph. Then for any $E \in E(H)$ there exists a set $S_E \subseteq V(H)$, which is not comparable with any edge of $H - \{E\}$. Clearly, $E \subseteq S_E$ or $S_E \subseteq E$. Denote by E^* the larger of E, S_E, and let

$$V(H^*) = V(H), E(H^*) = \{E^* : E \in E(H)\}.$$

We claim that H^* is a clutter. Assume indirectly that $E^* \subseteq F^*$. If $E^* = S_E$, $F^* = F$, then $S_E \subseteq F$, which is a contradiction since S_E is comparable only with E by definition. We get a similar contradiction, if $E^* = E$, $F^* = S_F$. If $E^* = E$, $F^* = F$, then $S_E \subseteq E$ and $E \subseteq F$, so $S_E \subseteq F$. If $E^* = S_E$, $F^* = S_F$, then $E \subseteq S_E \subseteq S_F$. We get a contradiction in both cases again.

Thus H^* is a clutter and so by Sperner's theorem

$$|E(H)| = |E(H^*)| \leq \binom{n}{\lfloor n/2 \rfloor}$$

[P. Erdős].

24. Following the hint, let

$$p = \left\lfloor \frac{n - k + 1}{2} \right\rfloor.$$

Then

$$p + k - 1 = \left\lfloor \frac{n + k - 1}{2} \right\rfloor$$

and hence, $p + (p + k - 1) = n$ or $n - 1$. Therefore $\binom{n}{p}, \ldots, \binom{n}{p+k-1}$ are the k largest binomial coefficients (if $n - k$ is even there is another such set: we could replace $\binom{n}{p}$ by $\binom{n}{p-k+2}$ but this is unimportant). Form the hypergraph M on $V(H)$ consisting of all $p, \ldots, (p+k-1)$-tuples as in the hint. Let $K_1, \ldots, K_t$ be a decomposition of $2^{V(H)}$ into disjoint symmetric chains as in 13.20. Obviously,

$$|E(M) \cap K_i| = \min(k, |K_i|);$$

on the other hand, $E(H)$ contains no chain of length k, whence

$$|E(H) \cap K_i| \leq \min(k, |K_i|).$$

Since this holds for every $i = 1, \ldots, t$, we conclude that

$$|E(H)| \leq |E(M)| = \binom{n}{p} + \ldots + \binom{n}{p + k - 1}.$$

[P. Erdős, *Bull. Amer. Math. Soc.* **51** (1945) 898–902.]

25. (a) Suppose, by way of contradiction, that the edges of H cannot be covered by one point. Let $E = \{x_1, \ldots, x_r\} \in E(H)$. Then x_i does not cover all edges, so we find an edge F_i such that $x_i \notin F_i$ $(i = 1, \ldots, r)$. Now $E, F_1, \ldots, F_r$ have no point in common; for such a point ought to be in E, but x_i does not cover F_i.

(b) We construct j edges $E_1, \ldots, E_j$ such that

$$|E_1 \cap \ldots \cap E_j| \leq r - (j - 1)(\tau(H) - 1)$$

for $j = 1, \ldots, k$. For $j = 1$, E_1 can be chosen arbitrarily. Assume that $E_1, \ldots, E_j$ have been selected $(j \leq k - 1)$. Since every other edge intersects $E_1 \cap \ldots \cap E_j$ this set must be of cardinality at least $\tau(H)$. Let $X \subseteq E_1 \cap \ldots \cap E_j$, $|X| = \tau(H) - 1$. Since X cannot cover all edges, there must be an edge E_{j+1} such that $E_{j+1} \cap X = \emptyset$. Thus

$$|E_1 \cap \ldots \cap E_{j+1}| \leq |(E_1 \cap \ldots \cap E_j) - X| =$$
$$= |E_1 \cap \ldots \cap E_j| - (\tau(H) - 1) \leq r - j(\tau(H) - 1),$$

i.e. E_{j+1} has the desired property.

Now $E_1 \cap \ldots \cap E_k \neq \emptyset$, i.e.

$$1 \leq |E_1 \cap \ldots \cap E_k| \leq r - (k - 1)(\tau(H) - 1),$$

whence the inequality of the problem follows immediately.

26. Take any edge F. If $|E \cap F| \geq 2$ for every edge E we can take $S = F$. Suppose that $|F' \cap F| = 1$ for some F', say $F' \cap F = \{x\}$. Let F'' be an edge of $H - x$ (such an edge exists since $\tau(H) \geq 2$), and set $S = F \cup F' \cup F''$. Clearly

$$|S| \leq 3r - 3.$$

On the other hand, let $E \in E(G)$. If $x \notin E$, then E must intersect F and F' in distinct points so $|E \cap S| \geq 2$. If $x \in E$, then E must meet F'' in a point different from x so $|E \cap S| \geq 2$ again.

27. Let $(x_1,\ldots,x_p)$ be any ordering of W, where W is a minimal set such that any two edges of H_W meet ($|W|=p$). Then there is at most one point x_i such that both $\{x_1,\ldots,x_i\}$ and $\{x_i,\ldots,x_p\}$ contain edges E, E' of H_W (clearly $E\cap E' = \{x_i\}$). For if $j\neq i$ and $\{x_1,\ldots,x_j\}$ contained F, $\{x_j,\ldots,x_p\}$ contained $F'(F,F'\in E(H_W))$, then assuming, e.g. $i<j$ we would have $E\cap F'=\emptyset$, a contradiction. So if we carry out the counting for every ordering of W, we count at most $p!$ points.

Now every element of W is counted many times. In fact, if $x\in W$, then there are edges E, $F\in E(H_W)$ such that $E\cap F=\{x\}$, by the minimality of W. Set $|E|=s$, $|F|=t$. Let us arrange the points of E, then put x, then put the points of F; finally "stick in" the remaining points arbitrarily. This can be done in

$$(t-1)!(s-1)!\binom{p}{s+t-1}(p-s-t+1)! = p!\frac{(s-1)!(t-1)!}{(s+t-1)!}$$

ways. Now $s,t\leq r$, hence

$$\frac{(s+t-1)!}{(s-1)!(t-1)!} = s\binom{s+t-1}{t-1} \leq r\binom{2r-1}{r-1}.$$

Also we can exchange the role of E and F. Thus every point in W is counted at least $\frac{2p!}{r\binom{2r-1}{r-1}}$ times. Hence the number of points in W is at most

$$p!/\frac{2p!}{r\binom{2r-1}{r-1}} = \frac{r}{2}\binom{2r-1}{r-1} = (2r-1)\binom{2r-3}{r-1}.$$

The second assertion follows by considering the following hypergraph. Let $|U|=2r-2$. With each partition $\{X,Y\}$ of U such that $|X|=|Y|=r-1$ we associate a new point v_{XY}; let V be the set of these points v_{XY} and $V(H)=U\cup V$. Moreover, let $E(H)$ consist of all r-tuples in U and also of all sets $X\cup\{v_{XY}\}$, $Y\cup\{v_{XY}\}$. It is easy to check that any two edges of H intersect but that no subhypergraph has this property. Thus the minimal W is $V(H)$, and

$$|V(H)| > |V| = \frac{1}{2}\binom{2r-2}{r-1} = \binom{2r-3}{r-1}.$$

[A sharpened version of a theorem of M. Calczynska-Karlowicz, *Bull. Acad. Polon. Sci.* Ser. Math. Astr. Phys. **12** (1964) 87–89.]

28. (a) Let x be any point of C. Then x is the endpoint of at most one A_i; for if A_i, A_j had a common endpoint then, since they are distinct, they ought to start in different directions and, as $|V(C)|\geq 2k$, they would be edge-disjoint.

Now observe that, since each A_j $(j=2,\ldots,t)$ has an edge in common with A_1, one of the endpoints of A_j is an inner point of A_1. Hence there are at most $k-1$ arcs A_j with $2\leq j\leq t$ and this implies that $t\leq k$.

(b) Let $\pi=(x_1,\ldots,x_n)$ be a cyclic permutation of $V(H)$. We consider a circuit C of length n and associate $x_1,\ldots,x_n$ with the edges of C (in the same cyclic order). Then if $E\in E(H)$ consists of consecutive points in π, then it corresponds to an arc A of C of length r. Any two of these arcs have an edge in common. Thus, by the preceding "lemma". there are at most r such arcs, i.e., *for any cyclic permutation π, there are at most r edges, which consist of consecutive point*. Since there are $(n-1)!$ cyclic permutations, this way we count not more than

$$r(n-1)!$$

edges. Let us see, how often a given edge E is counted here. To specify a cyclic permutation for which E consists of consecutive points, we have to order E and $V(H) - E$. Thus E is counted $r!(n-r)!$ times. So the number of edges is

$$m \le \frac{r(n-1)!}{r!(n-r)!} = \binom{n-1}{r-1}.$$

[P. Erdős–Chao Ko–R. Rado, *Quart. J. of Math.* Oxford, II. **12** (1961) 313–320, this proof is due to G. O. H. Katona, *J. Comb. Theory* **13** (1972) 183–184.]

29. (a) Let $F_1,\ldots,F_t$ be all the sets which are unions of ν disjoint edges of H. The fact that H is ν-critical means that

$$F_1 \cap \ldots \cap F_t = \emptyset.$$

By 13.25a, we find $r\nu + 1$ sets among $F_1,\ldots,F_t$, say $F_1,\ldots,F_{r\nu+1}$, whose intersection is void. Let $S = F_1 \cup \ldots \cup F_{r\nu+1}$. We claim that $|S \cap E| \ge 2$ for every edge E of H. In fact, $S \cap E$ meets every one of $F_1,\ldots,F_{r\nu+1}$ (otherwise E would be disjoint from F_i as a $(\nu+1)^{\text{st}}$ one). Since by their construction $F_1,\ldots,F_{r\nu+1}$ cannot be covered by one point it follows that $|S \cap E| \ge 2$. Now $|S| \le (r\nu+1)r\nu$. Thus the number of edges E is at most

$$\binom{|S|}{2}\binom{n}{r-2} \le \binom{(r\nu+1)r\nu}{2}\binom{n}{r-2} = O(n^{r-2}).$$

(b) We use induction on ν. If H is ν-critical, (a) settles the question. If H is not ν-critical, let $x \in V(H)$ be such that $\nu(H-x) < \nu(H)$. Then

$$|E(H-x)| \le \binom{n-2}{r-1} + \ldots + \binom{n-\nu}{r-1}$$

by the induction hypothesis and since x cannot be contained in more than $\binom{n-1}{r-1}$ edges, the assertion follows.

Equality holds for the hypergraph on n points consisting of those r-tuples, which meet a given ν-element set. [A. Hajnal–B. Rothschild, *J. Comb. Theory* **15** (1973) 359–362.]

30. Let $\{t_x\}$ $(x \in V(H))$ be an optimum fractional cover of H. Then

$$\sum_{x \in E} t_x \geq 1$$

holds for every edge E. Summing this for every edge of $H_i = H - x_1 - \ldots - x_i$, we obtain

$$\sum_{E \in E(H_i)} \sum_{x \in E} t_x \geq |E(H_i)|.$$

The left-hand side can be written as

$$\sum_x t_x \sum_{\substack{E \ni x \\ E \in E(H_i)}} 1 \leq (\sum_x t_x) d_{i+1} = \tau^* d_{i+1}.$$

So

(1) $$|E(H_i)| \leq \tau^* d_{i+1}.$$

Now suppose that the procedure stops after t steps, i.e. $x_1, \ldots, x_t$ cover all edges, $E(H_t) = \emptyset$. Then clearly

$$\tau(H) \leq t.$$

We know that

$$|E(H_i)| - |E(H_{i+1})| = d_{i+1},$$

so

$$t = \sum_{i=0}^{t-1} \frac{|E(H_i)| - |E(H_{i+1})|}{d_{i+1}} = \sum_{i=1}^{t-1} |E(H_i)| \left(\frac{1}{d_{i+1}} - \frac{1}{d_i}\right) + \frac{|E(H)|}{d}.$$

Since $d_i \geq d_{i+1}$ we can estimate $|E(H_i)|$ by (1) and obtain

$$t \leq \tau^* \left\{ \sum_{i=1}^{t-1} d_{i+1} \left(\frac{1}{d_{i+1}} - \frac{1}{d_i}\right) + 1 \right\} = \tau^* \left\{ \sum_{i=1}^{t-1} \frac{d_i - d_{i+1}}{d_i + 1} \right\}$$

$$\leq \tau^* \sum_{k=1}^{d} \frac{1}{k} < \tau^* (1 + \log d).$$

[L. Lovász, *Discrete Math.* **13** (1975) 383–390.]

31. (a) It suffices to prove the case $k = r - 1$. Furthermore, we only exclude trivial cases by assuming that $u > v$ and $w > r - 2$. By 1.42(i), we can write

$$\binom{u}{r} = \binom{w+1}{r} + \binom{w}{r-1}\binom{u-w}{1} + \binom{w-1}{r-2}\binom{u-w+1}{2} + \ldots,$$

and similarly,

$$\binom{v}{r} = \binom{w+1}{r} + \binom{w}{r-1}\binom{v-w}{1} + \binom{w-1}{r-2}\binom{u-w+1}{2} + \ldots.$$

Hence

$$0 = \binom{u}{r} - \binom{v}{r} - \binom{w}{r-1}$$

$$(1) \qquad = \binom{w}{r-1}(u-v-1) + \binom{w-1}{r-2}\left\{\binom{u-w+1}{2} - \binom{v-w+1}{2}\right\}$$

$$+ \binom{w-2}{r-3}\left\{\binom{u-w+2}{3} - \binom{v-w+2}{3}\right\} + \cdots$$

The binomial coefficient $\binom{x}{i}$ is positive and monotone increasing for $x > i-1$. Hence all terms but the first on the right hand side are positive, and so the first must be negative, i.e., $u < v+1$. Similarly,

$$\binom{u}{r-1} - \binom{v}{r-1} - \binom{w}{r-2}$$

$$= \binom{w}{r-2}(u-v-1) + \binom{w-1}{r-3}\left\{\binom{u-w+1}{2} - \binom{v-w+1}{2}\right\}$$

$$+ \binom{w-2}{r-4}\left\{\binom{u-w+2}{3} - \binom{v-w+2}{3}\right\} + \cdots$$

To get this sum from the right hand side of (1), we have to multiply the first (negative) term by $(r-1)/(w-r+2)$, and the others by $(r-2)/(w-r+2)$, $(r-3)/(w-r+2)$, ... Since the negative term gets larger multiplier than the positive terms, the sum decreases.

(b) We use induction on $|E(H)|$. It suffices to treat the case $k = r-1$. Let x be a point with minimum degree d. We may assume that there are no isolated points, i.e. $d \geq 1$; also we may suppose that there are at least two edges, so that $d < \binom{u}{r}$.

Let $H_1 = H - x$. Then $|E(H_1)| = \binom{u}{r} - d$. Let us write this number in the form $\binom{v}{r}$ ($v \geq r$); then, by the induction hypothesis, there are at least $\binom{v}{r-1}$ $(r-1)$-tuples covered by the edges of H_1. Note that none of these contains x. Let us define H_2 by $V(H_2) = V(H) - \{x\}$, $E(H_2) = \{E - \{x\} : x \in E \in E(H)\}$. Then $|E(H_2)| = d$, and H_2 is $(r-1)$-uniform. Let us write d in the form $\binom{w}{r-1}$, $w \geq r-1$, then by the induction hypothesis, there are at least $\binom{w}{r-2}$ $(r-2)$-tuples covered by the edges of H_2. Adding x to these $(r-2)$-tuples, we obtain at least $\binom{w}{r-2}$ $(r-1)$-tuples contained in edges of H. Since each of these $(r-1)$-tuples contains x, they are different from the ones found previously.

Thus the number of $(r-1)$-tuples contained in edges of H is at least

$$\binom{v}{r-1} + \binom{w}{r-2}.$$

Using (a) we will be done if we can show that $v \geq w$. To this end, observe that the average degree is

$$r\frac{|E(H)|}{|V(H)|} = r\frac{\binom{u}{r}}{|V(H)|}$$

and here, obviously, $|V(H)| \geq u$. Thus

$$\binom{w}{r-1} = d \leq r\frac{\binom{u}{r}}{u} = \binom{u-1}{r-1},$$

whence

$$w \leq u - 1.$$

Thus

$$\binom{v}{r} = \binom{u}{r} - \binom{w}{r-1} \geq \binom{u}{r} - \binom{u-1}{r-1} = \binom{u-1}{r}, \quad v \geq u-1 \geq w,$$

as stated. [J. B. Kruskal, in: *Math. Opt. Techniques*, Univ. of Calif. Press 1963, 251–278; G. Katona, in: *Theory of Graphs*, Akadémiai Kiadó 1966, 187–207. See these references for the exact optimum when u is not an integer.]

(c) Let H have m distinct edges, which are r-tuples and mutually intersecting. Let H' be defined by

$$V(H') = V(H)$$
$$E(H') = \{V(H) - E : E \in E(H)\}.$$

Then by assumption no edge of H can occur as a subset of any edge of H'. Let us write

$$m = \binom{u}{n-r} \quad (u \geq n - r, \text{ real}),$$

then by (b), there are at least $\binom{u}{r}$ r-tuples among the subsets of edges of H'. Hence

$$\binom{u}{n-r} + \binom{u}{r} \leq \binom{n}{r}.$$

Since

$$\binom{n-1}{n-r} + \binom{n-1}{r} = \binom{n}{r},$$

it follows that $u \leq n-1$ and thus

$$m = \binom{u}{n-r} \leq \binom{n-1}{n-r} = \binom{n-1}{r-1}.$$

[D. E. Daykin, *J. Comb. Th.* A **17** (1974) 254–255.]

32. (a) Observe that if $(x_1, \ldots, x_n)$ is any permutation of $V(H)$, then there is at most one index i for which every element of A_i has smaller index than every element of B_i. So we count at most $n!$ pairs (A_i, B_i).

Moreover, if i is any given index, then the number of permutations of $V(H)$ for which every element of A_i anticipates every element of B_i is

$$\binom{n}{p+q}p!q!(n-p-q)! = \frac{n!}{\binom{p+q}{p}}.$$

Hence, a given pair (A_i, B_i) is counted this number of times. Hence

$$m \leq n! \bigg/ \frac{n!}{\binom{p+q}{p}} = \binom{p+q}{p}.$$

(b) Let $E_1, \ldots, E_m$ be the edges of H. Since H is τ-critical, $H - \{E_i\}$ has a $(\tau(H) - 1)$-element point-cover T_i. Then, obviously, $E_i \cap T_i = \emptyset$ while $T_i \cap E_j \neq \emptyset$, if $i \neq j$. Thus (a) implies that

$$m \leq \binom{r + (\tau(H) - 1)}{r}.$$

[B. Bollobás; see B].

33. Let $V(H) = \{v_1, \ldots, v_n\}$. We color $v_1, v_2, \ldots, v_n$ by one of the colors red and blue, so that we do not color all points of any edge with the same color. Suppose that $v_1, \ldots, v_i$ are colored $(1 < i < n)$. If we cannot color v_{i+1} red, then there is an edge $E \subset \{v_1, \ldots, v_{i+1}\}$, $v_{i+1} \in E$ such that all the other points of E are red. Similarly, if v_{i+1} cannot be given color blue, there must be an edge $F \subset \{v_1, \ldots, v_{i+1}\}$, $v_{i+1} \in F$ all of whose points except v_{i+1} are blue. Now $E \cap F = \{v_{i+1}\}$, a contradiction. Thus we can color v_{i+1} either red or blue.

34. Suppose first that H has one point x_1 of degree 1 and all other points have degree 2. Let x_2 be a point adjacent to x_1, x_3 a point adjacent to x_2 or x_1, and generally, let x_{i+1} be point different from $x_1, \ldots, x_i$ and adjacent to one of them. Such a point x_{i+1} exists, because H is connected. Now going "back" we can color $x_n, x_{n-1}, \ldots, x_1$ successively with red and blue in the same way as in the previous solution.

Now suppose that every point of H has degree 2. We may assume that H is not a graph. Then it contains an edge E with $|E| \geq 3$. Let $x \in E$, $E' = E - \{x\}$ and $H' = H - \{E\} + \{E'\}$. Now if H' is connected, then, since x has degree 1 in H', by the first part of the solution H' is 1-chromatic. A 2-coloration of H' is a 2-coloration of H. If H' is not connected, then $H = H_1 \cup H_2$, $V(H_1) \cap V(H_2) = \{x\}$. 2-coloring H_1 and H_2 (which have points with degree 1) we can put these 2-colorations together after exchanging colors in H_1, if necessary.

35. Suppose that we have no 1-element edge (which is a trivial case). Also remove isolated points. We verify the assertions in the hint.

(a) Let $x \in V(H)$, $E \in E(H)$ be such that $x \in E$. Color $E - \{x\}$ red, $V(H) - E \cup \{x\}$ blue. Clearly there will be no red monochromatic edge. Since H is not 2-colorable, there must be a blue monochromatic edge F. F meets E in a single point and this point must be blue; hence $x \in F$. Thus x has degree at least 2.

(b) Let $x \in E_1, \ldots, E_d \in E(H)$, $d \geq 4$, and suppose indirectly that $|E_d| \geq 3$. Let $y_i \in E_i - \{x\}$ $(i = 2, \ldots, d-1)$. Since $d \geq 4$, there is at most one edge containing $y_2, \ldots, y_{d-1}$. So we can select a point $y_d \in E_d - \{x\}$ such that no edge contains $y_2, \ldots, y_d$. Now let us color $E_1 - \{x\}$, $y_2, \ldots, y_d$ red, all other points blue. It is easy to check that no monochromatic edge can occur.

This settles the case when there is a point x of degree at least 4; more generally, if there is a point x which is adjacent to 2-element edges only, then consider an edge E not containing x (such an edge clearly exists). E must contain the (single) point of $F - \{x\}$ for every edge $F \ni x$. This shows that E is unique and so $V(H) = E \cup \{x\}$, $E(H) = \{E\} \cup \{\{x, y\} : y \in E\}$.

So suppose that each point belongs to at least one edge of cardinality at least 3. Then there is no 2-element edge. For let $\{x, y\} \in E(H)$ and let E, F be $\geq$ 3-element edges containing x and y, respectively. Let u be the (unique) common point of E and F, and let $v \in E - \{u, x\}$, $w \in F - \{u, y\}$. By (a), there are other edges A, B containing v and w. Since A, B must meet $\{x, y\}$, we have $y \in A$, $x \in B$.

Now color every point in $E \cup F - \{u\} - \{y\}$ red, the remaining points blue. Then no red monochromatic edge occurs; for such an edge X must meet $\{x, y\}$ in a red point, i.e. in x; it must meet A in a red point, i.e. in v; but then $X = E$, which is not monochromatic. Also no blue monochromatic edge arises; for such an edge X must meet $\{x, y\}$ in a blue point, i.e. in y; it must meet E in a blue point, i.e. in u; but then $X = F$, which is not monochromatic. Thus H is 2-colorable, a contradiction.

Now we are prepared to prove (c). Let x, $y \in V(H)$, and select edges E, F such that $x \in E$, $y \in F$. We may assume that $x \notin F$ and $y \notin E$, otherwise we have nothing to prove. Let us color $E \cup F - \{x, y\}$ red, everything else blue. Then no monochromatic red edge can arise; for such an edge could only consist of one point of $E - \{x\}$ and one point of $F - \{y\}$, but 2-element edges are excluded. But then a blue monochromatic edge A arises, which must meet E in x and F in y, which proves (c).

It follows that every edge is a 3-tuple. For if $E \in E(H)$, $|E| \geq 4$, then let $x \in V(H) - E$. For each point y in E, there exists an edge containing x, y and these edges are, obviously, distinct. But then x has degree at least 4, a contradiction.

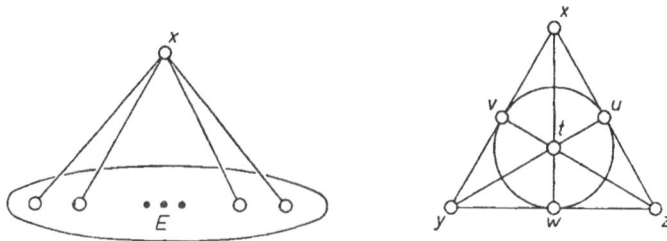

FIGURE 99

Now let E, F, G be three edges such that $E \cap F \not\subset G$. Let $E \cap F = \{x\}$, $F \cap G = \{y\}$, $E \cap G = \{z\}$. Then there are unique further points u, v, w of E, F, G. There is a unique edge $\{x,w,t\}$ containing x and w. There is a unique edge containing y and u, which must also contain t since this is the only point in which it can meet $\{x,w,t\}$. Similarly, $\{z,v,t\}$ and $\{u,v,w\}$ are edges and there are no other edges. This hypergraph (usually called the *Fano configuration* or 7-point-plane) is 3-chromatic. Thus we have found the hypergraphs shown in Fig. 99 as the only possibilities.

36. Suppose indirectly that H has a 2-coloration with red and blue. Let n_1 and n_2 denoted the numbers of red and blue points, respectively. Then the number of red–red–blue edges is

$$a\binom{n_1}{2},$$

where a is the number of edges containing a given pair. Similarly, there are

$$a\binom{n_2}{2}$$

blue-blue-red edges. Thus

(1) $$|E(H)| = a\left[\binom{n_1}{2} + \binom{n_2}{2}\right].$$

On the other hand, each edge contains exactly two pairs (x,y), where x is blue, y is red. Thus,

(2) $$2|E(H)| = a n_1 n_2.$$

From (1) and (2) we get

$$n_1 n_2 = n_1^2 + n_2^2 - n_1 - n_2 = n^2 - n - 2n_1 n_2$$

or, equivalently,

(3) $$3 n_1 n_2 = n^2 - n.$$

But we have

$$n_1 n_2 \leq \left(\frac{n_1 + n_2}{2}\right)^2 = \frac{n^2}{4},$$

so (3) implies that

$$\frac{n^2}{4} \geq n^2 - n \quad \text{or} \quad n \leq 4,$$

a contradiction. [L. Lovász, *Proc.* 4th *SE Conf. on Comb. Graph Theory and Computing*, Utilitas Math. 1973, 1–12.]

37. Consider the polyhedron P in n-dimensional space described by

$$\left\lfloor \frac{|E_j|}{2} \right\rfloor \leq \sum_{i=1}^{n} a_{ij} x_i \leq \left\lceil \frac{|E_j|}{2} \right\rceil \qquad (j = 1, \ldots, m),$$

$$0 \leq x_i \leq 1 \qquad (i = 1, \ldots, n).$$

P is non-empty since the vector $\left(\frac{1}{2}, \ldots, \frac{1}{2}\right)^T$ belongs to it. Also the matrix $\binom{A^T}{I}$ is totally unimodular and the constants are integers, so by the Hoffman–Kruskal theorem, P contains a lattice point. Let $(\xi_1, \ldots, \xi_n)^T$ be this lattice point. Color v_i red, if $\xi_i = 0$ and blue, otherwise. This coloration is a legitimate one, if H contains no 1-element edges. More exactly, there are at least $\lfloor |E_j|/2 \rfloor$ and at most $\lceil |E_j|/2 \rceil$ blue points in every edge E_j. This proves the "only if" part of the assertion.

Now suppose that each subhypergraph has a bicoloration as formulated. Then we show that every subdeterminant of the incidence matrix A is 0 or ± 1. Since a square submatrix of A is the incidence matrix of some partial subhypergraph which, obviously, has the same property it suffices to show that if $n = m$, then $\det A = 0$ or ± 1.

We claim that there exists a $W \subseteq V(H)$, $W \neq \emptyset$ such that $|W \cap E_1|, \ldots, |W \cap E_{n-1}|$ are even; in fact, consider the vectors

$$\mathbf{a}_j = \begin{pmatrix} a_{1j} \\ \vdots \\ a_{nj} \end{pmatrix}, \qquad \text{where } a_{ij} = \begin{cases} 1 & \text{if } v_i \in E_j, \\ 0 & \text{otherwise}, \end{cases}$$

$(i = 1, \ldots, n-1)$ over $GF(2)$. They do not generate the whole space, so there is a 01-vector

$$\mathbf{b} = \begin{pmatrix} b_1 \\ \vdots \\ b_n \end{pmatrix} \neq 0$$

orthogonal to all of them by 5.32f. Now set

$$W = \{v_i : b_i = 1\}.$$

Then $|W \cap E_j| \equiv \mathbf{b} \cdot \mathbf{a}_j \equiv 0 \pmod 2$ $(j = 1, \ldots, n-1)$. $|W \cap E_n|$ may be odd or even.

We may assume that $W = \{v_1, \ldots, v_i\}$. By the assumption $W = A_0 \cup A_1$, $A_0 \cap A_1 = \emptyset$ so that

$$\left\lfloor \frac{|E_j \cap W|}{2} \right\rfloor \leq |E_j \cap A_0| \leq \left\lceil \frac{|E_j \cap W|}{2} \right\rceil \qquad (j = 1, \ldots, n).$$

Note that, by the choice of W,

$$|E_j \cap A_0| = \frac{|E_j \cap W|}{2}$$

for $j = 1, \ldots, n-1$. Let, say, $A_0 = \{v_1, \ldots, v_s\}$. Then add the $2^{\text{nd}}, \ldots, s^{\text{th}}$ columns to the first one but subtract the $(s+1)^{\text{st}}, \ldots, t^{\text{th}}$ columns from it. This way we get 0's in all but the last entry of the first column, and get 0 or ± 1 in the last entry. If this last entry is 0, then $\det A_1 = 0$. Otherwise, we can expand by the first column and proceed by induction [A. Ghouila-Houri; see B].

38. (a) First we prove the statement of the hint. It suffices to consider the case $k = 2$. Let $E \in E(H)$, $E_1, E_2 \subset E$, $E_1 \cap E_2 = \emptyset$. Consider

$$H' = H - \{E\} + \{E_1, E_2\}.$$

We claim that H' has no odd circuits. For let C be any circuit of H'. If C contains none of E_1, E_2, then it corresponds to a circuit of H in the natural way and hence it is even. Suppose that it contains both of E_1, E_2. Then it is of the form

$$(E_1, x_1, F_1, \ldots, x_k, E_2, y_1, G_1, \ldots, y_l),$$

where $k \geq 2$, $l \geq 2$ as $E_1 \cap E_2 = \emptyset$. Now

$$(E, x_1, F_1, \ldots, x_k) \text{ and } (E, y_1, G_1, \ldots, y_l)$$

are circuits of H, hence k and l are even. Then $k + l$ is even, i.e. C is even.

Now replace each edge E by $\lfloor |E|/2 \rfloor$ disjoint pairs. The resulting graph is bipartite by the above result and thus it possesses a 2-coloration with red and blue. Now each edge E contains at least $\lfloor |E|/2 \rfloor$ red and at least $\lfloor |E|/2 \rfloor$ blue points, which proves the assertion.

(b) This follows trivially by (a) and 13.37 [C. Berge; see B].

39. Suppose indirectly that H is not balanced. Then it contains an odd circuit $(x_1, E_1, \ldots, x_{2k+1}, E_{k+1})$ which is not balanced. This means that there is no other incidence between $x_1, \ldots, x_{2k+1}$ and $E_1, \ldots, E_{2k+1}$ than is indicated by the definition of circuits. Hence A contains a $(2k+1) \times (2k+1)$ submatrix of the form

$$\begin{pmatrix} 1 & 1 & & & & & \\ & 1 & & & & 0 & \\ & & \ddots & & & & \\ & & & \ddots & & & \\ & 0 & & & 1 & 1 \\ 1 & & & & & 1 \end{pmatrix},$$

which has determinant 2, a contradiction [see B].

40. Suppose indirectly that there are balanced hypergraphs, which are not 2-chromatic and consider a minimum counterexample H. Define a graph G by $V(G) = V(H)$, $E(G) = \{E \in E(H) : |E| = 2\}$.

We claim that G is a connected graph. Suppose indirectly that $V(G) = V_1 \cup V_2$, $V_1 \cap V_2 = \emptyset$, $V_1, V_2 \neq \emptyset$ such that each $E \in E(G)$ is contained in V_1 or V_2. Consider H_{V_1} and H_{V_2} and remove the one-element edges of them. This way we obtain two balanced hypergraphs H_1, H_2.

By the minimality of H, H_1 and H_2 are 2-chromatic. Let us 2-color each of them by red and blue, we claim this is a good 2-coloration of H. Let $E \in E(H)$. If $|E| = 2$, then E is spanned by one of V_1 or V_2 and therefore it gets both colors. If $|E| \geq 3$, then $|E \cap V_i| \geq 2$ for one of $i = 1$, 2 and so, $E \cap V_i$ gets both colors in the 2-coloration of H_i. Thus we have obtained a 2-coloration of H, a contradiction.

Now G has no odd circuits, because H is balanced and since it is connected, it has an (essentially unique) 2-coloration with red and blue. We claim that this is a legitimate 2-coloration of H. Assume indirectly that there were a monochromatic edge E, $|E| > 1$. Since G is connected, it contains a path $P = (x_0, F_1, x_1, \ldots, F_k, x_k)$ with $x_0 \in E$, $x_k \in E$. We may assume that $x_i \notin E$ for $i = 1, \ldots, k-1$. The points $x_0, \ldots, x_k$ are alternately red and blue, hence k is even. Thus

$$(x_0, F_1, x_1, \ldots, x_k, E)$$

is an unbalanced odd circuit, a contradiction. Thus the 2-coloration we considered is a good 2-coloration of H, a contradiction again [C. Berge; see B, LP].

41. Let $E(H) = \{E_1, \ldots, E_m\}$ $(m \leq 2^{r-1})$ and color the points of H with red and blue at random, independently of each other and with probability $1/2$. Let A_i denote the event that E_i is monochromatic. Then

$$\mathsf{P}(A_i) = \frac{1}{2^{r-1}}$$

as there are 2^r ways to color E_i and 2 of these come into consideration. So the probability that a random coloration contains a monochromatic edge is

$$\mathsf{P}(A_1 + \ldots + A_m) < \sum_{i=1}^{m} \mathsf{P}(A_i) = \frac{m}{2^{r-1}} \leq 1,$$

(the first strict inequality follows from the fact that $A_i, \ldots, A_m$ are not mutually exclusive: all occur, when all points are red) [P. Erdős; see ES].

42. Let H be defined as in the hint. Then

$$|E_P| = \binom{|X|}{r} + \binom{|Y|}{r} \geq 2\binom{r^2}{r}$$

for any edge $E_P(P = \{X, Y\})$. Therefore,

$$t_x = \frac{1}{2\binom{r^2}{r}}$$

defines a fractional cover. Thus

$$\tau^*(H) \leq \frac{\binom{2r^2}{r}}{2\binom{r^2}{r}}.$$

(It would not be difficult to show that one has equality here.)

Now let T be any set of points of H covering every edge E_P. Then T is a collection of r-tuples, so $V(H_0) = S$, $E(H_0) = T$ defines an r-uniform hypergraph. The fact that T covers E_P means that P does not give a legitimate 2-coloration of H_0; so H_0 is not 2-chromatic. Thus if T is a minimum cover we have by 13.30,

$$|E(H_0)| = \tau(h) \le (1 + \log d(H))\tau^*(H).$$

Here, however,

$$\tau^*(H) = \frac{\binom{2r^2}{r}}{2\binom{r^2}{r}},$$

$$d(H) = 2^{r^2 - r},$$

and so

$$|E(H_0)| \le (1 + (r^2 - r)\log 2)\frac{\binom{2r^2}{r}}{2\binom{r^2}{r}} < r^2 \frac{\binom{2r^2}{r}}{\binom{r^2}{r}} =$$

$$= r^2 \cdot \frac{2r^2}{r^2} \cdot \frac{2r^2 - 1}{r^2 - 1} \cdots \frac{2r^2 - r + 1}{r^2 - r + 1} < c \cdot r^2 \cdot 2^r.$$

[P. Erdős; see ES].

43. Let us color the points of H at random again, as in the solution of 13.41. Observe that if $E_1, \ldots, E_p$ are disjoint from E_q, then $A_1, \ldots, A_p$ and any polynomial in them are independent of A_q. So if we form $L(H)$ and associate the event A_j with E_i, then this graph and the associated events satisfy the first condition of 2.18. Moreover, by the assumption, each point of $L(H)$ has degree at most 2^{r-3}, hence the second condition is satisfied as well. Therefore

$$\mathsf{P}(\overline{A}_1 \ldots \overline{A}_m) > 0,$$

i.e. there is a coloration in which no edge is monochromatic. [P. Erdős–L. Lovász, in: *Infinite and Finite Sets*, Coll. Math. Soc. J. Bolyai **10**, Bolyai–North-Holland (1974) 609–627.]

44. (a) Suppose that H is not 2-chromatic. Then by 13.43, there is an edge E which is intersected by more that 2^{r-3} other edges. Then one of the points of E belongs to more than $2^{r-3}/r$ of these. Thus, the maximum degree d of H is at least $2^{r-3}/r$.

Now let x be a point with maximum degree d. Then the edges containing x have no other point in common by the assumption and so, they cover at

$$1 + \frac{(r-1)2^{r-3}}{r} > 2^{r-4}$$

least points, a contradiction.

(b) For $E \in E(H)$, let $\varphi(E)$ denote the point of E which has largest degree in H (or one of these) and set $E' = E - \{\varphi(E)\}$,

$$H' = (V(H); \{E' : E \in E(H)\}).$$

Then H' must be at least 3-chromatic since any 2-coloration of H' would yield a 2-coloration of H. Thus by the solution of (a), there is a point x with degree $t > 2^{r-4}/(r-1)$ in H'. Let $E_1', \ldots, E_t'$ be the edges of H' containing x. Then, by definition, the degree of $\varphi(E_i)$ in H is at least the degree of x in H, which is at least the degree of x in H', which is equal to t. Moreover, the points $\varphi(E_1), \ldots, \varphi(E_t)$ are distinct, as $E_1, \ldots, E_t$ have no point in common other than x. Thus we have t points with degree at least t, which proves the assertion.

(c) Suppose indirectly that H has chromatic number at least 3. Then, by the preceding problem, there are $2^{r-4}/r$ points with degree $> 2^{r-4}/r$. If we count the edges adjacent to these points we get $4^{r-4}/r^2$, and each edge is counted not more than r times. Hence

$$|E(H)| > 4^{r-4}/r^3.$$

[ibid.]

45. (a) By 13.5b, we can select a pair $f(E)$ from each edge E of H so that the hypergraph (in fact, graph) $G = (V(H), \{f(E) : E \in E(H)\})$ is a forest. Hence G is 2-colorable and a 2-coloration of G gives a 2-coloration of H [M. Las Vergnas, L. Lovász; see B].

(b) Let H_2 be the triangle and suppose that we already have an r-uniform hypergraph H_r with the desired property. Define H_{r+1} as follows; for every edge $E \in E(H_r)$, take an $(r+1)$-element set E'; let the sets E' be disjoint from each other and from $V(H_r)$. Define

$$V(H_{r+1}) = \bigcup_{E \in E(H_r)} E' \cup V(H_r),$$

$$E(H_{r+1}) = \{E' : E \in E(H_r)\} \cup \{E \cup \{x\} : E \in E(H_r), x \in E'\}.$$

Then, obviously, H_{r+1} is $(r+1)$-uniform. Moreover, H_{r+1} is not 2-colorable. For let α be a 2-coloration of H_{r+1}. Then there is an edge $E \in E(H_r)$, which is monochromatic since H_r is not 2-colorable. Let, e.g. all points of E be red. Then for any $x \in E'$, $E \cup \{x\}$ is an edge of H_{r+1}, so x must be blue. But then E' is a monochromatic edge of H_{r+1}, a contradiction.

Let H' be any partial hypergraph of H_{r+1}, we claim that $|V(H')| \geq |E(H')|$. Suppose not, and let H' be a minimal counterexample. Then every point in H' has degree at least 2, since otherwise it could be removed. Therefore, if $x \in E'(E \in E(H_r))$ is a point of H', then $E' \in E(H')$, and also $E \cup \{y\} \in E(H')$ for each $y \in E'$.

Let A denote the set edges of H' of form E', $E \in E(H_r)$. For every $E \in E(H') - A$, let $E^* = E \cap V(H_r)$. Clearly $E^* \in E(H_r)$. Thus the hypergraph H'' defined by

$$V(H'') = V(H') \cap V(H_r),$$

$$E(H'') = \{E^* : E \in E(H') - A\}$$

(each set E^* is taken only once) is a partial hypergraph of H_r, and thus

$$|V(H'')| \geq |E(H'')|.$$

Also we have

$$|E(H')| = (r+1)|E(H'')|,$$
$$|V(H')| = |V(H'')| + r|E(H'')|$$

and so

$$|V(H')| \geq (r+1)|E(H'')| = |E(H')|.$$

This proves that H_{r+1} has the desired properties. We remark that every partial hypergraph H' of H_r even satisfies

$$|V(H')| \geq |E(H')| + 1$$

except for $H' = H_r$. [D. R. Woodall, in: *Combinatorics*, Proc. Conf. Southend-on-Sea, 1972, 322–340.]

(c) Assume that the incidence vectors of the edges of H do not span $\Rightarrow^n$. Then there exists a non-zero vector a orthogonal to the incidence vector of every edge. In other words, we can assign real numbers $a(v)$ to the vertices, not all zero, so that the sum of numbers assigned to the elements of any edge is zero. Let V^0, V^+ and V^- denote the set of vertices v for which $a(v) = 04$, $a(v) > 0$, and $a(v) < 0$, respectively.

Let H' be the partial hypergraph induced by V^0. By hypothesis, H' is 2-colorable; let $V_1 \cup V_2 = V^0$ be a 2-coloration of H'. Then $(V_1 \cup V^+) \cup (V_2 \cup V^-)$ is a 2-coloration of H. In fact, if an edge A is contained in V^0, then it must intersect both V_1 and V_2 by definition. If A is not contained in V^0, then a can add up to zero on A only if it contains both negative and positive entries. [P. D. Seymour, *Quart. J. Math. Oxford*, **25** (1974), 303–312.]

46. (a) Let H_r ba an r-uniform non-2-colorable hypergraph, without multiple edges, in which any two edges meet. We construct an $(r+1)$-uniform hypergraph H_{r+1} with similar properties. Let $|U| = r+1$, $U \cap V(H_r) = \emptyset$. Define H_{r+1} by

$$V(H_{r+1}) = V(H_1) \cup U,$$
$$E(H_{r+1}) = \{U\} \cup \{E \cup \{x\} : E \in E(H_r), x \in U\}.$$

It is easy to verify that H_{r+1}, has the desired properties. Also,

$$|E(H_{r+1})| = (r+1)|E(H_r)| + 1,$$

whence

$$|E(H_r)| = \lfloor (e-1)r! \rfloor > r!.$$

(b) Let H, H' be two hypergraphs. Define $H[H']$ as follows. Let

$$V(H[H']) = V(H) \times V(H'),$$
$$E(H[H']) = \{\{x_1\} \times F_1 \cup \{x_2\} \times F_2 \cup \ldots \cup \{x_r\} \times F_r :$$
$$: \{x_1, \ldots, x_r\} \in E(H), F_1, \ldots, F_r \in E(H')\}.$$

Then, if any two edges of H and of H' meet, the same holds for $H[H']$. In fact, if $\{x_1\} \times F_1 \cup \ldots \cup \{x_r\} \times F_r$ and $\{x_1'\} \times F_1' \cup \ldots \cup \{x_s'\} \times F_s'$ are two edges of $H[H']$, then $\{x_1, \ldots, x_r\}$ and $\{x_1', \ldots, x_s'\}$ meet; let $x_i = x_j'$, then also F_i and F_j' meet; let $y \in F_i \cap F_j'$, then (x_i, y) is a common point of the two edges of $H[H']$.

If H, H' are not 2-colorable, than neither is $H[H']$. For let us 2-color the points of $H[H']$. Then, for each $x \in V(H)$, one of the sets $\{x\} \times F (F \in E(H'))$ is monochromatic; choose such an edge F_x of H'. Now color x with the color of $\{x\} \times F_x$, then there will be an edge $\{x_1, \ldots, x_r\} \in E(H_r)$ which is monochromatic. Then the edge

$$\{\{x_1\} \times F_{x_1}, \ldots, \{x_r\} \times F_{x_r}\}$$

is monochromatic.

Let K denote the triangle and set

$$K_m = \underbrace{K[K[K \ldots [K] \ldots]]}_{m \text{ times}}.$$

Then, by the above, any two edges of K_m meet, but it is not 2-chromatic. It is $r = 2^m$-uniform and has $3^{2^m - 1}$ edges, as is easily seen by induction. Hence it has all the desired properties.

(c) The hypergraph constructed in the solution of 13.27 trivially has all the desired properties. [P. Erdős–L. Lovász, in: *Infinite and Finite Sets*, Coll. Math. Soc. J. Bolyai **10**, Bolyai–North-Holland (1974) 609–627.]

47. Consider such an r-uniform H without multiple edges, which is not 2-chromatic. Let x_1 be a point with maximum degree, then its degree is greater than $\frac{m}{r}$. For let $E \in E(H)$. Every other edge meets E, it meets itself in r points, hence

$$(m-1) + r \leq \sum_{F \in E(H)} |E \cap F| = \sum_{x \in E} d_H(x).$$

Thus

$$\max_{x \in E} d_H(x) \geq \frac{m}{r} - \frac{1}{r} + 1 > \frac{m}{r}.$$

Now suppose that $x_1, \ldots, x_i$ are chosen so that $\{x_1, \ldots, x_i\}$ is contained in more that $\frac{m}{r^i}$ edges $(i < r)$. Try to color $x_1, \ldots, x_i$ red, the rest blue. This cannot work, i.e. there must be an edge E_i such that $x_1, \ldots, x_i \notin E_i$. Any edge containing $\{x_1, \ldots, x_i\}$ meets E_i and thus, there is a point x_{i+1} which is contained in more than $\frac{m}{r^i}/r$ of them. Thus $\{x_1, \ldots, x_{i+1}\}$ is contained in more than $\frac{m}{r^{i+1}}$ edges.

We conclude that $\{x_1,\ldots,x_r\}$ is contained in more than m/r^r edges. Since $\{x_1,\ldots,x_r\}$ is, obviously, contained in at most one edge (it is either an edge or not), we have

$$\frac{m}{r^r} < 1, \qquad m < r^r$$

which proves the assertion. [ibid.]

48. Since each cover defines a fractional cover, the inequality

$$\tau^*(H) \leq \tau(H)$$

is trivial. Similarly

$$\nu(H) \leq \nu^*(H)$$

is obvious.

With the notation introduced in the hint, any real vector $\mathbf{x}$ such that

$$A\mathbf{x} \leq \mathbf{1}, \mathbf{x} \geq 0$$

defines a fractional matching; any real vector $\mathbf{y}$ such that

$$A^T\mathbf{y} \geq \mathbf{1}, \quad \mathbf{y} \geq 0$$

defines a fractional cover. Thus

$$\nu^*(H) = \max\{\mathbf{1}^T\mathbf{x}|A\mathbf{x} \leq \mathbf{1}, \mathbf{x} \geq 0\},$$
$$\tau^*(H) = \min\{\mathbf{1}^T\mathbf{y}|A^T\mathbf{y} \geq \mathbf{1}, \mathbf{y} \geq 0\},$$

and the relation

$$\nu^*(H) = \tau^*(H)$$

follows by the Duality theorem in linear programming.

49. (a) Let t be any k-cover of a graph G. Set

$$A = \{x : t(x) \geq k\}$$
$$B = \{x : t(x) = 0\}$$

and define

$$t'(x) = \begin{cases} 2 & \text{if } x \in A, \\ 0 & \text{if } x \in B, \\ 1 & \text{otherwise;} \end{cases}$$
$$t''(x) = t(x) - t'(x).$$

We claim that t' is a 2-cover and t'' is a $(k-2)$-cover. Let $(x,y) \in E(G)$. If x, $y \notin B$, then $t'(x)+t'(y) \geq 2$. If, say, $y \in B$, then $t(x)+t(y) \geq k$ implies that $t(x) \geq k$, $x \in A$, and $t'(x)+t'(y) \geq 2$ again. So t' is a 2-cover. On the other hand, $t''(x)+t''(y) \geq (t(x)-1)+t(y)-1 \geq k-2$ unless, e.g. $x \in A$. But then $t''(x)+t''(y) \geq t''(x) \geq k-2$ follows again.

(b) Let w be a k-matching, and suppose that k is even. Replace each edge E by $w(E)$ parallel edges. The resulting graph G' has degrees at most k. By adding new edges (loops are allowed) we can obtain a k-regular graph G''. By 7.40 G''

has a 2-factor; the rest of its edges form a $(k-2)$-factor. Removing the edges of $E(G'') - E(G')$, we obtain a decomposition of $E(G')$ into a 2-matching and a $(k-2)$-matching. These yield a decomposition of w into a 2-matching and a $(k-2)$-matching.

(c) Let t be an optimal fractional cover with rational weights (it is known from linear programming that such an optimal fractional cover exists). Let $2k$ be a common denominator of the weights in t. Then $2kt$ is a $(2k)$-cover. By (a), $2kt = t_1 + \ldots + t_k$, where $t_1, \ldots, t_k$ are 2-covers. Thus

$$t = \frac{1}{k}\left(\frac{t_1}{2} + \ldots + \frac{t_k}{2}\right),$$

where $\frac{t_1}{2}, \ldots, \frac{t_k}{2}$ are fractional covers. Since t is optimal, it follows that $\frac{t_1}{2}, \ldots, \frac{t_k}{2}$ must be optimal fractional covers, too. New, e.g. $\frac{t_1}{2}$ consists of halves of integers.

For matchings the assertion follows similarly. Also

$$2\tau^*(H) = \sum_{x \in V(H)} t_1(x)$$

is an integer.

(d) Let t be an optimal 2-cover. Set

$$A = \{x : t(x) = 0\}.$$

Obviously, A is independent. Moreover for any point y in $\Gamma(A)$, $t(y) \geq 2$, since there is an x in A adjacent to y and $t(x) + t(y) \geq 2$. Obviously, $t(y) \leq 2$ for every point, thus $t(y) = 2$.

Now $t(z) \geq 1$ for every $z \notin A \cup \Gamma(A)$; moreover, taking $t(z) = 1$ is sufficient to cover every edge of G twice. Thus $t(z) = 1$ by the optimality of t. So t is determined by A and

$$\sum_{x \in V(H)} t(x) = 2|\Gamma(A)| + |V(G) - \Gamma(A) - A| = |V(G)| + |\Gamma(A)| - |A|.$$

So, using (c),

$$2\tau^*(G) = \min_t \sum_x t(x) = |V(G)| + \min_{A \text{ indep.}} \{|\Gamma(A)| - |A|\}.$$

Thus (c) implies 7.37.

50. Let $\{U,V\}$ be a 2-coloration of G, and let t be a k-cover. For $1 \leq i \leq k$, define

$$t_i(x) = \begin{cases} 1 & \text{if } x \in U \text{ and } t(x) \geq i \\ & \text{or } x \in V \text{ and } t(x) \geq k-i+1, \\ 0 & \text{otherwise.} \end{cases}$$

We claim that $t_1 + \ldots + t_k = t$ and that every edge is covered by every t_i. The first assertion is trivial. Let $(x,y) \in E(G)$, $1 \leq i \leq k$. Since $t(x) + t(y) \geq k$, we have either $t(x) \geq i$ or $t(y) \geq k-i+1$. Accordingly, $t_i(x)=1$ or $t_i(y)=1$. The assertion concerning matchings follows exactly as in the previous solution only now we can use 7.10 instead of 7.40.

51. (a) Let t, t' be minimum fractional covers of G, H, respectively. Define

$$\bar{t}(x,y) = t(x)t'(y) \quad (x \in V(G), y \in V(H)),$$

then it is easy to check that t is a fractional cover of $G \otimes H$. Moreover,

$$\sum_{(x,y)\in V(G\otimes H)} \bar{t}(x,y) = \left(\sum_x t(x)\right)\left(\sum_y t'(y)\right) = \tau^*(G)\tau^*(H),$$

whence

$$\tau^*(G \otimes H) \leq \tau^*(G)\tau^*(H).$$

An analogous argument with fractional matchings yields

$$\nu^*(G \otimes H) \geq \nu^*(G)\nu^*(H),$$

and hence 13.48 implies the first relation.

Let S be a minimum point cover of $G\otimes H$. Let $F \in E(H)$ and

$$S_F = \{x \in V(G) : (x,y) \in S \text{ for some } y \in F\}.$$

Then, clearly, S_F covers all edges of G, whence $|S_F| \geq \tau(G)$. Set $\tau(G)=k$. Define, for $y \in V(H)$,

$$t(y) = |(V(G) \times \{y\}) \cap S|.$$

Then

$$\sum_{y\in F} t(y) \geq |S_F| \geq k,$$

i.e. t is a k-cover of H. Thus

$$\tau_k(H) \leq \sum_{y\in V(H)} t(y) = |S| = \tau(G \otimes H),$$

and thus

$$\tau(G \otimes H) \geq \tau_k(H) \geq k\tau^*(H).$$

This proves the upper bound in the second statement. The lower bound follows by a similar method as in the first step of the solution. The assertion concerning matching number follows analogously and therefore we omit details.

(b) Assume that $\tau(H)=\tau^*(H)$. Then by (a),

$$\tau(G)\tau(H) \geq \tau(G \otimes H) \geq \tau(G)\tau^*(H),$$

which proves that

$$\tau(G \otimes H) = \tau(G)\tau(H).$$

Conversely, assume that $\tau(H) > \tau^*(H)$. Then $\tau_n(H) < n \cdot \tau(H)$ for some n. Let $N = \tau_n(H)$ and define a hypergraph G on $1,\dots,N$ to consist of all $(N-n+1)$-

element subsets. Let $\{x_1, \ldots, x_N\}$ be a minimum n-cover of H (the same point may occur several times on this list) and

$$S = \{(i, x_i) : 1 \le i \le N\}.$$

Then S represents all edges of $G \otimes H$; for if $E \times F$ is an edge of $G \otimes H$ ($E \in E(G)$, $F \in E(H)$), then there are at least n points of S of the form (i, x_i), $x_i \in F$, and at most $n - 1$ of these have $i \notin E$ by the definition of G. Thus

$$\tau(G \otimes H) \le |S| = N.$$

Since clearly $\tau(G) = n$, we have

$$\tau(G \otimes H) = N = \tau_n(H) < n\tau(H) = \tau(G)\tau(H).$$

[C. Berge–M. Simonovits, in: *Hypergraph Seminar*, Lecture Notes in Math. 411, Springer, 1974, 21–33.]

    (c) By (a),

$$\tau^*(H^p) = \tau^*(H)^p,$$

thus

(1) $$\tau(H^p) \ge \tau^*(H)^p.$$

To estimate $\tau(H^p)$ from above we use 13.30. We have

$$d(H^p) = d(H)^p$$

by easy computation and thus,

(2)
$$\tau(H^p) \le (1 + \log d(H^p))\tau^*(H^p) =$$
$$= (1 + p\log d(H))\tau^*(H)^p.$$

By (1) and (2),

$$\tau^*(H) \le \sqrt[p]{\tau(H^p)} \le \sqrt[p]{1 + p\log d(H)}\,\tau^*(H).$$

Thus

$$\lim_{p \to \infty} \sqrt[p]{\tau(H^p)} = \tau^*(H).$$

[R. J. McEliece–E. C. Posner, *Ann. Math. Stat.* **42** (1971) 1706–1716.]

52. Suppose that H contains two edges E_1, E_2 with a common point x. Since it is τ-critical there exists a 1-cover t_i of $H - \{E_i\}$ such that

$$\sum_y t_i(y) = \tau(H) - 1.$$

Then define

$$t(y) = \begin{cases} 1/2, & \text{if } y = x, \\ (t_1(y) + t_2(y))/2, & \text{otherwise.} \end{cases}$$

Then, clearly, t is a fractional cover. Moreover, its size is

$$\sum_y t(y) = \frac{1}{2} + \frac{1}{2}\left(\sum_y t_1(y) + \sum_y t_2(y)\right) = \tau(H) - \frac{1}{2}.$$

Hence

$$\tau^*(H) \le \tau(H) - \frac{1}{2} < \tau(H).$$

[L. Lovász, in: *Hypergraph Seminar*, Lecture Notes in Math. 411, Springer, 1974, 111–126.]

53. Let $E_0 = \{v_1, \ldots, v_k\} \in E(H)$. Obviously, $k \ge 2$. Let w_i be a maximum matching of $H - v_i$ $(i = 1, \ldots, k)$; then, since H is ν-critical,

$$\sum_E w_i(E) = \nu(H) \qquad (i = 1, \ldots, k).$$

Set

$$w(E) = \begin{cases} 1/k, & \text{if } E = E_0, \\ (w_1(E) + \ldots + w_k(E))/k, & \text{if } E \ne E_0. \end{cases}$$

Then it is easy to check that w is a fractional matching; in fact, if $x \notin E_0$, then

$$\sum_{E \ni x} w(E) = \frac{1}{k}\left(\sum_{E \ni x} w_1(E) + \ldots + \sum_{E \ni x} w_k(E)\right) \le 1$$

and if $x = v_1$ (say), then

$$\sum_{E \ni v_1} w(E) = \frac{1}{k}\left(1 + \sum_{E \ni \nu} w_2(E) + \ldots + \sum_{E \ni v_1} w_k(E)\right) \le 1.$$

Thus $\nu^*(H)$ can be estimated as follows:

$$\nu^*(H) \ge \sum_E w(E) = \frac{1}{k}\left(1 + \sum_{E \ne E_0} (w_1(E) + \ldots + w_k(e))\right) =$$

$$= \frac{1}{k}(1 + k\nu(H)) = \nu(H) + \frac{1}{k} > \nu(H).$$

[ibid.]

54. Suppose indirectly that there are balanced hypergraphs with $\tau(H) > \nu(H)$ and consider a minimum example. Then

$$\tau(H - \{E\}) = \nu(H - \{E\})$$

for every $E \in E(H)$ and so,

$$\tau(H - \{E\}) = \nu(H - \{E\}) \le \nu(H) < \tau(H),$$

i.e. H is τ-critical. Since H cannot consist of disjoint edges, there are $E_1, E_2 \in E(H)$, $x \in V(H)$ with $x \in E_1 \cap E_2$.

Since $\tau(H - \{E_i\}) < \tau(H)$, we have a $(\tau(H) - 1)$-element point-cover T_i in $h - \{E_i\}$. Obviously, $E_i \cap T_i = \emptyset$. Set

$$W = (T_1 \triangle T_2) \cup \{x\}.$$

By 13.38, the subhypergraph H_W has chromatic number 2 (if 1-element edges are ignored). Let $\{A, B\}$ be bicoloration of it, $A \cap B = \emptyset$, $A \cup B = W$, $|A| \le |B|$. Since

$$|W| = |T_1 - T_2| + |T_2 - T_1| + 1 = 2\tau(H) - 2|T_1 \cap T_2| - 1,$$

we have $|A| \le \tau(H) - |T_1 \cap T_2| - 1$.

Now $A \cup (T_1 \cap T_2)$ covers every edge of H. For if an edge $E \in E(H)$, $E \ne E_1$, E_2 does not meet $T_1 \cap T_2$, then it meets both $T_1 - T_2$ and $T_2 - T_1$, i.e. it has at least two points in W and since $\{A, B\}$ is a bicoloration of H_W, $E \cap A \ne \emptyset$. Similarly, if $E = E_i$, then it has two points in W and we conclude as before. So $A \cup (T_1 \cap T_2)$ is a point-cover of H. But

$$|A \cup (T_1 \cap T_2)| \le \tau(H) - 1,$$

a contradiction [C. Berge; B, LP].

55. Obviously, (iii)$\Rightarrow$(i) and (iii)$\Rightarrow$(ii). Assume that (i) holds but (ii) fails, and consider a minimal such counterexample. Then for every vertex x,

$$\tau(H) > \nu(H), \qquad \tau(H - x) = \nu(H - x),$$

and since

$$\tau(H - x) \ge \tau(H) - 1,$$

we have

$$\nu(H - x) = \nu(H),$$

i.e. H is ν-critical. By 13.53, however, it follows that
$$\nu^*(H) > \nu(H),$$
a contradiction. So (i)$\Rightarrow$(iii).

Similarly, it follows that, if H is a minimal hypergraph, which satisfies (ii) but not (iii), then H is τ-critical (cf. 13.54), and so, by 13.52, $\tau(H) > \tau^*(H)$, a contradiction.

Thus (i)$\Leftrightarrow$(iii)$\Leftrightarrow$(ii). [L. Lovász; B, LP]

56. Let us first verify the assertion of the hint.

It suffices to show that, if we add a new edge, parallel to an old one to H, then the resulting hypergraph H_0 has property (iv). Let E be this new edge, parallel to $E_0 \in E(H)$.

It suffices to show that $d(H_0) = q(H_0)$ since this follows similarly for the partial hypergraphs. $d(H_0) \le q(H_0)$ is trivial.

If $d(H_0) = d(H) + 1$, then

$$q(H_0) \le q(H) + 1 = d(H) + 1 = d(H_0).$$

So suppose that $d(H_0) = d(H)$, i.e. no point of E_0 has maximum degree in H. Since H has property (iv), its edges can be $d(H)$-colored. Let E_0 have color red, say, and let $F_1,\ldots,F_t$ be the other red edges. Let $H' = H - \{F_1,\ldots,F_t\}$. Let us determine $d(H')$. If x has degree less than $d(H)$ in H, then it has degree less than $d(H)$ in H'; if x has maximal degree $d(H)$ in H', then it must be adjacent to a red edge, and since the points of E_0 have degree less than $d(H)$, $x \notin E_0$. Hence x has degree less than $d(H)$ in H'. Thus $d(H') < d(H)$, and, consequently, the edges of H' can be $(d(H)-1)$-colored. Color the edges $E, F_1,\ldots,F_t$ by a further color; we obtain a $d(H)$-coloration of the edges of H_0. This proves the assertion of the hint.

Now we turn to the solution of the problem.

(iv)$\Rightarrow$(i). Assume that $\nu(H) < \nu^*(H)$, and let w be a fractional matching of H of size larger than $\nu(H)$. Clearly we may assume that w is rational; let k be a common denominator of the weights $w(E)$. Replace each edge of H by $kw(E)$ parallel edges. The resulting hypergraph H_1 has $d(H_1) \leq k$ by the definition of w. Since it also satisfies (iv) by the lemma, we have

$$q(H_1) \leq k,$$

i.e. the edges of H_1 can be k-colored.

Since each color occurs at most $\nu(H_1) \leq \nu(H)$ times, it follows that

$$|E(H_1)| \leq k\nu(H).$$

Since by the construction of H_1

$$|E(H_1)| = k\nu^*(H),$$

it follows that

$$\nu^*(H) \leq \nu(H).$$

Since the converse inequality is trivial, (i) follows.

The relation (iii)$\Rightarrow$(iv) can be reduced to the relation (iv)$\Rightarrow$(iii) by constructing a hypergraph H^* as follows. Let $V(H^*)$ consist of all maximal (with respect to inclusion) 1-matchings of H; and, for every $E \in E(H)$, let the set E^* of all maximal 1-matchings of H containing E be an edge of H^*. We omit the details.

(iii)$\Rightarrow$(v). Suppose that H is normal, i.e. it satisfies (iii).

Let $E_1,\ldots,E_k \in E(H)$, $E_i \cap E_j \neq \emptyset$ $(1 \leq i,j \leq k)$. Then set $H' = (V(H), \{E_1,\ldots,E_k\})$. We have

$$\nu(H') = 1,$$

so we must have

$$\tau(H') = 1,$$

i.e. $E_1,\ldots,E_k$ must have a point in common. Thus H has the Helly-property.

Let G_1 be an induced subgraph of $\overline{L(H)}$. Then $G_1 \cong \overline{L(H')}$ for some partial hypergraph H' of H. We have

$$\chi(G_1) = \tau(H');$$

for a k-coloration of G_1 corresponds to a partition of $L(H')$ into k cliques and so, to a k-element point-cover of H' (since H' has the Helly-property). Similarly,

$$\omega(G_1) = \nu(H').$$

Since H is normal, we have

$$\chi(G_1) = \tau(H') = \nu(H') = \omega(G_1),$$

i.e. G is perfect.

The converse follows similarly [L. Lovász, B].

57. Suppose that G is perfect. Let $C_1, \ldots, C_N$ be the cliques of G. Take N points $p_1, \ldots, p_N$ and, with each $x \in V(G)$, associate an edge E_x such that

$$p_i \in E_x \Leftrightarrow x \in C_i.$$

Then

$$H = (\{p_1, \ldots, p_N\}; \{E_x : x \in V(G)\})$$

is a hypergraph such that $L(H) \cong G$. We claim that H satisfies property (iv) of normal hypergraphs. Let H' be any partial hypergraph of H. Then $L(H')$ is an induced subgraph of G, hence $L(H')$ is perfect, i.e.

$$\chi(L(H')) = \omega(L(H')).$$

Now observe that $\chi(L(H')) = q(H')$. Also, $\omega(L(H')) = d(H')$; for if $E_{x_1}, \ldots, E_{x_p}$ are edges of H' adjacent to the same point, then $x_1, \ldots, x_p$ are points of $L(H')$ forming a complete subgraph; hence $\omega(L(H')) \geq d(H')$. On the other hand, if $x_1, \ldots, x_p$ form a complete subgraph of $L(H')$, then there is a C_j such that $x_1, \ldots, x_p \in C_i$ and then p_j is adjacent to $E_{x_1}, \ldots, E_{x_p}$, hence $d(H') \geq \omega(L(H'))$. So

$$q(H') = \chi(L(H')) = \omega(L(H')) = d(H').$$

Thus H satisfies (iv) in 13.56 and so, it is normal. This implies that $\overline{L(H)} = \overline{G}$ is perfect.

If $\overline{G}$ is perfect, then the fact that G is perfect follows by interchanging the roles of G and $\overline{G}$. [ibid.]

§ 14. Ramsey Theory

1. (a) We use induction on $k+l$. For $k = 1$ or $l = 1$ the assertion is clear. Suppose that $k, l > 1$. Let $x \in V(G)$. Since

$$d_G(x) + d_{\overline{G}}(x) = |V(G)| - 1 = \binom{k+l}{k} - 1 = \binom{k+l-1}{k-1} + \binom{k+l-1}{k} - 1,$$

we have either

$$d_G(x) \geq \binom{k+l-1}{k-1} \quad \text{or} \quad d_{\overline{G}}(x) \geq \binom{k+l-1}{k}.$$

Assume, e.g. that the first inequality holds, and let G_1 be the subgraph induced by the neighborhood of x. Then, by the induction hypothesis, G_1 contains either a complete k-graph or $l+1$ independent points. In the second case we are done.

If G_1 contains a complete k-graph H, then $V(H) \cup \{x\}$ spans a complete $(k+1)$-graph.

(b) By (a),

$$R_2(a_1, a_2) \leq \binom{a_1 + a_2 - 2}{a_1 - 1}.$$

Now suppose that $k \geq 3$ and that $R_{k-1}(b_1, \ldots, b_{k-1})$ exists for each $b_1, \ldots, b_{k-1}$. Then we claim that

$$R_k(a_1, \ldots, a_k) \leq R_{k-1}(a_1, \ldots, a_{k-2}, R_2(a_{k-1}, a_k)).$$

In fact, consider a complete graph with $R_{k-1}(a_1, \ldots, a_{k-2}, R_2(a_{k-1}, a_k))$ points and k-color its edges. Identify the $(k-1)^{\text{st}}$ and k^{th} colors for the moment; then we know that either the ith color contains a complete a_i-graph for some $1 \leq i \leq k-2$ (in which case we are done) or the last color contains a complete $R_2(a_{k-1}, a_k)$-graph K. The edges of K are colored with $k-1$ and k and so, by definition, either color $k-1$ contains a complete a_{k-1}-graph or color k contains a complete a_k-graph. This completes the proof [see H, B, ES].

2. (a) I. We use induction on k. For $k = 2$, $R_2(3,3) \leq 6$ follows from 14.1a (here equality holds, as is shown by the pentagon).

Let x be any point of a k-colored complete graph on $\lfloor ek! \rfloor + 1$ points. There are $\lfloor ek! \rfloor$ edges adjacent to this point, which are split into k classes. Since

$$\lfloor ek! \rfloor = \left\lfloor \sum_{j=0}^{\infty} \frac{k!}{j!} \right\rfloor = \sum_{j=0}^{k} \frac{k!}{j!} = 1 + k \sum_{j=0}^{k-1} \frac{(k-1)!}{j!} = 1 + k \lfloor e(k-1)! \rfloor,$$

one of the k colors, let us call it red, contains at least $\lfloor e(k-1)! \rfloor + 1$ edges adjacent to x. Let S be the set of those points joined to x by a red edge.

If S spans a red edge this forms, together with x, a red triangle and we are finished. If S spans no red edge it spans a complete graph with $\lfloor e(k-1)! \rfloor + 1$ points, whose edges are $(k-1)$-colored; thus by the induction hypothesis, one of these $k-1$ colors contains a triangle and we are done again [see B].

II. To prove $R_3(3,3,3) = \lfloor e \cdot 3! \rfloor = 17$, we have to 3-color the edges of a complete 16-graph avoiding monochromatic triangles. Let x be any point, it easily follows that x must be adjacent to 5 edges of each color and those points joined to x by a red edge must span a green and a blue pentagon (Fig. 100). It is natural to assume that rotation through $120°$ maps red onto green and green onto blue, hence it suffices to specify the red edges. It is also quite natural to assume that rotation of each of the three pentagons by $72°$ keeps each color invariant. Some experimenting results in Fig. 101 (where the point x is not shown for the sake of clearness and the three pentagons are drawn in a geometrically different way to show this latter symmetry).

Remark: This coloration can be described as follows: Let P be a pentagon and $V(K_{16})$ the set of subsets of $V(P)$ with even cardinality. For $A, B \in V(K_{16})$, let (A, B) be colored

$$\begin{cases} \text{red} & \text{if } A \triangle B \text{ is an edge of } P, \\ \text{blue} & \text{if } A \triangle B \text{ is an edge of } \overline{P}, \\ \text{green} & \text{if } |A \triangle B| = 4. \end{cases}$$

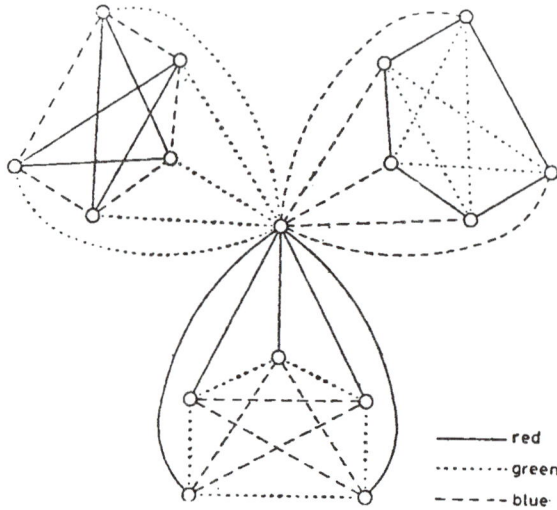

————— red

·············· green

— — — — blue

FIGURE 100

It easy to see that each edge gets one of the three colors. If $A, B, C \in V(K_{16})$, the

$$(A \triangle B) \triangle (B \triangle C) = A \triangle C;$$

hence it is seen that none of the three colors contains a triangle.

(b) The upper bound is trivial from 14.1a:

$$R_2(k, k) \leq \binom{2k-2}{k-1} < 2^{2k}.$$

To prove the lower bound we may suppose that $k \geq 4$. Let $n = \lceil 2^{k/2} \rceil$ and consider the hypergraph defined in the hint. Clearly this hypergraph is 2-colorable iff the edges of K_n can be 2-colored in such a way that no monochromatic complete k-graph arises, i.e. if $n < R_2(k, k)$.

The number of edges of H is, clearly, $\binom{n}{k}$ and H is $\binom{k}{2}$-uniform. So by 13.41, H is 2-colorable provided

$$\binom{n}{k} < 2^{\binom{k}{2}-1}.$$

But this is so, because

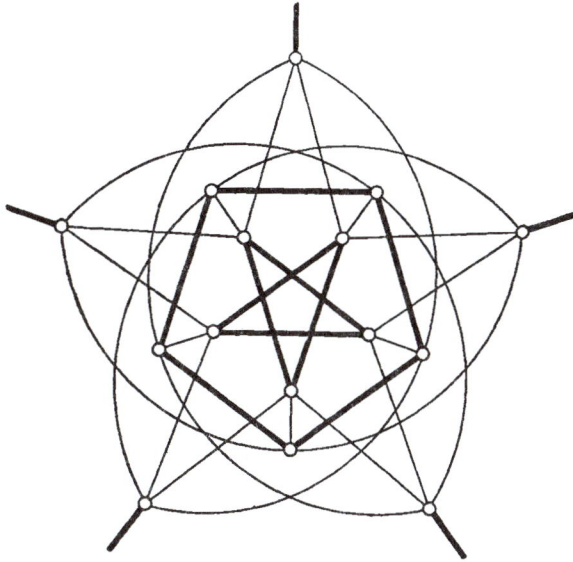

FIGURE 101

$$\binom{n}{k} < \frac{n^k}{2^{k-1}} < \frac{2^{\frac{k^2}{2}}}{2^{k-1}} = 2^{\binom{k}{2}-\frac{k}{2}+1} \le 2^{\binom{k}{2}-1}$$

[P. Erdős; see ES].

3. We use induction on r and $a_1 + \ldots + a_k$. For $r = 2$ we already know that $R_k^2(a_1, \ldots, a_k) = R_k(a_1, \ldots, a_k)$ exists. Suppose that $r > 2$ (as a matter of fact, $R_k^1(a_1, \ldots, a_k)$ trivially exists and we could start the induction with $r = 2$). Also, if $a_i = 1$ for any i, the assertion is trivial; so we may suppose that $a_i \ge 2$. Now let $x \in V(K_n^r)$. Color every $(r-1)$-tuples S in $V(K_n^r) - \{x\}$ with the color of $S \cup \{x\}$. Thus the edges of K_{n-1}^{r-1} are k-colored. So if

$$n - 1 \ge R_k^{r-1}(b_1, \ldots, b_k),$$

then, for some $1 \le i \le k$, there is a $V_i \subseteq V(K_n^r) - x, |V_i| = b_i$ such that all r-tuples of $V_i \cup \{x\}$ containing x have color i. Now if we choose here

$$b_i = R_k^r(a_1, \ldots, a_{i-1}, a_i - 1, a_{i+1}, \ldots, a_k)$$

(this number exists by the induction hypothesis on $a_1 + \ldots + a_k$) we find that either V_i has an a_j-subset all of whose k-tuples have color j for some $j \ne i$ (in which case we are done), or else, V_i contains an $(a_i - 1)$-subset V' all of whose r-tuples are colored i. Thus all k-subsets of $V' \cup \{x\}$ are colored i, and we are done again [see H, B].

(b) Let $m = R_k^r(a)$, $N = r + k\binom{m}{r}$ and $S = \{1, \ldots, N\}$. Trivially $N < k^{m^r}$ except for trivial cases. Assume that the set of all $(r+1)$-element subsets of S is k-colored. We prove that there is an a-element subset of S all of whose $(r+1)$-subsets have the same color. We follow the hint.

We define elements x_i and a "reservoir" U_i of points recursively such that

(i) $x_i, \ldots, x_i \notin U_i, x_{i+1}, \ldots, x_m \in U_i$,

(ii) for each subset $1 \le \nu_1 < \ldots < \nu_r \le i$, all $(r+1)$-tuples $(x_{\nu_1}, \ldots, x_{\nu_r}, y)$ with $y \in U_i$ have the same color, and moreover

(iii) $|U_i| \ge k^{\binom{m}{t} - \binom{i}{r}}$ $(i = 1, \ldots, m)$.

All these conditions clearly hold for $i = 1, \ldots, r-1$, if $x_1 = 1, \ldots, x_{r-1} = r-1$ and $U_i = \{i+1, \ldots, N\}$. Now assume that $x_1, \ldots, x_i, U_1, \ldots, U_i (r-1 \le i \le m-1)$ have been defined in accordance with (i), (ii) and (iii). Choose an arbitrary element of U_i for x_{i+1}. To define U_{i+1}, call two elements $y, y' \in U_i - \{x_{i+1}\}$ *equivalent*, if the colors of

$$(x_{\nu_1}, \ldots, x_{\nu_{r-1}}, x_{i+1}, y) \text{ and } (x_{\nu_1}, \ldots, x_{\nu_{r-1}}, x_{i+1}, y')$$

are the same for every $1 \le \nu_1 < \ldots < \nu_{r-1} \le i$. There are at most $k^{\binom{i}{r-1}}$ equivalence classes. Define U_{i+1} as the largest one of these. (i) and (ii) are trivially fulfilled. Also

$$|U_{i+1}| \ge \left\lceil \frac{|U_i| - 1}{k^{\binom{i}{r-1}}} \right\rceil \ge k^{\binom{m}{r} - \binom{i+1}{r}}.$$

Thus (iii) holds as well.

We now define a k-coloration of the r-subsets of $\{x_1, \ldots, x_m\}$ by coloring $(x_{\nu_1}, \ldots, x_{\nu_r})$ $(1 \le \nu_1 < \ldots < \nu_r \le m)$ with every common color of $(r+1)$-sets $(x_{\nu_1}, \ldots, x_{\nu_r}, y)$ $(y \in U_{\nu_r})$. By the choice of m, there is an a-subset A of $\{x_1, \ldots, x_m\}$ such that all r-tuples of A have the same color, say red. But then all $(r+1)$-subsets of A are red in the original coloration. In fact, let $(x_{\nu_1}, \ldots, x_{\nu_{r+1}})$ be an $(r+1)$-tuple of elements of A. Then by the choice of A, $(x_{\nu_1}, \ldots, x_{\nu_r})$ is red. But then, since $x_{\nu_{r+1}} \in U_{\nu_r}$, the above $(r+1)$-tuple is red by the definition of the coloration of r-tuples. This proves that $R_k^{r+1}(a) \le N$. [P. Erdős–R. Rado, *Proc. London Math. Soc.* Series **3**, 2 (1952) 417–439. Concerning the sharpness of this estimation, see P. Erdős–A. Hajnal–R. Rado, *Acta Math. Hung.* **16** (1965) 93–196.]

4. For $n = 3$ the assertion is trivial; suppose that $n > 3$. Let us remove a point x; we find, by induction, an $(n-1)$-circuit C in the rest, which is either monochromatic or has two points a, b such that one (a, b)-arc P is red, while the other one, Q is blue. The first case can be regarded as the special case of the latter when $a = b$.

We may assume that edge (a, x) is red. Let u be the point on Q next to a (if Q consists of a single point, u can be any neighbor of a). Then

$$P + (a, x) + (x, u) + (Q - a)$$

is a Hamiltonian circuit in the complete n-graph such that either the two (b, x)-arcs are red and blue, respectively, or the two (b, u) arcs are so [H. Raynaud; see B].

5. Let P, Q_1, Q_2 be as in the hint. By induction on k, we may assume that P has length $k - 1$. Let a, b; c_1, d_1; c_2, d_2 denote the endpoints of P, Q_1, Q_2, respectively. Suppose indirectly that there is a point $x \notin V(P) \cup V(Q_1) \cup V(Q_2)$. Since, from the structure of Q_1 and Q_2,

$$|V(Q_i) \cap V(P)| = \frac{1}{2}(|V(Q_i)| - 1) = |V(Q_I) - V(P)| - 1,$$

we have

$$|V(P) - V(Q_1) - V(Q_2)| = k + 2 - |V(Q_1) \cup V(Q_2) - V(P)| >$$
$$> k + 2 - |V(G) - V(P)| = 2k + 2 - \left\lfloor \frac{3k + 1}{2} \right\rfloor = \left\lfloor \frac{k}{2} \right\rfloor + 2.$$

Thus $Q_1 \cup Q_2$ misses at least $\left\lfloor \frac{k}{2} \right\rfloor + 3$ point of P. This implies that there is an edge $(u, v) \in E(P)$ such that $u, v \neq a, b$ and $u, v \notin V(Q_1 \cup Q_2)$.

One of the edges $(x, u), (x, v)$ must be blue; otherwise, replacing (u, v) by (u, x) and (x, v) we could increase the length of P. Let, say, (x, u) be blue. Then (c_1, u) must be red otherwise (c_1, u) and (x, u) could be added to Q_1. But then (c_1, v) must be blue by the same argument as above. Similarly, (c_2, v) is blue. Now

$$Q_1' = Q_1 + (c_1, v) + (c_2, v) + Q_2,$$
$$Q_2' = \{x\}$$

are two paths with total length greater than $|E(Q_1)| + |E(Q_2)|$ satisfying the same conditions as Q_1 and Q_2, a contradiction.

Thus $P \cup Q_1 \cup Q_2$ covers all points. Let us consider the edges (c_1, a), (c_2, a), (d_1, b), (d_2, b). By the maximality of P, these are blue. Hence, the circuit

$$C = Q_1 + (c_1, a) + (c_2, a) + Q_2 + (d_1, b) + (d_2, b)$$

is blue. The length of C is

$$2|V(G) - V(P)| = 2 \left\lfloor \frac{k + 1}{2} \right\rfloor.$$

If k is odd this is equal to $k+1$ and so, removing any edge of C we have a blue path of length k. Suppose that k is even. If there is a blue edge joining C to $V(G) - V(C)$, we find such a blue path again. So suppose that all edges connecting C to $V(G) - V(C)$ are red. Then a red path with length $2 \left\lfloor \frac{k+1}{2} \right\rfloor \geq k$ can trivially be

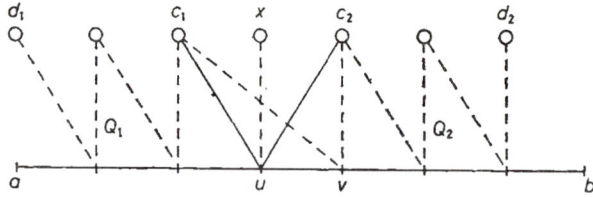

FIGURE 102

found (Fig. 102) [L. Gerencsér–A. Gyárfás, *Annales Univ. Sci. R. Eötvös* Sectio Math. X. (1967) 167–170].

6. (a) Let $(x_0, \ldots, x_{2k})$ be a red circuit. If (x_i, x_{i+2}) is red for some $0 \le i \le 2k$ (where $x_{2k+1} = x_0$, $x_{2k+2} = x_1$ etc.), then we trivially have a red $(2k)$-circuit. Thus we may assume that (x_i, x_{i+2}) is blue for each $0 \le i \le 2k$.

The circuit $(x_0, x_2, \ldots, x_{2k}, x_1, \ldots, x_{2k-1})$ is a blue $(2k+1)$-circuit. Therefore as above, we may assume that (x_i, x_{i+4}) is red for each $1 \le i \le 2k$. Now $(x_0, x_4, \ldots, x_{2k}, x_2, x_1)$ is a $(2k)$-circuit, which is red except possibly for its edge (x_{2k}, x_2). Thus we may assume that (x_{2k}, x_2) is blue and similarly, so is (x_i, x_{i+3}) for each $0 \le i \le 2k$. Now the circuit $(x_0, x_3, x_6, x_8, \ldots, x_{2k}, x_1, x_{2k-1}, x_{2k-3}, \ldots, x_5, x_2)$ is a blue $(2k)$-circuit.

(b) Let $(x_0, \ldots, x_{2k-1})$ be a red circuit, and suppose that there is no monochromatic $(2k-1)$-circuit. As before, (x_i, x_{i+2}) is blue for $1 \le i \le 2r$ (now setting $x_{2k+i} = x_i$).

We claim that $(x_i, x_{i+2\nu})$ is blue for each $1 \le i \le 2k$, $1 \le \nu \le k-1$. Suppose indirectly that $(x_0, x_{2\nu})$ is red (say). Then all edges of $(x_0, x_{2\nu}, x_{2\nu-1}, \ldots, x_1, x_{2\nu+2}, x_{2\nu+3}, \ldots, x_{2k-1})$ are red except possibly $(x_1, x_{2\nu+2})$. Thus $(x_1, x_{2\nu+2})$ must be blue. Similarly, the circuit

$$(x_0, x_1, \ldots, x_{2\nu-2}, x_{2k-1}, x_{2k-2}, \ldots, x_2)$$

implies that $(x_{2\nu_2}, x_{2k-1})$ is blue. Now the circuit

$$(x_{2k-1}, x_{2\nu-2}, x_{2\nu-4}, \ldots, x_0, x_{2k-2}, \ldots, x_{2\nu+2}, x_1, x_3, \ldots, x_{2k-3})$$

has length $2k-1$ and is blue, a contradiction. So $(x_i, x_{i+2\nu})$ is blue for every i, ν and thus, $\{x_1, \ldots, x_{2k-1}\}$, $\{x_0, \ldots, x_{2k-2}\}$ induce blue complete subgraphs.

(c) The case $m = 4$ can be verified directly; suppose that $m \ge 5$.

For the two graphs G_1, G_2 formed by the red and blue edges, respectively, we have

$$|E(G_1)| + |E(G_2)| + \binom{2m-1}{2},$$

whence, e.g.

$$|E(G_1)| \ge \frac{1}{2}\binom{2m-1}{2} = \frac{m-1}{2}(2m-1) > \frac{(m-1)(|V(G_1)| - 1)}{2}.$$

Thus by 8.26, G_1 contains an M-circuit for some $M \ge m$.

Consider a monochromatic M-circuit C with $M \geq m$, M minimal; let, e.g. C be red. If $M = m$, we are done. Suppose indirectly that $M > m$. By (a), (b), we also have a monochromatic $(M-1)$-circuit (contradicting the choice of M) or $M = 2k$ is even and we have two disjoint complete graphs K and H, both in blue, with $V(K) \cup V(H) \supseteq V(C)$, $|V(K)| \geq k$, $|V(H)| \geq k$. Choose here $|V(K)| + |V(H)|$ to be maximal.

If there are two independent blue edges connecting K to H, then obviously we will have a blue $(M-1)$-circuit. Thus we may suppose that all blue edges between K and H have a point in common $x \in V(K)$. If m is even we trivially form a red m-circuit from the edges connecting H to K so suppose that m is odd. Let $v \in V(G) - V(C)$. If v is joined by red edges to both $K - x$ and H, we can form a red m-circuit from these two red edges and the red (K, H)-edges. So suppose that for each $v \in V(G) - V(C)$ either all $(v, K-x)$-edges or all (v, H)-edges are blue. The latter cannot happen by the maximality of $|V(H)| + |V(K)|$ and the same reasoning implies that (v, x) is red.

If $|V(H)| \geq m$, we are trivially done so suppose that $|V(H)| \leq m-1$. Consider the sets $S = V(K-x)$, $T = V(G) - S - V(H)$. Then $|S| + |T| \geq (2m-1) - (m-1) = m$, $|S| \geq k \geq \frac{m-1}{2}$, and all edges spanned by S or joining S to T are blue. If $|S| > \frac{m-1}{2}$ here we trivially find a blue m-circuit. Similarly, we are finished, if $|S| = \frac{m-1}{2}$ but T induces at least one blue edge. Thus $|V(H)| = m-1$, $|S| = \frac{m-1}{2}$, $|T| = \frac{m+1}{2}$ and T induces a red complete graph. If a point of T is joined to at least two points of H by blue edges, we are finished again.

Thus we may suppose that at most one blue edge connects each $v \in T$ to H. Since $|V(H)| = m-1$, H contains $(m-3)/2$ points, which are connected to T by red edges only and a further point, which is connected to T by at most one blue edge. If M is the set of these $(m-1)/2$ points, we easily find a red m-gon in $V(T) \cup M$ (Fig. 103). [A. Bondy–P. Erdős, *J. Comb. Theory* **14** (1973) 46–54. For other Ramseyan results concerning circuits, see V. Rosta, *J. Comb. Theory* (B) **15** (1973) 94–104 and 105–120.]

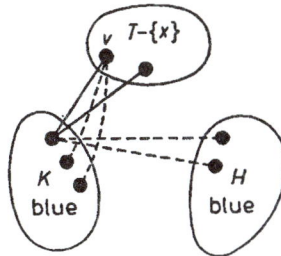

FIGURE 103

7. Suppose indirectly that the points of the plane can 3-colored with no two points at distance 1 having every same color. Let a in Fig. 17 in the hint have, say, color red. Then b, c must be blue and green (in some order). Therefore, f must be red. Similarly, g must be red. But then f, g form a pair at distance 1 with the same color, a contradiction [H. Hadwiger, *Elemente d. Math.* **16** (1961) 103–104].

8. Suppose that the triangle ABC in Fig. 18 in the hint has red vertices. Then, if any of D, E, F is red we are done; so suppose that D, E and F are blue. Now, if G or H is blue, then DEG or DFH is a monochromatic triangle; if G and H are red, then BGH is one [P. Erdős–R. L. Graham–P. Montgomery– B. L. Rothschild–J. Spencer–E. G. Straus, *J. Comb. Th.* **14** (1973) 341–363].

9. (a) Let us split the plane into strips of width $\frac{\sqrt{3}}{2}$, counting their lower edge with the strips but not the upper one. Color these strips alternately red and blue. It is easy to see that any regular triangle with sides 1 has to contain vertices from two adjacent strips and so, it must not be monochromatic.

(b) We may as well consider 1, $\sqrt{3}$, $\frac{\pi}{\sqrt{2}}$ as this only means a transformation of the unit by a scalar factor. We show that, if we 2-color the points of the plane, there will be a monochromatic regular triangle with side 1 or $\sqrt{3}$; in either case, the preceding problem will imply the existence of a monochromatic triangle with sides 1, $\sqrt{3}$, $\pi/\sqrt{2}$.

If not all points have the same color there must be a pair a, b of points at distance 2 colored differently; for any pair of points with different colors can be connected by a polygon with sides 2 and one of these sides must have differently colored endpoints.

So we may assume that a is red and b is blue in Fig. 19 in the hint. Also we may assume without loss of generality that c is red. If d or e is red we have a monochromatic regular triangle with side 1; if both d, e are blue we have a monochromatic regular triangle with side $\sqrt{3}$. [P. Erdős et al. in: *Infinite and Finite Sets*, Coll. Math. Soc. J. Bolyai **10** (1974) 529–583.]

10. (a) Let a, b the lengths of edges of the rectangle R. Let $v_0, \ldots, v_N$ be the vertices of a regular simplex in the N-dimensional space with side a, where $N = k^{k+1}$; let $w_0, \ldots, w_k$ be the vertices of a regular simplex in the k-dimensional space with side b. Consider the points (v_i, w_j) of the $(N+k)$-dimensional space $(0 \le i \le M, 0 \le j \le k)$. For each fixed i, the $k+1$ points (v_i, w_j) can be colored in $k^{k+1} = N$ ways; therefore, there are two indices $0 \le i_1 < i_2 \le N$ such that (v_{i_1}, w_j) and (v_{i_2}, w_j) have the same color for each $0 \le j \le k$. Also, there are two indices $0 \le j_1 < j_2 \le k$ such that (v_{i_1}, w_{j_1}) and (v_{i_1}, w_{j_2}) have the same color. Hence,

$$(v_{i_1}, w_{j_1}), (v_{i_1}, w_{j_2}), (v_{i_2}, w_{j_1}), (v_{i_2}, w_{j_2})$$

have the same color. These four points obviously form a rectangle congruent to R.

(b) Let $\mathbf{a}$, $\mathbf{b}$ be the two side vectors of R, then, as R is not rectangular, $\mathbf{ab} \ne 0$. We may suppose that $\mathbf{ab} = 1$, since this only means transformation of the unit by a scalar factor. We claim $n(4, R)$ does not exist. For any n, construct a 4-coloration

of the n-dimensional Euclidean space by giving $\mathbf{w}$ the color i ($0 \le i \le 3$), if $\lfloor \mathbf{w}^2 \rfloor \equiv i \pmod 4$. We prove that there exists no monochromatic set congruent to R.
Suppose that R' is such a set. Then

$$R' = \{\mathbf{w}, \mathbf{w} + \mathbf{a}', \mathbf{w} + \mathbf{b}', \mathbf{w} + \mathbf{a}' + \mathbf{b}'\}.$$

for some $\mathbf{w}, \mathbf{ab} \in E^n$. Since this is monochromatic, we have

$$\lfloor \mathbf{w}^2 \rfloor \equiv \lfloor (\mathbf{w} + \mathbf{a}')^2 \rfloor \equiv \lfloor (\mathbf{w} + \mathbf{b}')^2 \rfloor \equiv \lfloor (\mathbf{w} + \mathbf{a}' + \mathbf{b}')^2 \rfloor \pmod 4$$

and so,

$$\lfloor \mathbf{w}^2 \rfloor - \lfloor (\mathbf{w} + \mathbf{a}')^2 \rfloor - \lfloor (\mathbf{w} + \mathbf{b}')^2 \rfloor + \lfloor (\mathbf{w} + \mathbf{a}' + \mathbf{b}')^2 \rfloor \equiv 0 \pmod 4.$$

But

$$\lfloor \mathbf{w}^2 \rfloor - \lfloor (\mathbf{w} + \mathbf{a}')^2 \rfloor - \lfloor (\mathbf{w} + \mathbf{b}')^2 \rfloor + \lfloor (\mathbf{w} + \mathbf{a}' + \mathbf{b}')^2 \rfloor =$$
$$= \mathbf{w}^2 - (\mathbf{w} + \mathbf{a}')^2 - (\mathbf{w} + \mathbf{b}')^2 + (\mathbf{w} + \mathbf{a}' + \mathbf{b}')^2 + \Theta =$$
$$= 2\mathbf{a}'\mathbf{b}' + \Theta = 2 + \Theta,$$

where $-2 < \Theta < 2$, and so, this number cannot be divisible by 4. [Erdős et al. ibid.]

11. If the edges of a complete $(n+1)$-graph are k-colored, then, by 14.2a, we find a monochromatic triangle. Thus in the complete graph on $\{1, \ldots, n+1\}$ where, following the hint, we color the edge (i,j) with color ν, if $|i - j| \in A_\nu$, we find three points i, j, k such that $x = |i - j|$, $y = |j - k|$, $z = |i - k|$ belong to the same class. Let, say, $i < j < k$, then

$$x + y = z,$$

which proves the assertion. [I. Schur, *Jahresb. Deutschen Math.-Ver.* **25** (1916) 114.]

12. *First solution.* Suppose that $n_0(k, r - 1)$ exists (certainly so does $n_0(k, 1)$), then we prove that we can take

(1) $$n_0(k, r) = k^{n_0(k, r-1)} + n_0(k, r - 1).$$

For suppose that $n \ge n_0(k, r)$ and $\{1, \ldots, n\}$ is k-colored. Consider the colors of $i, i+1, \ldots, i + n_0(k, r-1) - 1$, for each $1 \le i \le k^{n_0(k, r-1)} + 1$. There are $k^{n_0(k, r-1)}$ sequences of colors, which can occur here, thus we have two numbers $i < j$ such that, for each $0 \le \nu \le n_0(k, r - 1) - 1$, the colors of $i + \nu$ and $j + \nu$ are the same.

By induction, we have a number a, $i \le a \le i + n_0(k, r - 1) - 1$, and integers $d_1, \ldots, d_{r-1}$ such that all numbers $a + d_{i_1} + \ldots + d_{i_\nu}$, $1 \le i_1 < \ldots < i_\nu \le r - 1$ have the same color. Set $d_r = j - i$, then, by the choice of i and j, all numbers

$$a + d_{i_1} + \ldots + d_{i_\nu}, 1 \le i_1 < \ldots < i_\nu \le r$$

have the same color.

Second solution. First we prove: If $A \subseteq \{1, \ldots, n\}$, $|A| \ge 2$, then there is a number $d \ge 1$ such that the set

$$A_d = \{a \in A : a + d \in A\}$$

has

$$|A_d| > \frac{|A|^2}{4n}.$$

In fact

$$\sum_{d=11}^{n-1} |A_d| = \binom{|A|}{2} \geq \frac{|A|^2}{4},$$

whence

$$\max_{1 \leq d \leq n-1} |A_d| \geq \frac{|A|^2}{4(n-1)} > \frac{|A|^2}{4n}.$$

Now let A be the largest color-class in a k-coloration of $\{1,\ldots,n\}$. Then we have a $d_1 \geq 1$ such that

$$|A_{d_1}| > \frac{|A|^2}{4n} \geq \frac{n}{4k^2} = \frac{4n}{(4k)^2} \quad \left(\text{as } |A| \geq \frac{n}{k} \right).$$

Similarly, we have a $d_2 \geq 1$ such that

$$A_{d_1,d_2} = |\{a \in A_{d_1} : a + d_2 \in A_{d_1}\}| = |\{a \in A :$$
$$a + d_1 \in A, a + d_2 \in A, a + d_1 + d_2 \in A\}| >$$
$$> \frac{n}{64k^4} = \frac{4n}{(4k)^4}$$

and repeating this r times we get numbers $d_1,\ldots,d_r$ such that

$$|A_{d_1,\ldots,d_r}| = |\{a \in A : a + d_{i_1} + \ldots + d_{i_\nu} \in A,$$
$$\text{for each } 1 \leq i_1 < \ldots < i_\nu \leq r\}| > \frac{4n}{(4k)^{2^r}}.$$

Any element of $A_{d_1,\ldots,d_r}$ satisfies our requirements. We need, however, $|A_{d_1,\ldots,d_i}| \geq 2$ for each $1 \leq i \leq r-1$. This is satisfied provided

$$n > \frac{1}{4}(4k)^{2^{r-1}}.$$

13. Let $n = \lfloor ek! \rfloor$. Assume that the non-empty subsets of $\{1,\ldots,n\}$ are k-colored with colors $1,\ldots,k$. Now color the pair $\{i,j\}$, $1 \leq i < j \leq n+1$ by the color of the interval $\{i,\ldots,j-1\}$. By 14.2a, there is a monochromatic triangle. If $1 \leq p < q < r \leq n+1$ are the vertices of this triangle, then $X = \{p,\ldots,q-1\}$, $Y = \{q,\ldots,r-1\}$ and $X \cup Y = \{p,\ldots,r-1\}$ have the same color.

14. (a) Suppose that $N(k,t)$ exists for every k ($t \geq 1$); we prove $N(k,t+1)$ exists for every k. Since $N(k,1)$ exists by the preceding problem, this will prove the assertion.

Let $a = N(k^2, t)$ and $N(k, t+1) = N(k^{2^a}, 1) + a$. Suppose that a coloration α of the set of all subsets of an $N(k, t+1)$-element set S is given. Let $T \subseteq S$, $|T| = a$. First we define a coloration α^* of the subsets of $S - T$, for which

$$\alpha^*(X) = (\alpha(X \cup Y) : Y \subseteq T).$$

This coloration uses k^{2^a} colors and so by the choice of $N(k, t+1)$, there exist two disjoint sets A_1, $B_1 \subseteq S - T$ such that $\alpha^*(A_1) = \alpha^*(B_1) = \alpha^*(A_1 \cup B_1)$, i.e.

(1) $$\alpha(A_1 \cup X) = \alpha(B_1 \cup X) = \alpha(A_1 \cup B_1 \cup X)$$

for each $X \subseteq T$. Now define a coloration α^{**} of the subsets of T by

$$\alpha^{**}(X) = (\alpha(X), \alpha(x \cup A_1)).$$

This coloration uses k^2 colors and hence by the definition of a, there are $2t$ disjoint sets A_2, $B_2, \ldots, A_{t+1}$, B_{t+1} such that for each sequence $2 \le i_1 < \ldots < i_\nu \le t+1$,

$$\alpha^{**}(C_{i_1} \cup \ldots \cup C_{i_\nu})$$

is the same for any choice of $C_j = A_j$ or B_j or $A_j \cup B_j$. Then A_1, $B_1, \ldots, A_{t+1}$, B_{t+1} satisfy the requirements of our problem.

(b) We use induction on r; for $r = 2$ the preceding problem contains the solution.

Assume that $b = n(k, r)$ exists; we define $n(k, r+1) = N(k, b)$ and prove it has the desired property. Let α be any k-coloration of all subsets of an $n(k, r+1)$-element set S. By (a), there exist $2b$ disjoint sets A_1, $B_1, \ldots, A_b$, B_b such that for each $1 \le i_1 < \ldots < i_\nu \le b$, all sets of the form

$$C_{i_1} \cup \ldots \cup C_{i_\nu} (C_j = A_j \text{ or } B_j \text{ or } A_j \cup B_j)$$

have the same color. Define a k-coloration α^* of the set $\{1, \ldots, b\}$ by

$$\alpha^*(\{i_1, \ldots, i_\nu\}) = \alpha(A_{i_1} \cup \ldots \cup A_{i_\nu}).$$

By the choice of b, there are disjoint subsets $I_1, \ldots, I_r \subseteq \{1, \ldots, b\}$ such that the color of any non-empty union of them is the same, say red. Now set

$$X_j = \bigcup_{i \in I_j} A_i, \quad Y_j = \bigcup_{i \in I_j} B_i.$$

Then $X_1, \ldots, X_r$, $Y_1, \ldots, Y_r$ are disjoint and any non-empty union of them is red. In fact, let V, $W \subseteq \{1, \ldots, r\}$ and set $I = \bigcup_{j \in V \cap W} I_j$, $I' = \bigcup_{j \in V - W} I_j$, $I'' = \bigcup_{j \in W - V} I_j$. Then we have

$$\alpha\left(\left(\bigcup_{j \in V} X_j\right) \cup \left(\bigcup_{j \in W} Y_j\right)\right) = \alpha\left(\left\{\bigcup_{i \in I}(A_i \cup B_i)\right\} \cup \left\{\bigcup_{i \in I'} A_i\right\} \cup \left\{\bigcup_{i \in I''} B_i\right\}\right).$$

By the choice of A_i and B_i this color is the same as

$$\alpha\left(\bigcup_{i\in I\cup I'\cup I''} A_i\right) = \alpha^*(I\cup I'\cup I'').$$

which is red.

Since the number of sets $X_1,\ldots,X_r$, $Y_1,\ldots,Y_r$ is $2r\geq r+1$, this proves the assertion. [R. L. Graham–B. L. Rothschild, *Bull. Amer. Math. Soc.* **75** (1969) 418–422; also see R. L. Graham–K. Leeb–B. L. Rothschild, *Adv. in Math.* **8** (1972) 417–433.]

15. Let $n=n(k,r)$, where $n(k,r)$ is e function in the preceding problem. Suppose that α is a k-coloration of $\{1,\ldots,n\}$. Then define a k-coloration α^* of the subsets of $S=\{1,\ldots,n\}$ by

$$\alpha^*(X) = \alpha(|X|)\,(X\subseteq S).$$

By 14.14, there are disjoint sets $X_1,\ldots,X_r\subseteq S$ such that all non-empty unions of any of them have the same color. Then the choice $d_i=|X_i|$ is an appropriate one. [J. Folkman–J. Sanders; see also R. L. Graham–K. Leeb–B. L. Rothschild, *Adv. in Math.* **8** (1972) 417–433.]

16. (a) Take $n_1(k,m)=k$; set $S=\{x_1,\ldots,x_k\}$. Consider the m-vectors

$$(m-2)\cdot\{x_1,\ldots,x_k\};\ (m-1)\{x_1\}+(m-2)\cdot\{x_2,\ldots,x_k\};$$
$$\ldots,(m-1)\cdot\{x_1,\ldots,x_{k-1}\}+(m-2)\cdot\{x_k\};\ (m-1)\cdot\{x_1,\ldots,x_k\}.$$

Some two of these, say

$$(m-1)\cdot\{x_1,\ldots,x_i\}+(m-2)\cdot\{x_{i+1},\ldots,x_k\}$$

and

$$(m-1)\cdot\{x_1,\ldots,x_j\}+(m-2)\cdot\{x_{j+1},\ldots,x_k\}$$

have the same color $(0\leq i<j\leq k)$. Let $X=\{x_{i+1},\ldots,x_j\}$, $\mathbf{b}=(m-1)\cdot\{x_1,\ldots,x_i\}+(m-2)\cdot\{x_{j+1},\ldots,x_k\}$, then the requirement is satisfied.

(b) We prove the existence of $n_r(k,m)$ by induction on r.

Set $a=k^{m^{r-1}}$, $k'=k^{m^a}$, $\hat{n}_{r+1}(k,m)=n_r(k',m)+a$. We prove that this choice is correct. Let $T\subset S$, $|T|=a$. Define a coloration of m^{S-T} by

$$\alpha^*(\mathbf{a}) = (\alpha(\mathbf{a}+\mathbf{d});\ \mathbf{d}\in m^T).$$

Here we use m^a-tuples of the original set of colors, so we use not more than $k^{m^a}=k'$ colors. Hence by the choice of $n_{r+1}(k,m)$, there exist r disjoint non-empty subsets $X_1,\ldots,X_r$ of $S-T$ and an m-vector $\mathbf{b}_1\in m^{S-X_1-\cdots-X_r}$ such that

$$\alpha^*\left(\sum_{i=1}^r a_i X_i + \mathbf{b}_1\right) = \alpha^*\left(\sum_{i=1}^r \min(a_i,m-2)X_i + \mathbf{b}_1\right)$$

holds for every $0\leq a_i\leq m-1$; i.e. we have

$$(1) \qquad \alpha\left(\mathbf{d} + \sum_{i=1}^{r} a_i X_i + \mathbf{b}_1\right) = \alpha\left(\mathbf{d} + \sum_{i=1}^{r} \min(a_i, m-2) X_i + \mathbf{b}_1\right)$$

for every $0 \le a_i \le m-1$ and $\mathbf{d} \in m^T$. Now define a coloration α^{**} of m^T by

$$\alpha^{**}(\mathbf{d}) = \left(\alpha\left(\mathbf{d} + \sum_{i=1}^{r} a_i X_i + \mathbf{b}_1\right) : 0 \le a_i \le m-2\right).$$

In this way an m^r-tuple of colors is a new color, so α^{**} uses at most $k^{m^r} = a$ colors. Since $|T| = a$, we find by (a) a set $X_{r+1} \subseteq T$ and an m-vector $\mathbf{b}_2 \in m^{T-X_{r+1}}$ such that

$$\alpha^{**}(mX_{r+1} + \mathbf{b}_2) = \alpha^{**}((m-2)X_{r+1} + \mathbf{b}_2),$$

i.e.

$$(2) \qquad \alpha\left((m-1)X_{r+1} + \sum_{i=1}^{r} a_i X_i + \mathbf{b}_1 + \mathbf{b}_2\right)$$

$$= \alpha\left((m-2)X_{r+1} + \sum_{i=1}^{r} a_i X_i + \mathbf{b}_1 + \mathbf{b}_2\right)$$

for any integers $0 \le a_i \le m-2$. Now let $\mathbf{b} = \mathbf{b}_1 + \mathbf{b}_2$ (this is an m-vector, as $\mathbf{b}_1$, $\mathbf{b}_2$ are disjoint). Then $X_1, \ldots, X_{r+1}$, $\mathbf{b}$ satisfy the requirement. Clearly $X_1, \ldots, X_{r+1}$ are disjoint from each other and $\mathbf{b}$. Let

$$\mathbf{a} = \sum_{i=1}^{r+1} a_i X_i + \mathbf{b},$$

then

$$\alpha(\mathbf{a}) = \alpha\left(\sum_{i=1}^{r} \min(a_i, m-2) X_i + a_{r+1} X_{r+1} + \mathbf{b}\right)$$

by (1) and

$$\alpha\left(a_{r+1} X_{r+1} + \sum_{i=1}^{r} \min(a_i, m-2) X_i + \mathbf{b}\right) =$$

$$= \alpha\left(\min(a_{r+1}, m-2) X_{r+1} + \sum_{i=1}^{r} \min(a_i, m-2) X_i + \mathbf{b}\right)$$

by (2). This completes the proof.

17. We use induction on m. For $m=1$ the answer follows by 14.14b (we may even take $\mathbf{b} = \emptyset$). Suppose that $N(k,r,m)$ exists and define $a = N(k,r,m)$,

$$N(k,r,m+1) = n_a(k,m+1),$$

where $n_a(k, m+1)$ is defined as in 14.16b. Then if $|S| \geq N(k, r, m+1)$ we find disjoint sets $Y_1, \ldots, Y_a \subseteq S$ and an $(m+1)$-vector $\mathbf{b}_1$ of points of $S - Y_1 - \ldots - Y_a$ such that

$$(1) \qquad \alpha \left(\sum_{i=1}^{a} a_i Y_i + \mathbf{b}_1 \right) = \alpha \left(\sum_{i=1}^{a} \min(a_i, m-1) Y_i + \mathbf{b}_1 \right)$$

holds for every $0 \leq a_i \leq m$. For each $\mathbf{j} \in m^{\{1, \ldots, a\}}$, define

$$\alpha^*(\mathbf{j}) = \alpha \left(\sum_{i=1}^{a} a_i Y_i + \mathbf{b}_1 \right),$$

where a_i is the number of times i occurs in $\mathbf{j}$. Now by the choice of a, there are disjoint subsets $V_1, \ldots, V_r \subseteq \{1, \ldots, a\}$ and an m-vector $\mathbf{k}$ of elements of $\{1, \ldots, a\} - V_1 - \ldots - V_r$ such that

$$\alpha^* \left(\sum_{j=1}^{r} b_j V_j + \mathbf{k} \right)$$

is the same color, say red, for every choice of $0 \leq b_j \leq m-1$. Define

$$X_j = \bigcup_{i \in V_j} Y_i, \mathbf{b} = \mathbf{b}_1 + \sum_{i=1}^{a} a_i Y_i,$$

where a_i is the number of times i occurs in $\mathbf{k}$. Clearly $\sum_{i=1}^{a} a_i Y_i$ is disjoint from X_j and $\mathbf{b}_1$, and so $X_1, \ldots, X_r, \mathbf{b}$ are mutually disjoint. Moreover, if $0 \leq a_i \leq m$, then

$$\alpha \left(\sum_{i=1}^{r} a_i X_i + \mathbf{b} \right) = \alpha \left(\sum_{i=1}^{r} \min(a_i, m-1) X_i + \mathbf{b} \right)$$

by (1); moreover, this is equal to

$$\alpha^* \left(\sum_{i=1}^{r} \min(a_i, m-1) V_i + \mathbf{k} \right),$$

which is red. This proves that $N(k, r, m+1)$ has the desired property. [A. Hales–R. I. Jewett, *Trans. Amer. Math. Soc.* **106** (1963) 222–229.]

18. Let $w = (m-1)N(k, 1, m)$, $|S| = N(k, 1, m)$. Let α be any k-coloration of $\{0, \ldots, w\}$. We define a k-coloration of m^S by setting

$$\alpha^*(\mathbf{a}) = \alpha(|\mathbf{a}|).$$

(here $|\mathbf{a}|$ denotes the sum of entries of $\mathbf{a}$). Since $0 \leq |\mathbf{a}| \leq (m-1)|S| = w$, this coloration is well defined. By the choice of $|S|$, there exists a set $X \subseteq S$ and an m-vector $\mathbf{b}$ on $S - X$ such that

$$\alpha^*(\mathbf{b}) = \alpha^*(\mathbf{b} + X) = \alpha^*(\mathbf{b} + 2X) = \ldots = \alpha^*(\mathbf{b} + (m-1)X),$$

i.e.

$$\alpha(|\mathbf{b}|) = \alpha(|\mathbf{b}| + |X|) = \ldots = \alpha(|\mathbf{b}| + (m-1)|X|).$$

Thus, $(|\mathbf{b}|, |\mathbf{b}|+|X|, \ldots, |\mathbf{b}|+(m-1)|X|)$ is a monochromatic arithmetic progression [cf. R. L. Graham–B. L. Rothschild, *Proc. Amer. Math. Soc.* **42** (1974) 385–386].

19. Suppose that $\{1, \ldots, N\}$ has a k-coloration α_N for which no subset has property P. Consider the colors $\alpha_N(1)$. Since we use only k colors, there will be one occurring infinitely many times, say

$$\alpha_{N_1^1}(1) = \alpha_{N_2^1}(1) = \ldots .$$

Denote this common color by $\alpha(1)$. Now consider the colors $\alpha_{N_i^1}(2)$. Again we will find an infinite subsequence $N_1^2, N_2^2, \ldots$ of $N_1^1, N_2^1, \ldots$ such that

$$\alpha_{N_1^2}(2) = \alpha_{N_2^2}(2) = \ldots = \alpha(2).$$

Now suppose that we have defined colors $\alpha(1), \ldots, \alpha(n)$, so that still infinitely many indices $N_1^n, N_2^n, \ldots$ exist with

$$\alpha_{N_i^n}(m) = \alpha(m) \qquad \text{for } 1 \leq m \leq n \text{ and } i = 1, 2, \ldots .$$

Then among the colors $\alpha_{N^n}(n+1)$, one color occurs infinitely often; denote this by $\alpha(n+1)$.

We claim the coloration α of all natural numbers is such that no finite subset has property P. Assume indirectly that a subset $S \subseteq \{1, \ldots, n\}$ has property P. Then

$$\alpha_{N^n}(m) + \alpha(m)$$

for each $m \in S$ and thus, S has property P in the coloration $\alpha_{N_1^n}$ too, a contradiction.

20. (a) Write each natural number m in the form $m = 5^{\alpha_m}(5\beta_m + \gamma_m)$, $1 \leq \gamma_m \leq 4$ and color m with color γ_m. Thus the set of all natural numbers is 4-colored. We claim that $x + y = 3z$ has no monochromatic solution. Suppose that it has one in the γth class, then

$$5^{\alpha_x}(5\beta_x + \gamma) + 5^{\alpha_y}(5\beta_y + \gamma) = 3 \cdot 5^{\alpha_z}(5\beta_z + \gamma).$$

Divide here by 5^α, where $\alpha = \min(\alpha_x, \alpha_y, \alpha_z)$, and consider the equation mod 5. Then we get

$$\xi_x \gamma + \xi_y \gamma \equiv 3\xi_z \gamma \pmod 5,$$

where $\xi_x = 1$, if $\alpha = \alpha_x$ and $\xi_x = 0$, if $\alpha < \alpha_x$ (similarly for ξ_y and ξ_z). Then

$$\xi_x + \xi_y \equiv 3\xi_z \pmod 5,$$

which, obviously, has a unique solution in ξ_x, ξ_y, $\xi_z = 0, 1$: $\xi_x = \xi_y = \xi_z = 0$. But this is a contradiction as α must be equal to one of α_x, α_y, α_z.

(b) We prove by induction on k that $x + 2y = z$ has a monochromatic solution. For $k = 1$, this is obvious. Suppose that it holds for $k - 1$ colors.

It follows as in 14.19 that there is a natural number N with the property that if we $(k-1)$-color the numbers $\{1, \ldots, N\}$, then one of the color classes will contain a solution of $x + 2y = z$.

Now suppose that the natural numbers are k-colored. By van der Waerden's theorem 14.18, one of the classes, the "red" class say, contains an arithmetic progression

$$\{a, a + d, a + 2d, \ldots, a + 2Nd\}.$$

Let us seek a monochromatic solution of (b) in the form $x = a$, $y = \kappa d$, $z = a + 2\kappa d$. Now x and z are red for each $1 \le \kappa \le N$, so if κd is red for any $1 \le \mathbf{x} \le N$ we are finished. Suppose that this is not the case, then we have the numbers $d, 2d, \ldots, Nd$ colored with $k - 1$ colors. Define a new coloration of $\{1, \ldots, N\}$ by giving κ the color of κd. Here we use $k - 1$ colors, hence by the definition of N, there are x_1, y_1, z_1 having the same color and satisfying $x_1 + 2y_1 = z$. Now $(dx_1) + 2(dy_1) = (dz_1)$ and dx_1, dy_1, dz_1 have the same color in the original k-coloration, which proves the assertion.

21. I. (i)$\Rightarrow$(ii). Consider a $(0, 1)$-solution of $(*)$; we may assume that it looks like

$$\sum_{i=1}^{r} a_i \cdot 1 + \sum_{i=r+1}^{m} a_i \cdot 0 = 0.$$

Now since the equation

$$(**) \qquad\qquad a_1 \cdot x - a_1 \cdot y + \left(\sum_{i=r+1}^{m} a_i \right) \cdot z = 0$$

also satisfies (i) a any monochromatic solution of $(**)$ yields a monochromatic solution of $(*)$ with $x_1 = x$, $x_2 = \ldots = x_r = y$, $x_{r+1} = \ldots = x_m = z$, we may assume that $(*)$ looks like

$$(***) \qquad\qquad Ax - Ay + Bz = 0.$$

We use induction on k. The case $k = 1$ is trivial, so suppose that $k > 1$ and (ii) holds for $k - 1$ colors. As in 14.19, we can find a natural number N such that if $\{1, \ldots, N\}$ is $(k-1)$-colored, $(***)$ has a solution in one of the color classes. Now suppose that all natural numbers are k-colored. By van der Waerden's theorem, we find a monochromatic (say, red) arithmetic progression

$$a, a + d, \ldots, a + BNd.$$

Then from

$$A(a + B\kappa d) - Aa + B(\kappa Ad) = 0$$

it follows that if κAd is red for some $1 \leq \kappa \leq N$, we are done. So suppose that κAd is never red; then by the choice of N, we find numbers x_1, y_1, z_1 such that $x_1 Ad$, $y_1 Ad$, $z_1 Ad$ have the same color and then

$$A(x_1 Ad) - A(y_1 Ad) + B(z_1 Ad) = 0.$$

II. (ii)$\Rightarrow$(i). Let p be a very large prime and write $t = p^{\alpha_t}(\beta_t p + \gamma_t)$ $(1 \leq \gamma_t \leq p-1)$. Let us color t with γ_t. Then by (ii), (*) has a monochromatic solution; i.e. there are non-negative integers $\alpha^{(1)}, \ldots, \alpha^{(m)}$, $\beta^{(1)}, \ldots, \beta^{(m)}$, $\gamma (1 \leq \gamma \leq p-1)$ such that

(1)
$$\sum_{i=1}^{m} a_i p^{\alpha^{(i)}} (\beta^{(i)} + \gamma) = 0.$$

Set $\alpha = \min(\alpha^{(1)}, \ldots, \alpha^{(m)})$ and divide here by p^{α}, then consider (1) mod p. We get

$$\sum_{i=1}^{m} a_i \xi_i \gamma \equiv 0 \pmod{p},$$

where $\xi_i = 1$ or 0 according as $\alpha_i = \alpha$ or $\alpha_i > \alpha$. Hence

$$\sum_{i=1}^{m} a_i \xi_i \equiv 0 \pmod{p}.$$

Since p is large (e.g. it suffices to take $p > |a_1| + \ldots + |a_m|$), we get

$$\sum_{i=1}^{m} a_i \xi_i = 0.$$

Since $\alpha = \alpha_i$ for at least one i, $\xi_1, \ldots, \xi_m$ is a non-trivial $(0, 1)$ solution of (*). This proves (i). [R. Rado, *Math. Zeitschr.* **36** (1933) 424–480.]

22. We specify $f(k)$ later. Let $1, \ldots, k$ be k colors and color the integers with them at random, independently of each other and giving each integer each color with probability $1/k$. Let A_j denote the event that $S + j$ does not meet one of the colors. Then

$$P(A_j) \leq \sum_{i=1}^{k} P(S + j \text{ does not meet color } i) = k \left(1 - \frac{1}{k}\right)^{f(k)}.$$

Now form a graph G on Z, the set of integers, by connecting j_1 to j_2, if $(S+j_1) \cap (S+j_2) \neq$. Then A_j is independent of

$$\{A_\nu : (j, \nu) \notin E(G)\}.$$

Finally, observe that the degree of each point in G is

$$d = |\{\nu : (0, \nu) \in E(G)\}| = |\nu : (S + \nu) \cap S \neq 0\}| = |\{s - s' : s, s' \in S\}| \leq f(k)^2.$$

Thus, if

(1)
$$k\left(1 - \frac{1}{k}\right)^{f(k)} \le \frac{1}{4f(k)^2},$$

we have

$$\mathsf{P}(A_j) \le \frac{1}{4d}$$

and thus, by 2.18,

(2)
$$\mathsf{P}(\overline{A}_0 \ldots \overline{A}_N) > 0$$

for any N. Of course, we cannot assert that $P\left(\prod_{-\infty}^{\infty} \overline{A}_j\right) > 0$, but the existence of a coloration such that all translated copies of S meet all colors still follows. One possibility would be to apply compactness theorems (cf. 14.19) but here we can construct this coloration directly.

Let r be the diameter of S ($r = \max S - \min S$) and $N > k^{r+1}$. Consider a k-coloration for which $\overline{A}_0 \ldots \overline{A}_N$ holds, i.e. $S+1, \ldots, S+N$ meet all colors. Since $N > k^{r+1}$, there are two segments $\{\nu, \nu+1, \ldots, \nu+r\}$ and $\{\mu, \mu+1, \ldots, \mu+r\}$ ($\nu < \mu$), which are colored similarly. Now, if we keep the coloration of $\nu, \nu+1, \ldots, \mu-1$ but extend this in both directions with period $\mu - \nu$ then, as is easily verified, $S+j$ will meet all colors for every integer j.

It only remains to find an $f(k)$ for which (1) holds. Certainly $f(k) = \lfloor ck \log k \rfloor$ will be appropriate for large enough c, because

$$\left(1 - \frac{1}{k}\right)^{f(k)} \approx \left(1 - \frac{1}{k}\right)^{cl \log k} < e^{-c \log k} = \frac{1}{k^c},$$

while

$$\frac{1}{4f(k)^2} = \frac{1}{4c^2 k^2 \log^2 k} > \frac{1}{k^{c-1}}.$$

[P. Erdős–L. Lovász, in: *Infinite and Finite Sets*, Coll. Math. Soc. J. Bolyai **10**, Bolyai–North-Holland, (1974) 609–627.]

23. (a) Let us denote the elements of $GF(q)$ by $0, 1, \ldots, q-1$ (the operations between them will play no role.) Each element of the n-dimensional affine space over $GF(q)$ can be represented as a vector $(a_1, \ldots, a_n)$, $0 \le a_i \le q-1$, which in turn can be regarded as a $(q-1)$-vector of elements of $\{1, \ldots, n\}$. Thus if these are k-colored and $n > N(k, r, q-1)$, where $N(k, r, q-1)$ is defined as in 14.17, then there will be disjoint non-empty sets $V_1, \ldots, V_r \subseteq \{1, \ldots, n\}$ and a q-vector $\mathbf{b}$ of points of $\{1, \ldots, n\} - V_1 - \ldots - V_r$ such that all vectors $\sum_{i=1}^{r} a_i V_i + \mathbf{b} (0 \le a_i \le a-1)$ have the same color; let, say,

$$V_i = \{k_i + 1, \ldots, k_{i+1}\} \quad (i = 1, \ldots, r; \quad 0 = k_1 < k_2 < \ldots < k_{r+1} = m)$$

and let j occur in $\mathbf{b}\,b_j$ times (clearly $b_j = 0$ for $1 \leq j \leq m$), then all vectors of the form

$$(\underbrace{a_1, \ldots, a_1}_{1}, \underbrace{a_2, \ldots, a_2}_{k_2}, \underbrace{a_3, \ldots,}_{k_3} \underbrace{a_r}_{k_{r+1}}, b_{m+1}, \ldots, b_n)$$

have the same color. But these vectors form an r-dimensional subspace.

(b) We prove the (formally) more general assertion formulated in the hint by simultaneous induction on $r_1, \ldots, r_k$ and k. For $k = 1$ we clearly can take

$$(1) \qquad \overline{N}(1; r_1; q-1) = r_1.$$

On the other hand, if $r_i = 0$ for some $1 \leq i \leq k$, say $r_k = 0$, then

$$(2) \qquad \overline{N}(k; r_1, \ldots, r_{k-1}, 0; q-1) = \overline{N}(k-1; r_1, \ldots, r_{k-1}, q-1)$$

will be an appropriate choice. For if any point has color k we are done; if color k does not occur we have the case with $k-1$ colors. So we may assume that $r_i > 0$ for $1 \leq i \leq k$. Then let

$$(3) \qquad \overline{N}(k; r_1, \ldots, r_{k-1}; q-1) = N(k; m, q-1),$$

where

$$m = 1 + \max_{1 \leq i \leq k} \overline{N}(k; r_1, \ldots, r_{i-1}, r_i - 1, r_{i+1}, \ldots, r_k; q-1)$$

and $N(K, m, q-1)$ is defined as in part (a). To prove that (3) is an appropriate choice let $n \geq \overline{N}(k; r_1, \ldots, r_k; q-1)$ and suppose that the points of the n-dimensional projective space over $GF(q)$ are k-colored. Let $\sum_0$ be a hyperplane (the "infinitely distant" hyperplane). Remove $\sum_0$, then the remaining affine space contains a monochromatic subspace Φ of dimension m by (3) and (a). Let Φ have, e.g. color 1. Let Φ_1 be the subspace of the projective space spanned by Φ and put $\Psi = \Phi_1 \cap \sum$. Then Ψ is an $(m-1)$-dimensional projective space, which is k-colored. Thus, by the choice of m, Ψ has a monochromatic subspace Δ of dimension $r_1 - 1$ in the first color or of dimension r_i in the ith color ($2 \leq i \leq k$). In the latter case we are done. If Δ is $(r_1 - 1)$-dimensional and has color 1, we can take any r_1-dimensional subspace Δ_1 of Φ_1 such that $\Delta_1 \cap \sum = \Delta$ and then δ_1 is an r_1-dimensional subspace with color 1. [R. L. Graham–B. L. Rothschild, *Trans. Amer. Math. Soc.* **159** (1971) 257–292.]

24. *First solution*: Let $n \geq R_{k+1}^{r-1}(r, \ldots, r)$, where R is the Ramsey number as in 14.3. Let $S = \{1, \ldots, n\}$, and denote by V the set of all $(r-1)$-element subsets of S. Define a hypergraph on V by letting the r $(r-1)$-tuples of each r-subset of S form an edge of H. Then clearly H is r-uniform, any two edges of H have at most one point in common, and the fact that H is not k-colorable is just the assertion of Ramsey's theorem.

Second solution: Let $n \geq N(k; 1, q-1)$, where q is a prime power. Let $V(H)$ be the set of points of the n-dimensional affine space over $GF(q)$ and let $E(H)$ consist of the lines of this affine space. Then any two edges of H have at most one points in common and it is q-uniform. The assertion of 14.23a is just that H is not k-colorable. This settles the problem for the case when r is a prime power.

Now let r be arbitrary and select a prime power $q \geq r$. Carry out the above construction for q and then select an arbitrary r-subset of each edge of the resulting hypergraph. These r-element sets constitute a hypergraph with the desired properties.

25. For each a_i, let $t(i)$ denote the maximum length of subsequences

$$a_i \leq a_{j_1} \leq \ldots \leq a_{j_{t-1}}$$

with

$$i < j_1 < \ldots < j_{t-1}.$$

Suppose that $\max t(i) \leq k$ (otherwise we are done). Then there is a τ such that $t(i) = \tau$ holds for at least $k+1$ indices i. Let

$$t(i_1) = \ldots = t(i_{k+1}) = \tau, \ i_1 < \ldots < i_{k+1}.$$

Then we claim that

$$a_{i_1} > a_{i_2} > \ldots > a_{i_{k+1}}.$$

Suppose indirectly that

$$a_{i_\nu} \leq a_{i_\mu} (\nu < \mu).$$

Then, since $t(a_{i_\mu}) = \tau$, we have a sequence

$$i_\mu < j_1 < \ldots < j_{\tau-1}$$

with

$$a_{i_\mu} \leq a_{j_1} \leq \ldots \leq a_{j_{\tau-1}}.$$

Then

$$a_{i_\nu} \leq a_{i_\mu} \leq a_{j_1} \leq \ldots \leq a_{j_{\tau-1}},$$

contradicting $t(a_{i_\nu}) = \tau$.

Remark: Define $i \prec j$ if $i < j$ and $a_i \leq a_j$. Then we have a partial order on the set $\{1, \ldots, n^2 + 1\}$. Thus the result would follow from 9.32.

26. (a) Consider $t(x)$ as in the hint. Observe that $\max t(x) = t(1)$. We use induction on n.

First we prove the assertion of the hint. Assume indirectly that $f(x) > 2^k$ and $t(x) \geq t(1) - k$. Then $x > 2^k$ and $k < n-1$. Let

$$x = a_1 < \ldots < a_{t(x)},$$
$$f(a_1) \leq \ldots \leq f(a_{t(x)}).$$

By the induction hypothesis, there exist

$$1 \leq b_0 < b_1 < \ldots < b_k \leq 2^k$$

with

$$f(b_0) \leq f(b_1) \leq \ldots \leq f(b_k).$$

Then

$$f(b_0) \le f(b_1) \le \ldots \le f(b_k) \le b_k \le 2^k < f(a_1) \le \ldots \le f(a_{t(x)}),$$

whence

$$t(1) \ge t(x) + k + 1,$$

a contradiction.

Now observe that if $t(x) = t(y)$, then $f(x) \ne f(y)$ (otherwise, the chain starting with $y(x < y)$ could be extended to x) and thus,

$$|\{x : t(x) = t(1) - k\}| \le 2^k \quad (0 \le k \le t(1) - 1).$$

Hence,

$$2^{n-1} = \sum_{k=0}^{t(1)-1} |\{x : t(x) = t(1) - k\}| \le \sum_{k=0}^{t(1)-1} 2^k = 2^{t(1)} - 1.$$

Hence

$$t(1) \ge n,$$

which was to be proved.

(b) Let

$$\varphi_0(2^k + l) = 2^k - l(k = 0, \ldots, n - 2; 0 \le l \le 2^k - 1).$$

Then $\varphi_0(i) \le i$ $(i = 1, \ldots, 2^{n-1} - 1)$ and φ_0 consists of $n - 1$ monotone decreasing intervals; any sequence

$$1 \le a_1 < \ldots < a_t \le 2^{n-1} - 1$$

with

$$f(a_1) \le \ldots \le f(a_t)$$

contains at most one number from each of these intervals. Hence $t \le n - 1$. [E. Harzheim, *Publ. Math. Debrecen* **14** (1967) 45–51.]

27. (a) Let $\eta(x,y)$ be the maximum length of a path $(x, y = y_1, y_2, \ldots, y_\eta)$ with

$$f(x, y_1) \le f(y_1, y_2) \le \ldots \le f(y_{\eta-1}, y_\eta)$$

and also, let $\varphi(x,y)$ be the maximum length of paths $(x, y = y_1, y_2, \ldots, y_\varphi)$ with

$$f(x, y_1) > f(y_1, y_2) > \ldots > f(y_{\varphi-1}, y_\varphi).$$

Observe that
(∗) for any two (adjacent) edges (x, y) and (y, z), either $\eta(x, y) > \eta(y, z)$ or $\varphi(x, y) > \varphi(y, z)$. In fact, let, e.g. $f(x, y) \le f(y, z)$ and $(y, z = z_1, z_2, \ldots, z_{\eta(y,z)})$ be a path with

$$f(y, z_1) \le f(z_1, z_2) \le \ldots \le f(z_{\eta(y,z)-1}, z_{\eta(y,z)}).$$

Then the path $(x, y, z_1, \ldots, z_{\eta(y,z)})$ shows that $\eta(x, y) \ge \eta(y, z_1) + 1$.

For $w \in V(T_n)$, let

$$P_w = \{(\eta(w,x), \varphi(w,x)) : (w,x) \in E(T_n)\}$$

and let $\overline{P}_w$ consist of all maximal elements of P_w (with respect to the partial ordering $(\eta, \varphi) \leq (\eta', \varphi')$ iff $\eta \leq \eta'$ and $\varphi \leq \varphi'$). Then we claim that

$$(**) \qquad\qquad\qquad \overline{P}_w \neq \overline{P}_v, \text{ if } w \neq v.$$

For suppose that, e.g., $(v,w) \in E(T_n)$. By the definition of $\overline{P}_v$, there is a pair $(\eta_0, \varphi_0) \in \overline{P}_v = \overline{P}_w$ where $\eta(v,w) \leq \eta_0$ and $\varphi(v,w) \leq \varphi_0$. Let $\eta_0 = \eta(w,x)$ and $\varphi_0 = \varphi(w,x)$. Then $\eta(v,w) \leq \eta(w,x)$ and $\varphi(v,w) \leq \varphi(w,x)$, which contradicts $(*)$.

All that remains is to count how many sets come into consideration as sets $\overline{P}_w$. If $\overline{P}_w = \{(\eta_1, \varphi_1), \ldots, (\eta_k, \varphi_k)\}$, then $\eta_1 1, \ldots, \eta_k$ are distinct (so are $\varphi_1, \ldots, \varphi_k$) and if, say, $\eta_1 < \ldots < \eta_k$, then $\varphi_1 > \ldots > \varphi_k$. Hence, the two sets $\{\eta_1, \ldots, \eta_k\}$ and $\{\varphi_1, \ldots, \varphi_k\}$ uniquely determine $\overline{P}_w$. For a fixed k, there are at most $\binom{p'}{k}$ ways to choose $\eta_1, \ldots, \eta_k$ and $\binom{q'}{k}$ ways to choose $\varphi_1, \ldots, \varphi_k$, where $p' = \max \eta(x,y)$ and $q' = \max \varphi(x,y)$. Thus the number of distinct sets $\overline{P}_w$ is

$$\leq \sum_{k=0}^{p'} \binom{p'}{k}\binom{q'}{k} = \sum_{k=0}^{p'} \binom{p'}{k}\binom{q'}{q'-k} = \binom{p'+q'}{p'}.$$

Thus

$$\binom{p'+q'}{p'} \geq n > \binom{p+q-2}{p-1},$$

whence either $p' \geq p$ or $q' \geq q$. This proves the assertion.

(b) With a slight change in notation, we are going to construct a transitive tournament T_n on $n = \binom{p+q}{p}$ points and a function

$$f : E(T_n) \to \mathbf{Z}$$

(which will be an injection) such that the maximum length of a monotone increasing path is p and the maximum length of a monotone decreasing path is q. The case $p = 1$ or $q = 1$ is trivial; so suppose that $p, q \geq 2$.

We have

$$\binom{p+q}{p} = \binom{p+q-1}{p} + \binom{p+q-1}{p-1}.$$

By induction on $p+q$, we construct a transitive tournament T_m on $m = \binom{p+q-1}{p}$ points and a tournament T_l on $l = \binom{p+q-1}{p-1}$ points, together with two functions $g : E(T_m) \to \mathbf{Z}$, $h : E(T_l) \to \mathbf{Z}$ such that the maximum length of monotone increasing [decreasing] paths in T_m and T_l are p and $p-1$ [$q-1$ and q], respectively. Let $V(T_n) = V(T_m) \cup V(T_l)$,

$$E(T_n) = E(T_m) \cup E(T_l) \cup (V(T_m) \times V(T_l))$$

and

$$f(x,y) = \begin{cases} 2g(x,y), & \text{if } (x,y) \in E(T_m), \\ 2h(x,y) + 1, & \text{if } (x,y) \in E(T_l), \\ -M(x,y) & \text{otherwise,} \end{cases}$$

where $M: E(T_n) \to \mathbf{Z}$ is an injection such that $M(x,y) > \max(|g(x,y)|, |h(x,y)|)$. It is easy to verify that T_n and f have the desired properties. [V. Chvátal–J. Komlós, *Canad. Math. Bull.* **14** (1971) 151–157.]

28. (a) If $0 \leq i \leq m-1$, let x_{i+1} be any element of

$$S - \{f(S - \{x_1, \ldots, x_\nu\}) : 0 \leq \nu \leq i\}$$

(which set is clearly non-empty), and let

$$T = S - \{x_1, \ldots, x_m\}.$$

Then by the choice of the x_i's,

$$f(S - \{x_1, \ldots, x_\nu\}) \in T \ (\nu = 0, \ldots, m-1)$$

and this holds trivially for $\nu = m$ too.

(b) We select $x_{2^n}, x_{2^n-1}, \ldots$ recursively. If $x_{2^n}, \ldots, x_{\mu+1}$ are already selected, then let x_μ be an element of $S - \{x_{2^n}, \ldots, x_{\mu+1}\}$, which occurs the least number of times in the form $f(S - \{x_{2^n}, \ldots, x_{\nu+1}\})\}$, $\mu \leq \nu \leq 2^n$.

To handle this procedure, let $\varphi(m)$ denote the number of indices $\nu \geq m+1$ such that

$$f(\{x_1, \ldots, x_\nu\}) \in \{x_1, \ldots, x_m\}.$$

Then we have

(1) $$\varphi(m) - \varphi(m-1) \leq \frac{1}{m}\varphi(m) - 1$$

since the indices ν counted in $\varphi(m)$ but not in $\varphi(m-1)$ are those with $\nu \geq m+1$ and

$$f(\{x_1, \ldots, x_\nu\}) = x_m$$

and by the choice of x_m, the number of such indices ν is at most $\varphi(m)/m$; and m is counted in $\varphi(m-1)$, but not in $\varphi(m)$.

We want to show that $\varphi(2^k) \geq (n-k)2^k$; more generally we show that

(2) $$\varphi(2^k + l) \geq (n-k)2^k + (n-k-2)l \ (0 \leq l \leq 2^k).$$

Note that for $l = 2^k$ this inequality is the same as

$$\varphi(2^{k+1}) \geq (n-k-1)2^{k+1}.$$

Inequality (2) is easily proved by induction on $n - 2^k - l$. If $n - 2^k - l = 0$, it is trivially true. Suppose that $2^k + l < n$. We may assume that $l < 2^k$ since otherwise we could consider the decomposition $2^k + 0 = 2^{k-1} + 2^{k-1}$. By (1)

$$\varphi(m-1) \geq 1 + \frac{m-1}{m}\varphi(m),$$

whence

$$\varphi(2^k + l) \geq 1 + \frac{2^k + l}{2^k + l + 1}\varphi(m).$$

By induction hypothesis

$$\varphi(m) = \varphi(2^k + l + 1) \geq (n - k)2^k + (n - k - 2)(l + 1) =$$
$$= (n - k)(2^k + l + 1) - 2(l + 1).$$

Thus

$$\varphi(2^k + l) \geq 1 + (2^k + l)(n - k) - \frac{2(l + 1)(2^k + l)}{2^k + l + 1} =$$
$$= (2^k + l)(n - k) + 1 - 2l - 2 + \frac{2l + 2}{2^k + l + 1}$$

and since the left-hand side is an integer,

$$\varphi(2^k + l) \geq (2^k + l)(n - k) + 1 - 2l - 2 + 1 =$$
$$= (n - k)2^k + (n - k - 2)l,$$

as stated. Thus, $\varphi(0) = \varphi(1) + 1 \geq n + 1$

(c) Let φ_0 be the mapping of $\{1, \ldots, 2^n\}$ into itself defined in 14.26b. We define a mapping f associating with each k-element subset X of $\{1, \ldots, 2^n - 1\}$ the $\varphi_0(k)^{\text{th}}$ largest element of X. This is possible, because $1 \leq \varphi_0(k) \leq k$.

Now suppose that $X_0 \subset X_1 \subset \ldots \subset X_n$ is a chain of subsets with $f(X_0) = \ldots = f(X_n) = y$ (one may suppose that $X_0 = \{y\}$, but this is irrelevant). Let $k_i = |X_i|$. Then, by the definition of f,

$$\varphi_0(k_i) = |X_i \cap \{y, \ldots, 2^n - 1\}|,$$

thus $\varphi_0(k_0) \leq \varphi_0(k_1) \leq \ldots \leq \varphi_0(k_n)$. But this is impossible by 14.26.

Remark: Part (b) gives, of course, a new proof of part (a) of 14.26; but the proof given there is much simpler. [E. Harzheim, *Publ. Math. Debrecen* **15** (1968) 19–22.]

29. Consider the complete graph K_N on the given points; orient its edges upward and color an edge color i, if it forms with the positive half of the x-axis an angle between $\frac{i}{n}\pi$ and $\frac{i+1}{n}\pi (i = 0, \ldots, n - 1)$ (we may assume that these angles themselves never occur or else we rotate the picture). Suppose that there exists no sequence $a_0, \ldots, a_n$ of points in which $\sphericalangle a_{i-1}a_i a_{i+1} > \left(1 - \frac{1}{n}\right)\pi$ for $i = 1, \ldots, k - 1$. Then the graph with edges of color i contains no directed path of length k; such a path would exactly mean an almost straight broken line. Therefore by 9.9 the graph formed by edges of color i is k-colorable. Since K_N is the union of n such graphs it follows by 9.3 that $\chi(K_N) \leq k^n$, a contradiction [P. Erdős–G. Szekeres, *Compositio Math.* **2** (1965) 463–470].

30. (a) Define a tournament on S with values associated with the edges as in the hint. By 14.27a there exists a monotone increasing path of length $p + 1$ or a monotone decreasing path of length $q+1$. Let, e.g. the first case occur. Then there are points $(x_0, y_0), \ldots, (x_{p+1}, y_{p+1})$ such that $x_0 < x_1 < \ldots < x_{p+1}$ and

$$\frac{y_0 - y_1}{x_0 - x_1} < \ldots < \frac{y_{p+1} - y_p}{x_{p+1} - x_p}.$$

Clearly these points form a $p+2$-gon convex from below.

(b) We construct a set S of $\binom{p+q}{p}$ points spanning no $(p+2)$-gon convex from below and no $(q+2)$-gon convex from above, by induction on $p+q$. The case $p = 1$ or $q = 1$ is trivial, so let $p, q \geq 2$. Let S_1 be a set of $\binom{p+q-1}{p-1}$ points spanning no $(p+1)$-gon convex from below and no $(q+2)$-gon convex from above. Also let S_2 be a set of $\binom{p+q-1}{p}$ points spanning no $(p+2)$-gon convex from below and no $(q+1)$-gon convex from above; both S_1, S_2 should consist of points in general position.

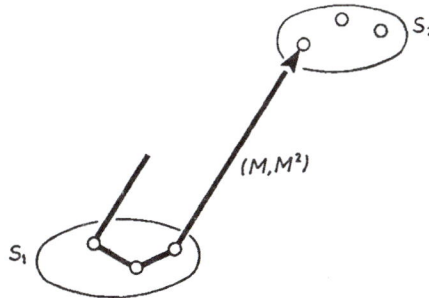

FIGURE 104

Let us place S_1, S_2 on the plane in such a way that the vector connecting some two points of them is (M, M^2), where M is very large (Fig. 104). Let us consider any polygon P convex from below; let $(x_1, y_1), \ldots, (x_t, y_t)$ be its points, $x_1 < \ldots < x_t$. If all these points belong to S_1 or S_2, then $t \leq p+1$ by definition. So suppose that $(x_1, y_1), \ldots, (x_i, y_i) \in S_1$, $(x_{i+1}, y_{i+1}), \ldots, (x_t, y_t) \in S_2$ (clearly any point of S_1 is to the left from any point of S_2 if M is large enough). Then

$$\frac{y_{i+1} - y_i}{x_{i+1} - x_i} \approx M,$$

so if M is larger than

$$\max\left\{ \frac{y' - y}{x' - x} : (x, y), (x', y') \in S_2 \right\},$$

then our polygon cannot be convex unless $i+1 = t$. Since $i \leq p$ by the choice of S_1 it follows that $t \leq p+1$. Similarly we conclude that any polygon convex from above spanned by S has at most $q+1$ vertices [P. Erdős–G. Szekeres; ibid.].

31. (a) We show first that 5 points always determine a convex quadrilateral. If the convex hull of them is a quadrilateral we are done, so suppose that this convex hull is a triangle abc. This contains two further points d, e. Suppose that the line de intersects the edges ab and ac of the triangle. Then $bcde$ is a convex quadrilateral (Fig. 105).

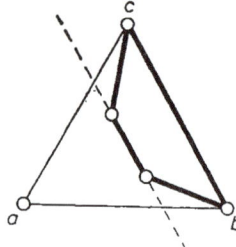

FIGURE 105

Now let us 2-color all quadruples of the given n points: let a quadruple be red, if it forms the vertices of convex quadrilateral and blue otherwise. If

$$(1) \qquad\qquad n \geq R_2^4(m, 5),$$

then we have either m points with all quadruples red or 5 points with all quadruples blue. The latter case is impossible by the above. Hence we have m points any 4 of which form a convex quadrilateral. Hence the m points form a convex m-gon.

(b) I. We show that (1) can be replaced by

$$n \geq \binom{2m - 4}{m - 2} + 1.$$

In fact, given any $\binom{2m-4}{m-2} + 1$ points in general position, they always span an m-gon convex from below or from above by 14.30a; this is a convex m-gon.

II. We construct a set of 2^{m-2} points spanning no convex m-gon. Let T_i be a set of points in general position such that $|T_i| = \binom{n-2}{i}$ and T_i spans no $(i+2)$-gon convex from above a no $(n-i)$-gon convex from below. Let T_i have diameter less than 1 $(i = 0, 1, \ldots, n-2)$. Also we may suppose that any line formed by two points of T_i forms an angle less than $45°$ with e x-axis (by applying an affinity perpendicular to the x-axis, if necessary).

Let P be a large regular $(4n - 4)$-gon with centre at the origin and let $v_0, \ldots, v_{n-2}$ be its vertices in the segment of $\pm 45°$ around the positive half of

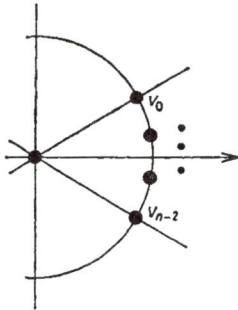

FIGURE 106

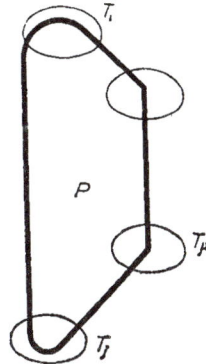

FIGURE 107

the x-axis (Fig. 106). Let us place T_i around v_i (i.e. let us translate one point of T_i into v_i). The resulting set T has

$$|T| = \sum_{i=0}^{n-2} |T_i| = \sum_{i=0}^{n-2} \binom{n-2}{i} = 2^{n-2}$$

elements. We claim that it contains no convex n-gon. Let P be any convex polygon spanned by T. If P is contained in a $T_i (0 \leq i \leq n-2)$, let a and b be the vertices of P with least and largest x-coordinate, respectively. The diagonal ab splits P into two polygons, convex from above and below, respectively. Hence by the definition of T_i, the number of vertices of these polygons is $\leq i+1$ and $\leq n-i-1$, respectively. Hence the number of vertices of P is $\leq (i+1) + (n-i-1) - 2 = n-2$.

Now suppose that P intersects T_i and T_j, $i < j$. Let i be the least and j be the largest index for which this holds. Now P can intersect T_μ, $i < \mu < j$ in at most one point, and its pieces in T_i and T_j are convex from above and below, respectively (Fig. 107). Hence P has at most

$$(i+1) + (j-i-1) + (n-j-1) = n-1$$

points.

(c) $K(4) \geq 5$ trivially; $K(4) \leq 5$ by the first paragraph of the solution. Thus $K(4) = 5$.

$K(5) \geq 9$ by (b). So what we have to show is that any 9 points (in general position) contain a convex pentagon. If the convex hull of the 9 points is a pentagon we are done. So we may suppose that it is a quadrilateral or a triangle.

$1°$ Suppose that the convex hull is a quadrilateral $xyzu$. Consider the 5 remaining points. If they form a convex pentagon we are done. Otherwise, there are 4 among them, which do not form a convex quadrilateral; let a point v be contained in a triangle abc (Fig. 108). Consider the angles avb, bvc, cva. One of these contains x and y. Then $xyavb$ is a convex pentagon.

$2°$ Suppose the convex hull is a triangle abc. Again we have two subcases:

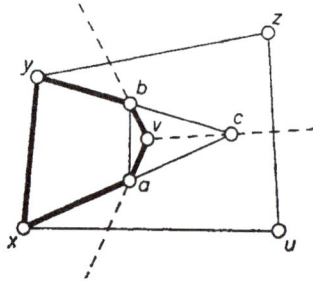

FIGURE 108

I. The convex hull of the remaining 6 points is a quadrilateral $xyzu$. Let p, q be two further points (inside $xyzu$). If the line pq intersects two adjacent edges of $xyzu$ we have a pentagon trivially (Fig. 109). So suppose that pq intersects the edges xy and uv. We may suppose that the intersection with xy is nearer p than q (Fig. 109). Now consider the semilines px, py, qz, qu. They divide the outside of $xyzuu$ into 4 regions. If the region $ypqz$ contains one of a, b, c we are done. Similarly if $xpqu$ contains one of a, b, c.

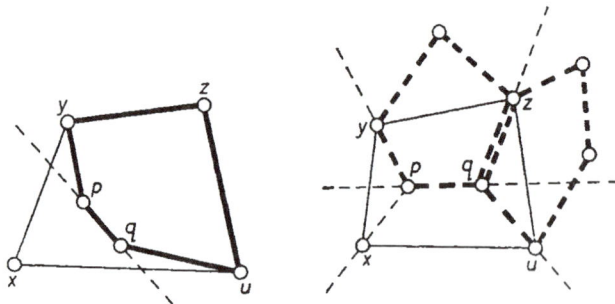

FIGURE 109

If $ypqz$ and $xpqu$ contain none of a, b, c, then either xpy or zqu contains two of those; then we find a convex pentagon again.

II. Finally, suppose that the convex hull of the 6 points inside abc is another triangle xyz. Again let p, q be two points inside xyz. We may assume that the line pq intersects the edges xy and xz. Consider the angles xpy, xqz. Each of these may contain at most one of a, b, c otherwise we find a convex pentagon. Hence one of a, b, c lies in the region determined by the semilines py, qz and the segment pq; but this yields a convex pentagon again (Fig. 110) [ibid.].

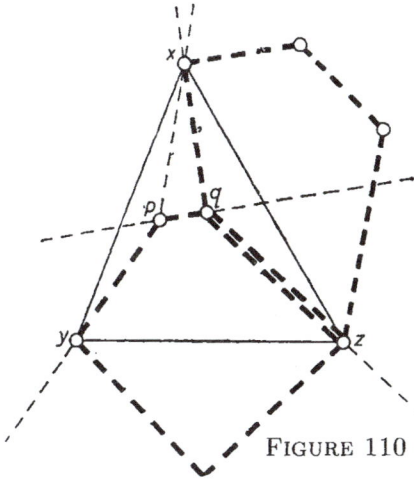

FIGURE 110

§ 15. Reconstruction

1. (a) Let $e_1, \ldots, e_k$ $(k \geq 4)$ be the edges adjacent to a given point $x \in V(G)$. Then $e_1, \ldots, e_k$ form a complete subgraph of $L(G)$. Moreover, they form a maximal complete subgraph (clique) of $L(G)$; for if $e \in E(G)$ is any other edge, then it does not contain x and thus, it meets at most two of $e_1, \ldots, e_k$.

Conversely, let $e_1, \ldots, e_k$ form a clique of $L(G)$, $k \geq 4$, we claim that they are the edges adjacent to one of the points of G. For let x be the common point of e_1 and e_2. If $e_3, \ldots, e_k$ also contain x, then we are finished, since, by the maximality of the complete subgraph of G formed by $e_1, \ldots, e_k$, no other edge can contain x.

Assume indirectly that e_3 does not contain x, then e_1, e_2, e_3 form a triangle. Since $k \geq 4$, we also have an e_4; obviously, e_4 cannot meet all of e_1, e_2, e_3, a contradiction.

The above assertions show that, if G has degrees at least 4, then we obtain G from $L(G)$ as follows: Let $C_1, \ldots, C_n$ be those cliques of $L(G)$ with at least 4 points; associate a point x_i with C_i and join x_i and x_j iff C_i and C_j have a point in common. The resulting graph G' will be isomorphic to G.

Thus, if G_1, G_2 are graphs with all degrees at least 4 and $L(G_1) \cong L(G_2)$ then, "reconstructing" G_1 and G_2 as above, we get isomorphic graphs.

(b) Let $S_1, \ldots, S_n$ be the stars of the points of G_1 and let $S_1', \ldots, S_m'$ be the stars of the points of G_2. Consider an isomorphism $\sigma : L(G_1) \to L(G_2)$. We may assume that $n \geq m$.

$\sigma(S_i)$ cannot be a star for each $i = 1, \ldots, n$. For if $\sigma(S_i)$ is a star, the $\sigma(S_i)$, $\sigma(S_j)$ $(i \neq j)$ are different stars (since $S_i \neq S_j$, except in the trivial case $G_1 \cong K_2$, which we may ignore) and so, it would follow that $m = n$, $\sigma(S_i) = S_i'$ $(i = 1, \ldots, n)$ with a suitable choice of indices. But then mapping the center of S_i onto the center of S_i' $(i = 1, \ldots, n)$, we get an isomorphism between G_1 and G_2.

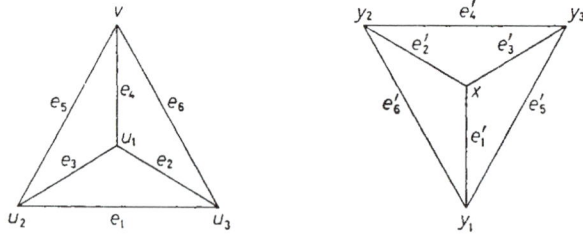

FIGURE 111

Thus $\sigma(S_1)$ is not a star (say). We claim that there are three edges e_1, e_2, e_3 in G_1 and corresponding edges e_1', e_2', e_3' in G_2, which form a triangle in one graph and a star in the other.

1° If $|S_1| = 1$. i.e. $S_1 = \{e_1\}$, then no endpoint of $e_1' = \varrho(e_1)$ has degree 1. So there are two edges e_2', $e_3' \neq e_1'$ starting from the two endpoints of e_1'. $e_2 = \sigma^{-1}(e_1')$ and $e_3 = \sigma^{-1}(e_3')$ meet e_1, which is only possible if they start from the same endpoint of e_1. But then e_2' and e_3' must meet, too, i.e. e_1', e_2', e_3' form a triangle.

2° If $S_1 = \{e_1, e_2\}$, then $e_1' = \sigma(e_1)$, $e_2' = \sigma(e_2)$ have a point x' in common. There must be a further edge e_3' starting from x'. $e_3 = \sigma^{-1}(e_3')$ meets both e_1 and e_2, hence they form a triangle.

3° If $S_1 = \{e_1, e_2, e_3\}$, then e_1', e_2', e_3' must form a triangle; otherwise, they would start from the same point and since $\sigma(S_i)$ is not a star, there would be a fourth edge meeting all of them, which is clearly impossible.

So we have three edges e_1, e_2, e_3 forming a triangle, so that the corresponding edges e_1', e_2', e_3' start from the same point x. $e_1 = (u_2, u_3)$, $e_2 = (u_1, u_3)$, $e_3 = (u_1, u_2)$, $e_i' = (x, y_i)$ $(i = 1, 2, 3)$.

The above pair yields a solution: $G_1 = (\{u_1, u_2, u_3\}, \{e_1, e_2, e_3\}) \not\cong G_2 = (\{x, y_1, y_2, y_3\}, \{e_1', e_2', e_3'\})$ but $L(G_1) \cong L(G_2)$ We claim there is no further such pair. If there was, we would have $|E(G_i)| \geq 4$.

Let $e_4 \in E(G_1)$, $e_4 \neq e_1, e_2, e_3$. We may assume that $e_4 = (u_1, v)$. Then $e_4' = \sigma(e_4)$ meets e_1 and e_3, hence $e_4' = (y_2, y_3)$. Since $G_1 \not\cong G_2$, we have a further edge $e_5 \in E(G)$. Again we may assume that it meets one of e_1, e_2, e_3, e_4. Obviously, $u_1 \notin e_5$.

If e_5 does not meet e_4, then it meets e_1 and e_2 (say). $e_5' = \sigma(e_5)$ must meet e_1' and e_2' but not e_3' and e_4', which is impossible. If e_5 meets e_4 but not e_1, then e_5' meets e_4' but not e_1', e_2', e_3', which is impossible again. Thus $e_5 = (u_2, v)$ or (u_3, v); say $e_5 = (u_2, v)$, the $\sigma(e_5) = e_5' = (y_1, y_3)$. Since $G_1 \not\cong G_2$, we must have a sixth edge e_6. As above, we get $e_6 = (u_3, v)$, $\sigma(e_6) = (y_1, y_2)$. Now $G_1 \not\cong G_2$ implies that there must be a seventh edge, but the same argument as above shows this is impossible (Fig. 111).

(c) Let $\sigma : L(G_1) \to L(G_2)$ be a non-trivial isomorphism. Again let $S_1, \ldots, S_n$ a $S_1', \ldots, S_m'$ be the stars of points in G_1 and G_2, respectively. If $\sigma(S_i)$ is a star of each $1 \le i \le n$, then we get, as before, that σ is a *trivial* isomorphism. Otherwise, we can follow the preceding solution but everywhere, where we argued "since $G_1 \ncong G_2$ we have a further edge" we must stop and list an example. So we get the pairs shown in Fig. 112. [H. Whitney, *Amer. J. Math.* **54** (1932) 160–168.]

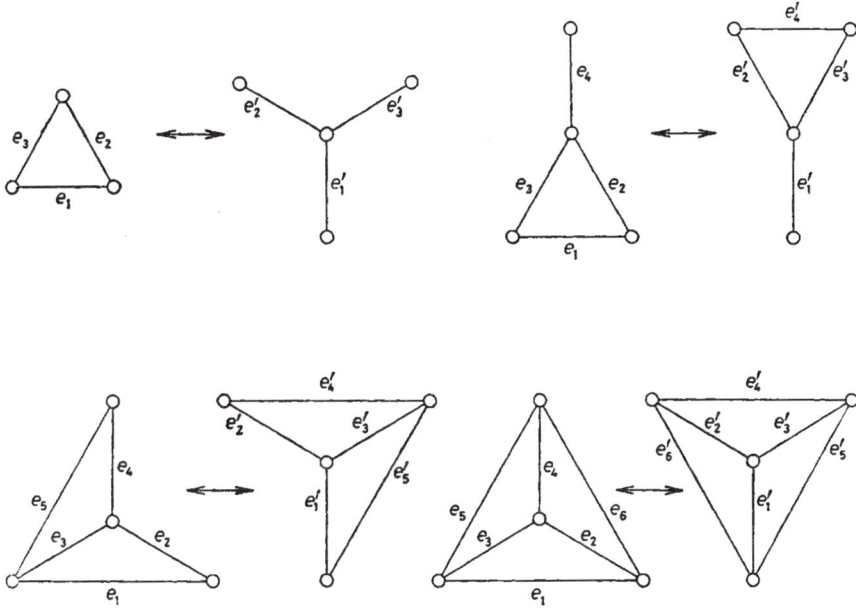

FIGURE 112

2. (a) Let $E, F \in E(K_n^r)$. Then the number of edges of K_n^r disjoint from E and F is

$$\binom{n - |E \cup F|}{r} = \binom{n - 2r + |E \cap F|}{r}.$$

Similarly, the number of edges of K_n^r disjoint from $\alpha(E)$ and $\alpha(F)$ is

$$\binom{n - 2r + |\alpha(E) \cap \alpha(F)|}{r}.$$

Thus

$$\binom{n - 2r + |E \cap F|}{r} = \binom{n - 2r + |\alpha(E) \cap \alpha(F)|}{r}.$$

Since $n - 2r \ge r$, this implies that $|E \cap F| = |\alpha(E) \cap \alpha(F)|$.

We want to show, as in 15.1, that the set of all r-tuples containing any given point x is mapped onto the set of r-tuples containing a point $\beta(x)$. This then defines a permutation β of $V(K_n^r)$, which induces α.

So let $x \in V(K_n^r)$, and let E, F be two edges with $E \cap F = \{x\}$. Then, by the above, $\alpha(E)$ and $\alpha(F)$ have a unique point $\beta(x)$ in common. We claim that, if any edge $A \in E(K_n^r)$ contains x, then $\alpha(A)$ contains $\beta(x)$ (and conversely).

Let $A \in E(K_n^r)$, $x \in A$, and let B be an r-tuple such that $B \cap A = A - E - F$. Then

(1)
$$|B \cap A| = |A - E - F| = |A| - |A \cap E| - |A \cap F| +$$
$$+ |A \cap E \cap F| = r + 1 - |A \cap E| - |A \cap F|.$$

Since

$$|\alpha(B) \cap \alpha(E)| = |B \cap E| = 0,$$
$$|\alpha(B) \cap \alpha(F)| = |B \cap F| = 0,$$

we have

$$|\alpha(B) \cap \alpha(A)| \le |\alpha(A) - \alpha(E) - \alpha(F)| = r - |\alpha(A) \cap \alpha(E)| -$$
$$- |\alpha(A) \cap \alpha(F)| + |\alpha(A) \cap \alpha(E) \cap \alpha(F)|.$$

Here $|\alpha(B) \cap \alpha(A)| = |B \cap A|$, $|\alpha(A) \cap \alpha(E)| = |A \cap E|$, $|\alpha(A) \cap \alpha(F)| = |A \cap F|$, thus (1) implies that

$$|\alpha(A) \cap \alpha(E) \cap \alpha(F)| \ge 1,$$

i.e. $\beta(x) \in \alpha(A)$.

(b) We sharpen the preceding argument. The number of r-tuples disjoint from two given r-tuples E, F with $|E \cap F| = r - 1$ is

$$\binom{n - r - 1}{r} > 0,$$

and is less than this if $|E \cap F| < r - 1$. Therefore, $|E \cap F| = r - 1$ iff $|\alpha(E) \cap \alpha(F)| = r - 1$.

Now suppose that we know $|E \cap F| = k \Leftrightarrow |\alpha(E) \cap \alpha(F)| = k$ for $k = 0, 1, \ldots, k_0$ (we certainly know this for $k = 0$). Then $|E \cap F| = k_0 + 1 \Leftrightarrow |E \cap F| > k_0$ and there is an r-tuple A with $|E \cap A| = r - 1$, $|F \cap A| = k_0 \Leftrightarrow |\alpha(E) \cap \alpha(F)| > k_0$ and there is an r-tuple A with $|\alpha(E) \cap \alpha(A)| = r - 1$, $|\alpha(F) \cap \alpha(A)| = k_0 \Leftrightarrow |\alpha(E) \cap \alpha(F)| = k_0 + 1$.

This proves that $|E \cap F| = |\alpha(E) \cap \alpha(F)|$ for any two r-tuples E, F.

Now let x be a point of K_n^r and E, F two r-tuples in K_n^r with $E \cap F = \{x\}$. Let $\alpha(E) \cap \alpha(F) = \{\beta(x)\}$, we show that $x \in A$ implies $\beta(x) \in \alpha(A)$ or, equivalently, $x \notin A$ implies $\beta(x) \notin \alpha(A)$.

Let $x \notin A$, and assume indirectly that $\beta(x) \in \alpha(A)$. Set $|A \cap E| = i$, $|A \cap F| = j$, $d = |V(K_n^r) - A - E - F|$. Then

$$d = n - (3r - 1 - i - j) \le i + j + 1 \le r.$$

Also,

$$|V(K_n^r) - A - \{x\}| = n - r - 1 \ge r.$$

So we can find an r-tuple B such that

$$V(K_n^r) - A - E - F \subseteq B \subseteq V(K_n^r) - A - \{x\}.$$

Consider $\alpha(B)$. Obviously, $\alpha(B) \cap \alpha(A) = \emptyset$, so $\beta(x) \notin \alpha(B)$. Moreover,

$$|\alpha(B) \cap \alpha(E)| = |B \cap E|$$
$$|\alpha(B) \cap \alpha(F)| = |B \cap F|,$$

thus

$$|\alpha(B) - \alpha(E) - \alpha(F)| = |B| - |B \cap E| - |B \cap F| = d.$$

Hence

$$|V(K_n^r) - \alpha(A) - \alpha(E) - \alpha(F)| \geq |\alpha(B) - \alpha(E) - \alpha(F)| = d.$$

But

$$|V(K_n^r) - \alpha(A) - \alpha(E) - \alpha(F)| = n - 3r + |\alpha(A) \cap \alpha(E)| +$$
$$+ |\alpha(A) \cap \alpha(F)| + |\alpha(E) \cap \alpha(F)| - |\alpha(A) \cap \alpha(E) \cap \alpha(F)| =$$
$$= n - 3r + i + j + 1 - 1 = d - 1,$$

a contradiction. [C. Berge, *Hypergraph Seminar*, Lecture Notes in Math. 411, Springer, 1974, 1–12.]

3. Let E, F be two r-tuples, $|E \cap F| = j$. Let us count the number of r-tuples adjacent to both E and F. We get
    (1) if $j < 2i - r$, there are no such r-tuples,
    (2) if $2i - r \leq j$, the number of r-tuples adjacent to both E and F is

$$\sum_{\nu = 2i-r}^{\min(i,j)} \binom{j}{\nu} \binom{r-j}{i-\nu}^2 \binom{n-2r+j}{r-2i+\nu} = f_j(n)$$

(we consider i and r fixed).
    The polynomials $f_j(n)$ are distinct for $j = 2i-r, 2i-r+1, \ldots, r-1$, since their leading terms are

$$\binom{r-j}{i-j}^2 \frac{n^{r-2i+j}}{(r-2i+j)!} \quad \text{if } 2i - r \leq j \leq i,$$

$$\binom{j}{i} \frac{n^{r-i}}{(r-i)!} \quad \text{if } i \leq j < r.$$

Thus, if n is large enough the numbers $f_j(n)$ are distinct for $j = 2i-r, \ldots, r-1$. This implies that we have characterization:
    $|E \cap F| = j$ iff *the number of r-tuples adjacent to both E and F is equal to* $f_j(r)$.
    Thus, if we define a graph on the points of $L_i(K_n^r)$ by joining two of them when the number of neighbors they have in common is $f_{i-1}(n)$ (note that $2i-r \leq i-1$), we get $L_{i-1}(K_n^r)$. α is therefore an automorphism of $L_{i-1}(K_n^r)$. Proceeding similarly, we get that α is an automorphism of $L_0(K_n^r)$. But $L_0(K_n^r) = \overline{L(K_n^r)}$, so α is an automorphism of $L(K_n)$. By 15.2, this implies that α is induced by a permutation of $V(K_n^r)$ [C. Berge; ibid.].

4. (a) Let $S_1, \ldots, S_n$ be the stars of points of G, and consider the hypergraph

$$H = (E(G), \{S_1, \ldots, S_n\}).$$

It is obvious that $L(H) \cong G$.

More generally, let $C_1, \ldots, C_m$ be complete graphs such that $C_1 \cup \ldots \cup C_m = G$. For each $x \in V(G)$, let U_x be the set of those C_i's containing x, and set

$$H = (\{C_1, \ldots, C_m\}; \{U_x : x \in V(G)\}).$$

Then $L(H) \cong G$. The natural correspondence is $\sigma(U_x) = x$ and it is trivial to verify that it is an isomorphism.

Conversely, suppose that $L(H) \cong G$ for a hypergraph H and let $\sigma : L(H) \to G$ be an isomorphism. For each $v \in V(H)$, let C_v be the subgraph of G spanned by the points $\sigma(E)$, where $v \in E \in E(H)$. Obviously, C_v is a complete subgraph, $\bigcup_{v \in V(H)} C_v = G$ and if U_x denotes the set of complete graphs C_v containing x, we have

$$H \cong (\{C_v : v \in V(H)\}; \{U_x : x \in V(G)\}).$$

(b) We prove by induction on n that G is the union of $\frac{n^2}{4}$ complete graphs. For $n = 2$ this is obvious. Let $n > 2$, and let x, y be two adjacent points. $G - x - y$ is the union of $t = \left[\frac{(n-2)^2}{4}\right] = \left[\frac{n^2}{4}\right] - n + 1$ complete graphs $C_1, \ldots, C_t$ and certain isolated points.

Let $z \in V(G) - x - y$. If z is adjacent to both of x, y, then let C_z be the triangle xyz. If z is adjacent to x or y only, then define C_z as the edge joining z to x, y. Otherwise, let $C_z = \{z\}$. Also let C_0 be the edge (x, y). Now adding the new complete graphs C_0 and C_z ($z \in V(G) - x - y$) we get

$$t + n - 2 + 1 = \left[\frac{n^2}{4}\right]$$

complete graphs whose union is G.

Thus we can cover the edges of G by $\left[\frac{n^2}{4}\right]$ complete graphs and this yields a hypergraph with line-graph G on $\frac{n^2}{4}$ points.

The result is sharp, as is shown by a complete bipartite graph $K_{\left[\frac{n}{2}\right] \cdot \left[\frac{n+1}{2}\right]}$ [P. Erdős–A. W. Goodman–L. Pósa, *Canad. J. Math.* **18** (1966) 106–112].

5. (a) Let x, y, z be three points of G such that, removing any combination of them, G remains connected (i.e. x, y, z are three endpoints of a spanning tree of G; if the only spanning tree of G is a path, then G is a path or circuit and therefore a line-graph).

Let $G - x \cong L(H_x)$, $G - y \cong L(H_y)$, ..., $G - x - y \cong L(H_{xy})$, ..., $G - x - y - z \cong L(H_{xyz})$. Then $L(H_{xyz}) \cong L(H_{xy} - e_z)$ for the edge e_z corresponding to z. Since $|E(H_{xyz})| > 6$, this means that the isomorphism between $L(H_{xyz})$ and $L(H_{xy} -$

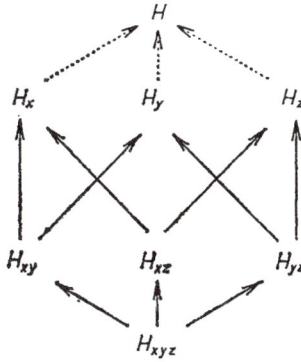

FIGURE 113

e_z) is induced by an embedding of H_{xyz} into H_{xy}. Similarly, we get a whole lot of embeddings (illustrated in Fig. 113), which commute with the isomorphisms between $L(H_{xy})$, etc. and the corresponding induced subgraphs of G (Fig. 113).

Since we also know that the embedding of H_{xyz} into H_x, H_y, H_z is unique, this diagram commutes. So we may identify all points mapped onto each other by embeddings, and form the union H. The edges of H correspond to the points of G in the natural way and it is easy to check that this correspondence is an isomorphism between $L(H)$ and G.

(b) Let $G_1, \ldots, G_k$ be those simple graphs which are not line-graphs but each proper induced subgraph of which is a line-graph. Since by the preceding problem they have at most nine points, their number is finite.

Now if G is any simple graph which is not a line-graph, then it contains a minimal induced subgraph, which is not a line-graph, and this is one of $G_1, \ldots, G_k$. Conversely, if G contains one of $G_1, \ldots, G_k$ then, obviously, it cannot be a line-graph.

6. (a) Suppose that the points a, b, c and d of $G = L(H)$ form a 3-star with center a. Then the edge a is adjacent to b, c and d; some two of them b and c say, start from the same endpoint of a. But then b and c are adjacent.

Now let abc, bcd be odd triangles, we claim a, b, c start from the same point of H. Suppose indirectly that a, b, c form a triangle of H. By definition, there is an edge x adjacent to one or three of a, b, c. But this is impossible since x must start from a point of the triangle formed by a, b, c in H and so, x is adjacent to two of a, b, c. Thus a starts from the common point of b, c. Similarly, d starts from the common point of b, c implying that a, d are adjacent.

(b) We may assume that G is connected since each component satisfies (i) and (ii) and if they are line-graphs, then so is G. Let $(a, b) \sim (c, d)$ be as in the hint. Suppose that $(a, b) \sim (c, d) \sim (e, f)$ are distinct edges of G, we show that $(a, b) \sim (e, f)$ (the other criteria for an equivalence relation are trivially fulfilled). Observe that $\{a, b, c, d\}$ spans a complete graph. Consider a and e (say), $a \neq e$. If $a = c$ or d, then a, e are trivially adjacent so suppose that $a \neq c$, d; similarly $e \neq$

c, d. Then acd is an odd triangle (if $b=c$ or d this follows immediately from the definition of equivalence, otherwise b is adjacent to three points of it), similarly cde is an odd triangle, whence a and e are adjacent by (ii). Now, if (a,b) and (e,f) have no common endpoint we immediately conclude that $(a,b) \sim (e,f)$. Suppose that $a=e$. If c or d is different from a, b, f, then it is adjacent to three points of the triangle abf, so this triangle is odd; if (c,d) is also an edge of this triangle, then it must be odd again. So $(a,b) \sim (e,f)$ follows again.

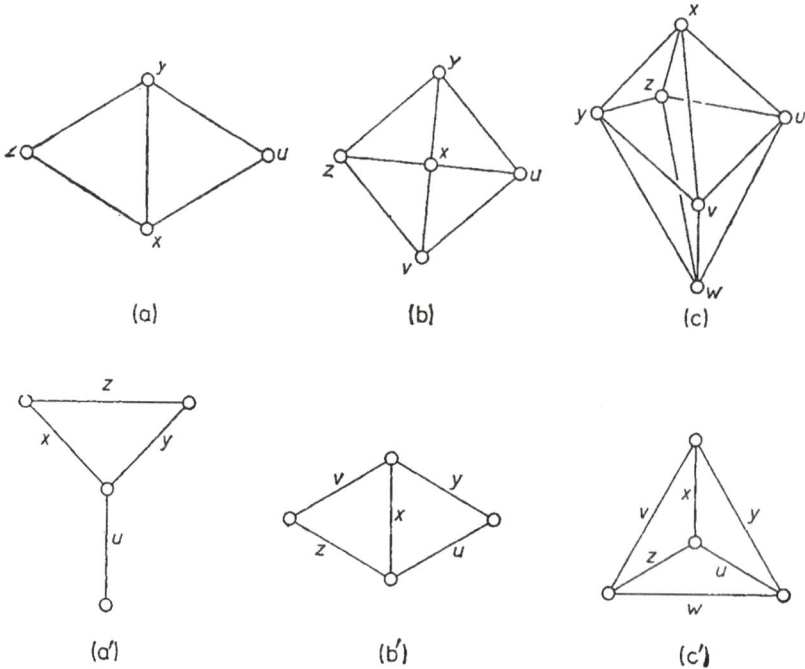

FIGURE 114

Let $C_1, \ldots, C_N$ be the equivalence classes of the relation $\sim$. They obviously form complete graphs. We show that each point of G is contained in at most two of these complete graphs unless G is one of the exceptional graphs shown in Fig. 114a,b,c.

Assume that $x \in V(G)$ is contained in three complete graphs i.e. there are three non-equivalent edges (x,y), (x,z) (x,u) adjacent to x. By (i) some two of y, z, u are adjacent; say $(y,z) \in E(G)$. Since xyz is not an odd triangle, u must be adjacent to one of y, x, say $(u,y) \in E(G)$, and then $(u,z) \notin E(G)$. If $|V(G)|=4$, then we get (a) in Fig. 114.

If G is larger, then there is a point v adjacent to one of x, y, z, u. It must be adjacent to an even number of points of the triangles xyz and xyu, therefore it must be adjacent to one of x, y. Let, say, $(v,x) \in E(G)$ (the roles of x and y have been symmetric as, obviously, (y,x), (y,z) and (y,u) are non-equivalent). By (i)

v is adjacent to one of z and u, say $(u,v) \in E(G)$. Again, because xyz and xyu are not odd triangles $(v,y) \in E(G)$ and $(v,z) \in E(G)$. If G has only five points, we get (b) in Fig. 114.

Obviously, (v,x) is non-equivalent to at least one of (x,z) and (x,u), say (v,x) is non-equivalent to (x,u), i.e. vxu cannot be an odd triangle; for if it were, then a point adjacent to an odd number of points of vxu would be adjacent to an odd number of points of one of the triangle vxz, xyz, xyu, as is easy to verify. Thus, the roles of y, z, u, v are symmetric.

Now suppose that G has a further point w, adjacent to one of the x, y, z, u, v. w cannot be adjacent to x; for if it were, then one of (y,v) as well as one of z, u ought to be a neighbor of w; let, say, (y,w), $(z,w) \in E(G)$, then xyz is odd, a contradiction. So w is adjacent to y (say) but not to x. Then it follows that it is adjacent to u, v, z also. So if G has 6 points, we have (c) in Fig. 114.

Observe that no two edges spanned by $x, \ldots, w$ can be equivalent, hence the roles of $x, \ldots, w$ are symmetric. Therefore, the argument of the last paragraph shows that no further edge can be adjacent to any of the points $x, \ldots, w$, i.e. indeed we only get the three graphs of Fig. 114.

Now it is easy to complete the proof. If G is one of the graphs of Fig. 114. then it is a line-graph of the corresponding graph in Fig. 114a', b', c'. Otherwise, G can be covered by edge-disjoint complete subgraphs so that each point is contained in exactly two of them (we add one-point graphs, if necessary). As in 15.40a, this implies that G is a line-graph. [A. M. van Roof–H. S. Wilf, *Acta Math. Hung.* **16** (1965) 263–269.]

(c) We claim that G is a line-graph of a simple graph iff it does not contain any of the graphs shown in Fig. 115 as an induced subgraph.

It is immediately clear that these graphs are not line-graphs since they violate (i) or (ii) of (a). Conversely, suppose that G is not a line-graph. Then it either contains the 3-star as an induced subgraph or it violates (ii), i.e. there are points a, b, c, d such that abc, bcd are odd triangles and a, d are non-adjacent. Let e and f be points adjacent to an odd number of points of abc and bcd, respectively. Then there are several cases to distinguish: $e = f$ or $e \neq f$; which points of the corresponding triangle are adjacent to e, f. The consideration of these cases is straightforward but lengthy; they yield the graphs shown in Fig. 115. We leave the details to the reader [L. W. Beineke; see B, H].

7. From 15.4 we know that a graph is the line-graph of a 3-uniform hypergraph without multiple edges iff its edges can be covered by complete subgraphs so that each point is contained in three of them and each edge is contained in at most two of them.

Now suppose that the graph in the hint (shown in Fig. 20) is a line-graph, i.e. it has such a decomposition into complete graphs. Obviously, (x_0, z_1) and (y_0, z_1) must be complete graphs by themselves. Since z_1 is contained in three complete graphs only, z_1, x_1, y_1 must be one. Similarly, u_1, x_1, y_1 must be one of the covering complete subgraphs. Thus (x_1, y_1) is already contained in two covering complete subgraphs, hence (x_1, z_2) and (y_1, z_2) must be covering complete sub-

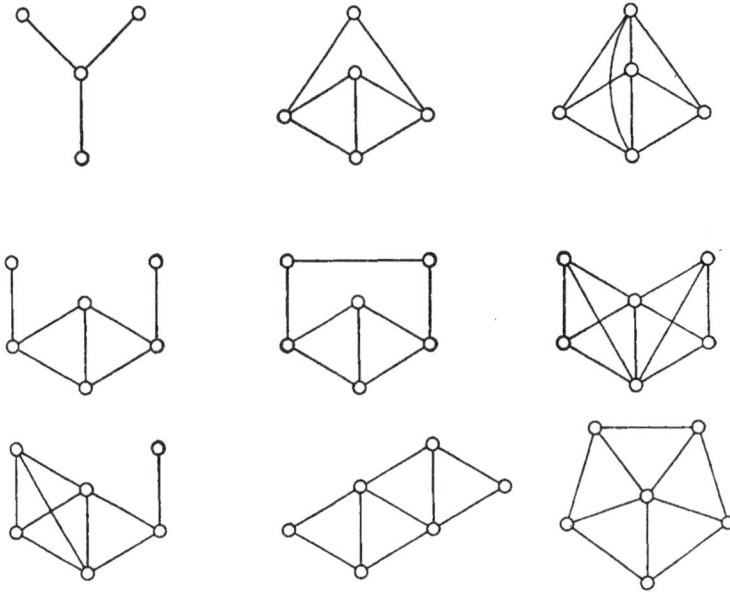

FIGURE 115

graphs by themselves. From here on, the argument periodically repeats till we get that (x_n, z_{n+1}) and (y_n, z_{n+1}) are covering complete subgraphs by themselves. Since, obviously, so are (x_{n+1}, z_{n+1}) and (y_{n+1}, z_{n+1}), z_{n+1} is contained in four of them, a contradiction.

Now remove a point from G. There are essentially four kinds of points: x_i (or y_i), z_i, u_i, v_i (or w_i). Figure 116 shows how to decompose $G - t$ into complete subgraphs, if $t = x_i$, z_i, u_i, v_i [L. Nickel].

8. (a) See Fig. 117.

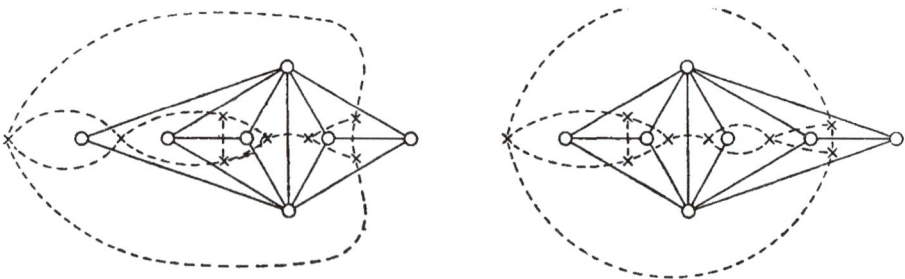

FIGURE 117

(b) This is just a re-formulation of the corollary, which has been drawn from 6.69.

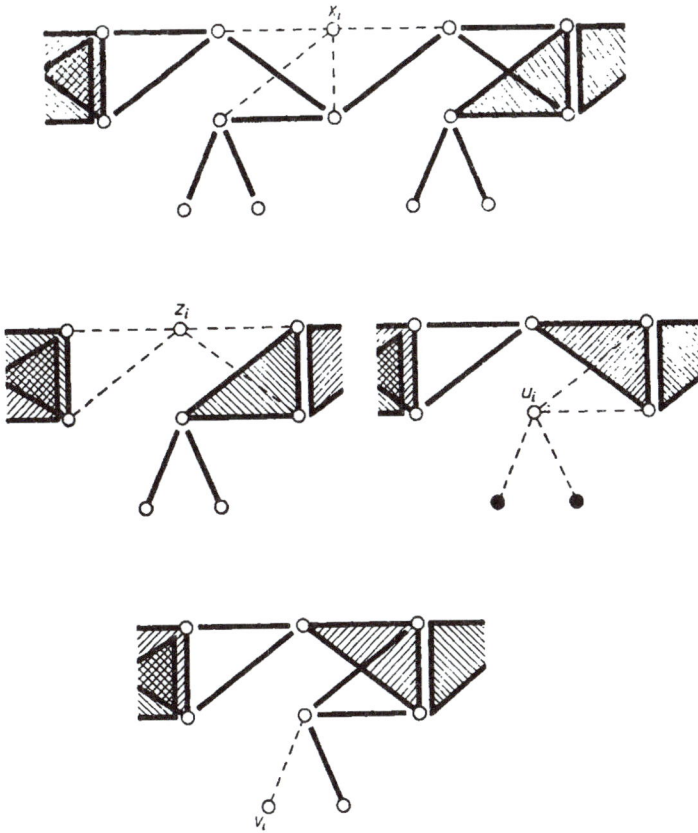

FIGURE 116

9. (a) Let $e_1 = (x, y)$, $e_2 = (x, z)$ be adjacent edges of G_1; we show that $\psi(e_1)$, $\psi(e_2)$ are adjacent as well. If e_1, e_2 are parallel, clearly so are $\psi(e_1)$, $\psi(e_2)$. Since $G_1 - x$ is 2-connected it contains a circuit C_x through y and z, if $y \neq z$. Then $\psi(C)$ is a circuit in G_2. Since $C + e_1$ contains no other circuit, neither does $\psi(C) + \psi(e_1)$. Therefore, $\psi(e_1)$ is not a chord of $\psi(C)$, i.e. it has an endpoint u not on $\psi(C)$. But $C + e_1 + e_2$ contains a circuit C' through e_1 and e_2. Then $\psi(C') \subseteq \psi(C) + \psi(e_1) + \psi(e_2)$ is a circuit in G_2 containing $\psi(e_1)$ and $\psi(e_2)$. Thus there is an edge $\neq \psi(e_1)$ of $\psi(C')$ adjacent to u. This edge must be $\psi(e_2)$. So $\psi(e_1)$ and $\psi(e_2)$ are adjacent.

(b) Now let $x \in V(G_1)$, and let $e_1, \ldots, e_k$ be all edges adjacent to x. Then $\psi(e_1), \ldots, \psi(e_k)$ intersect by pairs and no three of them form a triangle. Hence there is a point $\eta(x)$ contained in all of $\psi(e_1), \ldots, \psi(e_k)$. The point $\eta(x)$ is unique; for if a $u \neq \eta(x)$ also belonged to $\psi(e_1), \ldots, \psi(e_k)$, then $\psi(e_1), \ldots, \psi(e_k)$ would be parallel; thus $e_1, \ldots, e_k$ would be parallel, i.e. x would have a unique neighbor, which is impossible, as G_1 is 3-connected. Hence η is well-defined. Let $u \in V(G_2)$

and $f_1, \ldots, f_k$ be all edges adjacent to u. As above, there is a unique point $x \in V(G_1)$ contained in $\psi^{-1}(f_1), \ldots, \psi^{-1}(f_k)$. Then $\eta(x)$ is contained in $f_1, \ldots, f_k$ and hence $u = \eta(x)$. Therefore, $\eta(x)$ is onto. Similarly, it follows that it is one-to-one. Also, if $(x, y) = e \in E(G)$, then $\eta(x)$, $\eta(y)$ are endpoints of $\psi(e)$. Therefore η is an isomorphism. [H. Whitney, *Amer. J. Math.* **54** (1932) 160–168.]

10. If $r = 2$ the assertion is trivial. Assume that $r > 2$. Let v (v') be the point next to x_r (x_r'), which has degree at least 3. Remove x_r and all inner points of the path joining x_r to v from T, and remove x_r' and the inner points of the (x_r', v')-path from T'. This way we get two trees T_1, T_1' with endpoints $x_1, \ldots, x_{r-1}$ and $x_1', \ldots, x_{r-1}'$, respectively. Obviously, the distance between x_i and x_j is the same in T and T_1 $(1 \le i < j \le r-1)$ and similarly for T' and T_1'. Thus, by the induction hypothesis, $T_1 \cong T_1'$, and, in fact, there exists an isomorphism φ such that $\varphi(x_i) = x_i'$.

We show that $\varphi(v) = v'$. Consider the unique $(\varphi(v), v')$-path, and extend it to a path joining x_i' to x' $(1 \le i < j \le r-1)$. Then

$$d(x_i', x_j') + d(x_i', x_r') - d(x_j', x_r') = 2d(x_i', v'),$$

while

$$d(x_i, x_j) + d(x_i, x_r) - d(x_j, x_r) = 2d(x_i, v).$$

So by the assumption $d(x_i, v) = d(x_i', v')$, which implies that $v' = \varphi(v)$. It also follows that $d(x_r, v) = d(x_r', v')$; for

$$d(x_r, x_i) + d(x_r, x_j) - d(x_i, x_j) = 2d(x_r, v),$$
$$d(x_r', x_i') + d(x_r', x_j') - d(x_i', x_j') = 2d(x_r', v')$$

and the left-hand sides are equal by the assumption. Now φ extends to an isomorphism of T onto T' obviously. [E. A. Smolenskii, *Zhurnal Vychisl. Mat. i Matem. Fiz.* (1962) 371–372.]

11. (a) Let P_{ij} be the (unique) (x_i, x_j)-path in T. Let Q be the (unique) (P_{ij}, x_k)-path and denote by v its beginning point. Then, obviously

$$d(x_i, x_j) + d(x_j, x_k) - d(x_i, x_k) = 2d(x_j, v)$$

which proves (a).

Now consider the two paths P_{ij} and P_{kl}. They may have 0, 1 or at least 2 points in common. Suppose first that they do not meet, then let Q be the (P_{ij}, P_{kl})-path with endpoints $u \in V(P_{ij})$, $v \in V(P_{kl})$. Let R_i, R_j be the (x_i, u), (x_j, u)-paths and R_k, R_l be the (x_k, v), (x_l, v)-paths in T. Then

$$P_{ik} = R_i \cup Q \cup R_k, \qquad P_{jl} = R_j \cup Q \cup R_l$$
$$P_{jk} = R_j \cup Q \cup R_k, \qquad P_{il} = R_i \cup Q \cup R_l.$$

So

$$d_{ik} + d_{jl} = d_{il} + d_{jk} = d_{ij} + d_{kl} + 2d(u,v),$$

which proves assertion (b) in this case.

Now suppose that P_{ij} and P_{kl} have exactly one point v in common. Then considering the form of paths joining v to x_i, x_j, x_k, x_l we get, in the same way as above,

$$d_{ij} + d_{kl} = d_{ik} + d_{jl} = d_{il} + d_{jk}.$$

Finally, if P_{ij} and P_{kl} have more than one point in common, then either P_{ik} and P_{jl} or P_{il} and P_{jk} do not meet and we have the first case by exchanging indices.

(c) Suppose that a tree T' is constructed with endpoints $x_1, \ldots, x_{r-1}$ such that $d(x_i, x_j) = d_{ij}$ holds in T' for each $1 \le i < j \le r-1$.

Let x_i, x_j be two points of T' such that

$$d_{ri} + d_{rj} - d_{ij}$$

is minimal. Let v be the point on the (x_i, x_j)-path in T' for which

$$d(v, x_i) = \frac{1}{2}(d_{ri} + d_{ij} - d_{rj})$$

and join x_r to v by a path of length $\frac{1}{2}(d_{ri} + d_{rj} - d_{ij})$ (this is possible since these numbers are natural numbers by (a)). We claim that the resulting tree T satisfies $d(x_k, x_l) = d_{kl}$, for any $1 \le k < l \le r$. This is trivial unless $l = r$. It is also easy to see, when $k = i$ or $k = j$ from the definition of u, so suppose that $k \ne i$, j.

By (b), two of the numbers $d_{ij} + d_{rk}$, $d_{ir} + d_{jk}$, $d_{jr} + d_{ik}$ are equal and not less than the third. We claim that $d_{ij} + d_{rk}$ is not less than the other two. For suppose that

$$d_{ij} + d_{rk} < d_{ir} + d_{jk}.$$

Then

$$d_{rk} + d_{rj} - d_{jk} < d_{ri} + d_{rj} - d_{ij},$$

which contradicts the choice of i and j. So, for a suitable choice of indices i, j,

$$(1) \qquad d_{ij} + d_{kr} = d_{ir} + d_{jk} \ge d_{jr} + d_{ik}.$$

Now T itself satisfies (b), and, in fact, the (x_r, x_k)-path and (x_i, x_j)-path in T certainly intersect. Hence, by the solution of the preceding problem

$$(2) \qquad d(x_i, x_j) + d(x_k, x_r) = d(x_{i_0}, x_r) +$$
$$+ d(x_{j_0}, x_k) \ge d(x_{j_0}, x_r) + d(x_{i_0}, x_k),$$

where (i_0, j_0) is a permutation of (i, j). Since we have $d(x_i, x_j) = d_{ij}$, $d(x_{i_0}, x_r) = d_{i_0 r}$, etc., (1) and (2) imply that we may assume that $i_0 = i$, $j_0 = j$ and then (1) and (2) imply

$$d(x_k, x_r) = d_{kr}.$$

[K. A. Zaretskii, *Usp. Matem. Nauk.* **20** (1965) 6, 94–96.]

12. (a) The number $2l$ occurs in $\{a_i + a_j : 1 \le i < j \le n\}$ exactly

$$\sum_{\nu=0}^{l} \binom{k+1}{2\nu}\binom{k+1}{2l-2\nu}$$

times if l is odd and

$$\sum_{\nu=0}^{l} \binom{k+1}{2\nu}\binom{k+1}{2l-2\nu} - \frac{1}{2}\left[\binom{k+1}{l}^2 + \binom{k+1}{l}\right]$$

times if l is even. It occurs in $\{b_i + b_j : 1 \le i < j \le n\}$ exactly

$$\sum_{\nu=0}^{l-1} \binom{k+1}{2\nu+1}\binom{k+1}{2l-2\nu-1} - \frac{1}{2}\left[\binom{k+1}{l}^2 + \binom{k+1}{l}\right]$$

times if l is odd and

$$\sum_{\nu=0}^{l-1} \binom{k+1}{2\nu+1}\binom{k+1}{2l-2\nu-1}$$

times if l is even. Suppose that, e.g., l is odd, then one has to verify that

$$\sum_{\nu=0}^{l} \binom{k+1}{2\nu}\binom{k+1}{2l-2\nu} = \sum_{\nu=0}^{l-1} \binom{k+1}{2\nu+1}\binom{k+1}{2l-2\nu-1} - \frac{1}{2}\left[\binom{k+1}{l}^2 + \binom{k+1}{l}\right]$$

or, equivalently,

$$\sum_{\mu=0}^{2l}(-1)^\mu \binom{k+1}{\mu}\binom{k+1}{2l-\mu} = -\frac{1}{2}\left[\binom{k+1}{l}^2 + \binom{k+1}{l}\right],$$

which follows from identity (d) in 1.42.

(b) Suppose that $\{a_i + a_j : 1 \le i < j \le n\} = \{b_i + b_j : 1 \le i < j \le n\}$, but $\{a_i : 1 \le i \le n\} \ne \{b_i : 1 \le i \le n\}$. We show that n is a power of 2. We may assume that $a_i, b_i \ge 0$, since adding any constant to all a_i and b_i the condition remains valid. Set

$$f(x) = \sum_{i=1}^{n} x^{a_i}, \qquad g(x) = \sum_{i=1}^{n} x^{b_i}.$$

Then

$$f^2(x) = \sum_{i=1}^{n} x^{2a_i} + 2\sum_{1 \le i < j \le n} x^{a_i+a_j} = f(x^2) + 2\sum_{1 \le i < j \le n} x^{a_i+a_j},$$

$$g^2(x) = g(x^2) + 2\sum_{1 \le i < j \le n} x^{b_i+b_j}$$

and thus,

$$f^2(x) - g^2(x) = f(x^2) - g(x^2), \quad f(x) + g(x) = \frac{f(x^2) - g(x^2)}{f(x) - g(x)}.$$

We want to substitute $x = 1$ in the last identity, since, then we will have $2n$ on the left-hand side. But $x = 1$ is a root of $f(x) - g(x)$; so write

$$f(x) - g(x) = (1 - x)^{k+1} p(x) \quad (k \geq 0, \ p(1) \neq 0).$$

Then

$$\frac{f(x^2) - g(x^2)}{f(x) - g(x)} = (1 + x)^{k+1} \frac{p(x^2)}{p(x)}$$

and

$$2n = f(1) + g(1) = (1 + 1)^{k+1} \frac{p(1)}{p(1)} = 2^{k+1},$$

which proves the assertion. [J. Selfridge–E. G. Straus, *Pac. J. Math.* **8** (1958) 847–856.]

13. More generally, we prove that, if

$$h(s_1, \ldots, s_k) = \sum_{i=1}^{M} \alpha_i \binom{c_{i1}}{s_1} \cdots \binom{c_{ik}}{s_k}$$

(where $(c_{i1}, \ldots, c_{ik}) \neq (c_{j1}, \ldots, c_{jk})$ for $i \neq j$) vanishes for any choice of $s_1, \ldots, s_k \geq 0$, then $\alpha_1 = \ldots = \alpha_M = 0$. The assertion of the problem will follow by considering $h = f - g$.

Let $c_1, \ldots, c_p$ be the distinct values of c_{ik} $(1 \leq i \leq N)$. Then

$$h(s_1, \ldots, s_k) = \sum_{i=1}^{p} h_i(s_1, \ldots, s_{k-1}) \binom{c_i}{s_k},$$

where h_i is of the same form as h. Setting $x_i = h_i(s_1, \ldots, s_{k-1})$, we have

$$x_1 + \ldots + x_p = 0,$$

$$\binom{c_1}{1} x_1 + \ldots + \binom{c_p}{1} x_p = 0,$$

(1)

$$\vdots$$

$$\binom{c_1}{p-1} x_1 + \ldots + \binom{c_p}{p-1} x_p = 0.$$

We claim that the determinant of this system of equations is non-zero. We use the formula

$$\sum_{k=0}^{n} S(n, k) k! \binom{x}{k} = x^n,$$

where $S(n,k)$ are Stirling numbers of second kind (see 1.7). Multiply the $(i+1)^{\text{st}}$ row of the determinant of (1) by $i!$ and then add $S(j,i)$ times the $(i+1)^{\text{st}}$ row to the $(j+1)^{\text{st}}$ row $(j=1,\ldots,p-1;\ i=j-1,\ldots,0)$. Then we get

$$\begin{vmatrix} \binom{c_1}{0} & \cdots & \binom{c_p}{0} \\ \vdots & & \vdots \\ \binom{c_1}{p-1} & \cdots & \binom{c_p}{p-1} \end{vmatrix} = \frac{1}{0!\,1!\ldots(p-1)!} \begin{vmatrix} 1 & \cdots & 1 \\ c_1 & \cdots & c_p \\ \vdots & & \vdots \\ c_1^{p-1} & \cdots & c_p^{p-1} \end{vmatrix} =$$

$$= \prod_{1 \le i < j \le p} \frac{c_i - c_j}{i - j} \ne 0.$$

Thus (1) has only the trivial solution for $x_1,\ldots,x_p$, i.e. $h_i(s_1,\ldots,s_{k-1})=0$ for any choice of non-negative integers $s_1,\ldots,s_{k-1}$. Using induction on k ($k=0$ can be considered as true) we find that all coefficients $h_i(s_1,\ldots,s_{k-1})$ are 0. Thus so are all coefficients in $h(s_1,\ldots,s_k)$.

14. Form

$$\sum_{x \ne y} |E(G_i - x - y)|.$$

Here each edge is counted at each pair (x,y) different from its endpoints, hence

$$\sum_{x \ne y} |E(G_i - x - y)| = \binom{|V| - 2}{2} |E(G_i)|.$$

This implies that $|E(G_1)| = |E(G_2)|$. Now form

$$\sum_{y \ne x_0} |E(G_i - x_0 - y)|,$$

where x_0 is fixed. If e is an edge adjacent to x_0, then e is not counted here; otherwise, it is counted $|V|-3$ times. Thus,

$$|E(G_i)| - \frac{1}{|V| - 3} \sum_{y \ne x_0} |E(G_i - x_0 - y)|$$

is the degree of x_0 in G_i. Since this value is the same for $i=1$ and 2, $d_{G_1}(x_0) = d_{G_2}(x_0)$.

Finally, we have

$$|E(G_i)| - |E(G_i - x - y)| = d_{G_i}(x) + d_{G_i}(y) - \varepsilon_i(x,y),$$

where $\varepsilon_i(x,y)$ is the number of (x,y)-edges in G_i. Since the other terms do not depend on i, we have $\varepsilon_1(x,y)=\varepsilon_2(x,y)$, i.e. G_1, G_2 are identical.

15. (a) Let H_1, H_2 be two circuits with $V(H_1) = V(H_2) = V$. Then $H_1 - x$ is a path of length $|V|-2$ and so is $H_2 - x$. So $H_1 - x \cong H_2 - x$ for all $x \in V$, but, of course, H_1 and H_2 need not be identical. Thus the assertion is false.

(b) Let $G_i - x$ have $f(H, G_i - x)$ subgraphs isomorphic to H. Form

$$\sum_{x \in V} f(H, G_i - x).$$

Then each subgraph of G_i isomorphic to H is counted here $|V| - |V(H)|$ times, so

$$\sum_{x \in V} f(H, G_i - x) = (|V| - |V(H)|) f(H, G_i).$$

Since $|V| > |V(H)|$ by the assumption and the left-hand side is the same for $i = 1, 2$, we get

$$f(H, G_1) = f(H, G_2)$$

as stated. The same argument works for induced subgraphs.

(c) Suppose first that G_1 consists of an isolated point and a connected component H on $|V| - 1$ points. Then G_2 has an induced subgraph H' isomorphic with H, by the preceding assertion. Since we have, by the preceding assertion again, $|E(G_1)| = |E(G_2)|$, it follows that G_2 has no edge other than those of H', i.e. G_2 consists of H' and an isolated point. Thus $G_1 \cong G_2$.

Conversely, if G_2 has a component with $|V| - 1$ points then, similarly, $G_1 \cong G_2$. Thus, in every case, the number of components of G_1 and G_2 isomorphic with H, where $|V(H)| = |V| - 1$, is the same.

Suppose that this holds for each connected graph H with $|V(H)| = |V| - 1, \ldots, k+1$. Let $|V(H)| = k$. To get the number of components of G_i isomorphic with H we count all induced subgraphs of G_i isomorphic with H and subtract the number of induced subgraphs, isomorphic with H, of larger components of G_i. By the preceding problem and the induction hypothesis, this number will be the same for $i = 1$ and 2.

So G_1, G_2 have the same number of components isomorphic with any given H, hence $G_1 \cong G_2$. [P. J. Kelly, *Pacific J. Math.* **7** (1957) 961–968; for a survey on the reconstruction problem see J. A. Bondy–R. Hemminger, *J. Graph Theory* **1** (1977) 227–268.]

16. (a) First we remark that since

$$d_{T_1}(x) = |V| - 1 - |E(T_1 - x)| = |V| - 1 - |E(T_2 - x)| =$$
$$= d_{T_2}(x),$$

T_1 and T_2 have the same endpoints. Let W be the set of these endpoints.

If $|W| = 2$, then both T_1, T_2 are paths and thus $T_1 \cong T_2$. We will assume in the sequel that $|W| \geq 3$.

Let P be any maximum path in T_1 and x an endpoint not on P. Then by $T_2 - x \cong T_1 - x \supseteq P$ it follows that T_2 also contains a path of the same length. Hence the diameter of T_2 is at least as large as the diameter of T_1. This being true conversely as well, it follows they have the same diameter d.

(b) We discuss the case of even d. The odd case is simpler. Let c_i be the center of T_i. We may assume that $d > 2$ since, if T_i is a star, then it contains a point of degree $n-1$ and hence, the other tree is also a star.

$1°$ Suppose first that $T_i - x$ has diameter d for each endpoint x.

Let us determine the number $m_i(D)$ of branches of T_i attached to c_i, which are isomorphic to a specified tree D with a specified "root" (i.e. with a specified point to be attached to c_i). We prove by "backwards" induction that $m_1(D) = m_2(D)$. If $|V(D)| \geq |V|$ this is trivially true.

Let us count the number of all occurrences of branch D attached to the center of $T_i - x$, $x \in W$. Then by $T_1 - x \cong T_2 - x$, we obtain the same number. The branches of $T_i - x$ isomorphic to D arise in two ways; they are either branches of T_i themselves or arise from a larger branch of T_i by removing x. The numbers of branches D of the graphs $T_i - x$ arising from larger branches of T_i are the same for $i = 1$, 2 since T_1 and T_2 have the same larger branches. Therefore, the numbers of branches D of the graphs $T_i - x$, which are branches D of T_i are the same for $i = 1$, 2. Since each branch D of T_i is counted $|W| - l$ times, where l is the number of endpoints of D, this proves that $m_1(D) = m_2(D)$. Thus $m_1(D) = m_2(D)$ holds for any rooted tree D. This implies that $T_1 \cong T_2$.

$2°$ Let us assume that there is exactly one point x_0 such that the diameter of $T_1 - x_0 = T_2 - x_0$ is less than d. Then there must be exactly two branches B_1^i, B_2^i attached to c_i in T_i, which contain a $(d/2)$-path; there may be any number of other branches. B_1^i contains exactly one point at distance $d/2$ from c_i; B_2^i contains at least two such points.

$2°$(a) Suppose first that neither of B_1^1, B_1^2 is a single path. Then B_2^i is the largest branch in the trees $T_i - x$ $(x \in W - \{x_0\})$ relative to the center, which has at least two points at distance $d/2$ from the root and so, B_2^1 and B_2^2 are isomorphic rooted trees. Hence it follows as in $1°$ that T_1, T_2 have the same branches attached to their centers.

$2°$(b) Assume that B_1^1 is a single path. Then the endpoint x_1 of $T_1 - x_0$ adjacent to x_0 has the property that it is contained in all $(d-1)$-paths of $T_1 - x_0$. Hence $T_2 - x_0$ also contains an endpoint x_2, which occurs in all $(d-1)$-paths. Now x_0 must be adjacent to x_2 in T_2; since otherwise both x_2 and x_0 would be endpoints of T_2 contained in all d-paths, a contradiction. x_0 is connected to corresponding points of $T_1 - x_0 \cong T_2 - x_0$, i.e. $T_1 \cong T_2$.

$3°$ The case when $T_1 - x \cong T_2 - x$ has diameter $d-1$ for two points x (it clearly cannot have diameter $d-1$ for more than 2 endpoints) can be treated analogously [15.15; ibid.].

17. (a) Let H be any graph on n points and let $|G \to H|$ denote the number of those mapping $\varphi : V(G) \to V(H)$, which are one-to-one and map adjacent points onto adjacent points. Then $|G \to H|$ can be expressed by an inclusion-exclusion formula as follows. For each $S \subseteq E(G)$ let G_S denote the graph $(V(G), S)$. Then,

by a method similar to the one in 5.18,

$$|G \to H| = \sum_{S \subseteq E(G)} (-1)^{|S|} |G_S \to \overline{H}|.$$

In particular,

(1)
$$|G_1 \to G_2| = \sum_{S \subseteq E(G_1)} (-1)^{|S|} |G_{1,S} \to \overline{G}_2|,$$

(2)
$$|G_2 \to G_2| = \sum_{S \subseteq E(G_2)} (-1)^{|S|} |G_{2,S} \to \overline{G}_2|.$$

Here the graphs $G_{1,S}$ $(S \subset E(G_1))$ and $G_{2,S}(S \subset E(G_2))$ are pairwise isomorphic; this follows just like 15.14b. Moreover,

$$|G_{1,E(G_1)} \to \overline{G}_2| = |G_1 \to \overline{G}_2| = 0$$

since G_1 has more edges that $\overline{G}_2$. Similarly,

$$|G_{2,E(G_2)} \to \overline{G}_2| = 0.$$

Thus, the right-hand side of (1) and (2) have the same terms. This implies that

$$|G_1 \to G_2| = |G_2 \to G_2| > 0.$$

Thus G_1 has a one-to-one mapping into G_2. Since they have the same number of edges, $G_1 \cong G_2$. [L. Lovász, *J. Comb. Theory* **13** (1972) 309–310.]

(b) We may assume that $V(G_1) = V(G_2) = V$. Consider a random permutation π of V. Denote by $A_i(B_i)$ the event that e_i is mapped onto a pair of non-adjacent points in G_1 (in G_2). Define, as in 2.2a,

$$A_I = \prod_{i \in I} A_i, \qquad B_I = \prod_{i \in I} B_i,$$

$$\sigma_j = \sum_{|I|=j} \mathsf{P}(A_I), \qquad \sigma'_j = \sum_{|I|=j} \mathsf{P}(B_I).$$

Note that as in 15.14b or part (a) it follows that

$$\sigma_j = \sigma'_j \quad \text{for} \quad j = 0, 1, \ldots, m-1.$$

Denote by η and ξ the numbers of A_i's and B_i's that occur. Then by 2.7b,

$$\mathsf{E}(x^\eta) = \sum_{j=0}^{m} (x-1)^j \sigma_j, \; \mathsf{E}(x^\xi) = \sum_{j=0}^{m} (x-1)^j \sigma'_j.$$

Hence

(1)
$$\mathsf{E}(x^\eta) - \mathsf{E}(x^\xi) = (x-1)^m (\sigma_m - \sigma'_m).$$

Set here $x = -1$ and use the trivial relations

$$|\mathsf{E}((-1)^\eta)| \leq 1 \quad \text{and} \quad |\mathsf{E}((-1)^\xi)| \leq 1.$$

Then we get by (1)

$$|\sigma_m - \sigma'_m| \le \frac{1}{2^{m-1}} < \frac{1}{n!}$$

by hypothesis. But from the definition of the probability field, σ_m and σ'_m are rationals with denominator $n!$. Hence they must be equal.

Now, taking the constant terms in (1), we obtain

$$P(\eta = 0) = P(\xi = 0).$$

But, if π is the identity, then $\xi = 0$. So

$$P(\eta = 0) = P(\xi = 0) > 0,$$

i.e. there is a permutation, which maps each e_i onto an edge. This is then an isomorphism between G_1 and G_2. [W. Müller, *J. Comb. Theory* [B] **22** (1977) 281–283.]

18. We are going to show that, if $|Y| = k - 1$, then

$$|E(H_1 - X)| = |E(H_2 - X)|.$$

Let $X_1, \ldots, X_t$ be the k-tuples containing Y ($t = |V| - k + 1$). Then

(1)
$$\sum_{i=1}^{t} |E(H_1 - X_i)| = (|V| - k - r + 1)|E(H_1 - X)|,$$

which follows by a simple counting. Similarly,

(2)
$$\sum_{i=1}^{t} |E(H_2 - X_i)| = (|V| - k - r + 1)|E(H_2 - X)|.$$

Since the left-hand sides of (1) and (2) are equal and $|V| - k - r + 1 > 0$, (1) and (2) imply that $|E(H_1 - Y)| = |E(H_2 - Y)|$. Similarly this follows for each Y with $|Y| \le k$.

Let $T \subseteq V$, $|T| = r$. We want to calculate the multiplicity of T as an edge of H_i; i.e. we want to determine the number of those edges of H_i, which do not belong to $H_i - t$ for any $t \in T$. By inclusion-exclusion, this number is

$$\sum_{X \subseteq T} (-1)^{|X|} |E(H_i - X)|.$$

Since $k \ge r$, this number is the same for $i = 1$, 2. Hence H_1 and H_2 are identical [V. Faber, *Hypergraph Seminar*, Lecture Notes in Math. 411, Springer, 1974, 85–94].

19. Let $\nu_i(X)$ denote the multiplicity of $X \subseteq V$ as an edge of H_i ($i = 1$, 2). The assumption implies that

(1)
$$\nu_1(X - \{x\}) + \nu_1(X) = \nu_2(X) = \nu_2(X - \{x\}) + \nu_2(X),$$

since $\nu_i(X - \{x\}) + \nu_i(X)$ is the multiplicity of the edge $X - \{x\}$ in $H_i \setminus x$. Now assume that $H_1 \ne H_2$. Then there are sets $X \subseteq V$ with $\nu_1(X) \ne \nu_2(X)$. Let X be a minimal such set. Then by (1), $\nu_1(X - \{x\}) \neq \nu_2(X - \{x\})$ for any $x \in X$. This is only possible, if $X = \emptyset$. So we may assume that $\nu_1(\emptyset) > \nu_2(\emptyset)$.

We prove by induction on $|X|$ that

$$\nu_1(X) > \nu_2(X), \quad \text{if } |X| \text{ is even,}$$
$$\nu_1(X) < \nu_2(X), \quad \text{if } |X| \text{ is odd.}$$

This follows easily from (1). In particular,

$$\nu_1(X) > \nu_2(X) \geq 0, \quad \text{if } |X| \text{ is even,}$$

i.e. every even subset of V is an edge of H_1. Similarly, every odd subset of V is an edge of H_2 [see C. Berge, *Hypergraph Seminar*, Lecture Notes in Math. 411, Springer, 1974, 1–12].

20. (a) Let $\hom(A, B)$ and $\mathrm{epi}(A, B)$ denote the numbers of homomorphisms of A into B and epimorphisms of A onto B, respectively. Observe first:

(*) If $\mathrm{epi}(G_1, H) = \mathrm{epi}(G_2, H)$ for every H, then $G_1 \cong G_2$.

In fact, we have $\mathrm{epi}(G_1, G_2) = \mathrm{epi}(G_2, G_2) > 0$, $\mathrm{epi}(G_2, G_1) = \mathrm{epi}(G_1, G_1) > 0$, i.e. G_1, G_2 are epimorphic images of each other. Since they are finite, they must be isomorphic.

So we want to relate $\hom(G_i, H)$ and $\mathrm{epi}(G_i, H)$. Obviously, each epimorphism is a homomorphism and a homomorphism is an epimorphism iff every edge of H is in the image. So we can use the inclusion-exclusion formula and get

$$\mathrm{epi}(G_i, H) = \hom(G_i, H) - \sum_{e \in E(H)} \hom(G_i, H - e) +$$

$$+ \sum_{\{e_1, e_2\} \subseteq E(H)} \hom(G_i, H - e_1 - e_2) - \ldots = \sum_{S \subseteq E(H)} (-1)^{|S|} \hom(G_i, H - S).$$

Since the right-hand sides are the same for $i = 1, 2$ and any fixed H, we conclude that

$$\mathrm{epi}(G_1, H) = \mathrm{epi}(G_2, H)$$

holds for every graph H. By (*), this implies that $G_1 \cong G_2$.

(b) Let $\mathrm{mon}(A, B)$ denote the number of monomorphisms (= one-to-one homomorphisms) of A into B. Again,

(*) if $\mathrm{mon}(H, G_1) = \mathrm{mon}(H, G_2)$ for every H, then $G_1 = G_2$.

In fact, $\mathrm{mon}(G_2, G_1) = \mathrm{mon}(G_1, G_1) > 0$, $\mathrm{mon}(G_1, G_2) = \mathrm{mon}(G_2, G_2) > 0$, i.e. G_1, G_2 have monomorphisms into each other, whence $G_1 \cong G_2$ follows because they are finite.

To relate $\mathrm{mon}(H, G_i)$ and $\hom(H, G_i)$, observe that a homomorphism is a monomorphism if and only if it does not identify any pair of points. Let $\Theta((x_1, y_1), \ldots, (x_k, y_k))$ be the least equivalence relation such that x_i, y_i belong to the same class $(i = 1, \ldots, k)$ and denote, for any simple graph H and equivalence relation Θ on $V(H)$, by H/Θ the simple graph obtained by identifying the points in every given class of Θ and forgetting about multiplicities of edges in the resulting graph. Then we can express $\mathrm{mon}\,(H, G_i)$ by inclusion-exclusion.

$$\mathrm{mon}\,(H, G_i) = \mathrm{hom}\,(H, G_i) - \sum_{\{x_1, y_1\} \subseteq V(H)} \mathrm{hom}\,(H/_{\Theta(x_1, y_1)}, G_i) +$$

$$+ \sum_{\{x_1, y_1\} \neq \{x_2, y_2\} \subseteq V(H)} \mathrm{hom}\,(H/_{\Theta((x_1, y_1), (x_2, y_2))}, G_i) - \ldots =$$

$$\sum_{k=0}^{\binom{|V(H)|}{2}} (-1)^k \sum_{\{x_1, y_1\} \neq \ldots \neq \{x_k, y_k\} \subseteq V(H)} = \mathrm{hom}\,(H/_{\Theta((x_1, y_1), \ldots, (x_k, y_k))}, G_i).$$

Again, the right-hand side is the same for $i = 1$ and 2, so $\mathrm{mon}(H, G_1) = \mathrm{mon}(H, G_2)$ and $(*)$ proves the assertion. [L. Lovász, *Periodica Math. Hung.* **1** (1971) 145–156.]

21. Let $\varphi, \psi: H \to G_1$ be homomorphisms, and set

$$\xi(v) = (\varphi(v), \psi(v)) \quad (v \in V(H)).$$

Then, as is easily verified, ξ is a homomorphism of H into $G_1 \times G_1$; conversely, each homomorphism of H into $G_1 \times G_1$ arises uniquely in this way. Hence

$$(1) \qquad\qquad \mathrm{hom}\,(H, G_1 \times G_1) = (\,\mathrm{hom}\,(H, G_1))^2.$$

Now we have

$$(\,\mathrm{hom}\,(H, G_1))^2 = \mathrm{hom}\,(H, G_1 \times G_1) = \mathrm{hom}\,(H, G_2 \times G_2) = (\,\mathrm{hom}\,(H, G_2))^2$$

and, since $\mathrm{hom}\,(H, G_i) \geq 0$,

$$\mathrm{hom}\,(H, G_1) = \mathrm{hom}\,(H, G_2).$$

This holds for any H, so by 15.19b, we get $G_1 \cong G_2$. [ibid.].

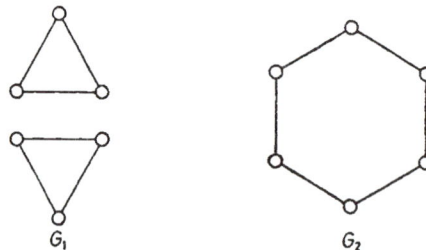

FIGURE 118

22. (a) We try $F = K_2$. Then observe that, if $v \in V(G_1)$, then $(v, x) \in V(G_1 \times F)$ has the same degree as v. Therefore, G_1 and G_2 must have the same degrees. A simple non-isomorphic pair with the same degree sequence is shown in Fig. 118. Forming the direct products with K_2 (we set $V(K_2) = \{1, 2\}$), we get two disjoint 6-cycles in each case, as shown in Fig. 119.

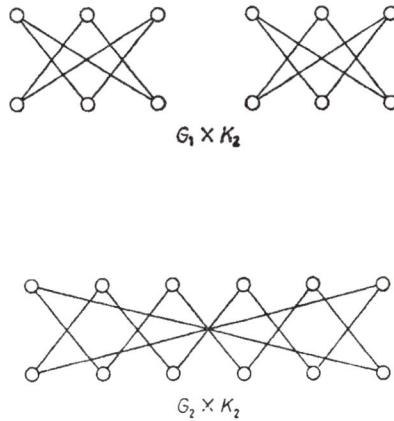

FIGURE 119

(b) Let H be any graph. For reasons similar to those in the preceding solution,

$$\hom(H, G_1)\hom(H, F) = \hom(H, G_1 \times F) =$$

(1)

$$= \hom(H, G_2 \times F) = \hom(H, G_2)\hom(H, F).$$

We want to show that

(2) $$\hom(H, G_1)\hom(H, F') = \hom(H, G_2)\hom(H, F').$$

If $\hom(H, F') = 0$, then (2) is trivially satisfied. If $\hom(H, F') > 0$ then, obviously, $\hom(H, F) > 0$, so (1) implies that $\hom(H, G_1) = \hom(H, G_2)$, whence (2) follows again.

(c) The relation $(G \cdot H)^0 = G^0 \times H^0$ formulated in the hint is easily verified. Then

$$G_1^0 \times H^0 = (G_1 \cdot H)^0 \cong (G_2 \cdot H)^0 = G_2^0 \times H^0.$$

Let E^0 denote the graph with one point and one loop. Then E^0 is a subgraph of H^0 and so by (b)

$$G_1^0 \cong G_1^0 \times E^0 \cong G_2^0 \times E^0 \cong G_2^0.$$

(d) The assertion formulated in the hint follows by an argument similar to the one in 15.20b except that $\Theta((x_1, y_1), \ldots, (x_k, y_k))$ must be interpreted as the least *congruence relation* containing the pairs $(x_1, y_1), \ldots, (x_k, y_k)$.

Now suppose that G_1, G_2, F are finite groups and

$$G_1 \times F \cong G_2 \times F.$$

Denoting the number of homomorphisms of the group H into the group G by $\hom(H, G)$ we have

$$\hom(H, G_1)\hom(H, F) = \hom(H, G_2)\hom(H, F),$$

for each finite group H. Since $\hom(H, F) \neq 0$, this yields

$$\hom(H, G_1) = \hom(H, G_2),$$

whence $G_1 \cong G_2$.

For finite rings the proof is the same. In fact, it is the same for any three algebras G_1, G_2, F provided F has a 1-element subalgebra. [For a generalization to categories see L. Lovász, *Acta Sci. Math.* **33** (1972) 319–322.]

Dictionary

of the combinatorial phrases and concepts used

Abel identities: see 1.44.

Access time: see *random walk*.

Adjacency matrix of a graph G with vertices $v_1, \ldots, v_n$: the matrix $A_G = (a_{ij})_{i,j=1}^n$, where A_{ij} is the number of (v_i, v_j)-edges.

Adjacent: see *graph*.

Arborescence: a digraph G with a specified vertex a called the *root* such that each point $x \neq a$ has indegree 1 and there is a unique (a, x)-path for each point x. Arborescences can be obtained by specifying a vertex a of a tree and then orienting each edge e such that the unique path connecting a to e ends at the tail of e. An *inverse* $\sim$ is the digraph obtained from an arborescence by inverting its edges.

Automorphism *of a [di]graph*: a permutation α of $V(G)$ such that the number of (x, y)-edges is the same as the number of $(\alpha(x), \alpha(y))$-edges $(x, y \in V(G))$. We also speak of the automorphisms of a graph G with colored edges. This means a permutation α such that the number of (x, y)-edges is the same as the number of $(\alpha(x), \alpha(y))$-edges with any given color. The set of all automorphisms of a [di]graph forms a permutation group $A(G)$.

Balanced circuit *in a hypergraph*: a circuit $(x_1, E_1, \ldots, x_k, E_k)$ such that either $k = 2$ or there is an incidence $x_i \in E_j$, where $j \neq i$, $i - 1$ and $(i, j) \neq (1, k)$.

$\sim$ **hypergraph:** a hypergraph in which every circuit of odd length is balanced.

totally $\sim$ **hypergraph:** every circuit is balanced.

Bell number: see *partition*.

Binomial coefficient $\binom{n}{k}$: the number of ways to select k elements out of n. One has

$$(1) \qquad \binom{n}{k} = \frac{n!}{k!(n-k)!} = \frac{n(n-1)\ldots(n-k+1)}{k!}, \qquad (0 \leq k \leq n)$$

and by definition

$$\binom{k}{0} = 1, \quad (k = 0, 1, \ldots).$$

Formula (1) defines $\binom{n}{k}$ for any real (or complex) value of n.

Bipartite *(2-chromatic) graph*: A graph G admitting a *bipartition* or *2-coloration* $\{A, B\}$ of its points such that each edge connects a point of A to a point of B (cf. *chromatic number*).

Block *of a graph* G: a cut-edge or a maximal 2-connected subgraph. Each edge is contained in a unique block. Blocks could also be defined as classes of the equivalence relation on $E(G)$ defined by "e and f are on a circuit or $e = f$". The blocks of a graph give a "cactus-like" structure; each point belonging to more than one block is a cutpoint, and the number of branches relative to any point is the same as the number of blocks containing this point (Fig. 120). Blocks containing one cutpoint are called *endblocks*.

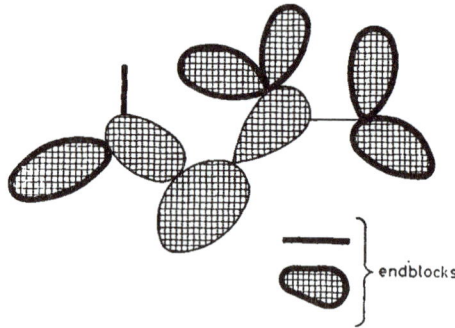

FIGURE 120

Bottleneck Theorem: see 6.71.

Branch *of a graph* G *relative to a point* x: a subgraph consisting of a component G_1 of $G - x$, the point x and all edges connecting x to G_1.

Bridge *of a subgraph* G_1: A (connected) subgraph B such that either B consists of a single edge with both endpoints in G_1 or B consists of a connected component of $G - V(G_1)$ together with all edges connecting this component to G_1 and their endpoints in G_1. The bridges of G_1 partition $E(G) - E(G_1)$; i.e. they could also be defined as classes determined by the equivalence relation "$e_1 \sim e_2$ iff $e_1 = e_2$ or there is a path connecting e_1 and e_2 disjoint from G_1".

Brooks' Theorem: see 9.13.

Brun's Sieve: see 2.13.

Burnside Lemma: see 3.23.

Catalan numbers see 1.33, 1.37–40.

Cayley Formula: see 4.2.

Characteristic polynomial *of a graph* G: The polynomial $p_G(\lambda) = \det(\lambda I - A_G)$, where A_G is the adjacency matrix of G. This clearly does not depend on the labelling of points. The roots of the characteristic polynomial, i.e. the eigenvalues of A_G, are called the *eigenvalues* of the graph G.

Chord *of a subgraph* $G_1 \subseteq G$: an edge $e \in E(G) - E(G_1)$ connecting two points of G_1.

Chromatic index *of a* [*hyper*]*graph* G: the least integer k for which the edges of G can be k-colored such that adjacent edges have different colors. We denote it by $q(G)$. Clearly $q(G) = \chi(L(G))$.

Chromatic number *of a graph* [*digraph, hypergraph*] G: the least integer k for which G has a "good k-coloration" (see below), we denote this number by $\chi(G)$. Clearly, $\chi(G) > 0$, if G is non-empty; $\chi(G) > 1$, if $E(G)$ is non-empty. $\chi(G) = \infty$ if G has a loop [or, when G is a hypergraph, it has an edge with at most 1 endpoint].

Chromatic polynomial $P_G(\lambda)$ *of a graph* G: the number of good λ-colorations of G ($\lambda = 0, 1, \ldots$). This turns out to be a polynomial in λ (for fixed G) and so, its definition can be extended to all real (or complex) values of λ. Note that two λ-colorations differing in the labelling of colors count as different.

Circuit *in a graph*: a walk $(x_1, e_1, \ldots, x_k, e_k, x_{k+1})$ such that $x_1, \ldots, x_k$ are distinct points, $e_1, \ldots, e_k$ are distinct edges and $x_1 = x_{k+1}$. If the graph is simple, we will denote it by $(x_1, \ldots, x_k)$.

$\sim$ *in a digraph*: a circuit in the graph obtained from G by replacing each oriented edge by an unoriented one with the same endpoints. Cf. *cycle*.

$\sim$ *in a hypergraph*: a sequence $(x_1, E_1, \ldots, x_k, E_k)$, where $x_1, \ldots, x_k$ are distinct points, $E_1, \ldots, E_k$ are distinct edges and $x_i \in E_i$ ($i = 1, \ldots, k$), $x_{i+1} \in E_i$ ($i = 1, \ldots, k-1$) and $x_1 \in E_k$. k is the *length* of this circuit.

Clique: a maximal complete subgraph of a graph.

Clutter: a hypergraph in which no edge contains another.

Collection: a set S together with an assignment of positive integers called *multiplicities* to the elements of S; anything not in S can be said to have multiplicity 0.

Coloration: a (legitimate, good) k-coloration of graph [digraph, hypergraph] is an assignment of "colors" (usually one of the integers $1, \ldots, k$) to the points such that every edge meets at least two different "colors".

k-**colorable** *graph* [*digraph, hypergraph*]: it has a good k-coloration.

Commute time: see *random walk*.

Complement *of a simple graph* G: the simple graph $\overline{G}$ defined by

$$V(\overline{G}) = V(G), \ E(\overline{G}) = \{(x,y) : x,y \in V(G), \ x \neq y, (x,y) \notin E(G)\}.$$

Clearly $\overline{(\overline{G})} = G$.

~ *of a simple digraph* G: the simple digraph $\overline{G}$ defined by

$$V(\overline{G}) = V(G), \quad E(\overline{G}) = V(G) \times V(G) - E(G).$$

Complete *graph*: a simple graph in which any two distinct points are adjacent. A complete graph on n points will be denoted by K_n.

~ *rooted d-ary tree*: see *tree*.

Complete bipartite graph: a simple graph whose points can be partitioned into two classes U, W such that two points are adjacent iff one of them belongs to U and the other to W. If $|U| = n$, and $|W| = m$, then the complete bipartite graph is denoted by $K_{n,m}$.

Component (*connected* ~) *of a graph* G: a maximal connected subgraph of G. Any two connected components of G are vertex-disjoint and each vertex (and edge) belongs to one of them. Their number is denoted by $c(G)$; $c_1(G)$ denotes the number of those connected components with an odd number of points.

strong ~ (*of a digraph*): maximal strongly connected subgraph.

Conductance *of a graph* G: the minimum of $\delta_G(S)/|S|$ over all non-empty sets $S \subseteq V$ with $|S| \leq |V|/2$, where $\delta_G(S)$ denotes the total number of edges joining S to its complement $V \setminus S$.

Connected *graph*: a graph that is not representable in the form $G_1 \cup G_2$, where G_1 and G_2 are vertex-disjoint non-empty graphs. Equivalently: any two points are connected by a path in the graph.

weakly ~ *digraph*: a digraph not representable as $G_1 \cup G_2$, where G_1, G_2 are vertex-disjoint non-empty digraphs.

strongly ~ *digraph*: a digraph in which any two points are connected by a (directed) path.

~ **hypergraph:** not representable as $H_1 \cup H_2$, where H_1, H_2 are vertex-disjoint non-empty hypergraphs. Note that if $\emptyset \in E(H)$, H is not connected.

k-**connected** *between a and b*: if less than k points ($\neq a, b$) and/or edges are removed, there still exists an (a,b)-path in the [di]graph (removal of edges is only needed when a, b are adjacent).

~ **[di]graph** G: on at least $k+1$ points, k-*connected* between any two points. Equivalent formulations: $|V(G)| \geq k+1$ and $G - X$ is [strongly] connected for any set $X \subset V(G)$, $|X| \leq k-1$. Or, for a graph: $|V(G)| \geq k+1$ and G cannot be represented as $G_1 \cup G_2$, where $V(G_1)$, $V(G_2) \neq V(G)$ and $|V(G_1) \cap V(G_2)| \leq$

$k - 1$. The complete graph K_n is thus $(n - 1)$-connected but non n-connected. Connected and 1-connected are equivalent for graphs with at least two points.

Contraction *of an edge* e *of a* [*di*]*graph*: removal of this edge and identification of its endpoints. Contracting a subgraph means contracting all edges of it (the order in which the contraction is made is irrelevant). Note that multiple edges may appear.

k-**cover** *of a* [*hyper*]*graph* G: a collection of points such that each edge contains at least k of them. (*Point-*)*cover* means the same as 1-cover; the minimum size of a point-cover is denoted by $\tau(G)$. A k-cover can also be regarded as a mapping $t : V(G) \to \{0, 1, \ldots\}$ such that $\sum_{x \in E} t(x) \geq k$ for each edge E.

fractional cover *of a* [*hyper*]*graph* G: an assignment of a non-negative real weight $t(x)$ to each point x such that $\sum_{x \in E} t(x) \geq 1$ for every edge E. The *size* of a fractional cover is $\sum_{x \in V(G)} t(x)$. The minimum size of a fractional cover is denoted by $\tau^*(G)$.

Cover time: see *random walk*.

Critical: a graph G is called (*edge-*)*critical* with respect to property P, or *critically having* property P if G has it but, on removing any edge, the resulting graph will not have property P. *Point-critical* is defined analogously.

α-*critical*: $\alpha(G - \{e\}) > \alpha(G)$, for each edge e.

χ-*critical*: $\chi(G - \{e\}) < \chi(G)$, for each edge e

τ-*critical*: $\tau(G - \{e\}) < \tau(G)$, for each edge e.

ν-*critical hypergraph*: $\nu(H - x) = \nu(H)$, for each $x \in V(G)$ (cf. problem 7.26 for an explanation of this name).

factor-critical graph: G has no 1-factor but $G - x$ has a 1-factor for each point x.

(a, b)-**cut:** a set F of edges representing (covering) all (a, b)-paths. The cut determined by $S \subseteq V(G)$ is the set of edges connecting S to $V(G) - S$. If C is a cut in the digraph G determined by S', the C^* is the cut determined by $V(G) - S$.

Cutset(*separating set*): a set of points [edges] in a connected graph whose removal results in a disconnected graph. A *cutpoint* [*cut-edge* or *isthmus*] is a point [edge] forming a cutset by itself.

Cycle *in digraph*: a walk $(x_1, e_1, \ldots, e_k, x_{k+1})$, where $x_1, \ldots, x_k$ are distinct and $x_{k+1} = x_1$.

Cycle index *of a permutation group* Γ: the polynomial

$$\frac{1}{|\Gamma|}p_\Gamma(x_1,\ldots,x_n) = \sum_{\pi\in\Gamma} x_1^{k_1(\pi)}\ldots x_n^{k_n(\pi)},$$

where n is the number of objects Γ acts on and $k_i(\pi)$ is the number of i-cycles in the cycle decomposition of the permutation π.

Degree *(valency) of a point* x *in a* [*hyper*]*graph* G: the number of edges containing x (in the case of graphs a loop is counted twice). The degree of x is denoted by $d_G(x)$. $d(G)$ denotes the maximum degree of G. A graph is *k-regular* if every point has degree k. The *indegree* [*outdegree*] of a point x in a digraph G is defined as the number of edges with their head [tail] in x; they are denoted by $d_G^-(x)$ and $d_G^+(x)$, respectively.

Diameter *of a graph* G: the maximum distance between points of G.

Digraph: a set $V(G)$ of *points* or *vertices*, a set $E(G)$ of *edges*, and an assignment of an ordered pair of points with each edge; the first and second elements of this pair are called the *tail* and the *head* of the edge, respectively. Two edges are *parallel*, if they have the same heads and tails. The digraph is *simple*, if it contains no two parallel edges. In this case, $E(G)$ can be considered as a subset of $V(G)\times V(G)$. If G is a graph and for each edge we declare one of its endpoints to be its head, the other one its tail, we obtain a digraph called an *orientation* of G. The phrase *oriented graph* is reserved for an oriented simple graph, i.e., for a digraph without loops in which at most one edge joins any two points. If $e = (x,y) \in E(G)$, we use either of the phrases: e is directed from x to y; y can be reached from x on e; e leaves x and enters y; e has *head* y and *tail* x. Cf. also *graph*.

Distance *between two points* x, y *in a graph* G: the minimum length of (x,y)-paths; if no path of G connects x to y, their distance is ∞. It is denoted by $d_G(x,y)-$ or shortly $d(x,y)$, if the graph G under consideration is well-determined.

Dual map *of a connected planar map* G: the map G^* constructed as follows. We select a point x_F in each of the faces F of G; these will be the vertices of G^*. Also we select a point p_e on each edge e of G. We connect each point p_e to the points x_F, $x_{F'}$, by Jordan curves J_e, J_e' interior to F and F', respectively, where F, F' are the two faces adjacent to e. If $F = F'$ (i.e. the same face of G bounds e from both sides), then J_e, J_e' should connect p_e to x_F such that they leave p_e on different sides of e (this happens, if e is a cutting edge). Moreover, let us choose J_e, J_e' such that the arcs J_e connecting x_F to points p_e on the boundary of F should have no point in common other than x_F. Set $e^* = J_e \cup J_e'$ and $E(G^*) = \{e^* : e \in E(G)\}$. Then G^* is also a planar map. If we consider G and G^* embedded in the sphere then the dual map is essentially uniquely defined, i.e. if $\hat{G}^*$ is another dual map of G, then there is a homeomorphism φ of the sphere onto itself such that $\varphi(x) = x$ for each $x \in V(G)$, $\varphi(e) = e$ for each $e \in E(G)$, $\varphi(V(G^*)) = V(\hat{G}^*)$

and if $\hat{e}^*$ is the edge corresponding to e^* in $\hat{G}^*$, then $\varphi(e^*)=\hat{e}^*$. The dual of G^* is G. The above construction and these last assertions involve much from plane topology that we accept here without proof (Fig. 121).

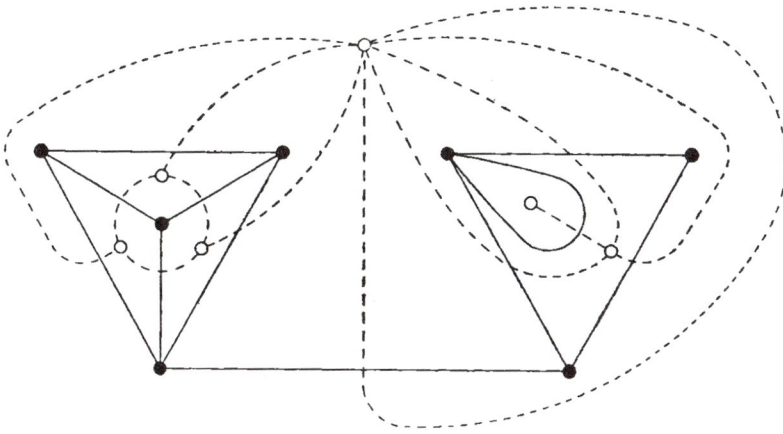

FIGURE 121

Ear-decomposition: see 6.28.

Edge: see *graph, digraph, hypergraph.*

Edge-cover *of a* [*hyper*]*graph*: a set of edges containing all points.

k-**edge-connected** between a and b: the removal of no more than $k-1$ edges results in a [di]graph containing an (a,b)-path.

$\sim$ [di]graph: k-edge connected between any two points. Equivalently: the removal of no more than $k-1$ edges results in a [strongly] connected [di]graph.

Edmonds' Matching Algorithm: see 7.34.

Elementary bipartite graph: see 7.7.

Empty [*hyper*]*graph*: has no points and no edges.

Endomorphism *of a* [*di*]*graph* G: a homomorphism of G into itself. The set of all endomorphisms of G with the composition as multiplication, form a semigroup, denoted by $\text{End}(G)$.

Endpoint of a graph G: a point with degree 1.

$\sim$ of an edge: *see graph.*

Erdös–de Bruijn Theorem: see 8.14.

Erdös–Ko–Rado Theorem: see 13.28.

Erdös–Stone Theorem: see 10.38.

Euler trail: a trail connecting all edges of a [di]graph.

Eulerian [*di*]*graph*: it has an Euler trail.

Euler's Formula: see 5.24.

Expansion rate: see *conductance*.

Face: see *planar map*.

f-factor of a graph: given a function f defined on $V(G)$ an f-factor is a subgraph G' with $V(G') = V(G)$ such that $d_{G'}(x) = f(x)$ for each point x. A *1-factor* is, therefore, a system of independent edges covering all points.

Ferrer's diagram: see 1.16.

(a,b)-**flow:** a non-negative real valued function f defined on the edges of a digraph such that, for each point $x_0 \neq a, b$,

$$\sum_{e=(x_0,y)\in E(G)} f(e) = \sum_{e=(y,x_0)\in E(G)} f(e)$$

(i.e. the amount of "water" entering x_0 is the same as the amount leaving it; "Kirchhoff's Law"). The *value* of an (a,b)-flow f is the "net gain" at the source a, i.e.,

$$w(f) = \sum_{e=(a,x)} f(e) - \sum_{e=(x,a)} f(e).$$

Forest: a graph without circuits. The components of a forest are trees.

Frucht's Theorem: see 12.5.

Gallai–Edmonds Structure Theorem: see 7.32.

Generating function *of a sequence* $\{a_n\}_{n=0}^{\infty}$: the function

$$f(x) = \sum_{n=0}^{\infty} a_n x^n.$$

exponential $\sim$: $f(x) = \sum_{n=0}^{\infty} \frac{a_n}{n!} x^n.$

Geometric lattice: a lattice L such that L is generated by its atoms and whenever x covers $x \wedge y$, $x \vee y$ covers y. The rank $r(x)$ is defined as the maximum length of $(0,x)$-chains less one. This function satisfies: $r(x) \geq 0$, $x > x' \rightarrow r(x) \geq r(x')$, if x covers y, then $r(y) \leq r(x) \leq r(y) + 1$, and $r(x \vee y) + r(x \wedge y) \leq r(x) + r(y)$. Examples of such lattices are the lattices of subspaces of any subset of a projective or affine plane.

Girth *of a graph* G: the length of its shortest circuit. The girth is 1 iff G has a loop and it is 2 iff G has multiple edges.

Graph: a graph G consists of a finite set $V(G)$ of *points (vertices)* and a finite set $E(G)$ of *edges*, and an assignment of an unordered pair of elements of $V(G)$ to each edge $e \in E(G)$, called the *endpoints* of e. We write $G = (V(G), E(G))$. An edge is said to *join* or *connect* its endpoints. If e connects x to y then e is called an (x,y)-*edge*. An edge with two identical endpoints is a *loop*. Two edges with the same pair of endpoints are *parallel* or *multiple*. The graph is *simple* if it has no loops or parallel edges. In this case we may consider $E(G)$ as a set of 2-subsets of $V(G)$. An edge and a point are *adjacent (incident)* if the point is an endpoint of the edge. Two edges are *adjacent* if they share an endpoint. Two points are *adjacent (neighboring)* if they are connected by an edge. The set of edges adjacent to a point is its *star*. The set of points adjacent to $X \subseteq V(G)$ is denoted by $\Gamma_G(X)$ or simply $\Gamma(X)$ if the graph in question is clear from the context. Graphs without loops are special hypergraphs (cf. also *digraph*).

Hajós' Construction: see 8.16.

Hamiltonian circuit [*cycle, path*]: a circuit [cycle, path] containing all points of a [di]graph.

Hamiltonian [*di*]*graph*: it has a Hamiltonian circuit [cycle].

Homomorphism *of* [*di*]*graph* G_1 *into* [*di*]*graph* G_2: a mapping $\varphi : V(G_1) \to V(G_2)$ such that if $(x,y) \in E(G_1)$, then $(\varphi(x), \varphi(y)) \in E(G_2)$.

Hypergraph (*set-system*): a hypergraph H consists of a finite set $V(H)$ of *points (vertices)*, a finite set $E(G)$ of edges and an assignment of a subset of $V(H)$ to each edge E, the set of *endpoints (elements)* of E. So we write $H = V(H), E(H))$. Two edges with the same endpoints are *parallel*. The hypergraph is *simple*, if it contains no parallel edges. In this case, $E(H)$ can be considered as a set of subsets of $V(H)$. An edge and a point are called *incident*, if the point is an endpoint of the edge. The hypergraph is r-*uniform*, if every edge has r endpoints. A 2-uniform hypergraph is a graph without loops. The *complete* r-*uniform hypergraph* on n vertices is a simple hypergraph having all r-subsets of its vertex set for its edges. It is denoted by K_n^r.

Identification *of two points* x, y *of a* [*di*]*graph* G results in a [di]graph G' with $V(G') = V(G) - \{x,y\} \cup \{\overline{xy}\}$, where $z = \overline{xy}$ denotes a new point, $E(G') = E(G)$ and each edge $e \in E(G)$ should have the same endpoints in G' as in G except if it has had x or y as endpoint it should have $\overline{xy}$ instead. Thus each (x,y)-edge becomes a loop in $\overline{xy}$.

Incidence matrix *of a graph* G with vertices $v_1, \ldots, v_n$ and edges $e_1, \ldots, e_m$: the matrix $B_G = (b_{ij})_{i=1 \, j=1}^{n \quad m}$, where $b_{ij} = 1$ if v_i and e_j are adjacent and 0 otherwise.

Incident: see *graph, hypergraph*.

Inclusion-exclusion Formula: see § 2.

Independent points *in a* [*di*]*graph*: no two are adjacent. The maximum number of independent points in G is denoted by $\alpha(G)$.

$\sim$ **edges** *in a* [*hyper,di*]*graph*: no two have an endpoint on common. The maximum number of independent edges in G is denoted by $\nu(G)$. Independent edges form a *matching*.

$\sim$ **paths** *in a* [*di*]*graph*: paths having no points in common except possibly their endpoints.

Induced subgraph: see *subgraph*.

Interval graph: a simple graph whose vertices are intervals on a line, and two vertices are adjacent iff they intersect as intervals.

Isolated point *of a graph*: a point adjacent to no edge.

Isomorphism *of G_1 onto G_2*: a one-to-one mapping φ of $V(G_1)$ onto $V(G_2)$ and a one-to-one mapping $\tilde{\varphi}$ of $E(G_1)$ onto $E(G_2)$ such that if x is an endpoint [head, tail] of e, then $\varphi(x)$ is an endpoint [head, tail] of $\tilde{\varphi}(e)$. If G_1 and G_2 are simple, then $\tilde{\varphi}$ plays no important role and we define an isomorphism as a one-to-one mapping φ of $V(G_1)$ onto $V(G_2)$ such that $(x,y) \in E(G_1) \leftrightarrow (\varphi(x),\varphi(y)) \in E(G_2)$.

Isthmus: see *cut-edge*.

Kernel: an independent set S of points in a digraph such that, for each $x \in V(G) - S$, there exists a $y \in S$ such that $(y,x) \in E(G)$.

König's Theorem: see 7.2.

Kuratowski's Theorem: see 5.38.

Laplacian of a graph G with vertices $v_1,\ldots,v_n$: the matrix $L_G = (\ell_{ij})_{i,j=1}^n$, where ℓ_{ij} is the negative of the number of (v_i,v_j)-edges if $i \neq j$, and ℓ_{ii} is the degree of vertex i. For a d-regular graph, $L_G = dI - A_G$.

Lattice: a partially ordered set such that any two elements x,y have a unique least upper bound $x \vee y$ (called their *join*) and a unique largest lower bound $x \wedge y$ (called their *meet*). All lattices we consider are finite. Every finite lattice has a unique smallest element 0 and a unique largest element 1. A smallest non-zero element is called an *atom*.

Line-graph *of a* [*hyper*]*graph G*: a simple graph $L(G)$ defined by
$$V(L(G)) = E(G),$$
$$E(l(G)) = \{(e,f) : e, f \in E(G), \ e, f, \text{ have an endpoint in common}\}.$$
$\sim$ *of a digraph G*: a simple digraph $L(G)$ defined by
$$V(L(G)) = E(G)$$
$$E(L(G)) = \{(e,f) : e, f \in E(G), \text{ the head of } e \text{ is the tail of } f\}.$$

Loop: see *graph.*

Matching: a *k-matching* in a [hyper]graph G is a collection of edges of G such that each point belongs to at most k of them (note that repetition of edges is allowed). A 1-matching is also called a *matching.* The maximum cardinality of a matching of G is denoted by $\nu(G)$; we set $\nu(G) = \infty$ if $\emptyset \in G$. A k-matching can be considered as a mapping $w : E(G) \to \{0, 1, \ldots\}$ such that $\sum_{E \ni x} w(E) \leq k$ for every point x ($w(E)$ is the multiplicity of E in the matching). A *perfect k-matching* is a k-matching such that each vertex belongs to exactly k members of it (note the difference between this and a k-factor: there an edge of G can occur at most once). A *fractional matching* is an assignment of a non-negative real weight $w(e)$ to each edge e such that $\sum_{E \ni x} w(e) \leq 1$ for every point x. The *size* of the fractional matching w is the value $\sum_{e \in E(G)} w(e)$; the minimum size of a fractional matching is denoted by $\nu^*(G)$.

Max-Flow-Min-Cut Theorem: see 6.74.

Maximum [minimum] means maximal [minimal] in cardinality.

Maximal [minimal] means maximal [minimal] with respect to inclusion.

Menger's Theorem: see 6.39.

Min-Path-Max-Potential Theorem: see 6.72.

Möbius function: see 2.22.

Möbius Inversion Formula: see 2.26.

Orientation: see *digraph.*

Parallel edges: see *graph, digraph, hypergraph.*

Partial hypergraph *of a hypergraph* H: a hypergraph H' with $V(H') \subseteq V(H)$, $E(H') \subseteq E(H)$.

Partition *of a set* S: a system $\{A_1, \ldots, A_k\}$ of disjoint non-empty subsets of S (called *classes* of the partition) such that $A_1 \cup \ldots \cup A_k = S$. The number B_n of partitions of an n-element set is called a *Bell number.*

$\sim$ *of a number* n: a collection $\{a_1, \ldots, a_k\}$ ($a_1 \geq \ldots \geq a_k$) of positive integers such that $a_1 + \ldots + a_k = n$.

Path *in a [di]graph:* a walk $(x_1, e_1, \ldots, e_k, x_{k+1})$, where $x_1, \ldots, x_{k+1}$ are disjoint points. It can be denoted by $(x_1, \ldots, x_{k+1})$, if the [di]graph is simple.

(X, Y)-**path:** a path in a [di]graph, connecting a point of X to a point of Y and having no other point in common with $X \cup Y$.

Perfect graph: a simple graph G such that each induced subgraph G' of G satisfies

$$\omega(G') = \chi(G').$$

Perfect Graph Theorem: see 13.57.

Permanent *of a matrix* $(a_{ij})_{i=1}^{n}{}_{j=1}^{n}$:

$$\operatorname{per} A = \sum_{\pi} a_{1,\pi(1)} \cdots a_{n,\pi(n)},$$

where π ranges over all permutations of $\{1,\ldots,n\}$.

Permutation *of a set* Ω: a one-to-one mapping of Ω onto itself. The number of permutations of an n-element set is $n! = 1 \cdot 2 \cdot \ldots \cdot n$. The identity permutation is denoted by 1. If γ is a permutation of Ω, $x \in \Omega$ and $\gamma(x) = x$, then x is called a *fixed point* of γ.

    cyclic $\sim$: identify two orderings $(x_1,\ldots,x_n)$, $(y_1,\ldots,y_n)$ of the same set S, of $y_1 = x_{k+1},\ldots,y_{n-k} = x_n$, $y_{n-k+1} = x_1,\ldots,y_n = x_k$ for some $1 \leq k \leq n$. A class of orderings identified in this way is a cyclic permutation.

    $\sim$ group: a group Γ such that to each $\gamma \in \Gamma$ a permutation $\tilde{\gamma} \in \Gamma$ of a finite set Ω is assigned such that $\widetilde{\gamma\delta}(x) = \tilde{\delta}(\tilde{\gamma}(x))$ $(\gamma,\delta \in \Gamma)$. If no confusion can arise we will assume the elements of Γ are permutations themselves, the same permutation occurring possibly several times. If $\tilde{\gamma} = \tilde{\delta}$ implies $\gamma = \delta$, or, equivalently, $\tilde{\gamma} \neq 1$ for $\gamma \neq 1$, the permutation group is called *effective*. If for any pair x, $y \in \Omega$ there is at most one $\gamma \in \Gamma$ such that $\gamma(x) = y$ the group is *transitive*. If for any pair x, $y \in \Omega$, there is at most one $\gamma \in \Gamma$ with $\gamma(x) = y$, or, equivalently, no $\gamma \in \Gamma$, $\gamma \neq 1$ has fixed points, the permutation group is *semiregular*. If it is both semiregular and transitive it is called *regular*. In this case $|\Gamma| = |\Omega|$ and the elements of Ω can be identified with the elements of Γ such that $\gamma(\delta) = \delta\gamma$ for all γ, $\delta \in \Gamma$.

Petersen graph: see Fig. 9 in problem 11.2.

Pfaffian *of a skew-symmetric matrix* $(a_{ij})_{i=1}^{2n}{}_{j=1}^{2n} = A$ (i.e. $a_{ii} = 0$, $a_{ij} = -a_{ji}$):

$$\operatorname{Pf} A = \sum \varepsilon_{i_1 j_1,\ldots,i_n j_n} a_{i_1 j_1},\ldots,a_{i_n j_n},$$

where $\varepsilon_{i_1 j_1,\ldots,i_n j_n}$ is the sign of the permutation

$$\begin{pmatrix} 1 & 2 & \ldots & 2n-1 & 2n \\ i_1 & j_1 & \ldots & i_n & j_n \end{pmatrix}$$

and the summation is taken over all partitions of $\{1,\ldots,2n\}$ of the form $\{\{i_1,j_1\},\ldots,\{i_n,j_n\}\}$. It is easy to see that the term corresponding to partition $\{\{i_1,j_1\},\ldots,\{i_n,j_n\}\}$ does not depend on the order of classes and/or the order of the two elements within a class.

Planar *map*: a graph whose vertices the points in the plane and whose edges are Jordan curves in the plane (ending in their corresponding graphical endpoints) with no point in common except their endpoints. A connected component of the set obtained by removing the edges and vertices of a planar map from the plane is called a *face* (region, country). The boundary of a face is always the union of certain edges; if the map G is 2-connected as a graph, the boundary of each face is a closed Jordan curve, consisting of the edges of a circuit in G.

$\sim$ *graph*: a graph G isomorphic to a planar map. Such a planar map is called an *embedding* of G in the plane.

Problem: see *graph, digraph, hypergraph*.

Pólya's Enumeration Method: see 3.26–30.

Product *of two simple* [*di*]*graphs*: we consider three kinds of products:

(weak) direct $\sim G_1 \times G_2$ of G_1 and G_2, defined by

$$V(G_1 \times G_2) = V(G_1) \times V(G_2),$$
$$E(G_1 \times G_2) = \{((x_1, x_2), (y_1, y_2)) : (x_1, x_2) \in E(G_1), (x_2, y_2) \in E(G_2)\}.$$

strong direct $\sim G_1 \cdot G_2$ of G_1 and G_2, defined by

$$V(G_1 \cdot G_2) = V(G_1) \times V(G_2),$$
$$E(G_1 \cdot G_2) =$$
$$= \{((x_1, x_2), (y_1, y_2) : (x_1, x_2) \in E(G_1) \text{ and } (x_2, y_2) \in E(G_2), \text{ or } x_1 = y_1$$
$$\text{and } (x_2, y_2) \in E(G_2) \text{ or } (x_1, y_1) \in E(G_1) \text{ and } x_2 = y_2\}.$$

Cartesian $\sim G_1 \oplus G_2$, defined by

$$V(G_1 \oplus G_2) = V(G_1) \times V(G_2),$$
$$E(G_1 \oplus G_2) = \{((x_1, x_2), (y_1, y_2) : x_1 = y_1 \text{ and } (x_2 = y_2 \in E(G_2);$$
$$\text{or } (x_1, y_1) \in E(G_1) \text{ and } x_2 = y_2)\}.$$

Thus $G_1 \cdot G_2 = (G_1 \times G_2) \cup (G_1 \oplus G_2)$ (see Fig. 122).

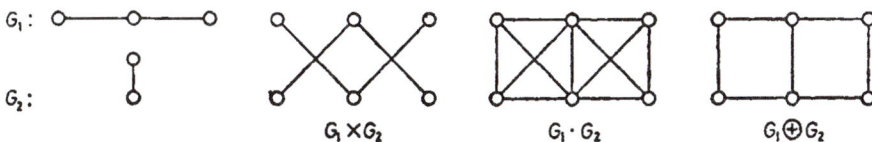

FIGURE 122

$\sim$ *of two hypergraphs* H_1, H_2: the hypergraph $H_1 \times H_2$ defined by

$$V(H_1 \times H_2) = V(H_1) \times V(H_2),$$
$$E(H_1 \times H_2) = \{E_1 \times E_2 : E_1 \in E(H_1), E_2 \in E(H_2)\}.$$

Prüfer code: see 4.5.

Pseudosymmetric *digraph*: see *symmetric*.

Ramsey's Theorem: see § 14.

Random walk *on a graph G*: an (infinite) sequence of random vertices $v_0, v_1, v_2, \ldots$, where v_0 is chosen from some given initial distribution (often concentrated on a single point) and for each $i \geq 0$, v_{i+1} is chosen from the uniform distribution on the neighbors of v_i. The *return time* (of vertex u) is the number of steps of the random walk starting at u before it returns to u (this is a random variable). The *access time* (from vertex u to vertex v) is the number of steps of the random walk starting at u before visiting v. The *commute time* (between vertices u and v) is the number of steps of the random walk starting at u before visiting v and returning to u. The *cover time* (from vertex u) is the number of steps of a random walk starting at u before every vertex is visited. The *mean return time of u*, the *mean access time* from u to v, the *mean commute time* between u and v, and the *mean cover time* from u are the expectations of the corresponding random variables. Thus the mean commute time between u and v is the sum of the access time from u to v and the access time from v to u.

Regular *graph* see *degree*; $\sim$ *group*: see *permutation group*.

Removal *of a set* $X \subseteq V(G)$ *from a [hyper-, di-]graph G*: removal of points in X together with all edges adjacent to them. The resulting [hyper, di-]graph is denoted by $G - X$; if $X = \{x\}$ we write simply $G - x$.

Return time: see *random walk*.

Restriction *of a hypergraph H onto* $X \subseteq V(H)$: the hypergraph H_X on the set X, for which $E(H_X)$ is the collection of sets $E \cap X$ $(E \in E(H))$. If $X = V(H) - Y$ then we adopt the further notation $H_X = H \setminus Y$ and $H_X = H - y$, if $Y = \{y\}$.

Rigid *graph*: has no proper endomorphism.

Selberg Sieve: see 2.14–17.

Semiregular *group*: see *permutation group*.

Separate: a set X of points and edges is said to be separate A and B $(A, B \subseteq V(G))$, if it represents (covers) all (A, B)-paths (cf. also *cutset*).

Sieve Formula: see § 2.

Simple *graph, digraph, hypergraph*: see *graph, digraph, hypergraph*.

Spanning subgraph *of G*: subgraph G' such that $V(G') = V(G)$.

Spectrum *of a graph* G: The spectrum (collection of eigenvalues) of the adjacency matrix A_G of G. Since A_G is symmetric, the eigenvalues of G (the elements of its spectrum) are real.

Sperner's Lemma: see 5.29.

Sperner's Theorem: see 13.21.

Splitting *a point x of a graph G into points $x_1,\ldots,x_k$*: we remove x, add $x_1,\ldots,x_k$ as new points and replace each (x,y)-edge $(y \in V(G) - \{x\})$ by an (x_i,y)-edge for exactly one i, $1 \leq i \leq k$.

Star: a tree with one point connected to all other points.

$\sim$ *of a vertex of a graph*: see *graph*.

Stationary distribution *for a random walk on a graph*: see 11.35.

Stirling cycle number $\begin{bmatrix} n \\ k \end{bmatrix}$: the number of permutations of an n-element set with exactly k cycles. The numbers $(-1)^{n-k} \begin{bmatrix} n \\ k \end{bmatrix}$ are also called *Stirling numbers of the first kind*, and denoted by $s(n,k)$.

Stirling partition number $\begin{Bmatrix} n \\ k \end{Bmatrix}$: the number of partitions of n objects into exactly k classes. These numbers are also called *Stirling numbers of the second kind*, and denoted by $S(n,k)$.

Subdivision *of a graph* G: a graph G' arising from G by replacing each edge e by a path P_e (of length ≥ 1), connecting the endpoints of e and having no other point in G, such that the paths P_e $(e \in E(G))$ are independent. We call the points of G *principal points* of G'.

Subgraph *of* G: a graph G' with $V(G') \subseteq V(G)$ and $E(G') \subseteq E(G)$. We write $G' \subseteq G$.

$\sim$ *of G induced by a set $X \subseteq V(G)$*: the graph $G[X]$ with $V(G[X]) = X$, $E(G[X]) = \{e \in E(G) : e \subseteq X\}$.

Substitution *of a graph G for a point x of a graph H* (supposing G and H are vertex-disjoint): we remove x from H, and replace each (x,y)-edge $(y \in V(H) - \{x\})$ by the $|V(G)|$ edges connecting y to the points of G.

Symmetric *digraph*: a simple digraph such that $(x,y) \in E(G)$, whenever $(y,x) \in E(G)$. *Pseudosymmetric*: $d^+(x) = d^-(x)$ at every point.

System of distinct representatives *of a hypergraph* H: a one-to-one mapping $\varrho: E(H) \to V(H)$ such that $\varrho(E) \in E$ for each $E \in E(H)$. If no confusion can arise, we also call the range $\varrho(E(H))$ a system of distinct representatives.

Tournament: a (simple) digraph T without loops in which exactly one of (x,y) and (y,x) is an edge for every pair $x \neq y$, $x, y \in V(T)$.

Trail: see *walk*.

Transitive *tournament*: a tournament T such that $(x,y) \in E(T)$ and $(y,z) \in E(T)$ imply $(x,z) \in E(T)$. The points of a transitive tournament have an ordering $(x_1,\ldots,x_n)$ such that $(x_i,x_j) \in E(G) \leftrightarrow i < j$.

$\sim$ *permutation group*: see *permutation group*.

Tree: a connected graph without circuits. It may also be defined as a connected graph such that removing any edge disconnects it; or as a circuit-free graph in which the introduction of any new edge will produce a circuit. A tree on n points has exactly $n-1$ edges and it always has at least two points of degree 1, provided $|V(G)| \geq 2$. A *rooted tree* is a tree with a specified vertex called its *root*. A *rooted d-ary tree* is a rooted tree such that the root has degree d and every non-root vertex has degree $d+1$ or 1. A rooted d-ary tree is *complete* if all its endpoints are at the same distance from the root.

Triangulation (*planar*): a planar map in which each face is a triangle.

$\sim$ *of a circuit C*: a graph consisting of this circuit and $n-3$ non-crossing "interior" diagonals (n is the length of C).

Turán's Theorem: see 10.34.

Tutte's Theorem: see 7.27.

Uniform: see *hypergraph*.

Valency: see *degree*.

Vertex: see *graph, digraph, hypergraph*.

Walk *in a [di]graph*: a sequence $(x_1,e_x,\ldots,x_k,e_k,x_{k+1})$ in which $x_1,\ldots,x_k$ are points and e_i is an (x_i,x_{i+1})-edge ($i=1,\ldots,k$). If the [di]graph is simple we may describe the walk by the sequence $(x_1,\ldots,x_{k+1})$. The walk is *open [closed]* iff $x_{k+1} \neq x_1$ [$x_{k+1}=x_1$]. The *length* of the walk is k above. A walk is a *trail* if no edge is used twice.

wheel: a graph obtained from a circuit by connecting all points to a new point (the "center" of the wheel).

Θ-**graph:** a graph consisting of three independent paths connecting two points.

Notation

A_G:	adjacency matrix of graph G.
$A(G)$:	automorphism group of G.
B_G:	incidence matrix of G.
B_n:	Bell number
$c(G)$:	number of components of G.
$c_1(G)$:	number of odd components (i.e. components with an odd number of points) of G.
$d_G(x)$:	degree of point x in G.
$d(G)$:	maximum degree in G.
$d_G(x,y)$:	distance between points x, y in G.
$d_G^+(x)$:	outdegree of point x in G.
$d_G^-(x)$:	indegree of point x in G.
$E(G)$:	set of edges of G.
$\mathrm{End}(G)$:	endomorphism semigroup of G.
$\exp(x)$:	e^x.
$n!!$:	n semifactorial: $n!! = n(n-2)\cdots = \displaystyle\prod_{1 \le k \le n,\ k \cong n \ (\mathrm{mod}\ 2)} k$.
I:	identity matrix.
$\mathbf{j}[J]$:	a vector [matrix] all entries of which are 1's.
$k_i(\pi)$:	number of i-cycles in permutation π.
K_n^r:	complete r-uniform hypergraph on n points. For $r=2$ the superscript is omitted.
$L(G)$:	line-graph of G.

$\left[\begin{smallmatrix} n \\ k \end{smallmatrix}\right]$:   Stirling cycle number.

$\left\{\begin{smallmatrix} n \\ k \end{smallmatrix}\right\}$:   Stirling partition number.

$O(f(n))$:   a function $g(n)$ such that $g(n)/f(n)$ is bounded.

$o(f(n))$:   a function $g(n)$ such that $g(n)/f(n) \to 0$ as $n \to \infty$.

$p_G(\lambda)$:   characteristic polynomial of G.

$P_G(\lambda)$:   chromatic polynomial of G.

$p_\Gamma(x_1, \ldots, x_n)$:   cycle index of permutation group Γ.

$\operatorname{per} A$:   permanent of matrix A.

$\operatorname{Pf} A$:   Pfaffian of matrix A.

$q(G)$:   chromatic index of G.

$V(G)$:   set of vertices of G.

$\mathbf{Z}$:   the set of integers.

$\delta_G(X)$:   number of $(X, V(G) - X)$-edges in G.

$\alpha(G)$:   maximum number of independent points.

$\Gamma_G(X)$:   set of points in graph G adjacent to at least one point of $X \subseteq V(G)$.

$\nu(G)$:   maximum number of independent edges in G (matching number).

$\nu^*(G)$:   maximum size of a fractional matching in G.

$\varrho(G)$:   minimum number of edges of G covering all points.

$\tau(G)$:   minimum number of points of G representing all edges.

$\tau^*(G)$:   minimum size of a fractional cover of G.

$\chi(G)$:   chromatic number of G.

$\omega(G)$:   maximum number of points in a clique (or in a complete subgraph) of G.

$d(x), d^+(x),$

$\Gamma(X)$, etc.:   abbreviations for $d_G(x)$, $d_G^+(x)$, $\Gamma_G(x)$ in the case when the graph G is understood.

$\lfloor x \rfloor$:   integer part of x: largest integer not greater than x.

$\lceil x \rceil$:   least integer not less than x.

$G - F$,   (where G is a graph (digraph, hypergraph) and $F \subseteq E(G)$) removal of all edges in F (but removing no points).

$G-f$, (where $f \in E(G)$) shorthand for $G - \{f\}$, when no confusion can arise.

$\overline{G}$: the complement of [di]graph G.

$G_1 \cup G_2$: the [di]graph defined by $V(G_1 \cup G_2) = V(G_1) \cup V(G_2)$, $E(G_1 \cup G_2) = E(G_1) \cup E(G_2)$ (here G_1, G_2 may have points or edges in common).

$G-X$: (where G is a graph [digraph, hypergraph] and $X \subseteq V(G)$) removal of points in X and all edges adjacent to them.

$G-x$: (where $x \in V(G)$) shorthand for $G - \{x\}$.

$H \setminus X$: (where H is a hypergraph and $X \subseteq V(H)$) the restriction of H onto $V(H) - X$.

$H \setminus x$: (where $x \in V(H)$) shorthand for $H \setminus \{x\}$.

$G[X]$: the subgraph of G induced by $X \subseteq V(G)$.

H_X: the restriction of hypergraph H onto $X \subseteq V(H)$.

G/F: (where $F \subseteq E(G)$) the graph arising from graph G by contracting all edges in F.

G/f: (where $f \in E(G)$) shorthand for $G/\{f\}$.

$G_1 \times G_2$: (weak) direct product of [di]graphs G_1, G_2.

$G_1 \cdot G_2$: strong direct product.

$G_1 \oplus G_2$: Cartesian product.

$H_1 \times H_2$: direct product of hypergraphs H_1, H_2.

Index

of the abbreviations of the textbooks and monographs

B : C. Berge, *Graphs and Hypergraphs*, North-Holland–American Elsevier, 1973

Biggs : N. Biggs, *Algebraic Graph Theory*, Cambridge Univ. Press, 1974.

ES : P. Erdős–J. Spencer, *Probabilistic Methods in Combinatorics*, Akadémiai Kiadó, Budapest, 1974.

Fe : W. Feller, *An Introduction to Probability Theory and its Applications*, 2nd ed., Wiley, New York–Chapman Hall, London, 1957.

FF : R. L. Ford–D. R. Fulkerson, *Flows in Networks*, Princeton Univ. Press, 1962.

H : F. Harary, *Graph Theory*, Addison–Wesley, 1969.

Hall : M. Hall, *Combinatorial Theory*, Blaisdell, 1967.

Hu : T. C. Hu, *Integer Programming and Network Flows*, Addison–Wesley, 1969.

K : D. König, *Theorie der endlichen und unendlichen Graphen*, Leipzig, 1936.

LP : L. Lovász–M. D. Plummer, *Matching Theory*, Akadémiai Kiadó–North Holland, 1986.

M : J. W. Moon, *Topics on Tournaments*, Holt, Rinehart and Wilson, 1968.

Mi : L. Mirsky, *Transversal Theory*, Academic Press, 1971.

O : O. Ore, *Theory of Graphs*, Amer. Math. Soc. Coll. Publ., 1962.

OF : O. Ore, *The Four Color Problem*, Academic Press, 1967.

R : J. Riordan, *An Introduction to Combinatorial Analysis*, Wiley, 1958.

Ré : A. Rényi, *Foundations of Probability*, Holden-Day, San Francisco–Cambridge–London–Amsterdam, 1970.

S : H. Sachs, *Einführung in die Theorie der endlichen Graphen*, Teil I–II., Teubner, 1970–1972.

St : R. P. Stanley, *Enumerative Combinatorics*, vol. 1, Wadworth & Brooks/Cole, Monterey. 1982.

W : K. Wagner, *Graphentheorie*, Biblographisches Inst. AG. 1970.

Wi : R. J. Wilson, *Introduction to Graph Theory*, Oliver & Boyd, 1972.

WV : H. Walther–H.-J. Voß, *Über Kreise in Graphen*, VEB Deutscher Verlag der Wiss., 1974.

Subject Index

Author Index
* Asterisks indicate references to books

Corrections to
Combinatorial Problems and Exercises

by László Lovász
(Second Edition, Akadémiai Kiadó, Budapest, 1993)

Problems

1.15 Second displayed formula correctly:

$$s(x) = \sum_{n=0}^{\infty} \frac{S_n}{n!} x^n.$$

1.28 In line 3, ϑ^k should be ϑ_k.

3.7 In displayed formula, y^y should be y^k.

3.29 Reference to 4.26 should be 3.26.

5.25 Assume that $n \geq 3$.

6.4 Assume that both G_1 and G_2 have at least two nodes.

Solutions

1.12(a) Last displayed formula correctly:

$$p(x) = \prod_{i=1}^{\infty} \left(\sum_{k_i=0}^{\infty} \frac{x^{ik_i}}{k_i!(i!)^{k_i}} \right) = \prod_{i=1}^{\infty} \exp\left(\frac{x^i}{i!} \right) = e^{e^x - 1}.$$

1.18 Last displayed formula correctly:

$$\gamma_d = \sum 2^{\beta_i(d)}.$$

1.28 In third displayed formula, replace ϑ_n^k by ϑ_ν^k.

1.30 End of (1) correctly: "... if it starts with a $[b]$". In first displayed formula, replace n_x by x_n.

2.2 In last displayed formula, replace $|A|$ by $|A_I|$.

3.2 Line 3: Replace "partitions" by "permutations".

3.8 First displayed formula correctly:

$$\frac{1}{n} \sum_{m=1}^{n} x_{n/(n,m)}^{(n,m)} = \frac{1}{n} \sum_{d|n} x_d^{n/d} \sum_{\substack{m \le n \\ (m,n)=n/d}} 1 = \frac{1}{n} \sum_{d|n} \varphi(d) x_d^{n/d}.$$

3.11 In first displayed formula, second line, and also in second displayed formula, summation should start with $k = 1$.

3.29 Last displayed formula correctly:

$$\sum_{n=0}^{\infty} a_n x^n = \frac{1}{|\Gamma|} \sum_{\gamma \in \Gamma} \sum_{n=0}^{\infty} q_n(\gamma) x^n = \frac{1}{|\Gamma|} \sum_{\gamma \in \Gamma} \prod_{j=1}^{|D|} r(x^j)^{k_j(\gamma)} = F(r(x), r(x^2), \dots)$$

3.31 Second line: replace R by D.

4.9 Third paragraph should start with "Second, suppose that no point of G' has..."

5.27 In line 13 of part (b), the formula should be: $2m - 2(m - n + 2) = 2n - 4 < 2n$.

6.13 In line 5, the formula should be: $x \in V(T) \setminus V(C)$.

9.11 In part (b), end of line 3 and beginning of line 4 should be: "... thus $\varepsilon(b) \le \varepsilon(a) - 1$. Conversely, $W_b + (b, a)$ is ...".

9.18 In line 12, the third case for $\alpha(z)$ should be "$\beta(x)$ if $z = x'$".

9.19 Second paragraph of (b) correctly: "Conversely, suppose that $\chi(G_0/e) \le k$ for every $e \in E(G_0)$. We show that G_0 can be embedded into a critically $(k+1)$-chromatic graph even as an induced subgraph. The assumption means that for each $e \in E(G_0)$, there exists a k-coloration α_e of $S = V(G_0)$ that associates the same color with the endpoints of e, but different colors with the endpoints of any other edge of G_0. Let P_e be the partition of S induced by α_e. By 9.8, we can find a graph G such that $S \subseteq V(G)$ and the k-colorations of G induce the partitions P_e ($e \in E(G_0)$) of S and no other partition. Set $G' = G \cup G_0$."

9.56 Insert at the end of the solution: "The converse is easy: assume that the graph has a 3-coloring with red, blue and green, and consider any node v with color red (say). The neighbors of v must be alternating between blue and green, and so their number must be even."

10.41 First displayed formula correctly:

$$\sum_{i=1}^{n} \binom{d_i}{2} \ge n \cdot \binom{\frac{n-1}{2}}{2} = \frac{n(n-1)(n-3)}{8}$$

11.11 End of second line correctly: "Thus $(\nu, i_\nu) \in E(G)$ ($\nu = 1, \dots, n$)."

11.14(a), Part III, second line: replace D by λ_1.

11.19 Formula (2) should start with $\left|\mathbf{w}^T A_G \mathbf{w}\right| = \ldots$ The seventh line below should start with "Thus, if $a_{ij} \neq 0$, then w_i and w_j have different signs." Last displayed formula in (a) correctly:

$$\sum_j a_{ij} w'_j = \sum_j a_{ij} w_j = \lambda_1 w_i = -\lambda_1 w'_i.$$

11.29 Second sentence correctly: "Since $\mathbf{1}$ is an eigenvector belonging to $\lambda_1 \neq \lambda_2$, it is orthogonal to $\mathbf{x}$, i.e., we have $\sum_i x_i = 0$." Last displayed formula should end with

$$\ldots = \frac{1}{k-1}(x_k - x_1)^2 > \frac{1}{Dn}.$$

12.13 First sentence correctly: "Let Q_n be the n-cube."

12.15 In last sentence, reference to Fig. 88 should be Fig. 91.

12.24 In last paragraph, reference to Fig. 92 should be Fig. 96.

13.2 Third paragraph, third line: $(h - \{E\}) \setminus E$ should be $(H - \{E\}) \setminus E$.

13.54 Second paragraph, second line: $h - \{E_i\})$ should be $H - \{E_i\})$.

Dictionary

Product: In **weak direct** $\sim$, second displayed line correctly:

$$E(G_1 \times G_2) = \{((x_1, x_2), (y_1, y_2)) : \ (x_1, y_1) \in E(G_1), (x_2, y_2) \in E(G_2)\}.$$

In **strong direct** $\sim$, second displayed line correctly:

$$E(G_1 \cdot G_2) = \{((x_1, x_2), (y_1, y_2) : \ (x_1, y_1) \in E(G_1) \text{ and } (x_2, y_2) \in E(G_2), \text{ or }$$
$$x_1 = y_1 \text{ and } (x_2, y_2) \in E(G_2) \text{ or } (x_1, y_1) \in E(G_1) \text{ and } x_2 = y_2\}.$$

Author Index

ISBN 978-0-8218-4262-1

9 780821 842621

CHEL/361.H

About this book

The main purpose of this book is to provide help in learning existing techniques in combinatorics. The most effective way of learning such techniques is to solve exercises and problems. This book presents all the material in the form of problems and series of problems (apart from some general comments at the beginning of each chapter). In the second part, a hint is given for each exercise, which contains the main idea necessary for the solution, but allows the reader to practice the techniques by completing the proof. In the third part, a full solution is provided for each problem.

This book will be useful to those students who intend to start research in graph theory, combinatorics or their applications, and for those researchers who feel that combinatorial techniques might help them with their work in other branches of mathematics, computer science, management science, electrical engineering and so on. For background, only the elements of linear algebra, group theory, probability and calculus are needed.